Lecture Notes in Networks and Systems

Volume 213

The series "Lecture Notes in Networks and Systems" publishes the latest developments in Networks and Systems—quickly, informally and with high quality. Original research reported in proceedings and post-proceedings represents the core of LNNS.

Volumes published in LNNS embrace all aspects and subfields of, as well as new challenges in, Networks and Systems.

The series contains proceedings and edited volumes in systems and networks, spanning the areas of Cyber-Physical Systems, Autonomous Systems, Sensor Networks, Control Systems, Energy Systems, Automotive Systems, Biological Systems, Vehicular Networking and Connected Vehicles, Aerospace Systems, Automation, Manufacturing, Smart Grids, Nonlinear Systems, Power Systems, Robotics, Social Systems, Economic Systems and other. Of particular value to both the contributors and the readership are the short publication timeframe and the world-wide distribution and exposure which enable both a wide and rapid dissemination of research output.

The series covers the theory, applications, and perspectives on the state of the art and future developments relevant to systems and networks, decision making, control, complex processes and related areas, as embedded in the fields of interdisciplinary and applied sciences, engineering, computer science, physics, economics, social, and life sciences, as well as the paradigms and methodologies behind them.

Indexed by SCOPUS, INSPEC, WTI Frankfurt eG, zbMATH, SCImago.

All books published in the series are submitted for consideration in Web of Science.

More information about this series at http://www.springer.com/series/15179

Jennifer S. Raj · Ram Palanisamy ·
Isidoros Perikos · Yong Shi
Editors

Intelligent Sustainable Systems

Proceedings of ICISS 2021

Springer

Editors
Jennifer S. Raj
Gnanmani College of Engineering
and Technology
Namakkal, Tamil Nadu, India

Isidoros Perikos
Department of Computer Engineering
and Informatics
University of Patras
Patra, Greece

Ram Palanisamy
Department of Business Administration,
The Gerald Schwartz, School of Business
St. Francis Xavier University
Nova Scotia, NS, Canada

Yong Shi
Department of Computer Science
Kennesaw State University
Kennesaw, GA, USA

ISSN 2367-3370 ISSN 2367-3389 (electronic)
Lecture Notes in Networks and Systems
ISBN 978-981-16-2421-6 ISBN 978-981-16-2422-3 (eBook)
https://doi.org/10.1007/978-981-16-2422-3

This Springer imprint is published by the registered company Springer Nature Singapore Pte Ltd.
The registered company address is: 152 Beach Road, #21-01/04 Gateway East, Singapore 189721, Singapore

We are honored to dedicate the proceedings of 4th ICISS 2021 to all the participants, organizers, and editors of 4th ICISS 2021.

Foreword

On behalf of the organizing committee, it is my pleasure to welcome you all to the fourth International Conference on Intelligent Sustainable Systems (ICISS 2021). The theme of the 2021 conference is *Intelligent Sustainable Systems for a More Sustainable Future*, a topic that is gaining significant research attraction from both academia and industries because of the influence of sustainability in digital age and develop sustainability integrated information management systems. Intelligent sustainable systems feature prominently to improvise and assist the transition towards sustainable societies and economies. The well-established research track of sustainable systems research in different ICT domains mandates further research exploration on intelligent sustainable systems and makes ICISS 2021 an excellent venue for exploring state-of-the-art sustainability-related foundations in intelligent information systems.

This year, ICISS has received 252 papers in different conference tracks, and based on the 3–4 expert reviews from the technical program committee, internal and external reviewers, and 65 papers were finally selected for the conference. The entire conference proceedings include papers from different tracks like Intelligent Systems, Sustainable Systems, and Applications. Each paper, regardless of track, has received at least three reviews, who have professional expertise in the particular research domain of the paper.

The entire success of 4th ICISS 2021 depends on the efforts and research expertise of the researchers in the sustainability domain, who have written and submitted their innovative research results on different research topics included in the conference. A deep appreciation to the conference program and review committee members, who have invested their valuable time in assessing the multiple research submissions made to the conference to maintain a high standard for this conference event. I extend my gratitude to the Springer publications for their exhaustive support in updating the conference production deadlines and details to the conference participants.

Dr. P. Ebby Darney
Principal
SCAD College of Engineering and Technology
Tirunelveli, Tamil Nadu, India

Preface

With a deep gratification, we are delighted to welcome you to the proceedings of the 4th International Conference on Intelligent Sustainable Systems (ICISS 2021) organized at SCAD College of Engineering and Technology, Tirunelveli, India on February 26–27, 2021. The major goal of this international conference is to gather the academicians, industrialists, researchers, and scholars together in a common platform to share their innovative research ideas and practical solutions toward the development of intelligent sustainable systems for a more sustainable future.

ICISS 2021 guarantees to be both invigorating and enlightening with an outstanding array of technical program chairs, keynote speakers, reviewers, and researchers across the globe. The conference delegates had a wide range of technical sessions based on different technical domains involved in the theme of conference. The conference program has included invited keynote sessions on developing a sustainable future, state-of-the-art research work presentations, and informative discussions with the distinguished keynote speakers by covering a wide range of topics in information systems and sustainability research.

We are pleased to thank the conference organization committee, conference program committee, and technical reviewers for working generously toward the success of the conference event. A special mention to the internal and external reviewers for working very hard in reviewing each and every paper received at the conference and for giving valuable suggestions to the authors for maintaining the quality of the conference. We are truly obliged to the authors, who have contributed their innovative research results to the conference. Special thanks go to the Springer Publications for their impeccable support and guidance throughout the publication process.

We wish the proceedings of ICISS 2021 will give an enjoyable and technical-rewarding experience for both attendees and readers.

Namakkal, India
Nova Scotia, Canada
Patra, Greece
Kennesaw, USA

Technical Program Chairs
Prof. Dr. Jennifer S. Raj
Prof. Dr. Ram Palanisamy
Dr. Isidoros Perikos
Prof. Dr. Yong Shi

Acknowledgments

The 4th edition of the International Conference on Intelligent Sustainable Systems (ICISS 2021) was organized by SCAD College of Engineering and Technology, Tirunelveli, India during 26–27 February, 2021. The conference has created a platform to bring together a huge number of developing researchers, scholars, and experts to leverage their latest and insightful research ideas, findings, and results. We are deeply grateful to our institution SCAD College of Engineering and Technology for their constant encouragement, guidance, and support in successfully organizing the fourth edition of ICISS. The computer science and engineering [CSE] department is honored by the Springer publications for publishing the proceedings of the conference event and we express our special gratitude to them and expect their support in our future endeavors.

We are deeply thankful to the organizing committee members for their valuable support and suggestion on many occasions of the conference event. We are pleased to thank the keynote speaker, Dr. Valentina Emilia, Balas for her ingenious and resourceful talk during the conference session. A successful conference of this magnitude is the significant result of formidable efforts made by various technical program/review/session chairs. We convey our gratitude to all the organizing chairs, faculty and non-faculty members, and reviewers for their outstanding efforts in delivering the right balance for ensuring the quality of the conference. We sincerely thank all the authors for expressing their potential and interest toward the conference. We would like to acknowledge all the members of the conference advisory committee and program committee for providing their excellent cooperation and guidance.

On behalf of the conference committee, we welcome you to Tirunelveli to experience the vibrancy of South India. We wish you to find an insightful research experience in 4th ICISS 2021.

Contents

Editors and Contributors

About the Editors

Dr. Jennifer S. Raj received Ph.D. degree from Anna University and Master's Degree in communication system from SRM University, India. Currently, she is working in the Department of ECE, Gnanamani College of Technology, Namakkal, India. She is a life member of ISTE, India. She has been serving as an organizing chair and a program chair of several international conferences and in the program committees of several international conferences. She is a book reviewer for Tata McGraw hill publication and publishes more than fifty research articles in journals and IEEE conferences. Her interests are in wireless healthcare informatics and body area sensor networks.

Prof. Ram Palanisamy is a professor of enterprise systems in the Business Administration Department at the Gerald Schwartz School of Business, St. Francis Xavier University. Dr. Palanisamy teaches courses on Foundations of Business Information Technology, Enterprise Systems using SAP, Systems Analysis and Design, SAP Implementation, Database Management Systems, and Electronic Business (Mobile Commerce). Before joining StFX, he taught courses in management at the Wayne State University (Detroit, USA), Universiti Telekom (Malaysia), and National Institute of Technology (NITT), Deemed University, India. His research interest includes enterprise systems (ES) implementation; ES acquisition; ES flexibility and ES success; knowledge management systems; and healthcare inter-professional collaboration.

Dr. Isidoros Perikos completed his Ph.D. in Computer Engineering and Informatics, Computer Engineering and Informatics Department at the University of Patras, Greece (2016), and M.Sc. in Computer Science and Technology, Computer Engineering and Informatics Department at University of Patras (2010). He has completed Engineering Diploma (five-year program, M. Eng.) in Computer Engineering and Informatics, Computer Engineering and Informatics Department at University of Patras (2008). His research interest includes semantic web and ontology engineering, web intelligence, natural language processing and understanding, human–computer interaction, and affective computing robotics. He has published in national and international journals and conferences.

Dr. Yong Shi is currently working as an associate/tenured professor of computer science, Kennesaw State University, and a director/coordinator of the Master of Computer Science. He is responsible for directing the Master of Computer Science program, reviewing applications for the Master of Computer Science. He has published more than 50 articles in national and international journals. He acted as an editor, a reviewer, an editorial board member, and a program committee member in many reputed journals and conferences. His research interest includes cloud computing, big data, and cybersecurity.

Contributors

Md. Abdul Momen Swazal Department of Electrical and Electronic Engineering, University of Asia Pacific, Dhaka, Bangladesh

Pawan Agarwal Graphic Era Hill University, Bhimtal, Uttarakhand, India

S. A. Akbar Academy of Scientific and Innovative Research (AcSIR), Ghaziabad, UP, India;
CSIR-Central Electronics Engineering Research Institute, Pilani, Rajasthan, India

K. S. Ananthu Department of Computer Science and IT, Amrita School of Arts and Sciences, Amrita Vishwa Vidyapeetham, Kochi, India

B. Angel Central Institute of Plastics Engineering and Technology, Chennai, India

D. Angel Sathyabama Institute of Science and Technology, Chennai, India

C. Anitha Department of CSE, NICHE, Kumaracoil, Tamil Nadu, India

Abu Shamim Mohammad Arif Computer Science and Engineering Discipline, Khulna University, Khulna, Bangladesh

Shweta Arora Graphic Era Hill University, Bhimtal, Uttarakhand, India

T. Arumuganathan ICAR-Sugarcane Breeding Institute, Coimbatore, India

M. Ashok Department of CSE, Rajalakshmi Institute of Technology, Chennai, Tamilnadu, India

J. Aswathy Academy of Scientific and Innovative Research (AcSIR), Ghaziabad, India;
CSIR-Central Electronics Engineering Research Institute, Pilani, India

V. Baby Shalini Information Technology, Kalasalingam Academy of Research and Education, Krishnankoil, Srivilliputhur, India

Anupam Kumar Bairagi Khulna University, Khulna, Bangladesh

Geetha Bala Department of Electronics and Communication Engineering, Prince Dr. K. Vasudevan College of Engineering & Technology, Chennai, India

R. Balasubramian Department of Computer Science, J.J College of Arts & Science (Autonomous), Pudukkottai, India

Shahana Bano Department of Computer Science and Engineering, Koneru Lakshmaiah Education Foundation, Vaddeswaram, Andhra Pradesh, India

Rohitash Kumar Banyal Department of Computer Science and Engineering, Rajasthan Technical University, Kota, India

Shailendra Baraniya Department of Electrical and Electronics Engineering, CMR Institute of Technology, Bengaluru, India

S. Barathi Kanna Department of Electronics and Communication Engineering, Amrita School of Engineering, Coimbatore, Amrita Vishwa Vidyapeetham, Coimbatore, India

Amlan Jyoti Baruah Department of Computer Science and Engineering, Assam Kaziranga University, Jorhat, Assam, India

Siddhartha Baruah Department of Computer Applications, Jorhat Engineering College, Jorhat, Assam, India

B. Bebeshko Faculty of Information Technologies, Kyiv National University of Trade and Economics, Kyiv, Ukraine

Pradeep Bedi Lingayas Vidyapeeth, Faridabad, India

Varsha Bhatia Amity University, Gurugram, Haryana, India

Nikhil Bhatt Akhilesh Das Gupta Institute of Technology and Management, Delhi, India

R. Bhavani Department of CSE/SRC, SASTRA Deemed University, Kumbakonam, India

V. Bhavani Department of Computer Science and Engineering, Koneru Lakshmaiah Education Foundation, Vaddeswaram, AP, India

Shishir Kumar Bhowmick Department of Electrical and Electronic Engineering, University of Asia Pacific, Dhaka, Bangladesh

Ankur Biswas Department of Computer Science & Engineering, Adamas University, Kolkata, India

Dibya Jyoti Bora School of Computing Science, Assam Kaziranga University, Jorhat, Assam, India

A. Bramendran Machine Intelligence Research Laboratory, Department of Electronics and Communication Engineering, Amrita School of Engineering, Coimbatore, Amrita Vishwa Vidyapeetham, India

Preksh Chadha Akhilesh Das Gupta Institute of Technology and Management, Delhi, India

M. H. Chaithra Department of Computer Science and Engineering, Visvesvaraya Technological University, Belagavi, Karnataka, India

Rekha Chakravarthi School of Electrical and Electronics, Sathyabama Institute of Science and Technology, Chennai, India

Guru Sai Chilukuri Department of Computer Science and Engineering, Koneru Lakshmaiah Education Foundation, Vaddeswaram, Andhra Pradesh, India

Tinnavalli Deepika Department of CSE, Koneru Lakshmaiah Education Foundation, Vaddeswaram, India

Rohan Desai Department of Information Technology, Terna Engineering College, Navi Mumbai, India

Usha Desai Department of Electronics and Communication Engineering, NMAM Institute of Technology, Nitte, Udupi, Karnataka, India

A. Desiatko Faculty of Information Technologies, Kyiv National University of Trade and Economics, Kyiv, Ukraine

Mikkili Divya Department of Electronics and Communication Engineering, Vignan's Nirula Institute of Technology and Science for Women, Guntur, AP, India

S. Divya Department of ECE, Sri Ramakrishna Engineering College, Coimbatore, India

Ekta Dixit Punjabi University, Patiala, Punjab, India

D. Evangeline M.S. Ramaiah Institute of Technolgy, Bangalore, India

Kondeth Fathima Amrita School of Arts and Science, Kochi, India

R. Femi Department of EEE, University College of Engineering, Nagercoil, India

R. Senthil Ganesh Department of ECE, Sri Krishna College of Engineering and Technology, Coimbatore, Tamilnadu, India

Mayank Garg Department of Computer Engineering & Application, GLA University, Mathura, India

K. Gayathri Department of Electronics and Communication Engineering, Sri Ramakrishna Engineering College, Coimbatore, India

Angelina Geetha Department of Computer Science and Engineering, Hindustan Institute of Technology and Science, Padur, Chennai, India

Deepika Ghanta Department of Computer Science and Engineering, Koneru Lakshmaiah Education Foundation, Vaddeswaram, Andhra Pradesh, India

S. Gokul Department of Electronics and Communication Engineering, Amrita School of Engineering, Coimbatore, Amrita Vishwa Vidyapeetham, Coimbatore, India

E. A. Gopalakrishnan Amrita School of Engineering, Center for Computational Engineering and Networking (CEN), Amrita Vishwa Vidyapeetham, Coimbatore, India

JyotiProkash Goswami Department of Computer Applications, Assam Engineering College, Guwahati, Assam, India

S. B. Goyal City University Malaysia, Petaling Jaya, Malaysia

Abhinay Gupta Department of Electrical and Electronics Engineering, KLE Dr. M. S. Sheshgiri College of Engineering & Technology, Belagavi, India

Megha Gupta Department of Computer Science, MSCW, University of Delhi, Delhi, India

S. Guruprasad Department of ISE, BMS Institute of Technology and Management, Bengaluru, India

K. Hemachandran Woxsen School of Business, Woxsen University, Hyderabad, India

S. G. Hiremath Department of Electronics and Communication Engineering, East West Institute of Technology, Bangalore, Karnataka, India

M. S. Hossain Lipu Department of Electrical, Electronic and Systems Engineering, Universiti Kebangsaan Malaysia, Bangi, Malaysia

G. K. Jakir Hussain Department of ECE, KPR Institute of Engineering and Technology, Coimbatore, Tamilnadu, India

Shoeb Hussain Department of Electrical Engineering, University of Kashmir, Srinagar, India

Vivek Jaglan Graphic Era Hill University, Dehradun, India

Deepa Elizabeth James Amrita School of Arts and Science, Amrita Vishwa Vidyapeetham, Kochi Campus, Kochi, India

Urooj Javeed Department of Electrical Engineering, University of Kashmir, Srinagar, India

S. Jayanthy Department of ECE, Sri Ramakrishna Engineering College, Coimbatore, India

P. Jayashree Department of Computer Technology, MIT Campus, Anna University, Chennai, India

Jayshree Department of Electronics & Communication Engineering, National Institute of Technology, Nagaland, India;
Indian Institute of Information Technology, Tiruchirappalli, India

Vandana Jindal DAV College, Bathinda, Punjab, India

Himanshu Joshi Graphic Era Hill University, Bhimtal, India

Kavita A. Joshi Graphic Era Hill University, Bhimtal, Uttarakhand, India

J. Judeson Antony Kovilpillai Department of ECE, Sri Ramakrishna Engineering College, Coimbatore, India

Mohammed Junaid Woxsen School of Business, Woxsen University, Hyderabad, India

Rakhee Kallimani Department of Electrical and Electronics Engineering, KLE Dr. M. S. Sheshgiri College of Engineering & Technology, Belagavi, India

Sowmya Kamath Department of Electronics and Communication Engineering, NMAM Institute of Technology, Nitte, Udupi, Karnataka, India

Sai Pavan Kamma Department of Computer Science and Engineering, Koneru Lakshmaiah Education Foundation, Vaddeswaram, Andhra Pradesh, India

G. S. Karthick Department of Computer Science, Bharathiar University, Coimbatore, Tamil Nadu, India

Durga Karthik Department of CSE/SRC, SASTRA Deemed University, Kumbakonam, India

S. Karthikeyan Department of Information Technology, V.H.N. Senthikumara Nadar College, Virudhunagar, Tamilnadu, India

T. Kathirvalavakumar Research Centre in Computer Science, V.H.N.Senthikumara Nadar College, Virudhunagar, Tamilnadu, India

Naman Kathuria Department of Information Technology, Netaji Subhas University of Technology, Delhi, India

Vaishali Khairnar Department of Information Technology, Terna Engineering College, Navi Mumbai, India

M. D. Reyad Hossain Khan Khulna University, Khulna, Bangladesh

Shoaib Khan Department of Electrical Engineering, University of Kashmir, Srinagar, India

O. Kharchenko Faculty of Information Technologies, Kyiv National University of Trade and Economics, Kyiv, Ukraine

K. Khorolska Faculty of Information Technologies, Kyiv National University of Trade and Economics, Kyiv, Ukraine

Akshata Koodagi Department of Electrical and Electronics Engineering, KLE Dr. M. S. Sheshgiri College of Engineering & Technology, Belagavi, India

J. Judeson Antony Kovilpillai Department of ECE, Sri Ramakrishna Engineering College, Coimbatore, India

Pambavasan Krishna Prasad Department of Computer Science and IT, Amrita School of Arts and Sciences, Amrita Vishwa Vidyapeetham, Kochi, India

S. Kukan Department of Electronics and Communication Engineering, Amrita School of Engineering, Coimbatore, Amrita Vishwa Vidyapeetham, Coimbatore, India

Jugnesh Kumar St. Andrews Institute of Technology & Management, Gurgaon, India

Krishanth Kumar Amrita School of Engineering, Center for Computational Engineering and Networking, Amrita Vishwa Vidyapeetham, Coimbatore, India

Navjot Kumar Academy of Scientific and Innovative Research (AcSIR), Ghaziabad, India;
CSIR-Central Electronics Engineering Research Institute, Pilani, India

R. Mathusoothana S. Kumar Department of IT, NICHE, Kumaracoil, Tamil Nadu, India

Sunny Kumar Department of Information Technology, Netaji Subhas University of Technology, Delhi, India

Tushar Kumar Graphic Era Hill University, Bhimtal, India

Sunita Kumawat Amity University, Gurugram, Haryana, India

K. Laila Department of Computer Technology, MIT Campus, Anna University, Chennai, India

V. Lazorenko Faculty of Information Technologies, Kyiv National University of Trade and Economics, Kyiv, Ukraine

K. Leena Department of Electronics and Communication Engineering, East West Institute of Technology, Bangalore, Karnataka, India

Monika Maan Department of Pharmacy, Banasthali University, Vanasthali, Rajasthan, India

Yamini Madan Amrita School of Engineering, Center for Computational Engineering and Networking (CEN), Amrita Vishwa Vidyapeetham, Coimbatore, India

R. Madhumathi Department of Computer Science and Engineering, Sri Ramakrishna Engineering College, Coimbatore, India

S. Mahalakshmi Department of ISE, BMS Institute of Technology and Management, Bengaluru, India

Sharder Shams Mahamud Computer Science and Engineering Discipline, Khulna University, Khulna, Bangladesh

Vaibhav Singh Makhloga Akhilesh Das Gupta Institute of Technology and Management, Delhi, India

Amarjit Malhotra Department of Information Technology, Netaji Subhas University of Technology, Delhi, India

Mudit Mangal Department of Computer Engineering & Application, GLA University, Mathura, India

T. Manivannan Department of Computer Applications, Alagappa University, Karaikudi, India

B. Mithra Department of Biomedical Engineering, Sri Ramakrishna Engineering College, Coimbatore, India

Sazib Mittro Department of Electrical and Electronic Engineering, University of Asia Pacific, Dhaka, Bangladesh

V. Mounika Department of Computer Science and Engineering, Koneru Lakshmaiah Education Foundation, Vaddeswaram, AP, India

N. Naga Lakshmi Department of Information Technology, Anurag Group of Institutions, Hyderabad, India

V. Nagaraj Department of ECE, Knowledge Institute of Technology, Salem, Tamilnadu, India

A. Nagarajan Department of Computer Applications, Alagappa University, Karaikudi, India

S. Nagarajan Department of Computer Science and IT, Amrita School of Arts and Sciences, Amrita Vishwa Vidyapeetham, Kochi, India

Sabout Nagaraju Department of Computer Science, PUCC, Pondicherry University, Lawspet, India

A. Nageswara Rao CSE Department, CMR Institute of Technology, Medchal, Hyderabad, India

Abdullah-Al Nahid Khulna University, Khulna, Bangladesh

N. M. Nandhitha School of Electrical and Electronics, Sathyabama Institute of Science and Technology, Chennai, India

Ronaq Nazir Department of Electrical Engineering, University of Kashmir, Srinagar, India

Gorsa Lakshmi Niharika Department of Computer Science and Engineering, Koneru Lakshmaiah Education Foundation, Vaddeswaram, Andhra Pradesh, India

S. Nithya Tanvi Nishitha Department of CSE, Koneru Lakshmaiah Education Foundation, Vaddeswaram, India

S. Niveda Department of Electronics and Communication Engineering, Sri Ramakrishna Engineering College, Coimbatore, India

Jayashree Padmanabhan Department of Computer Technology, MIT Campus, Anna University, Chennai, India

Krishna Pai Department of Electrical and Electronics Engineering, KLE Dr. M. S. Sheshgiri College of Engineering & Technology, Belagavi, India

Pallavi Department of ISE, BMS Institute of Technology and Management, Bengaluru, India

P. C. Panchariya Academy of Scientific and Innovative Research (AcSIR), Ghaziabad, UP, India;
CSIR-Central Electronics Engineering Research Institute, Pilani, Rajasthan, India

P. B. Pankajavalli Department of Computer Science, Bharathiar University, Coimbatore, Tamil Nadu, India

Amrita Verma Pargaien Graphic Era Hill University, Bhimtal, India

Saurabh Pargaien Graphic Era Hill University, Bhimtal, India

R. Parvathi Vellore Institute of Technology, Chennai, TamilNadu, India

Surendra Singh Patel Academy of Scientific and Innovative Research (AcSIR), Ghaziabad, UP, India;
CSIR-Central Electronics Engineering Research Institute, Pilani, Rajasthan, India

Debadatta Pati Department of Electronics & Communication Engineering, National Institute of Technology, Nagaland, India

Pravin S. Patil Department of Electronics and Telecommunication Engineering, SSVPSBS Deore College of Engineering, Dhule, India

Sarang D. Patil Department of Electronics and Telecommunication Engineering, Gangamai College of Engineering, Nagaon, India

Chandra Shaker Pittala Department of Electronics and Communication Engineering, MLR Institute of Technology, Hyderabad, India

E. Prabhu Department of Electronics and Communication Engineering, Amrita School of Engineering, Coimbatore, Amrita Vishwa Vidyapeetham, Coimbatore, India

M. Sandeep Prabhu Department of Electronics and Communication Engineering, Canara Engineering College, Benjanapadavu, Mangaluru, Karnataka, India

Saranya Prabu Department of Computer Technology, MIT Campus, Anna University, Chennai, India

Yerramreddy Lakshmi Pranathi Queen's University Belfast, Belfast, UK

S. Pratap Singh Marri Laxman Reddy Institute of Technology and Management, Dundigal, Hyderabad, India

C. Swetha Priya Lawspet, Pondicherry, India

N. Radha Vellore Institute of Technology, Chennai, TamilNadu, India

T. Raghavendra Gupta CSE Department, Hyderabad Institute of Technology and Management, Medchal, Hyderabad, India

N. Raghavendra Sai Department of Computer Science and Engineering, Koneru Lakshmaiah Education Foundation, Vaddeswaram, AP, India

Kartikay Raheja Akhilesh Das Gupta Institute of Technology and Management, Delhi, India

Kumar Rahul Department of Basic and Applied Sciences, NIFTEM, Sonipat, India

S. Pravinth Raja Presidency University, Bangalore, India

M. Rajasekar Department of Computer Applications, Hindustan Institute of Technology and Science, Padur, Chennai, India

G. M. Rajathi Department of ECE, Sri Ramakrishna Engineering College, Coimbatore, India

M. Rakshana Department of Biomedical Engineering, Sri Ramakrishna Engineering College, Coimbatore, India

P. Ramachandran Department of Computer Science, J.J College of Arts & Science (Autonomous), Pudukkottai, India

B. M. S. Rani Department of Electronics and Communication Engineering, Vignan's Nirula Institute of Technology and Science for Women, Guntur, AP, India

Yamini Ratawal Akhilesh Das Gupta Institute of Technology and Management, Delhi, India

Sathish Kumar Ravichandran Department of Computer Science and Engineering, Christ University, Bangalore, India

Md. Rezaul Karim Sohel Department of Electrical and Electronic Engineering, University of Asia Pacific, Dhaka, Bangladesh

Ritika St. Andrews Institute of Technology & Management, Gurgaon, India

Raul V. Rodriguez Woxsen School of Business, Woxsen University, Hyderabad, India

S. Emalda Roslin School of Electrical and Electronics, Sathyabama Institute of Science and Technology, Chennai, India

Abhishek Roy Department of Computer Science & Engineering, Adamas University, Kolkata, India

V Sai Krishna Academy of Scientific and Innovative Research (AcSIR), Ghaziabad, India;
CSIR-Central Electronics Engineering Research Institute, Pilani, India

M. S. Sangeeetha School of Electrical and Electronics, Sathyabama Institute of Science and Technology, Chennai, India

Harini Sankar Department of ECE, Sri Ramakrishna Engineering College, Coimbatore, India

C. Santhosh Kumar Machine Intelligence Research Laboratory, Department of Electronics and Communication Engineering, Amrita School of Engineering, Coimbatore, Amrita Vishwa Vidyapeetham, India

K. Santhosh Kumar Department of Computer Technology, MIT Campus, Anna University, Chennai, India

Hansika Sati Department of Agriculture and Environmental Sciences, NIFTEM, Sonipat, India

S. Savitha Department of Computer Science and Engineering, Christ University, Bangalore, India

Md. Sazal Miah School of Engineering and Technology, Asian Institute of Technology, Pathumthani, Thailand

Gopalakrishnan Seetharaman Indian Institute of Information Technology, Tiruchirappalli, India

Harkesh Sehrawat University Institute of Engineering and Technology, MD University, Rohtak, India

G. Thamarai Selvi Department of ECE, Sri Sai Ram Institute of Technology, Chennai, Tamilnadu, India

G. Shanmugarathinam Presidency University, Bangalore, India

Suman Sharma Department of Pharmacy, Banasthali University, Vanasthali, Rajasthan, India

Laxmi shaw Department of ECE, CBIT, Hyderabad, India

R. Shenbagalakshmi Department of Electrical Engineering, Sinhgad Institute of Technology, Lonavala, Maharashtra, India

N. Shenbagavadivu Department of Computer Applications, University College of Engineering, Anna University, Bharathidasan Institute of Technology Campus, Trichirapalli, India

Akshaya D. Shetty Department of Electronics and Communication Engineering, NMAM Institute of Technology, Nitte, Udupi, Karnataka, India

R. Shruthi Department of Computer Science and Engineering, Sri Ramakrishna Engineering College, Coimbatore, India

M. Sinthuja M.S. Ramaiah Institute of Technolgy, Bangalore, India

N. Siva Priya Department of Computer Applications, University College of Engineering, Anna University, Bharathidasan Institute of Technology Campus, Trichirapalli, India

A. Siva Sakthi Department of Biomedical Engineering, Sri Ramakrishna Engineering College, Coimbatore, India

S. A. Sivakumar Department of ECE, Ashoka Women's Engineering College, Kurnool, Andhra Pradesh, India

Vikas Siwach University Institute of Engineering and Technology, MD University, Rohtak, India

Rumaisa Sofi Department of Electrical Engineering, University of Kashmir, Srinagar, India

K. P. Soman Amrita School of Engineering, Center for Computational Engineering and Networking (CEN), Amrita Vishwa Vidyapeetham, Coimbatore, India

V. Sowmya Amrita School of Engineering, Center for Computational Engineering and Networking (CEN), Amrita Vishwa Vidyapeetham, Coimbatore, India

Jujavarapu Sravana Department of Electronics and Communication Engineering, Institute of Aeronautical Engineering, Hyderabad, India

T. Sree Renga Raja Department of EEE, University College of Engineering, Nagercoil, India

B. Sreedevi Department of CSE/SRC, SASTRA Deemed University, Kumbakonam, India

K. T. Sreekumar Machine Intelligence Research Laboratory, Department of Electronics and Communication Engineering, Amrita School of Engineering, Coimbatore, Amrita Vishwa Vidyapeetham, India

Swati Srivastava Department of Computer Engineering & Application, GLA University, Mathura, India

Patel Surendra Singh Academy of Scientific and Innovative Research (AcSIR), Ghaziabad, India;
CSIR-Central Electronics Engineering Research Institute, Pilani, India

Kavita Tewani Computer Science and Engineering Department, Institute of Technology, Nirma University, Ahmedabad, Gujarat, India

T. Thamaraikannan Department of ECE, Sri Ramakrishna Engineering College, Coimbatore, India

Rajat Toshniwal Woxsen School of Business, Woxsen University, Hyderabad, India

A. Udayavannan Department of Computer Technology, MIT Campus, Anna University, Chennai, India

Abdul Hasib Uddin Computer Science and Engineering Discipline, Khulna University, Khulna, Bangladesh

Sai Krishna Vaddadi Academy of Scientific and Innovative Research (AcSIR), Ghaziabad, UP, India;
CSIR-Central Electronics Engineering Research Institute, Pilani, Rajasthan, India

S. Vagdevi Department of Information Science and Engineering, Dayananda Sagar Academy of Technology & Management, Bangalore, India

Rajeev Ratna Vallabhuni Bayview Asset Management, LLC, Coral Gables, FL, USA

Kartik Vashisth Department of Information Technology, Netaji Subhas University of Technology, Delhi, India

N. Veena Department of ISE, BMS Institute of Technology and Management, Bengaluru, India

Iswarya Kannoth Veetil Amrita School of Engineering, Center for Computational Engineering and Networking (CEN), Amrita Vishwa Vidyapeetham, Coimbatore, India

Vallabhuni Vijay Department of Electronics and Communication Engineering, Institute of Aeronautical Engineering, Hyderabad, India

K. Vijayarekha School of EEE, SASTRA Deemed University, Tirumalaisamudram, India

E. R. Vimina Department of Computer Science and IT, Amrita School of Arts and Science, Amrita Vishwa Vidyapeetham, Kochi Campus, Kochi, India

S. S. Vishnu Priyan Department of Electronics and Communication Engineering, Amrita School of Engineering, Coimbatore, Amrita Vishwa Vidyapeetham, Coimbatore, India

V. Yaremych Faculty of Information Technologies, Kyiv National University of Trade and Economics, Kyiv, Ukraine

S. Zahid Nabi Dar Department of Electrical & Electronics Engineering, CMR Institute of Technology, Bengaluru, India

Deep Learning-Based Approach for Parkinson's Disease Detection Using Region of Interest

Yamini Madan, Iswarya Kannoth Veetil, V. Sowmya, E. A. Gopalakrishnan, and K. P. Soman

Abstract Deep Learning plays a major role in advancements in the healthcare domain, of which early disease diagnosis is one main application. With respect to the same, deep learning-based classification of brain MRI at the subject level is the requirement in the medical field. In this work, we have implemented an algorithm to identify the most discriminative range of MRI slices at the subject level to differentiate between Normal Cohorts (NC) and Parkinson's Disease (PD) subjects. We have also focused on handling data leakage and verified the model generalizability using Stratified k-fold cross-validation.

Keywords Parkinson's disease · Subject-level classification · T2 MRI · Data leakage · ROI

1 Introduction

Parkinson's Disease (PD) mostly affects people around the age of 65 years and above [11]. It is the second most common neurodegenerative disease, and is strongly associated with increased mortality rate in the recent years [10]. PD occurs due to the loss of dopamine-producing neurons in the substantia nigra region present in the midbrain which occurs due to the accumulation of iron in the nigral region [3, 12]. The disease is characterized by visible clinical symptoms such as restricted movement, stiffness and rest tremors in limbs which occur after the disease has progressed to the advanced stages [15]. Deep learning (DL) can help in identifying biomarkers for early disease detection. Morphological changes in the brain can be captured by several neuroimaging techniques like Magnetic Resonance Imaging (MRI) scan [2], Positron Emission Tomography (PET), Computed Tomography (CT) and Single-Photon Emission Computed Tomography (SPECT). However, MRI is the most preferred as it does not involve harmful radiations and is also effective in detecting

Y. Madan · I. K. Veetil · V. Sowmya (✉) · E. A. Gopalakrishnan · K. P. Soman
Amrita School of Engineering, Center for Computational Engineering and Networking (CEN), Amrita Vishwa Vidyapeetham, Coimbatore, India

J. S. Raj et al. (eds.), *Intelligent Sustainable Systems*, Lecture Notes in Networks and Systems 213, https://doi.org/10.1007/978-981-16-2422-3_1

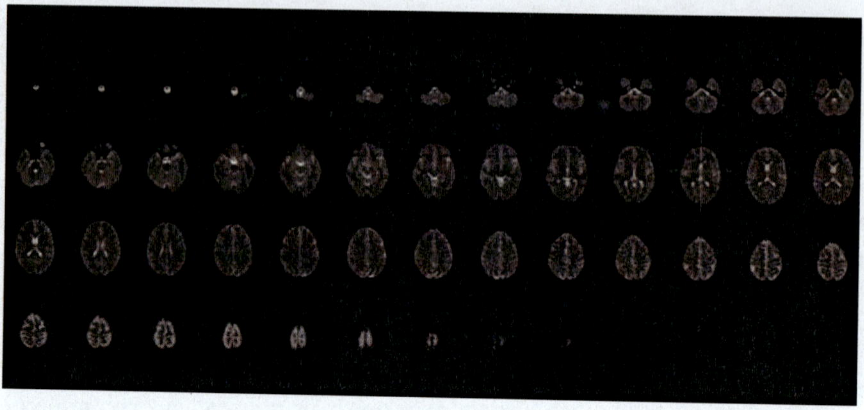

Fig. 1 T2 MRI of a subject consisting of 48 slices

PD [16]. As shown in Fig. 1, an MRI of every subject comprises multiple slices. However, there are some setbacks to the effective performance of the models [19] with high performance in the testing stages, yet lacking efficiency when deployed on real-time data or test data. This is called overfitting that could occur due to data leakage.

1.1 Subject-Level Classification

Classifying imaging results at the subject level is the requirement for disease diagnosis. The AI algorithm in use can classify a subject as affected by disease or undiseased (Normal Cohort). This will also help in easy interpretation of the results and does not require much technical expertise.

1.2 Region of Interest

In DL, the model's performance improves based on the availability of a large dataset for training. However, the quality of images and features to be learnt is the basic necessity for a good model performance. Since there are multiple slices in an MRI, certain slices could be less informative than the others. Extracting and identifying the slices which correspond to the midbrain is of importance and the Region of Interest in this study.

1.3 Model Generalizability

When we test the model performance on 1 split of test data, it gives an understanding of the model performance on that particular test split. This is influenced by factors such as classes of image in the split, quality of images, etc. On using cross-validation, the model is validated on k-folds of the data [1, 19] which gives a broader insight into the model's performance over unseen data as well.

1.4 Data Leakage

Data Leakage is a common problem in machine learning which occurs when a part of the training data is made accessible to test data or vice versa which will lead the model to 'memorize' features. In terms of detecting a disease using MRI image slices, when slices of a patient are present in both training and testing sets, data is said to be 'leaked'. Data Leakage can occur when data is split into train and test sets at the slice level, or when data augmentation is done before splitting the data. This could result in images of the same patient being found in both training and test sets.

2 Literature Survey

DL is widely used in image and signal processing. Jaswal et al. [7] have compared performance of Convolutional Neural Networks on different datasets and evaluated the performance using standard statistical metrics. There are several applications in the biomedical domain as well. In [5], the performance of DL architectures like RNN, GRU, CNN and LSTM in detecting three diseases—namely, Myocardial Infarction, Atrial defibrillation and arrhythmia are described. In [13], the DL architectures are compared to identify the optimal model for bio-metric recognition where a person is identified based on their ear morphology. Harini et al. [6], different CNN DL architectures are used in the detection of abnormalities from musculoskeletal radiograph images. Vijaya Kumar [18] uses CapsNet to identify the presence of brain tumour for the classification of brain cancer.

Sivaranjini and Sujatha [16] uses AlexNet to classify the presence of PD on T2 MRI images from the PPMI dataset at the slice level. However, whether model is evaluated at the slice level or subject level is ambiguous. In [14], an extended work of [16], ensemble learning model of three architectures—VGG16, VGG19 and DenseNet201 is used to detect PD using T2 MRI. Data augmentation has been used to handle class imbalance. However, the augmentation process is done before splitting the data into train and test sets that will result in data leakage. In [1], the model is evaluated at the subject level for T1 images using a 3D CNN to identify the presence of PD, and model generalizability is verified using k-fold cross-validation.

Classification is done at the subject level on T1 MRI images. There has been no initiative in evaluating DL architectures at the subject level on T2 MRI images with the view of verifying the model's generalizability to the best of our knowledge. Hence, in this work, our objectives are as follows:

(i) Identify the range of slices (ROI) that aid in the detection of PD
(ii) Classification of predicted results at the subject level
(iii) Model generalizability using Stratified k-fold cross-validation
(iv) Depict the effect of data leakage on model performance.

3 Dataset Description

Parkinson's Progressive Marker's Initiative (PPMI) is a well-known public repository that has imaging, biological samples and clinical and behavioural assessment data to aid in the identification of biomarkers of PD progression [8]. In this work, we have used 100 T2 weighted MRI images acquired with a field strength of 3T using the Siemens TrioTim scanner along the axial plane. The 100 subjects considered for the study belong to the age group of 62 ± 7.23 years, of which 50 belong to baseline order PD and 50 are Normal Cohorts (NCs). Each subject has 48 slices (see Fig. 1). The raw images are obtained in NifTi format.

4 Methodology

Figure 2 represents the block diagram of the methodology followed in the proposed work.

4.1 Data Preprocessing

There are three stages involved in preprocessing the T2 MRI images in the proposed method. Figure 3 shows a sample raw MRI image and the resultant image after each preprocessing stage.

(1) Bias Field Correction: Intensity values could be concentrated (or biased) in certain regions of the image. Using the N4ITK module of 3D Slicer software [4], the intensity distribution is normalized uniformly throughout the image, thereby correcting the bias field.

(2) Skull Stripping: The MRI image consists of both cranium and brain tissue. Since the model doesn't need features of the cranium in the identification of PD, the region corresponding to the skull is removed from the image using the FSL BET [9].

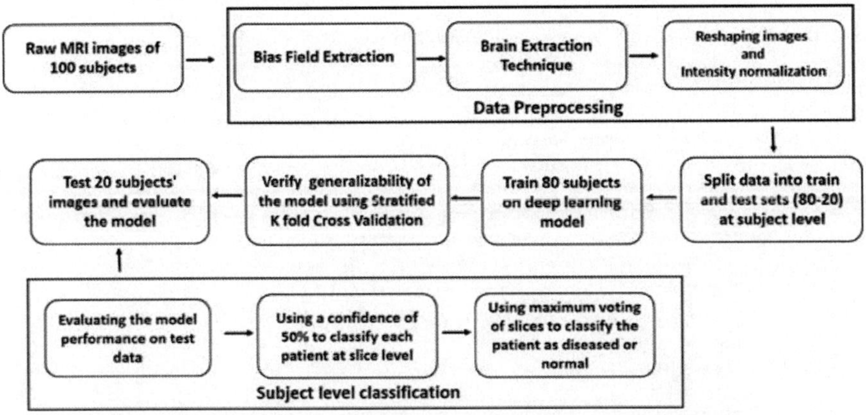

Fig. 2 Block diagram representing the flow of the methodology

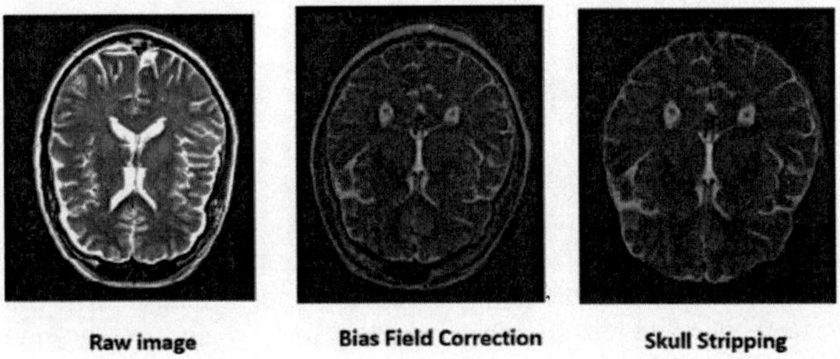

Fig. 3 Preprocessing of T2 MRI

(3) Intensity Normalization and Reshaping: The skull stripped images are then reshaped into RGB images by of the shape 224*224*3 by stacking each slice thrice. This is done for the input shape be compatible with the pretrained DL architectures available in the Keras Application Module.

4.2 Architecture

Data is split into training and test sets at the subject level with an 80:20 ratio resulting in 80 subjects in the training set and 20 subjects in the test set. The number of PD and NC subjects are balanced throughout. Figure 4 represents the split of the original data at the subject level into train and test sets. Further to determine the model generalizability stratified 5-fold cross-validation is performed.

Transfer learning of DL architectures—VGG19, DenseNet121 and Densenet201 are experimented with and their performance is recorded to find the best architecture

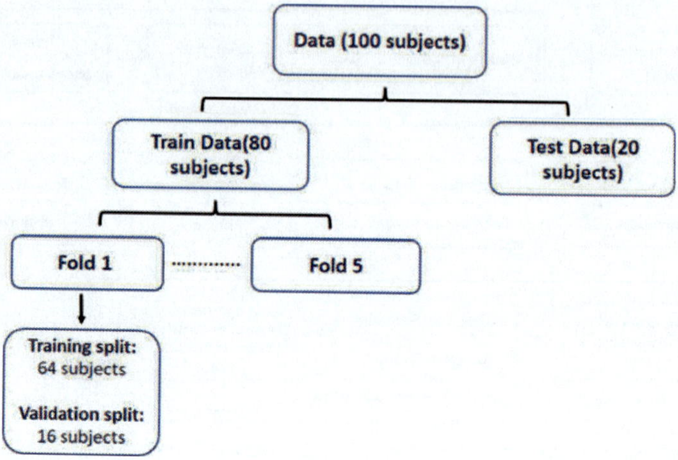

Fig. 4 Data split and Stratified 5-fold cross-validation

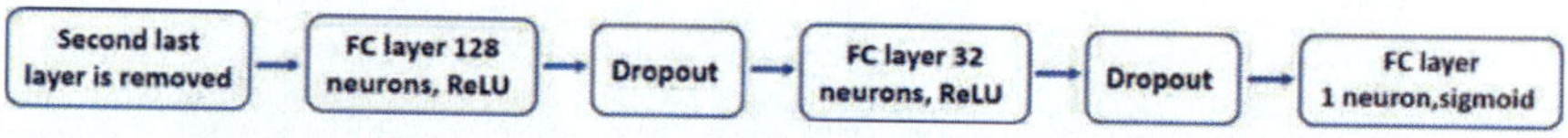

Fig. 5 Modifications applied to the three DL architectures

for the chosen dataset. Weights pretrained on Imagenet dataset will improve the model's performance and lead to a faster convergence when compared to training a model to learn the weights from scratch. The following architectural changes have been applied commonly to all the three architectures under consideration. The second last layer of the Keras model is removed and dense layers with Dropout are added. The final dense layer has one neuron with the sigmoid activation function to classify the presence of disease in an image. Dropout layer is added to reduce overfitting and is a regularization parameter [17]. Details regarding to architecture is shown in Fig. 5.

All layers of DenseNet201 and DenseNet121 are trained, while for VGG-19, only the last three layers of the architecture are trained to get a comparable performance. Over the best-performing DL architecture, we have experimented on the range of minimum number of slices that contains the necessary features for the model to learn to identify PD.

The number of hidden layers and neurons in the hidden layer, value of dropout, learning rate and number of epochs the model is trained for are the hyperparameters of the model that cannot be learned by the model and have to be specified before training. The tuned hyperparameter values for each of the model is specified in Table 1. Different optimizers like Stochastic Gradient, Adam, AdaBoost and Adagrad have been tried, and Adam was identified as performing the best in backpropagating the loss during training. Binary Cross-Entropy loss function is used as we are dealing with a binary classification problem. Addition of a dropout layer enables in training

Table 1 Data description

Modality	Magnetic Resonance Imaging (MRI)
Research group	Normal Cohort and Parkinson's Disease
Visit	Baseline
Acquisition plane	Axial
Acquisition type	2D
Field strength	3T
Slice thickness	3mm
Scanner manufacturing	Siemens Trio Tim scanner
Pixel spacing	X=0.9, Y=0.9
Weighting	T2

the model for more epochs by which the right features are learnt. This prevents the model from overfitting. Also, a smaller learning rate is required for the model to be able to identify the difference in brain features of an image having PD and NC. Larger learning rates tend to make the model to be biased towards one particular class. 1e-4 is the learning rate for DenseNet201 and 1e-5 is the learning rate for the other two architectures.

4.3 Evaluation Metrics

Sensitivity, Specificity, F1 score, Accuracy and confusion matrix are the evaluation metrics that have been used in this study. Sensitivity (or recall) measures the efficiency of detection of diseased patients and Sensitivity measures the efficiency in identifying the NCs. It is important for a model to minimize the number of misclassifications and, Sensitivity and Specificity help in monitoring the same. F1 score which is the harmonic mean of precision and recall, is an important metric in biomedical analysis as it measures the performance of the model in detecting diseased subjects precisely.

$$Sensitivity = \frac{TP}{TP + FN} \tag{1}$$

$$Specificity = \frac{TN}{TN + FP} \tag{2}$$

$$F1score = \frac{2 * precision * recall}{precision + recall} \tag{3}$$

$$Accuracy = \frac{TP + TN}{TP + TN + FP + FN} \tag{4}$$

5 Experiments and Results

5.1 Subject-Level Classification

The predicted result for a slice using the sigmoid activation from the final layer of the DL architecture is a probability value ranging between 0 and 1. With a confidence value of 50%, the predicted probability for every slice is classified into either of the classes (0 or 1) which is represented as a vector 'pred'. To generate a subject-level decision on the patient, maximum voting is applied which can be denoted as

$$y = argmax(pred[i]) \tag{5}$$

where the index i is the number of slices per subject and y is the resultant scalar value which is either 0 or 1 indicating the presence of disease in a subject. For the test data, weights trained on the training data are used on the same slices as in training slice set.

5.2 Region of Interest

This works aims to identify the range of slices of interest which could possibly contain features that differentiate a normal brain from a brain of a PD subject. From Fig. 1, it is clear that the first few slices of the MRI contain mostly black pixels and lack any information pertaining to the brain tissue. Slices towards the end do not have any information about the midbrain which has the features indicating PD. Hence, we hypothesize that slices in the middle range are of importance in this study.

On calculating the average pixel intensity of every slice of the 80 training subjects as shown in Fig. 6, it is noted that slice range between 15 and 42 has pixels of intensity above the mean average pixel intensity value represented by the red line, which is considered as a threshold. The slice range which satisfies this threshold is expected to be the Region of Interest. To validate this we have compared the performance of DenseNet201 on all the 48 slices and with just the selected ROI slices. On experimentation, it is observed that the slice range of 20–40 gave better results (see Table 2). Therefore instead of using all 48 slices, only 20 slices of each subject is used. This leads to better detection of PD as a lot of unwanted information gets removed. The model trained on 48 slices has high Specificity as it learns the features of an NC better and has low Specificity since 5 PD subjects were misclassified as a NC.

On the other hand, using 20 slices improves the performance. There is notable increase in the values of F1 average score and a reduced variation in the standard deviation. The Sensitivity and Specificity values are balanced, meaning the model is able to differentiate between NC and PD (Table 3 and Fig. 7).

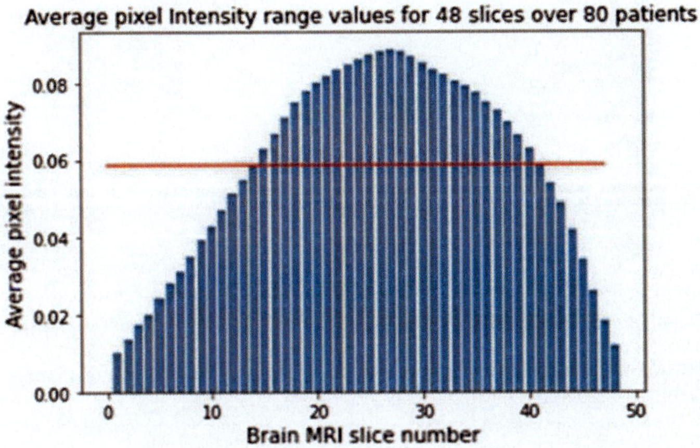

Fig. 6 Average Pixel Intensity for 48 slices of 80 subjects

Table 2 Hyperparameter tuning for the three DL architectures

Architecture	Learning rate	Dropout	Number of epochs
VGG19	1e-5	0.5	10
DenseNet121	1e-5	0.4	8
DenseNet201	1e-4	0.2	5

Table 3 Comparing the performance of DenseNet201 on tuning the number of slices

Number of slices per patient	Model performance over cross-validation				Test results	
	Specificity	Sensitivity	F1 avg	F1 std dev	Accuracy (%)	F1 score
48	**0.89**	0.66	0.708	0.02	55	0.69
20	0.7	**0.7**	**0.83**	**0.01**	**70**	**0.7**

5.3 *Model Generalizability*

To determine the generalizability of the model performance over unseen data, 5-fold stratified cross-validation is used resulting in each fold having 20% of the data (16 subjects) in validation set and remaining data for training (64 subjects). By using Stratified validation, the number of NC and PD patients in every fold is equally distributed (see Fig. 8) and hence gives us a good idea of the performance of the model on detecting both classes. The five-fold split is implemented with a fixed seed number so that the performance can be compared for the three different architectures. The metric used to evaluate the performance over each fold is F1 score. Hyperparameters

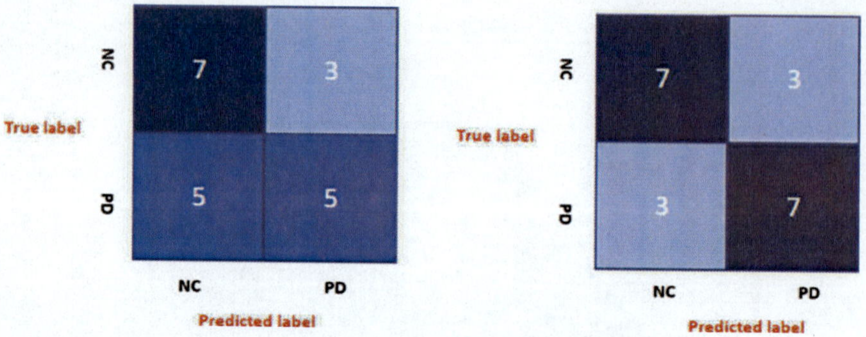

Fig. 7 Confusion matrix representing classification of test results using model trained on: **a** 48 slices of 80 patients **b** 20 slices of 80 patients

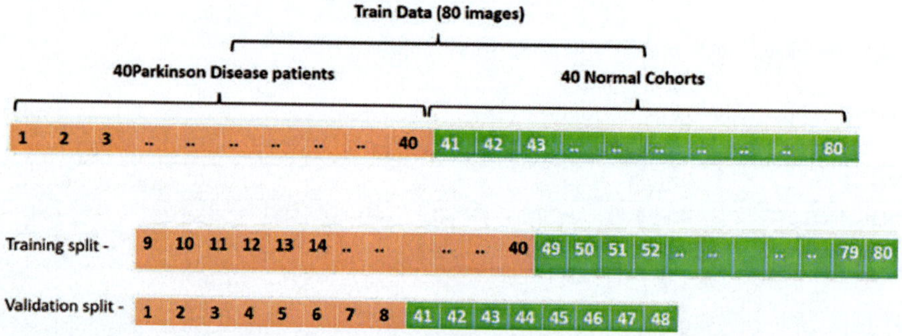

Fig. 8 Sample of Stratified five-fold cross-validation shown for one-fold

are tuned in a way so as to maximize the F1 average value across the folds and minimize the F1 standard deviation. By doing this the generalizability of the model will be achieved.

To identify the suitable architecture for the given dataset, 20 slices of 80 subjects have been trained on 3 DL architectures and their performances compared.Over the 5-folds during cross-validation, we can observe that Specificity of VGG-19 is higher than its Sensitivity which means that it identifies NCs better than PD patients. DenseNet201 maintains optimal trade-off between identifying diseased and non-diseased images, i.e. Sensitivity and Specificity are in a similar range. And the F1 average score of DenseNet201 is high and standard deviation is the least amongst the architectures. The metric values of cross-validation during training are specified in Table 4. DenseNet201 is the architecture that generalizes data the best amongst the three architectures while DenseNet121 performs the least.

While evaluating the models on the test data—20 slices of 20 subjects were tested using the weights learned by the model during training, it is observed that all the three models perform as well as the corresponding architecture's cross-validation results during training.

Table 4 Comparing model generalizability for training data on ROI slices

Architecture	Model performance over 5-fold CV			
	Specificity	Sensitivity	F1 avg	F1 std
VGG-19	**0.88**	0.74	0.84	0.06
DenseNet121	0.68	0.73	0.76	0.08
DenseNet201	0.81	**0.88**	**0.85**	**0.01**

Table 5 Comparing model performance over test data

Architecture	Model performance over test data			
	Specificity	Sensitivity	Accuracy (%)	F1 score
VGG-19	0.80	0.60	70	0.73
DenseNet121	0.60	0.90	70	0.75
DenseNet201	0.70	0.70	70	0.70

Table 6 Performance over test data

	F1 score	Accuracy (%)
With data leakage	**0.92**	**91.7**
Without data leakage (slice level)	0.87	84.1

5.4 *Effect of Data Leakage*

To depict the effect of data leakage on model performance, we have simulated a data split strategy at the slice level and subject level and compared the performance results of both the experiments on test set and external test data.

As seen from Tables 5 to 6, the model which is trained on data split into train and test sets at the slice level performs quite well when evaluated over test data, but performance drastically reduces when the model is deployed over new test data. On the other hand, although the model that is split at the subject level performs comparatively lower than the model with data leakage over test data, the performance of the model is observed to maintain a similar performance over test data as well as new data.

6 Conclusion

Initially, the data has been split into train and test set at the subject level which avoids data leakage. The slices that are most essential in identifying the presence of PD is our Region of Interest, which are found to be in the slice range of 20–40.

DenseNet201 is concluded to be the model that generalizes best amongst the three architectures. Training a model on 20 slices instead of 48 slices shows improved performance in detecting PD. The test results are classified at the patient level so that the easily interpretable model can be of assistance to the medical personnel. Also, the effect of data leakage on model performance has been simulated and compared with the performance of a model without data leakage.

Future scope of the current work is to use the best-performing architecture as a classifier in comparison of simple data augmentation techniques with adversarial techniques on NC images. Future direction involves performing hyperparameter tuning of the architecture using optimization techniques like Bayesian Hyperparameter Optimization.

References

1. Chakraborty, S., Aich, S., Kim, H.C.: Detection of Parkinson's disease from 3t t1 weighted MRI scans using 3d convolutional neural network. Diagnostics **10**(6), 402 (2020)
2. Chandy, A.: A review on IoT based medical imaging technology for healthcare applications. J. Innov. Image Process. **1**(01), 51–60 (2019). https://doi.org/10.36548/jiip.2019.1.006
3. Chen, Q., Chen, Y., Zhang, Y., Wang, F., Yu, H., Zhang, C., Jiang, Z., Luo, W.: Iron deposition in Parkinson's disease by quantitative susceptibility mapping. BMC Neurosci. **20**(1) (2019). https://doi.org/10.1186/s12868-019-0505-9
4. Fedorov, A., Beichel, R., Kalpathy-Cramer, J., Finet, J., Fillion-Robin, J.C., Pujol, S., Bauer, C., Jennings, D., Fennessy, F., Sonka, M., Buatti, J., Aylward, S., Miller, J.V., Pieper, S., Kikinis, R.: 3d slicer as an image computing platform for the quantitative imaging network. Magn. Reson. Imaging **30**(9), 1323–1341 (2012). https://doi.org/10.1016/j.mri.2012.05.001
5. Gopika, P., Sowmya, V., Gopalakrishnan, E., Soman, K.: Transferable approach for cardiac disease classification using deep learning. In: Deep Learning Techniques for Biomedical and Health Informatics, pp. 285–303. Elsevier (2020). https://doi.org/10.1016/b978-0-12-819061-6.00012-4
6. Harini, N., Ramji, B., Sriram, S., Sowmya, V., Soman, K.: Chapter five - musculoskeletal radiographs classification using deep learning. In: Das, H., Pradhan, C., Dey, N. (eds.) Deep Learning for Data Analytics, pp. 79–98. Academic (2020). https://doi.org/10.1016/B978-0-12-819764-6.00006-5
7. Jaswal, D., Vishvanathan, S., Kp, S.: Image classification using convolutional neural networks. Int. J. Sci. Eng. Res. **5**(6), 1661–1668 (2014)
8. Marek, K., Jennings, D., Lasch, S., Siderowf, A., Tanner, C., Simuni, T., Coffey, C., et al.: The Parkinson progression marker initiative (PPMI). Prog. Neurobiol. **95**(4), 629–635 (2011). https://doi.org/10.1016/j.pneurobio.2011.09.005
9. McCarthy, P.: Fsleyes (2020). https://doi.org/10.5281/ZENODO.3937147
10. Pinter, B., Diem-Zangerl, A., Wenning, G.K., Scherfler, C., Oberaigner, W., Seppi, K., Poewe, W.: Mortality in Parkinson's disease: a 38-year follow-up study. Mov. Disord. **30**(2), 266–269 (2014). https://doi.org/10.1002/mds.26060
11. Poewe, W., Seppi, K., Tanner, C.M., Halliday, G.M., Brundin, P., Volkmann, J., Schrag, A.-E., Lang, A.E.: Parkinson disease. Nat. Rev. Dis. Prim. (2017). https://doi.org/10.1038/nrdp.2017.13
12. Pyatigorskaya, N., Gallea, C., Garcia-Lorenzo, D., Vidailhet, M., Lehericy, S.: A review of the use of magnetic resonance imaging in Parkinson's disease. Ther. Adv. Neurol. Dis. **7**(4), 206–220 (2013). https://doi.org/10.1177/1756285613511507

13. Radhika, K., Devika, K., Aswathi, T., Sreevidya, P., Sowmya, V., Soman, K.P.: Performance analysis of NASNet on unconstrained ear recognition. In: Nature Inspired Computing for Data Science, pp. 57–82. Springer International Publishing (2019). https://doi.org/10.1007/978-3-030-33820-6-3
14. Rajanbabu, K., Veetil, I.K., S.V.G.E.A.S.K.P.: Ensemble of deep transfer learning models for Parkinson's disease classification
15. Shinde, S., Prasad, S., Saboo, Y., Kaushick, R., Saini, J., Pal, P.K., Ingalhalikar, M.: Predictive markers for Parkinson's disease using deep neural nets on neuromelanin sensitive MRI. NeuroImage: Clin. **22**, 101748 (2019). https://doi.org/10.1016/j.nicl.2019.101748
16. Sivaranjini, S., Sujatha, C.M.: Deep learning based diagnosis of Parkinson's disease using convolutional neural network. Multimed. Tools Appl. **79**(21–22), 15467–15479 (2019). https://doi.org/10.1007/s11042-019-7469-8
17. Srivastava, N., Hinton, G., Krizhevsky, A., Sutskever, I., Salakhutdinov, R.: Dropout: a simple way to prevent neural networks from overfitting. JMLR.org (2014)
18. Vijaya Kumar, T.: Classification of brain cancer type using machine learning **2019**(2) (2019). https://doi.org/10.36548/jaicn.2019.2.006
19. Wen, J., Thibeau-Sutre, E., Diaz-Melo, M., Samper-González, J., Routier, A., Bottani, S., Dormont, D., Durrleman, S., Burgos, N., Colliot, O.: Convolutional neural networks for classification of Alzheimer's disease: overview and reproducible evaluation. Med. Image Anal. **63**, 101694 (2020). https://doi.org/10.1016/j.media.2020.101694

Classification of Class-Imbalanced Diabetic Retinopathy Images Using the Synthetic Data Creation by Generative Models

Krishanth Kumar, V. Sowmya, E. A. Gopalakrishnan, and K. P. Soman

Abstract Diabetic retinopathy (DR) is a complication of diabetes which is due to the impairment of blood vessels of photosensitive cells in the eyes. This complication results in loss of eyesight if not diagnosed in the early stages. There are five stages of diabetic retinopathy: No DR, mild, moderate, severe, and proliferative. DR detection by traditional methods consumes a lot of time. An automatic and precise model would require an adequate amount of data for training which is not available. Publicly available dataset is highly imbalanced for other classes apart from No DR, especially proliferative and severe classes. In this paper, a model is created in two stages. The first stage involves the generation of synthetic data points using Deep Convolutional Generative Adversarial Network (DCGAN). The synthetic data are for highly imbalanced classes namely severe and proliferative. The second phase involves the augmented data classification using a CNN architecture.

Keywords Diabetic retinopathy · Deep convolutional generative adversarial · Convolutional neural network

1 Introduction

Diabetic retinopathy (DR) is an eye disease affiliated with diabetes. Around the world, most number of blindness cases among working-aged adults is caused by diabetic retinopathy [1]. There are five stages of Diabetic retinopathy: No DR, Mild, Moderate, Severe, and Proliferative. The grading is based on these indicators microaneurysm, blood vessel sausage, and retinal Detachment. In the case of NO DR, there is no sign of these indicators. In the case of mild, there are few microaneurysm as

K. Kumar (✉) · V. Sowmya · E. A. Gopalakrishnan · K. P. Soman
Amrita School of Engineering, Center for Computational Engineering and Networking, Amrita Vishwa Vidyapeetham, Coimbatore, India

V. Sowmya
e-mail: v_sowmya@cb.amrita.edu

E. A. Gopalakrishnan
e-mail: ea_gopalakrishnan@cb.amrita.edu

J. S. Raj et al. (eds.), *Intelligent Sustainable Systems*, Lecture Notes in Networks and Systems 213, https://doi.org/10.1007/978-981-16-2422-3_2

compared to moderate. In the case of severe, the blood vessel sausage comes into play and, for proliferative grading, retinal detachment is taken into account [2, 3]. The grading of DR is a tiring and laborious process. Whereas, with the innovation and enhancement in the technology, a significant difference can be achieved by automating the grading process [4–6].

The eyepacs dataset [7] is highly imbalanced to No DR due to the medical and privacy policies. Statistically, the diagnosis of a disease is very much similar to figure out a defect or malfunction which is an unlikely event that results in a highly imbalanced dataset. Traditional augmentation techniques in CNN have been applied to resolve the class imbalance of a dataset, detailed in [8–11]. But these traditional techniques namely, rotational, flipping and cropping and mirroring have overcome the problem of class imbalance only to a certain extent. The model performance is limited for augmented images as they are not varied [12]. In 2014 Generative Adversarial Network(GAN) was introduced [13]. GAN is predominantly used in the analysis of medical images, for example: Computerized Tomography scans, Magnetic Resonance images, X-Rays [14–16]. Apart from medical image analysis, GAN is also used for Hyperspectral image (HSI) classification [17], segmentation of speed bumps [18], identifying epiphytes in drone images [19]. Kaplan et al. [20], an unlearned DCGAN was used to generate varied fundus images, but the vessel trees of the fundus images could not be captured by the model clearly. Balasubramanian et al. [21], proposed a method for DR grading where in the augmented images for proliferative class were made equal to severe class which has the second least number of images. The images are trained and validated on the test dataset.

In the present work, we created synthetic images for proliferative and severe class such that it's equal to mild class. Based on the validation results and previous literature, the best set of images is selected for testing. Here the aim is to improve the performance of classification, especially for the proliferative and severe classes.

2 Methodology

Synthetic images are generated using DCGAN for severely imbalanced classes such that it is equal to the next classes with the least number of images. Therefore, synthetic images for proliferative class and severe is generated and augmented with real images such that the number of images (synthetic + Real) is equal to mild class. Synthetic images of 1735 are added to proliferative class and 1570 synthetic images are added to severe class. Once the images are augmented they are sent to the classifier for classification. The aforementioned process is shown in Fig. 1.

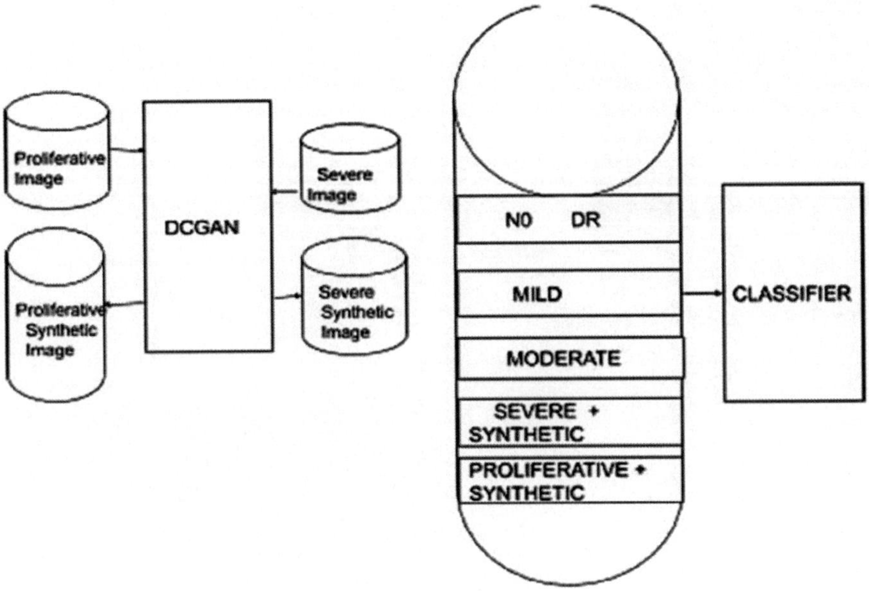

Fig. 1 Block diagram of the proposed methodology

Table 1 Class wise information of train and test. Eyepacs

Class	Train	Test
No DR	25,810	39,553
Mild	2443	3762
Moderate	5292	7861
Severe	873	1214
Proliferative	708	1208

2.1 *Dataset Description*

Kaggle Eyepacs dataset [7] is used for the current project. As we can see, the dataset is highly imbalanced. A total of 35126 images for training and a test dataset 53,576 is used. For experimentation, the train dataset is split into 80–20 and the resultant model is applied on the test dataset. Table 1 and Fig. 2 give the details of the dataset.

2.2 *Retinal Synthetic Image Generation*

There are two phases involved while training a GAN model. One is a generator model and the other is a discriminator model. The generator model generates the

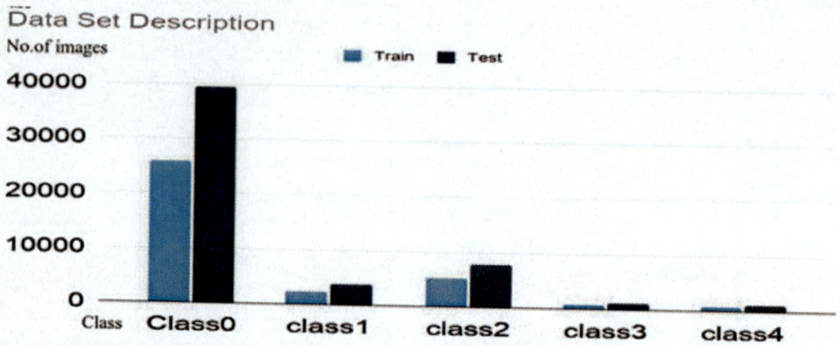

Fig. 2 Bar graph of eyepacs dataset

Fig. 3 Synthetic image of class 3 and class 4

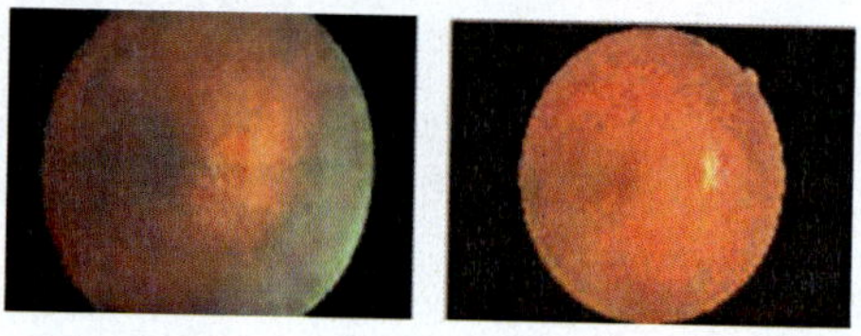

images from noise latent points. Discriminator(D) is set to forward-propagate with the task of classifying real and fake images. This is achieved by calculating their probabilities using either softmax or sigmoid function, depending upon the number of target classes. Also, the ideal probability of synthetic data is set as 0 while real as 1. The images from the generator are sent to the discriminator, but, now the generator is trained to create real images therefore the ideal probability is set to 1, accumulated loss is calculated and backpropagated to the generator (Fig. 3). The generator generates syntheic images as shown in Fig. 3.

The architecture of DCGAN [22] is mentioned in Fig. 4. In the generated model 4 transpose convolution layer and a convolution layer are used to generate an image of size $128 \times 128 \times 3$. The discriminator model has 5 convolutional layers and the activation function used is leaky ReLU. The discriminator is used to classify the image as Real or fake.

2.3 *CNN Classifier*

Shown in Fig. 5 is the classifier model [10], which gave an exemplar kappa score for the eyepacs dataset where in the loss function is categorical cross-entropy. The classifier is trained with augmented data and without augmented data. Based on the validation results of augmented data (proliferative and severe) the best model is selected. Further in one of the experiments cosine distances is calculated, the previous literature is validated and the best batch size is selected for classification on the test dataset.

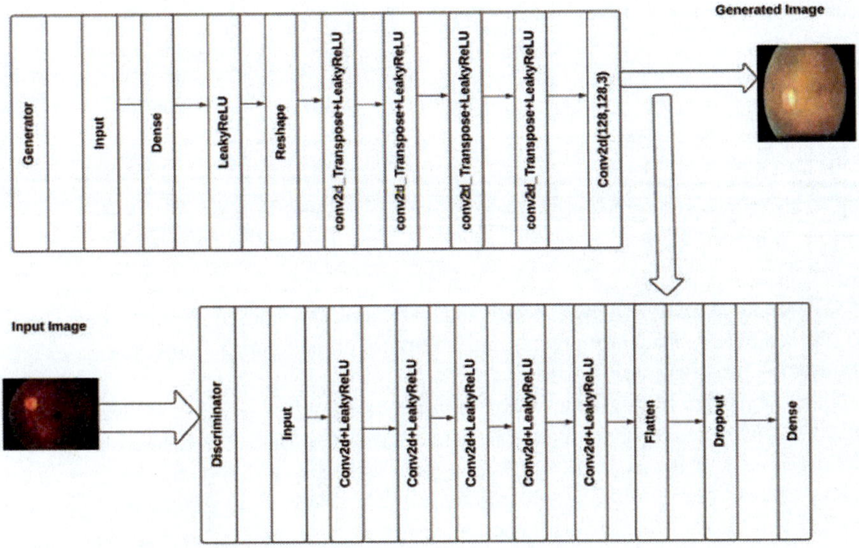

Fig. 4 The DCGAN architecture used in the present work

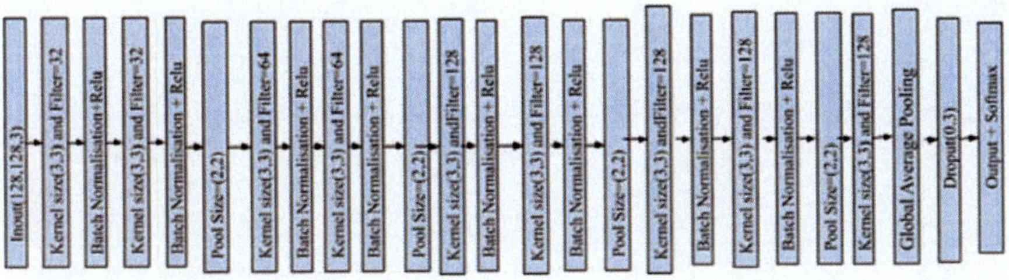

Fig. 5 CNN architecture for classifier used in the present work

In Fig. 5, the architecture of the classifier [10] is mentioned. The CNN has 9 convolution layers, 4 max pool layers and 8 Batch normalization layers are used. The activation function used is ReLU. The input to the classifier is $128 \times 128 \times 3$.

3 Experimental Results and Discussions

The synthetic images are augmented to the dataset and the train test split of 80–20 is done on the augmented dataset. The validation results are observed and the best model is implemented on the test set. In Table 2, we can observe the hyperparameters for classifier and DCGAN. The hyperparameters of the classifier are the same throughout the experiments and for DCGAN only the batch size is varied.

Table 2 Parameters for classifier and DCGAN

Parameters	Classifier	DCGAN
Batch size	15	16
Learning rate	1e-4	0.0002
Optimizer	Adam	Adam
Epoch	100	500
Loss function	Cross-entropy	Cross-entropy

There are a total of 7 experiments performed on the validation and test dataset for 5 classes of DR. Precision, recall and F1 are taken as metrics. Support suggests the number of images used for metrics.

The objective of this experiment is to check the effect of class 0 on class 3 and class 4 with data augmented and No data augmented. Based on the results obtained from Table 3, it is observed that, when class 0 is removed and trained, class 3 and class 4 have a very minute increase in their metrics. When the augmented images are added and trained, the performance metrics of class 3 and class 4 increases.

The objective of this experiment is to check the effect of augmented images. Based on the results of Table 4, when the images of class 3 and class 4 are increased to 2443 by adding synthetic points in the training, the validation metrics for the dataset increases.

Table 3 Classification results for DR grading with class 0 and without class 0

Class	No data augmented				No data augmented without class 0				Batch Size 16 without class 0, data augmented			
	Precision	Recall	F1	Support	Precision	Recall	F1	Support	Precision	Recall	F1	Support
0	0.75	0.88	0.81	5162	–	–	–	–	–	–	–	–
1	0.11	0.05	0.07	489	0.33	0.29	0.31	489	0.21	0.29	0.24	489
2	0.20	0.11	0.14	1058	0.58	0.69	0.63	1058	0.53	0.69	0.60	1058
3	**0.05**	**0.02**	**0.03**	**175**	**0.07**	**0.05**	**0.05**	**175**	**0.52**	**0.31**	**0.38**	**489**
4	**0.11**	**0.05**	**0.07**	**142**	**0.12**	**0.05**	**0.07**	**142**	**0.67**	**0.25**	**0.36**	**489**

Table 4 Classification results for DR grading with data and without data augmentation

Class	No data augmented				Data augmented			
	Precision	Recall	F1	Support	Precision	Recall	F1	Support
0	0.75	0.88	0.81	5162	0.75	0.89	0.81	5162
1	0.11	0.05	0.07	489	0.10	0.04	0.06	489
2	0.20	0.11	0.14	1058	0.21	0.12	0.15	1058
3	**0.05**	**0.02**	**0.03**	**175**	**0.82**	**0.65**	**0.73**	**489**
4	**0.11**	**0.05**	**0.07**	**142**	**0.91**	**0.71**	**0.80**	**489**

Table 5 Classification results for DR grading for different batch size

Class	Batch size 4				Batch size 8				Batch size 32			
	Precision	Recall	F1	Support	Precision	Recall	F1	Support	Precision	Recall	F1	Support
0	0.75	0.89	0.82	5162	0.74	0.88	0.81	5162	0.74	0.90	0.82	5162
1	0.12	0.04	0.06	489	0.09	0.04	0.05	489	0.09	0.02	0.04	489
2	0.23	0.13	0.17	1058	0.20	0.11	0.14	1058	0.21	0.11	0.14	1058
3	**0.86**	**0.65**	**0.74**	**489**	**0.79**	**0.63**	**0.70**	**489**	**0.87**	**0.66**	**0.75**	**489**
4	**0.88**	**0.75**	**0.81**	**489**	**0.84**	**0.72**	**0.78**	**489**	**0.89**	**0.70**	**0.79**	**489**

The objective of this experiment is to check the effect of batch size on the model. Based on the results obtained from Table 5, batch size 4 gives the best performance metrics on the validation dataset.

The objective of this experiment is to check as to how batch size affects the similarity between real and synthetic images. Literature survey from Table 6, shows that as the batch size the similarity between real and synthetic images increases with decrease in batch size. Based on the results obtained from Table 7, as the batch size is increased the difference cosine distances between Real and augmented images increases. It further validates that batch size 4 should be selected for future experiments.

The objective of this experiment is to check the previously trained model for batch size 4 on the test dataset. Based on the results obtained from Table 8, it can be observed that validation metrics and the test metrics are not in sync. The reason would be due to the validation dataset having more number of synthetic points during training as the split 80–20 split is random.

The objective of this experiment is to check the effect of synthetic images on validation split during training. Based on the results from Table 9, it can be inferred

Table 6 Results from "Analysis of Adversarial based Augmentation for Diabetic Retinopathy Disease Grading" [21]

Batch size	Average cosine distance (original)	Average cosine distance (augmented)	Difference
4	0.244492	0.243563	0.000929
8	0.244492	0.234786	0.009706
16	0.244492	0.292456	0.047964

Table 7 Hyperparameters and average cosine distances

Batch Size	Learning rate	Optimizer	Epoch	Loss function	Average cosine distance (original)	Average cosine distance (augmented)	Difference
60	0.0002	Adam	100	Cross-Entropy	0.402453	0.273985	0.13

Table 8 Classification results for DR grading for batch size 4 with class 0 and without class 0

Class	Batch size 4				Batch size 4 without class 0			
	Precision	Recall	F1	Support	Precision	Recall	F1	Support
0	0.75	0.89	0.81	39553	–	–	–	–
1	0.08	0.03	0.04	3762	0.35	0.23	0.28	3762
2	0.19	0.11	0.14	7861	0.57	0.78	0.66	7861
3	**0.03**	**0.01**	**0.01**	**1214**	**0.14**	**0.05**	**0.07**	**1214**
4	**0.05**	**0.01**	**0.02**	**1208**	**0.18**	**0.06**	**0.08**	**1208**

Table 9 Classification results for DR grading when real images used in validation

Class	Using real images in validation (validation results)				Using real images in validation (test results)			
	Precision	Recall	F1	Support	Precision	Recall	F1	Support
0	0.75	0.91	0.82	39553	0.74	0.90	0.82	39553
1	0.08	0.04	0.05	3762	0.08	0.03	0.05	3762
2	0.21	0.08	0.12	7861	0.21	0.08	0.11	7861
3	**0.09**	**0.04**	**0.05**	**1214**	**0.04**	**0.02**	**0.02**	**1214**
4	**0.05**	**0.01**	**0.01**	**1208**	**0.06**	**0.01**	**0.02**	**1208**

that during the validation when only real images are used the validation results are in sync with the test results. Therefore, selection of excess synthetic points in validation during the train-test split causes the spike in metrics of validation results.

The objective of this experiment is to check the effect of using 20%real and 20%synthetic points in validation and adding 1000 points of synthetic images to class 3 and class 4 in the test dataset. Based on the results from Table 10, adding 20%real and 20%synthetic images to the validation dataset gives better metrics. The synthetic points generated are classified as class 3 and class 4 based on the augmentation of synthetic points in the test dataset.

Table 10 Performance metrics of validation and test dataset on 20%real + 20%synthetic in validation

Class	Using real images in validation				Using 20%real + 20%synthetic images on valid dataset (validation metrics)				Using 20%real + 20%synthetic images on validation dataset and adding 1000 synthetic points to class 3 and class 4 (test metrics)			
	Precision	Recall	F1	Support	Precision	Recall	F1	Support	Precision	Recall	F1	Support
0	0.74	0.90	0.82	39553	0.74	0.88	0.81	5162	0.74	0.89	0.81	39553
1	0.08	0.03	0.05	3762	0.10	0.06	0.08	489	0.09	0.05	0.06	3762
2	0.21	0.08	0.11	7861	0.20	0.08	0.12	1058	0.19	0.08	0.11	7861
3	**0.04**	**0.02**	**0.02**	**1214**	**0.87**	**0.64**	**0.70**	**489**	**0.80**	**0.45**	**0.58**	**2214**
4	**0.06**	**0.01**	**0.02**	**1208**	**0.90**	**0.71**	**0.80**	**489**	**0.77**	**0.46**	**0.58**	**2208**

4 Conclusion

It can be concluded that, as the batch size is decreased, the image similarity increases between the synthetic and real images. Further during training, an appropriate ratio of synthetic and real images should be used. The model was able to classify the synthetic test images but the same could not be translated to real images. For the future works, a label of microaneurysm can be generated by image processing techniques from the training images and these labels which have ground truth information about microaneurysm can be transferred to the synthetic images using cycle GAN which is known for style transfer between images. This would ensure that synthetic images are similar to the real images.

References

1. Vasudevan, S., Senthilvel, S., Sureshbabu, J.: Study on risk factors associated with Diabetic Retinopathy among the patients with Type 2 Diabetes Mellitus in South India. Int. J. Ophthalmol. **17**(9), 1615–1619 (2017)
2. Wilkinson, C.P., et al.: Proposed international clinical diabetic retinopathy and diabetic macular edema disease severity scales. Ophthalmology **110**(9), 1677–1682 (2003)
3. Eye Institute of Corpus Christi. http://cceyemd.com/diabetes-eye-exams
4. Seoud, L., Chelbi, J., Cheriet, F.: Automatic grading of diabetic retinopathy on a public database. In: Ophthalmic Medical Image Analysis Second International Workshop Held in Conjunction with MICCAI (2015)
5. Gulshan, V., et al.: Development and validation of a deep learning algorithm for detection of diabetic retinopathy in retinal fundus photographs. JAMA **316**(22), 2402–2410 (2016)
6. Pratt, H., et al.: Convolutional neural networks for diabetic retinopathy. Proc. Comput. Sci. **90**, 200–205 (2016)
7. Kaggle diabetic retinopathy detection competition. https://www.kaggle.com/c/diabetic-retinopathydetection
8. Islam, S.M.S., Mahedi Hasan, Md., Abdullah, S.: Deep learning based early detection and grading of diabetic retinopathy using retinal fundus images (2018). arXiv:1812.10595
9. Li, X., et al.: Convolutional neural networks based transfer learning for diabetic retinopathy fundus image classification. In: 2017 10th International Congress on Image and Signal Processing, BioMedical Engineering and Informatics (CISP-BMEI). IEEE, pp. 1–11 (2017)
10. de La Torre, J., Puig, D., Valls, A.: Weighted kappa loss function for multi-class classification of ordinal data in deep learning. Pattern Recogn. Lett. **105**, 144–154 (2018)
11. Gao, J., Leung, C., Miao, C.: Diabetic retinopathy classification using an efficient convolutional neural network. In: 2019 IEEE International Conference on Agents (ICA). IEEE, pp. 80–85 (2019)
12. Zhou, Y., et al.: DR-GAN: conditional generative adversarial network for fine-grained lesion synthesis on diabetic retinopathy images (2019). arXiv:1912.04670
13. Goodfellow, I., et al.: Generative adversarial nets. In: Advances in Neural Information Processing Systems, pp. 2672–2680 (2014)
14. Bhattacharya, D., et al.: GAN-based novel approach for data augmentation with improved disease classification. In: Advancement of Machine Intelligence in Interactive Medical Image Analysis, pp. 229–239. Springer (2020)
15. Frid-Adar, M., et al.: Synthetic data augmentation using GAN for improved liver lesion classification. In: 2018 IEEE 15th International Symposium on Biomedical Imaging (ISBI 2018). IEEE, pp. 289–293 (2018)

16. Costa, P. et al.: Towards adversarial retinal image synthesis (2017). arXiv:1701.08974
17. Nirmal, S., Sowmya, V., Soman, K.P.: Open set domain adaptation for hyperspectral image classification using generative adversarial network. Lecture Notes in Networks and Systems (2020)
18. Patil, S.O., Sajith Variyar, V.V., Soman, K.P.: Speed bump segmentation an application of conditional generative adversarial network for self-driving vehicles. In: 2020 Fourth International Conference on Computing Methodologies and Communication (ICCMC), Erode, India, India (2020)
19. Shashank, A., Sajith Variyar, V.V., Sowmya, V., Soman, K.P., Sivanpillai, R., Brown, G.K.: Identifying epiphytes in drones photos with a conditional generative adversarial network (C-GAN). In: The International Archives of the Photogrammetry, Remote Sensing and Spatial Information Sciences, XLIV-M-2-2020, pp. 99–104 (2020). https://doi.org/10.5194/isprs-archives-XLIV-M-2-2020-99-2020
20. Kaplan, S., et al.: Evaluation of unconditioned deep generative synthesis of retinal images. In: International Conference on Advanced Concepts for Intelligent Vision Systems, pp. 262–273. Springer (2020)
21. Balasubramanian, R., Sowmya, V., Gopalakrishnan, E.A., Menon, V.K., Sajith Variyar, V.V., Soman, K.P.: Analysis of adversarial based augmentation for diabetic retinopathy disease grading. In: 2020 11th International Conference on Computing, Communication and Networking Technologies (ICCCNT) (2020)
22. Kora Venu, S.: Evaluation of deep convolutional generative adversarial networks for data augmentation of chest X-ray images (2020). ui.adsabs.harvard.edu

A Novel Leaf Fragment Dataset and ResNet for Small-Scale Image Analysis

Abdul Hasib Uddin, Sharder Shams Mahamud, and Abu Shamim Mohammad Arif

Abstract This paper introduces a dataset of leaf vein images from four different species (two from Monocotyledons and two from Dicotyledons). Then multiple instances of the dataset containing 64×64, 32×32, 16×16, 8×8, and 4×4 pixel single channel center-focused images were created, and a new Residual Neural Network (ResNet) based model has applied to each of the instances for cotyledon (seed leaf or embryonic leaf) group identification. Additionally, the same procedures were followed for plant species classification. The aim was to make use of only the vein patterns for the task. The results show that despite the difficulties of recognizing patterns from small-scale images, it is likely to efficiently categorize seed types and plant species by properly deploying residual blocks in neural network models. Also, the applied ResNet was compared against ResNet-152 V2 on the 64×64 pixel imageset in terms of species classification.

Keywords Leaf vein · Small-scale image · Cotyledon · Species · Classification · Residual network

1 Introduction

The number of plants is over extensive, with about 391,000 vascular species all around the world. Therefore, it is inconceivable and impractical for a specialist, to be able to recognize and categorize all the species. In summation, some plant species may have deep equivalency among each other, requiring much longer time and effort to categorize them. Hence, automatic plant identification is a growing and demanding problem that has acknowledged cumulative attention in recent years, particularly for identification based on leaf image analysis. The ultimate target of optimization, in this case, is to eliminate the use of human specialists handling huge estimated lists of plant species and to minimize classification time.

In recent times, in many exceptional appliances to machine learning, there is an inclination to replace classical techniques with deep learning algorithms. In deep

A. H. Uddin (✉) · S. S. Mahamud · A. S. M. Arif
Computer Science and Engineering Discipline, Khulna University, Khulna, Bangladesh

J. S. Raj et al. (eds.), *Intelligent Sustainable Systems*, Lecture Notes in Networks and Systems 213, https://doi.org/10.1007/978-981-16-2422-3_3

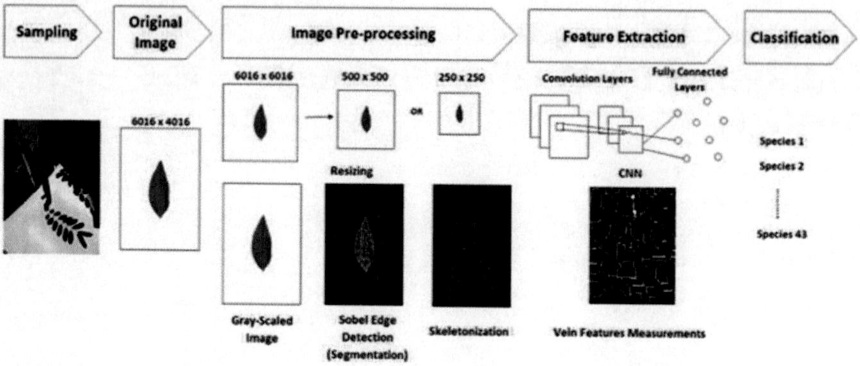

Fig. 1 Steps for utilizing vein morphometric for plant species classification [1]

learning methods, features are automatically extracted by the models, eliminating the need for handcrafted feature extraction. Moreover, classification outcomes are much improved than those earned with classical techniques.

Leaf shape is the most generally used feature for constructing an automated plant identification system. Other features than the shape, the leaf can supply some additional information, such as a vein structure, which is the key concept behind building the dataset.

Plant classification methods have been deployed by several researchers in recent years, which have accomplished great outcomes in the related fields. Various characteristics such as aspect ratio (ratio between length and width of the leaf), the ratio of the parameter to the diameter of the leaf, and vein characteristics were used to categorize the leaf to identify specific plants. Deep learning was used for plant species classification using leaf vein morphometric by Tan et al. [1]. Figure 1 presents the corresponding vein morphometric steps. Another work has been done by Lee et al., where deep learning was used to extract, and learn leaf features for plant classification [2]. Grinblat et al. have shown in his work how automatic classification of legumes can be done using leaf vein image features [3]. Example demonstration for image pre-processing and vein morphological plant classification are provided in Fig. 2. Walls utilized phylogenetic, standard ANOVA along with regression and introduced novel universal phylogenetic inspection on Angiospermae vein morphology, including characteristics of leaf functions [4]. Fractal-based approaches had investigated in terms of leaf complexity by Bruno et al. [5]. Hickey categorized dicot leaves based on their internal and external attributes [6]. Larese et al. proposed a scheme for differentiating legumes solely depending upon the leaf vein structures [7]. On the other hand, leaf category recognition by making use of the external features along with vein and color was performed with Probabilistic Neural Network by Kadir et al. [8]. Additionally, Aptoula and Yanikoglu made use of covariance-based morphology computation and an extended histogram of circular covariance to recognize plants [9]. Machado et al. applied differential equations for analyzing fractal patterns on leaves [10]. Price et al. discovered some significant leaf vein network properties

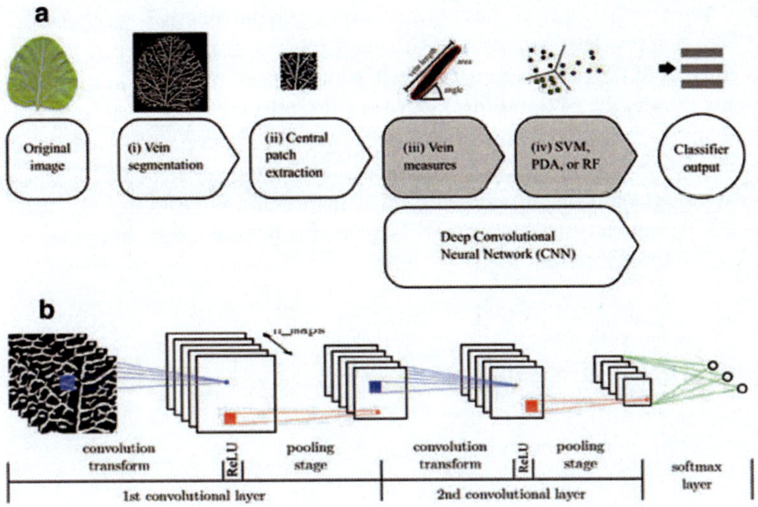

Fig. 2 **a** Image pre-processing steps for vein skeleton extraction [3]. **b** Deployed model for vein morphology-based plant categorization [3]

and strongly suggested their use while classifying plants based on their leaves [11]. A Machine Learning-based visual recognition system for vegetables and fruits had developed by Shakya [12]. Jacob exploited Capsule Network, which is competent in learning more powerful descriptions compared to CNN for image recognition [13]. Recently, a precision agriculture technique was proposed by Chandy et al. to identify different types of pests infesting coconut trees [14]. Additionally, Deep Learning methods were being applied by Vijayakumar et al. to recognize the mellowness of Dragon Fruit [15]. Lee et al. utilized leaf vein and shape for classifying leaf images [16]. On the other hand, Wu et al. developed an algorithm for plant leaf identification based on a probabilistic neural network [17].

Nonetheless, the availability of a full leaf image is not always possible, rather it can be sometimes very challenging. Additionally, working on images with high resolution is costly. Hence, in this manuscript, an experiment is conducted on small-scale images, where full leaf shape is not required and images do not contain high dimensions, which would otherwise require costly instruments.

A complex pattern recognition problem requires deeper neural networks. However, difficulties in training increase with network depth. Residual neural networks (ResNet) are capable of simplifying the complications introduced by the intense network depth [18]. For this reason, the ResNet-based method was developed for recognition purposes in this work.

This research presents a leaf image dataset to identify monocotyledonous and dicotyledonous plants which are much familiar and can be easily collected from the surroundings. Much care was given to collecting and preprocessing the dataset. While implementing Deep Residual Network, two different perspectives were considered—(i) cotyledon-type classification and (ii) plant species classification.

For the rest of the paper, Sect. 2 describes the dataset preparation steps. Section 3 represents the details of applying ResNet for identifying and classifying cotyledon type and plant species. The corresponding results are illustrated in Sect. 4. Additionally, some discussions are included in Sect. 5. In Sect. 6, a performance comparison is presented between the applied ResNet structure and the established ResNet-152 V2 model structure. Finally, in Sect. 7, research is concluded and some possible future contributions are outlined.

2 Dataset Preparation

The steps of the research are visualized in Fig. 3.

2.1 Data Collection

Images were collected from four different species of plants, two are from Monocotyledon plants (*Cocos nucifera, Eichhornia crassipes*) and the other two are from Dicotyledon plants (*Cucurbita moschata, Neolamarckia cadamba*).

Three different sets from each species were considered based on leaf age—(i) young, (ii) mid-aged, and (iii) old. Again, each set has two subsets based on the leaf side—(i) the front side and (ii) the reverse side. An example of the images from each species is included in Figs. 4 and 5. Hardware augmentation was performed on each specimen and captured video using a high-frequency camera.

The statistics of the initial imageset is provided in Table 1 (Column: Number of images (initial)). For *Cocos nucifera*, there a total of 26,947 images (4635 images for front-young, 4684 images for front-middle-aged, 4515 images for front-old, 5007 images for reverse-young, 3906 images for reverse-middle-aged, and 4200 images for reverse-old types). For *Eichhornia crassipes*, there was a sum of 20,908 images, of which, front-young, front-middle-aged, front-old, reverse-young, reverse-middle-aged, and reverse-old types correspond to 3765, 3623, 3925, 3227, 3273, and 3095 images, accordingly. Collectively, in the Monocotyledon group, there was a total of 47,855 images.

Now, in the Dicotyledon group, there were 33,672 images, of which, 13,964 images were from *Cucurbita moschata* and 19,708 images were from *Neolamarckia cadamba* species. Among the images from *Cucurbita moschata*, 2069 images for front-young, 1819 images for front-middle-aged, 3915 images for front-old, 2111 images for reverse-young, 1981 images for reverse-middle-aged, and 2069 images for reverse-old types. On the other hand, from among the images of *Neolamarckia*

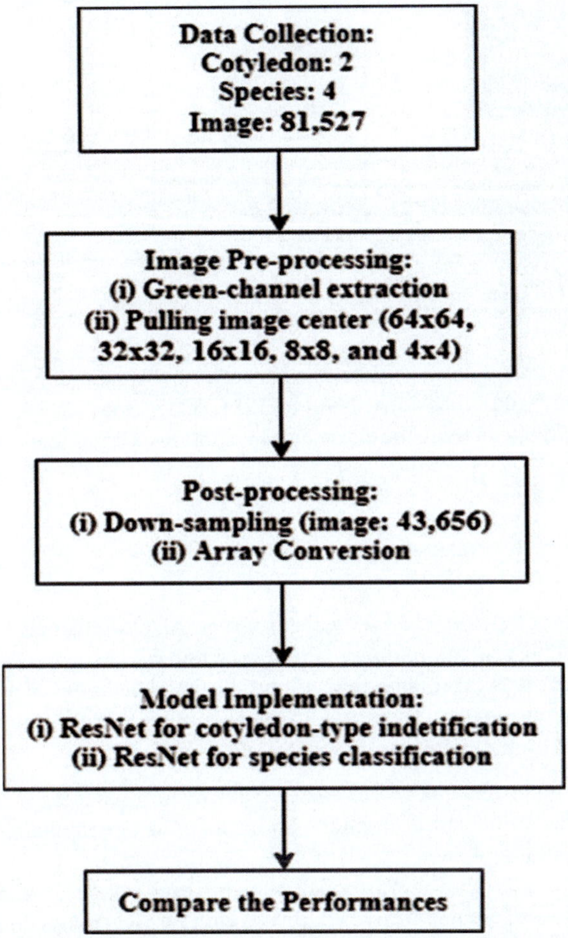

Fig. 3 Block schematic of the workflow

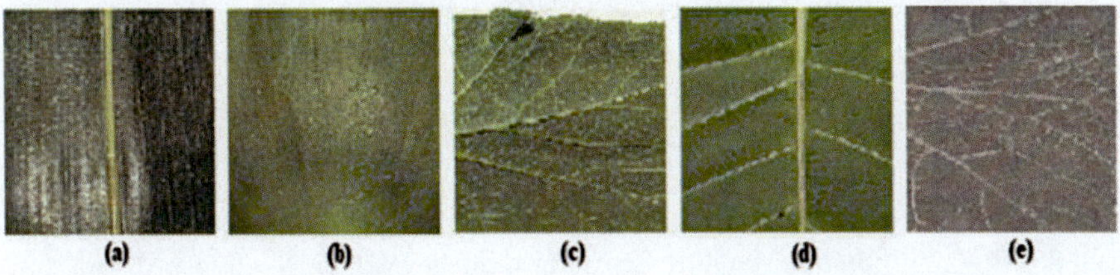

Fig. 4 Sample of **a** *Cocos nucifera* (Monocotyledon), **b** *Eichhornia crassipes* (Monocotyledon), **c** *Cucurbita moschata* (Dicotyledon), **d** *Neolamarckia cadamba* (Dicotyledon), **e** Green-channel extraction from color images

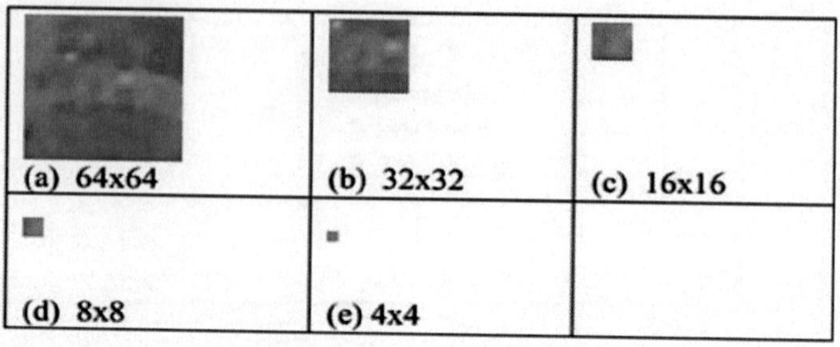

Fig. 5 Examples of center-focused images

cadamba, front-young, front-middle-aged, front-old, reverse-young, reverse-middle-aged, and reverse-old types correspond to 3531, 3318, 3367, 3135, 3248, and 3109 images, respectively. In total, the entire dataset contains 81,527 leaf fragment images.

2.2 *Image Pre-processing*

The dimensions of the initial images were different. Also, some images had leaf edge, which is against the primary intention. Point to be noted that for this experimentation, only vein pattern was taken as a feature and all other features, such as leaf shape, color, etc. were eliminated. Therefore, to achieve this goal, two-step preprocessing were performed on the corresponding dataset:

(i) First, the green channel was extracted from the original images, which contains most of the valuable information. An example of Green-channel extraction is provided in Fig. 4e.

(ii) Then, the center portion was pulled out from each image. In this step, multiple instances were created from each image with resolution 64 × 64, 32 × 32, 16 × 16, 8 × 8, and 4 × 4. The center point of an image was calculated from the height and width of each image. Only one segment was taken from an image. Example for each dimension is represented in Fig. 5.

2.3 *Post-processing*

The pre-processed images were unbalanced with different numbers of images from each group. Additionally, Machine Learning models work with numerical data. For these reasons, two steps of post-processing were carried out:

(i) The minimum number of images was collected for *Cucurbita moschata* mid-aged front-sided image group with 1819 samples. So, the dataset was down-sampled by selecting 1819 images from each group. After this step, each species contained 10914 images (21828 images for Monocotyledon and 21828

Table 1 Statistics for initial and final datasets

Cotyledon-type	Species	Leaf age	Leaf side	# images (initial)	# images (final)
Monocotyledon	*Cocos nucifera*	Young	Front	4635	1819
		Mid-aged		4684	
		Old		4515	
		Young	Reverse	5007	
		Mid-aged		3906	
		Old		4200	
	Total from *Cocos nucifera*			26947	10914
	Eichhornia crassipes	Young	Front	3765	1819
		Mid-aged		3623	
		Old		3925	
		Young	Reverse	3227	
		Mid-aged		3273	
		Old		3095	
	Total from *Eichhornia crassipes*			20908	10914
Total from Monocotyledon				47855	21828
Dicotyledon	*Cucurbita moschata*	Young	Front	2069	1819
		Mid-aged		1819	
		Old		3915	
		Young	Reverse	2111	
		Mid-aged		1981	
		Old		2069	
	Total from *Cucurbita moschata*			13964	10914
	Neolamarckia cadamba	Young	Front	3531	1819
		Mid-aged		3318	
		Old		3367	
		Young	Reverse	3135	
		Mid-aged		3248	
		Old		3109	
	Total from *Neolamarckia cadamba*			19708	10914
Total from Dicotyledon				33672	21828
Total images in dataset				81527	43656

images for Dicotyledon). The statistics are represented in Table 1 (Column: Number of images (final)).

(ii) Next, each image was converted into a NumPy array, flattened, and saved into a '*.csv' file. Each row in a file represents a single image, while the columns signify the flattened grayscale values for that image.

2.4 Filename-Format

Each group of images is saved in a single file. The name-format of each file is as follows:

<cotyledon-type>_<species-first-name>_<leaf-age>_<leaf-size>_<image-height>x<image-width>p.<file-format>

An example filename from the post-processed final dataset is,

'Dicot_Cucurbita_midAged_front_32 $\times$ 32p.csv', which contains 32-by-32 pixel images (flattened to 1024 numerical columns) for the images captured from the front-side, mid-aged leaves from Dicotyledon-type of *Cucurbita moschata* species plant.

3 Residual Neural Network for Cotyledon-Type Identification and Plant Species Classification

3.1 Dataset Formulation and Feature Description

The dataset, prepared as described in Sect. 2 was fed directly into the model after subdividing it into training–validation–test sets. The only feature that was utilized in this research was the vein texture from green channel images, which was the aim of this work. All other probable features were eliminated for the experimental purpose and no other segmentation processes were involved in this proposed work.

The imageset (43,656 images) was split into 80% training (34,968 images), 10% validation (4,344 images), and 10% testing (4,344 images) dataset. Thus, each cotyledon group and species had 21,828 and 10,914 images for training, accordingly. The test indices were selected randomly from each species.

3.2 Residual Block

The implementations of this paper are based on residual blocks. The corresponding structure of the deployed Residual network is presented in Fig. 6. Two-dimensional convolution layers were used with three different filters. For filter f1 and f3, a 1×1 kernel was integrated. On the other hand, a 3×3 kernel was used for the f2 filter. In all cases, the deployed stride was 1×1. There are three main stages of each residual block. All the stages are almost similar, except the first stage having a 2×2 max-pooling layer after the very first Conv2D layer. Between two subsequent Conv2D layers, batch normalization followed by the ReLU activation layer was added. At the end of each stage, the stage outcome is added with the original image and passed through a ReLU activation layer before starting the next stage.

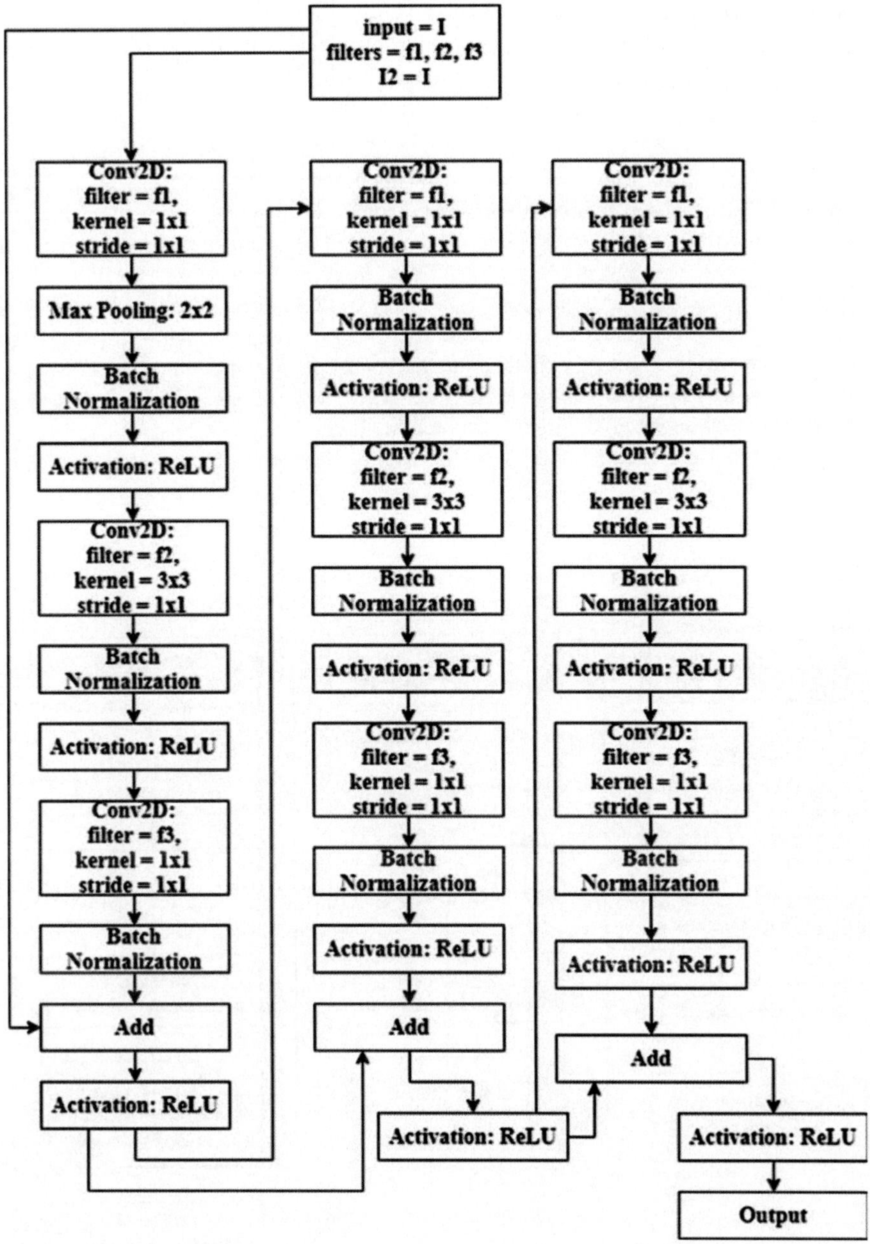

Fig. 6 A single Residual network block

3.3 Methodology

The block diagram for the applied model is given in Fig. 7. The model takes an input image and performs zero-padding of size 1 × 1. Then, each image passes through a 2D convolution layer, where a filter of size 64 with kernel size 5 × 5 and stride 1 × 1 is applied. After that, the model performs batch normalization, applies ReLU activation, and a 3 × 3 max-pooling with stride 1 × 1. Next, four residual blocks, each with three filter sizes are applied. The outcome of the residual blocks is passed through a 2 × 2 average pooling layer. Then, there are two dense layers, the first one with 4096 neurons and the second one with 2048 neurons. Now, in the case of a cotyledon identification scheme, the model applies an extra two dense layers (1024 and 512 neurons, respectively) before the final dense layer with 2 neurons. However, for species classification, after a dense layer with 2048 neurons, the model directly applies the final dense layer with 4 neurons. In all dense layers, ReLU is used as the activation function, except the final layers, where softmax is the activation function.

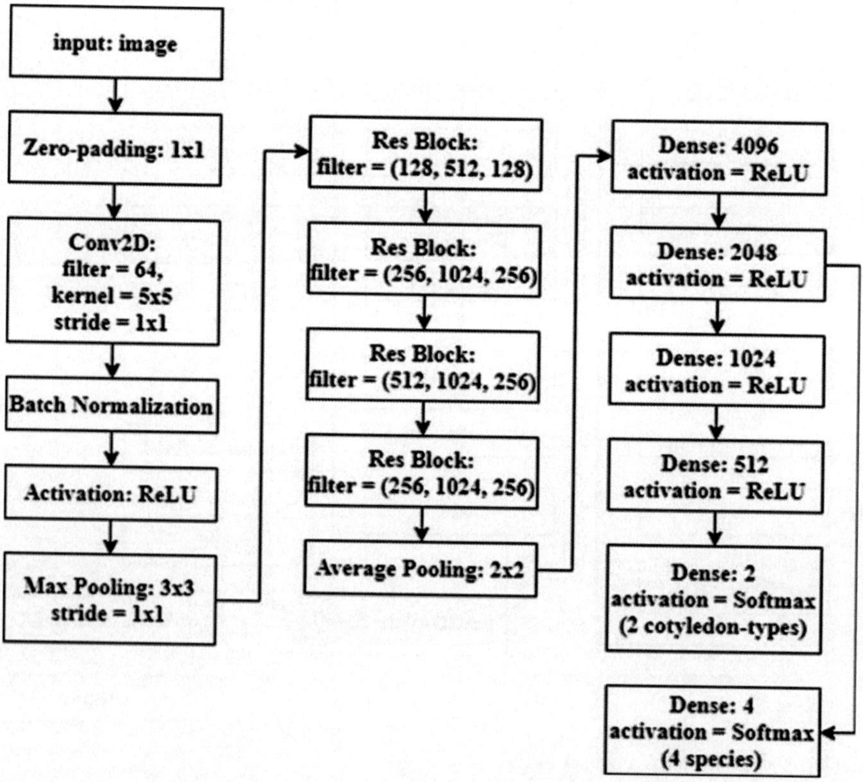

Fig. 7 Implemented model structure for cotyledon-type identification and species classification

The novelty part of this network is that it is, in reality, a Residual-Dense hybrid construction. For leaf vein analysis, a Residual network following several Dense layers was found efficient in classifying images. As for ResNet152 V2 [19], it is primarily based upon Convolution layers and there are no Dense layers embedded into it. However, this simple addition to the deployed 146-layer ResNet led to a significant difference in performance as demonstrated in Sect. 4.

As for kernel initializer, 'glorot_uniform' with seed 0 was used. Point to be noted that in the case of species low-resolution images, the pooling layers were omitted in Fig. 7 because each pixel is valuable in low dimension. The learning rate was set to 0.0001, Beta-1 regularizer was 0.9 and Beta-2 regularizer was 0.999. For cotyledon identification, binary cross-entropy was used and for species classification, categorical cross-entropy was applied as the loss function. In all cases, Adam optimizer was applied for optimization.

In the case of cotyledon identification, the batch size was 32. Similarly, for species classification, it was 8 (except for 8 × 8 pixel images in species classification, where the batch was 32). In all implementations, no epochs were 15.

Figure 8a illustrates the learning curves for the best implementation for cotyledon identification, which was gained from a 64 × 64 imageset. The training loss was significantly low at the start and decreased continuously towards nil. However, from the validation loss curve, we can see that it had a hard time learning the features.

On the other hand, Fig. 8b visualizes the learning curves for the best implementation in terms of plant species classification. The training loss, in this case, was much higher at the beginning. Nonetheless, it fell gradually as training continued. The validation loss curve demonstrates that the model learned more easily than cotyledon cases.

4 Results

4.1 Cotyledon-Type Identification

Table 2 summarizes the final results for cotyledon identification (results for species classification are provided in Table 3). For 64 × 64 and 32 × 32 pixel imagesets, the accuracy was 91.18% and 81.08%, respectively. However, with low-resolution images, accuracy dropped significantly. For 16 × 16 and 8 × 8 pixel imagesets, the accuracy was 63.14 and 65.35%, accordingly. The lowest performance gained for 4 × 4 pixel data instances was about 57.55% only.

The corresponding precision, recall, and F1-score for cotyledon identification as listed in Table 4 are 0.95, 0.87, and 0.91 for Monocotyledon and 0.86, 0.72, and 0.78 for Dicotyledon, respectively. Table 5 holds the confusion matrix for the 64 × 64 image-based cotyledon identification process. The model was able to identify 1,892 Monocotyledon and 2,069 Dicotyledon images correctly.

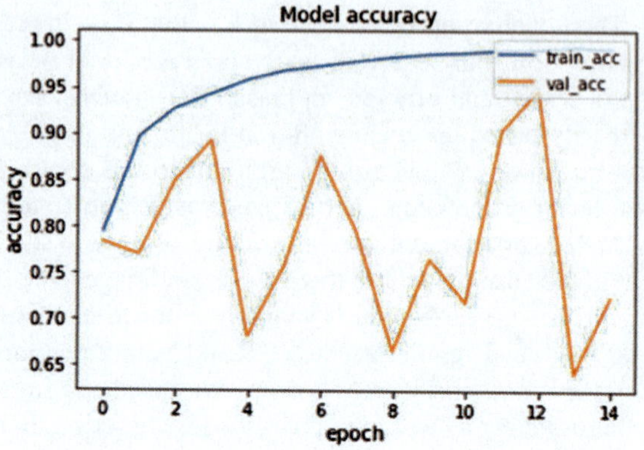

(a)

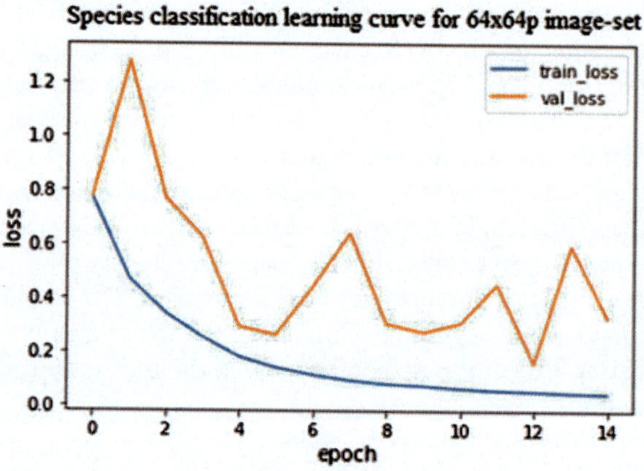

(b)

Fig. 8 Training versus validation loss curve for **a** cotyledon identification and **b** plant species classification using 64 × 64 images

Table 2 Results for cotyledon-type identification using Residual network

Implementation no.	Image dimension	Test accuracy (%)
1	64 × 64	**91.18**
2	32 × 32	81.08
3	16 × 16	63.14
4	8 × 8	65.35
5	4 × 4	59.48

Table 3 Results for plant species classification using Residual network. Corresponding test accuracy from ResNet-152 V2 [19] for 64 × 64p imageset in species classification is 69.57%

Implementation no.	Image dimension	Test accuracy (%)
1	64 × 64	**85.38**
2	32 × 32	71.16
3	16 × 16	61.42
4	8 × 8	47.86
5	4 × 4	34.00

Table 4 Elaborated results for the implementation with the highest (64 × 64 pixel imageset). Values in the first brackets represent the corresponding outcomes for ResNet-152 V2 in species classification

Species	Precision	Recall	F1 score
Cocos nucifera	0.81 (0.95)	0.95 (0.63)	0.88 (0.75)
Eichhornia crassipes	0.97 (0.77)	0.85 (0.77)	0.90 (0.77)
Monocotyledon	0.95	0.87	0.91
Cucurbita moschata	0.81 (0.56)	0.90 (0.85)	0.85 (0.68)
Neolamarckia cadamba	0.69 (0.66)	0.79 (0.53)	0.74 (0.59)
Dicotyledon	0.86	0.72	0.78

Table 5 Confusion matrix for cotyledon-type identification based on 64 × 64 pixel imageset (highest accuracy)

T = True P = Predicted	Monocotyledon (T)	Dicotyledon (T)
Monocotyledon (P)	**1892**	280
Dicotyledon (P)	103	**2069**

4.2 *Plant Species Classification*

The results for classifying plant species classification are unified in Table 3. Similar to cotyledon-type identification, the highest accuracy was achieved for images with resolution 64 × 64 pixel, and the lowest accuracy was for 4 × 4 pixel images. It can be observed that for 64 × 64 dimension images, the accuracy was as high as 85.38%. For 32 × 32 and 16 × 16 resolution imagesets, the performance was 71.16% and 61.42%. However, for 8 × 8 and 4 × 4 resolution dataset instances, the performances were only 47.86% and 43%, respectively.

For species classification, the precision, recall, and F1-score are 0.81, 0.95, 0.88 for *Cocos nucifera,* 0.97, 0.85, 0.90 for *Eichhornia crassipes,* 0.81, 0.90, 0.85 for *Cucurbita moschata*, and 0.86, 0.72, 0.78 for *Neolamarckia cadamba*, accordingly. The confusion matrix species classification for the highest accuracy implementation (64 × 64 imageset) is provided in Table 6. The corresponding model successfully classified 1,035 *Cocos nucifera*, 918 *Eichhornia crassipes*, 976 *Cucurbita moschata*, and 780 *Neolamarckia cadamba* images. However, the ResNet152 V2 was able to classify 681 *Cocos nucifera*, 835 *Eichhornia crassipes*, 927 *Cucurbita moschata*, and 579 *Neolamarckia cadamba* images, respectively. Therefore, for all species, applied ResNet performed better than ResNet152 V2 model.

5 Discussion

As the results depict, high-dimension imageset helps to gain significantly high performance easily. However, low-resolution images can be classified by using deeper neural networks. Additionally, performances in the cases of cotyledon-type identification were noticeably higher than species classification. It can be due to the number

Table 6 Confusion matrix for species classification based on 64 × 64 pixel imageset (highest accuracy). Values in the first brackets represent the corresponding outcomes for ResNet-152 V2 in species classification

T = True P = Predicted	*Cocos nucifera* (T)	*Eichhornia crassipes* (T)	*Cucurbita moschata* (T)	*Neolamarckia cadamba* (T)
Cocos nucifera (P)	**1035 (681)**	0 (8)	14 (267)	37 (130)
Eichhornia crassipes (P)	73 (8)	**918 (835)**	72 (173)	23 (70)
Cucurbita moschata (P)	21 (4)	22 (51)	**976 (927)**	67 (104)
Neolamarckia cadamba (P)	148 (25)	9 (189)	149 (293)	**780 (579)**

of images per class. While each class of cotyledon has as much as 21,828 images, each class of a plant species has only 10,914. Deeper network structures require a larger dataset. After subtracting the validation and testing dataset, each species only has 8,731 images solely for training, which is not enough in this case.

As for the change in dimension, Tables 2 and 3 clearly show that performance improves significantly with increasing resolutions. Nonetheless, the model was even able to achieve significant accuracy (34%) for 4 × 4 pixel species classification, which is much better than 25% (blindly classifying all data into one class).

Moreover, it can be decided that to help increasing model performance, either the dimension of the images can be increase or more data with low resolution might be added.

Finally, Fig. 8 signifies that although plant species or cotyledon types can be efficiently identified by using deep neural networks, like Residual Networks, finding the global optimum point can be challenging. This can be due to the low resolution of the images, or the lack of more features. As this experiment focuses only on center-focused, green channel leaf images, all other features, such as leaf shape, color, or other probable helpful features were deliberately avoided. Additionally, to keep pre-processing steps as simple as possible and to push the boundary of neural networks to classify challenging images, this research does not even conduct popular tactics, for example, boundary extraction. Hence, the performances shown in this manuscript are solely based upon a challenging dataset with only one (leaf texture) or arguably two (green channel leaf texture) features.

6 Comparison Between Applied ResNet and ResNet-152 V2

The paper compares the performances of the applied ResNet architecture against ResNet-152 V2, which was introduced in 2016 [17]. The ResNet-152 V2 was trained on the 64 × 64 pixel imageset for species classification with a similar setup as the applied ResNet model—batch size 32, learning rate 0.0001, optimizer Adam, 0.9 as the value of beta 1, and 0.999 for beta-2 regularizer.

The corresponding accuracy is 69.57%, where the accuracy for the applied model is 85.38% (Table 3). Table 4 enlists the respective Precision, Recall, and F1-score, whereas Table 5 contains the corresponding confusion matrix values for species classification on the 64 × 64 imageset. This comparative study demonstrates that the applied ResNet structure is more efficient than the state-of-the-art ResNet-152 V2 structure for leaf vein analysis and classification.

7 Conclusion and Possible Future Contributions

This research paper introduces a novel processed center-focused green channel image dataset for cotyledon-type identification and plant species classification. Moreover, multiple instances of the dataset were created based on image dimension and shown the effects of image dimension on Residual Neural Network. It can be seen that if more data are added, low resolution is not a problem. On the other hand, a high-resolution image eradicates the necessity for more data. Any one of the two strategies is worth to be followed as per the situation.

Multiple points in this paper can be enhanced in the future. For example, instead of down-sampling the dataset from each group, they can be either up-sampled using Random Sampling with Repeat, or more images in each category can be added. Moreover, higher resolution images can be extracted from the imageset and applied to neural network models for better performance. Also, red and blue channels can be extracted from the images to observe the impacts.

Author Contribution A.H.U. proposed the topic. S.S.M. collected all data and performed hardware augmentation. A.H.U. processed the dataset and performed implementations. A.H.U. and S.S.M. prepared the manuscript. A.S.M.A. supervised the entire process.

Data Link Mendeley data link: https://data.mendeley.com/datasets/ngds35s8vr/1.

References

1. Tan, J.W., Chang, S.-W., Kareem, S.B.A., Yap, H.J., Yong, K.-T.: Deep learning for plant species classification using leaf vein morphometric. IEEE/ACM Trans. Comput. Biol. Bioinform. (2018)
2. Lee, S.H., Chan, C.S., Mayo, S.J., Remagnino, P.: How deep learning extracts and learns leaf features for plant classification. Pattern Recogn. **71**, 1–13 (2017)
3. Grinblat, G.L., Uzal, L.C., Larese, M.G., Granitto, P.M.: Deep learning for plant identification using vein morphological patterns. Comput. Electron. Agric. **127**, 418–424 (2016)
4. Walls, R.L.: Angiosperm leaf vein patterns are linked to leaf functions in a global-scale data set. Am. J. Bot. **98**(2), 244–253 (2011)

5. Bruno, O.M., de Oliveira Plotze, R., Falvo, M., de Castro, M.: Fractal dimension applied to plant identification. Inf. Sci. **178**(12), 2722–2733 (2008)
6. Hickey, L.J.: Classification of the architecture of dicotyledonous leaves. Am. J. Bot. **60**(1), 17–33 (1973)
7. Larese, M.G., Namías, R., Craviotto, R.M., Arango, M.R., Gallo, C., Granitto, P.M.: Automatic classification of legumes using leaf vein image features. Pattern Recogn. **47**(1), 158–168 (2014)
8. Kadir, A., Nugroho, L.E., Susanto, A., Santosa, P.I.: Leaf classification using shape, color, and texture features (2013. arXiv:1401.4447)
9. Aptoula, E., Yanikoglu, B.: Morphological features for leaf based plant recognition. In: 2013 IEEE International Conference on Image Processing, pp. 1496–1499. IEEE (2013)
10. Machado, B.B., Casanova, D., Gonçalves, W.N., Bruno, O.M.: Partial differential equations and fractal analysis to plant leaf identification. J. Phys.: Conf. Ser. **410**(1), 012066 (2013)
11. Price, C.A., Wing, S., Weitz, J.S.: Scaling and structure of dicotyledonous leaf venation networks. Ecol. Lett. **15**(2), 87–95 (2012)
12. Shakya, S.: Analysis of artificial intelligence based image classification techniques. J. Innov. Image Process. (JIIP) **2**(01), 44–54 (2020)
13. Jacob, I.J.: Performance evaluation of caps-net based multitask learning architecture for text classification. J. Artif. Intell. **2**(01), 1–10 (2020)
14. Chandy, A.: Pest infestation identification in coconut trees using deep learning. J. Artif. Intell. Capsul. Netw. **1**(1), 10–18 (2019)
15. Vijayakumar, T., Vinothkanna, R.: Mellowness detection of dragon fruit using deep learning strategy. J. Innov. Image Process. (JIIP) **2**(01), 35–43 (2020)
16. Lee, K.-B., Hong, K.-S.: An implementation of leaf recognition system using leaf vein and shape. Int. J. Bio-Sci. Bio-Technol. **5**(2), 57–66 (2013)
17. Wu, S.G., Bao, F.S., Xu, E.Y., Wang, Y.-X., Chang, Y.-F., Xiang, Q.-L.: A leaf recognition algorithm for plant classification using probabilistic neural network. In: 2007 IEEE International Symposium on Signal Processing and Information Technology, pp. 11–16. IEEE (2007)
18. He, K., Zhang, X., Ren, S., Sun, J.: Deep residual learning for image recognition. In: Proceedings of the IEEE Conference on Computer Vision and Pattern Recognition, pp. 770–778 (2016)
19. He, K., Zhang, X., Ren, S., Sun, J.: Identity mappings in deep residual networks. In: European Conference on Computer Vision, pp. 630–645. Springer, Cham (2016)

Prediction of Covid 19 Cases Based on Weather Parameters

N. Radha and R. Parvathi

Abstract The aim of the paper is to study the relation between the weather parameters and the confirmed Covid 19 cases. The prime weather parameters are analysed with feature selection and an analysis is made on parameters which have more impact on confirmed cases. The problem is approached with linear regression, Decision tree regressor and Random Forest Regression. The metrics used for evaluation is mean squared error and R square score. The proposed experiment is to find the model that better makes prediction of confirmed cases with weather parameters as input. It also involves finding the important weather parameters which have relation with confirmed cases.

Keywords Feature selection · Regression · Decision tree regressor · Random forest regression · Weather factors · Transmission · Multiplication

1 Introduction

The study is made on various weather parameters such as average temperature, minimum temperature, maximum temperature, sea temperature and pressure, latitude, longitude, absolute humidity, relative humidity, precipitation, dew point and windspeed. Weather is made up of many factors which include temperature, humidity, solar radiation, pressure etc. The weather factors can be calculated to determine the atmospheric conditions and patterns which helps in weather predictions. The climatic conditions are found to have profound impact on a number of natural mechanisms and evolution of organisms.

The microorganisms have relation with weather conditions for transmission and multiplication. The analysis was made with feature engineering and with correlation. The models taken for analysis are linear regression, Random Forest Regression and

N. Radha (✉) · R. Parvathi
Vellore Institute of Technology, Chennai, TamilNadu, India
e-mail: radha.n2020@vitstudent.ac.in

R. Parvathi
e-mail: parvathi.r@vit.ac.in

J. S. Raj et al. (eds.), *Intelligent Sustainable Systems*, Lecture Notes in Networks and Systems 213, https://doi.org/10.1007/978-981-16-2422-3_4

Decision tree regression. The decision tree Regression shows better accuracy than the other two methods for prediction of confirmed cases.

The decision tree regression, generally, has better potency to predict cases when increasing the size of datasets. However, the random forest regression predicts the cases in much generalized way with averaging of the node values. In the experiment, the results from linear regression was more volatile than random forest regression and decision tree regressor. Decision tree regression predicts well within the set and random forest regressor can predict in a generalized way for inputs outside the training set as well (Fig. 1).

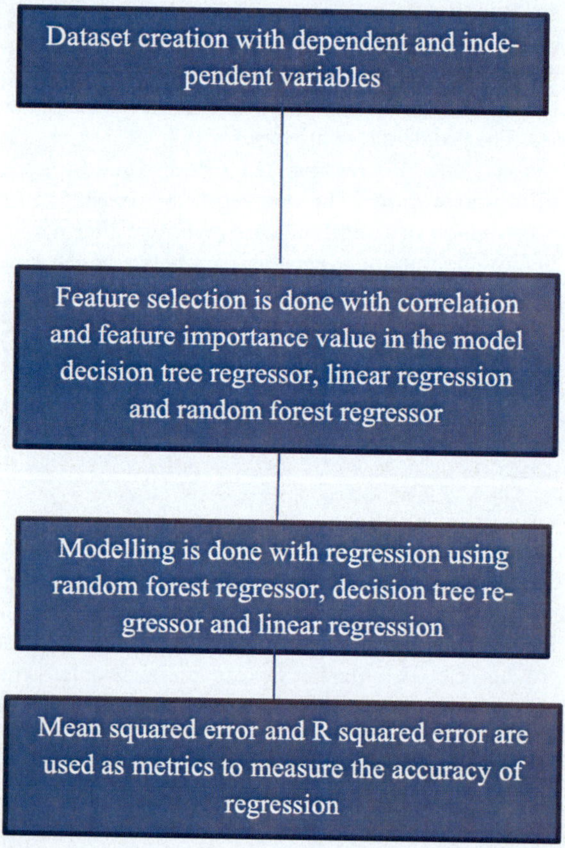

Fig. 1 Block diagram

2 Review of Related Literature Paper

The log linear model is used to find out the rate of growth of the Covid 19 confirmed cases. It is used to model the exponential increase or decrease of the occurrences of the infection over a period of time by modelling the log of the frequency of infections with reference to time. The function of time is linear.

$$Ln(z) = a + rt \tag{1}$$

z It is the count of cases
a it is the intercept
t time variable
r It is the increase rate [1].

The polynomial model which is of order k has the following equation.
Model can be represented in matrix format as

$$Y = Ta + e \tag{2}$$

T is the matrix of the independent variables and columns of T is not orthogonal.

Diffusion hierarchy models:

The sum total of COVID 19 cases from starting point to time t is found to be the cumulative sum of new coming cases with respect to time. It is represented as $F(t)$. It is not a decreasing function, irrespective of value of $f(t)$, because $f(t) \geq 0 \; \forall \; t$. The survival function describes the count of survivors from COVID-19, that is, the susceptible class, not yet infected by COVID-19, but are in treat of the disease and it is given by

$$G(t) = S - F(t) \tag{3}$$

S is the total population count [2].

Epidemic cases are studied previously with time series model and new statistical methods. The linear regression has greater significance to make predictions. It shows the high degree of accuracy in making predictions of covid cases based on travelling history and contacts [3].

A strong correlation factor is needed to determine the relation between the independent and dependent variables. A R squared value of 0.99 or 1 indicates a strong power of the model to make forecasts [4].

Ecological regression is important to bring out new innovations. It is an evolving research area. It helps to study the impact of air pollutant content on the transmission of Covid 19 cases. The ecological regression analysis does not have the capacity to explain biological mechanisms which tell the relationship that exist with different natural factors [5].

3 Methodology Used

3.1 Dataset Used

The dataset is from COVID 19 forecasting competition from Kaggle and NOAA GSOD dataset. It is in Kaggle repository. The data is upto March 18 2020. The factors under study are:

a. Longitude
b. Latitude
c. Wind speed
d. Precipitation
e. Relative humidity
f. Absolute humidity
g. Average temperature
h. Maximum temperature
i. Minimum temperature
j. Wind speed
k. Fog
l. Sea temperature and pressure
m. Dew point

The data set also has information related to country region with the fatalities and confirmed cases in the regions.

3.2 Correlation Study of the Factors

Removal of null values is done with the dataset. Columns with many null values are removed. The study of correlation is now made with the resulting dataset (Fig. 2).

The factors such as wind speed, latitude, longitude, temperatures, sea temperature and pressure and precipitation have more correlation with confirmed cases.

3.3 Mean Squared Error

This metric is used for valuation of accuracy in regression. It is the square of the sum of paired observations expressing error. The observations are to represent the same phenomena.

$$\mathrm{MSE} = \frac{1}{n}\sum_{t=1}^{n} e_t^2 \tag{4}$$

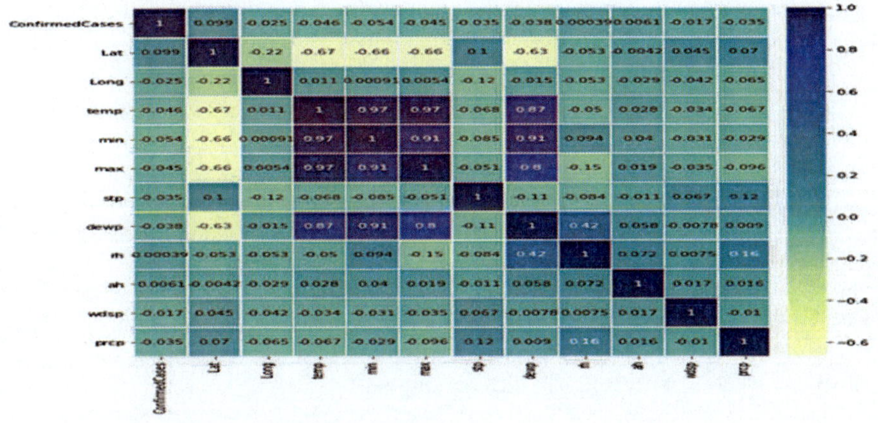

Fig. 2 Heatmap of the null value processed dataset

It is the squared sum of the deviations from predicted and true values of target values in the dataset divided by the total number of observations. Mean squared error and root mean squared error are used more as metrics in regression. Mean squared error and root mean squared error are enhancements of the MAE as metrics.

y: Predicted target values
x: True target values
e: $y - x$
n: The number of observations.

3.4 Models Used for Study

3.4.1 Linear Regression

It helps to create a model for a scalar response and many or one independent variables. When the number of independent factors if one it is called as simple linear regression. When the number of independent factors is more than one it is called as multiple linear regression.

3.4.2 Decision Tree Regressor

It is the model used to perform regression with decision trees. The depth of the tree parameter here helps to fine tune the quality of decision made. It creates a tree with decision nodes and leaf nodes.

3.4.3 Random Forest Regression

It is a supervised method of learning and uses ensemble learning for regression. It operates by constructing many decision tree regression trees in the training phase and outputs the mean of the classes as the output prediction.

3.5 *Analysis Performed*

The aim of the analysis is to determine the prime weather parameters that have impact in the spread of the Covid 19 cases. The study is made considering the confirmed cases and weather parameters in different regions.

4 Experiment Results

See Table 1.

4.1 *Linear Regression*

There is a great volatality in the results with linear regression. The mean squared error and R square error are high and low respectively.

Feature importance:

The average temperature and latitude are important features considering their priority in linear regression (Figs. 3 and 4).

Table 1 Experiment results

S. No	Regression results		
	Model fitted	Mean squared error	R square
1.	Linear Regression	10309.33	0.014
2.	Decision Tree Regression	106.504	0.99
3.	Random forest Regression	2148.04	0.95

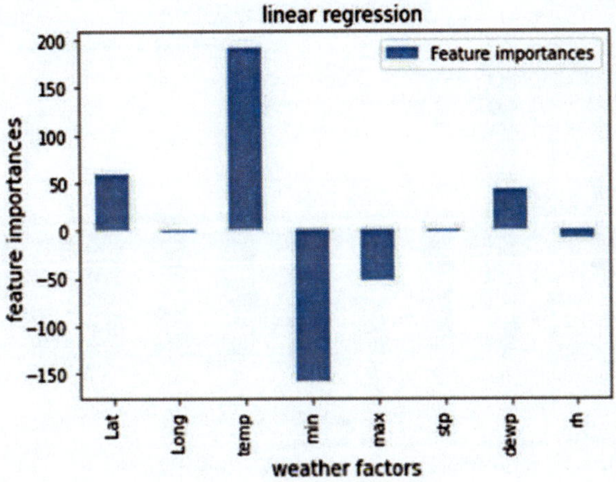

Fig. 3 Linear regression feature importances bar graph

	Feature importances
Lat	59.216647
Long	-0.656714
temp	190.856907
min	-158.324163
max	-52.504102
stp	-1.162626
dewp	43.275180
rh	-8.348996

Fig. 4 Linear regression feature importances

4.2 Decision Tree Regression

They are more constant and have low mean squared error and higher R squared score than the other models for the dataset.

Feature importance:

Latitude, longitude, maximum temperature, sea temperature and pressure and dew points are some important features under decision tree regression (Figs. 5 and 6).

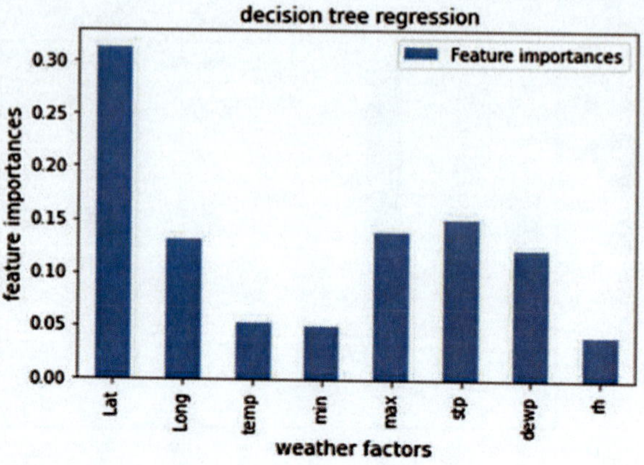

Fig. 5 Decision tree regression feature importances bar graph

	Feature importances
Lat	0.312434
Long	0.130546
temp	0.053371
min	0.050410
max	0.138115
stp	0.151462
dewp	0.122803
rh	0.040858

Fig. 6 Decision tree regression feature importances

4.3 Random Forest Regression

The mean squared error is higher when compared with decision tree regression but lower than linear regression. The R square score is lower than decision tree regressor but higher than linear regression.

Feature importance:

It has more similar feature importance as Decision tree regression. The maximum temperature, dewpoint, sea temperature and pressure, latitude and longitude are the important factors under random forest regression (Figs. 7 and 8).

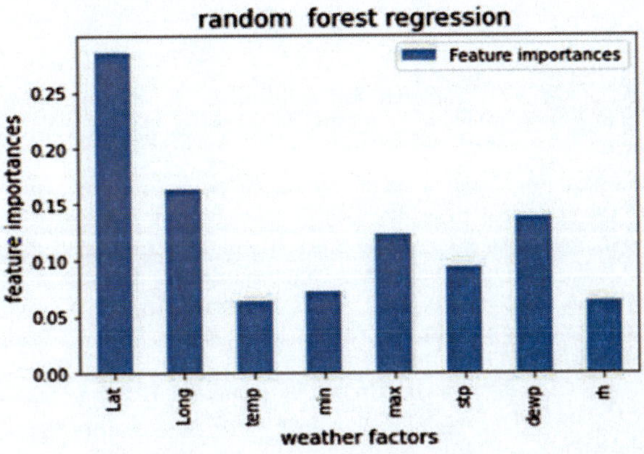

Fig. 7 Random forest regression feature importances bar graph

	Feature importances
Lat	0.285406
Long	0.162414
temp	0.063191
min	0.072122
max	0.122232
stp	0.093662
dewp	0.137513
rh	0.063461

Fig. 8 Random forest regression feature importances

5 Conclusion

The decision tree regression has better performance in terms of its metrics to make predictions. Therefore, the features which are considered as important under decision tree regression can be considered as important weather features having more impact in confirmed Covid 19 cases. Therefore, weather parameters having more impact on the expected cases are maximum temperature, dew point, sea temperature and pressure and geographical parameters are latitude then longitude of the region. However, to have a generalized approach for prediction outside the training set and known values, random forest regression can be used, though, its score is less than decision tree regression. The modelling can be improved with deep learning to perform feature engineering and to reduce the value of error.

References

1. Bhaskar, A., Ponnuraja, C., Srinivasan, R., Padmanaban, S.: Distribution and growth rate of COVID-19 outbreak in Tamil Nadu: A log-linear regression approach. Indian J. Public Health **64**(6), 188 (2020)
2. Ekum, M., Ogunsanya, A.: Application of hierarchical polynomial regression models to predict transmission of COVID-19 at global level. Int. J. Clin. Biostat. Biom **6**, 027 (2020)
3. Ogundokun, R.O., Lukman, A.F., Kibria, G.B., Awotunde, J.B., Aladeitan, B.B.: Predictive modelling of COVID-19 confirmed cases in Nigeria. Infect. Dis. Model. **5**, 543–548 (2020)
4. Rath, S., Tripathy, A., Tripathy, A.R.: Prediction of new active cases of coronavirus disease (COVID-19) pandemic using multiple linear regression model. Diabetes & Metab. Syndr.: Clin. Res. Rev. **14**(5), 1467–1474 (2020)
5. Wu, X., Nethery, R.C., Sabath, M.B., Braun, D., Dominici, F.: Air pollution and COVID-19 mortality in the United States: Strengths and limitations of an ecological regression analysis. Sci. Adv. **6**(45), eabd4049 (2020)

CloudML: Privacy-Assured Healthcare Machine Learning Model for Cloud Network

S. Savitha and Sathish Kumar Ravichandran

Abstract Cloud computing is the need of the twenty-first century with an exponential increase in the volume of data. Compared to any other technologies, the cloud has seen fastest adoption in the industry. The popularity of cloud is closely linked to the benefits it offers which ranges from a group of stakeholders to huge number of entrepreneurs. This enables some prominent features such as elasticity, scalability, high availability, and accessibility. So, the increase in popularity of the cloud is linked to the influx of data that involves big data with some specialized techniques and tools. Many data analysis applications use clustering techniques incorporated with machine learning to derive useful information by grouping similar data, especially in healthcare and medical department for predicting symptoms of diseases. However, the security of healthcare data with a machine learning model for classifying patient's information and genetic data is a major concern. So, to solve such problems, this paper proposes a Cloud-Machine Learning (CloudML) Model for encrypted heart disease datasets by employing a privacy preservation scheme in it. This model is designed in such a way that it does not vary in accuracy while clustering the datasets. The performance analysis of the model shows that the proposed approach yields significant results in terms of Communication Overhead, Storage Overhead, Runtime, Scalability, and Encryption Cost.

Keywords Healthcare datasets · Clustering · Encryption · Cloud privacy · Network performance · Machine learning

S. Savitha (✉) · S. K. Ravichandran
Department of Computer Science and Engineering, Christ University, Bangalore, India
e-mail: savitha.s@christuniversity.in

S. K. Ravichandran
e-mail: sathishkumar.r@christuniversity.in

J. S. Raj et al. (eds.), *Intelligent Sustainable Systems*, Lecture Notes in Networks and Systems 213, https://doi.org/10.1007/978-981-16-2422-3_5

1 Introduction

With the exponential growth of data in the Cloud Industry, there is a constant need to develop new tools and applications. Information processing is not an easy task and hence requires models designed artfully with appropriate services in built to the system. The services available in cloud have grown from IaaS (Infrastructure-as-a-Service) to FaaS (Function-a-as-Service) that particularly gave raise to the XaaS (Anything-as-a-Service) paradigm within the cloud network. These services enable hosting resources and infrastructure models to any part of the cloud world. Machine learning is also one such groundbreaking concept in the field of computer science that has invited huge number of developers and researchers, especially in the area of data analysis. Some of the domains where machine learning has stepped the foot are data science, artificial intelligence, prediction analysis, medicinal findings, cinema, and system rupture recognition which have all profited by progress in machine learning. However, data-owning participants often find it undesirable or impossible to share the information they collect. Companies may have obligations toward user privacy, aspire to monetize their data, or may need equal contribution to keep their data private to protect their platform. On real scale, medical and pharmacy industries have protection commitments under the Patient Privacy Act. With that, it is now an era to come up with a model to share information with progressive control over a more extensive scope of data sources. To unravel this quandary, cryptography is employed to empower the security features of machine learning paradigm. This leads to a new model, without uncovering a singular information that focuses on one another. It is understandable that numerous firms are keen in utilizing the existing security features than on developing a new one. Still, security is something which needs constant updation as the technology advances. So, the objective of this work is to develop a model with security-enabled cloud usage implemented on a machine learning platform. This will pave way to a versatile CloudML model for analyzing healthcare data. An abstract overview of the CloudML is shown in Fig. 1.

As unsupervised clustering algorithms are widely used to draw useful insights from the data, especially in the field of healthcare, the CloudML model implements a method to perform k-means on an encrypted dataset. So, to ensure the distance property among the datasets along with the security features, Homomorphic Encryption scheme is employed. This collaboration of homomorphic encryption scheme with k-means is deployed on the medical data to obtain a secured cluster of datasets. These clustered datasets are then made available to the doctors or researchers through the web interface which could enable them to draw insights on analyzing various factors causing a particular disease.

To develop this, a framework is formulated to gather various disjoint datasets that build up the k-means for analyzing the data. This further enables multi-party calculation with Paillier cryptosystem to actualize and secure the k-means grouping strategy for analyzing the information dispersed among various clusters. Paillier cryptosystem enables secure data storage in cloud by retaining the possibility to derive the ciphered data which can be deciphered later when needed. This enables no compromise on the integrity of the datasets in the proposed model.

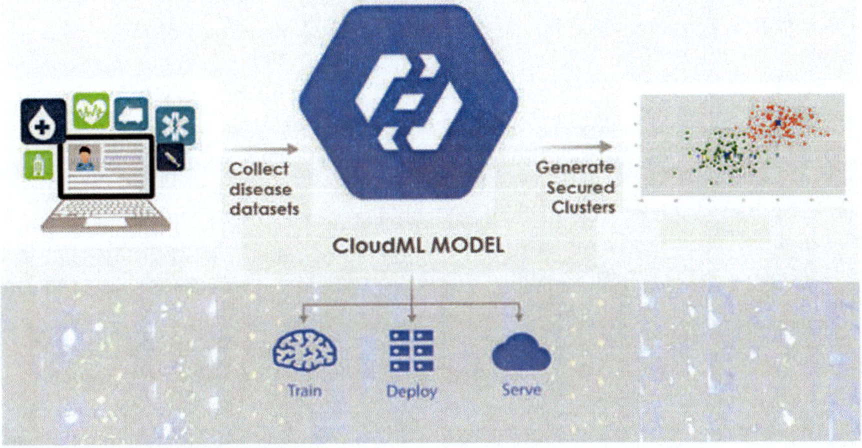

Fig. 1 CloudML abstract view

2 Related Work

Significant research has been done in the field of cloud security as it is crucial to store sensitive information without any third-party intervention and malicious attacks to the data. An analysis of various existing works that look into privacy preserving on the cloud has been done and is described below.

Vaidya and Clifton [10] presented a method for k-means clustering for privacy-preserving data mining algorithms. This methodology partitioned the data for a single entity in a vertical manner across multiple sites where each site had information about the subset of attributes for all the entities. It was found that clustering initiated the set of entities without disclosing any of the values where the clustering was centered. The advantage of this method is, given the mapping of points to clusters, each site was capable of calculating independent mean component that matched its respective attributes. This also ensured a reasonable privacy protection while limiting the communication costs. However, the main drawback of this approach was found to be the unawareness of the neighboring sites which generated less knowledge about the others sites' attributes.

Jagannathan and Wright [5] presented an efficient privacy-preserving protocol for K-means clustering which used both vertical and horizontal data partitioning. Since partition was done arbitrarily it was feasible to employ the protocols in this model for both the partitioning models. The major drawback of this system was that, although the intermediate cluster centers were not revealed, the algorithm leaked few credentials unintentionally with the intermediate cluster assignments.

Fahim et al. [4] presented a simple but effective clustering algorithm for k-means. This method used a simple data structure as a set or list for storing any information in the iteration, so that it could be used in the consecutive iterations. The experimental results showed that this scheme improved the k-means algorithm's computational speed with a sharp increase in magnitude of the total number of distance measurements and the overall time taken for the computation. With respect to the execution time of these k-means algorithms implemented along with the CLARA algorithm, it was found that all the algorithms performed almost on the same scale.

Bunn and Ostrovsky [1] presented a Two-Party k-means Clustering Protocol. This protocol was found to be appropriate in ensuring the privacy and performed better in the multi-party system. This method measured several k-means clustering iterations that had not disclosed any intermediate values of it. The downside of this approach was found to be the security assurance as it was based on cryptographic assumptions which denied the reversing such function criteria.

The work Vaidya et al. [11] gives a baseline for privacy-preserving protocols. It presented a protocol that used Naive Bayes classifier on both horizontal and vertical partitioned data. It also presented a method that bypasses the legal concern about sharing data. Nonetheless, one of the major drawbacks was that the cost and time for the classifier increased exponentially with the rise in data dimensionality, and Naive Bayes seemed to be inefficient in handling such large-scale datasets.

Doganay et al. [3] described about decentralized scenarios where data was partitioned vertically through multiple sites. It was found that, every site expected the clustering to be implemented without exposing the local server. The paper proposed a new clustering protocol based on additive secret sharing to protect the privacy of k-means. This new protocol was found to perform better than state-of-the-art scenario. However, increased communication and computation costs were the disadvantages of this methodology. The overall communication cost of the proposed algorithm was found to be lower up to a certain number of users (when using 1024-bit encryption up to 64 users), after which it increased exponentially.

Zhao et al. [15] proposed a MapReduce-based parallel k-means clustering algorithm which was found to be a powerful parallel programming technique for processing the datasets. This algorithm was found to be efficient in processing large datasets on commodity hardware with lower computation cost. However, this model achieved lower processing time when deployed in Hadoop environment, especially for MapReduce.

Sakuma and Kobayashi [6] implemented K-means clustering implemented with a new concept of privacy protection and user-centered privacy preservation. As this protocol functioned with large number of users, the system was found to be more scalable with confidentiality factors incorporated in it. It was found to be scalable up to one million users in real time, but there were issues with the asynchronous activity that led to low fault tolerance factor.

Cui et al. [2] discussed the problems of large-scale data processing using the K-means clustering algorithm. With MapReduce, a novel processing technique was introduced to eliminate the iteration dependence which led to excess computation. Experimental analysis on real scale with synthetic datasets proved that the optimized algorithm was found to perform better than the parallel K-means approach. The significant shortcoming of this approach is that it required regular checks to be carried out to confirm that every cluster has at most one center from C_i such that the collection of centers from sample i, where G_i is considered as the ith group of the center. This was found to be a time-consuming and compute-intensive task.

Zhang et al. [14] proposed High-Order PCM (HOPCM) algorithm for large data clustering that optimized the tensor space required for an objective function. A polynomial function was employed to enable the update option for the membership matrix and the clustering centers. This supported the BGV system's secure computing methodology. However, the inherent power of homographic encryption schemes of BGV was not clear.

Yuan and Tian [13] proposed an encryption scheme that deployed Learn with Error (LWE) hard problem in it. Since the encryption scheme was lightweight, the scheme achieved a high precision and clustering speed compared to the clustering in the k-means without enabling the privacy-protection criteria. Also, MapReduce architecture was incorporated into the design to support large-scale datasets that enabled parallel operations in the cloud computing platform as well. Since it was lightweighted, this model scaled up to 5 million users. Also, as LWE is an NP-hard problem, breaking the cluster within the given polynomial time was not possible.

Sharma et al. [7] discussed various tools to categorize and classify textual data in Mahout which was found to be a flexible machine learning library for data analysis. Mahout used TF-IDF weighting technique to create text vectors after which the clustering algorithms were applied on the vector sets. The clustering algorithms used for analysis were K-means, fuzzy K-means, Latent Dirichlet Allocation (LDA), and spectral clustering. Though fuzzy K-means clustering was found to execute the task in limited time for constant number of clusters, there are key areas that require assistance toward the future work. The work lags in a specific machine learning model with enhanced security features for the dataset.

Smys and Raj [8] proposed an IoT-based cloud model for monitoring the patients in the rural areas and small villages with the limited accessibility of the physicians and the healthcare centers. The model enabled timely suggestion of appropriate treatments for the patients with acceptable accuracy and data prediction. The patient's information is transmitted over the cloud from their smart devices for the analytical operations using Mahout [7]. Though the prediction and accuracy for analyzing the patients were acceptable security and privacy aspects for transmitting the patient's data have not been taken into consideration.

Suma [9] proposed a hybrid deep fuzzy hashing algorithm that retrieved the information by mapping similar information as correlated binary codes thereby training the underlying information using deep neural network and fuzzy logic from distributed cloud. The results of the proposed algorithm were compared against Support Vector Machine-Based Information Retrieval System and Deep Neural Network-Based Information Retrieval System based on the accuracy, specificity, sensitivity, and f-measure. Though the retrieval efficiency of the proposed algorithm was found to be beyond 95%, managing wide number of features and providing security to the models still remains unsolvable.

3 Proposed Approaches

The CloudML model has various levels of algorithms incorporated in it. These algorithms are linked with each module in an iterative manner to derive the analysis in a protected environment without any third-party intervention.

3.1 Unsupervised K-Means Clustering Algorithm

As the amount of data generated is growing exponential, particularly in the field of medicine, there is a need for an unlabeled processing model to solve intractable and abstract patterns from the datasets. Unsupervised K-means clustering is one such model that enables mining underlying patterns and the nature of distributed data in a detailed way. For the k-means clustering process, we first start by dividing the dataset into "k" clusters. Then the "k" cluster centers the datasets before starting with the development of the classifier that will be assigned to the datapoints of the corresponding cluster. Now, the datapoints are assigned in such a way to a cluster center so that the distance between the datapoint and the respective cluster is minimum. Once the initial assignment is complete, the CloudML model then computes the mean of each cluster and assigns that value as the cluster center for the next iteration. This process continues until either the cluster centers remain constant or if it has reached the maximum number of iterations. The proposed algorithm works in such a way that these cluster centers are calculated even if varied number of users own their own variety of datasets inside the cloud network.

3.2 Secured Multi-party Addition Algorithm

When there are multiple trustless users on the cloud network, it is always advisable to maintain the privacy for the computation of functions. So, this layer elaborates on secured multi-party addition algorithm that is employed in the cloud ML model.

Let there be k participants, $P_1, P_2, \ldots, P_k$, where each party P_i possesses a private input x_i. The algorithm works in such a way that every party P_i to end with a secret share L_i, the following equation must be satisfied.

$$\sum_{i=1}^{k} x_i = \prod_{i=1}^{k} l_i \tag{1}$$

In addition, an additive homomorphic encryption scheme named EncDec is employed on the private input to further enhance the security feature of the CloudML model. Also, KeyGen function creates a public encryption key as well as a private decryption key "d." The pseudocode for the algorithm is given below:

1. P_1 executes KeyGen to get the keys "d" and "e" and broadcasts "e" to all the participants in the network.
2. Now, P_1 calculates the encryption value of x_1 using "e" and sends it to P_2.
3. For $i = 2$ to $i = k - 1$, P_i calculates the following equation:

$$Enc_w(x_i).\prod Enc_e(x_j) \quad (2)$$

4. P_k takes the value of the product from $i = 1$ to $k - 1$ of Ence(x_i) provided to it by P_{k1} now, multiplies the value with the encryption of its input Ence(x_k).
5. Now, P_k randomly selects L_k (which is not an additional identity under Enc), to compute the inverse of L_k and then sends it to P_{k-1}.
6. Here For $i = k - 1$ to 2, P_i arbitrarily selects a non-identity L_i and its corresponding L_i inverse to put the value y_{i+1} that is received from P_{i+1} to the power of L_i inverse and sends back to P_{i-1}.
7. P_1 decrypts the message and finalizes the value as the share of L_1.

Algorithm 1: Secured Multi-Party Addition

Result: Shares l_i of the original secret values
Generate e and d for all processess P_i
$i = 2$;
while $i \neq k - 1$ **do**

Calculate $Enc(x_i), \prod_{j=1}^{i-1} Enc(x_j)$

Send the result to P_{i+1};

end
Randomly select l_k *and compute* l_k^{-1}

Send $y_k = \left(\prod_{j=1}^{i-1} Enc(x_i))^{l_k-1}\right)$ *to* P_{k-i}

$i = k - 1$;
While $i \neq 2$ **do**

Send $y_i = y_{i+1}^{l_i - I}$ *to* P_{i-1}

end
P_1 decrypts y_2 and sets the calculated value as it's share l_1
$l_1 = Dec_d y_2$

3.3 Pailier Homomorphic Encryption

The next layer of CloudML is Pailier Homomorphic Encryption. This algorithm functions in a procedural way where the participants can compute the functions on the data without granting access to the entire dataset inside the network. So, to allow operations on the encrypted datasets, Homomorphic Encryption is employed. The Paillier cryptosystem is incorporated in the model, which enables homomorphic encryption with ease. Such homomorphic additive encryption schemes will produce a non-deterministic encryption approach which makes our model more secured to illegal access. This proves that with any two ciphertexts, the model will calculate a new ciphertext. This encrypts the sum two available plaintexts curtailing the knowledge of the secret key (given the ciphertexts will not be decrypted). The core mathematical function is the multiplication operation of two ciphertexts. This is equal to the addition of the underlying plaintexts. This is formulated as follows:

$$\boldsymbol{E(m_1, r_1) \cdot E(m_2, r_2)(modn * n)} \tag{3}$$

This states decrypts on $m_1 + m_2$. Additionally, an encrypted plaintext (+) and an unencrypted plaintext (–) are multiplied with the power function. The formula for this given below:

$$\boldsymbol{pow(E(m_1, r_1) \cdot (modn * n), m_2)} \tag{4}$$

This denotes decrypts to $m_1 + m_2$.

4 Work Done

The CloudML system can manage parallel processes and participants in a cloud network without any compromise in performance. The processes are in built with sockets using ZeroMQ library to enable communication among the users in the network. For n participants, "n" number of processes are created and are assigned with a process number. This ranges from 0 to $n - 1$ enabling each process to know the appropriate socket to be communicated. The algorithm runs the required code in parallel to capture the order of messages sent and received. "Process 0" acts as the root process. Here, the generation of the public and private key is executed using the Paillier cryptosystem. Now, the root process distributes the public key to all the participants in the network. It is during this stage, the Multi-Party Addition function (getAddShares) is implemented which runs parallel on all of these processes. The inputs are added securely to each of these functions and the results are returned to the root process (Process 0). At each iteration, the instance of the k-means class is identified to calculate the local centroid of that particular cluster. Now, the getAddShares function computes the global centroids to obtain the current k-means object which will be updated with a new mean value.

5 Experimental Results and Discussion

Comparison of the performance of the CloudML model with Pailier Homomorphic Encryption with the existing system [13]. Learn with Error (LWE) hard problem is described in this section. There are significant outcomes in our proposed model with respect to Communication Overhead, Storage Overhead, Scalability, Encryption cost, and Runtime Analysis.

The metrics are presented in this section as follows:

1. The number of clusters (means) found in the dataset against the runtime in seconds.
2. Dimensionality of the data against the runtime in seconds.
3. The number of data points against runtime in seconds.
4. Monitoring the overhead due to the communication process.
5. Storage requirements varied with the number of data objects in millions.
6. Scaleup rate.
7. Time taken for encryption varied with number of data points.

Having all the variables constant, the parameters mentioned above are evaluated individually to obtain a clear analysis of the system. It is also observed that these variables have a collusive effect which keeps increasing. However, based on the changes observed on the individual effects of each variable, there are some conclusions obtained on the performance of the proposed system. Each parameter is individually discussed in the following sub-sections.

5.1 Communication Overhead

In the existing system, the overhead in communication occurs due to the interaction between the nodes after each round of clustering. Thus, the aggregate cost on communication for per-round of clustering with 4nK vector elements will take up to 8 bytes. In the proposed system, with the use of the multi-party additive concept with the k-means algorithm, there was a noticeable reduction in the communication overhead up to 4 bytes and thus there is an improvement in the performance as seen in Fig. 2.

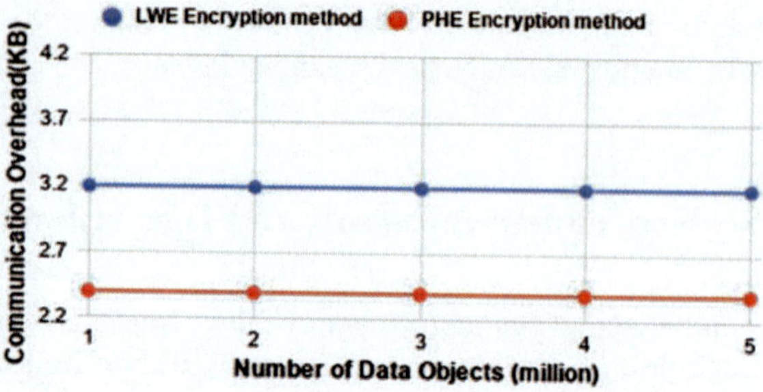

Fig. 2 Comparison of communication overhead

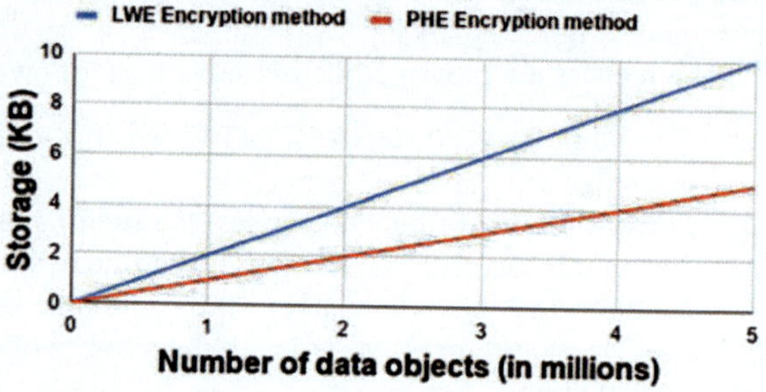

Fig. 3 Comparison of storage overhead

5.2 *Storage Overhead*

The method followed in the existing system led to an immense increase in overhead in terms of storage space as each data object as well as the cluster center is represented as vectors of "n"-dimensions. As the encryption of these vectors is done by converting them into twice the original dimension, this resulted in increase in the storage cost to be four times than that of the unencrypted k-means algorithm. However, in the proposed schema, the storage overhead is focused specifically as we do not require the 2n-dimensional vector and only the n-dimensional vector is taken into consideration. Therefore, the proposed model ensures that the storage cost is reduced as seen in Fig. 3.

5.3 *Scalability*

The evaluation of scalability of the existing scheme is done for "scaleup" as mentioned in (Xu et al., [12]. "Scaleup" is a measure to evaluate how the system will perform when given m-times larger resources thereby requiring the system to finish a scaled up job in lesser time than the actual job. Thus, if the actual job time is TOG, the rate of scaleup is calculated as the percentage of the job finished in TOG over a hundred percentage. When evaluating the existing system and proposed system on

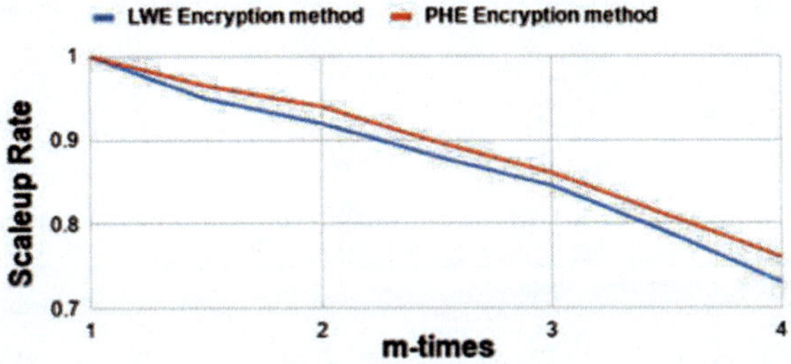

Fig. 4 Comparison of scaleup rate

this metric scale, the actual job value is set for performing the clustering process for 1 million data objects against two individual nodes. So, as seen in Fig. 4 it is evident that the scalability of the proposed system is higher than the existing approach.

5.4 Encryption Cost

In the existing system scheme, the owner of the data created the keys to perform the encryption as well as the updation of the cluster centroids. The generation of keys will only occur once at the beginning of the process while selecting the two invertible matrices. The encryption of "n" elements will require the owner to perform 4n operations which increases the encryption cost in the existing scheme. But, in the proposed system, we do the operations on n-dimensional vectors, thus the encryption cost is directly dependent on the time taken for encryption using the Paillier Homographic Encryption scheme. As shown in Fig. 5, it is clearly observable that the encryption cost has significantly reduced when compared to the existing model.

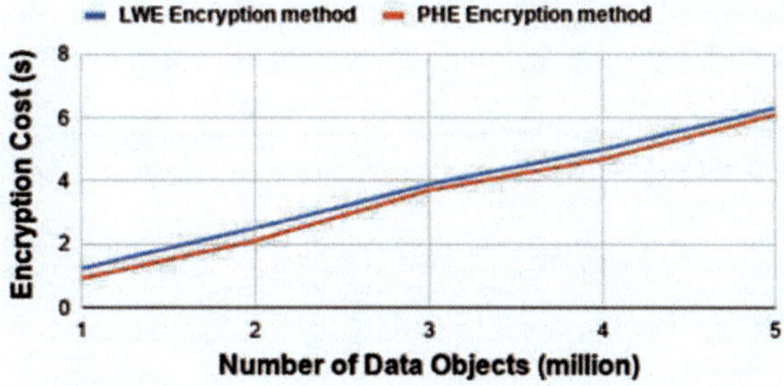

Fig. 5 Comparison of encryption cost

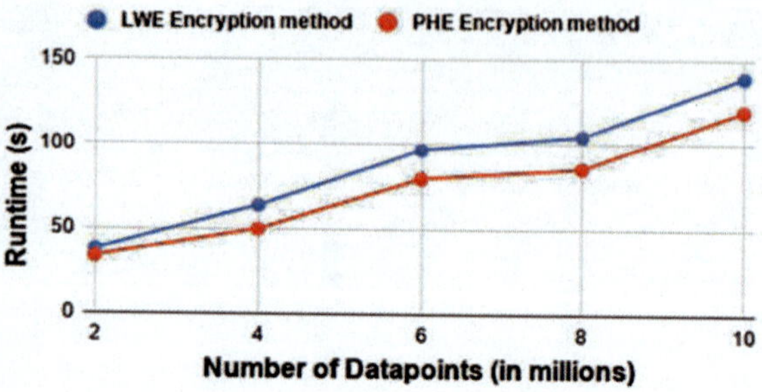

Fig. 6 Comparison of runtime

5.5 *Runtime Analysis on Data Points*

The runtime of the proposed system is compared with the existing system by increasing the number of data points. For data points lesser than 2 million, the runtimes of both the systems seem to converge. However, with increase in the number of data points, there is a considerable difference observed in the runtimes as shown in Fig. 6.

5.6 *Runtime Analysis on Data Dimensionality*

When comparing the dimensionality on the number of clusters, while considering the data points for up to 2.2 million data, it is evident that the existing as well as the proposed system shows varying performances with neither performing better than the other. So, there is a considerable drop in the runtime when the data point objects reach around 3 million in number. As the existing system requires storage which is double the dimensionality of the data, the processing speed tends to be lower. The variation in runtime is shown in Fig. 7.

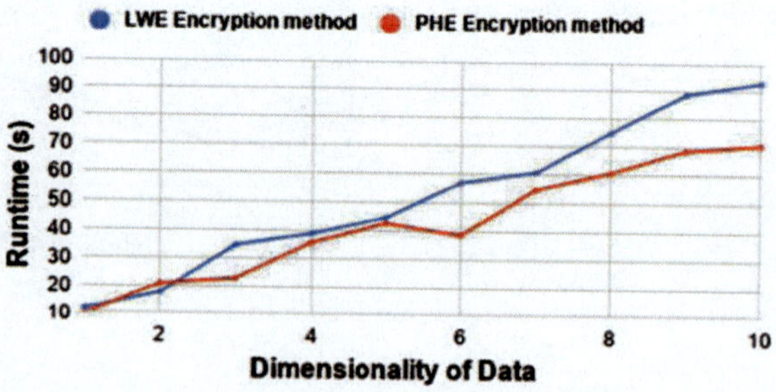

Fig. 7 Variation of runtime with dimensionality

6 Limitations and Future Work

The CloudML model that is developed is specific to identifying the severity of the heart-related diseases. As a continuation, this work could be extended to work with data related to other diseases from the MIMIC dataset including clinical notes, vital signals, and other measurements that would enable this model to identify and cluster disease datasets. This system prevented external attacks, ensured data confidentiality, and integrity. The primary work of this model is to stop the intruder from accessing and modifying the medical data without the knowledge of the participants in the cloud network. Along with this, an additional security layer is enabled in the login module for the patients as well as the doctors for preventing illegal entry to the patient's database. As a part of the future work, this model could be improved on increasing the security to specifically address certain attacks, namely, DDoS (distributed denial of service) and man-in-the-middle.

7 Concluding Remarks

The CloudML model is specially designed to develop a scalable and secured deployment of a multi-party machine learning algorithm for cloud network. The model implemented and explored the horizontally partitioned k-means clustering algorithm by balancing the performance as well as the security features of the cloud network. The proposed system is implemented in Python using the Paillier Homomorphic Encryption libraries that displayed the user interface for the patients to enter their symptoms, heart rate, and pressure, so that the doctors can view to which cluster the patient belongs to. Thus, the treatment will be carried out based on the severity of the disease. The evaluation metrics portrays that the developed model used performance-oriented algorithms, but still there is a substantial room for further development in usability of the current model. Having that as the concluding remarks, the future work could be focused toward including hybrid machine learning algorithm with computational primitives as a major factor for the cloud network.

References

1. Bunn, P., Ostrovsky, R.: Secure two party k-means clustering. In: Proceedings of the 14th ACM Conference on Computer and Communications Security, pp. 486–497. ACM (2007)
2. Cui, X., Zhu, P., Yang, X., Li, K., Ji, C.: Optimized big data k-means clustering using map reduce. J. Super Comput. **70**(3), 1249–1259 (2014)
3. Doganay, M.C., Pedersen, T.B., Saygin, Y., Savas, E., Levi, A.: Distributed privacy preserving k-means clustering with additive secret sharing. In: Proceedings of the 2008 International Workshop on Privacy and Anonymity in Information Society, pp. 3–11. ACM (2008)
4. Fahim, A.M., Salem, A.M., Af Torkey, F., Ramadan, M.A.: An efficient enhanced k-means clustering algorithm. J. Zhejiang Univ. Sci. **7**(10), 1626–1633 (2006)

5. Jagannathan, G., Wright, R.N.: Privacy preserving distributed k-means clustering over arbitrarily partitioned data. In: Proceedings of the Eleventh ACMSIGKDD International Conference on Knowledge Discovery in Data Mining, pp. 593–599 (2005)
6. Sakuma, J., Kobayashi, S.: Large scale k-means clustering with user centric privacy -preservation. Knowl. Inf. Syst. **25**(2), 253–279 (2010)
7. Sharma, I., Tiwari, R., Rana, H.S., Anand, A.: Analysis of mahout big data clustering algorithms. In: Singh, R., Choudhury, S., Gehlot, A. (eds.) Intelligent Communication, Control and Devices. Advances in Intelligent Systems and Computing, vol. 624. Springer, Singapore (2018)
8. Smys, S., Raj, J.S.: Internet of things and big data analytics for health care with cloud computing. J. Inf. Technol. Digit. World **01**(01), 9–18 (2019)
9. Suma, V.: A novel information retrieval system for distributed cloud using hybrid deep fuzzy hashing algorithm. J. Inf. Technol. Digit. World **02**(03), 151–160 (2020)
10. Vaidya, J., Clifton, C.: Privacy preserving k-means clustering over vertically partitioned data. In: Proceedings of the ninth ACMSIGKDD international conference on Knowledge discovery and data mining, pp. 206–215. ACM (2003)
11. Vaidya, J., Kantarcıoglu, M., Clifton, C.: Privacy preserving naive bayes classification. VLDB J. **17**(4), 879–898 (2008)
12. Xu, X., Jager, J., Kriegel, H.P.: A fast parallel clustering algorithm for large spatial databases. In: High Performance Data Mining, pp. 263–290. Springer (1999)
13. Yuan, J., Tian, Y.: Practical privacy preserving map reduce based k-means clustering over large scale dataset. IEEE Trans. Cloud Comput. (2017)
14. Zhang, Q., Yang, L.T., Chen, Z., Li, P.: Pphopcm: Privacy preserving high order possibilistic c-means algorithm for big data clustering with cloud computing. IEEE Trans. BigData (2017)
15. Zhao, W., Ma, H., He, Q.: Parallel k-means clustering based on map reduce. In: IEEE International Conference on Cloud Computing, pp. 674–679. Springer (2009)

Performance Evaluation of Hierarchical Clustering Protocols in WSN Using MATLAB

Sarang D. Patil and Pravin S. Patil

Abstract To improve the lifespan of WSN (wireless sensor networks), efficient optimization of network resources is necessary. Optimizing the network's energy utilization is the most important issue which is being addressed by most of the researchers these days. Depending on the structure, WSNs are divided into flat WSN and hierarchical or clustered WSN. In energy optimization for large-scale WSN, clustering proved an effective approach. Many researchers proposed many clustering protocols to date. Based on the energy associated with the sensor nodes, researchers divided networks into two categories: homogeneous networks and heterogeneous networks. In this paper, we tested LEACH protocol in a homogeneous environment and heterogeneous environment, centralized LEACH, SEP, DEEC, and developed DEEC protocols under different scenarios such as change in the sink position and change in the area. We evaluated and compared them on performance metrics such as network lifetime, throughput, and energy consumption.

Keywords LEACH · DEEC · SEP · DDEEC · Performance evaluation · Hierarchical clustering

1 Introduction

Tiny sensor nodes deployed in the field of interest to sense the information comprise a network called wireless sensor network [1]. The sensor node is capable of sensing, storing, transmitting, and receiving the data. These nodes are constraint in energy and processing capabilities due to their tiny size [2]. Therefore, to improve network lifetime energy of the node must be used optimally. Energy optimization to prolong network lifetime is the primary objective of researchers in this area [1, 3]. Scalability

S. D. Patil (✉)
Department of Electronics and Telecommunication Engineering, Gangamai College of Engineering, Nagaon, India

P. S. Patil
Department of Electronics and Telecommunication Engineering, SSVPSBS Deore College of Engineering, Dhule, India

J. S. Raj et al. (eds.), *Intelligent Sustainable Systems*, Lecture Notes in Networks and Systems 213, https://doi.org/10.1007/978-981-16-2422-3_6

and reliability are also important design goals that are focused on by researchers. Network protocols play a significant role in energy optimization. Many routing protocols were defined to optimize the energy consumption of the WSN [4].

WSN networks fall into two categories such as homogeneous networks and heterogeneous networks, depending on the availability of their resources [5, 6]. In homogeneous networks, sensor nodes bear identical energy, transmission capability, and processing capacity. However, in heterogeneous networks, case is precisely different. In heterogeneous networks, the node has different energy levels [6].

Routing protocols are categorized as clustering protocols, flat protocols, and location-based protocols according to the adopted network structure [7]. Among the routing protocols, clustering protocols have taken the lead in energy optimization and scalability [8]. Clustering protocols play a significant role in energy optimization when nodes are deployed randomly. Clustering protocols are proactive routing protocols which divide the network area into different clusters. Clustering protocols are further subdivided into hierarchical clustering and chain-based clustering [7, 9].

In hierarchical clustering, network area is subdivided into a small area known as a cluster. In every cluster, the network chooses any node as cluster head. All the nodes in a cluster transmit data to CH. These cluster heads perform routing, data aggregation functions. CH transmits aggregated data to the base station (BS) and hence hierarchy is maintained in clustering protocols. Therefore, these protocols are called hierarchical clustering protocols [6]. Figure 1 shows the networks' hierarchical structure. The network consists of three virtual layers. Nodes in the same virtual layer perform the same task for the assigned period. Generally, BS is only one for the entire network, and it is assumed that BS has unlimited resources like energy and processing power as compared to CHs and normal nodes [6, 10].

For a homogeneous network, many clustering protocols are developed, where all nodes carry same energy. LEACH protocol (low-energy adaptive clustering hierarchy) is the most popular and widely accepted protocol for homogeneous network [11]. Whereas, in heterogeneous networks, nodes with two or more energy levels are considered while designing protocols [12]. In this paper, we simulate and analyze well-referred hierarchical clustering protocols [13–15]. These protocols are centralized or distributed depending on the process of selecting cluster heads, the location of each node in the cluster, and their residual energy used to decide one of them as

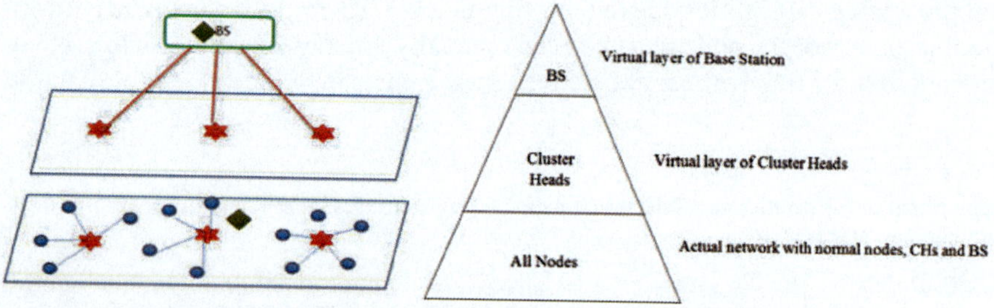

Fig. 1 Hierarchical structure in clustering protocol

a cluster head. The performance of LEACH [11, 16], LEACH centralized [16, 17], SEP (stable election protocol) [18], DEEC (distributed energy-efficient clustering) [19], and developed DEEC (developed distributed energy-efficient clustering) [20] protocols are analyzed and compared under the influence of different conditions such as different network area, change in the position of base station. In Sect. 2, we will explain radio model used. In Sect. 3, we will give brief overview of clustering protocols selected. In Sect. 4, we discuss performance metrics and scenarios. Sections 5 and 6 contain results and conclusions, respectively.

Out of these protocols, we have simulated LEACH in both homogeneous environments and heterogeneous environments [21], SEP, DEEC, and developed DEEC protocols. We tested them on parameters like network lifetime, the throughput of the network, and average energy consumption.

2 Radio Model

Majority of the protocols utilize a radio model, as shown in Fig. 2 [10, 21]. In this radio model, a sensor node comprises seven important blocks which includes transmission as well as a reception antenna, transmission as well as reception circuits, transmission amplifier, sensor, and data processor. System energy is consumed in three operations: data aggregation, data reception, and data transmission. System's energy equation is given as

$$E_{Tx} = E_{\mathrm{elect}} \times k + \left(\varepsilon_{fs} \times k \times d^2\right) \; if \;\, d \le d_0 \tag{1}$$

$$E_{Tx} = E_{\mathrm{elect}} \times k + \left(\varepsilon_{mp} \times k \times d^4\right) \; if \;\, d > d_0 \tag{2}$$

where E_{Tx} is transmission energy per bit, E_{elect} energy required by circuit per bit, ε_{fs} is amplifier energy per bit, ε_{mp} is amplifier energy per bit, number of transmitted bits are k and the distance between CH and node is d.

d_0 is calculated as

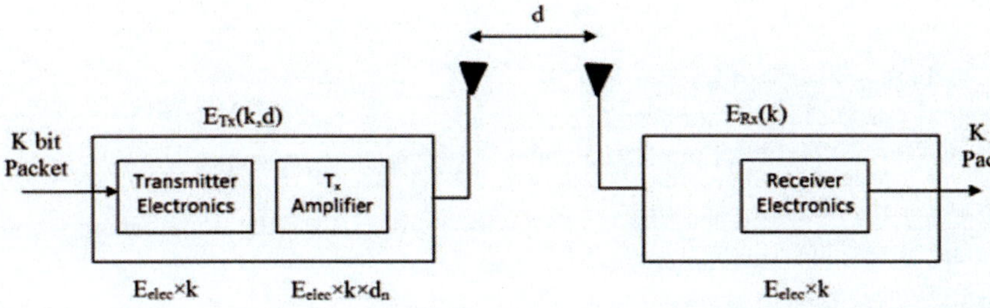

Fig. 2 Radio model

$$d_0 = \sqrt{\varepsilon_{fs}/\varepsilon_{mp}} \quad (3)$$

To receive message, energy required by receiver is

$$E_{Rx} = E_{\text{elect}} \times k \quad (4)$$

Also, energy is consumed in data aggregation. At CH level, data aggregation is performed to remove redundancy of similar data packets. Aggregated data packets reduce number of transmissions to BS.

3 Overview of Clustering Protocols

3.1 LEACH (Low-Energy Adaptive Clustering Hierarchy)

LEACH is a hierarchical clustering protocol used in homogeneous networks [11]. In LEACH, cluster formation is dynamic. In LEACH, cluster heads (CH) are elected randomly among the cluster members. The cluster head role rotates in the network to distribute the energy load evenly. In a cluster, data gathered at the cluster head has redundancy in it. The aggregation process removes this redundancy. CH performs data aggregation and sends aggregated data to BS to reduce transmission load. To reduce intra-cluster and inter-cluster collision, LEACH uses TDMA/CDMA MAC [11].

The operation of LEACH is performed in two phases: setup phase and steady-state phase. The election of CH is dependent on predetermined fraction p and the number of times node became CH previously. In the setup phase, the predetermined fraction of cluster members is elected as CH. The process of CH election is as follows [11]:

- The node who wants to become a CH chooses any number between 0 and 1.
- Formula for calculation of threshold value $t(n)$ of the node is

$$t(n) = \begin{cases} \frac{p}{1-p\times\left(r \bmod \frac{1}{p}\right)} \, if n \in G \\ 0, \, else \end{cases} \quad (5)$$

  - where p is desired fraction of CH;
  - r is current round number;
  - G represents the set of elected nodes in previous $1/p$ rounds.

- If the value of $t(n)$ is less than the selected number, then the node becomes CH at any point during next $1/p$ round.

When node becomes CH, it broadcasts its advertisement using CSMA MAC protocol. All non-CH members receive this advertisement and decide to which CH they should join. After deciding to which cluster it should join, non-CH informs CH about its membership in the cluster using CSMA MAC protocol. Once all the members join the cluster, the CH creates a TDMA schedule based on the number of nodes in a cluster. This schedule is broadcasted to all cluster members. Members send their data to CH in their specified time slot during the steady-state phase [11, 16].

3.2 LEACH in Heterogeneous Environment

In heterogeneous environment, m fractions of advanced nodes have α times higher energy than normal nodes. As the advanced nodes have more energy, they get more chance to become a CH in heterogeneous environment [21]. The total energy in two-level heterogeneous environment is increased by $(1 + \alpha m)$ times [21]. We observed that LEACH took some advantage of this increased energy, but in unstable region of operation CH selection process becomes unstable and no CH is selected. Due to this these advanced nodes remained idle in unstable region.

3.3 LEACH-C

Irrespective of some advantages, LEACH does not guarantee the placement and number of CH [22]. Using the central control algorithm, CH may be spread evenly in the entire network. LEACH centralized algorithm [17] works on this foundation.

This protocol uses a centralized clustering algorithm to form clusters and place CH accordingly. In LEACH-C protocol, BS needs the information of energy level and position of each node to calculate CH. The base station then calculates the average energy of the network and if the node is having less energy than average energy, then it cannot become CH. Due to this energy load balancing is achieved. After forming clusters and assigning their CH base station, broadcast a CH id message to all nodes. If the head id matches with its own id, the node is a CH. Otherwise, it acts as a normal node and it determines its TDMA schedule for data transmission from the received message. The steady-state phase of LEACH centralized is the same as that of LEACH [17].

3.4 SEP

SEP used energy heterogeneity in the sensor network. It uses normal node having normal energy, and advanced having α times more energy than normal nodes. In the network, advanced nodes are of m fraction than normal nodes. The spatial density of the network n SEP is the same as LEACH. Due to energy heterogeneity, the total energy of the network is increased by $(1 + \alpha m)$ times. To increase stable region, to keep the optimal value of CH constant, and to take advantage of energy heterogeneity, SEP elects advanced node as CH more than normal node [18].

The weighted probabilities for advanced nodes and normal nodes are

$$\text{For advanced node } P_{\text{adv}} = \frac{P_{\text{opt}}}{1 + a \cdot m} \times (1 + a) \tag{7}$$

$$\text{For normal node } P_{\text{nrm}} = \frac{P_{\text{opt}}}{1 + a \cdot m} \tag{8}$$

where m is the fraction of advanced nodes having a times higher energy than normal nodes. Cluster head probability P_{opt} can be given as

$$P_{\text{opt}} = \frac{k_{\text{opt}}}{n} \tag{9}$$

where k_{opt} is calculated as

$$k_{\text{opt}} = \sqrt{\frac{n}{2\pi}}\sqrt{\frac{\varepsilon_{fs}}{\varepsilon_{\text{amp}}}} \times \frac{M}{d_{toBS}^2} \tag{10}$$

So the threshold for selecting CH in SEP has two threshold values. Threshold values for normal node and for advanced node are given as

$$t(s_{\text{nrm}}) = \begin{cases} \frac{p_{\text{nrm}}}{1 - p_{\text{nrm}} \times \left(r mod \frac{1}{p_{\text{nrm}}}\right)} if s_{\text{nrm}} \in G' \\ 0, otherwise \end{cases} \tag{11}$$

$$t(s_{\text{adv}}) = \begin{cases} \frac{p_{\text{adv}}}{1 - p_{\text{adv}} \times \left(r mod \frac{1}{p_{\text{adv}}}\right)} if s_{\text{adv}} \in G'' \\ 0, otherwise \end{cases} \tag{12}$$

where G' is the set of the normal node which has not become a CH during last $1/p_{\text{nrm}}$ rounds. G'' is the set of the advanced node which has not become a CH during last $1/p_{\text{adv}}$ rounds [18].

3.5 DEEC

Performance of SEP in multi-level heterogeneous environment degrades as it is designed for two-level heterogeneity only. Also, it works poorly when there is heterogeneity as a result of the operation of the sensor network because its election probability and rotational epoch are directly related only to the initial energy [23].

To select cluster heads, DEEC uses the initial and residual energy levels of the nodes. By estimating the optimal value of network lifetime DEEC bypasses the need for global knowledge of network. This estimated optimal value is used to calculate the value ideal energy usage of each node in one round [19].

The weighted probability is calculated as

$$p_i = \frac{E_i(r)}{\overline{E}(r)} \tag{13}$$

where $\overline{E}(r)$ denote the average energy at round r of the network, $E_i(r)$ denotes residual energy of node i at round r.

The average energy of rth round $\overline{E}(r)$ is estimated as follows:

$$\overline{E}(r) = \frac{1}{N} E_{\text{total}} \left(1 - \frac{r}{R}\right) \tag{14}$$

The total number of rounds R of the system is estimated as

$$R = \frac{E_{\text{total}}}{E_{\text{round}}} \tag{15}$$

The E_{total} is calculated as

$$E_{\text{total}} = NE_0(1 + am) \tag{16}$$

And E_{round} is given by

$$E_{\text{round}} = L\left(2NE_{\text{elec}} + NE_{DA} + k\varepsilon_{mp} d_{toBS}^4 + N\varepsilon_{fs} d_{toCH}^2\right) \tag{17}$$

The average distance of the node to base station d_{toBS} is given by

$$d_{toBS} = 0.765\frac{M}{2} \tag{18}$$

The average distance of a node to cluster head d_{toCH} is given by

$$d_{toCH} = \frac{M}{\sqrt{2\pi k_{\text{opt}}}} \tag{19}$$

And k_{opt} is given by

$$k_{\text{opt}} = \sqrt{\frac{n}{2\pi}} \sqrt{\frac{\varepsilon_{fs}}{\varepsilon_{\text{amp}}}} \times \frac{M}{d_{toBS}^2} \tag{20}$$

The weighed probabilities for normal and advanced nodes are, respectively,

For normal node

$$P_{i_{nrm}} = \frac{P_{opt}}{1 + a \cdot m} \frac{E_i(r)}{\overline{E}(r)} \tag{21}$$

For advanced node

$$P_{i_{adv}} = \frac{P_{opt}(1 + a)}{1 + a \cdot m} \frac{E_i(r)}{\overline{E}(r)} \tag{22}$$

The probability threshold value used by s_i to decide itself as CH is as follows [19]:

$$t(s_i) = \begin{cases} \frac{P_i}{1 - P_i \times \left(r mod \frac{1}{P_i}\right)} ifn \in G \\ 0, otherwise \end{cases} \tag{23}$$

3.6 DDEEC

DDEEC implements same strategy as DEEC to estimate networks' average energy. The only difference between DEEC and DDEEC is that the normal node and the advanced node CH uses different expressions of probability. In DEEC, advanced nodes, which are having more energy, become CH more often than normal nodes. DEEC continues to force advance nodes to become a CH even if advanced node has same residual energy as normal node. Due to this advanced node energy drained faster than normal nodes. To avoid this unbalanced condition, some changes are made in DEEC equation for calculating the value optimal number of cluster heads [20]. Due to this advanced nodes are less punished. DDEEC introduces threshold limit of residual energy which is given below:

$$TH_{REV} = E_0 \left(1 + \frac{aE_{disNN}}{E_{disNN} - E_{disAN}} \right) \tag{24}$$

Advanced nodes and normal nodes use same probability to become CH once they cross the threshold residual energy limit. Therefore, CH selection is balanced and more efficient. Threshold limit of residual energy TH is given below:

$$TH_{REV} \simeq 0.7E_0$$

Average probability p_i for CH selection used in DDEEC is as follows as:

$$p_i = \begin{cases} \frac{p_{\mathrm{opt}} E_i(r)}{(1+am)\overline{E}(r)} \textit{for normal nodes}, E_i(r) > TH_{\mathrm{REV}} \\ \frac{p_{\mathrm{opt}} E_i(r)(1+a)}{(1+am)\overline{E}(r)} \textit{for advanced node}, E_i(r) > TH_{\mathrm{REV}} \\ c\frac{p_{\mathrm{opt}} E_i(r)(1+a)}{(1+am)\overline{E}(r)} \textit{for both normal advanced nodes } E_i(r) \le TH_{\mathrm{REV}} \end{cases} \tag{25}$$

where c is a real positive variable with nearest value of 0.02 which controls directly the clusters' head number.

4 Simulation Scenario and Performance Metrics

Performance of selected protocols is simulated using MATLAB. Simulation parameters are listed in Table 1. We have considered different scenarios for analysis purpose. To explain energy dissipation, we considered first-order radio model as shown in Fig. 2. In all three scenarios, we made a comparison among selected protocols on network lifetime, throughput, and average energy consumption [24]. While analysis we varied only one parameter at a time and kept all other parameters constant for all selected protocols. Nodes are randomly scattered in the network area [25]. Heterogeniety level of routing protocol is adjusted as per their proposed model to get realistic results [11, 12, 16, 17, 18, 19, 20, 21]. In the first scenario, we kept the area of the network $100 \times 100\ \mathrm{m}^2$. BS is kept at the center of the network. In the second scenario, we change the area of network field. We increase the area from $100 \times 100\ \mathrm{m}^2$ to $300 \times 300\ \mathrm{m}^2$. In the third scenario, we have changed the position of the base station while keeping the area of the network field and all other parameters constant. We kept base station at the center of the network field (50,50), at the corner of the network field (0,0), and outside of the network field. Before the discussion of

Table 1 The simulation parameters

Parameters	Value
Number of nodes	100
Optimal CH probability P_{opt}	0.1
Initial energy E_0	0.5 J
Energy dissipation at *Tx* & *Rx* Circuits E_{elec}	5 nJ/bit
Energy dissipation by transmitter and receiver amplifier to achieve acceptable communication $\in_{amp}$	100 pJ/bit/m^2
Message size	4000 bits
a fraction of energy of advance node than normal node	1.5
m fraction of number of advanced nodes than normal node	0.2

simulation results, performance metrics should be discussed. Following performance metrics are used to discuss results.

- **Stable region**: The interval between start of the network and death of the first node in the network (here we considered interval in terms of number of rounds).
- **Unstable region**: The interval between between the death of the first node and the last node of the network (here we considered interval in terms of number of rounds).
- **Network lifetime**: The interval between the network start to death of last sensor node in the network.
- **FND First node Dead**: Round number at which first node of network is dead.
- **HND Half Node Dead**: Round number at which 50% of nodes of network are dead.
- **LND Lasts node Dead**: Round number at which last node of the network is dead.
- **Throughput**: Total number of data packets successfully sent to the BS.

Other parameters for simulation purpose are given in Table 1.

5 Simulation Results

5.1 *Scenario 1*

In this scenario, we kept area constant at 100 × 100 m^2, BS at the center, and all other parameters as per given in Table 1 and checked the performance of the selected protocols.

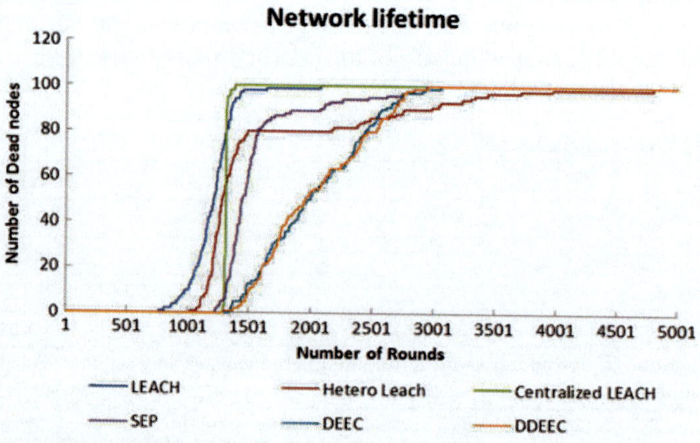

Fig. 3 Network lifetime comparison when area is 100 × 100 m^2

Table 2 Network lifetime comparison when area is $100 \times 100\ m^2$

Protocol	Number of round when node is dead					
	(FND) 1st	10th	20th	(HND) 50th	90th	(LND) 100th
LEACH	852	999	1103	1194	1316	2604
Hetero LEACH	1022	1162	1203	1280	2751	4798
Centralized LEACH	1294	1303	1305	1310	1313	1383
SEP	1224	1315	1354	1436	2072	2939
DEEC	1342	1478	1605	1964	2564	3055
DDEEC	1283	1497	1605	1950	2676	2953

Figure 3 and Table 2 clearly show that stable region of DEEC is highest among all protocols. Its first node died at 1342 round. As the LEACH centralized is having centralized control, its energy load distribution is very good and hence unstable region of LEACH centralized is the lowest. Stable region of DEEC is 57.51% more than LEACH in homogeneous and 31.31% more than LEACH in heterogeneous network. The unstable region of leach protocol in the heterogeneous environment is highest because leach protocol is unable to take advantage of heterogeneity.

Figure 4 and Table 3 show that DDEEC outperforms all other protocols. In DDEEC, number of packets received to BS is 4.75 times more than LEACH and 3.8 times more than LEACH in heterogeneous environment.

Figure 5 and Table 4 show that LEACH centralized utilizes energy of the nodes very efficiently. This is because it is centrally controlled. SEP, DEEC, and DDEEC efficiently used energy heterogeneity. During stable region, SEP utilized highest amount of average energy. In unstable region of operation, DDEEC utilized remaining energy of the network most efficiently. In unstable region of operation, DDEEC used 35.83% of entire energy and sent 35,907 packets to BS.

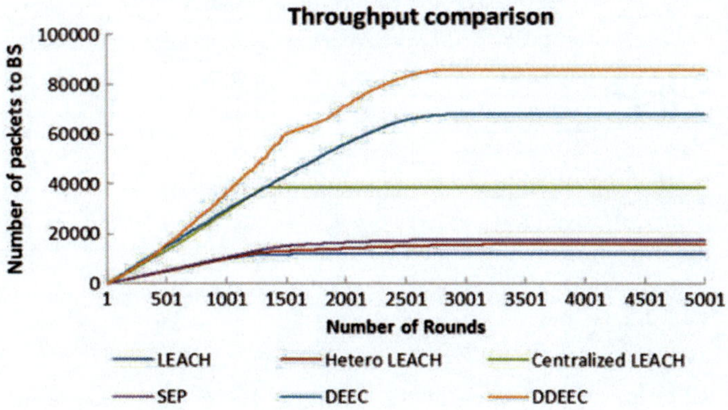

Fig. 4 Throughput comparison when area is $100 \times 100\ m^2$

Table 3 Throughput comparison when area is 100 * 100 m^2

Protocol	Packets received at BS till the death of node					
	(FND) 1st	10th	20th	(HND) 50th	90th	(LND) 100th
LEACH	8522	9910	10,810	11,440	11,780	11,996
Hetero LEACH	10,222	11,568	11,919	12,425	15,355	16,091
Centralized LEACH	37,953	38,166	38,205	38,310	38,368	38,564
SEP	12,895	13,796	14,147	14,758	16,748	17,674
DEEC	38,878	42,754	46,204	55,284	66,259	68,134
DDEEC	49,076	59,935	61,882	69,608	84,983	85,775

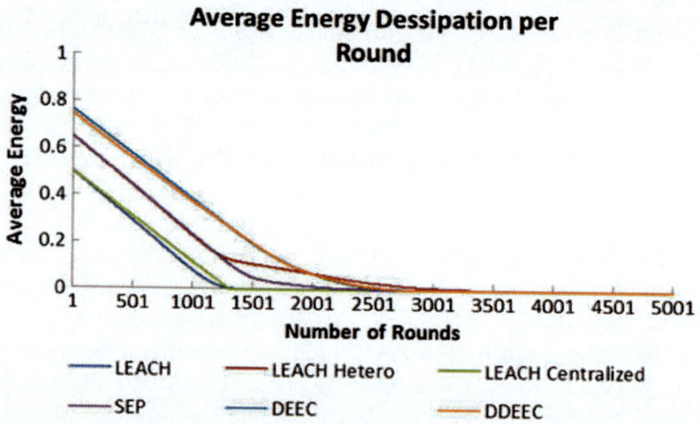

Fig. 5 Average energy dissipation of energy per round

Table 4 Percentage utilization of total energy per round

	1st (FND)	10th	50th (HND)	90th
LEACH	72.08527	83.84869	96.672252	99.561211
Hetero LEACH	66.64037	75.1333	80.00762	97.60694
Centralized LEACH	99.3922	99.8176	99.8732	99.8812
SEP	78.86797	84.43164	90.41793	98.63212
DEEC	67.42234	73.87572	90.58802	99.28389
DDEEC	63.86931	73.53724	90.01521	99.70874

5.2 *Scenario 2*

In second scenario, we vary area of the network, BS at the center, and all other parameters as per given in Table 1 and checked the performance of the selected protocols.

Figure 6 and Table 5 show that as the area of the network increases, stable region of operation of all protocol decreases. This is because as the area is increased keeping all other parameters constant density of the network decreases which results in increase in required transmission and reception power due to higher distance between nodes. Figure 6 and Table 5 also show that LEACH centralized is least scalable among all selected protocols.

Figure 7 and Table 6 show that as the area increases throughput, i.e., number of packet received at BS decreases. As the area is increased, the throughput of LEACH is decreased maximum by 8% and it remains in same range till area becomes 300 * 300 m^2. LEACH in heterogeneous environment performs poorly after area is increased above 141 * 141 m^2. Similarly, LEACH centralized and DDEEC performed very good till 141 * 141 m^2. Performance of DEEC degrades drastically as the area increased.

Figure 8 and Table 7 show that as the area increases average energy consumed during stable region of operation decreases. As the area is increased, the average energy consumed during stable region of operation of LEACH is decreased maximum by 23% till area becomes 225 * 225 m^2. The average energy consumed during stable

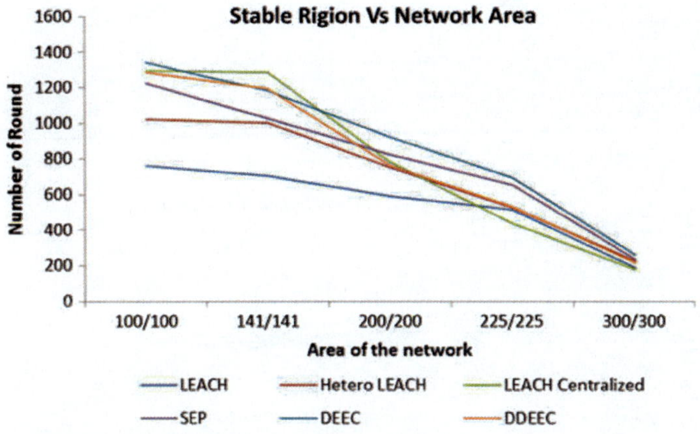

Fig. 6 Stable region of operation with respect to network area

Table 5 Number of round (FND) with respect to area

Area	Number of round (FND)					
	LEACH	LEACH hetero	LEACH centralized	SEP	DEEC	DDEEC
100 × 100	852	1022	1294	1224	1342	1283
141 × 141	703	998	1283	1024	1179	1194
200 × 200	592	753	782	822	922	762
225 × 225	512	523	434	654	694	531
300 × 300	186	221	174	232	261	215

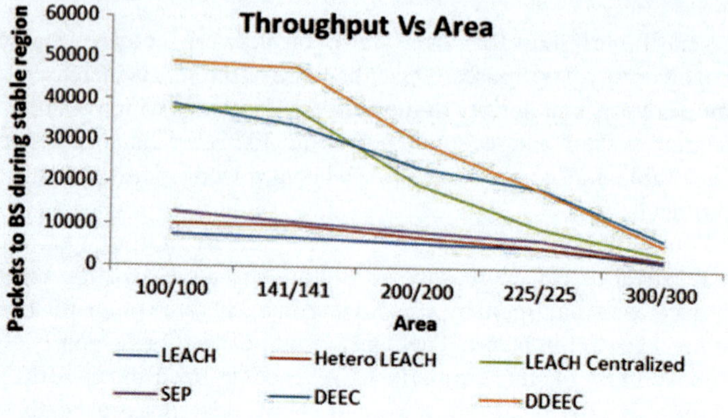

Fig. 7 Throughput versus area (FND)

Table 6 Throughput versus area (FND)

	LEACH	Hetero LEACH	LEACH centralized	SEP	DEEC	DDEEC
100 × 100	8522	10,222	37,953	12,895	38,878	49,076
141 × 141	7030	9978	37,778	10,797	33,816	47,819
200 × 200	5919	7524	19,993	8664	24,011	30,373
225 × 225	5123	5238	9634	6866	19,169	18,899
300 × 300	1866	2219	3522	2442	7160	5923

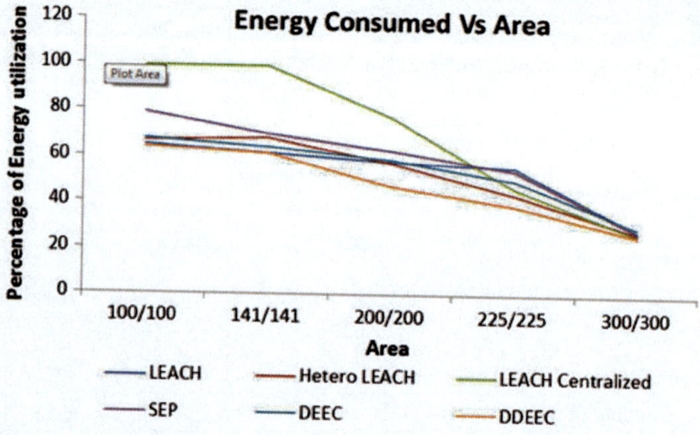

Fig. 8 Energy consumed versus area

region of operation of LEACH in heterogeneous environment, DEEC, and LEACH centralized is decreased after 141 * 141 m^2. Performance of SEP and DDEEC remains satisfactory till 225 * 225 m^2.

Table 7 Energy consumed area

	LEACH	Hetero LEACH	LEACH centralized	SEP	DEEC	DDEEC
100 × 100	72.08527	66.64037	99.3922	78.86797	67.42234	63.86931
141 × 141	61.52859	67.93784	98.999	69.64159	63.81977	61.314
200 × 200	57.89191	57.23296	77.3268	62.70408	58.73196	47.30934
225 × 225	55.46885	43.77035	46.46372	54.51355	49.96575	38.7614
300 × 300	28.16689	27.37623	26.483	27.90753	29.51112	26.19047

5.3 *Scenario 3*

In real time, it is not necessary that the sink is always positioned at the center of the area and hence in third scenario we kept area constant at 100 * 100 m^2 and change the position of the sink. We kept sink at the center of the area, at corner of the area, and outside the area. These three conditions resemble to real-time position of the sink.

Table 8 shows that in homogeneous and heterogeneous environments, if the position of the sink is within the area then the stability period of the LEACH did not change so much. In a heterogeneous environment, if the sink is positioned outside the area, then the LEACH stability period is decreased by 22%.

Performance of LEACH centralized, SEP, DEEC, and DDEEC is highly dependant on the position of the sink. If the sink is positioned outside the area, their stability period is decreased by 24%, 23%, 34%, and 55% respectively.

As per Table 9, throughput of LEACH centralized, DEEC, and developed DEEC are degraded by 65%, 68%, and 51%, respectively, when BS is positioned at the corner of the area. It is further degraded when BS is positioned outside the area.

Table 8 Network lifetime at different base stations

Protocol	BS at center		BS at corner		Outside BS	
	FND	LND	FND	LND	FND	LND
LEACH	852	2624	858	2899	793	1538
Hetero LEACH	1022	4798	944	4099	789	4681
LEACH centralized	1294	1383	1062	1475	977	1178
SEP	1224	2939	1028	5000	937	3422
DEEC	1342	3055	884	2730	877	2536
DDEEC	1283	2953	889	2848	578	2554

Table 9 Throughput at different base stations

Protocol	Position of base station					
	BS at center (50,50)		BS at corner (0,0)		Outside BS (50,150)	
	FND	LND	FND	LND	FND	LND
LEACH	8522	11,996	8580	11,405	7939	10,108
Hetero LEACH	10,222	16,091	9439	14,656	7884	13,792
LEACH centralized	37,953	38,564	13,162	14,516	7851	8505
SEP	12,895	17,674	10,826	17,258	9803	15,152
DEEC	38,878	68,134	12,389	28,916	8784	18,364
DDEEC	49,076	85,775	24,019	53,513	10,876	39,097

6 Conclusion

In WSN, to optimize the energy consumption of the network, many protocols are developed under the roof of clustering technique. This work is dedicated to evaluate the performance of LEACH, LEACH centralized, SEP, DEEC, and DDEEC protocols. Simulation results show that LEACH centralized uses network energy most efficiently. But, due to its centralized structure, it is least tolerant to change in area or position of the sink. DEEC and DDEEC protocols use their heterogeneity effectively to achieve more throughput. But, they are also intolerant to change in area and change in the sink position. From comparative evaluation, we found that while designing the clustering protocol, many points should be considered such as cluster size, optimal calculation of CH, rotation of CH, and position of the sink.

References

1. Latif, K., et al.: Performance analysis of hierarchical routing protocols in wireless sensor networks. In: 2012 Seventh International Conference on Broadband, Wireless Computing, Communication and Applications, pp. 620–625 (2012)
2. Akyildiz, I.F., Su, W., Sankarasubramaniam, Y., Cayirci, E.: Wireless sensor networks: a survey. Comput. Netw. **38**(4), 393–422 (2002)
3. Tyagi, S., Kumar, N.: A systematic review on clustering and routing techniques based upon LEACH protocol for wireless sensor networks. J. Netw. Comput. Appl. **36**(2), 623–645 (2013)
4. Akhtar, M.N., Ali, A., Ali, Z., Hashmi, M.A., Atif, M.: Cluster based routing protocols for wireless sensor networks: an overview. Int. J. Adv. Comput. Sci. Appl. **9**(12), 389–396 (2018)
5. Alnuaimi, M., Shuaib, K., Al Nuaimi, K., Abdel-Hafez, M.: Performance analysis of clustering protocols in WSN. In: 6th Joint IFIP Wireless and Mobile Networking Conference (WMNC), pp. 1–6 (2013)
6. Ameer Ahmed Abbasi and Mohamed Younis: A survey on clustering algorithms for wireless sensor networks. Comput. Commun. **30**, 2826–2841 (2007)
7. Biradar, R.V., Patil, V.C., Sawant, S.R., Mudholkar, R.R.: Classification and comparison of routing protocols in wireless sensor networks. Spec. Issue Ubiquitous Comput. Secur. Syst. **4**(2), 704–711 (2009)

8. Singh, S.K., Kumar, P., Singh, J.P.: A survey on successors of LEACH protocol. IEEE Access **5**, 4298–4328 (2017)
9. Jain, N., Sinha, P., Gupta, S.K.: Clustering protocols in wireless sensor networks: a survey. Int. J. Appl. Inf. Syst. (IJAIS) **5**(2), 41–50 (2013)
10. Mamalis, B., Gavalas, D., Konstantopoulos, C., Pantziou, G.: Clustering in wireless sensor networks. In: RFID and Sensor Networks: Architectures, Protocols, Security and Integrations, pp. 324–353 (2009)
11. Heinzelman, W.R., Chandrakasan, A., Balakrishnan, H.: Energy-efficient communication protocol for wireless microsensor networks
12. Han, G., Jiang, X., Qian, A., Rodrigues, J.J.P.C., Cheng, L.: A comparative study of routing protocols of heterogeneous wireless sensor networks. Sci. World J. **2014** (2014)
13. Sharma, T., Kumar, B., Tomar, G.S.: Performance comparison of LEACH, SEP and DEEC protocol in wireless sensor network. In: Proceedings of the International Conference on Advances in Computer Science and Electronics Engineering, pp. 10–15 (2012)
14. Kumari, S., Sharma, J., Bhattacharya, P.P.: Performance evaluation of SEP, LEACH and ZSEP under the influence of network area. Mody Univ. Int. J. Comput. Eng. Res. **1**(2), 85–91 (2017)
15. Iqbal, S., Shagrithaya, S.B., Sandeep, G.G.P., Mahesh, B.S.: Performance analysis of stable election protocol and its extensions in WSN. In: 2014 IEEE International Conference on Advanced Communications, Control and Computing Technologies, pp. 744–748 (2014)
16. Heinzelman, W.B.: Application-specific protocol architectures for wireless networks. Doctoral dissertation, Department of Electrical Engineering and Computer Science, Massachusetts Institute of Technology (2000)
17. Heinzelman, W.B., Chandrakasan, A.P., Balakrishnan, H.: An application-specific protocol architecture for wireless microsensor networks
18. Smaragdakis, G., Matta, I., Bestavros, A.: SEP: a stable election protocol for clustered heterogeneous wireless sensor networks. Boston University Computer Science Department (2004)
19. Qing, L., Zhu, Q., Wang, M.: Design of a distributed energy-efficient clustering algorithm for heterogeneous wireless sensor networks. Comput. Commun. **29**(12), 2230–2237 (2006)
20. Elbhiri, B., Saadane, R., Aboutajdine, D., et al.: Developed Distributed Energy-Efficient Clustering (DDEEC) for heterogeneous wireless sensor networks. In: 2010 5th International Symposium On I/V Communications and Mobile Network, pp. 1–4 (2010)
21. Sujee, R., Kannammal, K.E.: Behavior of LEACH protocol in heterogeneous and homogeneous environment. In: 2015 International Conference on Computer Communication and Informatics (ICCCI), pp. 1–8 (2015)
22. Kumar, A., Saini, R.K., Brijbhushan: Comparison and review of LEACH LEACH-C and PEGASIS routing protocols. Int. J. Adv. Res. Electron. Commun. Eng. (IJARECE) **5**(6), 1660–1665 (2016)
23. Thapa, R., Singh, H., Sharma, A.: A comparative analysis of LEACH and SEP using NS2. In: 2017 8th International Conference on Computing, Communication and Networking Technologies (ICCCNT), pp. 1–4 (2017)
24. Singh, S., Malik, A., Kumar, R.: Energy efficient heterogeneous DEEC protocol for enhancing lifetime in WSNs. Eng. Sci. Technol. Int. J. **2017**, 345–353 (2017)
25. Shen, J., Wang, A., Wang, C., Ren, Y., Wang, J.: Performance comparison of typical and improved LEACH protocols in wireless sensor network. In: 2015 First International Conference on Computational Intelligence Theory, Systems and Applications (CCITSA), pp. 187–192 (2015)

Speech Recognition Using Artificial Neural Network

Shoeb Hussain, Ronaq Nazir, Urooj Javeed, Shoaib Khan, and Rumaisa Sofi

Abstract Speech recognition can be an important tool in today's society for hand-free or voice-driven implementation. Using simple commands or triggers, it is possible for speech impaired human beings to communicate with increased ease of understanding. With the advent of various soft computing methods, a large class of nonlinearities can be handled. Artificial Neural Networks (ANN) have been applied in finding the solution for speech recognition. A lot of work is going on in this regard and mostly positive results have been achieved. Now, the research is being done to minimize the rate of error in obtaining the solution. In this paper, a comprehensive study of use of artificial neural networks in speech recognition is studied and proposes methods for training of the neural network so that an appropriate neural output can be obtained which is as close to the desired output. The paper demonstrates that ANN can indeed form the basis for a general-purpose speech recognition and neural network offers clear advantages over conventional methods. MATLAB simulation has been carried out to validate the results.

Keywords Speech recognition · Artificial neural network · Hearing impaired · ANFIS

1 Introduction

Voice recognition is gaining more and more attention. Being able to communicate for human beings is very important. Human beings learn to communicate through speech from a very early point in their lives. Developing an interface for communication of speech impaired beings is challenging and at the same time exciting. The biological organs involving speech generation exhibit nonlinear properties and as such the vocalizations can vary in terms of accent, pronunciation, articulation, roughness, volume, pitch, etc. [1]. Speech recognition or Automatic Speech Recognition (ASR)

S. Hussain (✉) · R. Nazir · U. Javeed · S. Khan · R. Sofi
Department of Electrical Engineering, University of Kashmir, Srinagar, India
e-mail: shoeb@uok.edu.in

J. S. Raj et al. (eds.), *Intelligent Sustainable Systems*, Lecture Notes in Networks and Systems 213, https://doi.org/10.1007/978-981-16-2422-3_7

is the process by which a machine identifies voice. Speech recognition will radically change the interaction between the humans and the computers. Moreover, speech recognition takes this one step further by using this application in order to identify, verify, and perceive basic commands.

Though great progress has been made in technology, computers are nowhere near the level of human brain performance at speech recognition. The brain's impressive superiority has motivated research on brain-like performance of computers for speech recognition [2]. Technology companies are recognizing interests in speech recognition technologies and are working toward making voice recognition a standard for most products. Commercial applications of computer-aided voice recognition were first introduced in the medical and legal fields. Speech recognition has definite potential for reducing pilot workload, but this potential was not realized consistently [3, 4]. ASR in the field of computer gaming and simulation is becoming more widespread. Disabled people are another part of population that benefit from speech recognition programs.

The advancement here is making the interaction between human and the machine easy and natural, and that is the main reason behind the implementation of artificial intelligence in speech recognition so the machine interprets the speech. Artificial Neural Networks (ANN) being a set of parallel computing devices emulate a human brain in many aspects [5]. Neural network offers a promising and futuristic approach for speech recognition. ANNs are biologically inspired tools that are quite capable in information processing. ANNs are nothing but the crude electronic models based on neural structure of brain.

This paper proposes the idea of using ANN using different training algorithms for speech recognition. ANN because of its advanced and highly intuitive formulation results in better response. The idea is to train the system through backpropagation algorithm and with increased input–output dependence. Simulation results confer that the proposed idea results in production of human-like speech.

The paper has been arranged in five sections. Following the Introduction, theory of ANN has been explained with respect to Speech Recognition in Sect. 2. Methodology has been presented in Sect. 3, simulation results in Sect. 4, and conclusion in Sect. 5.

2 Theory

2.1 Artificial Neural Networks from the Viewpoint of Speech Recognition

A human brain while more deep is different and somewhat slow from a typical computer in processing. A processor is able to execute instructions, commands, and process programs at incredible speeds. The human beings process information through nodes called neurons. These neurons pass electronic or chemical signals and communicate through synapses. The connections are dynamic and change each time any information is processed making a human being adaptive and a great learner. Neural networks can be trained to emulate the behavior of a human brain while processing information. The advantage of adding adaptability to speed makes neural networks extremely powerful AI technique.

The problem with speech recognition is, however, the audio spectrum which represents a quite unpredictable data. Training the neural network for such a large and unpredictable data is quite challenging. The proposed model of ANN-based speech recognition system sets a platform for improved speech recognition system using intelligent system.

The most common form of ANN architecture is arranged in mostly three layers only. These layers are the input layer, hidden layer, and an output layer. ANN requires training and for that a target matrix is required. Using a simple logical computation, the ANN is able to perform necessary computations. The input layer, being the first to receive the information through a combination of different inputs, forwards them to a set of interconnected neurons, processing the information through an activation function in order to determine the necessary output. The output layer sends the output through a set of neurons after passing through weighed gain matrices so as to reduce the error between output and target set. Implementation of this model uses the following mathematical equation:

$$y = F\{\Sigma x_k w_k + b\} \quad (1)$$

where w_k represents the synaptic weights, x_k represent the input, b represents the bias, and F represents the activation function.

2.2 Training

For an ANN architecture to be able to work efficiently and in a correct manner, it has to adjust its weights in a manner that fits the desired output profile for any given set of input matrix. In order to be able to do so, the training process of an ANN is very much important. The ANN should be able to replicate the actions of a human brain, have the capability to learn, and be responsive to any new information. This involves the iteration of training data from input matrix into the ANN architecture.

An optimization algorithm is used for emulating the learning behavior in the ANN. Various optimization algorithms have been identified in literature and each algorithm has its unique characteristic that improves the efficiency of ANN through either reduced memory requirements or greater speed or even in terms of higher precision. In this paper, speech recognition has been implemented through Levenberg–Marquardt algorithm to work specifically with loss functions [6].

Consider a loss function as sum of squared errors as below:

$$f = \Sigma e_i^2 \quad (2)$$

The Jacobian matrix of the loss function containing the derivatives of the errors with respect to the parameters is given as

$$J_{ij} = \partial e_i / \partial w_i \tag{3}$$

For $i = 1, \ldots m$ and $j = 1, \ldots n$ where n is the parameters in the neural network and w_i represents the weights. The gradient vector of the error function is computed as

$$\nabla f = 2\, J^T e \tag{4}$$

where e represents a row matrix containing only error terms. The updated weights can be thus presented as

$$w^{(i+1)} = [w^{(i)} - J^{(i)T} \cdot J^{(i)} + \lambda^{(i)} I]^{-1} [2 J^{(i)T} . e^{(i)}] \tag{5}$$

where λ is the damping factor.

2.3 *Proposed Model*

In order to model and train the input–output relationship in an ANN, it has been observed that conventionally, the outcomes rely upon the mathematical accurateness of the framework and its precision. A system with an incomplete mathematical model, it becomes difficult to analyze or predict the behavior of output [7]. Out of many techniques, artificial neural network technique has been chosen because of its better human–computer interfacing, ability to work with incomplete knowledge (less human dependency), parallel processing ability, ability to learn and model nonlinear models, and ability to characterize models with unpredictable output. The data associated with speech recognition is somewhat ill manufactured. The advantage that ANN offers in processing this ill-defined data is what drives the present work focusing on building a real-world system.

Figure 1 shows the flowchart illustrating the steps involved in speech recognition using ANN. So as to implement voice recognition through ANN, recording, analyzing, and manipulating audio files are required. Real-time recording approach has been used. The neural network is trained using supervised learning where both input and the target output are fed. Figure 2 shows the training methodology of ANN.

3 Methodology

All together for a neural system to perform proficiently and accurately it must have the option to adjust weighted functions so as to match the output and desired response [8]. The training of this system includes an input training matrix and an output

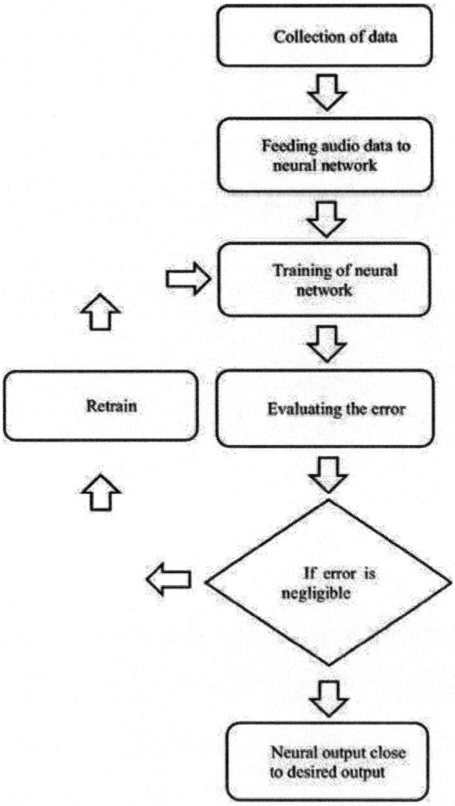

Fig. 1 Flowchart of Audio Recognition using AI

target matrix. The training in this fashion continues and the weights are accordingly adjusted in the system so as to reduce errors. The proposed model adopted a hardware for feeding the voice and sampling it in MATLAB. The inputs use a combination of number of voice samples. ANN processes the input and outputs the speech in response to command input, e.g., the input may be "What is your name," output response may be "Shoeb".

3.1 Different Approaches for Reducing the Error

Backpropagation is more suited for a neural network that is meant for pattern recognition. The algorithm is more common and advantageous. It is a form of supervised learning. It is driven by training dataset. For the feedforward neural network, backpropagation training algorithm is chosen where there is a backpropagation of errors until the derivative of the loss function is minimum. Mean Squared Error (MSE) is

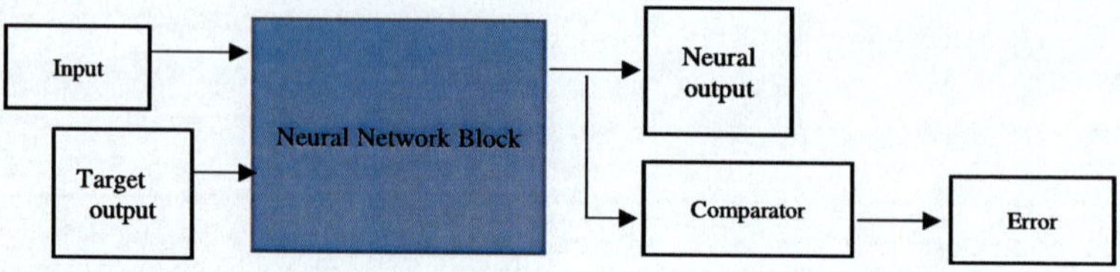

Fig. 2 Training of Artificial Neural Network

the average of the squared error that is used as the loss function for least squares regression [9–11].

3.2 Changing the Combination and Number of Inputs

In a data-driven model, the accessibility of an adequate number of information is an unquestionable requirement for accomplishing a decent model. By and large, it has been seen that the accessible information is not adequate for obtaining an accurate model. In such a situation, generation of more data and providing more variants of input data can yield better and more accurate results.

3.3 Increasing the Number of Hidden Neurons

Increasing the number of hidden layers might improve the accuracy or might not, it really depends on the complexity of the problem that one is trying to solve. When a neural network has too few hidden neurons, it does not have the capacity to learn enough of the underlying patterns. At increasing number of hidden neurons, the number of hidden neurons does not help too much. Determining the suitable number of neurons is mostly a hit-and-trial method. If acceptable range of neuron is not designated, model will become over-fitting and these cause wrong results. This implies ANN model offers higher result at model points; however, it conjointly offers worse results at check points [12, 13].

3.4 ANFIS (Adaptive Neuro-Fuzzy Inference System)

Combining the advantages of fuzzy logic with neural network makes the system more reliable. Fuzzy logic and neural network complement each other very well in acquiring a response of an ill-defined system [14, 15]. The disadvantage that ANFIS, however, offers is limited number of input and data handling capability.

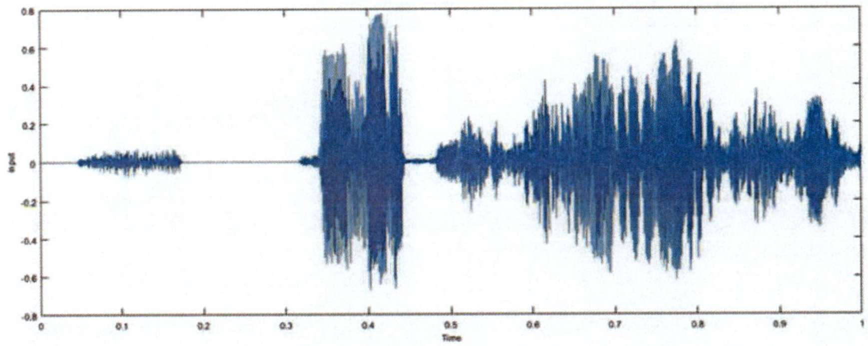

Fig. 3 Frequency spectrum of the input Audio waveform

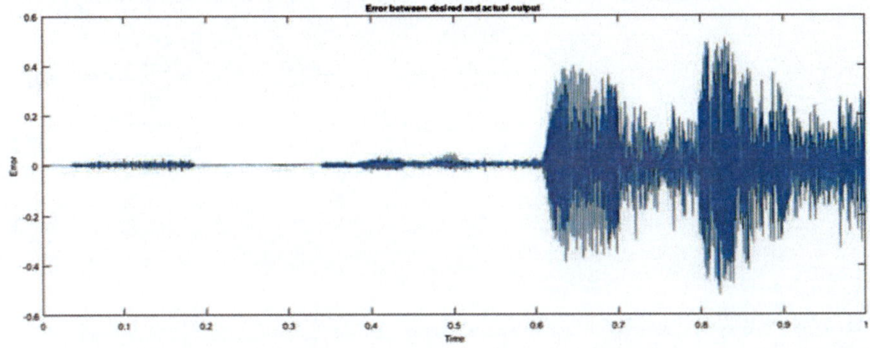

Fig. 4 Error between desired output and actual input using different neural network strategies

4 Simulation Results

Simulation of the proposed system is carried out using MATLAB. Neural network is trained using different training algorithms using different number of neurons. Input audio waveform is first sampled using in-built MATLAB commands. Figure 3 shows the frequency spectrum of the input audio waveform.

Figures 4, 5, 6, 7, and 8 show the error obtained between the actual audio input and the trained audio output resultant from the neural network. It can be observed from the results that using a recurrent neural network, the input audio waveform can be exactly obtained from the neural network using just an activation signal or command. The error in Fig. 4d, e is of the order of 10^{-4}. The neural network is able to reproduce the speech. A DSP-based hardware kit is further used to obtain the audio output through a speaker.

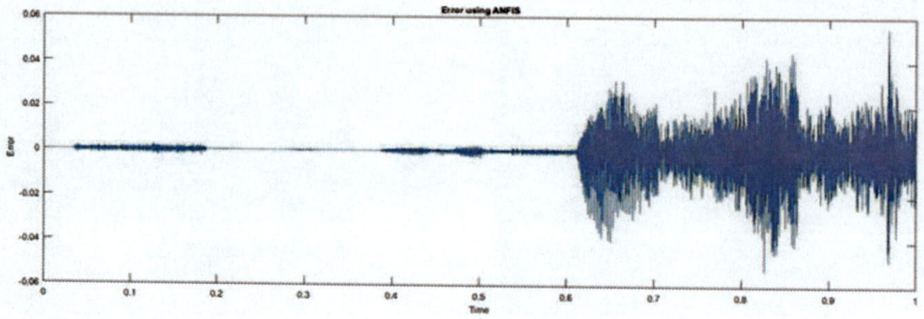

Fig. 5 Error between desired output and actual input using different neural network strategies

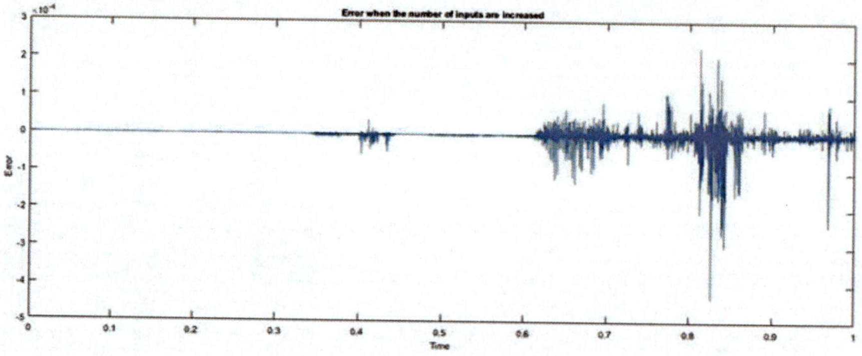

Fig. 6 Error between desired output and actual input using different neural network strategies

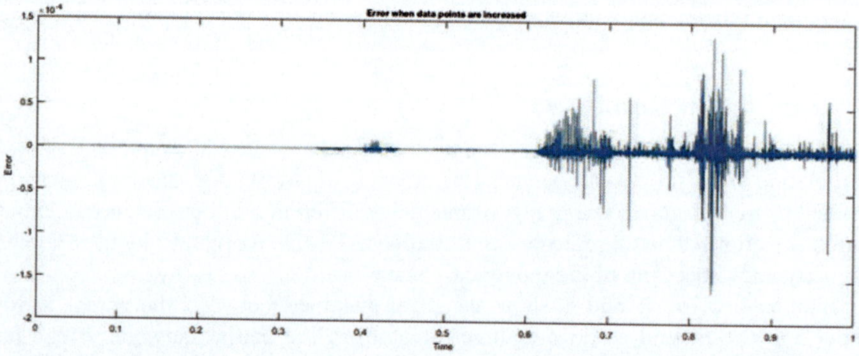

Fig. 7 Error between desired output and actual input using different neural network strategies

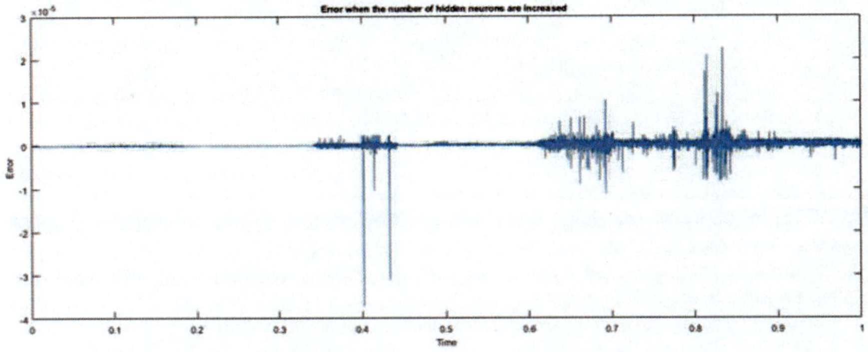

Fig. 8 Error between desired output and actual input using different neural network strategies

5 Conclusion

Motivated by the advantageous features of neural network, the paper presents a model of voice recognition based on artificial neural network. Different types of methodologies for obtaining an appropriate neural output same as the desired output have been presented. MATLAB simulation was carried out to validate the performance of the proposed model.

It can be concluded that the error obtained is because of the limitations of neural network. An idea to bring out the basic implementation of ANN in automatic speech recognition is presented through validated results from MATLAB to form the basis for future research. An improved response time and response validation is required in speech recognition. The proposed model adopted is simple offering an ease in implementation with reduced response time.

References

1. Kamble, B.C.: Speech recognition using artificial neural network - a review. Int. J. Comput. Commun. Instrum. Eng. (2016)
2. Tebelskis, J.: Speech recognition using Neural Networks. Doctoral Dissertation, School of Computer Science, Carnegie Mellon University, Pittsburgh, Pennsylvania (1995)
3. Gupta, A., Joshi, A.: Speech recognition using artificial neural network. In: International Conference on Communication and Signal Processing (ICCSP), Chennai, pp. 0068–0071 (2018)
4. Murat, G.D., Sazli, H.: Speech recognition with artificial neural networks. Digit. Signal Process. **20**(3), 763–768 (2010)
5. Alhawiti, K. M.: Advances in artificial intelligence using speech recognition. Int. J. Comput. Inf. Eng. (2015)
6. Singh, M., Sreejeth, M., Hussain, S.: Implementation of Levenberg-Marquadrt algorithm for control of induction motor drive. In: 3rd IEEE International Conference on Recent Trends in Electronics, Information & Communication Technology (RTEICT), Bangalore, India, pp. 865–869 (2018)

7. Bose, B. K.: Modern Power Electronics and AC Drives. Prentice Hall (2002)
8. Al Smadi, K., Al Issa, H., Trrad, I., Al Smadi, T.: Artificial intelligence for speech recognition based on neural networks. J. Signal Inf. Process. (2015)
9. Smys, S., Chen, J.I.Z., Shakya, S.: Survey on neural network architectures with deep learning. J. Soft Comput. Paradig. (JSCP) **2**(03), 186–194 (2020)
10. Gupta, A., Joshi, A.: Speech recognition using artificial neural network. In: 2018 International Conference on Communication and Signal Processing (ICCSP), Chennai, pp. 0068–0071 (2018). https://doi.org/10.1109/ICCSP.2018.8524333
11. Graves, A., Jaitly, N., Mohamed, A.: Hybrid speech recognition with deep bidirectional LSTM. ASRU, pp. 273–278 (2013)
12. Hraskoa, R., Pacheco, A.G.C., Krohlinga, R.A.: Time series prediction using restricted Boltzmann machines and backpropagation
13. Generating sequences with recurrent neural networks, June 2014 (2014)
14. Buragohain, M.: Adaptive Network based Fuzzy Inference System (ANFIS) as a tool for system identification with special emphasis on training data minimization. (Doctoral Dissertation, Indian Institute of Technology Guwahati, India (2018)
15. Hussain, S., Bazaz, M.A.: ANFIS implementation on a three phase vector controlled induction motor with efficiency optimisation. In: International Conference on Circuits, Systems, Communication and Information Technology Applications (CSCITA) 04/2014, pp. 391–396 (2014)

Multi-objective Optimization for Dimension Reduction for Large Datasets

Pradeep Bedi, S. B. Goyal, Jugnesh Kumar, and Ritika

Abstract In recent advancement of computational techniques, there is an exponential increase in amount of data. Learning on such large amount of data is a major area of concern with application of machine learning algorithms. Therefore, it is considered to be a complicated task to handle and perform computation on such large, complex, and heterogenous dataset. In this paper, a brief discussion about different dimension reduction or feature selection algorithms is given. A brief review about contribution of researchers for designing feature selection algorithms for large dataset is given. By analyzing exisitng problems, this paper is motivated to design a hybrid, robust, flexible, and dynamic feature selection model for classification of large datasets. For this, multi-objective optimized feature selection is proposed with an objective to minimize the error rate and execution time as well maximize accuracy of problem and to generate solution with high probability.

Keywords Machine learning · Big data · Feature selection · Feature dimensionality reduction · Multi-objective optimization

1 Introduction

In last few decades, computational techniques have seen a number of advancements, thereby increasing the data related to them [1]. It is not easy to make use of such huge amount of data without appropriate processing techniques. This further makes the system complex and inaccurate [2]. These result in a number of challenges in terms of storage of data, their analysis, and maintaining them in a safe and secure way. Analyzing such huge databases is the primary challenge faced by modern day

P. Bedi
Lingayas Vidyapeeth, Faridabad, India

S. B. Goyal (✉)
City University Malaysia, Petaling Jaya 46100, Malaysia

J. Kumar · Ritika
St. Andrews Institute of Technology & Management, Gurgaon, India

J. S. Raj et al. (eds.), *Intelligent Sustainable Systems*, Lecture Notes in Networks and Systems 213, https://doi.org/10.1007/978-981-16-2422-3_8

algorithms related to machine learning and data mining. At the very first instant, the data looks a bit easier to analyze due to its volume but practically it is a very complex process. This might be because of the hybrid and heterogeneous features involved in it [3–5]. It has wide areas of applications like marketing, bioinformatics, financial businesses, or medicine. The reason behind this is that of the large amount of data stored in it. If tools related to machine learning are to be applied for huge problems of data, the algorithms need to be redesigned and included in suitable environments. Therefore, several challenges and issues arised on such large amount of data with a huge number of classes and features. So, there is requirement of dimensionality reduction techniques and algorithms for efficiently computing on large datasets without decreasing performance [6].

Some of the advantages of applying dimensionality reduction over dataset are

- Less space is needed to store data.
- Less computational time required to process reduced dimentional data.
- Some algorithm, either classification or regression, does not perform accurately on high-dimentional data.
- Dimension reduction will help to reduce redundant features which takes unnecessary time and space.
- Easily to visualize and observe pattern of data.

In this paper, a brief discussion of dimension reduction is given along with their issues and types. Further, in this paper, a brief literature review is presented on different dimention reduction techniques given. Finally, an architecture is proposed for dimension reduction which is focused toward its adaptibility for any type of data mining or machine learning algorithms.

Big datasets consisting of multilevel variables are termed as big data. The growth rate related to them is very high. The most prominent aspect of big data is volume. Computer science and data processing fields have seen a large number of advancements in the past couple of years, which has further resulted in the rapid growth of attributes and records. This is one of the primary challenges in the field of data mining or in data science. Big data consists of large size of data which form datasets of multiple dimensions. Analyzing dataset of multiple dimensions and searching for a pattern within them is not an easy task. Depending upon a person's interest in the processes, data of high dimensions could be obtained from appropriate sources. A number of variables are responsible for the progress of a process. The variables may be measurable in some case while not in others. In order to obtain data from simulation results in an accurate way, it is required for us to deal with data of higher dimensions [7]. Figure 1 represents the dimention reduction process.

Dimension reduction is the process in which "N" features vector (high dimensional) is reduced to "n" feature vector (low dimensional), where "N" is greater than "n." Conversion of feature space from high dimensional to low dimension must be such that their conversion retains some useful information from original data. This reduction is required due to undesirable reasons such as curse of dimensionality. Dimension reduction is commonly used in applications such as visual or audio signal processing, bioinformatics, etc. [8, 9].

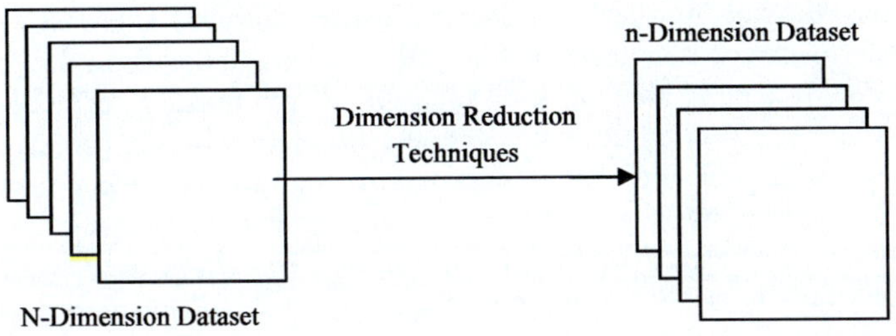

Fig. 1 Dimension reduction

2 Literature Review

The concept of genetic algorithm was put forward by Sıddiqi [10] in his work. Time-series datasets that were available publicly were utilized to carry out the experiments. Forecasting was also performed through feature sets through the use of a model known as Long Short-Term Memory (LSTM). A feature selection based on the Distributed Fuzzy Rough Set (DFRS) was proposed by Kong [11]. Parallel computing was done through DFRS feature selection by the separating and assigning tasks to a large number of nodes. Another algorithm termed as Multiple Relevant Feature Ensemble Selection (MRFES) was put forward by Ding in his research work [12]. This algorithm considered MCCM or Multilayer Co-evolutionary Consensus Map Reduce as its basis. To cope up with the feature ensemble selection of huge datasets, an MCCM model is constructed that is effective enough and has a number of sources for features. By utilizing the optimization techniques related to particle swarm and gray wolf the binary wrapper feature selection was put forward by Hasnony [13]. The optimized solutions are then figured out through the matrices of Euclidean separation with the K-nearest neighbor classifier. To avoid the algorithm between the locked state and the local optimal problem, a tent chaotic map is utilized. For the conversion of search space that is most appropriate for feature selection, a sigmoid function is employed. It makes the continuous vector a binary vector. To group large dimensional data streams, a dynamic feature mask was proposed by Fahy [14]. The masking is usually done with irrelevant or redundant features while the relevant ones are clustered. The mask gets improvised accordingly, if the features undergo changes. The mask is removed from irrelevant features and masking is applied to those with least relevance. For the section of diverse and informative genes belonging to the data related to gene expression, an important clustering and selection method was put forward by Yang [15] in his work. A hybrid genetic algorithm with wrapper-embedded feature approach (HGAWE) was put forward by Liang [16] in his research work. This algorithm aids in the combination of embedded regularization approaches (local) with the genetic algorithm (global). Feature selection algorithm's performance was analyzed by Zaffar and a student dataset was used to present the same [17]. An effective feature selection method of lightweight was proposed by Fong in his research work [18]. The accuracy in analysis with an optimum processing time was obtained through Accelerated Particle Swarm Optimization (APSO). For evaluation of performance, a collection of huge data with high degree of dimensionality is considered. In the research work by Lin [19], an optimization technique was put forward where two sets of variables were used to define weight of lists related to feature rank and final feature rank of the label. An algorithm related to feature selection was put forward by Peralta [20] in his research work considering evolutionary computation as the basis. To extract feature subsets from large datasets, it employs MapReduce paradigm. Three prominent classifiers are employed to evaluate feature selection method (Naive Bayes, Logistic Regression, and SVM).

3 Research Challenges

Curse of Dimensionality: One of the common issues in the data engineering field is curse of dimensionality which means large feature dimension will lead to increase in computational complexity as well decrease the efficiency level. Inappropriate dimension of features will compromise the efficiency level of big data during process of classification and analysis. High-dimensional set of data results in problems such as few samples with many features, redundancy, missing samples, no relation among features. If high-dimensional data with less similarity will be used, then it will result in more error and make it difficult for machine learning algorithms to give accurate results. So, to compensate the curse of dimentionality, it is required to feed the data mining or machine learning algorithms with low-dimensional with related or similar feature sets. To overcome abovementioned challenges that come with higher dimensional data, there is this need of reducing the dimensions of the data that is planned to be analyzed and visualized [4].

Overfitting Problem: Overfitting problem is one of the major issues in feature selection and dimension reduction techniques. Although traditional dimension reduction techniques, for high-dimension data, can identify relationship among available feature sets but cannot overcome from the influence of overfitting problem on the final selection of results which is still a great challenge.

Missing Value: In today's changing scenario, contineous changing data patterns, its size, volume, and format had lead to missing data problem. Machine learning training for missing data is also a challenging task.

4 Feature Selection and Extraction

Dimensionality reduction is performed by either feature selection or feature extraction. Feature selection is process by which some features are omitted from the available set of features that are not required for pattern analysis of data. In other terms, it is said that redundant and irrelevant features are ignored. Whereas in feature extraction process, the whole information or contents and maps are only useful content for pattern recognition. Among all processing techniques related to data, feature selection is the primary one. It is often employed to figure out features that are correlated and remove the uncorrelated ones from a set of features. A classifier who is searching for appropriate correlations is often disturbed by random or noisy features. The task of a classifier becomes much more challenging when he goes through a redundant feature as it makes the task much more complex without providing the classifier any useful information [3]. Methods of feature selection can be fundamentally categorized into three aspects [3, 5, 6]:

- Filtering Methods: These methods involve measures that are data related like crowding and separability. For instance, ANOVA, Pearson Correlation.
- Wrapper Methods: These methods depend on algorithm related to learning and are a part of functions related to fitness. For instance, recursive feature elimination, backward and forward selection [5].
- Embedded Methods: In the construction of the classifier, feature subsets are built in, for instance, Ridge and lasso regression.

As per the methods related to embedded and wrapper, a classifier or learning model dependency is observed, while in the filter method, measures related to data-like crowding and separability are observed. In large data environments, algorithms related to filter data method are much more efficient and beneficial. Feature ranking is primarily applicable in content of large scale and is a part of feature selection (Table 1).

The primary objective behind feature selection is the identification of original entity characteristics where m < d. In order to make classifications more accurate and minimize the time, a large number of classification problems employ feature selection. A complete set of functions can contain a lot of noise. The data is summarized or described with less information. This is useful for viewing dimensional data or for simplifying data that can then be used in a supervised learning process. Many of these reduction methods are suitable for classification and regression. The most common algorithms are Principal Component Analysis (PCA), Linear Discrimination Analysis (LDA), Autoencoders, etc. Some of the feature extraction techniques are discussed below.

Table 1 Comparison of commonly used feature selection methods

Interact with classifiers	Filter method	Wrapper method	Embedded method
Computational cost	NO	YES	YES
Accuracy	LOW	High	Depends on data
Robustness	LOW	High	High
Risk of overfitting	NO	YES	YES

4.1 Principle Component Analysis (PCA)

PCA is a commonly used dimensionality reduction algorithm that could give us a less dimensional approximation to the original dataset, while preserving potential variability. It is a type of pattern recognition in the data. PCA is a powerful tool for data analysis [7]. Once the data profiles and the compressed data have been found, the number of dimensions can be reduced without much information loss. The PCA phases are as follows:

- Get data and standardize: The range of contineous variables are standardized such that all variables contribute equally in analysis. If standardization is not performed then large range variable will dominate over lower range variables.
- Calculates the covariance matrix: In this step, covariance matrix is calculated. This step is essential part for dimension reduction as it is required to know relationship among variables as well as to reduce redundant information for further analysis.
- Calculates the eigenvalues of the covariance matrix: This step is performed to extract the principal components. As principle components represent the compressed and informatiove variable from initial variables.
- Select the components and create a feature vector.
- Get the new record.

4.2 Linear Discriminant Analysis (LDA)

In the pre-processing stage, the dimensional reduction for machine learning applications and classification of models is done through LDA method. A dataset is projected into a very trivial space and separation of good class is applied to minimize calculation cost and avoid over-regulation. LDA primarily finds its use in cases of unequal class frequencies and the examination of processes has been done on test data generated randomly. The relationship among class variation for a given record and the obtained class variance is maximized through the LDA method. Hence, large separation is guaranteed. The steps involved in LDA method are mentioned below:

- For various classes related to a dataset, the 2-D average vectors are calculated.
- The dispersion matrices are calculated.
- The eigenvalues ($\lambda_1, \lambda_2, \ldots, \lambda_d$) and eigenvectors ($e_1, e_2, \ldots, e_d$) related to the diffusion matrices are calculated.
- After lowering of eigenvalues and eigenvectors, eigenvector classification is done. K for which the value of eigenvalue is very large is selected, a 2-D array d * k is made with an objective to form w (eigenvectors in w are represented by its columns).
- In order to convert samples into subspace, the eigenvector d * k array is utilized. This can be simply represented by the matrix: Y = X * W (X = matrix n * d, n is the number of samples, and y = sample n * k converted into subspace).

4.3 Correlation Analysis (CA)

Pearson Correlation Analysis: It is mostly used to correlate the statistical measurement or relationship degree among variables and features. Pearson correlation coefficient ρ is calculated by the formula as given below:

$$\rho = \frac{n\sum x_i y_i - \sum x_i \sum y_i}{\sqrt{\sum x_i^2 - (\sum x_i)^2}\sqrt{n\sum y_i^2 - (\sum y_i)^2}} \tag{1}$$

where

x_i and y_i Individual variables (feature values).
n number of observations.

Spearman Correlation Analysis: It is considered to be as a non-parametric test that is used to measure the degree of association of data values. Spearman correlation coefficient σ is calculated by the formula mentioned below:

$$\sigma = 1 - (\frac{6\sum d_i^2}{n(n^2 - 1)}) \tag{2}$$

where

d_i Difference between the ranks of two independent variables, x and y.
n umber of observations.

Kendall Correlation Analysis: It is considered to be as a non-parametric test that is used to measure the strength of dependence of data values. Kendall correlation coefficient τ is calculated by the formula as given below:

$$\tau = \frac{\tau = (n_c - n_d)}{\frac{1}{2}n(n-1)} \tag{3}$$

where

n_c Number of concordant.

Concordant function is evaluated by signum function on variables such as

$$\text{signum}(x_2 - x_1) = \text{signum}(y_2 - y_1) \tag{4}$$

n_d Number of discordant.

Discordant function is also evaluated by signum function on variables such as

$$\text{signum}(x_2 - x_1) = -\text{signum}(y_2 - y_1) \tag{5}$$

n number of observations.

4.4 Autoencoders (AE)

AEs are unsupervised deep learning [6] neural networks with backpropagation learning process. These are used to represent high-dimentional data into low-dimentional feature vector. It generates intermediate feature vector of low dimension and reconstructs similar output vector as input. Its characteristics is similar to PCA but PCA is suitable only for linear transformation whereas it can be applied on both linear and nonlinear transformations of data. AEs have three layers: input layer, hidden layer, and output layer with activation function. The funcioning of AE is performed in two steps, one is to represent internal features as in Eq. 6 and another is to evaluated output value as in Eq. (7).

$$h(x) = f(W_e x_e + b_e) \tag{6}$$

$$X' = f(W_o h(x) + b_o) \tag{7}$$

where

x_e input value to encoder,
b_e bias value of nodes in encoder,
W_e weight value of encoder, $h(x)$ = output calculated by encider,
W_o output weight value,
b_0 output bias value,
X' final output.

To minimize the error between input feature vector and output feature vector, optimization function is used as stated in Eq. (8).

$$error = \text{argmin} \sum_{i=1}^{n} \|X_i - X_i'\| \tag{8}$$

where

X_i desired output value,
X'_i estimated output value.

4.5 Swarm Intelligence

Swarm intelligence (SI) is an artificial intelligence technique having self-organized decentralized behavior. These systems are composed of similar variables population interacting with locally. The swarm intelligence is based on the learning behavior from nature, especially biological systems. Some most common examples by which swarm intelligence is inspired are such as ant colonies, animal behavior, biological behavior, etc. These algorithms are based on collective population in search of best solution for specific problem. Each population represents a local best solution for the given problem. So, SI algorithm will improve the result by selecting best solution among local solutions until a stopping condition is not reached. So, these techniques would help in improvement of computational time and yield accurate clustering of similar features from high-dimensional data.

It is seen that swarm intelligence is efficient to solve NP-hard problems [9]. So, using swarm intelligence technique will result in a feature set that can satisfy the condition termed as fitness function, which can be further used in machine learning algorithms. Swarm-based feature selection or dimension reduction technique searches for functions in the search space based on a search strategy such as a global or local search method. Some of the best swarm-based dimension reduction and optimization techniques used in recent research works are such as particle swarm optimization (PSO), ant colony optimization (ACO), grasshopper optimization (GHO), crow search optimization (CSO), gray wolf optimization (GWO), etc.

5 Proposed Methodology

The act of obtaining a completely new dataset which does not have any redundant or unnecessary feature is actually known as feature selection. This also means that original data pattern is maintained in its original form and any useful information is not lost. When the amount of samples is large and the features are in large number, methods of feature selection are of vital importance. As dimensions can be successfully reduced through these methods, these are widely used among the users. It is actually a means to discover features that can very easily describe initial dataset in a precise and brief manner. This work is focused to design a proposed model for multi-objective feature selection problem for dimension reduction of large dataset processing problems. Many existing research works are focused on how many features to consider for a classifier, but has not considered for time-series problems (regression) in which starting time for computation is most important. By analyzing such problems, this research work is motivated to design a hybrid, robust, flexible, and dynamic model for any type of data and their problem domain. This multi-objective optimized feature selection is proposed with an objective to minimize the error rate and execution time as well maximize accuracy of problem and to generate solution with high probability. Optimal control for dynamic feature selection must satisfy following objectives such as maximization of accuracy; minimization of time; minimization of overall; computational cost; and minimization of error rate.

For these objectives, Multi-objective Optimization Algorithm (MOOA) is proposed. For a multi-objective optimization problem, the aim is to find a feature vector, $X = \{x_1, x_2 \ldots x_n\}$ which will satisfy the constraints (Eq. 9)

$$\begin{aligned}&\{f_1(x) \geq 0, \quad i = 1, 2, 3, \ldots, P\}\\&\{f_2(x) \geq 0, \quad j = 1, 2, 3, \ldots, Q\}\end{aligned} \tag{9}$$

and will minimize the vector function (Eq. 10).

$$\begin{aligned}&F(x) = \{f_1(x), f_2(x), \ldots, f_m(x)\}\\&X = \{x_1, x_2, \ldots, x_n\} \in \Omega\end{aligned} \tag{10}$$

where X = feature vector, $f_1(x)$ and $f_2(x)$= objective functions having maximum limits of P and Q, respectively. The set Ω denotes the feasible region, m = number of objective functions to be minimized. The flowchart of proposed swarm multi-objective optimized dimension reduction methodology is illustrated in Fig. 2.

Algorithm: Proposed Multi-objective Optimization for dimension reduction

---

```
Start
Select input parameters: population size (P), no. of population, maximum
iteration (i_max), local and global constant variables, objective functions, lower
and upper bounds of objective functions.
Generate population of size (x = x1, x2, x3...... xdim)
For i = 1 to i_max
        For j = 1 to P
        Fitness(population);
        If fitness_val < switch probability
           Update the position of population;
        End
           Fitness(population);
        If Updated Fitness(population) < Best Fitness(population)
           Return Best Fitness(population) and update position;
        End
        End
End
Output best population found
End
```

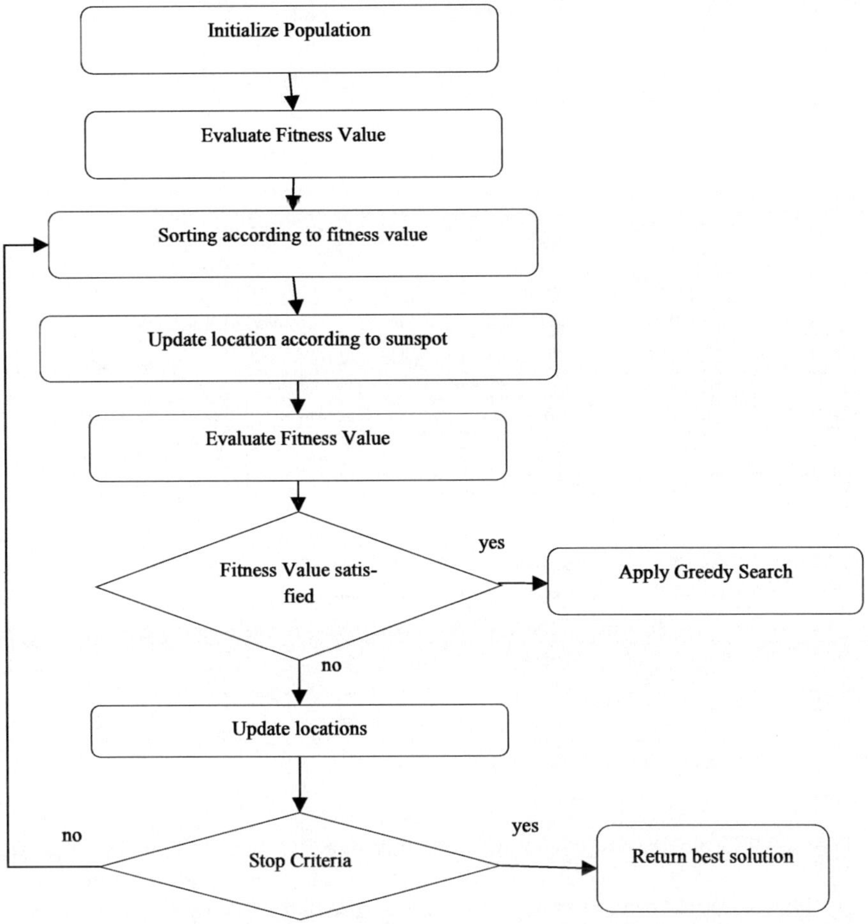

Fig. 2 Flowchart of multi-objective swarm optimization-based dimention reduction

6 Discussion

In this section, different contributions of researchers are summarized. After surveying different expertise works in field of dimension reduction, many quantitative parameters are identified such as given in Table 1. As discussed in Table 1, Sect. 2, different contributions of researchers for dimension reduction are focused on. Most of the researchers had focused their work only on improvement of accuracy level. Apart from accuracy there are several conditions and parameters that an algorithm has to meet. So, there is the requirement of integration of some solutions with existing algorithms to improve their efficiency. The best solution that comes to mind is multi-objective optimization as it can handle multiple problem domain at a time and thus reduces the computational complexity. So, this paper explores the application of

Table 2 Comparison of proposed method with existing methods

Features	Proposed	Khan et al. [8]	El-Hasnony et al. [13]	Liu et al. [16]	Gu et al. [21]	Emary et al. [22]
Swarm intelligence	Yes	No	Yes	Yes	Yes	Yes
Objective function	Multiple	Single	Single	Single	Single	Single
Type of data	Unsupervised and supervised	Supervised	Supervised	Supervised	Supervised	Supervised
Accuracy	Improved	Improved	Improved	Improved	Improved	Improved
Computational time	Reduced	Increased	Average	Increased	Average	Average
Overfitting	No	No	Yes	No	No	No
Cost optimization	Yes	No	No	No	No	No

swarm optimization with multiple objective functions to enhance the efficiency of the existing algorithms. Some of the theoretical comparisons of the proposed methodology are discussed in Table 2 to show the effectiveness of the proposed methodology when implemented in the real-time scenario.

7 Conclusion

Application of dimension reduction and feature selection technique will reduce the overfitting problem of classifiers for large dataset. The concentration of this paper is to provide an analytical review on current research challenges for dimension reduction on large data. After literature review it is seen that bio-inspired (swarm intelligence, genetic algorithm, etc.) is most popular method for finding relevant features as compared to traditional feature selection methods. In this paper, a multi-objective optimized feature selection technique for large dataset or big data analysis is proposed which can improve accuracy and computational time by selecting optimal and relevant features from large datasets.

References

1. Manyika, J., Chui, M., Brown, B., Bughin, J., Dobbs, R., Roxburgh, C., Byers, A.H.: Big Data: The Next Frontier for Innovation, Competition, and Productivity. (2011)
2. Tuo, Q., Zhao, H., Hu, Q.: Hierarchical feature selection with subtree based graph regularization. Knowledge-Based Syst. **163**, 996–1008 (2019)

3. Bolón-Canedo, V., Sánchez-Maroño, N., Alonso-Betanzos, A.: Recent advances and emerging challenges of feature selection in the context of big data. Knowl. Based Syst. **86**, 33–45 (2015)
4. Morán-Fernández, L., Bolón-Canedo, V., Alonso-Betanzos, A.: Centralized vs distributed feature selection methods based on data complexity measures. Knowl. Based Syst. **117**, 27–45 (2017)
5. Kohavi, R., John, G.H.: Wrappers for feature subset selection. Artif. Intell. **97**, 273–324 (1997)
6. Guyon, I., Elisseeff, A.: An introduction to variable and feature selection. J. Mach. Learn. Res. **3**, 1157–1182 (2003)
7. Singh, K., Kaur, L., Maini, R.: Comparison of principle component analysis and stacked autoencoder on NSL-KDD dataset. In: Computational Methods and Data Engineering, Advances in Intelligent Systems and Computing, vol. 1227, pp. 223–241 (2020)
8. Khan, M.A., Arshad, H., Nisar, W., Javed, M.Y., Sharif, M.: An ıntegrated design of fuzzy C-means and NCA-based multi-properties feature reduction for brain tumor recognition. In: Signal and Image Processing Techniques for the Development of Intelligent Healthcare Systems. pp. 1–28 (2020)
9. Brezoˇcnik, L., Fister, I., Podgorelec, V.: Swarm intelligence algorithms for feature selection: a review. Appl. Sci. **8**, 1521 (2018)
10. Siddiqi, U.F., Sait, S.M., Kaynak, O.: Genetic algorithm for the mutual ınformation-based feature selection in univariate time series data. IEEE Access. **8**, 9597–9609 (2020)
11. Kong, L., et al.: Distributed feature selection for big data using fuzzy rough sets. IEEE Trans. Fuzzy Syst. **28**, 846–857 (2020)
12. Ding, W., Lin, C., Pedrycz, W.: Multiple relevant feature ensemble selection based on multilayer co-evolutionary consensus MapReduce. IEEE Trans. Cybern. **50**, 425–439 (2020)
13. El-Hasnony, M., Barakat, S.I., Elhoseny, M., Mostafa, R.R.: Improved feature selection model for big data analytics. IEEE Access. **8**, 6698967004 (2020)
14. Fahy, C., Yang, S.: Dynamic feature selection for clustering high dimensional data streams. IEEE Access **7**, 127128–127140 (2019)
15. Yang, Y., Yin, P., Luo, Z., Gu, W., Chen, R., Wu, Q.: Informative feature clustering and selection for gene expression data. IEEE Access. **7**, 169174–169184 (2019)
16. Liu, X., Liang, Y., Wang, S., Yang, Z., Ye, H.: A hybrid genetic algorithm with wrapper-embedded approaches for feature selection. IEEE Access. **6**, 22863–22874 (2018)
17. Zaffar, M., Hashmani, M.A., Savita, K.S.: Performance analysis of feature selection algorithm for educational data mining. In: IEEE Conference on Big Data and Analytics (ICBDA), pp. 7–12 (2017)
18. Fong, S., Wong, R., Vasilakos, A.: Accelerated PSO swarm search feature selection for data stream mining big data. IEEE Trans. Serv. Comput. **9**, 33–45 (2016)
19. Lin, Y., Hu, Q., Zhang, J., Wu, X.: Multi-label feature selection with streaming labels. Inf. Sci. **372**, 256–275 (2016)
20. Peralta D., del Río, S., Ramírez-Gallego, S., Triguero, I., Benitez, J.M., Herrera, F.: Evolutionary feature selection for big data classification: a mapreduce approach. Math. Probl. Eng. Hindawi. (2015)
21. Gu, S., Cheng, R., Jin, Y.: Feature selection for high-dimensional classification using a competitive swarm optimizer. Soft Comput. **22**, 811–822 (2018)
22. Emary, E., Zawbaa, H.M., Grosan, C., Hassenian, A.E.: Feature subset selection approach by gray-wolf optimization. In: Afro-European Conference for İndustrial Advancement, pp. 1–13. Springer, Cham. (2015)

Modified Leader Algorithm for Under-Sampling the Imbalanced Dataset for Classification

S. Karthikeyan and T. Kathirvalavakumar

Abstract Data classification with a standard classifier automates the manual classification process in many fields. In a two-class dataset, when the number of samples in one class is more in number than the other class, namely, imbalanced, then the performance of a classifier gets degraded due to the limited availability of the training instances in a particular class. To overcome the problems with the imbalanced datasets, a new under-sampling method has been proposed with the baseline idea of an incremental clustering technique. Clusters are formed from the sum of features of the instances instead of finding distance between patterns. Representatives of the clusters are average of the instances of the cluster. Proposed algorithm has the ability to solve the problems than the existing under-sampling approaches with k-means algorithm and leader algorithm. The results produced through the proposed algorithm work better during the classification with good accuracy and reduced misclassification rate in both major and minor classes.

Keywords Imbalanced data · Under-sampling · Incremental clustering · Leader algorithm · Classification

1 Introduction

Samples are the vital component in a data classification problem. More number of training samples in all the classes helps the classifier to correctly classify the data. In some datasets, the distribution of samples in the classes is not balanced and is referred to as imbalanced datasets. Imbalanced class distribution problem is a special category where the number of samples in one class is more in numbers called majority class and less in numbers in another class called minority class. The classification accuracy

S. Karthikeyan
Department of Information Technology, V.H.N. Senthikumara Nadar College, Virudhunagar, Tamilnadu, India
e-mail: karthikeyan@vhnsnc.edu.in

T. Kathirvalavakumar (✉)
Research Centre in Computer Science, V.H.N.Senthikumara Nadar College, Virudhunagar, Tamilnadu, India

J. S. Raj et al. (eds.), *Intelligent Sustainable Systems*, Lecture Notes in Networks and Systems 213, https://doi.org/10.1007/978-981-16-2422-3_9

may not be a deciding factor with the imbalanced class distribution problems [1]. When performing classification with an imbalanced dataset, the focus of a classifier is toward the class having more number of instances and ignores the class having less number of instances. The classifier shows a better accuracy but the misclassification rate in minority class is high. To solve the problems with the imbalanced datasets, re-sampling strategy is adapted. Samples can be re-sampled through over-sampling or under-sampling or hybrid sampling. In over-sampling [2], the number of samples in the minority class is increased, in under-sampling [3], instances in the majority class get down-sampled, but in hybrid sampling [4] majority class samples are down-sampled and the minority class samples are over-sampled to achieve better classification accuracy.

Many approaches are available in the literature to solve the problems with the skewed class distribution. One popular approach for under-sampling is cluster-based under-sampling [5]. Clustering is an unsupervised method of grouping data based on the similarity constraints. Most cluster-based algorithms are available under the categories of partitional, hierarchical, and incremental.

Cluster-based synthetic data generation method with an evolutionary algorithm, namely, partitional clustering algorithm produces better results than the state-of-the-art ensemble algorithms [6]. Improvement in the classification performance is achieved in both minority and majority classes, when the subsets are formed after consolidating the values of the clusters [7]. Under-sampling experiment with k-means algorithm down-samples the majority class data by generating k-clusters by keeping the value of k equal to the count of the minority class data, and the observation shows that this strategy works for small-to-moderate imbalanced datasets and not for the extreme imbalanced datasets [8]. This limitation is solved by forming clusters using incremental clustering algorithms where the number of clusters is formed based on threshold. The quality of clustering is ensured by finding a large cluster whose diameter does not exceed a given user-defined diameter threshold. This measure prevents dissimilar patterns from being forced to stay in the same cluster and ensure that only good quality clusters are formed [9]. Another improvement in the under-sampling through k-means is proposed in [10] by analyzing the weak and noisy instances using the most influential features, which is determined using correlation-based feature subset selection algorithm.

Using the hierarchical clustering algorithm, under-sampling the majority class data using ensemble approach removes the negative samples which are dissimilar with the positive samples [11]. With automatic clustering and under-sampling algorithm [12], the variance is used to determine the samples from different clusters by checking whether the cluster is being further divided or not. This approach learnt the local spatial distribution of the data and clusters the majority class data hierarchically. With the incremental clustering algorithm, namely, mean shift clustering algorithm [13] forms clusters by iteratively assigning each data to its closest cluster centroid. With modified global k-means algorithm [14], the majority class data is under-sampled and considered for training the classifier which produces better results.

From the literature, it is observed that when data changes over time the incremental learning is needed. It is noted that very few research works are done on incremental clustering to deal the imbalanced data. To solve the problem with skewed data distribution, an efficient incremental clustering algorithm has been proposed to handle the data with less misclassification in both major and minor classes. Remaining sections of this paper are organized as follows: Sect. 2 describes the proposed algorithm and Sect. 3 discusses the experimental results.

2 Proposed Work

The proposed modified leader algorithm is an improved version of the leader algorithm [15], which is proposed for under-sampling the majority class data. Existing approaches of under-sampling balance the majority class data to the size of the minority class data [8] but this is eliminated in the proposed method. The proposed method sums all the features of each pattern in the majority class, and arranges the patterns in ascending order based on their sum. Clusters are formed, based on the sum values, in such a way that the difference between the sums of any two patterns of a cluster to be within the decided threshold value. After arranging the patterns in ascending order of their sum, the pattern with less sum value is considered as the first pattern of a cluster. Subsequent patterns from the arranged sum are included into the cluster one after the other until the difference of the new pattern sum and the sum of the first pattern of the cluster is lying below the threshold value. If the difference in the sum value crosses the threshold value, then the newly arrived pattern is the first pattern of the next new cluster. This procedure is repeated until all the patterns of the majority class are processed. After clustering the data, find the average of all the patterns inside each cluster by finding average of each feature individually. This newly calculated pattern of each cluster becomes clusters representative and these are the new training patterns of the majority class. Samples obtained through this procedure represent the properties of the majority class.

2.1 Algorithm

1. Calculate sum for each pattern by adding its features.
2. Arrange the pattern in ascending order based on its feature sum.
3. The pattern of a first sum in the order is a first pattern of a cluster.
4. Include the consecutive patterns inside the cluster one by one until the difference between the new pattern sum and the first pattern sum of the cluster is lesser than the threshold.
5. When the difference > threshold, corresponding pattern is the first pattern of a new cluster.

6. Repeat steps 4 and 5 repeatedly until all the patterns in the arranged order are processed.
7. Now representatives of each generated clusters are the average of patterns inside each cluster.
8. The cluster representatives are the members of majority class.

3 Experimental Results

The working process of the proposed experiment is shown in Fig. 1. To demonstrate the working of the proposed algorithm, it is tested under 13 imbalanced datasets collected from the Keel repository [16]. The total number of instances and its imbalanced ratio [17] are given in Table 1. The imbalance ratio of a dataset is calculated by dividing the total number of samples in the majority class by the total number of samples in the minority class. Imbalance ratio of the datasets used in this experiment ranges from 1.9% to 22.81%. Before under-sampling the majority class data, the imbalanced dataset is divided into two categories such as training dataset and testing dataset. The testing dataset contains 10% of data randomly drawn from the majority class and 10% from the minority class. Rest of the samples in the majority and minority classes are taken as the training dataset. The majority class samples of the training dataset are processed through the proposed under-sampling algorithm. Classifiers involved in this work include C4.5 and Bagging. C4.5 is a tree-based classifier which is widely used in the imbalanced datasets and Bagging classifier is an ensemble-based classifier which classifies the dataset by forming subsets and aggregate different predictions to form a concluding prediction. As C4.5 and Bagging classifiers are widely used in the literature to classify the imbalanced datasets, these classifiers are used in the proposed work.

To demonstrate the efficiency of the proposed experiment, the majority class data is under-sampled with two different algorithms, Leader and k-means. Leader

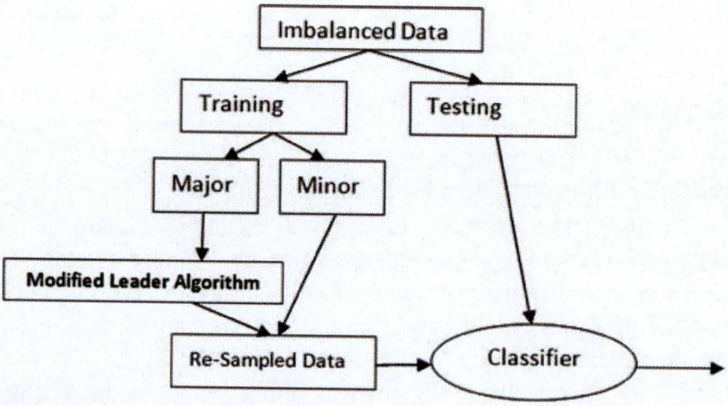

Fig. 1 Proposed work

Table 1 Dataset and its ımbalance ratio

Dataset	# of instances (major:minor)	Proportion major:minor	Imbalance ratio
Abalone9-18	731 (689:42)	0.942:0.057	16.68
Glass0	214 (144:70)	0.672:0.327	3.19
Glass1	214 (138:76)	0.644:0.355	10.39
Glass2	214 (197:17)	0.920:0.079	15.47
Glass4	214 (201:13)	0.939:0.060	22.81
Haberman	306 (225:81)	0.735:0.264	2.68
Iris0	150 (100:50)	0.666:0.333	2
New-Thyroid1	215 (180:35)	0.837:0.162	5.14
Pima	768 (500:268)	0.651:0.348	1.9
Vehicle1	846 (629:217)	0.743:0.256	2.52
Vehicle3	846 (634:212)	0.749:0.250	2.52
Yeast1vs7	459 (429:30)	0.934:0.065	13.87
Yeast0-5-6-7-9vs4	528 (477:51)	0.903:0.096	9.35

algorithm is an incremental clustering algorithm, which works on the basis of the threshold parameters, so the number of samples obtained from the majority class is not known until its convergence. K-means is a partitional clustering algorithm, with this algorithm the majority class samples are under-sampled in the ratio 1:1, that is, majority class samples under-sampled to the sample size of the minority class. The under-sampled majority class data is combined with the training instances of minority class data to form a new training dataset. The classification accuracy of the classifier is calculated using the obtained true positive (TP), true negative (TN), false positive (FP), and the false negatives (FN) of the confusion matrix. The classification accuracy is calculated using the formula

$$\text{Classification Accuracy} = \frac{\text{TP} + \text{TN}}{\text{TP} + \text{TN} + \text{FP} + \text{FN}} \tag{1}$$

The optimal threshold value and the classification accuracy of the proposed method under C4.5 and Bagging classifiers are shown in Table 2. The threshold value is selected by doing more trials. From Table 2, it is observed that Bagging shows better classification accuracy in eight datasets than C4.5 and C4.5 classifier shows better classification accuracy in one dataset than Bagging. Both classifiers produce same classification accuracy in four datasets. Regarding the misclassification rate of the majority class, the proposed work under C4.5 classifier suffers a lot when compared with the Bagging classifier.

The accuracy of the proposed work, k-means and Leader algorithm, is compared using C4.5 and Bagging classifiers and is shown in Figs. 2 and 3. In most datasets, the classification accuracy of the proposed work is higher than the data generated through

Table 2 Classification accuracy and misclassification rate (MR) under C4.5 and Bagging

Dataset	Threshold	C4.5	Bagging	MR under C4.5 major:minor	MR under Bagging major:minor
Abalone9-18	0.01	94.5	**97.26**	4/69:**0/4**	**2/69:0/4**
Glass0	0.01	80.95	**85.71**	4/14:**0/7**	3/14:**0/7**
Glass1	0.001	**100**	**100**	**0/14:0/8**	**0/14:0/8**
Glass2	0.001	90.9	**100**	2/20:**0/2**	**0/20:0/2**
Glass4	0.001	**100**	**100**	**0/20:0/1**	**0/20:0/1**
Haberman	0.25	**83.87**	77.41	**5/23:0/8**	7/23:**0/8**
Iris0	0.5	**100**	**100**	**0/10:0/5**	**0/10:0/5**
New-Thyroid1	0.5	**100**	**100**	**0/18:0/7**	**0/18:0/7**
Pima	5	67.53	**74**	25/50:**0/27**	**20/50:0/27**
Vehicle1	0.5	95.29	**96.47**	4/63:**0/22**	**3/63:0/22**
Vehicle3	0.5	97.61	**100**	2/63:**0/21**	**0/63:0/21**
Yeast1vs7	0.05	71.73	**73.91**	13/43:**0/3**	12/43:**0/3**
Yeast0-5-6-7-9vs4	0.02	83.01	**90.56**	9/48:**0/5**	**5/48:0/5**

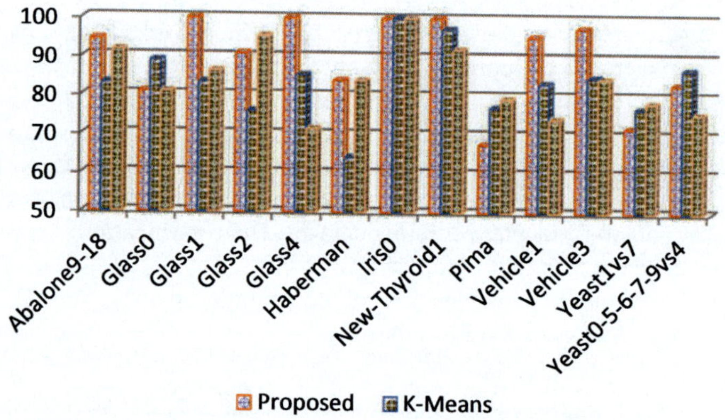

Fig. 2 Comparison of classification accuracy using C4.5

k-means algorithm and Leader algorithm. On comparing the classification accuracy of the proposed work with k-means and leader algorithms, it is observed that better results are obtained in all the datasets except Glass1, Glass2, Pima, Yeast1vs7, and Yeast0567vs4 under C4.5 classifier and lesser accuracy is obtained in the datasets Glass0, Haberman, Pima, and Yeast1vs7 with Bagging classifier.

The instances misclassified in the test dataset is shown in Tables 3 and 4. With C4.5 and Bagging classifiers the proposed algorithm produces slightly more misclassification than either k-means or leader in the majority class of the datasets Glass0, Glass2, Haberman, Pima, and Yeast1vs7 datasets. But no misclassification is observed under

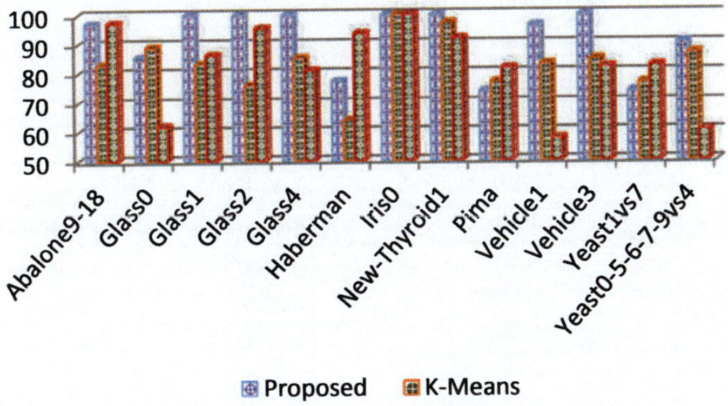

Fig. 3 Comparison of classification accuracy using Bagging

Table 3 Misclassification rate under C4.5

Dataset	Proposed work major:minor	K-means major:minor	Leader major:minor
Abalone9-18	**4/69:0/4**	27/69:**0/4**	6/69:**0/4**
Glass0	4/14:**0/7**	**2/14:0/7**	4/14:**0/7**
Glass1	**0/14:0/8**	2/14:**0/8**	3/14:**0/8**
Glass2	2/20:**0/2**	13/20:**0/2**	**1/20:0/2**
Glass4	**0/20:0/1**	2/20:**0/1**	6/20:**0/1**
Haberman	5/23:**0/8**	**4/23**:4/8	5/23:**0/8**
Iris0	**0/10:0/5**	**0/10:0/5**	**0/10:0/5**
New-Thyroid1	**0/18:0/7**	**0/18:0/7**	2/18:**0/7**
Pima	25/50:**0/27**	12/50:6/27	**11/50**:5/27
Vehicle1	**4/63:0/22**	9/63:5/22	20/63:2/22
Vehicle3	**2/63:0/21**	16/63:**0/21**	11/63:2/21
Yeast1vs7	13/43:**0/3**	13/43:1/3	**10/43:0/3**
Yeast0-5-6-7-9vs4	**9/48:0/5**	21/48:**0/5**	12/48:1/5

the proposed work in the minority class of all datasets. It is also observed that on comparing the proposed work with the Leader algorithm, marginal misclassification difference is observed in the majority class of three datasets Haberman, Pima, and Yeast1vs7.

The performance of an imbalanced datasets is evaluated using the statistical evaluation metrics other than classification accuracy to validate the proposed work. Here Area Under the Receiver Operating Characteristics (AUC) score and Brier score are considered for the evaluation. AUC score provides a cumulative measure of the performance across every possible classification threshold. The AUC score for

Table 4 Misclassification rate under Bagging

Dataset	Proposed work major:minor	K-means major:minor	Leader major:minor
Abalone9-18	**2/69:0/4**	32/69:0/4	**2/69:0/4**
Glass0	3/14:**0/7**	**0/14:0/7**	8/14:**0/7**
Glass1	**0/14:0/8**	6/14:**0/8**	3/14:**0/8**
Glass2	**0/20:0/2**	16/20:**0/2**	1/20:**0/2**
Glass4	**0/20:0/1**	4/20:1/1	4/20:**0/1**
Haberman	7/23:**0/8**	8/23:**0/8**	**2/23:0/8**
Iris0	**0/10:0/5**	**0/10:0/5**	**0/10:0/5**
New-Thyroid1	**0/18:0/7**	1/18:**0/7**	2/18:**0/7**
Pima	20/50:**0/27**	**12/50**:4/27	14/50:**0/27**
Vehicle1	**3/63:0/22**	21/63:3/22	32/63:4/22
Vehicle3	**0/63:0/21**	17/63:1/21	13/63:2/21
Yeast1vs7	12/43:**0/3**	17/43:1/3	**8/43:0/3**
Yeast0-5-6-7-9vs4	**5/48:0/5**	14/48:**0/5**	21/48:**0/5**

the proposed algorithm, k-means algorithm, and Leader algorithm under C4.5 and Bagging are shown in Tables 5 and 6.

From Tables 5 and 6, it is observed that from the AUC score of the proposed work with k-means under C4.5 classifier shows better results in ten datasets, same result in two datasets, and minor difference is observed in the remaining one dataset. On comparing the results with Leader algorithm, better result is obtained in eight datasets,

Table 5 AUC score under C4.5

Dataset	Proposed	K-means	Leader
Abalone9-18	**0.971**	0.841	0.957
Glass0	**0.857**	**0.857**	0.714
Glass1	**1**	0.964	0.893
Glass2	**0.975**	0.775	**0.975**
Glass4	**1**	0.775	0.85
Haberman	0.891	0.87	**0.913**
Iris0	**1**	**1**	**1**
New-Thyroid1	**1**	0.986	94.4
Pima	0.75	**0.91**	0.83
Vehicle1	**0.976**	0.897	0.802
Vehicle3	**0.984**	0.889	0.889
Yeast1vs7	0.849	0.709	**0.884**
Yeast0-5-6-7-9vs4	**0.896**	0.75	0.854

Table 6 AUC score under Bagging

Dataset	Proposed	K-means	Leader
Abalone9-18	**1**	0.96	**1**
Glass0	0.903	**1**	0.857
Glass1	**1**	**1**	0.995
Glass2	**1**	0.65	0.975
Glass4	**1**	**1**	**1**
Haberman	0.918	0.995	**1**
Iris0	**1**	**1**	**1**
New-Thyroid1	**1**	**1**	94.4
Pima	0.83	**0.974**	0.956
Vehicle1	0.998	**0.999**	0.938
Vehicle3	**1**	0.986	0.992
Yeast1vs7	0.884	0.729	**0.915**
Yeast0-5-6-7-9vs4	**1**	0.992	0.971

same result in two dataset, and poor performance in three datasets. With Bagging classifier, comparison of k-means algorithm with the proposed work performs better in five datasets, same result in four datasets, and lesser difference is observed in four datasets. With Leader algorithm, better results are observed in seven datasets, same result in three datasets, and lesser performance in three datasets.

Another statistical evaluation metric used in our work is Brier score. Brier score is used to measure the accuracy of the probabilistic predictions. The score 0 indicates a perfect classification and 1 indicates a worst performance. The results of the brier score evaluation under C4.5 and Bagging are shown in Tables 7 and 8.

From Tables 7 and 8, it is clear that the performance of the proposed work under C4.5 and Bagging is better in all the datasets except Glass0, Haberman, Pima, and Yeast1vs7. The time taken to perform under-sampling on the majority class data using incremental clustering algorithms is shown in Table 9. The time taken by the proposed work is higher than the Leader algorithm because of the data arrangement in order, but on comparing the results obtained through the proposed work, the time constraints get wiped out.

4 Conclusion

The performance of the re-sampled data generated using the proposed method shows better classification accuracy than the re-samplings generated through k-means and Leader algorithms. The selection of the threshold value plays a major role in forming clusters which leads to better classification. The proposed under-sampling algorithm helps the classifier to have a balanced focus on both classes. From the experiments,

Table 7 Brier score under C4.5

Dataset	Proposed	K-means	Leader
Abalone9-18	**0.055**	0.301	**0.055**
Glass0	0.19	**0.143**	0.286
Glass1	**0**	0.045	0.136
Glass2	**0.091**	0.409	0.045
Glass4	**0**	0.143	0.19
Haberman	**0.161**	0.226	**0.161**
Iris0	**0**	**0**	**0**
New-Thyroid1	**0**	**0**	0.08
Pima	0.312	**0.091**	0.234
Vehicle1	**0.047**	0.129	0.318
Vehicle3	**0.024**	0.131	0.179
Yeast1vs7	0.283	0.522	**0.217**
Yeast0-5-6-7-9vs4	**0.151**	0.415	0.245

Table 8 Brier score under Bagging

Dataset	Proposed	K-means	Leader
Abalone9-18	**0.026**	0.177	0.031
Glass0	0.168	**0.039**	0.274
Glass1	**0.02**	0.057	0.106
Glass2	**0.022**	0.38	0.047
Glass4	**0.007**	0.174	0.119
Haberman	0.172	0.113	**0.086**
Iris0	**0**	**0**	**0**
New-Thyroid1	**0.001**	**0.001**	0.073
Pima	0.245	**0.083**	0.156
Vehicle1	**0.031**	0.094	0.25
Vehicle3	**0.02**	0.113	0.122
Yeast1vs7	0.2	0.407	**0.134**
Yeast0-5-6-7-9vs4	**0.071**	0.199	0.219

it is clear that under-sampling with the proposed method along with the usage of Bagging classifier yields a better classification result, AUC score, Brier score, and reduced misclassification rate in both majority and minority classes. The novelty of the proposed work is forming clusters based on the sum of features of each pattern instead of finding distance calculation between patterns, and the leaders of the clusters are the average of the patterns inside the cluster instead of the pattern with lesser sum value.

Table 9 Under-sampling time using proposed and Leader algorithms

Dataset	Proposed (ms)	Leader (ms)
Abalone9-18	11,074	**2524**
Glass0	2390	**2096**
Glass1	**2089**	2093
Glass2	2746	**1799**
Glass4	3596	**2851**
Haberman	3459	**915**
Iris0	**1477**	1580
New-Thyroid1	2846	**2033**
Pima	5765	**5474**
Vehicle1	13,562	**7527**
Vehicle3	11,245	**5183**
Yeast1vs7	6556	**5417**
Yeast0-5-6-7-9vs4	7578	**5332**

References

1. Wenyu, H., Baili, Z.: Study of sampling techniques and algorithms in data stream environments. In: 9th International Conference on Fuzzy Systems and Knowledge Discovery (FSKD 2012), pp. 1028–1034 (2012).
2. Abdi, L., Hashemi, S.: To combat multi-class imbalanced problems by means of over-sampling techniques. IEEE Trans. Knowl. Data Eng. **28**, 238–251 (2016). https://doi.org/10.1109/TKDE.2015.2458858
3. Drummond, C., Holte, R.C.: C4.5, class imbalance, and cost sensitivity: why under-sampling beats over-sampling. In: Workshop on Learning from Imbalanced Datasets II, ICML, pp. 1–8 (2003). 10.1.1.68.6858
4. Seiffert, C., Hulse, J. Van, Raton, B.: Hybrid sampling for ımbalanced data. In: IEEE International Conference on Information Reuse and Integration, pp. 202–207, Las Vegas (2008).
5. Yen, S., Lee, Y.: Cluster-Based Sampling Approaches to Imbalanced Data Distributions, pp. 427–436. Springer LNCS (2006)
6. Lim, P., Goh, C.K., Tan, K.C.: Evolutionary cluster-based synthetic oversampling ensemble (eco-ensemble) for ımbalance learning. IEEE Trans. Cybern. **47**, 2850–2861 (2017). doi:https://doi.org/10.1109/TCYB.2016.2579658
7. Zhang, Y.P., Zhang, L.N., Wang, Y.C.: Cluster-based majority under-sampling approaches for class imbalance learning. In: 2010 2nd IEEE International Conference on Information and Financial Engineering, ICIFE 2010, pp. 400–404 (2010). doi:https://doi.org/10.1109/ICIFE.2010.5609385
8. Lin, W.C., Tsai, C.F., Hu, Y.H., Jhang, J.S.: Clustering-based undersampling in class-imbalanced data. Inf. Sci. (Ny). **409–410**, 17–26 (2017). doi:https://doi.org/10.1016/j.ins.2017.05.008
9. Nayini, S.E.Y., Geravand, S., Maroosi, A.: A novel threshold-based clustering method to solve K-means weaknesses. In: 2017 International Conference on Energy, Communication, Data Analytics and Soft Computing (ICECDS 2017), pp. 47–52 (2018). doi:https://doi.org/10.1109/ICECDS.2017.8389496

10. Kumar, N.S., Rao, K.N., Govardhan, A., Reddy, K.S., Mahmood, A.M.: Undersampled K-means approach for handling imbalanced distributed data. Prog. Artif. Intell. **3**, 29–38 (2014). doi:https://doi.org/10.1007/s13748-014-0045-6
11. Soltani, S., Sadri, J., Torshizi, H.A.: Feature selection and ensemble hierarchical cluster-based under-sampling approach for extremely imbalanced datasets: application to gene classification. In: 2011 1st International eConference on Computer and Knowledge Engineering (ICCKE 2011), pp. 166–171 (2011). doi:https://doi.org/10.1109/ICCKE.2011.6413345
12. Deng, X., Xu, Y., Chen, L., Zhong, W., Jolfaei, A., Zheng, X.: Dynamic clustering method for imbalanced learning based on AdaBoost. J. Supercomput. (2020). doi:https://doi.org/10.1007/s11227-020-03211-3
13. Comaniciu, D., Meet, P.: Mean shift analysis and applications. In: Proceedings of the Seventh IEEE International Conference on Computer Vision 2, pp. 1197–1203 (1999). doi:https://doi.org/10.1109/iccv.1999.790416
14. Moniruzzaman, M., Bagirov, A., Gondal, I.: Partial undersampling of ımbalanced data for cyber threats detection. In: ACM International Conference Proceedings Series, pp. 2–5 (2020). doi:https://doi.org/10.1145/3373017.3373026
15. Vijaya, P.A., Murty, M.N., Subramanian, D.K.: An efficient incremental protein sequence clustering algorithm. In: IEEE Region 10 International Conference TENCON 1, pp. 409–413 (2003). doi:https://doi.org/10.1109/tencon.2003.1273355
16. Small Scale Data Set. https://sci2s.ugr.es/keel/
17. Yen, L: Cluster-based under-sampling approaches for imbalanced data distributions. Expert Syst. Appl. **36**, 5718–5727 (2009). doi:https://doi.org/10.1016/j.eswa.2008.06.108

Weather Divergence of Season Through Regression Analytics

Shahana Bano, Gorsa Lakshmi Niharika, Tinnavalli Deepika, S. Nithya Tanvi Nishitha, and Yerramreddy Lakshmi Pranathi

Abstract Weather forecasting is carried out using different external measuring devices of weather predictions based on previous data. These devices or techniques obtain an obvious report of the prediction of weather. In our perspective, research is performed through many previous datasets in the weather forecast. Incorporating a model of multilinear regression and quantile regression allows visualizing the previous data to analyze the weather prediction. The dataset of precipitation levels is trained as a multilinear regression and gives the final level of estimation between rainfall in 2019 and rainfall in 2020. An analysis of a dependent variable and independent variables is multilinear regression. Through this study of approach algorithms, the seasons can be distinguished depending on the visualization.

Keywords Regression · Prediction · Weather · Season · Precipitation levels · Multilinear · Quantile

1 Introduction

Using the datasets, an investigation is performed for a specific place in the United States and a study of recent rainfall levels in the environment is performed. The weather forecast reports of the future are measured using machine learning algorithms.

By this study, prediction of weather is done through periodic assessment of weather satellites. Currently, improved technology forecasts the weather by scanning and measuring gadgets, radar systems, weather balloons, etc. Prediction of any data is only carried out through several iterations of past or current data preparation. Here in our research, a comparison report of past weather data is considered for a

S. Bano · G. L. Niharika (✉) · T. Deepika · S. N. T. Nishitha
Department of CSE, Koneru Lakshmaiah Education Foundation, Vaddeswaram, India

S. Bano
e-mail: shahanabano@icloud.com

Y. L. Pranathi
Queen's University Belfast, Belfast, UK

J. S. Raj et al. (eds.), *Intelligent Sustainable Systems*, Lecture Notes in Networks and Systems 213, https://doi.org/10.1007/978-981-16-2422-3_10

particular part of the United States. The dependent variable is the weather; it often varies with the external influences that have occurred through humans or on their own.

According to the World Meteorological Organization, a regular report on the local atmosphere, seas, soil, and all water supplies is always necessary. Therefore, consider all changes in the atmosphere, temperature changes, and precipitation values were sufficient to visualize our results. The Python language is used for representing the visualization of Boulder data temperature and precipitation [6] values. Many data clusters can be taken to visualize the precipitation values and temperatures. With the assistance of multilinear regression methods [8], meteorological changes have been well examined during our study.

Multilinear regression [13] is a study of dependent and independent variables. Here in our dataset, we have precipitation and temperature functions as dependent variables to modify climate change. If the rainfall levels are higher, then it creates a sign of precipitation. The visualization of climate data is related to temperature and precipitation levels. Finally, consider the July–December 2019 weather comparison report and January–June 2020, which describes the weather environment analysis based on the visualized graphical images obtained at the end of the code operation. The weather is changing factor dependent on temperature and precipitation values. The use of the weather datasets of a particular area indicates how the place can affect the climate.

Regression analysis models and prediction architects indicate where it increases and falls. A comparison report was prepared through the regression analysis report. This comparison report takes into account past data forecasting the season in which they are predicting the weather [1] in 2020 January to June through past reports of 2019 January to June.

2 Literature Survey

The regression analysis visualizes the data and proceeds to check where data is getting decreased, increased, and variable to be neglected. Choosing the best algorithm beyond all the existing algorithms is effectively explained by Anila, M., Pradeepini, G. [2]. Linear regression [12] builds an analysis on the individual dependent variable. Whereas temperature and precipitation are considered as the dependent variables. A single dependent variable is called linear regression. In the weather changing, the parameter is the only temperature that cannot make difference in it.

Linear regression [11] algorithm cannot provide a well-defined approach since there are two dependent variables in code. If one is missing in prediction parameter, there will be a loss of analytical and statistical approach. For that reason, many of the regression algorithms make sense of the prediction parameter. So, reading a large dataset like weather needs a very high effective algorithm. In our dataset, consider two dependent variables which affect the weather.

Multilinear regression statistics were initialized in our dataset evaluation for analyzing both temperature and precipitation, respectively. It always explains the data with a single independent and two or more independent variables. Here the relation to multilinear regression is bonding to the dependent variable which is the weather that day and the independent variable is temperature, precipitation where these two are independent of each other [7]. The final evaluation of data gives high accuracy of prediction which makes a spectacle of analytics.

Though multilinear regression effectively performs in variating seasons by comparing temperature and precipitation in the final output graph, the results are identified in a descriptive way of an attribute that varies the weather like temperature, precipitation, and so on. So, this parameter of approach varies the season from periodic metrics. Comparing and analyzing every dependent attribute plays an important role to change the seasonal climate [16].

3 Methodology

3.1 Importing All the Packages

To compare and run the model, it is required to import all the packages which excel in the statistical estimation of data exploration of machine learning. For all the classification of regressions, and for the comparison of datasets, a Scikit-Learn package is required to import inside the code. It supports the usage of NumPy and SciPy libraries. NumPy and pandas alone deal with the mathematical and statistical approach of output. To categorize the way of output, it describes the comparison parameter to both multilinear regression and quantile regression algorithms.

Here for statistical calculations, it lists the temperature and precipitation values of each day from July to December 2019, January to June 2020, which were utilized for generating Pandas, NumPy, and matplotlib. For the overall approach in the visualization concept, matplotlib is used. Visualization in a graph leads to a view of analysis.

Whereas in quantile regression parameter, the statsmodel API is used to assist the statistical estimation of data. Here, extensive data for the period is used to compare and estimate each dependent parameter that must be imported using statsmodel API.

3.2 Importing Dataset

It is required to import the dataset taken from a weather forecast of Boulder data temperature and precipitation values for the day to ensure that our system is working. The differences existing against each dependent class of weather are calculated in regression algorithms. The values are incorporated in the dataset from July to

December 2019 and from January to June 2020. These periods of data were taken in training our model.

3.3 Multilinear Regression

Multilinear regression is also a common resemblance of linear regression but here in multilinear regression, it predicts the values between one dependent variable and two or more independent variables. These independent variables are affecting the weather in our study. Here, the two independent variables are considered such as temperature and precipitation values of a place.

The regression algorithm of linear regressions varies with the data containing one dependent and one independent variable. In contrast to the linear regression algorithm, multilinear regression is used to summarize the whole data with a perfect accuracy.

Multilinear regression formulated by

$$y = \beta_0 + \beta_i x_i + \cdots + \beta_n x_n + \varepsilon \quad (1)$$

Here in this equation ε is error value, y is the dependent variable, x is independent variable, and β_i is a parameter.

3.4 Quantile Regression

Although the implementation of regressions leads to statistical way, the dependent attribute or parameter used in the case of study provides an impact to the analysis. So, in the same way, quantile regression [15] is the parameter programming where it fits the sample of data of p samples to do the regressions and for every quantile value the dataset randomly checks with day to form the parameter analysis. Here in our research, weather is parameter where temperature and precipitation are two independent attributes of each other but dependent in change of weather. Multilinear regression provides statistics by comparing both attributes and its impact on weather. For this way of approach, our study deals with the effectiveness of two algorithms.

A Quantile regression is the analysis estimated by calculating the median of a target. Here quantile regression deals with every dependent variable constraint and compares its dependency through every quantile. Traditional linear regression is contrast to this regression and quantile regression is not permitted to display the calculation of median, but it finds the dependency on quantile of data for feature variables (Eq. 2).

The formula for quantile regression includes similar structure of linear regression:

$$Q_\tau(y_i) = \beta_0(\tau)x_{i1} + \cdots + \beta_p(\tau)x_{ip} \; i = 1, 2, 3, \ldots, n \quad (2)$$

The absolute median value for quantile regression is calculated by

$$\text{median} = \frac{1}{n}\sum_{i=1}^{n} \rho_\tau(y_i - \beta_0(\tau)x_{i1} + \cdots + \beta_p(\tau)x_{ip}) \quad (3)$$

Here in this formula of equation, ρ is a function to check the error depending values on quantile and later transformation of stage gives dependency score on each dependent variable for every random value. For any consequence of data, transforming the data can lead to change the regression dependencies.

3.5 Covariance and Correlation

Here covariance and correlation [9] deal with the theoretical method to how two parameters are related to each other and in what way. Whereas covariance provides with how parameters relate to each other and how their relationship approach is worked out. The covariances are in two types such as positive covariance and negative covariance. Positive covariance is when one value increases, so it may also increase the other value. If one value increases, then the other value definitely decreases in negative covariance.

The formula for identifying covariance:

$$\mathrm{C\,(X, Y)} = \frac{\sum_{i=1}^{n} (X_i - \overline{X})(Y_i - \overline{Y})}{n - 1} \quad (4)$$

In Eq. 4, x and y are values of two comparing attributes for calculating the relation among them.

Correlations are the perfect way to determine the perfect inverse correlation and perfect positive correlation. And these values of analysis are in between −1 and 1 to integrate the relation among the variables.

3.6 Algorithm

Step 1: Start.

Step 2: The packages should be imported mainly Scikit-Learn, NumPy, Pandas, and all the essential packages.

Step 3: Import dataset required for training.

Step 4: Preprocessing the data.

Step 5: Testing the data with required dependent and independent variables.

Step 6: Fit these values in our model of algorithm that is multilinear and quantile regressions.

Step 7: Implements the confidence score and prediction intervals to predict the head.

Step 8: Displays the summary report and visualizes the graph for our approach.

Step 9: Repeat the steps 1, 2, 3, 4, 5, 6, 7.

Step 10: End.

4 Flowchart

Our model flow is to interpret the data by importing the necessary packages and libraries required for the dataset to be visualized. A well-defined analysis of the graph is given here by the Scikit-Learn package. The weather data is analyzed in the later stage by applying multilinear regressions and algorithms of quantile regressions. The statistics available between them are obtained by covariance and correlations. A precise calculation of each dataset is the final approach to our dataset. The accuracy obtained using the algorithm of multilinear regression is highly efficient. So, for better summarizing of seasonal variance in performance, one can use the multilinear regression algorithm. The main varying dependent variable in dataset can be relationally compared with other independent variables using quantile regressions (Fig. 1).

5 Results

Temperature and precipitation comparison is an approach which deals with data analysis. Through all the data obtained from July to December 2019 (Fig. 2), we are able to take an objective approach to weather prediction. The way of comparing temperatures and precipitation is hereby declared vice versa in Fig. 7. The ultimate objective of prediction through this analogy is seasonal weather recognition.

Scatterplot of temperature is also one of the visualization relation of attributes where it allows to plot the values (Fig. 3).

Multilinear regression methodologies are trained with dataset from July to December 2019 for dealing with temperature and precipitation in Fig. 4. From this visualization, obtained values are

Mean squared error = 56.223….
Root mean squared error = 7.498..
R2_score = 0.502…

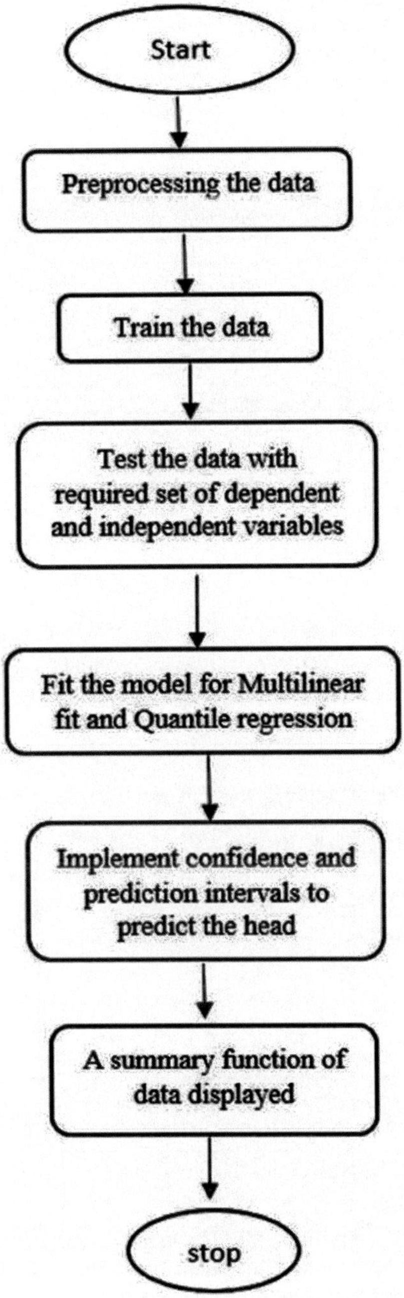

Fig. 1 Overview of the process

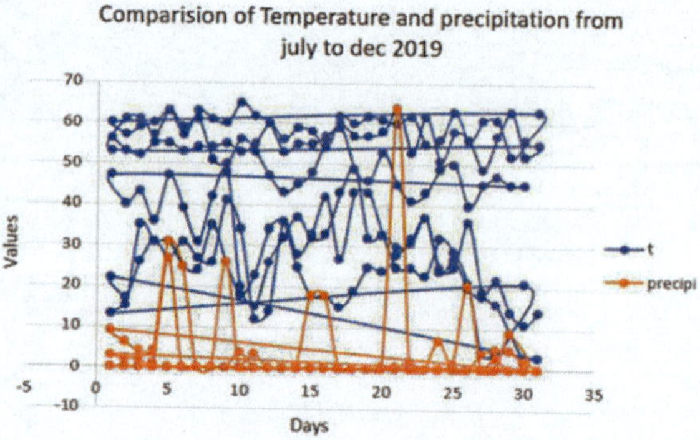

Fig. 2 Comparison of temperature and precipitation from July to December 2019

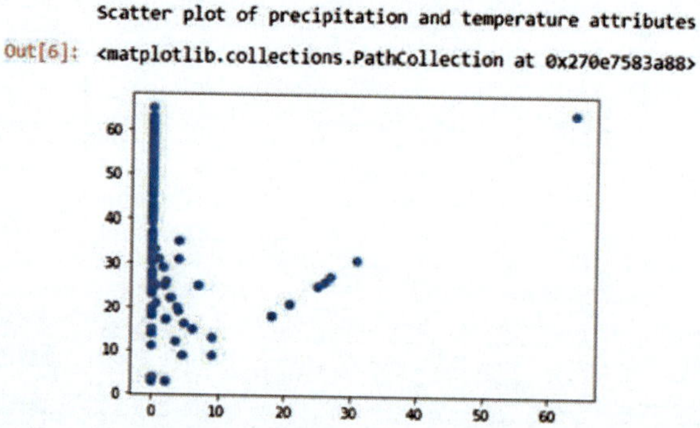

Fig. 3 Scatterplot of attributes

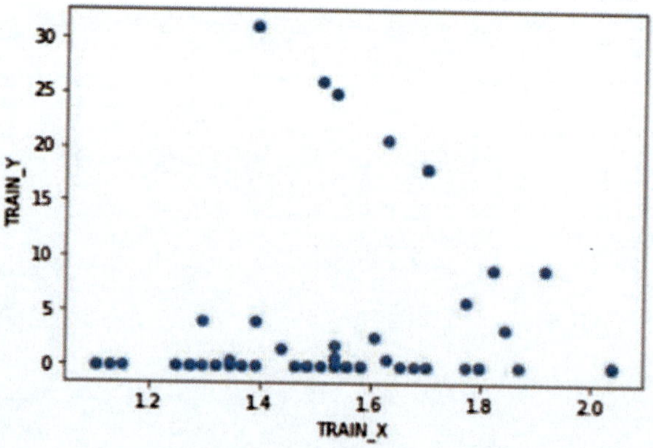

Fig. 4 Multiple linear regression result for July to December 2019

Accuracy is more important in every algorithm and it is achieved to use multilinear regression. It has given an effective accuracy where we get 1.0 after so many iterations of code (Fig. 5).

Figure 6 from the variance obtained between attributes relates how the relation between attributes is in and in which way they are moving with relation. Through

Fig. 5 Accuracy of multilinear regression algorithm

```
accuracy
1.0

classification
              precision    recall  f1-score   support

         low       1.00      1.00      1.00        56

    accuracy                           1.00        56
   macro avg       1.00      1.00      1.00        56
weighted avg       1.00      1.00      1.00        56
```

Fig. 6 Covariance and correlation with respect to day

```
Co-Variance =                day          t    precipi
day        78.825731  -25.219173  -2.972862
t         -25.219173  276.038845 -16.822767
precipi   -2.972862  -16.822767  43.994098
Co-Relation =              day          t    precipi
day        1.000000 -0.170967 -0.050483
t         -0.170967  1.000000 -0.152656
precipi  -0.050483 -0.152656  1.000000
```

our data of availability, co-variance and co-relation is obtained between temperature, precipitation with respect to day or date (Figs. 7 and 8).

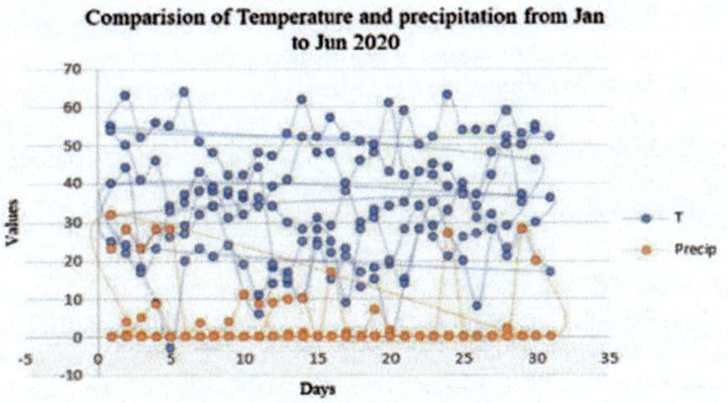

Fig. 7 Comparison of temperature and precipitation from January to June 2020

Fig. 8 Quantile regression dependent to temperature

```
                         QuantReg Regression Results
==============================================================================
Dep. Variable:                      t   Pseudo R-squared:              0.04208
Model:                       QuantReg   Bandwidth:                       14.65
Method:                 Least Squares   Sparsity:                        53.75
Date:                Sun, 06 Sep 2020   No. Observations:                  184
Time:                        13:05:37   Df Residuals:                      182
                                        Df Model:                            1
==============================================================================
                 coef    std err          t      P>|t|      [0.025      0.975]
------------------------------------------------------------------------------
Intercept     45.0000      2.044     22.018      0.000      40.967      49.033
precipi       -0.8000      0.300     -2.671      0.008      -1.391      -0.209
==============================================================================
```

6 Conclusion

This model of analysis makes to understand the relationship between dependent and independent variables. Quantile regression excels in dealing with the attribute relations and their dependencies. A multilinear regression comparison graph makes one analyze the seasonal differences between two parts of a year. By this regression algorithm, the data is more efficient with accuracy. Through this work of research, it can be concluded that the regression allows one to understand the visualization of a climate easily. Finally, in our work, the seasonal change is identified.

7 Future Scope

Weather divergence can be extended for further purpose to implement in any unknown places to differentiate its climate changes. It can also be implemented as an Android application.

References

1. Amirullah, I., Harun, N., Pallu, M.S., Arwan, A.: Statistic approach versus artificial intelligence for rainfall prediction based on data series. Int. J. Eng. Technol. (IJET). **5**(2), 1962–1969 (2013)
2. Anila, M., Pradeepini, G.: Study of prediction algorithms for selecting appropriate classifier in machine learning. J. Adv. Res. Dyn. Control Syst. **9**(Special Issue 18), 257–268 (2017)
3. Anisha, P.R., Vijaya Babu, B.: A germane prognosis paradigm for climate and weather research. Int. J. Control Theory Appl. **9**(34), 5769 (2016)
4. Chandrashaker Reddy, P., Suresh Babu, A.: Survey on weather prediction using big data analysis. In: Second International Conference on Electrical, Computer and Communication, 2017.
5. Das, B., Nair, B., Reddy, V.K. et al. Evaluation of multiple linear, neural network and penalised regression models for prediction of rice yield based on weather parameters for west coast of India. Int. J. Biometeorol. **62**, 1809–1822 (2018). doi:https://doi.org/10.1007/s00484-018-1583-6
6. Gholizadeh, M.H., Darand, M.: Forecasting precipitation with artificial neural networks (case study: Tehran). J. Appl. Sci. **9**(9), 1786–1790 (2009)
7. Heinemann, G.T., Nordmian, D.A., Plant, E.C.: The relationship between summer weather and summer loads - a regression analysis. In: IEEE Transactions on Power Apparatus and Systems, vol. PAS-85, no. 11, pp. 1144–1154. (1966). doi:https://doi.org/10.1109/TPAS.1966.291535
8. Kaya Uyanık, G., Güler, N.: A study on multiple linear regression analysis. Procedia Social Behav. Sci. **106**, 234–240 (2013). doi:https://doi.org/10.1016/j.sbspro.2013.12.027
9. Kumar, B.S.C., Sadasivan, K., Ramesh, K.: Correlation between compressive strength and split tensile strength of GGBS and MK based geopolymer concrete using regression analysis. J. Mech. Cont. Math. Sci. **14**(1), 21–36 (2019). doi:https://doi.org/10.26782/jmcms.2019.02.00002
10. Lavanya, K., Reddy, L.S.S., Reddy, B.E: Distributed based serial regression multiple imputation for high dimensional multivariate data in multicore environment of cloud. Int. J. Ambient Comput. Intell. **10**(2), 63–79 (2019). doi:https://doi.org/10.4018/IJACI.2019040105

11. Li, H., Pi, D., Wu, Y., Chen, C.: Integrative method based on linear regression for the prediction of zinc-binding sites in proteins. IEEE Access **5**, 14647–14656 (2017)
12. Pavuluri, B.L., Vejendla, R.S., Jithendra, P., Deepika, T., Bano, S.: Forecasting meteorological analysis using machine learning algorithms. In: 2020 International Conference on Smart Electronics and Communication (ICOSEC), Trichy, India, pp. 456–461 (2020). doi:https://doi.org/10.1109/ICOSEC49089.2020.9215440
13. Sreehari, E., Srivastava, S.: Prediction of climate variable using multiple linear regression. In: 4th International Conference on Computing Communication and Automation (ICCCA), 2018.
14. Sutawinaya, I.P., Astawa, I.N.G.A., Hariyanti, N.K.D.: Comparison of adaline and multiple linear regression methods for rainfall forecasting. In: The 2nd International Joint Conference on Science and Technology (IJCST) 2017, doi:https://doi.org/10.1088/1742-6596/953/1/012046
15. Yu, K., Lu, Z., Stander, J.: Quantile regression: applications and current research areas. J. R. Stat. Soc. Ser. D (The Statistician). **52**, (2003). doi:https://doi.org/10.1111/1467-9884.00363
16. Zibners, L.M., Bonsu, B.K., Hayes, J.R., Cohen, D.M.: Local weather effects on emergency department visits. Pediatr. Emerg. Care. **22**(2), 104–106 (2006)

A Case Study of Energy Audit in Hospital

Abhinay Gupta, Rakhee Kallimani, Krishna Pai, and Akshata Koodagi

Abstract Energy Audit is an important tool in transforming the fortunes of any organization. An energy audit is a systematic approach of surveying and performing analysis of energy flow for energy conservation in any sector. For every commercial firm in India, an energy audit is mandatory as per Energy Conservation Act 2001. As a result, among industrial consumers, energy conservation is getting more attention due to the realization that energy saved will be economically beneficial for them. Some rural areas in India are power deficient for domestic usage even after getting connected to the grid. This problem can be addressed by Energy conservation. Conservation of energy in the various industrial and commercial sectors will assist us by transferring the saved capacity to destitute individuals, and this will also lead to the reduction in setting up a new generation. The significant need of the energy audit is to moderate energy and fossil products for a group of people yet to come life. The energy auditing has been performed at a leading private hospital in the city to estimate the annual energy consumption. The energy audit focuses mainly on the equipment consumption, air-conditioning system, lighting system, etc. The annual power consumption of the hospital was 91,83,870 kWh. The paper suggests recommendations for energy savings which if implemented will lead to a saving of 5,51,032 kWh energy. Economic and efficient energy conservation measures implementation is subject to budgetary constraints and the effects of such measures have been recognized by reducing energy cost, with the added benefits of environmental safety.

Keywords Energy conservation · Energy audit · Data analysis · Management · Energy efficiency

A. Gupta (✉) · R. Kallimani · K. Pai (✉) · A. Koodagi
Department of Electrical and Electronics Engineering, KLE Dr. M. S. Sheshgiri College of Engineering & Technology, Belagavi, India

J. S. Raj et al. (eds.), *Intelligent Sustainable Systems*, Lecture Notes in Networks and Systems 213, https://doi.org/10.1007/978-981-16-2422-3_11

1 Introduction

The energy audit is a mandatory activity to monitor and control energy and utility costs as per Indian Energy Conservation Act 2001. To determine the energy saving and analyze the energy consumption of the consumers, the efficient tool in Energy management is Energy auditing. The auditing is done in different sectors such as residential, industrial, agricultural, and commercial. According to [1] the Ministry of Power in 2018–2019, the energy consumption in the residential sector was 24.76%, industrial sector was 41.16%, agricultural sector was 17.69%, and commercial sector was 8.24%. As shown in Fig. 1, the industrial sector covers the maximum consumption of energy out of all sectors, this motivates the researchers to provide recommendations for the reduction of energy consumption without affecting the Quality of Service (QoS) [2]. India's energy use is expected to double by 2040, and this leads to the growth in opportunities in less energy-intensive industry sectors where energy intensity could be more than half. Hence, the researchers are focusing on reducing the energy intensity with the objective of providing a policy framework and procedure to measure, monitor, and verify energy efficiency in individual sectors by the active participation of all stakeholders and implement energy conservation programs abiding by the act [3].

The industry sector covers aerospace, education, healthcare, transport, food, etc. We have done energy auditing in the field of the healthcare industry focusing on hospitals with an intention of providing a technical report signifying the potential energy saving [4].

In this study, the energy audit was carried out through a Walk-through (Preliminary) and Detailed audit. The preliminary audit focuses mainly on reduction of losses, overall energy savings, and strengthening of the energy management system. It also does not include any kind of advancements for which huge capital investment

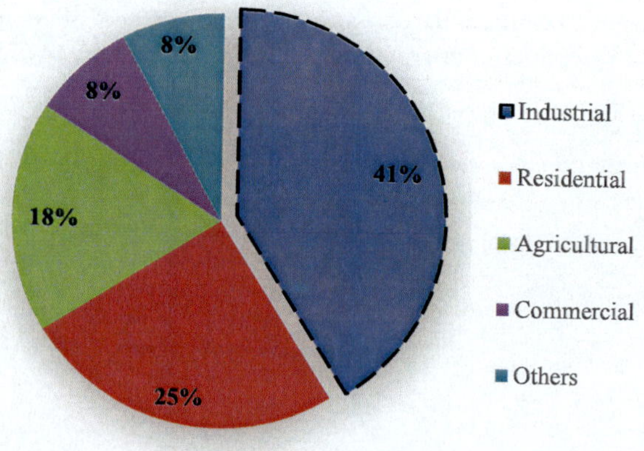

Fig. 1 Energy consumption in various sectors [1]

is required. The preliminary audit is aimed at the identification of common energy wastages, the rough estimation of losses and various saving potentials, by analyzing electricity bills, walk-through inspection for opportunities of energy saving, and determination of the various areas that need more attention in the detailed analysis [5]. The detailed energy audit needs approximately 12 months, aiming to perform comprehensive recording and study of energy usage and the pattern of consumption area-wise during the given time span. The Electricity bills and Transformer loading were collected from April 2016 to Mar 2017. Thereafter, the load was divided into the following five categories: illumination, cooling, computer–UPS–Printer, rotary equipment, and others. Hence, the monthly average consumption sums up to 7,65,322.5 kWh costing up to ₹ 53,01,695.83 (INR). Various equipment like PQ analyzer, Data logger, and Temperature monitoring system were used to obtain the data for the calculations of various losses and hence summarized in this paper.

In this study, we discuss the utilization of the energy-efficient equipment, conservation pattern, and scope of energy conservation of a Leading Private Hospital in the city consisting of four store buildings with nearly 1000 beds [6, 7]. We analyzed the utilization and provided an energy auditing report recommending reducing the energy demand to a reasonable minimum cost.

2 Methodology

According to the Energy Conservation Act 2001, the energy audit is described as "the verification, monitoring, and analysis of the use of energy including submission of technical reports containing recommendation for improving energy efficiency with cost–benefit analysis and an action plan to reduce energy consumption" [1, 4]. The energy audit is an organized strategy for decision making in the vicinity of energy management as an effective and efficient solution. It stabilizes the total energy inputs with its use and serves to add all the streams in a facility.

Energy, labor, and materials are the most contributing operational costs in industries. Energy is the only operational cost that can be managed. This results in the reduction of total operational costs without compromising the QoS. An Energy Audit plays a vital role in preventive maintenance by incorporating a positive approach in reducing the cost of energy and controlling the quality programs [8].

The flow diagram (Fig. 2) shows the complete method for Energy Auditing. Consider the first step to be planning and organizing the whole case study by having an informal meeting with the plant managerial members [9]. By following up the same Primary data and the energy flow charts are to be gathered for further processes. Having been doing a survey would help us to get more accurate output upon completion of the case study. Conducts of detailed trials and analysis on how the energy consumption is taking place are a must [10]. Various analyses for the identification of energy conservation, cost reduction, and development opportunities can be carried out. Finally, proper documentation and presentation to the management board would help to suggest the required recommendation for further well-being of the

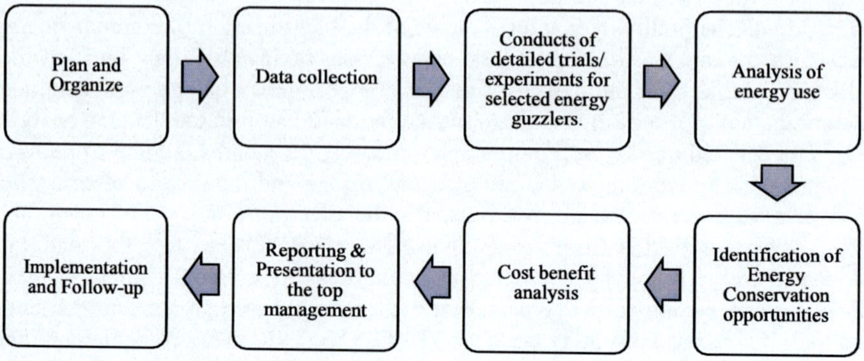

Fig. 2 Methodology

auditing area [11]. Once the recommendations are accepted, further implementation and follow-up are a must for continuous growth and development (Fig. 3 and Tables 1 and 2).

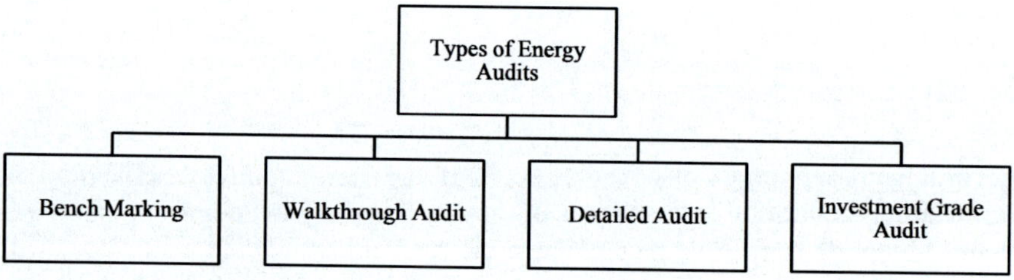

Fig. 3 Types of energy audit

Table 1 Walk-through phase details for energy auditing for an estimated time for completion to be ~8 days

Building size	Data collection formats	Monitoring systems	Forms checklist	Calculation tools	Expected results
Drafted the size of the building	Substation details and one-year electricity bill	None	Basic	Simple spreadsheets	A brief report detailing the limitations in the system and along with an area to attend for detailed auditing

Table 2 Detailed phase details for energy auditing for an estimated time for completion to be ~90 days

Building size	Data collection formats	Monitoring systems	Forms checklist	Calculation tools	Expected results
A detailed plan of the building including exterior elevation views	Connected load, transformers with specifications, and one-year electricity bill	Transformer loading, power quality analyzer	Detailed	Simple spreadsheet and programming using Jupyter notebooks (Pandas, NumPy, Matplotlib)	Summarized report briefing recommendation/limitations with technical and economical evaluation

3 Data Analysis

3.1 Data Cleaning

The statistics gathered from the hospital premises used to be in Simple Spreadsheet format. Using Python 3.8.5 and Jupyter Notebook tool, the data cleansing procedure was carried on the spreadsheet. Thereafter, the cleaned spreadsheet was once then transformed to csv (Comma-separated values) format file. The data analysis libraries like Pandas, NumPy, and Matplotlib have been used and the entire evaluation was once carried out, subsequently acquiring a number of output graphs. These graphs were analyzed in detail and required recommendations have been suggested.

3.2 Connected Load and Electricity Bill Analysis

In the hospital, the connected load has been divided into five categories. The five categories are illumination, cooling, computer–UPS–Printer, rotary equipment, and others. The total connected load is calculated as 541.492 kW which is categorized as below and the same is shown in the pie chart (Fig. 4).

3.3 Maximum Contract Demand Calculations

The maximum contract demand of the hospital campus is 1600 kVA. From the electricity bill, it is observed that for the period Apr 2016–Mar 2017, the penalty was paid for exceeding the contract demand with ₹ 180/kVA for sanctioned demand and ₹ 360/kVA for exceeded contract demand. Thus, we can avoid paying penalties

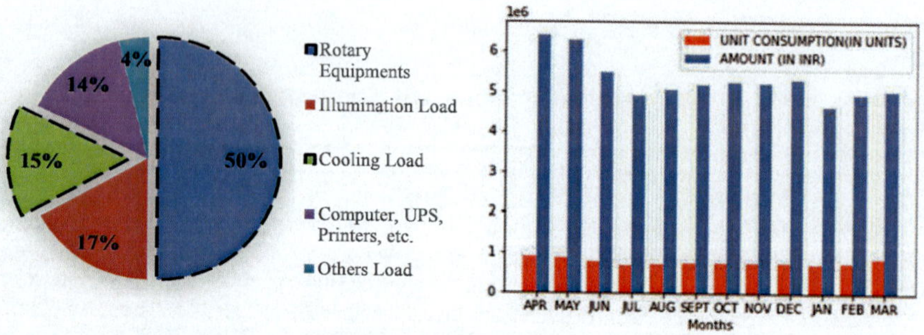

Fig. 4 **a** Load distribution and **b** Analysis of unit consumption

Table 3 Savings on recommended contract demand

Contract demand (kVA)	Recorded M.D (kVA)	Penalty in lakhs per year (₹ INR)	Savings in lakhs per year (₹ INR)
Present (1600)	1782	8,06,600	
Recommended (2200)	2200		6,98,600[a]

by increasing the maximum contract demand. The analysis carried is as shown in (Table 3).

Minimum recorded demand (Apr 2016–Mar 2017) = 1499 kVA
Maximum Recorded Demand (Apr 2016–Mar 2017) = 2087 kVA
Average exceeded contract demand = 1782 kVA
New contract demand = 2200 kVA.
Fixed charges for 1600 kVA = 1600 * 180 = ₹ 2,88,000 (INR)
Fixed charges for 2200 kVA = 2200 * 180 = ₹ 3,96,000 (INR)
Extra charges for 2200 kVA = 600 * 180 = ₹ 1,08,000 (INR)

3.4 *Transformer Analysis*

There were a total of three transformers of 1600 kVA each, 3300/440 V. A data logger was installed in these transformers, which has collected the data over a year from April 2016 to March 2017. Figure 5a explains the average voltage versus the date of collection. Where V1, V2, and V3 are average voltages of the transformers T1, T2, and T3, respectively. As per the observation, average voltage V1 and V2 show less variation when compared to average voltage V3. This drastic variation in V3 is due to load variation.

Figure 5b represents transformer load versus date of collection. Where P1, P2, and P3 are the average power of the transformers T1, T2, and T3, respectively. That shows that the load on T3 is very less as compared to load on transformers T2 and T1. Due to this unbalanced condition, the voltage variation is seen in Fig. 5a. The

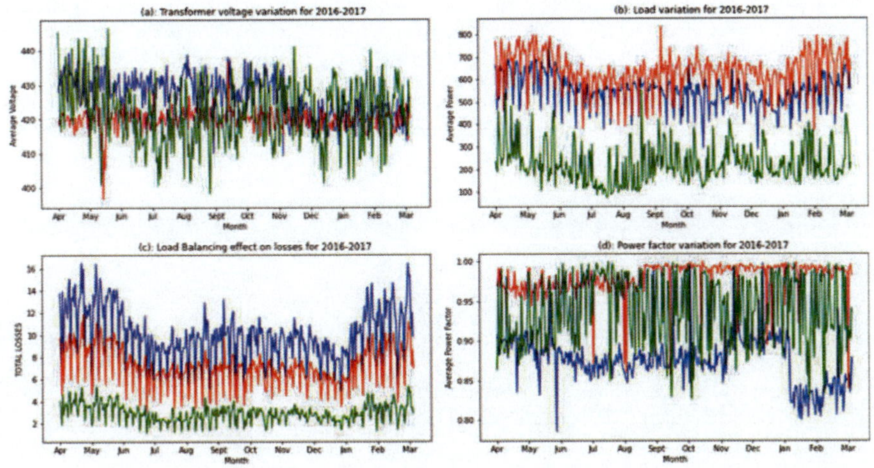

Fig. 5 **a** Transformer voltage variation for 2016–2017, **b** load variation, **c** load balancing effect on losses, and **d** power factor variation

load variation shown in Fig. 5c depicts that potential saving is possible by balancing the load on transformers. Figure 5d shows power factor variation with respect to the date of collection and it is ranging from 0.785 to 1.

3.5 Illumination Load Analysis

In the investigated building, analysis was carried out in two areas [12, 13]. Approximately, 1000 tube lights of 36 W with choke coil were in use as shown in Table 4. The table details the energy consumption of these tube lights in each area amounting to ₹ 40,84,989.48 (INR). The analysis with 28 W Electronic Choke Tube lights leads to an annual saving of ₹ 24,96,382.46 (INR) with an initial investment of ₹ 5,43,500.00 (INR) within a payback period of 3 months (Approx.).

It was observed during the inspection of the hospital that there were areas in which the light intensity was high during daytime, yet the illumination facilities were operating leading to unnecessary power consumption. Hence, we recommended an automatic control of illumination to be used using an Infrared technology-based Proximity Infrared Sensor (PIR). The recommendation is for corridors, entrances, and staircases where the illumination can be controlled. We have assumed a minimum of 6 h per day as saving and calculated energy saving as shown in Table 5. Assume 400 LEDs are to be connected to the PIR sensor [14].

Table 4 Illumination load analysis

Area	Number of tube lights to be replaced with 28 W	Present details of consumption with 36 W		Present details of consumption with 28 W		Initial investment (₹) for replacement of 36 W to 28 W	Total savings
		Energy consumed (kWh)	Amount (₹)[a]	Energy consumed (kWh)	Amount (₹)[a]		
1	332	191,949.12	1,247,669.28	74,646.88	485,204.72	166,000	762,464.56
2	755	436,510.8	2,837,320.2	169,754.2	1,103,402.3	377,500	1,733,917.9
Total	1087	6,28,459.92	40,84,989.48	2,44,401.08	15,88,607.02	5,43,500.00	24,96,382.46

[a]1 unit = ₹ 6.5
28 w electronic ballast = ₹ 500/- each.
Payback Year of 28 w Electronic Ballast = 3 Months Approx.
Energy Consumption = Wattage of each Bulb * Number of Days * Number of Hours * Number of Bulbs
Energy Cost = Wattage of each Bulb * Number of Days * Number of Hours * Number of Bulbs * Cost of 1 Unit

Table 5 PIR sensor calculation

Number of LEDs	Total wattage	Savings in units/year	Savings in cost/year	Total investment	Payback period
400	4.4 kW	9,636[a] kWh	₹ 62,634	₹ 40,000	8 months

[a]By considering 6 h per day and 365 days in a year

3.6 *Cooling Load Analysis*

In the hospital, a chiller plant has been used to provide the required cooling to the entire building. This consumes a maximum part of the energy and needs the attention of the auditor to recommend the appropriate saving technique. As per [12, 15], an increase in 1-degree temperature of refrigerant will improve the efficiency of the chiller compressor by 1–2%, also the reduction in power consumption by approximately 0.75 kW for 1 Tonne of refrigeration.

In the chiller plant, the refrigerant temperature is 8° throughout the day. But as the temperature at night time is less compared to daytime, the refrigerant temperature is increased from 8 to 9° for approximately 6 h in the night time. The savings achieved are as follows:

In the current scenario, the rating of the chiller plant is 300 TR.
Savings = 0.75 * 300 = 225 kW
Savings in cost = 225 kW * 6 h. * ₹ 6.5 * 180 days (except summer season)
= ₹ 15,79,500 (INR).

3.7 *PQ Analyzer*

Harmonic data is analyzed from the PQ analyzer Fluke 435 [16]. The collected data was analyzed for various voltage and current harmonic distortions. The analysis as shown in Fig. 6 urges the need for the third and the fifth harmonic to be filtered to obtain a stable system. Percentage harmonics of each phase are shown in Table 6.

Figure 6b shows BN phase experiences dominant Voltage Harmonic components ranging from ~0.45 to ~0.68 V; Fig. 6c shows C phase variation of Current Harmonic components ranging from ~4 to ~9 A. The third Current Harmonic component results in a rise in the current flowing through the neutral phase.

Due to the presence of nonlinear elements in the system, there is a generation of harmonic effect leading to distortion caused by the excess current in and out of the appliances. In order to reduce the distortion, filters are designed. We have designed a passive filter using Eq. (1). Filtering out these harmonic components will avoid the malfunctioning of the nearby equipment (Table 7).

$$f = \frac{1}{2\pi\sqrt{LC}} \tag{1}$$

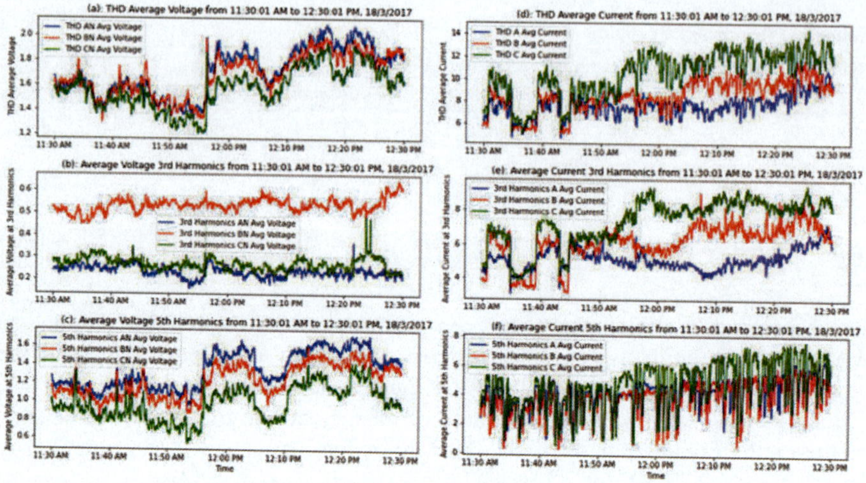

Fig. 6 Voltage and current distortion at THD, third, and fifth harmonic distortion

Table 6 Harmonic percentage

Phase\order	3rd (%)	5th (%)	THD (%)
R	5.25	4	7.875
Y	6	3.75	7.9
B	7.5	4.875	10.5

Table 7 Filter design

Sl.No	Order	Design frequency (Hz)	C (assumed)	L (calculated)
1	3rd	147	56 μF	20.92 mH
2	5th	245	56 μF	7.53 mH

where
f = Frequency (Hertz)
C = Capacitance (Farad)
L = Inductance (Henry)
The following results were obtained:

Table 8 Savings and payback

Recommended measures	Investment (in ₹)	Savings (in ₹)	Net savings (in ₹)	Payback period
Illumination	5,43,500.00	24,96,382.46	19,52,882.46	3 months
PIR sensors	40,000	62,634	22,634	8 months
Chiller plant		15,79,500	15,79,500	
Transformer		2,15,939	2,15,939	
Change in contract demand		6,98,600	6,98,600	
Total	5,83,500	50,53,055.46	42,75,555.46	

4 Discussion

The analysis of the energy bill for the year 2016–2017 was carried out and found the need of managing the energy in a systematic approach. The analysis during the audit recommends increasing the contract demand from 1600 to 2200 kVA which saves 1% of the billing amount without investment. In the investigated two areas, we recommended the use of 28 W electronic choke tube lights instead of 36 W tube lights with choke coils, and automatic control of light using a PIR sensor for selective areas which saves 3% of the annual bill. It is found that annually 0.3% saving is possible by loading all the transformers equally so that losses can be reduced. It is recommended to increase the temperature by 1° for approximately 6 h in the chiller plant to save 1.7% of losses annually [17]. The executive summary of the saving is as shown in Table 8.

The harmonic analysis was carried out by using PQ analyzer Fluke 435. It is found that third and fifth harmonics are present in the system. By designing a suitable filter, the performance of various rotatory equipment can be improved. Filters like LCL, LCL-th, and FR can be used along with LR reactors or FRE rejection filters: having 7% for fifth, andseventh harmonic dominant and 14% for third harmonic dominant [18].

5 Conclusion

The energy audit is an important approach for defining and pursuing a robust energy management program. The significance of energy conservation in regular daily existence was examined, which can assist each person with pondering the appropriate use of energy. Observing from the client's perspective, it is more important to understand the estimation of energy. Hence, the hospital which is taken for investigation has a huge area where there were many opportunities to save energy. We performed a detailed energy audit for the various connected loads. We analyzed the effectiveness of energy consumption in these areas by identifying the loss areas, and suitable recommendations were suggested. The analysis shows that an increase in the contract demand saves 1% of billing, 3% of billing can be saved in the illumination areas, 0.3% by balancing the transformer load, and 1.7% of saving can be achieved by increasing the temperature by 1° in the cooling system. In this paper, it is observed that if all the recommendation is implemented, this will result in a total annual saving of ₹ 50,53,055.46 (INR) with an investment of ₹ 5,83,500 (INR) with the payback period of 2 months (approx.) This systematic approach can be extended to any industrial or commercial sector in the future.

References

1. Ministry of Power: The Energy Conservation Act, Gaz. India, vol. 60, no. 2, p. 22 (2001). Available from: http://powermin.nic.in/sites/default/files/uploads/ecact2001.pdf
2. Elangovan, M., Ravichandran, A.T., Chellamuthu, P.: Need of energy audit in everyday life. **2**(11), (2020)
3. Chaphekar, S.N., Mohite, R.A., Dharme A.A.: Energy monitoring by energy audit and supply side management. International Conference on Energy Systems and Applications, ICESA 2015, no. ICESA, pp. 178–183 (2016). doi:https://doi.org/10.1109/ICESA.2015.7503335
4. Dongellini, M., Marinosci, C., Morini, G.L.: Energy audit of an industrial site: a case study. Energy Procedia. **45**, 424–433 (2014). doi:https://doi.org/10.1016/j.egypro.2014.01.046
5. Rayhana, R., Khan, M.A.U., Hassan, T., Datta, R., Chowdhury, A.H.: Electric and lighting energy audit: a case study of selective commercial buildings in Dhaka. In: 2015 IEEE International WIE Conference on Electrical and Computer Engineering WIECON-ECE 2015, pp. 301–304 (2016) doi:https://doi.org/10.1109/WIECON-ECE.2015.7443923
6. Nourdine, B., Saad, A.: Energy consumption in hospitals. In: 2020 International Conference on Electrical and Information Technologies ICEIT 2020, (2020). doi:https://doi.org/10.1109/ICEIT48248.2020.9113177
7. Gupta, S., Kamra, R., Swaroopa, M., Sharma, A.: Energy audit and energy conservation for a hostel of an engineering institute. In: 2018 2nd IEEE International Conference on Power Electronics, Intelligent Control and Energy Systems ICPEICES 2018, pp. 8–12 (2018). doi:https://doi.org/10.1109/ICPEICES.2018.8897298
8. Mendis, N.N.R., Perera, N.: Energy audit: a case study. In: 2006 International Conference on Information and Automation, ICIA 2006, pp. 45–50 (2006) doi:https://doi.org/10.1109/ICINFA.2006.374149
9. G. of I. Ministry of Petroleum & Natural Gas: Petroleum Conservation Research Association, © Petroleum Conservation Research Association, India. http://www.pcra.org/usefullinks/display/1
10. Sharma, R.: Energy audit of residential buildings to gain. In: 2015 International Conference on Energy Systems and Applications, no. ICESA, pp. 718–722 (2015)
11. Kumar, A., Ranjan, S., Singh, M.B.K., Kumari, P., Ramesh, L.: Electrical energy audit in residential house. Procedia Technol. **21**, 625–630 (2015). doi:https://doi.org/10.1016/j.protcy.2015.10.074
12. Goyal, P., Shiva Kumar, B., Sudhakar, K.: "Energy audit: A case study of energy centre and Hostel of MANIT, Bhopal". In: 2013 International Conference on Green Computing, Communication and Conservation of Energy, (ICGCE) 2013, pp. 644–648,vol. 1 (2013). doi:https://doi.org/10.1109/ICGCE.2013.6823515
13. Jadhav, V., Jadhav, R., Magar, P., Kharat, S., Bagwan, S.U.: Energy conservation through energy audit. In: International Conference on Trends in Electronics and Informatics, ICEI 2017, vol. 2018, pp. 481–485 (2018). doi:https://doi.org/10.1109/ICOEI.2017.8300974

14. Saravanakumar, R., Chetan, N., Chakaravarthy, P., Karthickkeyan, Rakesh, S., Ramkiran: M energy audit report. In: Third International Conference on Advances in Electrical, Electronics, Information, Communication and Bio-Informatics, AEEICB 2017, pp. 482–485 (2017). doi:https://doi.org/10.1109/AEEICB.2017.7972360
15. BEE Energy Audit: Energy efficiency in electrical utilities. Bureau of Energy Efficiency, New Delhi, India (2005)
16. Fluke Corporation: Fluke 434-II and 435-II power quality and energy analysers, Fluke 434-II and 435-II power quality and energy analysers, 2020. https://www.fluke.com/en-in/product/electrical-testing/power-quality/434-435
17. Stamenić, M., Jankes, G., Tanasić, N., Trninić, M., Simonović, T.: Energy audit as a tool for improving overal energy efficiency in Serbian industrial sector. In: 2nd International Symposium On Environment Friendly Energies And Applications, EFEA 2012, pp. 118–122 (2012). doi:https://doi.org/10.1109/EFEA.2012.6294075
18. Circutor: Filtering solutions for improving energy efficiency, Solutions. Available from: http://circutor.com/docs/Soluciones Filtrado_EN_Cat.pdf

The Comparison Analysis of Cluster and Non-cluster Based Routing Protocols for WAPMS (Wireless Air Pollution Monitoring System)

Ekta Dixit and Vandana Jindal

Abstract With increasing industries, the air pollution monitoring system has become the main challenge of intelligent cities. Several methods have been developed in APMS, and WSNs seem like the solution to sensors, because of less cost, and dense deployment. Thus, the deployment of the sensors is significant for better performance while assuring less economic cost. WSN is related to the set of distributed sensors for monitoring the PHY situations and arranging the gathered data at the central position. In this article, a comparative analysis is defined in routing protocols for WAPMS to study and defined the advantages and disadvantages. The conclusion showed the prior protocols into two different classes such as cluster and non-cluster based routing protocols. The cluster-based methods have designed the PHY network into virtual classes, whereas the non-cluster based protocols both use broadcasting tools for data communication cluster-based routing protocol, compared to LEACH; TLLEACH is based on the two-level cluster heads (CHs) with less communication distance, and less hop is necessary to transfer for distance to the base station. It reduces the average energy consumption. Their network parameters are calculated to E2D and energy in the network.

Keywords Wireless sensor network air pollution monitoring system (WAPMS) · Wireless sensor network (WSN) · Cluster-based protocols · Non-cluster based routing protocol

1 Introduction

WSNs have been measured as one of the vital skills in the twenty-first century [1]. It is capable of monitoring and regulating physical environments from remote locations with suitable accuracy. Wireless Sensor Network in the air pollution monitoring system is worldwide used in several applications such as controlling, surveillance,

E. Dixit (✉)
Punjabi University, Patiala, Punjab, India

V. Jindal
DAV College, Bathinda, Punjab, India

J. S. Raj et al. (eds.), *Intelligent Sustainable Systems*, Lecture Notes in Networks and Systems 213, https://doi.org/10.1007/978-981-16-2422-3_12

medical diagnosis, disaster maintenance, emergency inquiry, and home applications [2]. These networks are very famous from traditional Wireless Networks (WNs) because of their valuable features such as severe Energy Consumption (EC), Dense Node Deployment (DND), and un-reliability of Sensor Nodes (SNs) [3]. This paper introduced a WAPMS for urbanization and comparative analysis with several routing protocols such as cluster and non-cluster based protocols. APM is one of the vital concepts influencing the quality of life and health of the increasingly urban population of industrial sectors. Air is polluted in several cities by emissions from sources like cars and trucks, power plants, and manufacturing devices. If the particles and gases from those operations accumulated in the air with high concentration, it may be dangerous for the health of persons as well as the surroundings. However, meteorological situations complicate the quality of air problems in the region. To secure the health situation, the public requires timely data about the air quality and other factors that influenced it. An air-quality forecast permits residents to decrease the exposure if the concentration of the pollutant is high. Pollution monitoring is essential for the protection of the environment. Pollution may occur due to air pollutants that contaminate the surroundings and water. APM is measured as a complex job, but it is very significant. Normally, the data loggers were used to gather data intermittently, and the APM was more time-consuming and expensive [4]. Normally, the APM unit in MAURITIUS lacks assets and makes use of massive devices. These massive devices optimize the tractability of the system and create it not easy to ensure more timely and flexible. However, APM (air pollution monitoring) in WSN depends on some requirements as follows: (i) Develop a structural design to define the hops and their interface. (ii) Gather air pollution recordings from the region of interest. (iii) Collaborate between thousands of nodes to gather information and transfer it to the gateway hop, whereas optimizing the number of copies. (iv) Using suitable data aggregation to decrease energy consumption during the communication of huge information among a huge number of hops (Fig. 1) [27].

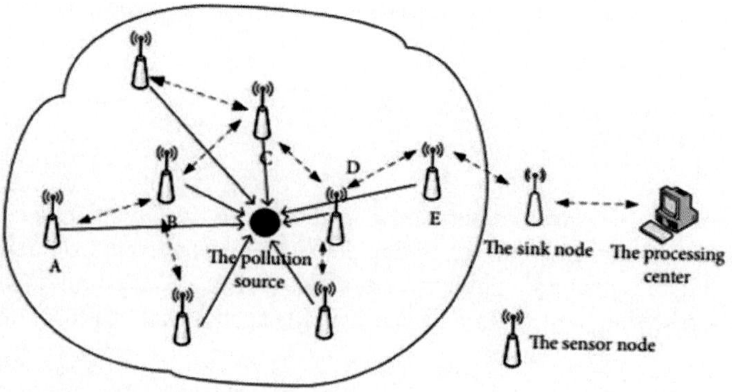

Fig. 1 The air pollution monitoring network system

Hence, monitoring [27] of the inside and outside air pollution involved the significant stages i.e., gathering of the data, collecting, and transmits the gathered information. The air pollution monitoring in wireless sensor networks is mainly used in actual applications for gathering and assessing information more smartly performed without human interference. Currently, WSN is engaged in commercial, engineering, and equipped applications [6]. In Sect. 3, a comparative analysis of various clustering routing methods has been done. Hence, various routing protocols in WSN have been compared in Tables 2, 3, and 4. In a cluster-based routing protocol, compared to LEACH and TL-LEACH is based on the two-level cluster heads (CHs) with less communication distance, and less hop is necessary to transfer for distance to the Base Station (BS). It reduces the average EC TEEN is the re-active protocol to a large number of modifications in sensory attributes that are more appropriate for the re-active system and time-critical situations. In a non-cluster-based protocol, PCRP has been simulated by the IEEE 802.11 MAC protocol. It is intended for wireless local area networks and it may not involve energy-efficient methods. Scalable source protocol may not always generate the direct route of topological data. Hence, it may not choose receiver hop in a suitable way that may result in communication delay. It is valuable for query-based applications. Thus, cluster-based routing protocols have better performance metrics in WSN.

Sections are elaborating as follows: Sect. 2 has included a description of the significance of the APMS in the networks. Moreover, a comparative analysis of the various protocols is given in this section. Section 3 provided a detailed survey of several articles that includes protocols, research gaps or problems, proposed methods, and proposed parameters. In Sect. 4, the comparative analysis between the cluster and non-cluster based methods are discussed and namely LEACH, TL-LEACH, TTDD, TEEN, PEGASIS, GBRP, PARP, SSRP, and PPRP, etc. The conclusion and further work have given in Sect. 5.

2 Main Significance of Air Pollution Monitoring System in WSN

The main significance of air pollution monitoring is examining the level of pollution related to ambient air–quality standards [7]. Standards are used as a measurement to set the objective for reducing air pollution. Thus, robust monitoring aims to guard against the high level of the events by alarm rate to individuals and perform an action. Air pollution has become the main crisis among human lives, but developed countries are facing significant challenges. Main cities like China are first in air pollution, and also at the top level of the list in the economic sector. China comes in the top-list among the most populated areas in Asia. The given Fig. 2, demonstrates the numerical level of the particle matter of the main cities.

However, India is at the peak rate of pollution set by WHO and USEPA as compared to other cities. However, the main source of the pollution occurred from

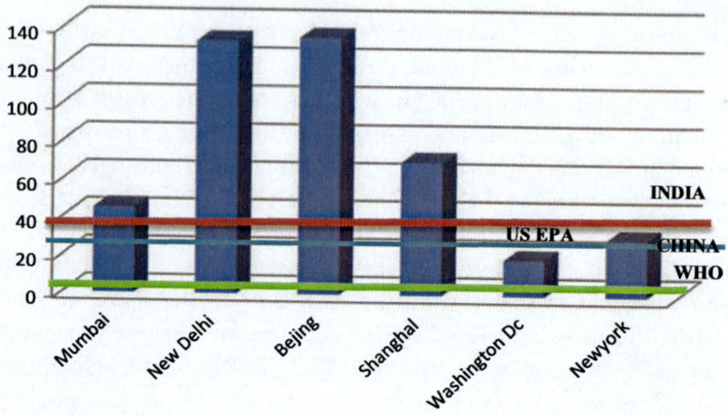

Fig. 2 Particle matter in cities in 2014

Table 1 Air quality levels in main cities [8]

Cities	SO_2			NO_2			Particle matters		
	Avg	IND	Res	Avg	IND	Res	Avg	IND	Res
New Delhi	5.3	0	0	57.6	1	1	222	−1	−1
Mumbai	5.4	0	0	34.9	0	1	119	1	+1
Kolkata	13.5	0	0	66.1	1	+1	116	1	−1
Chennai	12.0	0	0	19	0	0	64	1	1
Pune	32.3	0	0	57.8	0	0	112.6	1	+1

the burning of fuels, and biomass in industries. Hence, Table 1 demonstrates the pollution-level in various cities of India. Thus, the level of SO_2 (Sulfur Dioxide) is the standard range, a high concentration level of NO_2 (Nitrogen Dioxide) is analyzed. And, PM_{10} (particulate matter) is at peak level in all the cities and critical levels in Delhi and Kolkata [8]. It is moderate if the pollutant level is more than 50% of the standard value. In the given Table 1, given the air quality levels in main cities. The parameter values are as expected; low value = 0, High value = +1, Medium value = 1, and critical = −1.

3 Literature Review

This section summarized that the WAPMS routing protocols, and measured the pollution-related metrics. Various articles have been surveyed on the APM schemes in WSN. Saini, J. et al. [9] reviewed the custom of the wireless models for the establishment of the Cyber-Physical Schemes (CPS) for the real-time monitoring scheme. Moreover, it presented the assessment of microcontrollers utilized for the system design and issues in the establishment of real-time analysis schemes. This

Table 2 Comparison between cluster-based routing protocols with various performances metric

Protocol name	Energy-efficiency	Delivery	Delay	Scalability	Load balancing	Complexity
LEACH [2000]	VL [17]	VS	VS	VL	M	L
TEEN [2001]	VH [18]	S	S	L	L	H
TTDD [2019]	VL [19]	VL	VL	L	G	L
TTLEACH [2014]	L [20]	S	S	Mo	B	L
PEGASIS [2002]	L [21]	VL	VL	VL	M	H

Table 3 The summary of advantages and disadvantages of cluster-based routing protocols in WAPMS

Method name	Advantages	Disadvantages
LEACH [17] (2000)	• Each node has an equal chance to become Ch (Cluster heads) and Uses TDMA so it keeps Chs from un-required collisions	• Single—hop com-munication and can't use in huge-scale net-works
TEEN [18] (2001)	• Transmission can be con-trolled by varying thresholds and time critical applications	• Thresholds are not matching the SN will be commu-nicated and information may be lost if chs are not able to transmit to each other
TTDD [19] (2019)	• Event detecting WSNs among irregu-lar data traffic and resolve the moving problems of destination in the large scale network	• High latency and low energy-effi-ciency
TT-LEACH [20] (2014)	• Random revolution of local Cluster BSs and two level hierarchies	• Chs dies very early-stage due to loss of energy
PEGASIS [21] (2002)	• Energy load separated and mini-mize the overhead	• Maximum delay and WSN is not very scal-able

research presented a new goal in the field of Internal Air Pollution Monitoring (IAPM). Bathiya, B. et al. [5] developed low-cost multiple sensors hop for air pollution capacity and to improve the WSN protocols for information gathering, and data aggregation protocol. In remote locations, data gathering becomes the main problem. Other communication problems, gathering information at a stable location was time-consuming. Therefore, the proposed work aims to develop a sensor network at a low-cost sensor hop. The main goal of the research was to develop environmental monitoring in WSN. They had used Xbee radio units that depend on

Table 4 The summary of advantages and disadvantages of non-cluster-based routing protocols in WAPMS

Method name	Advantages	Disadvantages
GBRP [22] (2005)	• This protocol uses broadcast to transfer data and minimum communication overhead	• The query-based application uses Only and doesn't identified packet delay completely
PARP [23] (2020)	• This protocol can transmit power control and built many too many multicast routing trees	• Traffic load and maximum over-head
SSRP [24] (2006)	• Reactive protocol and it uses proactive protocol structure for virtual-ring construction	• This protocol doesn't create a minimum path and consider de-laying
PCRP [25] (2007)	• It uses power-aware and cross layer mechanism	• It uses hidden terminal issue and IEEE 802.11 MAC Proto-col design with WLANs

the application, radio units may change. Generally, in network design, tree-building methods were developed to manage current child relations and sleep schedules to manage the Parent–Child Relation (PC-R) and Sleep Scheduling (SS) approach. They experimented with a gas sensor and analyzed the temperature change from 25 °C and 43 °C. Mohamed, R. E. et al. [10] categorized the uses of WSN depending on various features to define main protocols policy problems. However, the power efficiency of the current proactive routing protocol was considered from various angles. The major challenge was finding a way to cover the region of interest and forward monitored information to the base station. They introduced the set stage and violation power, fairness restraint of active routing approaches. Though, the power was discussed during the setup stage and damage of power, fairness restraint of dynamic routing methods. They achieved maximum wireless sensor network performance such as network connectivity and coverage. In this research, they studied the wireless sensor network lifetime mainly depends on power usage. The main goal of the research can be explained as they introduced the classification of WSN uses measuring all the major design facets and interrelated system characteristics. Sun, J. et al. 2017 [11] proposed research on improving air quality monitoring (AQM) in the network that prevents the issues of traditional monitoring schemes. In this research, they built an AQM scheme using WSN that contains the sensor hops, gateway, mobile hops, and monitoring center. Also, a solar-energy supply mode was constructed to improve the network lifespan. Furthermore, to assure the transmission consistency and accuracy, the global scheme for the Mobile Communication (MC) and Global Positioning Scheme (GPS) were unified into the network. The analyzing data of the sensor node parameters such as temp values are 101, 110, and 106 °C, humidity values are 27.4, 29, and 28%, PM2.5 values are 143, 155, and 182 ug/m^3, carbon oxide values are 620, 650, 610, and 640 ppm and O3 with ug/m^3 values are 88, 89, 90, and 89, etc. Kaivonen, S. et al. [12] proposed research was done at Uppsala Campus in Sweden. The project was named green IoT has established a testbed with static and wireless sensors for air quality sensor (AQS) in the main center of the city (Uppsala). To balance the network coverage of fixed sensor nodes, they assessed to organize the air quality sensors on communal vehicles. Movable devices permit dimensions to be engaged at various sites in the city without the restraints and additional charges of connecting the devices on city structures. They were proficient to improve the network coverage of the sensing region considerably.

To summarize, one of the major issues in existing work was health-related issues due to the increase in air pollution in various regions and countries, which has gained more consideration of government and humanity. Prevention of air pollution was difficult due to biochemical, organic, radiological particles. Hence, it becomes necessary to monitor these parameters. Thus monitoring the air pollution in WSN has various advantages over the traditional methods. Monitoring stations are utilized to examine and gather the real pollutant region from the road traffic particle emission. Hence, the APM (Air Pollution Monitoring) is done in a large infrastructure environment for better maintenance. Other more methods used in the research are real-time monitoring systems, indoor and outdoor monitoring systems. Centralized monitoring, identify the pollution level through Google map and air quality monitoring system.

4 Category of the Routing Protocols Used in Air Pollution Monitoring System in WSN

The RP (Routing Protocol) is the procedure for choosing an appropriate route for traveling from start to sink. The information packets are sensed by the sensor hops in the WSN that are forwarded to the BS, which interconnects the sensor network that is gathered and assessed. The current routing protocols of wireless sensor networks are categorized into cluster and non-cluster based routing methods.

4.1 Cluster-Based Routing Protocols in WAPMS

The routing in WAPMS is a main challenging approach since of features that distinguish these sensor networks from the other sensor networks. Cluster-based routing is significant for the network applications in which large sensor hops are deployed for sensing applications. If sensors initiate to connect and involve in data communication, then there may be network congestion and collisions that result in the limited power of the network. Clustered nodes may address these problems. In these networks, nodes are partitioned into several smaller groups which are known as clusters [13]. Every cluster has a controller which is referred to as CH (Cluster Head), SN (Sensor Nodes). The clusters may be fixed and variable. It may be of similar or different size. WSN enables the effective use of limited energy of sensor hops and improves the network lifetime [14]. Existing protocols in WSN are classified as LEACH, TTDD, TL-LEACH, PEGASIS, and TEEN [15].

4.2 Non-cluster Based Protocols in WAPMS

In this routing protocol, all deployed hops may directly connect to the base station. One problem with the cluster-based routing protocol is data redundancy at the base station because the closest hops sense and forward similar information [16]. Another problem is that the power-constrained hops forward information far from the base station that may cause power depletion at hops and reduce the network lifetime. These protocols may use broadcasting and flooding methods for the transmission purpose. They may not build the physical network into virtual sets. Some of the non-cluster based routing protocols are classified as GBRP, PARP, SSRP, and PCRP.

Here Table 2 discusses the comparison between cluster-based protocols with several performance metrics such as energy, delivery, delay, scalability, load balancing, and complexity of the routing protocol. The full form of the short forms is VL = very low, VS = very small, M = medium, L = low, VH = Very high, S = small, G = good, Mo = Moderate, B = bad, and H = high.

Tables 3 and 4 summarizes the advantages and disadvantages in cluster-based and non-cluster based routing protocols in WAPMS.

Figure 3i shows the network performance parameter with average E2D vs. no. of sensor nodes (GBRP) protocol. It is considered that the rate of delay rises with the increase no. of actors and inversely proportional to the no. of sensor nodes. The comparison between the number of actors/nodes with Packet delivery Ratio. This procedure increases the packet delivery ratio as defined in Fig. 3ii. PARP protocol is not reliable for DN (dense network) as the RTS (routing table size) maximizes the size of the network. The power control routing protocol is replicated using IEEE 802.15.4 MAC protocol, i.e., normally developed for energy-constrained and DN sensor networks. The Value of average energy dissipation, delay, and delivery ratio for the power control protocol is studied under a protocol (IEEE 802.15.4 MAC) and showed in Fig. 3iii, iv, and v resp. It may consider that a power control protocol does not achieve a well-advanced protocol (IEEE 802.15.4 MAC).

Figure 4i shows the considered that scalable source protocol consumes high energy to send the data. This protocol also not verified which actor is chosen as a sink to transfer the information, and it can lead to an incorrect collection of actors i.e., isolated from the action field. It may consider that scalable source protocol damages the average E2D metric as defined in Fig. 4ii.

Figure 5i shows the energy over the interval time. LEACH and TL-LEACH protocol utilize up to all the energy available in the WSN. The improved LEACH protocol defines more steady energy consumption that at the last it may bring to maximize the network lifetime. It might be taken back to the fact that though in the LEACH protocol there is only a level before transferring to the BS, in the improved version of the LEACH structure 2-levels may be used for minimizing the distances of broadcast and so regarding intakes. Figure 5ii defines the EC with the TTDD protocol. It creates two assumptions. Initially, for the individual curve, the EC maximizes slowly but sub- linearly as the no. of destination increases. Since more destinations flood more local queries and broadcast nodes are complex in data transferring, both

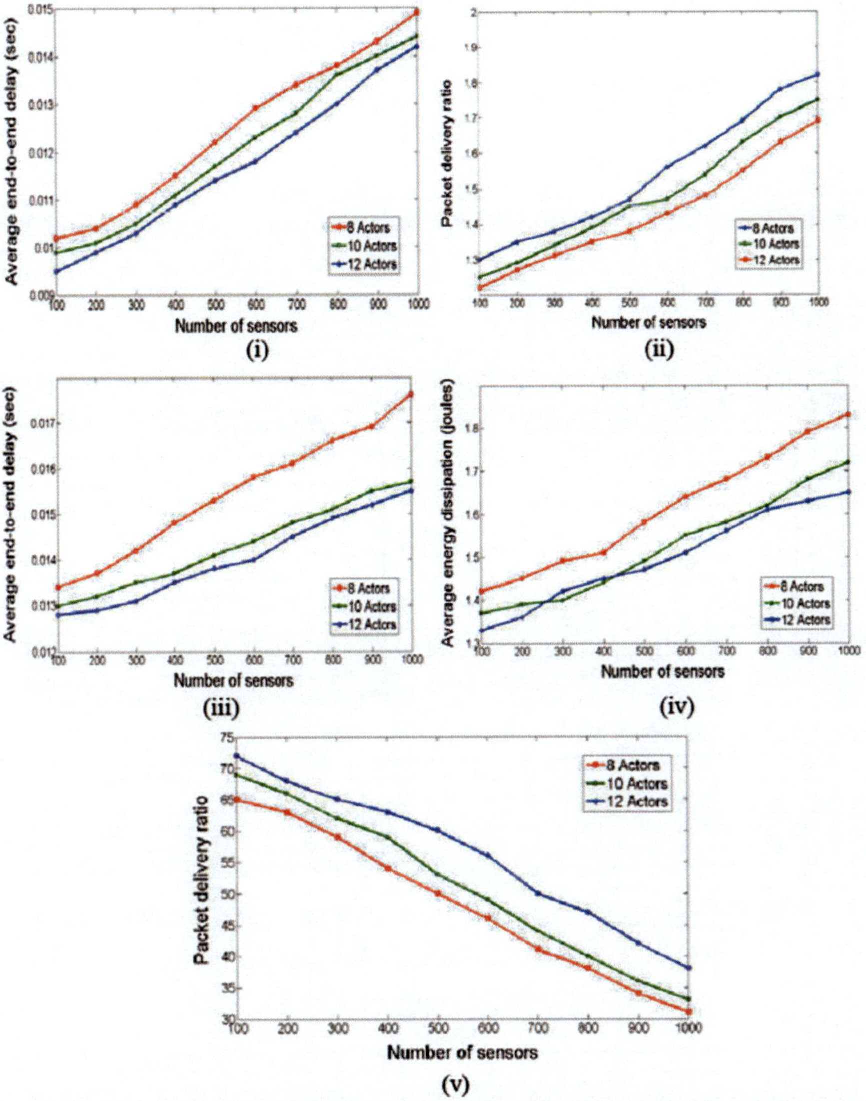

Fig. 3 **(i)** Average delay vs no. of sensors (GBRP) [22] **(ii)** Packet delivery Ratio vs. no. of sensors (PARP) [23] **(iii)** Average delay vs. no. of sensors(PCRP) [24] **(iv)** Average energy vs no. of sensors (PCRP), and **(v)** Packet Delivery Ratio vs no. of sensors (PCRP) [24] non-cluster based routing protocol [16]

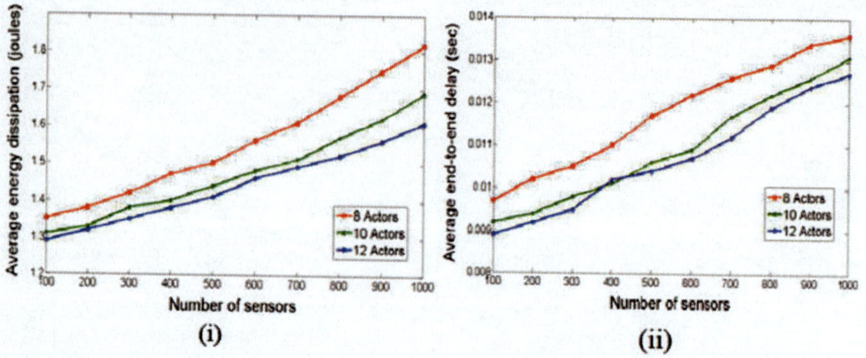

Fig. 4 **(i)** Average energy (SSRP) and **(ii)** Delay vs. no. of sensors (SSRP) [25] non-cluster based routing protocol [16]

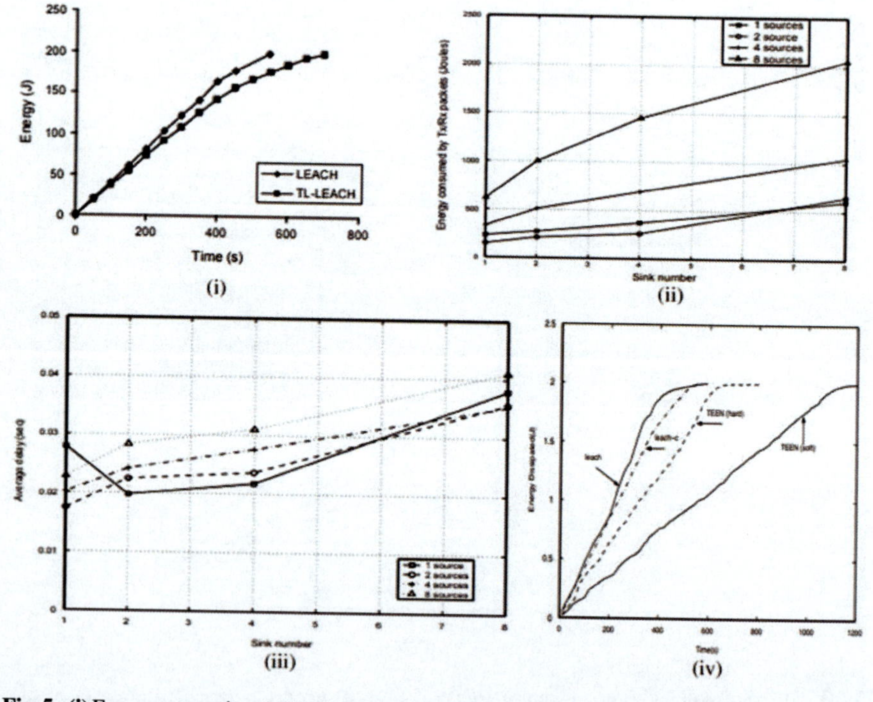

Fig. 5 **(i)** Energy comparison LEACH [28] and TL-LEACH [26, 29] **(ii)** Energy (TTDD) **(iii)** Delay (TTDD) protocol [19, 29], and **(iv)** Energy with TEEN and LEACH comparison [18] cluster-based routing protocol

consume high energy. Figure 5iii defines the delay metric, which values from 0.02 to 0.08 s. They incline to maximize when there are high destinations or sources. Maximum sources create high DPs (data packets), and more destinations required more LCF (local query flooding). Figure 5iv shows the comparison between two protocols such as LEACH and TEEN. The TEEN protocol performs much better than LEACH. If the CF (cluster formation) depends on the LEACH, the evaluation of the TEEN method is probable to be similarly better.

5 Conclusion and Future Scope

This work has concluded a comparison between the categories of cluster and non-cluster based routing protocols used in air pollution monitoring in the WSN network. These types of protocols are classified into two classes such as cluster and non-cluster based protocols. The simulation has been done for each routing protocol under the same environments. Individual routing protocol have been studied in isolated for the delay, delivery rate, and energy consumption changing the number of sensor nodes. It is considered that between the cluster-based routing protocol hierarchy, energy-efficiency and reliable protocol outperforms others. Also, non- cluster-based routing protocols have superior performance. The comparative analysis between cluster (LEACH, TEEN, TTDD, and TL-LEACH) and non-cluster based protocols (GBRP, SSRP, PCRP, and PARP) performs well for the network parameters. Therefore, it may infer that cluster-based routing protocols are more famous among the research communal of WAPMS. It will introduce the novel or hybrid routing protocol in WAPMS. It will improve the network lifetime and the overhead rate as compared to the other network parameters with wireless air pollution monitoring system.

References

1. Wei, Y., Heidemann, J., Estrin, D.: An energy-efficient MAC protocol for wireless sensor networks. In: Proceedings. Twenty-First Annual Joint Conference of the IEEE Computer and Communications Societies, IEEE, vol. 3, pp. 1567–1576 (2016)
2. Ying, Y.W., Lo, K.M., Mak, T., Leung, K.S., Leung, Y., Meng, M.L.: A survey of wireless sensor network based air pollution monitoring systems. Sensors **15**(12), 31392–31427 (2015)
3. Kadri, A., Yaacoub, E., Mushtaha, M., Abu-Dayya, A.: Wireless sensor network for real-time air pollution monitoring. In: 2013 1st International Conference on Communications, Signal Processing, and their Applications (ICCSPA), IEEE, pp. 1–5 (2013)
4. Abraham, S., Li, X.: A cost-effective wireless sensor network system for ındoor air quality monitoring applications. In: FNC/MobiSPC, pp. 165–171 (2014)
5. Bathiya, B., Srivastava, S., Mishra, B.: Air pollution monitoring using wireless sensor network. In: 2016 IEEE İnternational WIE Conference on Electrical and Computer Engineering (WIECON-ECE), IEEE, pp. 112–117 (2016)
6. Bhushan, B., Sahoo, G.: Routing protocols in wireless sensor networks. In: Computational İntelligence in Sensor Networks, pp. 215–248. Springer, Berlin, Heidelberg. (2019)

7. Patil, D., Thanuja, T.C., Melinamath, B.C.: Air pollution monitoring system using wireless sensor network (WSN). In: Data Management, Analytics and Innovation, pp. 391–400. Springer, Singapore. (2019).
8. Roopashree, J., Raghunath, C.R., Ravikumar, D. Low power EMC optimized wireless sensor network for air pollution monitoring system. Int. J. Recent Innovation Trends Comput. Commun. **3**(6), 3532–3537 (2015)
9. Saini, J., Dutta, M., Marques, G.: A comprehensive review on indoor air quality monitoring systems for enhanced public health. Sustainable Environ. Res. **30**(1), 6 (2020)
10. Mohamed, R.E., Saleh, A.I., Abdelrazzak, M., Samra, A.S.: Survey on wireless sensor network applications and energy efficient routing protocols. Wireless Pers. Commun. **101**(2), 1019–1055 (2018)
11. Sun, J., Zhang, Z., Shen, S., Zou, Z.: An Improved Air Quality Monitoring System Based on Wireless Sensor Networks. In: Proceedings of the 2017 2nd International Conference on Communication and Information Systems, pp. 31–36 (2017)
12. Kaivonen, S., Ngai, E.C.-H.: Real-time air pollution monitoring with sensors on city bus. Digital Commun. Networks **6**(1), 23–30 (2020)
13. Wei, C., Yang, J., Gao, Y., Zhang, Z.: Cluster-based routing protocols in wireless sensor networks: A survey. In: Proceedings of 2011 International Conference on Computer Science and Network Technology, IEEE, vol. 3, pp. 1659–1663 (2011)
14. Sumathi, J., Velusamy, R.L.: A review on distributed cluster based routing approaches in mobile wireless sensor networks. J. Ambient Intell. Hum. Comput. Springer (2020)
15. Hussain, I., Ahmed, Z.I., Saikia, D.K., Sarma, N.: A QoS-aware dynamic bandwidth allocation scheme for multi-hop WiFi-based long distance networks. EURASIP J. Wireless Commun. Networking **2015**(1), 160 (2015)
16. Kakarla, J., Majhi, B., Battula, R.B.: Comparative analysis of routing protocols in wireless sensor–actor networks: a review. Int. J. Wireless Inf. Networks **22**(3), 220–239 (2015)
17. Heinzelman, W.R., Chandrakasan, A., Balakrishnan, H.: Energy-efficient communication protocol for wireless microsensor networks. In: Proceedings of the 33rd Annual Hawaii International Conference on System Sciences, IEEE, p. 10 (2000)
18. Manjeshwar, A., Agrawal, D.P.: TEEN: a routing protocol for enhanced efficiency in wireless sensor networks. In: Proceedings 15th International Parallel and Distributed Processing Symposium, IPDPS, vol. 1, p. 189 (2001)
19. Yarinezhad, R.: Reducing delay and prolonging the lifetime of wireless sensor network using efficient routing protocol based on mobile sink and virtual infrastructure. Ad Hoc Netw. **84**, 42–55 (2019)
20. Braman, A., Umapathi, G.R.: A comparative study on advances in LEACH routing protocol for wireless sensor networks: a survey. **3**(2), 5883–5890 (2014)
21. Lindsey, S., Raghavendra, C., Sivalingam, K.M.: Data gathering algorithms in sensor networks using energy metrics. IEEE Trans. Parallel Distrib. Syst. **13**(9), 924–935 (2002)
22. Durresi, A., Paruchuri, V.: Geometric broadcast protocol for sensor and actor networks. In: 19th International Conference on Advanced Information Networking and Applications (AINA'05) Volume 1 (AINA papers), IEEE, vol. 1, pp. 343–348 (2005)
23. Somauroo, A., Bassoo, V.: Energy-efficient genetic algorithm variants of PEGASIS for 3D wireless sensor networks. Appl. Comput. Inf. (2020)
24. Fuhrmann, T.: Scalable routing in sensor actuator networks with churn. In: 2006 3rd Annual IEEE Communications Society on Sensor and Ad Hoc Communications and Networks, IEEE, vol. 1, pp. 30–39 (2006)
25. Zhou, Y., Ngai, E.C.-H., Lyu, M.R., Liu, J.: POWER-SPEED: a power-controlled real-time data transport protocol for wireless sensor-actuator networks. In: 2007 IEEE Wireless Communications and Networking Conference, IEEE, pp. 3736–3740 (2007)
26. Zafar, S., Bashir, A., Chaudhry, S.A.: Mobility-aware hierarchical clustering in mobile wireless sensor networks. IEEE Access **7**, 20394–20403 (2019)
27. Luo, X., Yang, J.: A survey on pollution monitoring using sensor networks in environment protection. J. Sensors **2019** (2019)

28. Loscri, V., Morabito, G., Marano, S.: A two-levels hierarchy for low-energy adaptive clustering hierarchy (TL-LEACH). In: IEEE Vehicular Technology Conference, IEEE, 1999, vol. 62, no. 3, p. 1809, (2005)
29. Luo, H., Ye, F., Cheng, J., Lu, S., Zhang, L.: Ttdd: a two-tier data dissemination model for large-scale wireless sensor networks. J. Mobile Networks Appl. (MONET) (2003)

Frequent Itemset Mining Algorithms—A Literature Survey

M. Sinthuja, D. Evangeline, S. Pravinth Raja, and G. Shanmugarathinam

Abstract A prominent subfield of data mining is Frequent Itemset Mining which explores mysterious and hidden patterns in the transaction database. However, as the volume of data increases, the mining of hidden patterns of the frequent itemset is more time-consuming. Moreover, dominant memory consumption is required in mining where the hidden pattern of the frequent itemset computation is complicated through the algorithm. Therefore, a powerful algorithm is needed to mine the hidden patterns of the frequent itemset within a more precise execution time and with lower consumption of memory while the size of data increases over the period. This study article focuses on the pros and cons of FP-growth, LP-growth, FIU-tree, IFP-growth algorithm for frequent pattern discovery, and more efficient frequent pattern mining algorithms can be further carried out.

Keywords Data mining · Frequent itemset mining · FP-tree · IFP-tree · LP-tree · FIUT

1 Introduction

In the present scenario of technological boom, storage of data is very significant. Business, research, medicine, government offices, supermarkets deal with different data to be assessed and audited. The audit process is so repetitive that a manual process with the assistance of an individual or technological expertise cannot be carried out, as data grows exponentially, though legacy/classical techniques cannot help effectively with this situation. A curious and active area of research is Data Mining, which had more attention in the last decades [9]. In the IT industry and society, data mining is very effective due to the extensive availability of an enormous amount of data. Domain data mining is the process of exploring productive information from a huge

M. Sinthuja (✉) · D. Evangeline
M.S. Ramaiah Institute of Technolgy, Bangalore, India

S. P. Raja · G. Shanmugarathinam
Presidency University, Bangalore, India

J. S. Raj et al. (eds.), *Intelligent Sustainable Systems*, Lecture Notes in Networks and Systems 213, https://doi.org/10.1007/978-981-16-2422-3_13

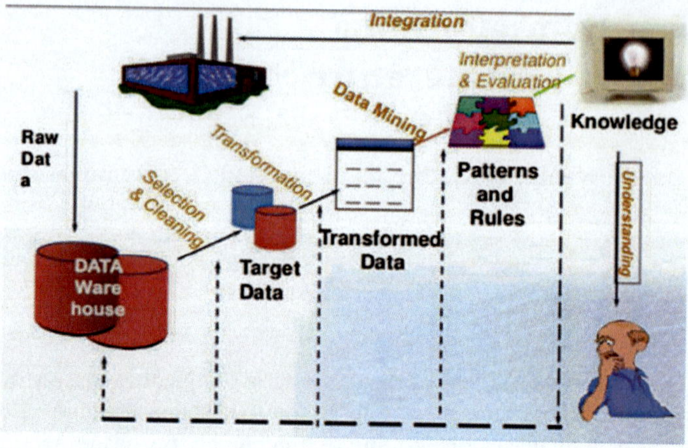

Fig. 1 Data mining: a KDD process data mining the core of knowledge

volume of data [2, 1]. The main aim is to explore knowledge that supports deciding on a critical situation.

Medicine, insurance, finance, education, fraud detection, retail sector, etc., are applications of data mining techniques. Data mining's major benefit is the generation of agile and efficient algorithms that can carry massive volumes of information. The data mining method is shown in Fig. 1.

Frequent item sets, association rules, and correlations that someone chooses to accomplish seem to be the most realistic data mining activity, instead of the most investigation [10, 8–12]. Frequent patterns are perhaps the frequent occurrence of itemsets in a dataset. In order to find association and interrelation between the dataset that are relevant, and other available mining tasks are also classified and clustered [14]. Market basket analysis is an example of association rule mining. Since the first phrase is the most extensively used data mining tool in various domains of frequent pattern mining. Frequent itemset mining is recognized as an intermediate stage for mining tasks such as classification and clustering. Through life science data analysis [1, 7, 3, 6], frequent pattern mining is implemented. Frequent pattern mining is applied to medical and biological data for a better diagnosis of disease.

2 Frequent Itemset Mining Algorithms—A Literature Survey

A new tree structure for the implementation of frequent itemset mining technique was proposed and named Linear Prefixtree (LP-Tree) [10]. The structure of the LP-tree in array form is to curtail the usage of pointers between nodes. LP-tree is extremely efficient in contrast to conventional methods. The reason behind this is a complication of pointers to ignore various arrays that are required to preserve the transaction. In the case of the FP-tree structure proposed by [5], individual nodes are utilized for storing every item in one transaction. When it comes to FP-tree individual arrays are used for storing each transaction. The simplicity of the array is explored while adding the transactions and reach through LP-tree. The authors have tried to form the specific structure of frequent pattern trees in linear prefix trees by adopting arrays and pointers. BNL acts as a ledger that contains information about the branched nodes. The modification will happen in BNL if any of the new branches are created in LP-tree excluding the root node.

The format of LP-tree is as follows:

$$\text{LP - tree} = \{\text{Headerlist, BNL, LPN}_1, \text{LPN}_2 \ldots \text{LPN}_n\}$$

The LP-tree header list is the same as the FP-tree, consisting of arranged objects, counting items, and pointing to the location of the array in which the items are arranged in a given LPN. Whereas the FP-tree consists of the items sorted, the count of the items and the pointer to indicate the initial node of the item are addressed. The database is scanned to discover each item's count. Generate the header list and categorize the object based on the item's count in decreasing order. Prune the rare items in the second scan and organize the items based on the count in decreasing order. The completed transactions are inserted into a newly created LPN in the form of an array. The length of an array constitutes to cumulative addition of all completed transactions.

The format of LPN of n items is illustrated as follows:

$$\text{LPN} = \{(\text{Parent\_Link}), (\text{i}_1,\text{S,L,b}), (\text{i}_2,\text{S,L,b}), \ldots\ldots, (\text{i}_\text{n}, \text{S, L, b})\}$$

The header of the LPN is Parent-Link that connects to its parent node(root), in, S, L, b is an nth item, 'S' denotes the frequency/support of the item, 'L' denotes the connector to the related item's next node, and 'b' is the flag that denotes the branch node's presence. If the first transaction item is not in the root, then the Parent connection of the newly created LPN will be directly linked to the root.

At a time when one child can be stored in an array, if nodes have more than one child then it is represented by an alternate format called BNL. If more than one child has a node, then it is considered to be a branch node. The children of each branch node are represented by different child nodes and are connected to their parent node. BNL acts as a ledger that contains child's data.

2.1 *Building of LP-Tree*

All the transactions have to be connected to the tree at the moment of the second scan. The intention is to discover the first search item of the newly inserted transaction for the root children. If the children are found, then the support item will be increased, or else a new LPN will be generated. And the contained transaction objects will also be added where the header node points to the root node. At the same time, the BNL root affiliates will be increased to one. If the whole item in the current transaction is the same as the previously added transaction, then the item support is increased to one item in the current LPN. A new LPN is created and the current LPN is marked as a parent by the header if all of the items are similar to a preexisting LPN where the pending items are different. The authors have attempted to cut out the complexity of pointers here; however, at the time of making LP-tree as BNL, those pointers are included. Every node of the FP-tree has the following details.

Item name, support, parent pointer, child pointer, node link.

In FP-tree, it is seen that each item is represented by each node where most of the items are added multiple times in FP-tree. On the contrary, LP-tree requires only one LPN to store a set of items where a multitude of arrays are used to store the multiple items of the transaction from the database. The process of mining is identical to FP-tree, mining travel inverse from bottom to top (root), i.e., from end item of the header list.

To find the parent of current LPNs, LP-growth performs a single traversal with the assistance of header pointers. The authors have specified that BNL can be deleted after the formation of LP-tree as traversing from lower level to higher level requires only the header point of every LPN. The transaction storage method of LPN is an advantage; there is a responsibility where a single transaction would be stored in a different LPN despite its capacity to store it in a single LPN.

Further, BNL can be used to track the deviation on the path. To ignore all these complexities, the integration method is proposed to keep a freshly inserted transaction with the present one and also BNL is pruned to release the additional memory. Arrays of LPN are used to save the transactions, and all the drawbacks of the array are applicable. To insert a transaction continuous free memory is required. By adopting this data structure, complexity of pointers is removed drastically and also execution time and consumption of memory are lower than FP-tree.

An innovative approach called Frequent Items Ultrametric Trees (FIUT) is initiated for generating frequent itemset [3]. The efficiency is enhanced in FIU-tree by acquiring frequent itemsets. Two important steps are

1. First step

The database is scanned twice to calculate the frequency of all 1-itemset in the first scan.

Pruning is performed in the second scan, where the remaining items in every transaction are grouped to separate clusters. The entire transaction with k *very prevalent* items is organized into one cluster. Complete transaction in one cluster has a similar sum of frequent items in it.

2. Second step

The main transaction database does not accept the transactions enclosed in the clusters. In this step, the FIU-tree is developed and the mining process is carried out.

Initially, for the construction of the k-FIU-tree, the highest number of items from the transaction category is considered, where k is from M down to 2 and M represents the maximum value of k among the transactions. For the mining of frequent *m*-itemsets, tree construction is initiated with m items.

2.2 Construction of k-FIU-Tree is as Described as Below

Construction of tree is initiated at the beginning of $k = m$ by generating a root node labeled as null. Insert the first transaction of the k group into the tree. Here, the root node acts as a parent to the first item of the transaction. Insert the pending items regularly to the next of the previously inserted item. To insert pending transactions, check whether the initial item is the root's children if so then children are to be as the parents of the following item in the transaction, in the same way, compare the next items. If any mismatch is found generate a fresh node as the child of the present item and insert all the items pending as fresh branch. The number of transactions from the set should be included as a count in the leaf node. As soon as adding entire k-itemsets, the k-frequent itemset will be mined. FIU tree is said to be the balanced tree where all single leaves are stacked at a similar level. The process of mining is initiated by analyzing the count value of the leaf link FIU tree. An itemset is said to be a frequent k-itemset only if the denotation of 'count/D' is higher than or equivalent to minimum support. Create (k-1)-FIU tree after checking all the leaf nodes. The k-FIU-tree will be changed to (k-1)-FIU tree and sum every original k-1 itemsets from its group. Likewise, iterations in the above-mentioned procedure are performed until all the groups become free.

In recent years, numerous frequent mining techniques have been explored. In this article, the author has proposed an alternative method of mining known as FUI-tree mining. While creating and mining frequent patterns, search space is reduced by hypothesizing k-itemset at a time. In this concept, construction of trees and the process of mining perform together. Instantly, after the creation of the k-FIU-tree establishment of k-itemsets will be initiated.

With the above approach of FP-Tree, it is not required to create a conditional database and conditional FP-trees for all available itemsets. At a point, a single k-FUI-tree will be available in memory, however, concerning the FP-tree complete set of the tree should have to be added into memory. Similarly, a single k-FUI-tree will be available in memory which decreases the execution time as the itemsets are not sorted before adding into the tree about the transactions and the entire items will be in etymology order.

To analyze the performance of this approach, the author has used two distinct transactional databases. Various experimentations were performed in contrast to the new approach with the existing FP-tree method. The execution time of FP-growth increases to 13 s from 7 s, whereas the FIU tree takes 4 to 6 s on processing 5000 to 10,000 transactions with the support of 0.14 to 0.08. In another experimental analysis, while fixing the minimum support of 0.02% for a transaction set 9000–25,000 FIU tree consumes execution time of 55–89 s. When it comes to FP-tree execution time is from 98 to 178 s. From all the experimentation, it is shown that FP-growth consumes higher execution time when compared to the FUI-tree with any of the transactional database. The execution time between two methods is differing from 5 s to 75 s, respectively. When fixing the minimum support to the lower-level execution time of FP-growth increases extremely, whereas in the FIU tree there is only a minor increment in execution time. From the experimental result, it is shown that FIUT mining approaches are efficient in distinct support threshold and a distinct number of datasets.

For mining association rules, an improved frequent pattern growth algorithm (IFP-growth) [5] is proposed. In this article, the author has compared the proposed algorithm with FP-tree and proved that the proposed algorithm is better in performance as well as consumption of memory is lower. The author of this article has proposed an advanced IFP-growth algorithm to enhance the conduct of the FP-growth algorithm. The address-table structure is employed in the IFP-growth algorithm to reduce the difficulty of finding each node in an FP-tree. The Hybrid FP-tree mining approach is used to diminish the requirement for the reconstruction of conditional FP-tree.

The process of reaching the conditional FP-tree and FP-tree interrupts the conduct of mining FP-growth. In the phase of building FP-tree, before inserting a fresh transaction with n items, the algorithm searches n times in FP-tree. It has to find 'm-i' times to insert each item where 'm' stands for the sum of items in the transactional database, 'i' is the location of the present item in the transaction. For the insertion of the first item, the algorithm has to execute $m-1$ times. In addition to the second item, it has to run $m-2$ times, and so on. Hence in the case with maximum odds, the author calculates the difficulty in the construction of a new path as $(m + (m - 1) + \cdots + (m - (n - 1)))$. In order to minimize the complexities in the tree construction process, the author used an appropriate data structure called an address table. An address table consists of group of items and pointers.

The respective node of that item is connected by the item pointer at the next level of the FP-tree. For accessing the child node of the FP-tree, the address table is used. Per node verifies its address table to verify the existence of a child in the tree. The author has included one address table with each node in this notion, exceptional to the last FP-tree object. In the address table, all the frequent items and initial level pointers are also included.

In the second level, all nodes will have separate address tables which contains all items succeeding present item in the header table. The address table remains persistent, whenever the items occur more frequently in FP-tree. This approach occupies huge memory. In the process of mining, numerous methods have been used to reduce the cost of the establishment for a conditional pattern based. At the time of implementation, infrequent items are not eliminated instead they use another technique called tree-level value evaluated by dividing the items in the header table into two.

In this analysis, the author has compared the IFP-growth algorithm with the existing algorithms of FP-growth. It is inferred that the execution time and memory of the FP-tree are more expensive than other algorithms. It is observed that there is diminutive difference in consumption of memory of FP-tree and IFP. At the time of mining frequent itemset of every node in FP-tree memory consumption of conditional FP-tree is temporary. When it comes to IFP memory space required to construct an address table for each node is not momentary.

3 Problems and Definitions

In this article, various existing frequent itemset mining algorithms are discussed in detail. All the algorithms have their pros and cons. In this section, few techniques suggested to overcome the limitations of the existing algorithms are provided. Among the approaches discussed in Sect. 2 for mining, frequent itemset Linear Prefix tree (LP-tree) is the simple method to mine frequent itemsets. Numerous complications faced by the algorithm are repeated formation of LP-tree. To overthrow the disadvantage, after the creation of LP-tree some other approaches can be adopted for the process of mining. In the FIUT algorithm, it is observed that this technique is efficient in contrast to the algorithm FP-tree. But it scans the database thrice where the alternative methods require only two database scans. Further, it includes processing such as the creation of a cluster, decomposing of trees, etc., in contrast to alternative methods. Some of the groups will be more if the transactions are very less which results in high computational time. In this situation, the algorithm will not provide better output. In this case, at the time of the mining process, it is not required to hold the entire FP-tree on primary memory. Modern clustering techniques can be applied to speed up the execution time. To hold an address table for every node excess memory is used by the IFP-growth algorithm in the third method which leads to a shortage of memory to save those supplementary data. The address table consists of the item name and pointer to its child.

In order to overcome the drawback, a newly modified address table may be used to minimize memory usage, such as changing the item's name to item codes or storing it for children exclusively.

4 Conclusion

A brief overview of FP-growth, LP-growth, FIU-tree, IFP-growth algorithms was discussed in this article. A familiar technique of data mining is association rule mining. It is not only used to find significant relationships from the massive amount of data, but also used in a database to differentiate between a variety of classes. The first step of Association rule mining is Frequent itemset mining. This article exposes that all the techniques have their strengths and weaknesses. However, the techniques specified above have huge improvements from the classic FP-tree algorithm. By contributing new approaches with these existing methods, the researcher can improve the efficiency of the algorithms. Modifications can be done to these algorithms effectively by reducing the execution time and usage of memory. Eventually, this survey will provide a schematic of the current work and provide a similar perception of frequent itemset mining to the researcher.

References

1. Aggarwal, C.C.:Data Mining: The Textbook. Springer (2015)
2. Aggarwal, C.C., Han, J.: Frequent Pattern Mining. Springer (2004)
3. Anand, J.V.: A methodology of atmospheric deterioration forecasting and evaluation through data mining and business intelligence. J. Ubiquitous Comput. Commun. Technol. **2**(02), 79–87 (2020)
4. Atluri, G., Gupta, R., Fang, G., et al.: Association analysis techniques for bioinformatics problems. In: International Conference on Bioinformatics and Computational Biology, pp. 1–13. Springer (2009)
5. Han, J., Pei, J., Yin, Y.: Mining frequent patterns without candidate generation. In: SIGMOD '00 Proceedings of the 2000 ACM SIGMOD International Conference on Management of Data, pp.1–12 (2000)
6. Kumar, T.S.: Data mining based marketing decision support system using hybrid machine learning algorithm. J. Artif. Intell. **2**(3), 185–193 (2020)
7. Lin, K.-C., Liao, I.E., Chen, Z.-S.: An improved frequent pattern growth method for mining association rules, Expert Syst. Appl. **38**, 5154–5161 (2011)
8. Naulaerts, S., Meysman, P., Bittremieux, W., et al.: A primer to frequent itemset mining for bioinformatics. Brief. Bioinform. **16**(2), 216–231 (2015)
9. Pei, J., Yin, Y.: Mining frequent patterns without candidate generation. ACM SIGMOD Record **29**(2), 1–12 (2000)
10. Pyun, G., Yun, U., Ryu, K.H.: Efficient frequent pattern mining based on linear Prefix tree. Knowledge Based Syst. **55**,125–139 (2014)
11. Rajaraman, A., Ullman, J.D., Ullman, J.D., Ullman, J.D.: Mining of Massive Datasets, 1. Cambridge University Press, Cambridge (2012)
12. Sinthuja, M., Puviarasan, N., Aruna, P.: Comparison of candidate itemset generation and non candidate itemset generation algorithms in mining frequent patterns. Int. J. Recent Innovation Trends Comput. Commun. **5**, 192–197 (2017)
13. Sinthuja, M., Puviarasan, N., Aruna, P.: Comparative analysis of association rule mining algorithms in mining frequent patterns. Int. J. Adv. Res. Comput. Sci. **8**, (2017)
14. Tanbeer, S.K., Ahmed, Jeong, B.S., Lee, Y.: Efficient single pass frequent pattern mining using a prefix-tree. Inf. Sci. **179**(5), 559–583 (2008)
15. Tsay, Y.-J., Hsu, T.-J., Yu, J.-R.: FIUT: a new method for mining frequent itemsets. Inf. Sci. **179**, 1724–1737(2009)

Impact of Segmentation Techniques for Condition Monitoring of Electrical Equipments from Thermal Images

M. S. Sangeeetha, N. M. Nandhitha, S. Emalda Roslin, and Rekha Chakravarthi

Abstract Infrared Thermography is used in condition monitoring of electrical equipment. Thermal images or thermographs are acquired using an IR camera and the hotspot/coldspot temperature is calculated. Image segmentation is the most important step in isolating the hotspot/coldspot. A segmentation technique must retain the full size and shape of the anomaly while completely removing the unwanted regions. Though various segmentation techniques are cited in the literature, these segmentation techniques could not detect anomalies of irregular shapes. In this paper, an Improved Active Contour Modeling technique is proposed to isolate the Region of Interest. The performance of the proposed technique is compared with that of the conventional segmentation techniques. IACM removes the undesirable regions and is successful in detecting the Region of Interest of any shape and size.

Keywords Condition monitoring · Infrared thermography · Segmentation · Multilevel thresholding · Improved active contour modeling

1 Introduction

Condition monitoring of electrical equipment is a major concern in India. Infrared Thermography is a suitable technique for condition monitoring as it is a non-invasive, non-hazardous, and non-contact technique and it does not need shut down of electrical equipment [1–6]. It uses an IR camera that captures the heat pattern and converts it into a thermograph. Hotspots represent high-temperature region which in turn corresponds to maximum current flow and coldspot corresponds to low-temperature region implying no current in that area. Hotspots/coldspots are the Regions of Interest. These regions are isolated; temperature is determined and is mapped periodically. Any abrupt variation demands inspection (Fig. 1).

As abnormal or abrupt variation indicates an anomaly, it is necessary to segment thermographs for identifying these regions (hotspots/coldspots). Segmentation of

M. S. Sangeeetha · N. M. Nandhitha (✉) · S. E. Roslin · R. Chakravarthi
School of Electrical and Electronics, Sathyabama Institute of Science and Technology, Old Mamallapuram Road, Chennai 600119, India

J. S. Raj et al. (eds.), *Intelligent Sustainable Systems*, Lecture Notes in Networks and Systems 213, https://doi.org/10.1007/978-981-16-2422-3_14

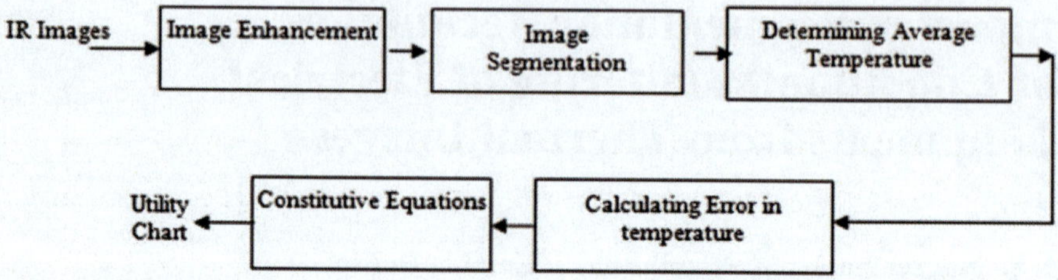

Fig. 1 Condition monitoring of electrical equipment

thermograph involves hotspot isolation [7, 8], feature extraction, and temperature detection (error detection).

2 Literature Survey

Novizon et al. [9] proposed a method in which a thermal image of the arrester is taken and processed to calculate the whole-body temperature. Here, the network is trained by using multi-layer backpropagation method having humidity, ambient temperature, and minimum, maximum, and whole-body temperature as input parameters. LABVIEW software is used to train the network. The surface temperature of the arrester is correlated with the leakage current to determine the condition of the surge arrester. The authors have obtained an accuracy of up to 1.5% error.

Manjiri et al. [10] used thermal image processing to do condition monitoring in electrical equipment. The author used the HSI model by taking the hue region instead of the grayscale image for extracting the exact hotspot region. The hotspot is easily extracted from the hue region using Sobel, Prewitt, Robert, and Otsu method. The MSE and PSNR are obtained for various images and tabulated. The author concludes that Sobel and Otsu method gives better results.

Olivier et al. [11] proposed a deep learning technique to determine the condition of the machine. Using this method, the condition of the machine and oil level is predicted. John et al. [12] proposed a method for condition monitoring of reciprocating compressor valves. Here, vibration-based condition monitoring method was developed to find the valve wear by using time–frequency analysis combined with image-based recognition techniques. Here, the operating data such as vibration, cylinder pressure, and shaft position are processed using time–frequency domain approach and the resulting diagrams are processed for feature extraction. 90% accuracy is achieved.

Keerthi et al. [13] used a convolutional neural network on the thermal image to detect the faults in rotating machinery. Here, thermal images of the machine are obtained and converted into the grayscale image, and the unwanted noise is removed. A novel 3D full convolution network is applied to the enhanced image for the screening stage. The data obtained from the screening stage is distinguished with the CNN model to obtain final detection results. Here, 90% accuracy is obtained in condition monitoring of rotating machines.

Though considerable research is carried out in this area, segmentation is still a challenging task as it is very difficult to obtain the true size of the hotspot (owing to the irregular shape of the hotspot) and to remove the undesirable regions completely. Hence, it is necessary to develop image segmentation techniques that are best suited for isolating the hotspots in real-time thermographs. In order to carry out the research, a set of real-time thermographs are acquired.

3 Research Database

FLIR T335 IR camera is used for acquiring thermal images. A total of 110 images are acquired of electronic and electrical equipment. Table 1 provides the metadata of thermal images acquired by varying the emissivity and distance. The acquired images are pseudocolor images of size m × n × 3.

Thermal images of CPU acquired at emissivity 0.1, 0.2, 0.3, and 0.4 are shown in Table 1. Figures 2, 3, 4, and 5 show the thermographs acquired at different emissivities and distances of UPS, USB, and PCB. The hotspot is the high-temperature region and is shown with high intensity.

Table 1 Metadata of thermal images

Specimen/device	images	Operational conditions	Real-time temperature in degree centigrade (T_{act}) (°C)
CPU	10	Emissivity	46
UPS	51	Emissivity, distance	41
USB	26	Emissivity, distance	31
PCB	33	Emissivity	28

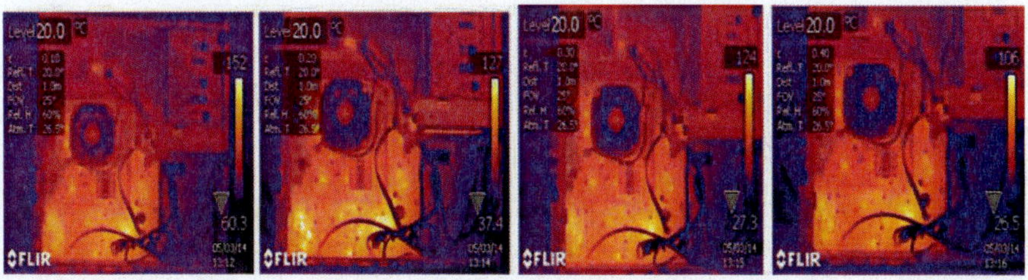

Fig. 2 Thermographs of CPU (emissivity-0.1, 0.2,0.3, and 0.4)

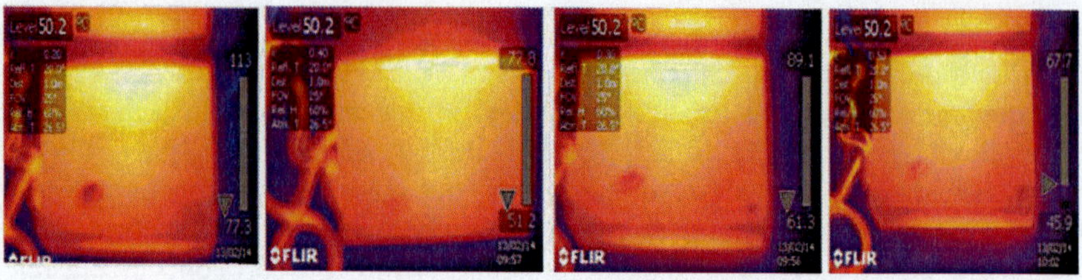

Fig. 3 Thermographs of UPS (emissivity-0.1 and 1; distance = 1 m, 2 m)

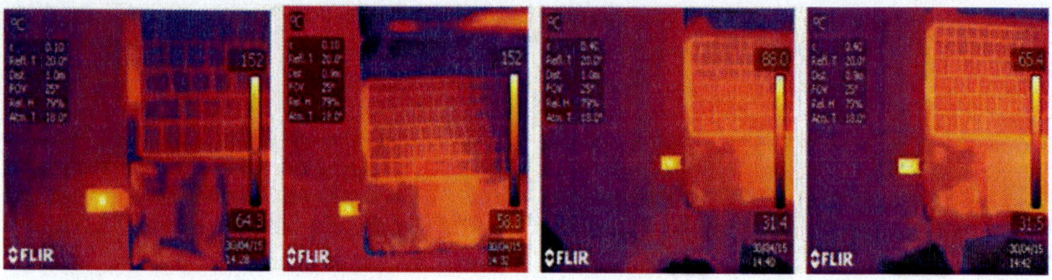

Fig. 4 Thermographs of USB (emissivity-0.1 and 0.4; distance = 1 m, 0.9 m)

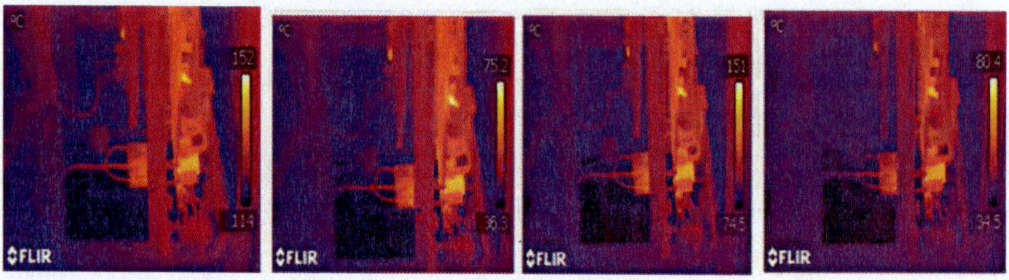

Fig. 5 Thermographs of PCB (emissivity-0.1, 0.9, 0.2, and 1)

4 Image Segmentation Techniques for Thermographs

In discontinuity-based segmentation, edges, points, and lines of the hotspot are detected and a morphological image processing technique is then applied for detecting the hotspot. The shape and size of the structuring elements chosen for morphological image processing are dependent on the hotspot size and shape and cannot be standardized. The similarity between the pixels namely intensity, brightness, contrast, etc. results in a heuristic choice of threshold, which is a major limitation of similarity-based segmentation techniques. Radon transform and Hough transform are the most commonly used transforms for image segmentation in similarity-based segmentation. The above techniques work well for hotspots mimicking the regular solid shapes. However, in real-time images, the Region of Interest is irregular in shape and hence the conventional transforms cannot be used for segmentation.

In such cases, the snake algorithm also known as Active Contour Modeling is used for the segmentation of Region of Interest. In the snake algorithm, the mask is initially generated and the mask is applied to the image. Gradients greater than the threshold are retained (Region of Interest). Hence, the performance of the snake algorithm is dependent on mask generation. If the mask generated is accurate, then the segmentation results in accurate detection of Region of Interest. In this method, a mask is generated using the Euclidean-based segmentation and it is used further. In the Euclidean-based segmentation, the thermograph is not converted into grayscale and the pseudo color thermograph is used as such. The mask is then used for segmenting the images.

In order to compare the performance of the proposed technique, region growing and multilevel thresholding techniques are used for segmenting the images. In region growing, seed pixel intensity is obtained from the Region of Interest and all the pixels with similar intensity are retained. The retained pixels indicate the anomaly.

In the case of multilevel thresholding, threshold values for each bin are fixed from the histogram of the original image. The intensities are fixed in such a way that the Region of Interest is highlighted (Tables 2, 3, 4, 5, and 6).

Four different sets of thermographs namely thermographs of CPU, USB drive, PCB, and UPS are considered. From the subjective analysis, it is understood that Region growing and multilevel thresholding cannot isolate the anomaly regions in certain thermographs. On the other hand, Improved Active Contour Modeling has isolated the anomaly in all the thermographs. Having segmented the Region of Interest, hotspot temperature and hence the error in temperature are calculated and tabulated. Tables 7,8,9, and 10 and Figs. 6,7,8, and9 provide an error in temperature for CPU, USB drive, PCB, and UPS images for three different types of segmentation.

From Tables 7, 8, 9 and 10 and Figs. 6, 7, 8 and 9, it is found that of the three segmentation techniques, Improved Active Contour Modeling provides the least error. It implies that the Region of Interest is segmented to its true size and the undesirable regions are completely removed. IACM provides accurate segmentation as it uses the generated mask for isolating the Region of Interest without a limitation on its shape Error in temperature for different segmentation techniques.

5 Conclusion and Future Work

Of the various segmentation techniques, used for isolating the Region of Interest, Improved Active Contour Modeling is the best-suited technique. Advantages of IACM are flexibility in developing the mask, automatic generation of the mask from the original image (i.e., each thermal image has an image-specific mask generated from the original thermograph), and segmentation is independent of the shape of the anomaly. Owing to the above facts, IACM can detect anomalies or abnormalities irrespective of the size and shape. Also, the proposed technique is independent of the size of the input image and its dimensionality (grayscale, RGB color model).

Table 2 Segmentation techniques—algorithm, advantages, limitations

Segmentation technique	Algorithm	Advantages	Limitations
Region growing	Seed pixel intensity is chosen Dissimilarity matrix is calculated dissimilarity < threshold is retained as ROI Other intensities are made zero	Computationally less complex	Undesirable regions are present No standard method for choosing the seed pixel and threshold
Multilevel thresholding	Histogram is obtained The number of levels is decided The range of intensities are determined The original images intensity is modified accordingly	Hotspots of any shape can be obtained	Undesirable regions are present Thresholds are dependent on the color palette of the thermographs
IACM	Masks are generated from pseudocolor thermographs [14, 15] Mask is slid over the thermograph and gradients are obtained Gradients within a specific threshold are retained	Hotspots of any shape and size can be obtained	Computationally complex and is suited for post-processing (not suited for online processing)

Table 3 Impact of segmentation techniques on isolating anomaly for CPU thermographs

Input image	Segmented image		
	Region growing	Multilevel thresholding	Improved Active contour modeling
IR_1679			
IR_1677			

Table 4 Impact of segmentation techniques on isolating anomaly for PCB thermograph

Input image	Segmented image		
	Region growing	Multilevel thresholding	Improved Active contour modeling
IR_1217			
IR_1219			

However, the work stopped at developing the software. Hardware can be developed that accepts the input thermographs and provides the segmented image as the output. Also, user-friendly GUI can also be developed. Code optimization can also be explored.

Table 5 Impact of segmentation techniques on isolating anomaly for UPS thermograms

Input image	Segmented image		
	Region growing	Multilevel thresholding	Improved Active contour modeling
IR_1403		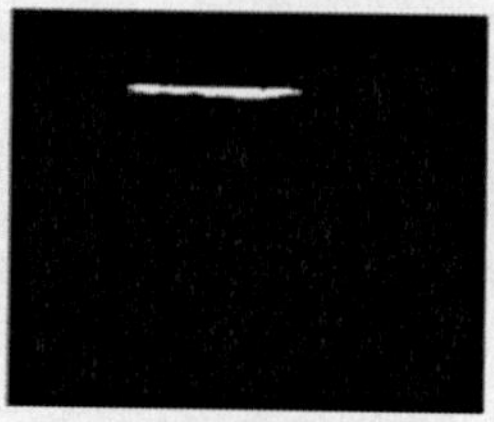	
IR_1405			

Table 6 Impact of segmentation techniques on isolating anomaly for USB thermograms

Input image	Segmented image		
	Region growing	Multilevel thresholding	Improved Active contour modeling
IR_2226			
IR_2228			

Table 7 Error in temperature for CPU thermographs

S. No	Input image	Error in temperature (°C)		
		Region growing	Multilevel thresholding	IACM
1	IR_1673	94.6248	94.6248	74.148
2	IR_1675	69.4497	69.4497	60.8134
3	IR_1677	66.3230	66.0892	55.7759
4	IR_1679	48.6353	48.6353	40.3908
5	IR_1681	31.7325	32.4682	23.5508
6	IR_1683	25.0741	25.1992	16.7532
7	IR_1685	10.7971	10.7971	5.0372
8	IR_1687	5.4190	Output not obtained	0.2167
9	IR_1689	Output not obtained	Output not obtained	−1.5472
10	IR_1691	Output not obtained	Output not obtained	−5.3676

Table 8 Error in temperature for USB drive thermographs

S. No	Input image	Error in temperature (°C)		
		Region growing	Multilevel contrast enhancement	IACM
1	IR_2226	97.6784	97.6784	84.5212
2	IR_2228	98.0012	97.6784	80.6878
3	IR_2230	98.1510	97.9143	85.5805
5	IR_2234	75.9095	77.2102	64.7030
6	IR_2236	79.4612	75.9095	64.9468
7	IR_2240	81.0781	69.4019	66.8899
12	IR_2258	−13.1288	Output not obtained	−2.3349
13	IR_2260	−13.0819	Output not obtained	−4.7517
14	IR_2262	−14.0529	Output not obtained	−0.8404
15	IR_2264	2.1095	Output not obtained	0.0068

Table 9 Error in temperature for PCB thermographs

S. No	Input image	Error in temperature (°C)	
		Region growing	IACM
1	IR_1217	71.0000	63.6735
2	IR_1219	70.0000	68.7824
3	IR_1221	70.0000	65.9084
4	IR_1223	52.0000	50.4483
5	IR_1227	17.0000	15.7227
6	IR_1229	13.4000	12.6115

Table 10 Error in temperature for UPS thermographs

Input image	Error in temperature (°C)	
	Region growing	IACM
IR_1403	16.3913	7.6471
IR_1405	13.8976	8.8637
IR_1407	9.6782	7.7401
IR_1409	11.4570	9.8637
IR_1411	15.5871	6.5975
IR_1413	17.7366	6.7654
IR_1415	20.8324	7.8583
IR_1417	24.3797	7.0538
IR_1419	12.8452	7.4764
IR_1421	10.8557	6.8635
IR_1423	8.9484	6.6661
IR_1425	7.2480	6.7538
IR_1441	24.7004	16.3932
IR_1443	29.3564	13.2985
IR_1447	29.4155	16.7063

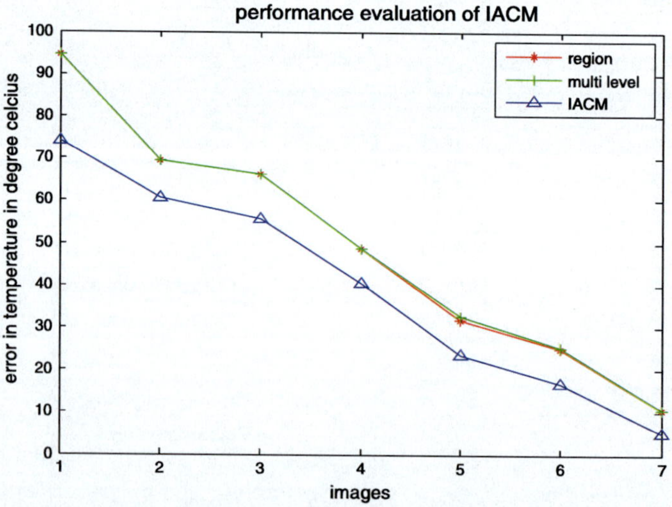

Fig. 6 Error in temperature for different segmentation techniques for CPU thermographs

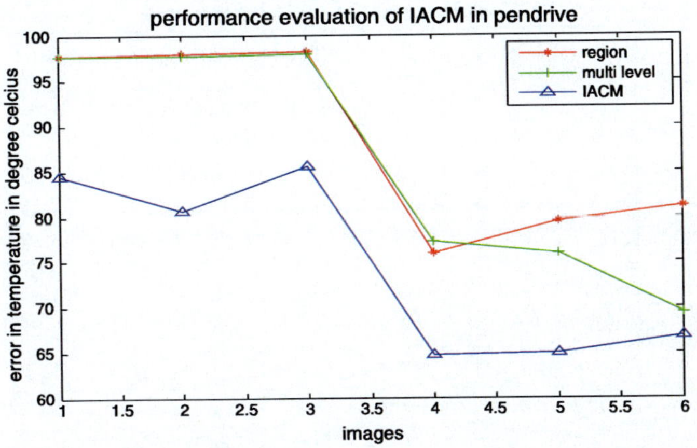

Fig. 7 Error in temperature for different segmentation techniques for USB drive thermographs

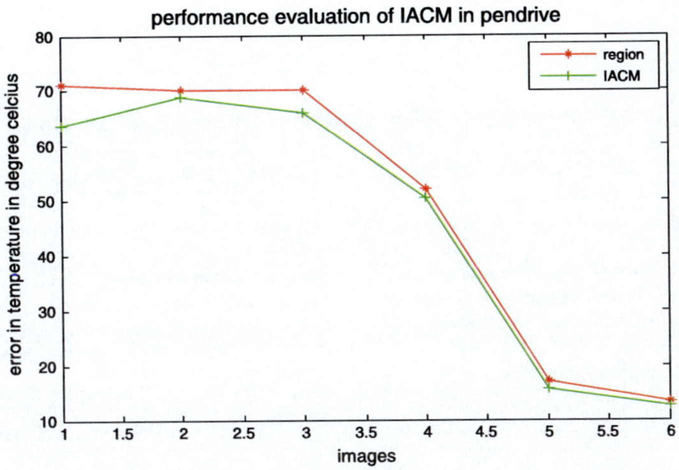

Fig. 8 Error in temperature for different segmentation techniques for PCB thermographs

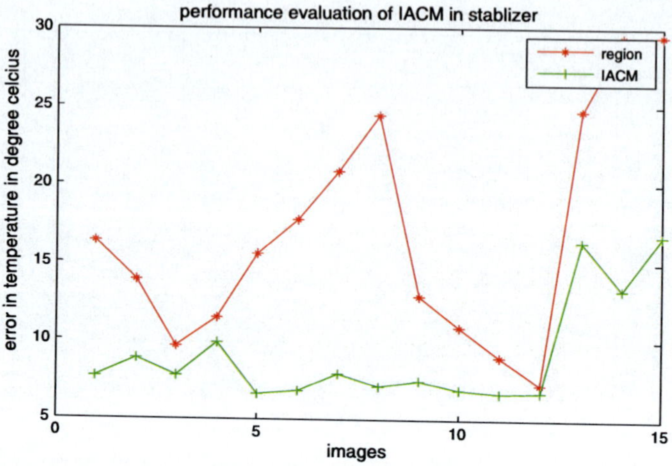

Fig. 9 Error in temperature for different segmentation techniques for UPS thermographs

References

1. Maldague, X.P.: Introduction to NDT by active infrared thermography. Mater. Eval. **60**, 1060–1073 (2016)
2. Sangeetha, M.S., Nandhitha, N.M.: Multilevel thresholding technique for contrast enhancement in thermal images to facilitate accurate image segmentation. Indian J. Sci. Technol. 1–7 (2016)
3. Maldague, X.: Applications of infrared thermography in non destructive evaluation, trends in optical nondestructive testing (invited chapter), 591–609 (2000)
4. Cielo, P., Lewak, R., Maldague, X., Lamontagne, M.: Thermal methods of NDE. Can. Soc. Nondestr. Test. J. **7**(2), 30–49 (1986)
5. Vavilov, V., Maldague, X., Dufort, B., Ivanov, A.: Adaptive thermal tomography algorithm. In: Proceedings SPIE: Thermosense XV, Allen, L.R. (ed.) (SPIE: Society of Photo-Optical Instrumentation Engineers), vol. 1933, pp. 166–173 (1993)
6. Bilodeau, G.-A., Ghali, R., Desgent, S., Langlois, P.: Where is the rat? tracking in low contrast thermographic images. In: IEEE Computer Society Conference on Computer Vision and Pattern Recognition Workshops (CVPRW), pp. 55–60 (2011)
7. Manoharan, S.: Embedded imaging system based behavior analysis of dairy cow. J. Electron. **2**, 148–154 (2020)
8. Chandy, A.: RGBD analysis for finding the different stages of maturity of fruits in farming. J. Innov. Image Process. 111–121 (2019)
9. Novizon, N., Abdul-Malek, Z., Bashir, N., Ghafar, N.: Thermal image and leakage current diagnostic as a tool for testing and condition monitoring of ZnO surge arrester. Jurnal Teknologi. 27–32 (2013)
10. Manjiri, A., Shweta, A.: Condition monitoring of electrical equipment using thermal image processing. Int. J. Res. Publ. Eng. Technol. **3**(4), 45–49 (2017)
11. Janssens, O., Van de Walle, R., Loccufier, M., Hoecke, S.: Deep learning for infrared thermal image based machine health monitoring. IEEE/ASME Trans. Mechatron. 151–159 (2017)
12. Trout, J.N., Kolodziej, J.R.: Reciprocating compressor valve condition monitoring using image-based pattern recognition. In: Annual Conference of the Prognostics and Health Management Society, pp. 1–10 (2016)
13. Keerthi, M., Rajavignesh: Machine health monitoring using infrared thermal image by convolution neural network. Int. J. Eng. Res. Technol. **6**(7), 1–5 (2018)

14. Sangeetha, M., Nandhitha, N.M.: Study on the impact of distance and emissivity measurement for condition monitoring of electronic circuit boards. 1330–1333 (2017)
15. Sangeetha, M., Nandhitha, N.M.: Improved active contour modelling for isolating different hues in infrared thermograms. Russ. J. Nondestr. Test. 142–147 (2017)

A Detailed Survey Study on Various Issues and Techniques for Security and Privacy of Healthcare Records

M. H. Chaithra and S. Vagdevi

Abstract It is noticed that the exponential data growth in the healthcare domain is manageable by the application of Big Data architecture and techniques. Various Machine Learning (ML) and Big Data techniques are influencing healthcare. It is essential to propose a secure and smart healthcare information system with the latest security mechanism. Similarly, provisions have to be made to secure classified healthcare records in the cloud. Cryptosystems, service-oriented architecture, secure multiparty computation, and secret share schemes are some of the security mechanism methods. Evaluation of a classification model in a cloud computing environment is considered in this paper for privacy preserving. For the success of healthcare organizations, a detailed survey study about privacy and security aspects is also dealt with in this paper. This has resulted in machine learning-based secured data processing of healthcare records in the cloud environment.

Keywords Privacy and security · Healthcare · Big data · Machine learning · Cloud computing · Masking encryption · Activity monitoring · Granular access control · Dynamic data encryption · End-point validation

1 Introduction

Hadoop is one of the prominent frameworks used to manage and analyze unstructured data. The effective and efficient utilization of huge data is important for valuable insight into big data. Big data refers to volume, velocity, and variety of data. Typically, a healthcare information system includes: a cloud environment, EHR, security layer, big data analytics, and information delivery. Controlling unauthorized access to health records and protecting patient's profiles and records are the

M. H. Chaithra (✉)
Department of Computer Science and Engineering, Visvesvaraya Technological University, Belagavi, Karnataka, India

S. Vagdevi
Department of Information Science and Engineering, Dayananda Sagar Academy of Technology & Management, Bangalore, India

J. S. Raj et al. (eds.), *Intelligent Sustainable Systems*, Lecture Notes in Networks and Systems 213, https://doi.org/10.1007/978-981-16-2422-3_15

primary concerns in this system. Users usually outsource big data over the cloud since it provides powerful storage and computational processing. Data collection, storage, and computation over the cloud are more preferred in the context of big data analytics.

The challenges in this context are complex computations for classification and developing an encryption-based privacy-preserving model for classification. A secured privacy-preserving model in the cloud environment is proposed in this paper as an outcome of the survey.

To bring in revolutionary changes in the healthcare ecosystem, various organizations and institutions need to devote huge time and resources in understanding the process and realize the benefits. It is known that the world's information is doubling every two years which also includes duplicates. This means big data has big opportunities for research. Various challenges in this context are: accuracy, privacy concerns, consistency, and facility. Once these issues are addressed, then numerous applications of big data can be explored such as: public sector, healthcare industry, education and learning, the insurance industry, transportation sector, industrial and natural resources, banking, fraud detection, and entertainment.

2 Related Works

The below table presents a comprehensive survey on various approaches used to secure healthcare records in big data and cloud computing environments from published research articles (Table 1).

3 Big Data Applications—An Analysis

We can find justifiable evidence about the importance of big data in numerous applications among various fields, ranging from healthcare to fraud detection [13]. Effective from 2017, huge efforts are being involved in the healthcare sectors. Similarly, different areas have different levels of applications of big data. Various tools and techniques used for big data analytics are discussed below [11]. NoSQL technique uses Cassandra, Hbase, and ZooKeeper. MapReduce technique uses Hive, Pig, Flume, and Oozie, and Storage technique uses HDFS.

The Healthcare domain majorly involves: patient care, real-time patient monitoring, predictive analysis of diseases, and improving treatment methods. Clinical trials, medical imaging, maintaining patient records electronically, and recording patient's behavior and preferences are covered in patient care [12]. Temporal data, patient surveillance, the entire IT system of the hospital, and customized services are included in real-time patient monitoring. The predictive analysis includes predicting health risks and anticipating early diagnosis of diseases. Finally, measures need to be defined clearly for improving the treatment methods. The article selection process,

Table 1 Remarks on previous research findings

Sl.no	References	Objectives	Methodology used	Conclusion	Future work	Remarks
1	Kaur et al. [6]	A design of smart and secure healthcare information system using machine learning and advanced security mechanism has been proposed to handle big data of healthcare records	Predictive analysis of diseases	An efficient model for securing healthcare records in the cloud environment is proposed	Fuzzy logic and information theory can enhance precision in disease diagnosis	It is inferred that it is essential to secure big data in a cloud environment especially in the healthcare sector. A model in this direction was proposed by the authors in 2018
2	Li et al. [9]	Design of privacy-preserving outsourced classification in cloud computing environment for healthcare records (POCC)	Homomorphic encryption, Naïve Baye's classifiers	Big data in healthcare records can be encrypted using different public keys	Reduce the cost of communication and computation	This paper presents a design of POCC and justifies the encryption mechanism using different public keys
3	Abouelmehdi et al. [1]	Security and privacy challenges in big data over the cloud for managing healthcare records	Big data security life cycle is proposed using authentication, masking, encryption, access control, and monitoring and audit control	Differentiation of security and privacy	Disease-based classification can be proposed	This paper provides comprehensive knowledge about privacy and security aspects, and the necessity of dealing with big data

(continued)

Table 1 (continued)

Sl.no	References	Objectives	Methodology used	Conclusion	Future work	Remarks
4	Wang [14]	Security by classification using fuzzy C-means clustering and SVM	Support vector machine and C-means clustering	Cloud-based image processing of patient's health records is proposed	Pixel enhancement and Gaussian filtering can be applied to images	This paper presents the classification of patient's health records using image processing techniques

discussed in [6], is evident that huge research works are already carried out by many researchers in their articles. From the Google Scholar database, one can have a detailed look over many research outcomes in published articles.

4 A Secured Big Data and ML-Based Framework

Based on the detailed survey on the Google Scholar database, a layered architecture encompassing big data and ML is proposed as below [5]. Data encryption involves the conversion of original information into the encrypted form to avoid unauthorized access as shown in Fig. 1. Data Encryption Standards (DES) are defined along with Advanced Encryption Standards (AES) to address security issues. It is necessary to monitor the activities of the authorized users. This will define multiple permission levels for authorized users, controlled access to surveillance issues, and execution of queries by various stack holders. So uniqueness of the proposed layered architecture can be appreciated by the deployment of the surveillance systems. This model can be proposed for healthcare and many other allied domains [8].

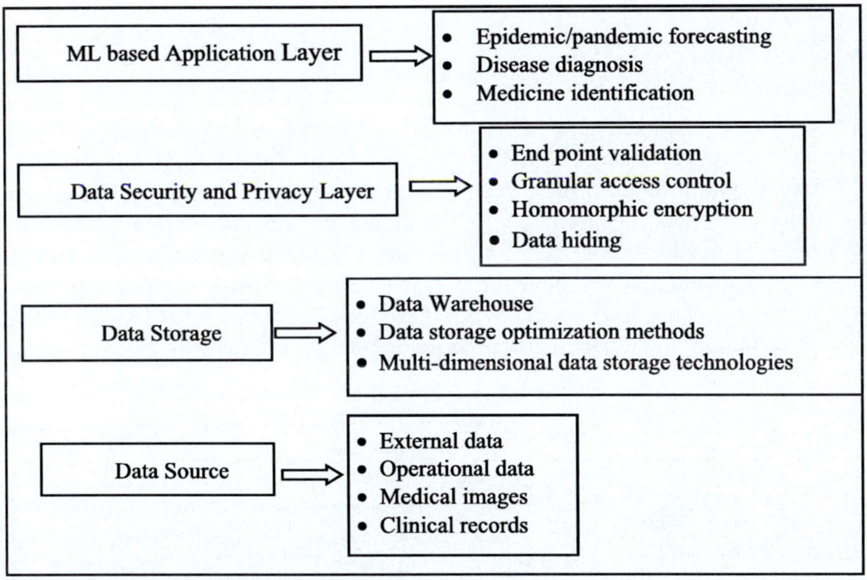

Fig. 1 Layered architecture for ML-based healthcare system

5 Classification on Privacy Preserving

The task of classification is considered as one of the main fundamental procedures in mining and ML projects. In the context of medical health records, if historical data of patients are recorded, and the new patient exhibits the same symptoms, then predicting the diagnosis steps is comfortable. But the information of the previous patient is temporal, which is sensitive to age, gender, and past and present medical symptoms. Therefore, a privacy-preserving medical diagnosis is essential. Kruse et al. [7] proposed two methods for classifying EGC signals namely linear branching based and neural network based. These were dependent on homomorphic encryption. Bhadani et al. [3] found a secured Naïve Bayesian decision support system based on the patient. They proposed a homomorphic aggregation scheme based on encryption. Barni et al. [2] developed a framework on a homomorphic encryption scheme as a privacy-preserving technique. This facilitates to realize the classifier models of decision. But these schemes are not suggested due to large computation and communication costs. Liu et al. [10] proposed risk minimization ML problems and two solutions to support dynamic differential privacy. They used dual variable permutation and primal variable perturbation.

6 Big Data Life Cycle

A typical flow of modules involved in the big data life cycle is shown in Fig. 2 particularly in the healthcare domain. Collection of data from various sources, ETL, and registration of them in local host are a primary responsibility [1]. These data are subjected to data transformation, filtering, and classification based on the need for action. Typically, the.*csv* files are read from a cloud environment, and by maintaining privacy and security, patient records are classified using suitable classification techniques [4]. A huge repository of these read data helps in testing future live data after training the model. ML-based prediction algorithms which are proposed

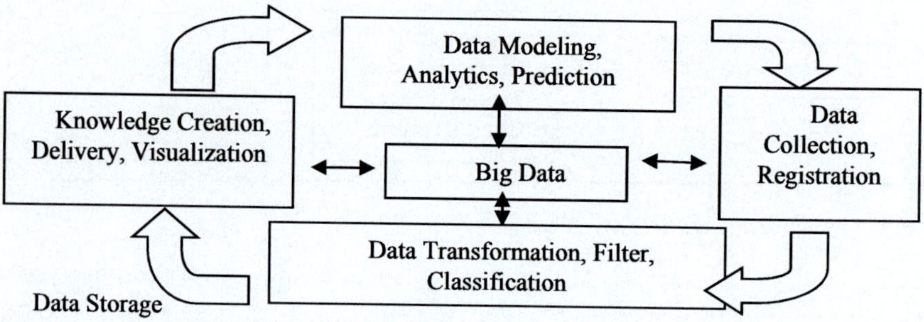

Fig. 2 Big data life cycle for secured healthcare

based on probability theory are helpful in decision making. The obtained results will be customized to the user's need to analyze and predict future treatment actions. Throughout the process of this life cycle, data storage plays a significant role in supplying required information from various sources. Several computational techniques and tools are heavily dependent on historical data for the success of the developed model. Many research articles which are published earlier are suggesting the same flow of modules in any typical big data model [15]. In the past decade, several researchers have confirmed the same flow in the healthcare domain.

7 Cloud-Based Framework with Image Processing

In a typical scenario, most of the patient's records are at high risk if security issues are not addressed. A multi-level security cover is proposed by [14] which covers image-based patient records and their security. The majority of health-related records are in image form. Some of the image processing techniques including the Convolutional Neural Network (CNN) are been used in recent times. Particularly with image data, CNN provides promising results in securing sensitive data. As shown in Fig. 3 above, the cloud dishes out a huge repository of patient's records for processing and decision making. These records are scrutinized and processed as per the requirements of the treatment.

Hospitals with radiology laboratories mainly concentrate on lab records which are images. These images require fine-tuned reading and decision making. Many ML-based prediction algorithms use probability theory to predict the most probable causes from images and help in proper decision making. In most cases, the classical architecture of cloud computing usually exposes customers' data to various and serious security threats. In this regard, a secured environment for data processing is proposed with components such as Client, *CloudSec,* and Cloud provider. *CloudSec* component encrypts all health data through HTTPS/SSL protocol and uses a segmentation approach to keep medical images in a secured manner.

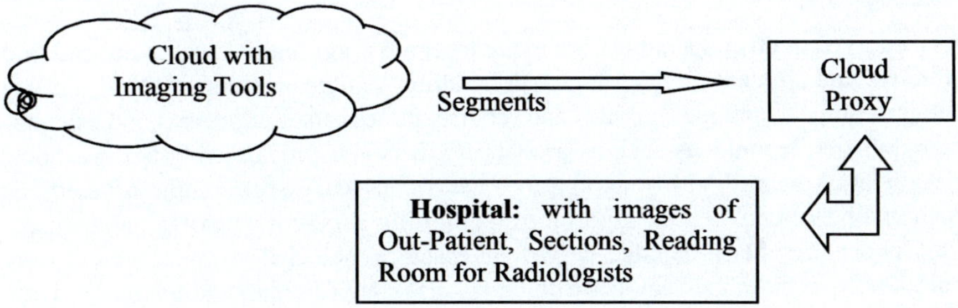

Fig. 3 Cloud-based framework for healthcare domain using image processing

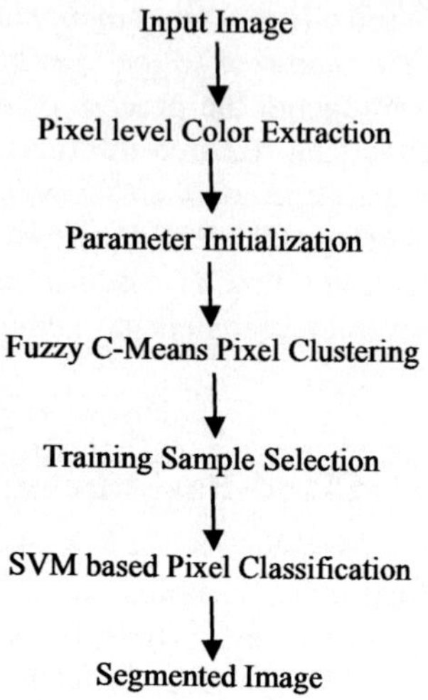

Fig. 4 Data protection through image processing

As shown in Fig. 4, images are processed to obtain segmented forms for better decision making. Fuzzy C-means is a computational technique under soft computing, which helps in identifying and segmenting various features of colored images of the input file. As an alternative to Fuzzy C-means, one can use Artificial Neural Network (ANN) as well as Genetic Algorithms (GA). All these three techniques are alternative to each other under soft computing. Initially, pixel-level decomposition of the input image is required for decomposing various segments of the image. As mentioned earlier, using a bias in ANN, the desired outcomes can be expected by running the model for several epochs. Both the training and testing modules are subjected to preprocessed images. The segmented images provide clarity in decision making for further treatment on the patients.

8 Conclusion

As discussed in this paper, there are many research opportunities to explore and find effective and efficient methodologies in handling healthcare records. Using various computational techniques, privacy and security issues can be addressed in the healthcare domain. Techniques such as probability theory to predict using ML methods, classification using SVM, mapping to trained data using soft computing methods such as fuzzy logic, genetic algorithms, and artificial neural networks can be used. This paper presents a detailed survey on research outcomes from the previously published papers on the healthcare domain. Image-based segmentation and decision making thereon are considered to be one of the most secured data analyses from this survey. We have seen many cloud-based services in a huge range of applications. But, healthcare seems to be one of the challenging and emerging areas, where the application of the latest computational techniques is essential. Securing privacy and security in such a vast domain of records increases the complexities in solving problems. This paper attempts to address these issues in a comprehensive manner.

References

1. Abouelmehdi, K., Beni-Hessane, A., Khaloufi, H.: Big healthcare data: preserving security and privacy. J. Big Data. **5**(1) 2018
2. Barni, M., Failla, P., Lazzeretti, R.: Efficient privacy-preserving classification of ECG signals. In: First IEEE International Workshop on, Information Forensics and Security, et al.: WIFS 2009, IEEE, pp. 91–95 (2009)
3. Bhadani, A.K., Jothimani, D.: Big data: challenges, opportunities and realities. In: Singh, M.K., Kumar, D.G. (eds.) Effective Big Data Management and Opportunities for Implementation, pp. 1–24 (2016)
4. Bost, R., Popa, R.A., Tu, S., Goldwasser, S.: Machine learning classification over encrypted data. NDSS (2015)
5. Gandomi, A., Haider, M.: Beyond the hype: big data concepts, methods, and analytics. Int. J. Inf. Manage. **35**(2), 137–144 (2015)
6. Kaur, P., Sharmab, M., Mittal, M.: Big data and machine learning based secure healthcare framework. In: International Conference on Computational Intelligence and Data Science (ICCIDS 2018), Procedia Comput. Sci. **132**, 1049–1059 (2018)
7. Kruse, C.S., Goswami, R., Raval, Y., Marawi, S.: Challenges and opportunities of big data in healthcare: a systematic review. JMIR Med Inf. **4**(4) (2016)
8. Lee, I.: Big data: dimensions, evolution, impact, and challenges. Bus. Horiz. **60**(3), 293–303 (2017)
9. Li, P., Li, J., Huang, Z., Gao, C.-Z., Chen, W.-B., Chen, K.: Privacy-preserving outsourced classification in cloud computing. Cluster Comput. **21**, 277–286 (2018)
10. Liu, X., Lu, R., Ma, J., et al.: Privacy-preserving patient-centric clinical decision support system on naive Bayesian classification. IEEE J. Biomed. Health Inf. **20**(2), 655–668 (2016)
11. Marwan, M., Kartit, A., Ouahmane, H.: Security enhancement in healthcare cloud using machine learning. In: The First International Conference On Intelligent Computing in Data Sciences. Procedia Comput. Sci. **127**, 388–397 (2018)
12. Ozgur, C., Kleckner, M., Li, Y: Selection of statistical software for solving big data problems: a guide for businesses, students, and universities. Sage Open. 1–12 (2015)
13. Raj, J.S.: A novel information processing in IoT based real time health care monitoring system. J. Electron. **2**(03), 188–196 (2020)
14. Wang, H.: IoT based clinical sensor data management and transfer using blockchain technology. J. ISMAC **2**(03), 154–159 (2020)
15. Zhang, T., Zhu, Q.: Dynamic differential privacy for ADMMbased distributed classification learning. IEEE Trans. Inf. Forensics Secur. **12**(1), 172–187 (2017)

Performance Analysis of Different Classification Algorithms for Bank Loan Sectors

K. Hemachandran, Raul V. Rodriguez, Rajat Toshniwal, Mohammed Junaid, and Laxmi shaw

Abstract The proposed research work aims to develop a novel algorithm to make predictions for various financial institutions to safeguard themselves from fraudsters, and at the same time to ease the pre-sanction process for availing loan and its related verification process. Currently, in the post pandemic world, the proposed algorithm is very essential for the financial institutions, as the rate of loan procurement by individuals has been unprecedently increased and at the same time the chances of loan default has also been increased. For all these cases, first, the bank requires to analyze their Credit Information Bureau India Limited [CIBIL] score and check whether they had done loan repayments within an appropriate time period. Data mining plays a key role to solve such problems and also different algorithms are available in the machine learning domain. Among that, K-nearest neighbor, decision tree, support vector machine, and logistic regression models are taken into consideration for performing data classification with good accuracy. In the present work, the performance of each algorithm is analyzed. The experiments were carried out using python. The accuracy of classifiers will be analyzed by using the following metrics such as Jaccard index, F1-score, and Log loss. This helps to find the best algorithm for classification and the potential of customer, which is much higher than the data mining classification algorithm, and thus it proves to be very helpful for bank officers for sanctioning loan.

K. Hemachandran (✉) · R. V. Rodriguez · R. Toshniwal · M. Junaid
Woxsen School of Business, Woxsen University, Hyderabad, India
e-mail: hemachandran.k@woxsen.edu.in

R. V. Rodriguez
e-mail: raul.rodriguez@woxsen.edu.in

R. Toshniwal
e-mail: rajat.toshniwal@woxsen.edu.in

M. Junaid
e-mail: mohamed.junaid@pgdm.woxsen.edu.in

L. shaw
Department of ECE, CBIT, Hyderabad, India

J. S. Raj et al. (eds.), *Intelligent Sustainable Systems*, Lecture Notes in Networks and Systems 213, https://doi.org/10.1007/978-981-16-2422-3_16

Keywords Decision tree · F1 score · Jaccard similarity score · KNN · Logistic regression · Log loss · Metrics · Support vector machine

1 Introduction

This paper is an explanation for the utilization of various statistical tools in python for the prediction of EMI payback or payment default made by an individual, when the loan is given to them [1, 3]. In addition, this paper describes the usage and provides the details of an algorithm that has been developed, and the dataset is then tested, and further the prediction details are evaluated with the level of accuracy obtained by using different statistical tools like K-nearest neighbor, decision tree, support vector machine, and logistic regression. Here, not just the statistical tools, but also the neural networks like support vector machine has been used, and the working of these tools are also described. To test the accuracy of these tools, various metrics such as Jaccard Index, F1-Score, and Log Loss are used [19]. The main use of these metrics is to perform classification based on probabilities.

2 Build Data Model

To test this algorithm, the experimental data was retrieved from the website of UCI, and at the same time the actual inputs that are asked by financial institutions [6, 18] were taken into consideration rather than considering any other inputs to test the accuracy rate of the algorithm, so this could be soon used by the financial institutions. Henceforth, it is required to import the packages shown in Fig. 1 in python for the functioning of the algorithm.

a. Itertools (Itertools is a package in python, which is used to implement the number of iterators building blocks. Iterator is used to create a continuous loop. This helps the algorithm to take the inputs on their own from the dataset)

```
In [41]: import itertools
         import numpy as np
         import matplotlib.pyplot as plt
         from matplotlib.ticker import NullFormatter
         import pandas as pd
         import numpy as np
         import matplotlib.ticker as ticker
         from sklearn import preprocessing
         %matplotlib inline
```

Fig. 1 Importing libraries

b. NumPy (NumPy is a package in python, which is used to perform various mathematical functions and work with arrays. This package helps in working with linear algebra, Fourier transform, and matrices.)
c. Matplotlib (Matplotlib is a package in python, which is used for plotting the graphs. This package helps to ease the data analysis process. Extension of Matplotlib that helps Matplotlib to be used as MATLAB is Pyplot, it is used in creation of 2D graphics.)
d. Pandas (Pandas is a package in python, which is used for performing data analysis and data manipulation. Pandas provide 2D table object in the form of data frame. This package mainly works with the tabular data.)
e. Sklearn (Sklearn is a package in python that provides various supervised and unsupervised algorithms. It is used in classification, regression, clustering, and dimension reduction process and it also has various tools for machine learning)
f. Seaborn (Seaborn is a package in python that provides statistical graphs. It builds on top of Matplotlib and integrates with panda data structure).

The dataset used consists of various inputs like loan amount, loan status, term, effective date, due date, age, education, and gender. Then, the traininging dataset will be loaded with the help of pandas library [13]. The size of the dataset can be checked by using (df.head). The count of the training dataset is 346. The dataset loaded is required to be formatted for date and time using (pd.to_datetime(df['due date'])). The changes made in the dataset can be verified by using (df.head). Now it is required to visualize and pre-process the dataset loaded [9, 12, 26, 27]. Further, it is required to check the count of loan paid off and loan defaulted using (df['loan_status'].value_counts()) [4, 16]. Now, the data should be pre-processed before using the statistical tools. The loan status of the individuals are obatined on the basis of gender and principal using seaborn, and the graphical representation can be used for easy visualization and analysis. Now, it is required to check the loan status of the individuals on the basis of gender and age using seaborn, and in this regard Figs. 2, 3, and 4 shows the graphical representation that can be used for easy visualization and analysis [8].

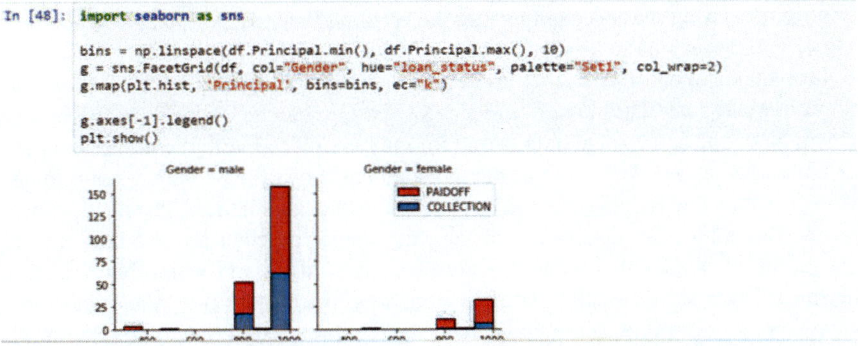

Fig. 2 Data visualizations-principal

```
In [49]: bins = np.linspace(df.age.min(), df.age.max(), 10)
         g = sns.FacetGrid(df, col="Gender", hue="loan_status", palette="Set1", col_wrap=2)
         g.map(plt.hist, 'age', bins=bins, ec="k")

         g.axes[-1].legend()
         plt.show()
```

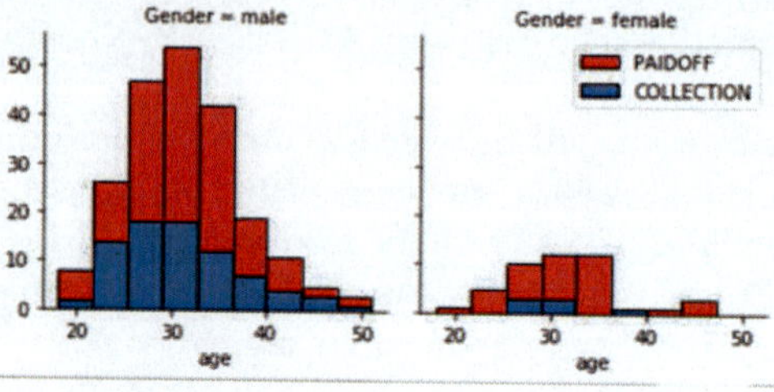

Fig. 3 Data visualization-age perspective

```
In [50]: df['dayofweek'] = df['effective_date'].dt.dayofweek
         bins = np.linspace(df.dayofweek.min(), df.dayofweek.max(), 10)
         g = sns.FacetGrid(df, col="Gender", hue="loan_status", palette="Set1", col_wrap=2)
         g.map(plt.hist, 'dayofweek', bins=bins, ec="k")
         g.axes[-1].legend()
         plt.show()
```

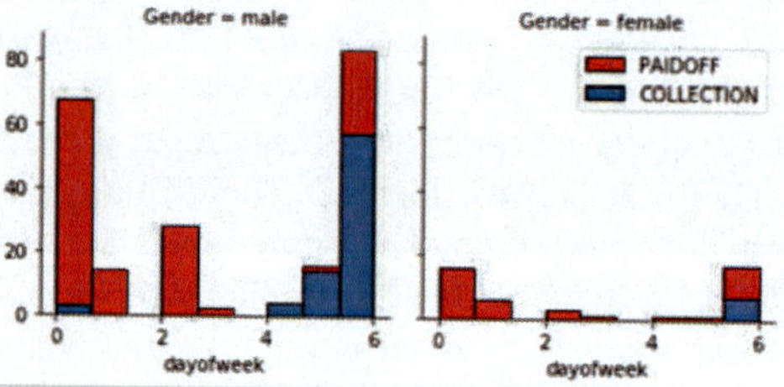

Fig. 4 Data visualization—day of week perspective

Now, the loan status of the individuals are checked on the basis of gender and effective date using seaborn, and further the graphical representation can be used for easy visualization and analysis.

Next, the categorical data such as gender and loan status are required to convert into numerical data. The numerical value of the two can be obtained in the binary format. Now, the categorical data is required to be replaced with the numerical data present in the dataset. The result of the replacement shows the percentage of females paying the loan and default, and males paying the loan and default. The result shows that loan defaulter are mostly male. Various categorical data are available in the dataset, and it is difficult to analyze the categorical data, and when there is large dataset, it becomes difficult to convert each categorical data into numerical data. To reduce this difficulty, "One Hot Encoding" can be used in the conversion of all categorical data into numerical data. This converts the categorical data into binary variables, and append them to the dataset. For attaining accuracy, the large dataset is converted into smaller sets. Here, sets of 5 are created to obtain the accuracy. The variables are required to be assigned for further process. X is termed as the input and Y is termed as the output (Loan Status). Moreover, it is required to normalize the data. In such situation, data standardization gives data zero mean and unit variance. After data standardization, various statistical tools are used to predict the loan status and find out the best suitable model for prediction. Here, the test data cannot be used to check the accuracy, instead the traininging data is splitted into two sets as traininging dataset and test dataset.

3 Classification Algorithms

3.1 Implementation of K-Nearest Neighbor (KNN)

A simple, easy-to-implement supervised machine learning algorithmic rule which can be accustomed to solve classification and regression challenges is the K-Nearest Neighbor [KNN] algorithm. KNN may be a straight forward algorithmic rule that stores out all the cases and classifies new cases supported by a similarity measure (e.g., distance functions) [7, 11, 17]. KNN has been employed by applying math estimation and pattern recognition that are already available from the starting of 1970's as a non-parametric technique. KNN test is performed by using Sklearn. First, the dataset should be splitted into traininging and test data. We can take the traininging data to be 80% of the dataset, and the rest 20% could be used as test data. This process is called training test split, shown in Fig. 5. KNN is used to optimize the K value, where K value is the parameter for performing the KNN function [2]. Now we need to import metrics to evaluate the algorithm. After traininging the dataset, testing of the dataset is to be done to check the accuracy [14, 15, 23]. We can measure the accuracy of both training and test data with the help of metrics, and the metrics used is F1_score shown in Fig. 6. It is statistical analysis of binary classification.

```
In [60]: from sklearn.model_selection import train_test_split
         X_train, X_test, y_train, y_test = train_test_split( X, y, test_size=0.2, random_state=4)
         print ('Train set:', X_train.shape,  y_train.shape)
         print ('Test set:', X_test.shape,  y_test.shape)

         Train set: (276, 8) (276,)
         Test set: (70, 8) (70,)
```

Fig. 5 Splitting of traininging and test set for KNN algorithm

```
In [64]: # accuracy evaluation

         from sklearn import metrics
         print("Train set Accuracy: ", metrics.accuracy_score(y_train, neigh.predict(X_train)))
         print("Test set Accuracy: ", metrics.accuracy_score(y_test, yhat))

         Train set Accuracy:  0.8152173913043478
         Test set Accuracy:  0.7571428571428571
```

Fig. 6 Accuracy of traininging and test set for KNN algorithm

3.2 *Implementation of Decision Tree*

A decision tree could be a flowchart-like structure during which every internal node represents a "test" on associate in Nursing attribute (e.g., whether or not a coin flip comes up heads or tails), every branch represents the result of the check, and every leaf node represents a category label (decision taken when computing all attributes). Decision trees square measure want to solve each classification and regression issues within the variety of trees that may be incrementally updated by reading the dataset into smaller datasets (numerical and categorical), wherever the results square measure delineated within the leaf nodes. Decision Trees area unit is a sort of supervised Machine Learning (that is, you justify what the input is and what the corresponding output is within the coaching data) wherever the info is unendingly split consistently with an exact parameter [21]. We can perform the Decision Tree test using Sklearn. First, we need to split the training dataset into training and test data. We can take the training data to be 70% of the dataset, and the rest 30% could be used as test data. This process is called training test split. Now we need to create a model for decision tree. Here the set is created of size 5. Now we need to import metrics to evaluate the algorithm. After traininging the dataset, testing of the dataset is to be done to check the accuracy shown in Figs. 7 and 8.

```
In [67]: from sklearn.model_selection import train_test_split

         X_trainset, X_testset, y_trainset, y_testset = train_test_split(X, y, test_size=0.3, random_state=3)

In [68]: # modelling
         loanTree = DecisionTreeClassifier(criterion="entropy", max_depth = 6)

         loanTree.fit(X_trainset,y_trainset)

Out[68]: DecisionTreeClassifier(class_weight=None, criterion='entropy', max_depth=6,
                     max_features=None, max_leaf_nodes=None,
                     min_impurity_decrease=0.0, min_impurity_split=None,
                     min_samples_leaf=1, min_samples_split=2,
                     min_weight_fraction_leaf=0.0, presort=False, random_state=None,
                     splitter='best')
```

Fig. 7 Splitting of training and test set for decision tree algorithm

```
In [70]: # Evaluation
         from sklearn import metrics
         print("DecisionTrees's Accuracy: ", metrics.accuracy_score(y_testset, predTree))

         DecisionTrees's Accuracy:  0.7019230769230769
```

Fig. 8 Accuracy–Decision Tree Algorithm

3.3 Implementation of Support Vector Machine (SVM)

The SVM kernel performs by taking a low dimensional input area and transforms it into a better dimensional area, i.e., it converts 2 dimensional to 3 or more dimensional. It is principally helpful in non-linear separation downside. Support vectors area unit is the information point that lies nearest to the choice surface (or hyperplane). They're the information points most tough to classify. They need direct referring to the optimum location. We can perform the SVM test using Sklearn. First, we need to split the training dataset into training and test data. We can take the training data to be 80% of the dataset, and the rest 20% could be used as test data. This process is called training test split. Now we need to create a model for SVM [27]. We use Support Vector Classifier (SVC) for classifying the dataset into various sets. Now we need to import metrics to evaluate the algorithm. After training the dataset, testing of the dataset is to be done to check the accuracy. We can measure the accuracy of both training and test data with the help of metrics, and the metrics used is f1_score shown in Figs. 9 and 10. It is the statistical analysis of binary classification.

```
In [72]: # modelling SVM

         from sklearn import svm
         clf = svm.SVC(kernel='rbf')
         clf.fit(X_train, y_train)

         /opt/conda/envs/Python36/lib/python3.6/site-packages/sklearn/svm/base.py:196: FutureWarning: The default value of gamma will ch
         ange from 'auto' to 'scale' in version 0.22 to account better for unscaled features. Set gamma explicitly to 'auto' or 'scale'
         to avoid this warning.
           "avoid this warning.", FutureWarning)

Out[72]: SVC(C=1.0, cache_size=200, class_weight=None, coef0=0.0,
           decision_function_shape='ovr', degree=3, gamma='auto_deprecated',
           kernel='rbf', max_iter=-1, probability=False, random_state=None,
           shrinking=True, tol=0.001, verbose=False)
```

Fig. 9 Splitting of training and test set for SVM algorithm

```
In [74]: # Evaluation
         from sklearn.metrics import f1_score
         f1_score(y_test, yhat, average='weighted')

Out[74]: 0.7540496161185816
```

Fig. 10 Accuracy–support vector machine algorithm

3.4 *Implementation of Logistic Regression*

Logistic regression may be an applied mathematics model that in its basic kind uses a provision operate to model a binary variable, though more advanced extensions exist. In regression analysis, logistic regression (or logit regression) estimates the parameters of a logistic model (a variety of binary regression). Logistic regression analysis is employed to look at the association of (categorical or continuous) independent variable(s) with one dependent variable. This can be in distinction to linear regression analysis during which the variable may be a continuous variable. Logistic Regression test is performed by using Sklearn. First, the training dataset should be splitted into training and test data. The training data utilizes 80% of the dataset, and the rest 20% could be used as test data. This process is called training test split. Here, it is intended to use confusion matrix that could be a tabular outline of a variety of correct and incorrect predictions that are created with the classifier [20] accuracy of both training and test data with the help of metrics, and the metrics used here is jaccard_similarity_score shown in Figs. 11 and 12. It measures the similarity between the different finite sample sets.

Now, after traininging the algorithm with training and test data, with different statistical features, the actual test dataset is used for testing the accuracy of the algorithm, and finding out the best statistical feature that can be used for predictions.

```
In [75]: # modelling

         from sklearn.linear_model import LogisticRegression
         from sklearn.metrics import confusion_matrix
         LR = LogisticRegression(C=0.01, solver='liblinear').fit(X_train,y_train)
         LR

Out[75]: LogisticRegression(C=0.01, class_weight=None, dual=False, fit_intercept=True,
                   intercept_scaling=1, max_iter=100, multi_class='warn',
                   n_jobs=None, penalty='l2', random_state=None, solver='liblinear',
                   tol=0.0001, verbose=0, warm_start=False)
```

Fig. 11 Splitting of training and test set for logistic regression algorithm

```
In [77]: # Evaluation
         from sklearn.metrics import jaccard_similarity_score
         jaccard_similarity_score(y_test, yhat)

Out[77]: 0.6857142857142857
```

Fig. 12 Accuracy–logistic regression algorithm

4 Performance Metrics

4.1 *Jaccard Similarity Score*

The Jaccard Index is conjointly referred to as the Jaccard similarity coefficient, maybe a statistics employed in understanding the similarities between sample sets [22]. The live emphasizes the similarity between finite sample sets and is formally made public as a result of the dimensions of the intersection divided by the dimensions of the union of the sample sets. Convolutional Neural Networks, that area unit ordinarily tasked with image identification applications, apply the Jaccard Index measurements as to how of conceptualizing accuracy of object detection [5]. As an example, if a pc vision algorithmic program is tasked with detective work faces from a picture, the Jaccard index is ready to quantify the similarities between the computer's identification of faces and those of the coaching knowledge. The formula to calculate precision and recall are shown in Eqs. 1 and 2.

$$\text{Precision} = \text{True Positive}/(\text{True Positive} + \text{False Positive}) \tag{1}$$

$$\text{Recall} = \text{True Positive}/(\text{True Positive} + \text{False Negative}) \tag{2}$$

4.2 *F1 Score*

In applied math analysis of binary classification, the F-score or F-measure could be a measure of a test's accuracy. It is calculated from the exactness and recall of the check, wherever the exactness is that the range of properly-known positive results divided by the quantity of all positive results, as well as those not known properly, and therefore, the recall is that the range of properly-known positive results divided by the quantity of all samples that ought to are known as positive. The F1 score is the mean value of the exactness and recall shown in Eq. 2. A lot of generic F score applies extra weights, valuing one in all exactness or recall over the opposite.

$$\text{F1 Score} = 2\left(\frac{\text{Precision} \times \text{Recall}}{\text{Precision} + \text{Recall}}\right) \tag{3}$$

4.3 Log Loss

Log loss, aka logistic loss or cross-entropy loss. This can be the loss perform employed in (multinomial) supplying regression and extensions of it like neural networks, outlined because of the negative log-likelihood of a supplying model that returns y_pred chances for its coaching information y_true [10]. The bolder the possibilities, the higher are your Log Loss nearer to zero. It is the measure of uncertainty (you could decide it entropy), thus a low Log Loss means that a low uncertainty/entropy of your model.

Log loss can be determined for each column in the informational index utilizing the Log loss equation Eq. 4.

$$L = (\mathrm{y} \times \log(\mathrm{y}') + (1 - \mathrm{y}) + \log(1 - \mathrm{y}') \quad (4)$$

The condition just estimates how far each anticipated likelihood is from the real name. A normal of the log loss from all the columns gives the perfect incentive for the log loss (Eq. 5).

$$\mathrm{Lloss} = \frac{1}{n} + \sum (\mathrm{y} \times \log(\mathrm{y}') + (1 + \mathrm{y}) + \log(1 - \mathrm{y}')) \quad (5)$$

A good model should have smaller log loss value.

Rather than testing the statistical features over one metrics, we will use three metrics to increase the level of accuracy. For performing the test, we need to import function (jaccard_similarity_score, f1_score, and log_loss) from Sklearn. We need to load the test dataset for evaluation of the algorithm. We need to change the categorical data in test dataset to numerical data using "One Hot Encoding". We need to assign the variables X to all the inputs, and Y to the output (Loan Status) that is to be predicted [24]. Now we can test the accuracy of the various statistical features with the various metrics imported.

5 Result Analysis

Here, the metrics Log_loss can only be used on logistic regression, while the others can be used on all the statistical features. While performing the evaluation on various metrics, we find difference in the level of accuracy by each metrics. Here are the results of the various statistical features and the metrics evaluation of those features.

It can be observed that, SVM has the highest accuracy with both the metrics, KNN has the least accuracy with Jaccard metrics, and Logistic Regression has least accuracy with F1 Score metrics.

SVM is the proposed statistical feature for loan prediction evaluation, having the highest accuracy of 79.6% and 75.8% with the prediction of loan status (Table 1).

Table 1 Accuracy result

Algorithm	Jaccard	F1 Score	LogLoss
K-nearest neighbor	0.703704	0.686067	N/A
Decision tree	0.722222	0.736682	N/A
Support vector machine	0.796296	0.758350	N/A
Logistic regression	0.740741	0.660427	0.567215

6 Conclusion

In this paper, the performance of different classificication algorithms were anlyzed based on three performance metrics Jaccard, F1 score, and log loss. The proposed algorithms are used to predict the loan repayment capability behavior of a customer in a cost effective way. The bank officers need to determine whether to approve loan for the applicant or not. This proposed methodology will protect the bank from further misuse, fraud applications, etc., by identifying the customers, whose repayment capability status is risky, especially in the banking sector. The experiment has proved that the classification accuracy of SVM is high compared with other classification algorithms.

References

1. Aafer, Y., Du, W., Yin, H.: DroidAPIMiner: mining API-Level features for robust malware detection in android. In: Security and Privacy in Communication Networks, pp 86–103 (2013)
2. Apilado, V.P., Waner, D.C., Dauten, I.J.: Evaluative techniques in consumer finance—experimental results and policy implications for financial institutions. J. Financ. Quant. Anal. **9**(2), 275–283 (1974)
3. Arun, K., Ishan, G., Sanmeet, K.: Loan approval prediction based on machine learning approch. IOSR J. Comput. Eng. NCRTCSIT **2016**, 18–21 (2016)
4. Boyle, M., Crook, J.N., Hamilton, R., Thomas, L.C.: Methods for credit scaling applied to slow payers. In: Proc. Conf Credit Scoring and Credit Control (eds L. C. Thomas, I. N.Crook and D. B. Edelman), pp. 75–90 (1992)
5. Chambers, J.M.: Computational Methods for Data Analysis. Applied Statistics, Wiley, **1**(2), 1–10 (1977)
6. Chen, M.C., Huang, S.H.: Credit scoring and rejected instances reassigning through evolutionary computation techniques. Expert Syst. Appl. **24**(4), 433–441 (2003)
7. Hand, D.J., Vinciotti, V.: Choosing k for two-class nearest neighbor classifiers with unbalanced classes. Pattern Recognit. Lett. **24**(9–10), 1555–1562 (2003)
8. Hanumantha Rao, K., Srinivas, G., Damodhar, A., Vikar Krishna, M.: Implementation of anomaly detection technique using machine learning algorithms. Int. J. Comput. Sci. Telecommun. **2**(3), 25–30 (2011)
9. He, Y., Han, J., Zeng, S.: Classification algorithm based on improved ID3 in bank loan application. Inf. Eng. Appl. 1124–1130 (2012)
10. Huang, L., Zhou, C.G., Zhou, Y.-Q., Wang, Z.: Research on data mining algorithms for automotive customers' behavior prediction problem. In: 2008 Seventh International Conference on Machine Learning and Applications (2008). doi:https://doi.org/10.1109/ICMLA.2008.23

11. Islam, M.J., Wu, Q.M.J., Ahmadi, M., Sid-Ahmed, M.A.: Investigating the performance of Naive-Bayes classifiers and K- nearest neighbor classifiers. In: International Conference on Convergence Information Technology (ICCIT 2007), pp. 1541–1546 (2007)
12. Kaishe, Q., Wenli, C., Junhong, W.: The ID3 algorithm an improved algorithm. Comput. Eng. Appl. **39**(25), 104–107 (2003)
13. Keerthi, S., Gilbert, E.: Convergence of a generalized SMO algorithm for SVM classifier design. Mach. Learn. **46**, 351–360 (2002)
14. Li, F.: The hybrid credit scoring strategies based on KNN classifier. In: Sixth International Conference on Fuzzy Systems and Knowledge Discovery, Tianjin, 2009, pp. 330–334 (2009)
15. Marinakis, Y., Marinaki, M., Doumpos, M., et al.: Optimization of nearest neighbor classifiers via metaheuristic algorithms for credit risk assessment. J. Global Optim. **42**, 279–293 (2008)
16. Otgler, Y.E.: A credit scoring model for commercial loans. J. Money Credit Bank. **2**(4), 435–445 (1970)
17. Paredes, R., Vidal, E.: A class-dependent weighted dissimilarity measure for nearest neighbor classification problems. Pattern Recogn. Lett. **21**(12), 1027–1036 (2000)
18. Ram, B., Rama Satish, A.: Improved of K-nearest neighbor techniques in credit scoring. Int. J. Dev. Comput. Sci. Technol. **1**(2), (2013)
19. Sahay, B.S., Ranjan, J.: Real time business intelligence in supply chain analytics. Inf. Manage. Comput. Secur. **16**(1), 28–48 (2008)
20. Sutrisno, H., Halim, S.: Credit scoring refinement using optimized logistic regression. In: 2017 International Conference on Soft Computing, Intelligent System and Information Technology (ICSIIT), Denpasar, pp. 26–31 (2017)
21. Wei, G., Yingjie, S., Mu, Y.X.: Commercial bank credit risk evaluation method based on decision tree algorithm. In: 2015 Seventh International Conference on Measuring Technology and Mechatronics Automation, Nanchang, pp. 285–288 (2015)
22. White, C.: The role of business intelligence in knowledge management. Bus. Intell. Network. (2005)
23. Xia, Li.: ID3 classification algorithm application in bank customers erosion. J. Comput. Technol. Dev. **19**(3), (2009)
24. You, H.: A knowledge management approach for real-time business intelligence. In: 2nd International Workshop on Intelligent Systems and Applications (2010). doi:https://doi.org/10.1109/IWISA.2010.5473385
25. Zhang, X., Zhou, Z.: Credit Scoring model based on kernel density estimation and support vector machine for group feature selection. In: 2018 International Conference on Advances in Computing, Communications and Informatics (ICACCI), Bangalore, pp. 1829–1836 (2018)
26. Zhu, H., Zhong, Y.: Based on improved ID3 information gain feature selection method. Comput. Eng. **36**(8), (2010)
27. Zou, Y., Fan, C.: Based on the attribute importance ID3 algorithm. Comput. Appl. **28**, 145–149 (2008)

Universal Shift Register Designed at Low Supply Voltages in 20 nm FinFET Using Multiplexer

Rajeev Ratna Vallabhuni, Jujavarapu Sravana, Chandra Shaker Pittala, Mikkili Divya, B. M. S. Rani, and Vallabhuni Vijay

Abstract Shift registers are utilized in personal computer systems as an element of ability, including RAM and numerous types of registers. Besides, automatic framework tasks including splitting, duplicating, and so forth are used to convert the identical information into sequential data or vice versa. This article provides a comprehensive shift register that can play a consistent and consistent shift activity. When the device limit is reduced after direct guidance, the short channel acts as power loss, surface scattering, and sinking speed. The recently planned relocation register uses CMOS, which will consume more power. The proposed registry, which takes advantage of the main innovations of FinFET, has the power to eliminate or alleviate the difficulties mentioned above. Relative performance studies are completed more rapidly than other standard schemes, taking into account necessary performance measures such as power delay product (PDP) and element energy delay product (EDP) measurements.

Keywords FinFET · Flip-flops · Multiplexers · Sequential circuits · Shift registers

R. R. Vallabhuni
Bayview Asset Management, LLC, Coral Gables, FL, USA
e-mail: rajeevratna@ieee.org

J. Sravana · V. Vijay (✉)
Department of Electronics and Communication Engineering, Institute of Aeronautical Engineering, Hyderabad 500043, India
e-mail: v.vijay@iare.ac.in

C. S. Pittala
Department of Electronics and Communication Engineering, MLR Institute of Technology, Hyderabad 500043, India

M. Divya · B. M. S. Rani
Department of Electronics and Communication Engineering, Vignan's Nirula Institute of Technology and Science for Women, Guntur, AP, India

J. S. Raj et al. (eds.), *Intelligent Sustainable Systems*, Lecture Notes in Networks and Systems 213, https://doi.org/10.1007/978-981-16-2422-3_17

1 Introduction

The shift registers are used to change serial data to parallel data and these shift registers are also used as coding keyboards and some to perform arithmetic operations and counter operations, but the thin film transistor circuit has the drawback of more power consumption that is due to operating voltage and wide circuit area [1–9]. Reports within side the literature consists of target research of the unique forms of shift registers in CMOS technology in terms of design, application, and performance. The technologies available for designing registers for replacement, along with their advantages and disadvantages, are also discussed [10–19].

This paper has an advanced design which uses D flip-flop in 4-bit universal shift register which have S_0 and S_1 inputs to operate the operations of the registers. In a D flip-flop, when input clock is high, the D flip-flop will work and presents the output with a given input. Similarly, to reset the D flip-flop, a "clear" input is used, as shown in schematic Fig. 2, which depicts a D flip-flop with a clear input. Therefore, the D flip-flop reflects the input according to the "clock" and "erase" activated. The switch register receives serial data as input and performs left and right shift operations on it. The switching process is carried out once in each case, i.e., right shift or left shift can be carried out. This is solved by a 4 to 1 multiplexer circuit. There are 4 to 1 multiplexers for each shift register bit [1, 20–23].

In this paper, the first part is an introduction to existing methods, proposed methods and methods defined in the chain, as well as the needs, advantages, and uses of the chain. The second section provides information about FinFET properties and modeling, including FinFET operations, symbols, benefits, and use of shift registers. In addition, this paper discusses the complete details of the proposed circuit design and the implementation of the advanced scheme, which consists of the details of the development of the structure, a brief description of the circuit, the schematic diagram, the truth table, and a complete description of the work. Full details of the results of the advance project simulation are described in the fourth section, which consists of the following subsections. This includes a description of the tools and environment used, the transition response to the proposed schema, simulation diagrams, table results, cumulative benefits, and applications. Finally, the fifth part consists of the final section, which contains the goals and results of the virtuoso. Lastly, references are mentioned.

2 Structural Design and Aspects in FinFETs

The FinFET is a type of metal oxide multipole area impact transistor (MOSFET). In the case of the FinFET, the body is formed via way of means of a thin silicon film wrapped over a conductive conduit. The name FinFET comes from the reality that its structure seems like a string of fins. The tool line duration is determined via way of means of the thickness of the tool. The length line of MOSFET must be between

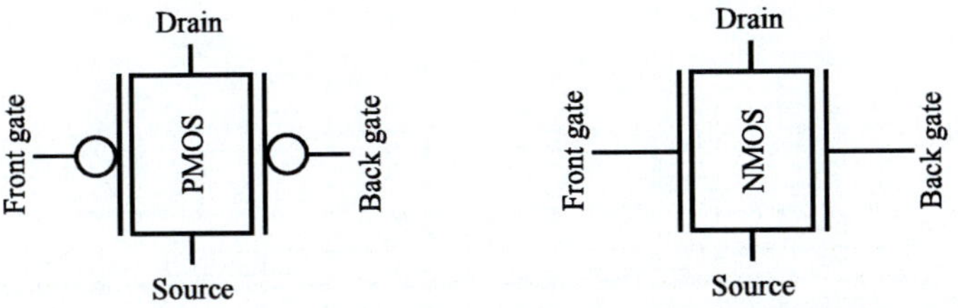

Fig. 1 FinFET symbols

the source connection and drain. In single gate transistor design, non planar dual gate transistor is based on either bulk silicon–on–insulator (SoI) or silicon wafer. There are two types of FinFET's [24–29]:

1. SoI FinFET
2. Bulk FinFET

Since the MOSFET was produced, the device line length has shrunk continuously to make the device compact and fast. The following MOSFET-related parameters underscore the need for smaller and more compact components and explain why MOSFETs are not the right choice. The shorter part of the sealing electrode is called length, and the longer part is called width. As the MOSFET channel length decreases, the short-channel effect increases. The short-channel effect is associated with two physical phenomena [30–35]:

a. The predicament imposed on electron float characteristics within the channel.
b. The change of the edge voltage because of the shortening channel duration (Fig. 1).

The development of nanoscale tool generation overcomes the scaling boundaries of MOSFET, that is, it has advantages like very low voltage, low power energy efficient and reduces leakage power and short line effects energy acquisition from cellular and biomedical applications are used in 6T SRAM cells for better performance and also used in developing high-speed DRAM cells. The FinFET is also used in the development of the highly efficient FinFET flash cells with a gate length of 20 nm technology [36, 37].

3 Design and Realization of 4-Bit Universal Shift Register

Prefabricated shift registers use CMOS which evaporates more power than the proposed scheme. You will design energy-efficient switch registers using superior FinFET technology to reduce the above challenges. Analysis of various parameters such as the effect of temperature on total power, total energy consumption, calculation of average constant power, etc., must be validated using FinFET technology.

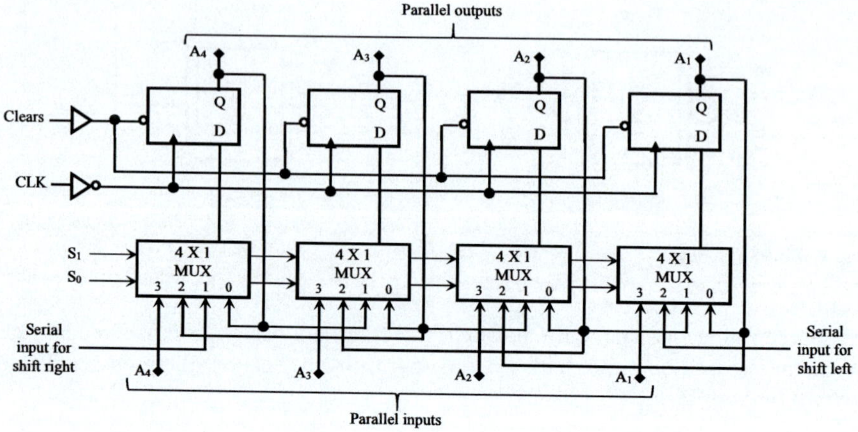

Fig. 2 4-bit universal shift register

The universal shift register in Fig. 2 uses four flip-flops and four 4×1 multiplexers, clock , and clear input. This design has two selection lines S_0 and S_1 which are used to select shift register mode in multiplexers [38, 39].

There are three modes to change the data that is a parallel register can transmit or acquire data in a parallel way. Serial registers can transmit or obtain data serially by using shift left and shift right modes these modes are done by universal shift register, but in case of few applications, obtain data or transmit data serial and parallel way. We use a selector switches S_0 and S_1 to shift the data, Table 1 shows the modes of shifts.

The register does not accept any data in locked mode, that is, (S_1, $S_0 = 0$). The register content is not affected by variations in inputs. As long as the $S_0 = 0$ and $S_1 = 0$, there is no effect on the output. For example, set $A_4A_3A_2A_1 = 1010$ and then set the clock cycles. See Table 2.

Table 1 Modes of shifts

Clock cycle	A_1	A_2	A_3	A_4	I_1	I_2	I_3	I_4
Cycle 1	0	1	0	1	0	0	0	0
Initial value	0	1	0	1	0	0	0	0

Table 2 Shifting operations

Operating mode	S_1	S_2
Parallel loading	1	1
Left shift	1	0
Right shift	0	1
lock	0	0

Table 3 Cycles of shifting operations

Clock mode	Right shift mode	I_4	I_3	I_2	I_1
	Initial value	0	0	0	0
Cycle 7	0	0	0	1	0
Cycle 6	0	0	1	0	0
Cycle 5	1	1	0	0	1
Cycle 4	0	0	0	1	1
Cycle 3	0	0	1	1	0
Cycle 2	1	1	1	0	0
Cycle 1	1	1	0	0	0

Table 4 Mode of shifts

Clock mode	I_1	I_2	I_3	I_4	A_1	A_2	A_3	A_4
Cycle 1	0	1	0	1	0	1	0	1
Initial value	0	0	0	0	0	1	0	1

For the right shift ($S_1S_2 = 01$), mode inputs are taken from I_4 to I_1. You can take this component with the aid of using setting the value of the right shift transfer in line with the sequence 1100100 as you cycle the clock as proven in Table 3. See that the signals circulate from I_4 to I_1. In the shift-left mode ($S_1S_0 = 10$), the register works in a comparable way, besides that the signals circulate from Q_0 to Q_3.

In parallel operation for select lines $S_1 = 1$ and $S_0 = 1$, data is taken from inputs A_1-A_4. For values of $A_4 = 1$, $A_3 = 0$, $A2 = 1$, and $A_1 = 0$ will results $I_4I_3I_2I_2 = 1010$ after cycling the clock as shown in Table 4.

The universal shift register is an integrated logic circuit that works in three different data transfer modes. Parallel registers can send or receive data in parallel. Serial registers can send or receive data one by one or by shifting it left or right. However, in some applications, universal shift registers receive data one by one and send them in parallel.

Inputs A_0, A_1, A_2, and A_3 are connected via port 11, so inputs A_0, A_1, A_2, and A_3 can only be accessed if $S_1S_0 = 11$. Feedback I_0, I_1, I_2, and I_3 occurs via port 00 If $S_1S_0 = 00$, output Q from trigger D is returned to the trigger input, leading to a change in register content. Port 01 is used for right shift. In mode $S_1S_0 = 01$ only port 01 is active and takes its value from the previous larger trigger and transfers it to the trigger, which is connected to the 4×1 mux output. Finally, port 10 is used for left shift. Since this is the only active port when $S_1S_0 = 10$, it assigns the output of a less significant trigger to the one connected to its 4×1 mux output. As a result of this interconnection model, where each register block is an exact copy of another block, the selector switch can regulate the behavior of all multiplexers at the same time. We call this combination of behavior the universal mode register behavior [37–41].

4 Simulation Results of 4-Bit Universal Shift Register

The short-channel effect of the shift register is reduced. The shift register has low power consumption. The main applications of the registers are: they are used in digital electronic devices like computers as temporary data storage, data transfer applications, data manipulation, and also used as a counter (Fig. 3).

The Fig. 4 shows the graph between Vdd versus delay. And x-axis shows as vdd and y-axis shows delay.

The Fig. 5 shows the graph between Vdd versus power. And x-axis shows as vdd and y-axis shows power.

The Fig. 6 shows the graph between vdd versus power delay product (PDP), and x-axis shows as Vdd and y-axis shows PDP.

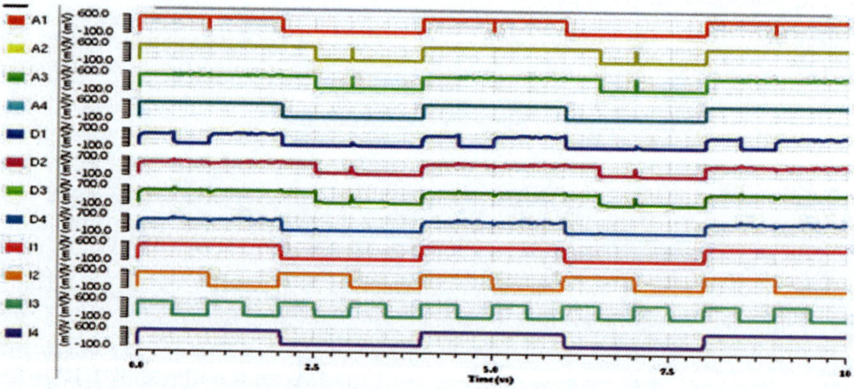

Fig. 3 Transient response of the proposed circuit

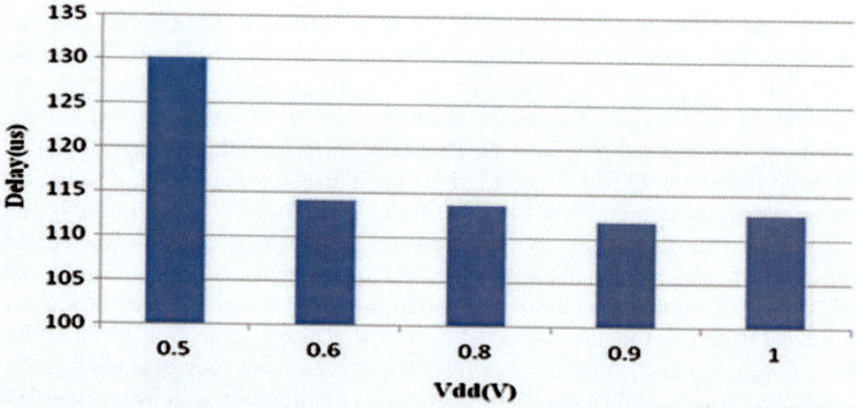

Fig. 4 Vdd versus Delay

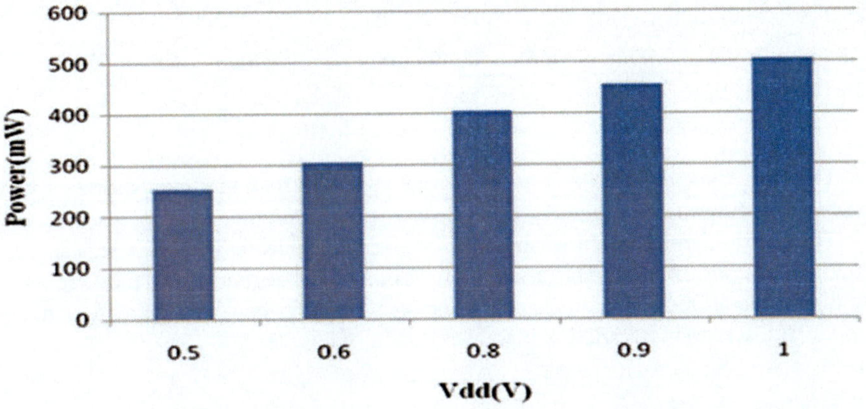

Fig. 5 Vdd versus Power consumption

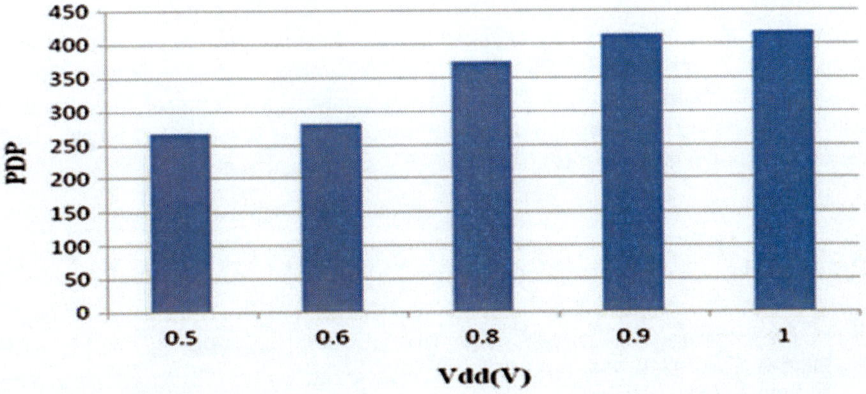

Fig. 6 Vdd versus PDP

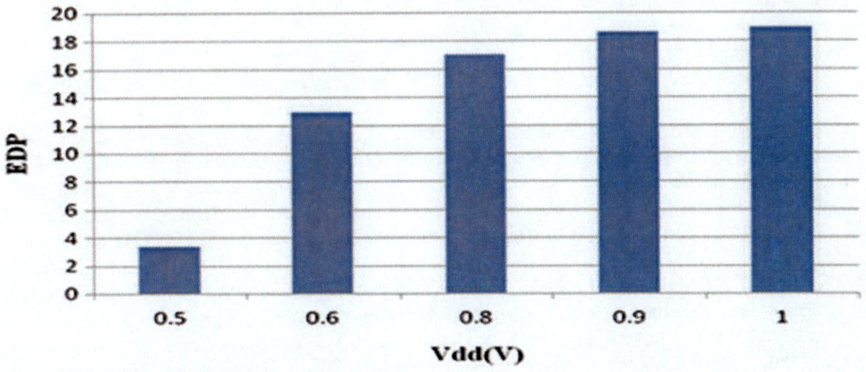

Fig. 7 Vdd versus EDP (FinFET)

Table 5 Performance metrics

Device	Power (mW)	Delay (nS)	PDP (nJ)	EDP ($\times 10^{-18}$Js)
FinFET based USR	454.4	5.57	2.52	14.1

The Fig. 7 shows the graph between vdd versus EDP, and x-axis shows as vdd and y-axis shows EDP (Table 5).

The short-channel effect of the shift register is reduced. The shift register has low power consumption. The main applications of the registers are: they are used in digital electronic devices like computers as temporary data storage, data transfer applications, data manipulation, and also used as a counter.

5 Conclusion

A universal shift register using D flip-flops and 4:1 multiplexers is designed and simulated in 20 nm FinFET technology. The 4-bit universal shift registers are used to added features like high speed, low power parameter have been calculated in this paper. Total power consumption of the proposed model is 454.4 mW. The proposed circuit has shown great results above the 0.5 V supply voltage.

References

1. Kim, Y.B.: Challenges for nanoscale MOSFETs and emerging nanoelectronics. Trans. Electr. Electron. Mater. **11**(3), 93–105 (2010)
2. Krishna, V.V.S.V., Monisha, A., Sadulla, Sk., Prathiba, J.: Design and implementation of an automatic beverages vending machine and its performance evaluation using Xilinx ISE and Cadence. In: 2013 Fourth International Conference on Computing, Communications and Networking Technologies (ICCCNT), pp. 1–6. IEEE (2013)
3. Ratna, V.R., Saritha, M., Saipreethi, N., Vijay, V., Pittala, C.S., Divya, M., Sadulla, S.: High speed energy efficient multiplier using 20 nm FinFET technology. In: Proceedings of the International Conference on IoT Based Control Networks and Intelligent Systems (ICICNIS 2020), pp. 1–8 (2020)
4. Khadir, M., Chaitanya, K., Sushma, S., Preethi, V., Vijay, V.: Design of carry select adder based on a compact carry look ahead unit using 18 nm Finfet technology. J. Crit. Rev. **7**(6), 1164–1171 (2020)
5. Vijay, V., Pittala, C.S., Siva Nagaraju, V., China Venkateswarlu, S., Sadulla, S.: High performance 2:1, 4:1 and 8:1 binary and ternary multiplexer realization using CNTFET technology. J. Crit. Rev. **7**(6), 1159–1163 (2020)
6. Kurra, A.K., Sadulla, S.: Analysis of physical unclonable functions (PUFS) for secure key generation on smartcard chips. J. Adv. Res. Dyn. Control Syst. **9**, 1735–1745 (2017)
7. Vallabhuni, R.R., Koteswaramma, K.C., Sadgurbabu, B., Gowthamireddy, A.: Comparative validation of SRAM cells designed using 18 nm FinFET for memory storing applications. In: Proceedings of the 2nd International Conference on IoT, Social, Mobile, Analytics & Cloud in Computational Vision & Bio-Engineering (ISMAC-CVB 2020), pp. 1–10 (2020)

8. Seo, J., Song, S.-J., Kim, D., Nam, H.: Robust low power DC-type shift register circuit capable of compensating threshold voltage shift of oxide TFTs. Displays (2017)
9. Vijay, V.: Second generation Differential Current Conveyor (DCCII) and its applications. Vignan's Foundation for Science, Technology & Research (Deemed to be University), Guntur (2017)
10. Purkayastha, T., De, D., Chattopadhyay, T.: Universal shift register implementation using quantum dot cellular automata. Ain Shams Eng. J. (2016)
11. Buynoski, M.S., An, J.X., Wang, H., Yu, B., Advanced micro devices Inc. Double spacer FinFET formation. U.S. Patent 6,709,982 (2004)
12. Vallabhuni, R.R., Lakshmanachari, S., Avanthi, G., Vijay, V.: Smart cart shopping system with an RFID interface for human assistance. In: Proceedings of the Third International Conference on Intelligent Sustainable Systems [ICISS 2020], Palladam, India, 4–5 December 2020, pp. 497–501 (2020)
13. Vijay, V., Siva Nagaraju, V., Sai Greeshma, M., Revanth Reddy, B., Suresh Kumar, U., Surekha, C.: A Simple and Enhanced Low-Light Image Enhancement Process Using Effective Illumination Mapping Approach. Lecture Notes in Computational Vision and Biomechanics, pp. 975–984. Springer, Cham, Switzerland (2019)
14. Pittala, C.S., Parameswaran, V., Srikanth, M., Vijay, V., Siva Nagaraju, V., Venkateswarlu, S.C., Shaik, S., Vallabhuni, R.R.: Realization and comparative analysis of thermometer code based 4-bit encoder using 18 nm FinFET technology for analog to digital converters. In: Advanced Intelligent Systems and Computing (AISC) (2020)
15. Vijay, V., Srinivasulu, A.: Grounded resistor and capacitor based square wave generator using CMOS DCCII. In: Proceedings of the 2016 IEEE International Conference on Inventive Computation Technologies (IEEE ICICT-2016), Coimbatore, India, 26–27 August 2016, pp. 79–82 (2016)
16. Pittala, C.S., Karthik, R., Krishna, O.K.S., Bhavana, A.: Design of low threshold full adder cell using CNTFET. Int. J. Appl. Eng. Res. **12**(12), 3411–3415 (2017)
17. Vijay, V., Pittala, C.S., Sadulla, S., Manoja, P., Abhinaya, R., Rachana, M., Nikhil, N.: Design and performance evaluation of energy efficient 8-bit ALU at ultra low supply voltages using FinFET with 20 nm technology. In: Nandan, D., Mohanty, B.K., Kumar, S., Arya, R.K. (eds.) VLSI Architecture for Signal, Speech, and Image Processing. CRC Press (2021)
18. Saritha, P., Vinitha, J., Sravya, S., Vijay, V., Mahesh, E.: 4-bit vedic multiplier with 18 nm FinFET technology. In: 2020 International Conference on Electronics and Sustainable Communication Systems (ICESC), Coimbatore, India, pp. 1079–1084 (2020)
19. Vijay, V., Prathiba, J., Niranjan Reddy, S., Praveen Kumar, P.: A review of the 0.09 μm standard full adders. Int. J. VLSI Des. Commun. Syst. 3(3), 119 (2012)
20. Venkateswarlu, S.C., Kumar, N.U., Kumar, N.S., Karthik, A., Vijay, V.: Implementation of area optimized low power multiplication and accumulation. Int. J. Innov. Technol. Explor. Eng. (IJITEE) **9**(9), 2278–3075 (2019)
21. Singh, H., Meenalakshmi, M., Akashe, S.: Power efficient shift register usingFinFET technology. In: 2016 International Conference on Emerging Trends in Electrical Electronics & Sustainable Energy Systems (ICETEESES) (2016)
22. Sreeja, M., Vijay, V.: A unique approach to provide security for women by using smart device. Eur. J. Mol. Clin. Med. **7**(1), 3669–3683 (2020)
23. Sabbaghi-Nadooshan, R., Kianpour, M.: A novel QCA implementation of MUX-based universal shift register. J. Comput. Electron. **13**(1), 198–210 (2014)
24. Rani, B.M.S., Mikkili, D., Vallabhuni, R.R., Pittala, C.S., Vallabhuni, V., Bobbillapati, S., Bhavani Naga Prasanna, H.: Retinal vascular disease detection from retinal fundus images using machine learning. Australia patent 2020101450
25. Al Mamun, M.S., Mandal, I., Hasanuzzaman, M.: Design of universal shift register using reversible logic. Int. J. Eng. Technol. **2**(9), 1620–1625 (2012)
26. Ashok Babu, P., Siva Nagaraju, V., Mariserla, R., Vallabhuni, R.R.: Realization of 8 x 4 barrel shifter with 4-bit binary to gray converter using FinFET for Low power digital applications. J. Phys.: Conf. Ser. **1714**(1), 012028. https://doi.org/10.1088/1742-6596/1714/1/012028. IOP Publishing

27. Vijay, V., Prathiba, J., Niranjan Reddy, S., Srivalli, Ch., Subbarami Reddy, B.: Performance evaluation of the CMOS Full adders in TDK 90 nm technology. Int. J. Syst. Algorithms Appl. **2**(1), 7 (2012)
28. Siva Nagaraju, V., Ashok Babu, P., Vallabhuni, R.R., Mariserla, R.: Design and implementation of low power 32-bit comparator. In: Proceedings of the International Conference on IoT Based Control Networks and Intelligent Systems (ICICNIS 2020), pp. 1–8 (2020)
29. Vijay, V., Srinivasulu, A.: Tunable resistor and grounded capacitor based square wave generator using CMOS DCCII. Int. J. Control Theory Appl. **8**, 1–11 (2015)
30. Vallabhuni, R.R., Sravya, D.V.L., Sree Shalini, M., Uma Maheshwararao, G.: Design of comparator using 18 nm FinFET technology for analog to digital converters. In: 2020 7th International Conference on Smart Structures and Systems (ICSSS), Chennai, India, 23–24 July 2020, pp. 318–323 (2020)
31. Vijay, V., Srinivasulu, A.: A square wave generator using single CMOS DCCII. In: Proceedings of the 2013 IEEE International SoC Design Conference (IEEE ISoCC-2013), Busan, South Korea, 17–19 November 2013, pp. 322–325 (2013)
32. Shaik, S., Kurra, A.K., Surendar, A.: High secure buffer based physical unclonable functions (PUF's) for device authentication. Telkomnika **17**(1) (2019)
33. Shaik, S., Kurra, A.K., Surendar, A.: Statistical analysis of arbiter physical unclonable functions using reliable and secure transmission gates. Int. J. Simul.–Syst. Sci. Technol. **19**(4) (2018)
34. Vallabhuni, R.R., Shruthi, P., Kavya, G., Siri Chandana, S.: 6Transistor SRAM cell designed using 18 nm FinFET technology. In: Proceedings of the Third International Conference on Intelligent Sustainable Systems [ICISS 2020], Palladam, India, 4–5 December 2020, pp. 1181–1186 (2020)
35. Vijay, V., Srinivasulu, A.: A low power waveform generator using DCCII with grounded capacitor. Int. J. Publ. Sect. Perform. Manag. **5**, 134–145 (2019)
36. Vallabhuni, R.R., Yamini, G., Vinitha, T., Sanath Reddy, S.: Performance analysis: D-Latch modules designed using 18 nm FinFET technology. In: 2020 International Conference on Smart Electronics and Communication (ICOSEC), Tholurpatti, India, 10–12 September 2020, pp. 1171–1176 (2020)
37. Vijay, V., Srinivasulu, A.: A novel square wave generator using second generation differential current conveyor. Arab. J. Sci. Eng. **42**(12), 4983–4990 (2017)
38. Vijay, V., Srinivasulu, A.: A DCCII based square wave generator with grounded capacitor. In: Proceedings of the 2016 IEEE International Conference on Circuits, Power and Computing Technologies (IEEE ICCPCT-2016), Kumaracoil, India, 18–19 March 2016, pp. 1–4 (2016)
39. Vallabhuni, R.R., Sravana, J., Saikumar, M., Sai Sriharsha, M., Roja Rani, D.: An advanced computing architecture for binary to thermometer decoder using 18 nm FinFET. In: 2020 Third International Conference on Smart Systems and Inventive Technology (ICSSIT), Tirunelveli, India, 20–22 August 2020, pp. 510–515 (2020)
40. Nagalakshmi, K., Srinivasulu, A., Ravariu, C., Vijay, V., Krishna, V.V.: A novel simple schmitt trigger circuit using CDTA and its application as a square-triangular waveform generator. J. Mod. Technol. Eng. **3**, 205–216 (2018)
41. Vallabhuni, R.R., Karthik, A., Sai Kumar, CH.V., Varun, B., Veerendra, P., Nayak, S.: Comparative analysis of 8-bit manchester carry chain adder using FinFET at 18 nm technology. In: Proceedings of the Third International Conference on Intelligent Sustainable Systems [ICISS 2020], Palladam, India, 4–5 December 2020, pp. 1158–1162 (2020)

Predicting Graphical User Personality by Facebook Data Mining

V. Mounika, N. Raghavendra Sai, N. Naga Lakshmi, and V. Bhavani

Abstract Flexible applications can benefit from having customer character models as an essential element. The wide range of spaces in which it is found to be crucial are: the type of reformist learning aid, e-commerce, therapeutic associations or referral structures, etc. The most reliable strategy is used to obtain the customer character that is attached to the customer reference. At the end of the day, from one point of view, it is fascinating to make the customer's lifestyle as impalpable as it is in the current conditions, but without going around and coming to terms with the unshakable quality of the model. The power components of the past, which are of human interests, online media turned into a surprising valuation target, provide a valid information to examine and display the customer behavior. For instance, Customer collaboration with Facebook produces computerized print paths, including action logs, "Preferences", and printed and visual information posted by customers, which is extensively collected and extracted for business purposes and is related to an access point valuable information for customers and analysts. Ongoing examinations have shown that, the salient points acquired with this information show critical connections with customer segment, behavior, and psycho-social attributes. This article examines the adaptation of the aroused customer's personality display using the methods for a reasonable arrangement with Facebook data characteristics.

Keywords Graphical distribution · Weekly app activity · Facebook · Graph color clustering technique

---

V. Mounika (✉) · N. Raghavendra Sai · V. Bhavani
Department of Computer Science and Engineering, Koneru Lakshmaiah Education Foundation, Vaddeswaram, AP, India
e-mail: vmounika@kluniversity.in

N. Raghavendra Sai
e-mail: nallagatlaraghavendra@kluniversity.in

V. Bhavani
e-mail: vasanthabhavani@kluniversity.in

N. Naga Lakshmi
Department of Information Technology, Anurag Group of Institutions, Hyderabad, India
e-mail: nagalakshmiit@cvsr.ac.in

J. S. Raj et al. (eds.), *Intelligent Sustainable Systems*, Lecture Notes in Networks and Systems 213, https://doi.org/10.1007/978-981-16-2422-3_18

1 Introduction

Social connections have come to be utilized in a generalized way and standard structures for the dissemination of knowledge also its speaks of brainstorm of social exchanges. Clients' obligations and activities impart considerable awareness of direct encounters, feelings, and lonely interests. Reflecting thereon personality, which surprisingly sees everyone among us, impacts many parts of human leadership, mental cycles, and therefore, the accumulation of sentimental responses, there's an enormous portal hospitable adding new character-based characters to the user interfaces. Modified structures utilized in spaces, e.g., E-learning, data separation, cooperation and web business, could greatly enjoy an interface that changes interaction (e.g., inspirational methodology, introduction styles, interaction modes and proposals) as shown. From the character of the customer [1], moving beyond customer exchanges is simply an early stage in clarifying customer leadership from a personality standpoint. Our guess is that clients are relied on a relative basis to get essential lead models as they interact through virtual housing partnerships, that these models are mined most of the time with a selected extraordinary goal of anticipating a character's tilt of the customer. The aim of characterizing is the evaluation of the connection with the customer within the context of social connects. Customer demonstration is vital when it involves versatile structures. Counting on the goal of every adaptive condition, it'll usually be essential to deal with and use some data on salient points, propensities, needs, practices, etc., of the customer.

As for the credits, which can have different purposes, the character is a fascinating segment. This research work will represent a person's character as an outsized number of salient points that activate an individual's propensity for title; this taste is constant over time and under conditions. Knowing the character of a given individual provides clues on how they might most emphatically answer various conditions. Seeing a customer's character can help understand, for instance, their sensitive needs in various contexts. Therefore, versatile applications can enjoy having customer character models to switch their course within the same way. There's a good range of spaces where this will be valuable, namely care advancement grades, e-learning, online business, clinical thinking, recommendation frameworks, among others. For instance, when it involves web activity, the items that are offered to a customer can change counting on their impulse-seeking character. Another domain of application would be preparation, as character influences how surrogate students learn and use their knowledge and changes the way substitute students perform their homework. Resolved clarifying structures it's imperative to know the surrogate character to propose the foremost sensitive companies in each environment as required (for example, just in case he has only 10 min available and is complicit within the situation, it's going to not be correct to propose him to create a capricious mission, no matter whether it isn't that awful if it's a critical neurotic-anxiety purpose). Therefore, the client's motivating nature can contribute during a general sense to the client's visualization of adaptable structures. The foremost commonly used framework for obtaining this information is to say that the customer completes the reviews.

However, even customers may find this task tedious, as most character reviews invoke multiple requests to react to due request regarding the request for an accurate customer profile [2, 3]. On the one hand, we believe that the customer's character must be obtained within the most discreet way possible, being things because it is, but without compromising the reliability of the model developed. I recently did some evaluations associated with the custom client or the autoloader showing earnings. A few of works have struggled to interpret some characteristics of the character, who would really like to be considered within the part of the bloody ending. Interpersonal organizations are used extensively and became primary modes of knowledge dispersion and facilitators of social collaborations.

Client engagements and exercises provide a meaningful insight into unique behaviors, perceptions, feelings, and interests. Taking under consideration the examination of the prophetic character that particularly identifiers of human behavior perspectives, character with exercises. Ask the character's customer to collaborate with Facebook information. The Facebook register to form different individual preferences for exercises. For the proposed framework, some investigations are proposed on characters contemplated altogether respects, information from Facebook for a prophetic examination, the character of unique individual exercises. Review information for likes, comments on any options that Facebook information awaits for customer determination. It took several factors from Facebook to play a number one role in setting the proper tests for our specific exams. Facebook profiles and exercises provide important insights into the customer's character, revealing genuineness, as against the prophetic examination of the character admired or extended within the frame. Our exploration has two interconnected destinations.

2 Related Work

Information extraction methods recognize a vital limitation in autonomous relationship plans between the character and variety of customer information obtained from various sources. In general, two approaches have been reached for taking into account the likeable recruits of the affiliate customer character. The main approach uses artificial intelligence data collection to collect coal network tested models to speak about the thoughts and character of the customer. The character related to highlights with semantic signs is developed below.

Various faith-breaking and inquiry procedures were used to create skillful character models across the five character assessments using the phonetic highlights from a dataset that included only thousands of reference pieces from early brain research students. [4]. Gadget for etymological evaluation. The revealed accuracies of the classifiers were 54–62% for all credits. Both Naïve Bayes and SMO are used to indicate all the 4 of the 5 character ratings excluding the n-gram highlights from a corpus of individual online journals on the web. Their results elevate the importance of the highlight choice example by expanding the accuracy of the classifiers, yielding 83–93% for the choice of altered elements. We should mention the capabilities of the datasets used in these assessments that correspond distinctly to ours, especially the remarkable correction procedures and the sources from which they were collected.

Not many ratings, that use an incredibly more prominent number of positions from the indistinguishable dataset under our rating, have a truly phenomenal goal of ours to express themselves to analyze associations between character attributes [5] and Facebook enhancement information and the relationship between credits. The likes information is very precious to facebook. These appraisals were not proposed to investigate the rich phonetic examples that happen in the utilization of language on social affiliations, which is, bringing together motivation behind this assessment.

3 Exam Dataset

My Personality experience gave a representation of 250 Facebook client with around 10,000 explanations utilized in our evaluation.

A large number of 725 characteristics associated with the model could be divided into five social events. The characteristics were obtained from both theoretical work and careful work in past evaluations [2]. One segment of these arrays addresses the characteristics obtained from the Facebook dataset, such as explicit demographics, activity limits, claims, and self-centered association data. Ultimately, the social settings, where individuals are acquainted permit us with search for ordinary examples for the five character attributes.

Another characteristic course of action was controlled by applying a large amount of typical language to take care of the procedures, and the unique word representation plans offered an explanation for being relevant to the topic under our assessment.

4 My Personality Information

The character rating reflected in the customer's Facebook exercises adds to a broad level of functionality [3, 6]. Undoubtedly, the most instinctively expected records are quantifiable data on customer activity (e.g., number of inclinations, states, parties, brands, and events). The characteristics of the fragments, for example, age and gender, tended to be taken into account, as their effect is known to be shown in the environment under consideration., for example, thickness, business, and brokerage, provide more information on the client's direct social tool for assessing personality through a couple of assessments.

Some evaluations have highlighted the gigantic correlation between the character and the signals pronounced or adapted by the etymologist (Status, emphasis control, and incredible lexical variety). Additionally, a survey was conducted on language-related words online, visit abbreviations, emojis, obscenities, and jargon to address the specific use of language displayed by social alliance customers.

POS tag parameters

A further phonetic evaluation using the NLTK tool stash-related Brown corpus was profiled to obtain an extended arrangement of the phonetic characteristics, which examines the usual numbers and word ratios to obtain unambiguous highlighting commands (e.g., intensifiers, descriptors, development words, and pronouns). Most of the work has been done to understand how these etymological signs relate to the five character attributes.

Afinn parameters

Afinn's union attributes a much more prominent breadth to character patterns [7], unequivocally, the consideration of character registers identified with toilet-positive minds. Specifically, the normal and summative valence of fiery and non-stacked words with sentiments in the customer's expresses, the control of restless words and the estimation of words with unequivocal valence.

H4Lvd boundaries

General Inquirer is a book information content appraisal doohickey, consolidating 182 classes of word names that join four sources, including HIV-4 and Lasswell, while regarding the word reference utilized in our notable course of action. A word is planned utilizing a strength level scale, which is a mix of a few valence classes, for example, positive versus negative, strong versus fragile, and dynamic versus idle.

5 Results

A few assessments have shown that restricting the features utilized for the arrangement of those that are unequivocally dissected improves the apparently clever execution. Thus, Pearson's offshoot rating was pushed for the whole features plan. Here is a conversation of quantifiably immense prize affiliations (*, ** and *** are utilized to show p-values at 0.05, 0.01, and 0.001 autonomously). As theorized, the main association impact was found for the proportion of accomplices and extroversion (r = 0.4 ***), which is as per the aftereffects of a few experts. An unbelievably precise and valuable pointer of character characteristics was simply the appraisal of the limits focused connection [7]. Monster relationship among transitivity and extraversion (r = 0.29 ***) were found in a path tantamount to transitivity and delicacy (r = 0.21 ***).

The outcomes demonstrated that extraversion is additionally identified with thickness, business, and financier with a coefficient of 0.31 ***, 0.28 ***, and 0.27 ***, independently. This illustration of consistency found for extroversion had no fundamental relationship to the remainder of the character's credits. The likelihood of the showed issue has been dependably identified in other Facebook activity limits [8], being unequivocal the connection among extroversion and proportion of names (r = 0.27 ***), relatively like those that exist between proportions of events and mindfulness. (r = 0.28 *). Sexual bearing was ordinarily expected to be a factor of greatness, anyway a lone association fundamental for neuroticism was discovered (r = 0.22 ***).

Obviously, the phonetic signs keeping an eye on Afinn's energized words were related to character qualities just as responsiveness to experiences; the excess four characters exhibited a relationship with the mean valence of the words (r between 0.14 ** and 0.21 **). A fundamental example was seen between the standards and the normal number of words with valence +2 (r = 0.24 **).

6 Classification Models Assessment

To help our models, we've done a ton of foreplay to investigate how extraordinary they are when review character credits. We tried obvious and acclaimed approach figurings, paying little mind to the Support Vector Machines (SVMs) [9] and their generally valuable and progressed groupings, Simple Minimal Optimization (SMO) and Boost tallies (MultiBoostAB and AdaBoostM1), showed an incredible favorable position forthright of exactness, this route going with the conversation is restricted to them. Without a doubt, the vast majority of the appraisals had lovely degrees of exactness when utilizing all capacities (producers' deliberate precision was set at 75%).

We then proceeded to refine the structures of the technique considering a better study of the characteristics that depend on the Pearson association coefficient. Choosing to interlace features by updating the relationship to one quality and limiting the relationship to multiple features limited the feature ratio to 5–16 and resulted in an accuracy improvement of up to 78%. The improvement was spread across multiple assessments, such as rule-based counts and decision trees a normal result, as these counts seem to work best to ensure the most information gathering capabilities.

There was an amazing partition in those attributes related with each character quality, certifying the intricacy and multidimensional pieces of the subject viable [10]. What was additionally maybe enlightening for our models was the choice of highlights inside Express Party affected too. The holes in the Afinn and H4Lvd gatherings could be clarified by the current semantic piece between word classes.

Appraisal sorting out work ascribed huge execution gains to the SMO classifier as demonstrated in Table 1, improving the consistency of the chose features.

Table 1 Classifier SMO measures for incorporate decision using situating computations

Trait/Measure	TP rate	FP rate	Exactness	Recall	ROC area
OPE	0.948	0.1	0.948	0.948	0.924
CON	0.92	0.08	0.92	0.92	0.92
EXT	0.928	0.092	0.928	0.928	0.918
AGR	0.86	0.144	0.86	0.86	0.858
NEU	0.864	0.162	0.864	0.864	0.851

A nearer assessment of the information uncovered that the best attributes of the coordination shifted among the characteristics entered: Openness to encounter: geospace, number of parts, incorporates, H4Lvd (deficiency, riches, edification individuals, delicate abilities) [11]. Phenomenal Guarantee: H4Lvd (accumulation of influence, comfort to control, dependence on others, deficiency towards others, descriptors of social relations) Extroversion: business, network size, characteristics, modifiers, movement words, H4Lvd (gain of influence, decline like cycle, different relationships).

Affectability: geo-district. Afinn number of words with valence +3/+5, transitivity, trademark, "a", H4Lvd (credits, perceives obvious degrees of attributes, disrespect).

Neuroticism: sexual direction, network size, number of social issues, number of names, "a", H4Lvd (big conclusions, affirmation, gratitude, desire for help, joy of a tendency, confidence, interest, and obligation). We give colossal proportions of information to social objectors like Facebook. Using this document to see your review of yourself and your partners, your names, age, region, sexuality, strong belief, political views is something we offer as compensation for being important to a relationship association like Facebook (Fig. 2) [12]. Such a large number and then talk about: Why does Facebook need this information? How can you deal with the information you accumulate? Is it possible to be part of a relationship association without providing this information? Would you have the ability to monitor it via the security settings? Let the substitutes read the Terms of Service with Facebook and discuss if they think about practically any terms they accept by joining (Fig. 2). What if you're not on Facebook?

Activity: In This Activity, Bar Chart Represent Facebook User Activities like Upload photos, Posted links, Upload Videos and Statuses to represent in a Bar chart in Day wise and Month wise in Drill Manner (Fig. 1).

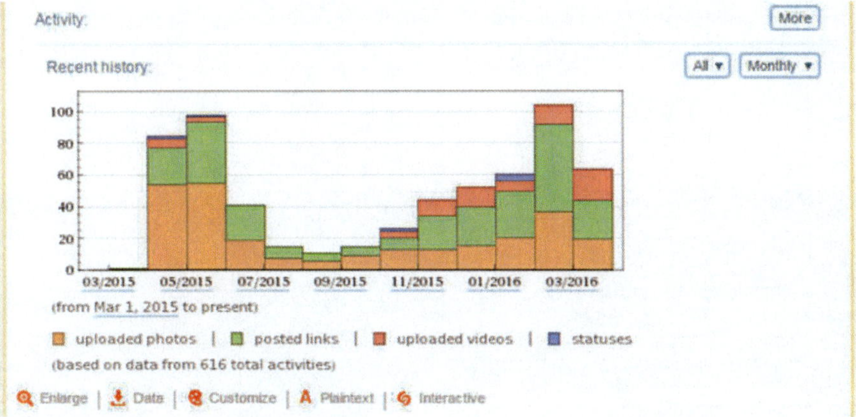

Fig. 1 Activity bar chart of user activities

1. Types: In this activity, pie chart represent face book user activities like upload photos, posted links, upload videos and statuses to represent in a pie chart in day wise and month wise in count manner (Fig. 2).
2. Weekly Distribution: In this activity, weekly distribution chart represent face book user activities like upload photos, posted links, upload videos and statuses to represent in a pie chart in day wise and month wise in distribution manner (Fig. 3).
3. Activity Chart Represent Facebook User Activities like nametests.com, JForFun.Com, MixFun.eu, and Udemy to represent in a chart in time wise manner (Fig. 4).
4. Graph Color Clustering of User Friends Data: This graph represents friend group network cluster by graph color clustering technique (Fig. 5).

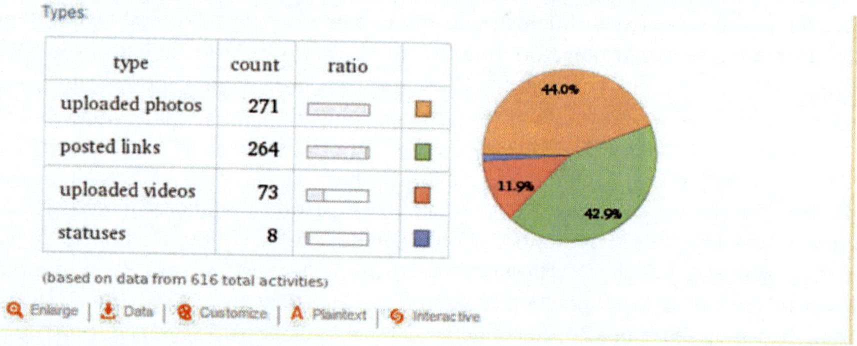

Fig. 2 Activity measure of users

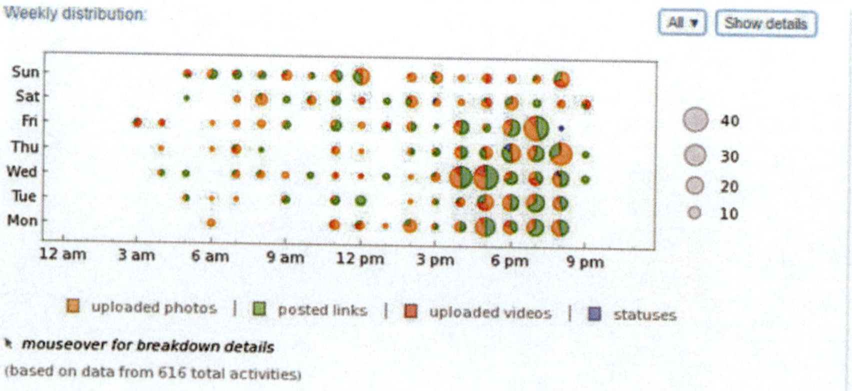

Fig. 3 Weekly distribution of user activities

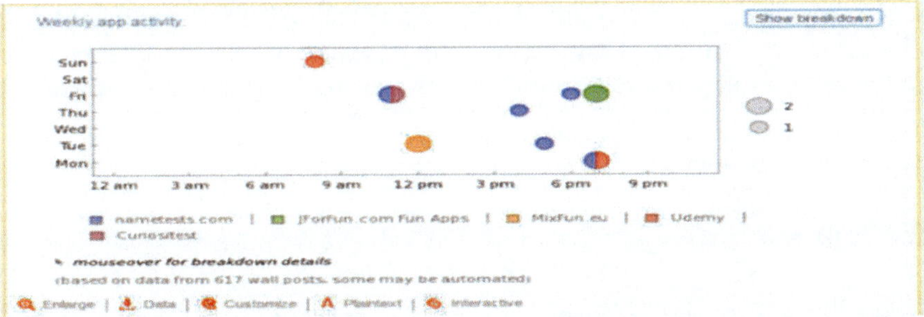

Fig. 4 Activity chart representation

Fig. 5 Graph color clustering of user friends data

7 Conclusion

The continuation of our assessment of Facebook is depends on character is the claim that includes the accuracy of the classifiers may be up to date. Dispensing enthusiastic information from colossal information efforts is just the beginning of our main goal of importance and a possible clarification of the chosen social affiliation style.

Difficulties in delivering dominant and equivalent performance results with broad advantage for more important datasets may require thinking about additional highlights, more wonderful information consisting of search and question strategies. Please encourage our future review efforts to develop character models with more theoretical highlights as of now addressed only quantitatively (e.g., pages, social events, occasions, likes). The future evaluationof the importance and affectivity of the different markers in the credits of the characters are waived. Our extracted goal is for the malevolent soul to express the appropriateness of sensitive models to the extent that the exploratory situation is based on meaningful assessments for selected domains.

References

1. Potharaju, S.P., Sreedevi, M.: A novel cluster of quarter feature selection based on symmetrical uncertainty. Gazi Univ. J. Sci. (2018)
2. Sharmila, P., Danapaquiame, N., Subhapriya, R., Janakiram, A., Amudhavel, J.: Secure data process in distributed cloud computing bioscience biotechnology research communications (2018)
3. Sai, M.K., Sivaramakrishna, N., Teja, P.V.N.S.R., Prakash, K.B.: A hybrid approach for enhancing security in Iot using RSA algorithm helix (2019)
4. Potharaju, S.P., Sreedevi, M.: An unsupervised approach for selection of candidate feature set using filter based techniques. Gazi Univ. J. Sci. (2018)
5. Danapaquiame, N., Balaji, V., Gayathri, R., Kodhai, E., Sambasivam, G.: Frequent item set using abundant data on hadoop clusters in big data bioscience biotechnology research communications (2018)
6. Ravinder, R.P., Sucharita, V.: A framework to automate cloud based service attacks detection and prevention. Int. J. Adv. Comput. Sci. Appl. (2019)
7. Sai, R.N., Rajesh, S.K.: A novel technique to classify the network data by using OCC with SVM. Int. J. Adv. Technol. (2018)
8. Muthukumar, V., Bhalaji, N.: MOOCVERSITY-deep learning based dropout prediction in MOOCs over weeks. J. Soft Comput. Paradig. (JSCP) **2**(03), 140–152 (2020)
9. Suma, V.: Data mining based prediction of demand in indian market for refurbished electronics. J. Soft Comput. Paradig. **3**, 153–159 (2020)
10. Sharma, N., Yalla, P.: Classifying natural language text as controlled and uncontrolled for UML diagrams. Int. J. Adv. Comput. Sci. Appl. (2017)
11. Raghavendra, S.N., Jogendra, K.M., Smitha, C.Ch.: A secured and effective load monitoring and scheduling migration VM in cloud computing. In: IOP Conference Series: Materials Science and Engineering, December 2020, vol. 981 (2020). ISSN- 1757-899X
12. RaghavendraSai, N., Satya Rajesh, K.: An efficient los scheme for network data analysis. J. Adv. Res. Dyn. Control Syst. (JARDCS) **10**(9) (2018). ISSN:1943-023X

Deep Learning in Precision Medicine

Kavita Tewani

Abstract Healthcare is considered as one of the prime sectors in any country. To improve the life style and medical health of the citizens, countries around the globe invests into this sector in order to give better medical facilities. With the advancement of the applications of Artificial Intelligence in interdisciplinary domain, healthcare system is now amalgamated with advance AI domains like Deep Learning, Machine Learning, Big Data, etc. The paper summarizes the applications of Deep Learning in several medical sectors and discusses various algorithms adopted by researchers to include the power of Deep Learning in current medical system.

Keywords Precision medicine · Deep learning · CNN · ANN

1 Introduction

Healthcare domain has attracted a lot of researchers these days due the challenges that medical industry brings on to the field of human healthcare. The traditional way of working with the healthcare data/challenges is time-consuming and economically expensive. Also, the past data that practitioners receive has become an important tool to come up with the smart technologies in a better way. The data available from past is nowadays used for customized prescription based on the features recorded, which provides advanced treatment methods to the patients. Various fields of Artificial Intelligence such as Big data, Machine Learning, Data Mining, Deep Learning, etc., have proved their worth in the healthcare data. The term **precision medicine** is coined for using machine intelligence to improve the medical industry by introducing the data-based decision into the process. Instead of relying on the general medicine based on the common physical traits, the precision medicine helps in recommending the personalized medicine varying from one individual to another and give customized healthcare advices. The healthcare industries are collecting the huge amount of the

K. Tewani (✉)
Computer Science and Engineering Department, Institute of Technology, Nirma University, Ahmedabad, Gujarat, India

J. S. Raj et al. (eds.), *Intelligent Sustainable Systems*, Lecture Notes in Networks and Systems 213, https://doi.org/10.1007/978-981-16-2422-3_19

data of patients every day, consider the smart devices that keeps track of personal healthcare in order to understand the patient's daily activities.

One of the promising subdomain of Artificial Intelligence, Deep Learning, is gaining the recognition for its ability to learn with the knowledge that it gathers using experience. The algorithms take away the need for a supervisor to formally specify what is required to be done [1]. Unlike machine learning, deep learning mimics the human brain by storing the information in the form of neuron (artificial neuron). The neurons are connected to each other in order to transfer the information, where the information is stored in the form of weight. Deep learning is favored over machine learning for its tendency to work on large amount of data without selecting the features explicitly. It consists of several layers of neurons starting from the input layer where data is fetched to process, followed by a stack of hidden layers which performs the mathematical transformation of data. Each hidden layer makes data more and more meaningful, after the hidden layer, there is an output layer, which gives the desired output. This powerful model makes deep learning as an obvious choice to work in the challenging field which require complex computations such as computer vision [2–4], speech recognition [5, 6], and natural language processing [7, 8].

2 Deep Learning in Medical Imaging

Dealing with medical images is not new in the field of medical research, the practitioners often use the MRI scans, CT scans into consideration in order to know the patient's disease status. To understand these images requires expertise and time to extract the information from them. One of the applications of deep learning lies in extracting the information from these images without human intervention. The images are preprocessed and then represented in the form of a pixel matrix for better understanding of the data and this matrix is given as an input to the input layer of the network architecture.

The processing of the image is majorly divided into medical image classification and medical image segmentation.

Medical Image Classification—In case of image classification, the model should be able to give the results, say whether an individual suffers from a particular disease or not. To prepare a model for such kind of problem, the image is first trained in a supervised manner, where the network learns to classify the images of infected and non-infected patients. Upon significant learning, the model is given a new image where it should be able to classify the never seen input image based on the previous learning experience. To create such model for classification, Convolutional Neural Network (CNN) is considered as the best option to deal with image dataset (Fig. 1).

Another deep learning model used for classification on images are MTANN (Massive training Artificial Neural Network) [10]. The architecture of MTANN is similar to the CNN, it consists of input layer, some hidden layers followed by output layer. However, in addition to the functionalities given by CNN layers, MTANN consists of an additional layer called scoring layer. This layer is added after every

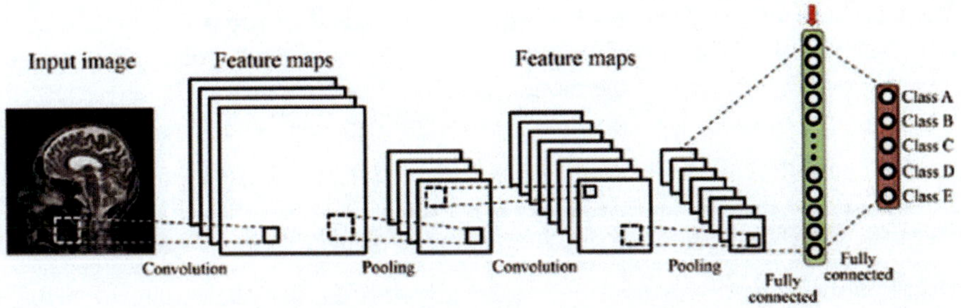

Fig. 1 CNN applied on the MRI scan image, which gives five categories as output. It shows two convolutional layers each followed by pooling layer and two fully connected layers [9]

output layer and gives a single score of likelihood based on the likelihood maps. So in MTANN, convolution happens outside the network and the output of MTANN is an image instead of a class (Fig. 2).

For the purpose of image segmentation, again CNN are used. The basic difference between classification and segmentation of image is that, in classification, the objects are grouped together into a single class, whereas in segmentation, each object in the image is labeled to identify to which class it belongs to. It assigns the label to every pixel in the image. The automatic segmentation of image use supervised voxel classification, where each classifier is trained to assign a class label to each voxel [12].

One of the recent advances in Deep Learning is GAN, which stands for Generative Adversarial Network. GANs have been used for generating synthetic medical images. They have been used for synthetic data augmentation which is useful to make the performance of CNN better for image classification task. One of the challenges of image dataset is the lack of annotated images for creating a model. Manual annotations consume so much of time and having precise annotation itself is a challenge. The GAN model consists of two networks, one network generates fake images while

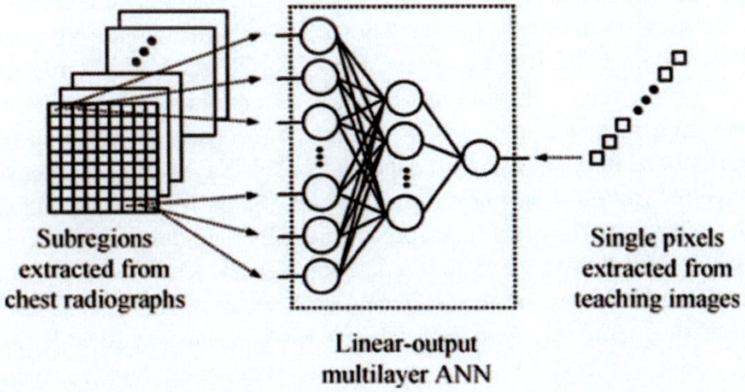

Fig. 2 Architecture of MTANN with linear output multilayer ANN [11]

the other network discriminates between the real and fake images generated repeatedly. The use of GANs in medical imaging to generate synthetic data has proved to increase the performance of the model significantly [13].

3 Deep Learning in Disease Detection

Deep learning algorithms have been used to detect the diabetic retinopathy based on images [14], where CNN is applied to learn to recognize the local features, in proposed algorithms, they have used Inception-v3 architecture to create the model. Deep learning has also shown its worth for detecting the critical diseases like Alzheimer [15], where they have proposed early diagnosis of the disease using stacked auto-encoder followed by a softmax output layer which required less labeled data. Hu et al. [16] surveyed the recent studies for applying deep learning to detect different types of cancer (Breast Cancer, Lung Cancer, Skin Cancer, Prostate Cancer, Brain Cancer, and Colon Cancer). They showcased the work done in detecting cancer using image dataset where the researchers have worked on these datasets with four popular deep learning architectures: CNN (Convolutional Neural Network), FCN (Fully Convolutional Networks), auto-encoders, and deep belief network. The paper also concludes that, out of more than 80 papers surveyed, 63 of them have used CNN which shows the power of this network for image dataset. In [17], the authors have worked on identifying metastatic breast cancer, they have followed supervised learning methods for patch based classification. For the purpose of training the model, 400 whole slide images (WSI) images from Camelyon16 dataset was used. Recently, Google's DeepMind in collaboration with Moorfields Eye Hospital has successfully created an AI system that give diagnosis recommendation for more than 50 eye diseases [18].

With the outbreak of Covid-19, which has infected millions throughout the world and with high spread rate challenged the researchers to come up with the speedy detection of the disease so that it can be treated in early stages. Scientist around the globe are working insensitively for early detection of virus spread. Various ways of detecting the virus spread like rapid test, RT-PCR, CT scan, and X-ray imaging is done. However, the result of rapid test has less accuracy and RT-PCR results takes much time. Using the power of AI with X-ray images for early detection and diagnosis of virus is proposed in many research papers [19–21]. Nayak et al. [19] have worked on 286 images of frontal-view chest X-ray images and performed data augmentation to increase the dataset. They have used pre-trained CNN models, i.e., AlexNet, VGG-16, GoogleNet, MobileNet-V2, Squeezenet, ResNet-34, ResNet-50, and Inceotion-V3 and obtained the best result in ResNet-34 with accuracy of 98.33%. Ozturk et al. [22] proposed the methodology to detect the Covid-19 using raw chest X-ray images, they used DarkNet-19 model as a classifier for real-time object detection using YOLO (you only look once). The paper proposed the algorithm with 17 convolutional layers with different filtering in each layer and have got accuracy of 98.08% for binary classification (Covid infected or not infected) on chest X-ray images. They proposed

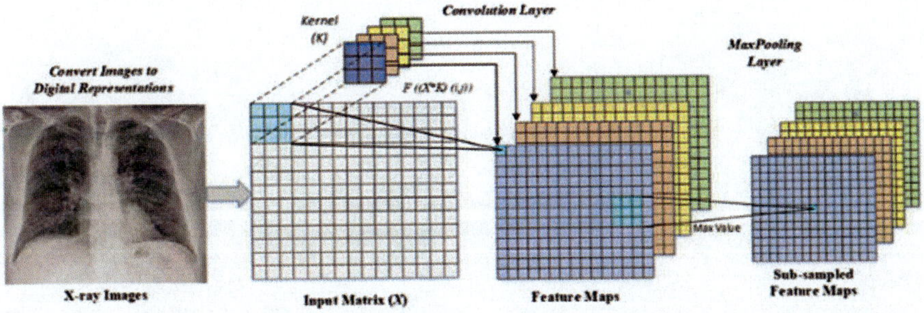

Fig. 3 Architecture of DarkCovidNet as proposed in [22]

DarkCovidNet architecture based on DarkNet where the number of layers were reduced compare to the original architecture (Fig. 3).

4 Deep Learning in Drug Discovery and Design

Discovery of a new drug is one of the biggest challenges in the field of medical science, as per London School of Hygiene and tropical medicine, the average cost of developing a new drug can cost up to $1.3 billion and years of trials before the drug come to the market [23]. This raises the demand of using the Artificial Intelligence in drug design and discovery to efficiently deliver new drug to the market with better performance. The role of AI in drug discovery is not new, but the challenges kept by the current situation demands to reduce the time and increase the efficiency of the whole process. One of the advantages of using Deep Learning in drug discovery is that it can deal with complex patterns of the structure and hence is an obvious choice where the protein structures are dealt with [24]. The article also mentioned the role of machine learning to understand the relationship molecular descriptors used to obtain the QSAR (Quantitative structure activity relationship) model. They mentioned the advantages of using deep learning architectures such as hierarchical architectures, deep architectures, and deep multitask neural network to identify the general features for drug discovery process including identifying protein target for efficient binding of the ligand [25]. Protein-ligand binding prediction is one of the crucial task in augmentation of drug discovery process. Stepniewska-Dziubinska et al. [25] has discussed the process by considering the 3D structures of the molecule and have used 3D convolutional model for feature map. In [26], authors have reviewed the role of Deep Learning in computational chemistry and chemoinformatics. They have also discussed the mainstream architecture like CNN, RNN, and generative networks for the purpose of small molecule drug design and development (Fig. 4).

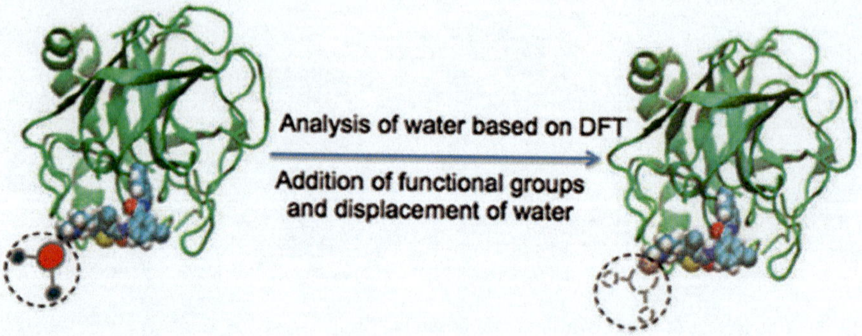

Fig. 4 Protein-ligand binding process for classical density functional theory (DFT) [27]

5 Deep Learning in Personalized Healthcare

Managing the electronic health record (EHR) manually is a tedious task and highly irregular in nature. However, the patient's historical data is important for the healthcare practitioners to take some crucial decision. So these historical data play an important role, but need high amount of preprocessing, with the huge amount of imbalance data available to the researchers, it becomes coherent to use the Deep Learning algorithms to fetch vital information from these data. Capturing the local important features from the irregular EHR using CNN is discussed in [28], they identified similarity scores to perform disease predictions and patient clustering. The paper also summarizes the various methodologies like supervised metric learning, logistic regression, multi-task learning method, etc., for the purpose of prediction and recommendation of patient's personalized healthcare. Naylor [29] lists out the reasons like availability of digital data, improved algorithms to work on image dataset, various open sources' use for adopting Artificial Intelligence and Deep Learning for medicine and healthcare, and discussed the evolution of these algorithms in healthcare system. Esteva et al. [30] shows the extensive use of CNN on image dataset for the purpose of object detection, classification, and segmentation, it also shows how this algorithm has outperformed the human experts and helped them to get rid of laborious task which takes immense time and man power. It also focusses on the use of Natural Language Processing (NLP) for analyzing and inferring the healthcare data by using image captioning, text generation, and machine translation. Auto-encoders (for unsupervised learning) are also used nowadays to predict the specific kind of diagnosis by compressing and reconstructing unlabeled data. To keep track of patient's visit, clinical voice assistants can be used, to record the patient's conversation speech to text can be used both of them are applications of NLP. Talking about Reinforcement Learning (RL), where the system learns from the feedback given from the environment, is also useful in Robotic–Assisted Surgery (RAS), where a surgeon guides the robot and help it with the instructions to perform the actions.

6 Summary

Table 1 summarized the various Deep Learning methodologies proposed by researchers to help and automate the task of medical science.

Table 1 Summary of deep learning methodologies used in medical domain

Paper	Methodologies used	Results/Conclusion	Remarks
Overview of deep learning in medical imaging [10]	Summarized and compared MTANN (Massive training ANN) and CNN models for 2D and 3D input image. Illustrated the working of models for lung nodules	Mentioned the various approaches of Machine Learning and Deep Learning in computer vision and medical images	Stated that segmented of images with complex background is still a challenge to classify the object of interest
Deep Learning for Multi-task Medical Image Segmentation in Multiple Modalities [12]	Trained a single CNN for different segmentation tasks over brain MRI, breast MRI and cardiac MRI	The network was trained with 25000 mini-batches and received significant results for all the three kinds of input images	Demonstrated the working of single CNN architecture for the purpose of multi class classification
GAN-based synthetic medical image augmentation for increased CNN performance in liver lesion classification [13]	Synthesized liver lesions images using GANs and used generated images on CNN for improved classification results	Have received better results (+5%) using synthetic data augmentation over the classic data augmentation approach	Used Generative Adversarial Networks (GANs) to generate medical images which can be used for data augmentation
Development and Validation of a Deep Learning Algorithm for Detection of Diabetic Retinopathy in Retinal Fundus Photographs [14]	Trained Deep CNN (DCNN) over image dataset with 128, 175 images to detect Diabetic retinopathy	Achieved more than 90% sensitivity on various input dataset (EyePACS-1, RDR and Messidor-2)	Using optimized neural network to detect the diabetic retinopathy they have achieved high sensitivity and specificity, however the algorithm's success is to be tested on clinical settings
Deep learning for image-based cancer detection and diagnosis – a Survey [16]	Focused on CNN, FCN (Fully convolutional networks), Auto-encoder and Deep Belief network for cancer detection and diagnosis	Summarized the cancer detection and diagnosis techniques using deep learning for breast cancer, lung cancer, skin cancer, brain cancer	Use of different deep learning were discussed for various kinds of cancer

(continued)

Table 1 (continued)

Paper	Methodologies used	Results/Conclusion	Remarks
Application of deep learning techniques for detection of COVID-19 cases using chest X-ray images: A comprehensive study [19]	Evaluated 8 pre-trained CNN models on publicly available chest X-ray images for early detection of Covid-19 infection	Obtained 98.33% accuracy using Resnet-34 model on frontal-view chest X-ray image dataset. The data was split into 70% training and 30% testing	Applied 8 different CNN models on images of size 224 × 224 and found that Resnet-34 architecture outperformed rest of the models for classifying chest X-ray images
Automated detection of COVID-19 cases using deep neural networks with X-ray images [22]	Implemented Darknet model using YOLO (you only look once) for the real time object detection system on radiological images	Achieved the result of 87.02% for the 3 class problem (No-findings, Covid-19 and Pneumonia) and 98.08% for binary classification (No-findings and Covid-19)	Designed DarkCovidNet architecture on the basis of Darknet-19 model for YOLO and hence some of the layers were taken from the already available architecture. The proposed model has 17 convolutional layers
Development and evaluation of a deep learning model for protein–ligand binding affinity prediction [25]	Utilized 3D structures of protein-ligand for the neural network and trained the model the over PBDbind dataset. The model used was Deep CNN with last layer as convolutional layer for flattening	After training the model for 14 epochs, the Pearson's correlation coefficient (R) of 0.70 and Standard Deviation (SD) of 1.61 is achieved	Used structure-based ligand discovery for drug discovery process using deep neural network called Pafnucy
Deep Patient Similarity Learning for Personalized Healthcare [28]	Utilized patient's electronic health record (EHR) for disease prediction and patient clustering using CNN to capture local information from the EHR	Represented the data in the form of 2D data and applied CNN on the matrix of more than 100,000 patients' record. The model received the accuracy of 85.34% to predict obesity and 91.78% to detect COPD	The patients' historical record can be used in proper way which is then used to find the distance between the patient pair. This is utilized to create patient cluster and used to draw the similarities between the records

(continued)

Table 1 (continued)

Paper	Methodologies used	Results/Conclusion	Remarks
A guide to deep learning in healthcare [30]	Discussed deep learning methods in NLP (Natural language processing), CV (Computer Vision) and Reinforcement learning for healthcare	Focused on CNN for image-level diagnosis, RNN models for analyzing textual and speech records	The paper summarized the use of general deep learning model in the field of healthcare to make the day to day computational of managing the patient's record in convenient way

References

1. Goodfellow, I., et al.: Deep Learning, vol. 1, no. 2. MIT Press, Cambridge (2016)
2. Voulodimos, A., et al.: Deep learning for computer vision: a brief review. Comput. Intell. Neurosci. (2018)
3. Ponti, M.A., et al.: Everything you wanted to know about deep learning for computer vision but were afraid to ask. In: 2017 30th SIBGRAPI Conference on Graphics, Patterns and Images Tutorials (SIBGRAPI-T). IEEE (2017)
4. Zeiler, M.D.: Hierarchical convolutional deep learning in computer vision. Dissertation, New York University (2013)
5. Deng, L., Hinton, G., Kingsbury, B.: New types of deep neural network learning for speech recognition and related applications: an overview. In: 2013 IEEE International Conference on Acoustics, Speech and Signal Processing. IEEE (2013)
6. Amodei, D., et al.: Deep speech 2: End-to-end speech recognition in english and mandarin. In: International Conference on Machine Learning (2016)
7. Deng, L., Liu, Y. (eds.): Deep Learning in Natural Language Processing. Springer (2018)
8. Young, T., et al.: Recent trends in deep learning based natural language processing. IEEE Comput. Intell. Mag. **13**(3), 55–75 (2018)
9. Peña-Solórzano, C.A., et al.: Findings from machine learning in clinical medical imaging applications–Lessons for translation to the forensic setting. Forensic Sci. Int. 110538 (2020)
10. Suzuki, K.: Overview of deep learning in medical imaging. Radiol. Phys. Technol. **10**(3), 257–273 (2017)
11. Suzuki, K., et al.: Image-processing technique for suppressing ribs in chest radiographs by means of massive training artificial neural network (MTANN). IEEE Trans. Med. Imaging **25**(4), 406–416 (2006)
12. Moeskops, P., et al.: Deep learning for multi-task medical image segmentation in multiple modalities. In: International Conference on Medical Image Computing and Computer-Assisted Intervention. Springer, Cham (2016)
13. Frid-Adar, M., et al.: GAN-based synthetic medical image augmentation for increased CNN performance in liver lesion classification. Neurocomputing **321**, 321–331 (2018)
14. Gulshan, V., et al.: Development and validation of a deep learning algorithm for detection of diabetic retinopathy in retinal fundus photographs. JAMA **316**(22), 2402–2410 (2016)
15. Liu, S., et al.: Early diagnosis of Alzheimer's disease with deep learning. In: 2014 IEEE 11th International Symposium on Biomedical Imaging (ISBI). IEEE (2014)
16. Hu, Z., et al.: Deep learning for image-based cancer detection and diagnosis — a survey. Pattern Recogn. **83**, 134–149 (2018)
17. Wang, D., et al.: Deep learning for identifying metastatic breast cancer (2016). arXiv:1606.05718

18. https://deepmind.com/blog/article/moorfields-major-milestone
19. Nayak, S.R., et al.: Application of deep learning techniques for detection of COVID-19 cases using chest X-ray images: a comprehensive study. Biomed. Signal Process. Control 102365 (2020)
20. Hemdan, E.E.-D., Shouman, M.A., Karar, M.E.: Covidx-net: a framework of deep learning classifiers to diagnose covid-19 in x-ray images (2020). arXiv:2003.11055
21. Alazab, M., et al.: COVID-19 prediction and detection using deep learning. Int. J. Comput. Inf. Syst. Ind. Manag. Appl. **12**, 168–181 (2020)
22. Ozturk, T., et al.: Automated detection of COVID-19 cases using deep neural networks with X-ray images. Comput. Biol. Med. 103792 (2020)
23. https://www.lshtm.ac.uk/newsevents/news/2020/average-cost-developing-new-drug-could-be-15-billion-lesspharmaceutical#:~:text=The%20study%20estimated%20that%20the,as%20high%20as%20%242.8%20billion
24. Gawehn, E., Hiss, J.A., Schneider, G.: Deep learning in drug discovery. Mol. Inf. **35**(1), 3–14 (2016)
25. Stepniewska-Dziubinska, M.M., Zielenkiewicz, P., Siedlecki, P.: Development and evaluation of a deep learning model for protein–ligand binding affinity prediction. Bioinformatics **34**(21), 3666–3674 (2018)
26. Jing, Y., et al.: Deep learning for drug design: an artificial intelligence paradigm for drug discovery in the big data era. AAPS J. **20**(3), 58 (2018)
27. Cui, D., et al.: The role of interfacial water in protein–ligand binding: insights from the indirect solvent mediated potential of mean force. J. Chem. Theory Comput. **14**(2), 512–526 (2018)
28. Suo, Q., et al.: Deep patient similarity learning for personalized healthcare. IEEE Trans. Nanobiosci. **17**(3), 219–227 (2018)
29. Naylor, C.D.: On the prospects for a (deep) learning health care system. JAMA **320**(11), 1099–1100 (2018)
30. Esteva, A., et al.: A guide to deep learning in healthcare. Nat. Med. **25**(1), 24–29 (2019)

Improved Stress Prediction Using Differential Boosting Particle Swarm Optimization-Based Support Vector Machine Classifier

P. B. Pankajavalli and G. S. Karthick

Abstract The emerging technology and stress-causing events in the life span of humans impose an effective stress prediction system. The intervention of early detection of stress will effectively reduce the stress and enriches the quality of life. With the assistance of a machine learning approach, a new stress prediction model is developed and it gives personalized as well as an adaptive stress prediction model. The process of learning uses physiological signals, which effectively identifies the stress status of the user. The identification and selection of the best features is a vital step of data preprocessing in case of dimensionality reduction. The performance of the classification model is degraded when the model is trained without reducing the dimensionality of the dataset which may result in poor performance. Hence, identifying and selecting the best features can improve the performance of the classifiers. The proposed stress prediction model with Differential Boosting Particle Swarm Optimization retrieves the best features and the Support Vector Machine (SVM) classifies the subjects into three categories, namely, stress, normal, and relax. The proposed DBPSO-based SVM is compared with the existing approaches and it is evaluated using six performance metrics. From the experimental results, the proposed model attains high accuracy with a low classification error rate.

Keywords Stress prediction · Machine learning · Differential evolution · Particle swarm optimization · Feature selection · Classification

1 Introduction

The experience of stress in the life of humans will reflect in the life's quality, work, health, and mood. The prevalence of stress causes about fifty disorders and signs, which is reported by the American Institute for Stress [1]. The most

P. B. Pankajavalli (✉) · G. S. Karthick (✉)
Department of Computer Science, Bharathiar University, Coimbatore, Tamil Nadu, India
e-mail: pankajavalli@buc.edu.in

G. S. Karthick
e-mail: karthickgs@outlook.com

J. S. Raj et al. (eds.), *Intelligent Sustainable Systems*, Lecture Notes in Networks and Systems 213, https://doi.org/10.1007/978-981-16-2422-3_20

important stress-triggered disorders are depression, heart attack, headaches, and the instabilities of the immune system. The stress is categorized by considering the sequence of incidence and duration, whereby the stressor is segregated into five categories [2]. They are Brief Naturalistic, Acute Stressor, Chronic Stressor, Stressful Events Sequences, and Distant Stressor. The clinical assessment, standardized questionnaires, and interviews are employed in the traditional ways of evaluating the stress. The circumstances and the stress episodes have made the traditional approaches complicated and ineffective [3].

The real-time scientific approaches are incorporated in the clinical health framework for the identification and monitoring of the stressful condition of humans. The shortcomings in the traditional system are rectified by recently developed health services and stress forecasting framework. The computer-based framework uses general health measurements and forecasts the level of stress, whereas it can easily adapt to the rapid changes in clinical measurement [4]. Mobile health (m-health), smart health (s-health), and electronic health (e-health) are the diverse forms of health care systems that incorporate several algorithms for the effective processing and predicting the real-time data [5].

In this article, an effective differential evaluation (DE) and particle swarm optimization (PSO) is incorporated with the support vector machine, which improves the prediction of stress. The clinical measurements like the rate of respiration, electromyography, physiological signals, and GSR are used to retrieve the classification level of stress. The proposed technique can be discriminated from the previous research contributions, wherein previous work used the mutation strategy DE/(rand/1), but in this proposed technique, DE/(rand/2) is used as a mutation strategy along with the updation of PSO. Another contribution is that the best particles are compared using sigmoid function. The proposed algorithm is evaluated using the performance metrics, namely, accuracy, precision, error rate, sensitivity, specificity, and Matthew's correlation coefficients (MCC).

Machine learning approaches were applied to stress prediction data to determine the effective classification model [6, 7]. The support vector machine (SVM), k-nearest neighbor, logistic regression, decision tree, and random forest are the widely used classification approaches [8–10]. The article is emphasized as follows: the DBPSO-based SVM for predicting the stress is given in Sect. 2, the evaluation of the existing and proposed approach is depicted in Sect. 3, and the stress predicting is concluded along with future scope in Sect. 4.

2 Related Works

Nowadays, people have to deal with diverse stress triggering situations, where high degree of stress will disrupt and obstruct the safety of normal lives. Predicting the stress in the earlier stage can prevent numerous health issues related with stress. When an individual is stressed, there is a prominent shift in numerous bio-signals, namely, optical, electrical, thermal, impedance, etc., by incorporating these signal's level of stress can be predicted. With the assistance of biological signal stress can be effectively predicted by different machine learning techniques. Various researchers had developed machine learning-based stress prediction models. These models help in the identification of stress at the earlier stages. Considerable literature survey has been carried out which encompasses varied research process includes datasets, feature selection, machine learning algorithm used in stress detection were presented in Table 1 along with its advantages and disadvantages.

3 Differential Boosting Particle Swarm Optimization-Based Support Vector Machine (DBPSO-SVM)

The main aim of the DBPSO-based SVM is to create a classifier to attain better classification accuracy. The DBPSO is initiated with the estimation of the velocity and looks for the superior position, which updates the current position. The DE crossover is incorporated with the selection process and global best is accomplished by the iterative procedure. From this procedure, a subset of optimized feature is retrieved. The entire process is persistent while waiting for the SVM convergence.

The proposed hybrid approach is applied for updating the strategy of mutation and it is compared with the strategy of new crossover that applies the sigmoidal activation function to retrieve the best solution. The velocity is updated between the range 0 and 1 and the process is calculated as

$$v_{x,G+1} = w \times v_{x,G} + c_1 \times k_{p1} \times (pbest_{ID} - I_x) + c_2 \times k_{p2} \times (gbest_{ID} - I_x) \quad (1)$$

$$I_{n,x} = \begin{cases} 1 \; if \; rand(.) < s(v_{ID}) \\ 0 \; Otherwise \end{cases}, where \; s(v_{ID}) = \frac{1}{1 + e^{-v_{ID}}} \quad (2)$$

where v_x is the velocity of the particle x in the dimension of DI. The c_1, c_2, k_{p1}, and k_{p2} denotes the acceleration constant. The inertia of weight is signified by w and it specifies the influence of preceding velocity. The values of random numbers lie between 0 and 1. The best and global best position of the particle is denoted by the $pbest_{ID}$ and $gbest_{ID}$, respectively. The activation function is utilized to scale the velocity range and it lies between 0 and 1. The constant C_1 and C_2 are assigned with the value 2.05.

DBPSO is utilized in minimizing the redundancy and appropriate feature selection process. The enrichment of fitness function and the investigation of relative significance relevance, as well as the duplicity, are attained by the addition of a new weight-based constraint. It is equated as follows:

$$Objective \; Function = \omega \times \frac{nnz(temp)}{length(temp)} + (1 - \omega) \times (1 - p) \quad (3)$$

Table 1 Literature study on stress and emotion detection systems using physiological parameters

Machine learning algorithm	Features selected	Dataset	Class labels	Advantages	Disadvantages	Inference
Logistic regression [11]	Respiratory signal	DEAP dataset	Positive, negative	Provides better accuracy than existing works due to its data driven nature	The performance is low and does not possibly applied to real-time systems	73% for negative subjects and 80% for positive subjects on DEAP dataset
SVM [12]	HR, rMSSD, pNN50, AVSCL, RD, SDNN, AVNN	International affective system dataset	Low, medium and high	The emotion detection has been validated among laboratory and wearable oriented sensor types	Results are not validated properly and only limited number of subjects was considered for evaluation	66% of accuracy rate achieved for negative and 70% of accuracy rate for positive samples
kNN, SVM [13]	Skin temperature, skin conductance, blood pressure	Real-time dataset	Fear, anger, sad, happy, surprise and neutral	Automatic weighting technique allows the real-time system to adopt new data	Samples considered were small and it suffers from generalization	80% of accuracy was achieved by kNN and 84% of higher accuracy was achieved by SVM
kNN, SVM [14]	EOG, EMG, respiration rate, skin temperature, blood pressure, GSR	DEAP dataset	Low, high	Results show that the dimensionality has been reduced significantly	Better generalization can be achieved using chaotic features and validation of results may be improved using Poincare analysis	kNN obtained better results when compared to SVM

(continued)

Table 1 (continued)

Machine learning algorithm	Features selected	Dataset	Class labels	Advantages	Disadvantages	Inference
SVM (linear kernel), neural networks [15]	Physiological signals	DEAP dataset	Positive, negative	The method used for feature selection yields better results	The system cannot be extended for real-time applications	The accuracy rate of 65% has been achieved, which turns as a major drawback of this work
Logistic regression, SVM [16]	EEG signal data	SJTU EEG dataset	Positive, neutral, negative	In this research work, high level experimentation with respect various classifiers	Validation has to be performed on more datasets	81% of accuracy rate has been achieved
SVM with linear kernel [17]	EEG signal data	Real-time data	Happy, sad	Varied time lengths has been considered for experimentation and validated accurately	The testing can be extended on predicting the emotions rather than happiness and sadness	When compared to other literatures highest accuracy rate of 94% has been achieved

where weight is denoted by ω that lies in the range of [0,1] and the samples are stored in the temporary variable temp that estimates the relevance among the label and class of the feature. The SVM classifier output is stored in p and the redundancy of the retrieved feature is evaluated. The objective function increases relevance and reduces redundancy. In DBPSO, the most relevant features were retrieved and the feature values would be $\omega > 1 - \omega$ and from the experimental analysis, it has been identified that the best features were obtained when $\omega = 0.8$.

Algorithm: Differential Boosting Particle Swarm Optimization Based Support Vector Machine (DBPSO Based SVM)

Input: Size of the Population (P), Generation Training Dataset, Rate of Crossover (rc), c_1, c_2 and w.

Output: Stress Classification

Initialization

For Generation=1

Iterate over P

Estimation of Random Integers

For every particle x, estimate the velocity

$$v_{x,G+1} = w \times v_{x,G} + c_1 \times k_{p1} \times (pbest_{ID} - I_x) + c_2 \times k_{p2} \times (gbest_{ID} - I_x)$$

Estimation and Updation of Best Positions

$$I_x = \begin{cases} 1 \; if \; rand\;() < s(v_{ID}) \\ 0 \; Otherwise \end{cases}, \quad where \;\; s(v_{ID}) = \frac{1}{1+e^{-v_{ID}}}$$

Apply DE Mutation Process DE/rand/2

Then

$$I_{x,g+1} = I_{r1,g} + F.\left(I_{r2,g} - I_{r3,g}\right) + F.\left(I_{r4,g} - I_{r5,g}\right)$$

Estimation of New Fitness Value

$$ft_i = I_{x,g+1} + v_{x,G+1}$$

Apply Crossover

$$C_{x,g+1} = \begin{cases} ft_i \; if \; rand(0,1) \le rc \\ l_i \; Otherwise \end{cases}$$

Pass the Best Feature Set to the SVM Classification Process

End

End

4 Results and Discussion

The results were acquired from the proposed DBPSO and the existing classification approaches namely support vector machine (SVM), k-nearest neighbor, logistic regression, decision tree, and random forest are discussed in this section.

4.1 Dataset Description

The dataset used in this article is taken from the PhysioNet repository and it is generated from human's wearable devices. The sample is available with 4132 subjects and 23 features. In the phase of preprocessing, subjects with missing values are eliminated, and hence, 529 subjects are eliminated. In terms of training, 3000 samples were considered and 1132 samples were taken for testing the model. From the feature selection process, significant features are retrieved and considered in the classification process. Three thousand six hundred and three subjects with twenty three features are taken as an input and the state of the person is identified using the classification approach. The target class contains three labels, namely, stress, normal, and relaxed state, which specify the mental state of the human.

4.2 Validation Technique

The k-fold cross-validation approach is applied in this dataset and it divides the dataset into k equal parts. In the machine learning classification process, the training process utilizes the $k - 1$ parts and the remaining part is used for verifying the performance of the classification. The performance measure is estimated averagely by the k-fold cross-validation.

4.3 Performance Metrics

The proposed and existing classification approaches are evaluated using the accuracy, sensitivity, specificity, Precision, MCC, AUC, and Error Rate. The comparison is attained by investigating the acquired values of the performance metrics.

Accuracy: Accuracy states the approximation of the definite value from the classified instances and it is the estimation of the true value and also it is the identification of both TP and TN values among the number of the evaluated classes.

$$Acc = \frac{True\ Positive + True\ Negative}{True\ Positive + True\ Negative + False\ Positive + False\ Negative}$$

Sensitivity: Sensitivity is the ratio of true negative values out of all the samples in the dataset and it has no definite condition.

$$Sensitivity = \frac{True\ Positive}{True\ Positive + False\ Negative}$$

Specificity: Specificity is the proportion of positive values that are appropriately identified and it is indicated as the rate of hit rate, true positive, and recall. It is calculated as

$$Specificity = \frac{True\ Negative}{True\ Negative + False\ Positive}$$

MCC: MCC is applied in the machine learning approaches as an evaluation of the excellence of classification. It is considered as the correlation coefficient among the forecasted and observed classification values. It is calculated as

$$MCC\ Value = \frac{TP \times TN - FP \times FN}{\sqrt{(TP + FP)(TP + FN)(TN + FP)(TN + FN)}}$$

AUC: The area under curve (AUC) denotes the separability ratio of the classifier to classify the instances correctly. The yielding of higher AUC value shows the potentiality of the classifier in discriminating the stress subjects as stress subjects and abnormal subjects as abnormal subjects.

Error Rate: The incidence of error is due to the data transformation, which is due to several factors, namely, noise, distortion, and interference. The error rate is a ratio of performance rate and it is the proportion of sequences that are incorrectly classified and attained by the decision-making model. It is calculated as

$$Error\ rate = \frac{FP + FN}{TP + TN + FP + FN}$$

4.4 Result Analysis

The proposed DBPSO algorithm identified 9 features as optimal features and then the selected optimal features passed on to the classifiers. The acquired results of the classification approaches are compared and investigated in this section. The acquired values during experimentation are given in Table 2 and the error rate is given in Table 3.

From the experimental results presented in Table 2 and Fig. 1, it can be determined that SVM-Linear Kernel is the best classifier which yields <93, 89, 96, 86, 93, 92> for <ACC, SEN, SPEC, MCC, AUC, F1-SCORE> concerning the features obtained by DBPSO. Table 2 shows the error rate of various classifiers with the features selected by DBPSO and it has been inferred that SVM achieves less error rate when compared to other classifiers.

From the analysis, it can be identified that the DBPSO-based SVM improves the overall stress prediction accuracy when compared to other research contributions identified in the literature, and those identified contributions were validated on various benchmark stress datasets. The comparison of this research work and other research contributions were statistically proven as well as graphically presented in Table 4 and Fig. 2, respectively.

Table 2 Performance evaluation metrics using DBPSO with DT, LR, kNN, SVM, and RF over physionet dataset

Stress prediction model	Hyper-parameters	ACC	SEN	SPEC	MCC	AUC	F1-score
DBPSO + DT	Entropy	81	71	90	64	81	78
	Gini index	78	53	100	61	76	69
DBPSO + LR	$C = 1, R = L1$	91	92	90	83	91	91
	$C = 1, R = L2$	90	92	92	80	90	89
DBPSO + kNN	$k = 2$	85	92	78	71	85	85
	$k = 3$	83	96	72	70	84	84
	$k = 11$	83	78	87	66	83	81
DBPSO + SVM	$C = 0.1$, linear	93	89	96	86	93	92
	$C = 0.1$, RBF	91	89	93	83	91	90
	$C = 1$, linear	91	96	87	84	92	91
	$C = 10$, RBF	91	92	96	83	91	91
DBPSO + RF	Entropy	81	71	90	64	81	78
	Gini index	80	71	87	60	79	76

Table 3 Analysis of classifiers error rate

Performance metric	Algorithms				
	DBPSO + DT	DBPSO + LR	DBPSO + kNN	DBPSO + SVM	DBPSO + RF
Error rate	5.6	7.1	5.4	**4.1**	4.9

5 Conclusion

From the above experimental results and discussion, it can be clearly stated that physiological data collected through sensing devices were highly prominent for predicting the stress level of the individual. In order to identify the best features, DBPSO algorithm is developed and then the various machine learning classifiers are applied. Among the classifier models, SVM with linear kernel is found to be the effective classifier for predicting the stress with respect to the features selected by DBPSO. Therefore, DBPSO-based SVM model attains high accuracy and reduced error rate when compared with the existing approach. In the future, the more contextual information of the user will be included in the prediction of stress.

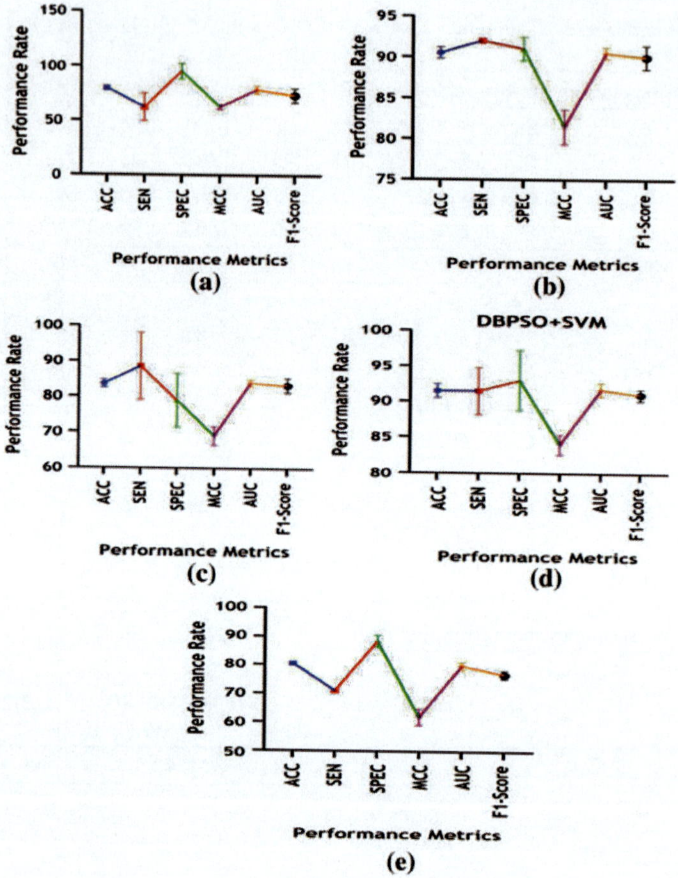

Fig. 1 Performance of classifiers based on multiple tuning parameters depicted using line plot **a** DT **b** LR **c** kNN **d** SVM **e** RF

Table 4 Performance analysis of DBPSO based SVM with other research contributions

Research contribution	Method used	ACC	SEN	SPEC
Pankajavalli and Karthick [5]	SVM	89	98	76
Sano and Picard [18]	Naive Bayes	71	75	90
Keshan et al. [19]	PCA +SVM (RBF, Linear), KNN	75	92	82
Ghaderi et al. [20]	LR	90	77	90
Lee and Chung [21]	Bagged & complex tree	80	75	96
Garcia-Ceja et al. [22]	GMM	70	72	88
Proposed	**DBPSO based SVM**	**93**	**89**	**96**

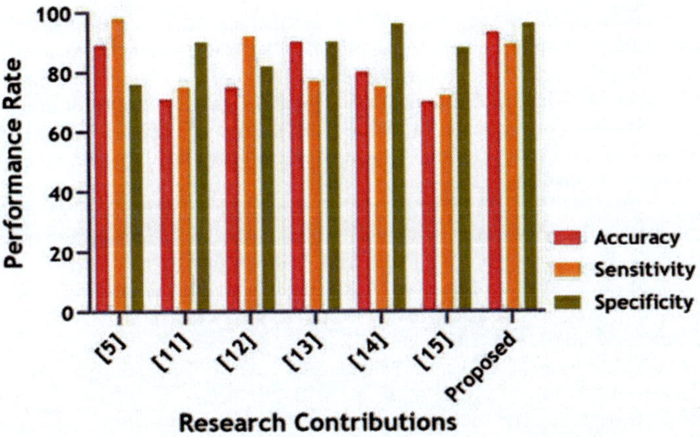

Fig. 2 Comparison of DBPSO based SVM with other research contributions

Acknowledgements The authors thank the Department of Science and Technology–Interdisciplinary Cyber Physical System (DST-ICPS), New Delhi (DST/ICPS/IoTR/2019/8), for the financial support to this research work.

References

1. Rachakonda, L., Kothari, A., Mohanty, S.P., Kougianos, E., Ganapathiraju, M.: Stress-log: an IoT-based smart system to monitor stress-eating. In: 2019 IEEE International Conference on Consumer Electronics (ICCE), January 2019, pp. 1–6. IEEE (2019)
2. Karthick, G.S., Pankajavalli, P.B.: A review on human healthcare Internet of Things: a technical perspective. SN Comput. Sci. **1**, 198 (2020). https://doi.org/10.1007/s42979-020-00205-z
3. Castro, D., Coral, W., Rodriguez, C., Cabra, J., Colorado, J.: Wearable-based human activity recognition using an iot approach. J. Sens. Actuator Netw. **6**(4), 28 (2017)
4. Din, I.U., Guizani, M., Rodrigues, J.J., Hassan, S., Korotaev, V.V.: Machine learning in the Internet of Things: designed techniques for smart cities. Futur. Gener. Comput. Syst. **100**, 826–843 (2019)
5. Pankajavalli, P.B., Karthick, G.S.: A unified framework for stress forecasting using machine learning algorithms. In: Chillarige, R., Distefano, S., Rawat, S. (eds.) Advances in Computational Intelligence and Informatics. ICACII 2019. Lecture Notes in Networks and Systems, vol. 119. Springer, Singapore (2020). https://doi.org/10.1007/978-981-15-3338-9_24
6. Mohapatra, A.G., Khanna, A., Gupta, D., Mohanty, M., de Albuquerque, V.H.C.: An experimental approach to evaluate machine learning models for the estimation of load distribution on suspension bridge using FBG sensors and IoT. Comput. Intell. (2020)
7. Bansal, A., Ahirwar, M.K., Shukla, P.K.: A survey on classification algorithms used in healthcare environment of the Internet of Things. Int. J. Comput. Sci. Eng. **6**(7), 883–887 (2018)
8. Al-Makhadmeh, Z., Tolba, A.: Utilizing IoT wearable medical device for heart disease prediction using higher order Boltzmann model: a classification approach. Measurement **147**, (2019)

9. Pandey, P.S.: Machine learning and IoT for prediction and detection of stress. In: 2017 17th International Conference on Computational Science and Its Applications (ICCSA), July 2017, pp. 1–5. IEEE (2017)
10. Samie, F., Bauer, L., Henkel, J.: From cloud down to things: an overview of machine learning in internet of things. IEEE Internet of Things J. **6**(3), 4921–4934 (2019)
11. Zhang, Q., Chen, X., Zhan, Q., Yang, T., Xia, S.: Respiration-based emotion recognition with deep learning. Comput. Ind. **92**, 84–90 (2017)
12. Ragot, M., Martin, N., Em, S., Pallamin, N., Diverrez, J.M.: Emotion recognition using physiological signals: laboratory vs. wearable sensors, advances in human factors in wearable technologies and game design. Adv. Intell. Syst. Comput. **608**, 15–23 (2018)
13. Khezria, M., Firoozabadi, M., Sharafata, A.R.: Reliable emotion recognition system based on dynamic adaptive fusion of forehead biopotentials and physiological signals. Comput. Methods Prog. Biomed. **122**, 149–164 (2015)
14. Mohammadi, Z., Frounchi, J., Amiri, M.: Wavelet-based emotion recognition system using EEG signal. Neural Comput. Appl. **28**, 1985–1990 (2017)
15. Kumar, N., Khaund, K., Hazarika, S.M.: Bispectral analysis of EEG for emotion recognition. Proc. Comput. Sci. **84**, 31–35 (2016)
16. Chai, X., Wang, Q., Zhao, Y., Liu, X., Bai, O., Li, Y.: Unsupervised domain adaptation techniques based on auto-encoder for non-stationary EEG-based emotion recognition. Comput. Biol. Med. **79**, 205–214 (2016)
17. Li, M., Lu, B.L.: Emotion classification based on gamma-band EEG. In: Proceedings of 31st Annual International Conference of the IEEE EMBS (2009)
18. Sano, A., Picard, R.W.: Stress recognition using wearable sensors and mobile phones. In: Proceedings of the 2013 Humaine Association Conference on Affective Computing and Intelligent Interaction, pp. 671676 (2013)
19. Keshan, N., Parimi, P.V., Bichindaritz, I.: Machine learning for stress detection from ECG signals in automobile drivers. In: Proceedings of the 2015 IEEE International Conference on Big Data (Big Data), pp. 2661–2669 (2015)
20. Ghaderi, A., Frounchi, J., Farnam, A.: Machine learning-based signal processing using physiological signals for stress detection. In: 2015 22nd Iranian Conference on Biomedical Engineering (ICBME), p. 9398 (2015)
21. Lee, B.G., Chung, W.Y.: Wearable Glove-type driver stress detection using a motion sensor. IEEE Trans. Intell. Transp. Syst. **18**(7), 18351844 (2017)
22. Garcia-Ceja, E., Osmani, V., Mayora, O.: Automatic stress detection in working environments from smartphones accelerometer data: a first step. IEEE J. Biomed. Heal. Inf. **20**(4), 10531060 (2016)

Closed-Loop Control of Solar Fed High Gain Converter Using Optimized Algorithm for BLDC Drive Application

R. Femi, T. Sree Renga Raja, and R. Shenbagalakshmi

Abstract The detailed analysis of closed loop control of high gain converter with Particle Swarm Optimization (PSO)-based PID controller fed BLDC motor was carried out using an alternative energy source. The main motive of using the advanced topology of the converter is to get high voltage transfer gain and to acquire maximum power from the solar array. In this work, PSO algorithm is introduced for PID controller. Traditional controllers cannot provide effective dynamic performance, hence PSO-PID control algorithm is proposed for tuning PID gain parameters. Moreover, the VSI adapts fundamental switching frequency to avoid power loss due its high switching frequency and the speed of Brushless DC (BLDC) drive is controlled through variable dc-link voltage of VSI. No additional control circuits are added to regulate the speed of the motor drive, thereby preventing additional power loss in the system. The battery storage provides additional support for an uninterrupted power supply to operate the BLDC motor in the absence of solar irradiance. By implementing the closed-loop control of this converter configuration using an optimized algorithm, the overall efficiency of the system can be increased. Using the MATLAB platform, device output is validated.

Keywords Energy resources · PO-SL luo converter · PSO-PID controller · BLDC drive · Optimization

1 Introduction

Due to its wide variety of applications, the use of renewable energy sources has increased demand in various fields over the past few decades. In the case of non-renewable energy supplies, a significant amount of carbon radiation is produced by the combustion of fossil fuels, which strongly influences the greenhouse effect and

---

R. Femi (✉) · T. Sree Renga Raja
Department of EEE, University College of Engineering, Nagercoil 629004, India

R. Shenbagalakshmi
Department of Electrical Engineering, Sinhgad Institute of Technology, Lonavala, Maharashtra, India

J. S. Raj et al. (eds.), *Intelligent Sustainable Systems*, Lecture Notes in Networks and Systems 213, https://doi.org/10.1007/978-981-16-2422-3_21

induces global warming. By incorporating certain eco-friendly energy sources such as solar, hydro, geothermal, tidal, etc., which are exceedingly present in a state of affairs. Sunlight energy features to reach forward to a chance of having these issues directly to observe a better outcome for the requisite energy regarding distinct benefits over other resources. The power from solar panel is not stable and fluctuates based on dynamic nature of environmental conditions. Because of its non-linear characteristics, the solar panel cannot be directly connected to the load, so high voltage transfer gain converters [1] are used for the necessary applications. For the suggested work, different kinds of literature surveys are performed. As compared to linear mode power supplies with less ripple current, compact size, and high transfer gain ratio, these advanced switch-mode power converters have many advantages. For conventional converters such as boost and buck converters, the steady-state analysis [2, 3] is obtained directly and this approach is not suitable for linear invariant systems. To overcome this drawback, the state-space averaging technique is used to analyze the system stability. Advanced DC-DC converters are of higher-order system and investigate their typical performance, which is rather complex, hence this work is considered as a reduced-order model [4, 5] for mathematical analysis and modelling of high gain. Also, these types of converters have complicated circuits in non-linear modes with varying parameter issues. PID controllers do not provide satisfactory performance, hence high gain converter used optimized control technique for regulating the output voltage. In the work, the PSO-based PID controller is designed to tune the parameters K_p, k_i and K_d value. A bidirectional converter is a fundamental mediator of connecting the resource and battery which regulates two-way power flow, hence it is used for two different combined energy sources [6, 7]. One of the major advantages of using a bidirectional converter is to improve overall efficiency and system performance at a low cost.

In general, DC motors are employed for commercial purposes, despite the fact that they are expensive and have a low efficiency. On the other hand, Induction Motor (IM) [8] is used for this type of system due to its ruggedness, low cost, high efficiency, and easy availability in markets. Induction Motor is not favourable for solar fed application due to its complicated control circuit and gets overheat under low voltage conditions. BLDC drive [9] is used to address these disadvantages, which are highly powerful, reliable, low EMI problems, and better performance in a broad range of speed control. The key contribution of the work is the stability of the converter when the minimum irradiance level is obtained using an optimized algorithm that smoothly regulates the speed of the BLDC motor with less fluctuation on the load side and continuous power supply with the help of battery converter. In order to run a 1HP AC motor, the proposed work uses solar as energy resources, shows better performance, and achieves high efficiency under starting and loading conditions.

2 Overview of DC-DC Converter Fed BLDC Drive with PSO-PID Controller

By altering the typical topology with simplistic structure, a cost-effective and PSO-PID controller is used for the efficient operation of advanced dc-dc converter fed Brushless DC drive using photovoltaic panel, which is designed and shown in Fig. 1. For high power applications, advanced DC-DC converter becomes more popular and is well suitable for non-linear characteristics of DC output from PV array. In order to obtain the stability for the high gain DC-DC converter, state feedback control is designed using a PSO-PID controller. From the PO-SL Luo converter, the well-regulated and controlled DC output voltage is given to the Brushless DC motor through the DC-link capacitor via VSI. Based on the voltage across the DC-link capacitor, the speed of the motor can be controlled, thereby reducing switching losses, which is attained using the fundamental frequency of VSI by means of electronic commutation [10]. Moreover, the purpose of using a bidirectional converter is to transfer power between two DC sources in both directions to provide an uninterrupted power supply for the required application. Its significant role is to control the flow of current between common DC bus and energy storage device like a battery to grant continuous and steady output to the Voltage Source Inverter (VSI). In this work, during normal operating condition, the bidirectional converter allows transferring the power to the battery for storing purpose, while under poor irradiance, the battery discharges the power and flow through a bidirectional converter to load side. In the absence of solar irradiance, the bidirectional converter is connected to the battery, providing essential power to drive the BLDC motor without losing the entire performance.

Sunlight energy is the most prominent energy source compared to other alternative sources. It was the only energy source always available and does not cause any pollution to the environment. The development of solar cell systems is still continuing and utilization of solar energy through solar cell systems can be used to the fullest.

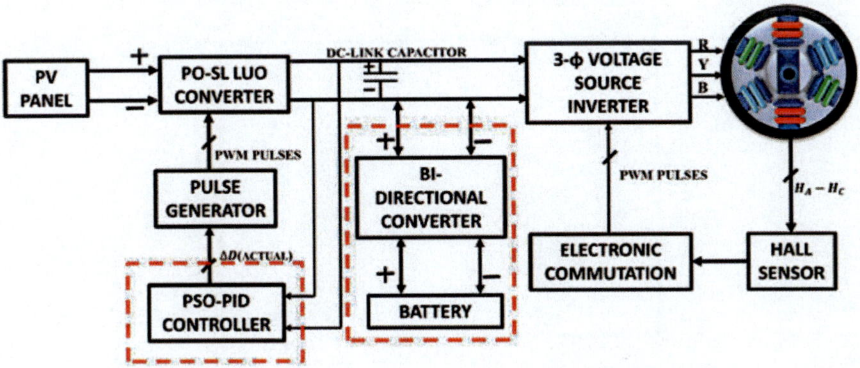

Fig. 1 Block diagram representation of PO-SL Luo converter fed BLDC drive

Table 1 Specifications of solar PV module-MTK 320 W/24v

Parameters	Range
Peak power (Pmp)	320.0 W
OpenCircuit Voltage (Voc)	45.0 V
Short circuit current (Isc)	9.172 A
Peak power voltage (Vmp)	36.26 V
Peak power current (Imp)	8.40 A
Number of cells in series	72

However, the solar panel requires battery storage for its generated electrical energy which is used when the sun does not luminesce. The mathematical model of the solar panel is discussed in [11, 12].

3 Designing and Modelling of High Gain DC-DC Converter

The equivalent circuit of high gain PO-SL Luo converter is shown in Fig. 4. It is one of the recently developed DC-DC converters in which the voltage lift technique is applied to design the Luo converter. The Positive Output super-lift is obtained from the elementary Luo converter, in which the voltage lift technique is employed by including one capacitor and one diode at the output side of the circuit to get high voltage gain and amplify the voltage stage by stage in arithmetic series. Here V_{PV} is the DC input voltage from the solar panel, S_L is an n-channel MOSFET switch, C and C_1 denote the capacitors, L_1, L_2, and L_3 indicates three inductors, in that, L_3 provides extra boosting output voltage even under overloading condition. Then two free-wheeling diodes D_1 and D_2 are used for circulating the current across capacitances, C_0 denotes the output Dc-link capacitor (Fig. 2).

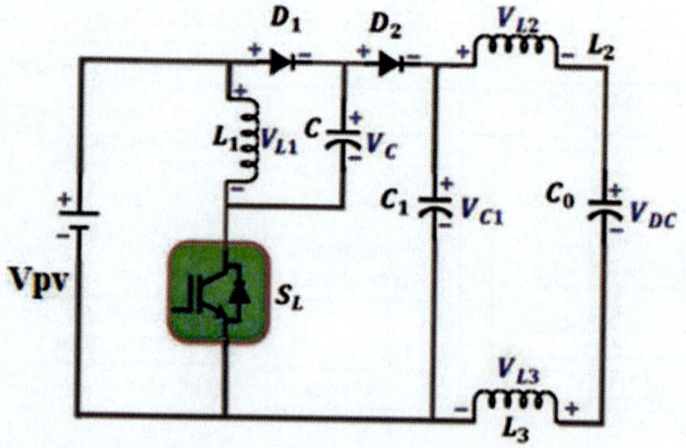

Fig. 2 Circuit diagram of high gain DC-DC Converter

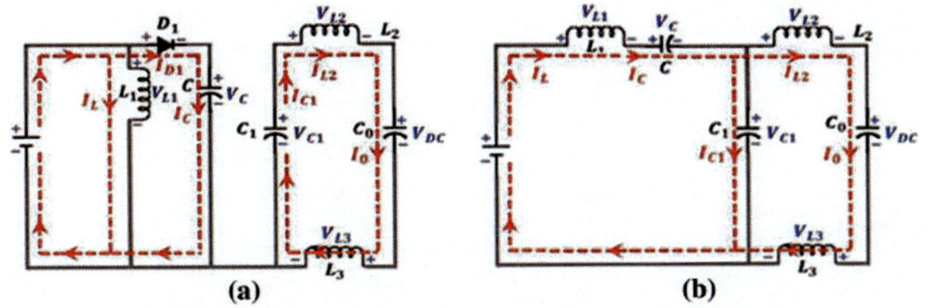

Fig. 3 **a** Switch On and **b** switch off condition of Positive Output-Luo converter

Figure 3a represents the mode 1 operation and Fig. 3(b) represents the mode 2 operation. During mode 1 operation, the current circulates through inductor L_1 and capacitor C which is powered by an input source. During this instant, the inductors at the output side, L_2 and L_3 discharges the current through the DC-link capacitor. During mode 2 operation, inductor current has decreased the value of ($V_{DC} - 2V_{PV}$). The peak to peak ripple current of inductor value is defined as

$$\Delta I_L = \frac{V_{DC} - 2V_{PV}}{L} T_{off} \quad (1)$$

where V_{DC} is the output voltage across the DC-link capacitor which is given to VSI, V_{PV} is the solar output voltage and fed to the converter T_{off} is given by (1 − d) T, where T_{off} is the turn-off period, d denotes duty cycle, total time period is denoted as T of the converter. The range of ripple value of the capacitor is considered to be 10% to 30% of output current. Similarly, the peak to peak capacitor value is defined as

$$\Delta V_{cs} = \frac{V_{DC(1-d)}}{f_{sw} R C_2} \quad (2)$$

where ΔV_{cs} is the ripple voltage of the capacitor, f_{sw} is the switching frequency of capacitor, and ripple value assumed for the capacitor is 1% to 2% of its output voltage. The amplification is achieved by adding n number of diode and n number of the capacitor at the output side of the converter. Now voltage gain is expressed as

$$G = \frac{1}{\lambda_{1......n}} \left[\frac{2-d}{1-d}\right]$$

where $\lambda_{1......n}$ is the co-efficient of voltage lift circuit. By using these formulas, the designing parameters are calculated.

By using Eqs. 1 and 2, inductor L and capacitance C values are calculated and illustrated in Table 2. To decrease the computational complexity of the closed loop

Table 2 Design parameter of high gain converter

Parameter value	Rating value
Input voltage (V_{pv})	30 V
Output Voltage (V_{out})	110 V
Inductor (L_1)	8.141×10^{-6}
Inductor (L_2,L_3)	3.569×10^{-5}
Capacitor (C_1,C_0)	9.2 μF, 22 μF
Switching Frequency (f_{sw})	50 kHz
Output power (P_o)	746 W
Input power (P_{in})	800 W

control proposed converter, a reduced order state model was developed. The equivalent circuit for reduced-order model calculation of PO-SL converter is given in [11]. This type of PWM converter, employed switches to generate the pulses for a sequence of turn-on and turn-off at a particular switching frequency f_{sw}. The state-space modelling equation is stated as

At mode 1:

$$\frac{di_L}{dT_{on}} = \frac{V_{pv}}{L} \tag{3}$$

$$\frac{dV_{DC}}{dT_{on}} = -\frac{V_{DC}}{R_o C_1} \tag{4}$$

At mode 2:

$$\frac{di_{L_1}}{dT_{off}} = \frac{2V_{pv}}{L} - \frac{V_{DC}}{L} \tag{5}$$

$$\frac{dV_{DC}}{dT_{off}} = \frac{1}{C_1} i_L - \frac{1}{R_0 C_1} V_{DC} \tag{6}$$

where T_{on} and T_{off} are an on-time and off-time period of switches. For reduced-order model R_0 is assumed at the load side. In general, state-space equation is derived as

$$\dot{x}(t) = a_1 x(t) + b_1 V_{pv} \text{ when s} = 1 \tag{7}$$

$$\dot{x}(t) = a_2 x(t) + b_2 V_{pv} \text{ when } s = 0 \tag{8}$$

where $\dot{x}(t)$ is the input variable, a_1, b_1 and a_2, b_2 are the co-efficient matrices, s = 0 shows the turn-off state and s = 1 shows the turn-on state of the converter.

$$a_1 = \begin{bmatrix} 0 & 0 \\ 0 & -\frac{d}{R_0 C_1} \end{bmatrix} \tag{9}$$

$$a_2 = \begin{bmatrix} 0 & \frac{d-1}{L} \\ \frac{1-d}{C_1} & \frac{d-1}{R_0 C_1} \end{bmatrix} \quad (10)$$

$$b_1 = \begin{bmatrix} \frac{d}{L} \\ 0 \end{bmatrix} \quad (11)$$

$$b_2 = \begin{bmatrix} \frac{2-2d}{L} \\ 0 \end{bmatrix} \quad (12)$$

The output voltage equation of the PO-SL Luo converter is defined as

$$y(t) = \begin{bmatrix} 0 & 1 \end{bmatrix} x(t) \quad (13)$$

The estimation of output DC-link capacitor value [9] using the fundamental frequency of VSI is related to the minimum and maximum speed value of BLDC drive. The formula for finding the C_0 is defined as

$$C_{0maxi} = I_0/6 \times \omega_{maxi} \times \Delta V_{DC} \quad (14)$$

$$C_{0mini} = I_0/6 \times \omega_{mini} \times \Delta V_{DC} \quad (15)$$

The C_{0maxi} and C_{0mini} are the capacitance value of the maximum and minimum speed of BLDC motor , ω_{maxi} and ω_{mini} are the maximum and minimum fundamental frequency.ΔV_{DC} is the maximum ripple output voltage. Usually, the highest value of the capacitor is chosen for designing a variable DC-link capacitor to complete the load demand.

4 Control of BLDC Drive-by Electronic Commutation

It is a most prominent electronic commutated device in which internal input is generated by rotor position sensors that give the information about the rotor's position for proper stator switching. Instead of sinusoidal back-emf, BLDC motor requires back-emf of trapezoidal shape which displaces 120° electrical degrees to each other. The rotor position is sensed by a sensor named hall sensor H_A, H_B, and H_C embedded into the stator. For every 60 degree, three specified combination of hall effect signals are created for each rotor position. The hall sensors are energized when the magnetic poles are nearer to it; the sensor signal becomes 0 or 1. These signals trigger PWM pulses from the pulse generator and give them to the switches of VSI. The main role of electronic commutation is by eliminating high switching losses by employing VSI at the fundamental frequency. The combination of Hall signals is tabulated in [9].

5 Control Algorithm

5.1 PSO Algorithm

PSO was first developed by Kennedy and Eberhart [13] and they introduced computational intelligence by availing similarity of social communication, and not as an individual mental skill. It was influenced by the living creature's behaviour, i.e., when the bird migrates for food searching. PSO is a method of searching food grains by a swarm and each food grain defines the objective function at its present location. Each swarm moves to seek for grains in a search space and acquire information about the past events in its current position in order to choose the optimum position. Among various control algorithms, PSO is a heuristic method of searching the optimized value to obtain better system performance [14]. Each particle swarm consists of three-dimensional vectors such as current position $\overrightarrow{x_i}$, the previous best position represents as $\overrightarrow{P_{best_i}}$ and denotes the velocity as $\overrightarrow{V_i}$.The step by step procedure for tuning the parameters of PID controllers using the PSO algorithm [15] is described as

Step 1. To assign the initial parameter that is needed for the tuning process, i.e. K_p, k_i and K_d values

Step 2. Define input values
Number particles = 50
Search space D = 3(K_p, k_i and K_d)
Length of iteration m = 30
Weight inertia ratio w = 0.4
Acceleration constant = 2.05
Initial velocity as $V_o = 0.1x_o$(10% of initial position x_o)

Step 3 Initialize each particle having position and velocity as x and v and noted as first iteration.

Step 4. Evaluate the fitness function which shows the best particle and named as p

Step 5. Now personal best as $P^m_{best_i} = x^m_i$ and global best as $G^m_{best} = x^m_p$,i denoted individual particle.

Step 6. Calculate weight inertia ratio using the formula as $= \frac{(w_{max} - m) \times w_{max} . w_{min}}{maximumiteration}$.

Step 7. Update velocity and position

$$v^{m+1}_{i,j} = w \times v^m_{i,j} + C_1 \times rand() \times \left(P^m_{best_{i,j}} - x^m_{i,j}\right) + C_2 \times rand() + G^m_{best_j} - x^m_{i,j}$$

$$x^{m+1}_{i,j} = x^m_{i,j} + v^{m+1}_{i,j}$$

Step 8. Again calculate the fitness function and find the best particle and noted as p_1

Step 9. Update personal best value, global best value, and check the condition if new fitness function value is lesser than the previous fitness value then

$P_{best_i}^{m+1} = x_i^{m+1}$ otherwise $P_{best_i}^{m+1} = P_{best_i}^{m}$

Similarly, check the new best fitness function p_1 is lesser than best fitness function p then $G_{best}^{m+1} = P_{best_{p_1}}^{m+1}$ and assign $p = p_1$ otherwise $G_{best}^{m+1} = G_{best}^{m}$

Step.10 Check the matrix length if it is less than final length value and go to step 6 otherwise displayed the optimal value.

6 Simulation Validation and Discussion of Proposed System

The PO-SL Luo converter is modelled and simulated with a PSO-PID controller. The design parameters are related to the time domain specification such as rise time(t_r), settling time(t_s), peak overshoot(M_p), and ripple voltage(V_p) are shown in the tabulation. Simulation results shown in Fig. 4 are evidence of achieving the systems better performance of output voltage, output current, and inductor current iL_1, iL_2, and iL_3. From the tabulation III and simulation results shown in Fig. 4, it is easily understood that for the variation of solar input values such as 29 V-30 V-31 V, the corresponding output voltage settles down faster and maintains a constant value of 110 V without any peak overshoot or undershoot, and ripples voltage is also negligible which shows the improved performance (Table 3).

The output voltage of the converter is increased up to 110 V, which supports the performance of the converter by limiting power device stresses. The steady-state behaviour of the system is analyzed at minimum irradiance of 200 W/m^2, which generates a minimum value of 24 V. Furthermore, the irradiance level of solar increases to 800 W/m^2, however, the converter is maintaining the required constant

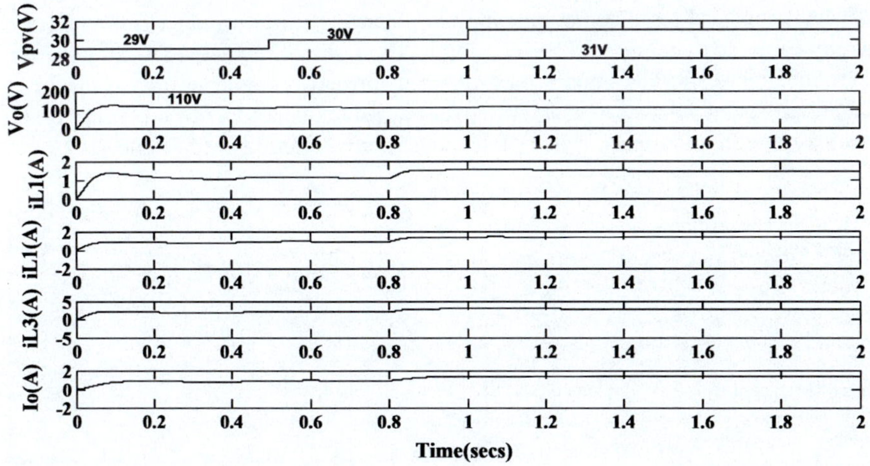

Fig. 4 Output response of PO-SL Luo converter of input voltage variation using PSO-PID controller method input voltage, output voltage, Inductor current iL_1, iL_2, and iL_3 and output current

Table 3 Performance parameters of PO-SL Luo converter using pso-pid controller

Sl. No	Tuned parameters of high gain Luo converter using PSO-PID technique	K_p	k_i	K_d	
1		13.1818	1.2138 × 10^4	0.058	
2	Performance parameter of high gain Luo converter	Settling time (t_s) (secs)	Rise time (t_r) (secs)	Peak overshoot (M_p) (%)	Ripple voltage (V_p)(V)
		0.09	0.07	0	0

output voltage of 110 V without any oscillation. This constant output voltage is achieved by tuning the parameter of the PSO-PID controller, which enhances the entire performance of the system. The obtained closed-loop transfer function of the system is

$$G(s) = \frac{1.583e^6}{S^3 + 123S^2 + 3.644e^5 S}$$

The tuned value for the above transfer function using the traditional ZN-PID controller is $K_p = 9.258$, $k_i = 1.1254 \times 10^4$ and $K_d = 0.\ 0054$. With the help of these reference values the PSO-PID choosing the bounding velocity to obtain the optimal value.

Figure 5 shows the speed curve of the BLDC motor under loaded conditions. BLDC motor is capable of running at a rated speed of 2500 rpm. Based on solar radiation, the speed can vary, but in this case, the converter is intended to prepare constant and hard-regulated output voltages. Hence, the performance is not affected at any cause of dynamic weather condition. The optimal regulated DC output from the converter results in a smooth start-up, when the load is connected across the drive, the motor speed value accepts a limited droop and fluctuates by 0.4 secs, after that it settles smoothly and reaches its steady-state at the required speed (Table 4).

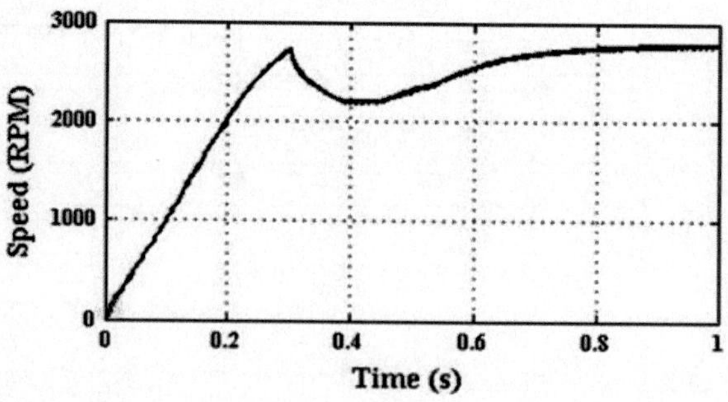

Fig. 5 Speed waveform at loaded condition

Table 4 Efficiency calculation of proposed work

Sl. No	Solar insolation level (w/m^2)	Output power $P_{pv}(W)$	Motor power $P_m(W)$	Efficiency (%)
1.	200	110	82	74.2
2	400	320	260	81.4
3	600	520	440	84
4	800	690	600	87
5	1000	800	746	92.6

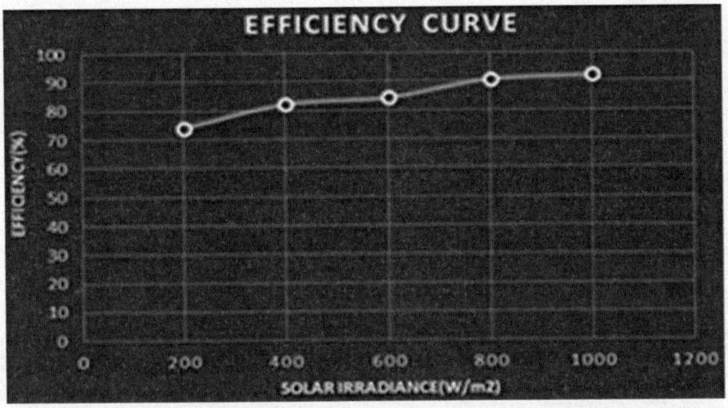

Fig. 6 Efficiency Curve under various solar irradiances

By analyzing the efficiency curve, the performance of solar energy fed BLDC drive is certainly understood. The efficiency curve is plotted for various irradiance levels and obtained maximum efficiency at highest power of 746W. The efficiency of 74% is attained at a lower irradiance level of 200 W/m^2 and achieved 92% of efficiency at a higher solar irradiance level. Under the dynamic weather conditions, it is the obvious efficient working converter that enhanced the efficiency of the entire system (Fig. 6).

7 Conclusion

The closed-loop control is designed for the PO-SL Luo converter in the continuous-time domain using a PSO-PID controller. The simulation results are obtained using the MATLAB platform and are simultaneously verified in mathematical calculation. The simulation results and analysis show that the PSO-PID controller is thus designed to attain reliable output regulation and dynamic performance. Because of these dynamic characteristics, the BLDC motor can perform extremely under variable load conditions. Moreover, even under poor weather conditions, the bidirectional converter connected battery, serves as supporting source for continuing the power supply to the load. Therefore, the proposed work attained its objective in the desired manner in terms of stability, robustness, smooth starting and running BLDC motor, and efficiency.

References

1. He, Y., Luo, F.L.: Analysis of PO-SL Luo Converters with voltage-lift circuit. IEE Proc.-Electr. Power Appl. **152**(5) (2005)
2. Lakshmi, S., Renga Raja, S.: Closed loop control of soft switched interleaved buck converter. Int. J. Power Electron. Drive System (IJPEDS) **2**(3), 313–324 (2012)
3. Narayan Kamat, S., Lakshmi, S.: Design and analysis of positive output self lift Luo converter. In: 2018 Fourth International Conference on Computing Communication Control and Automation (ICCUBEA), IEEE 2018
4. Arunkumar, N., Sivakumaran, T.S., Ramashkumar, K., Shenbagalakshmi, R.: Analysis, modeling and simulation of state feedback control for positive output super lift Luo converter. Circuits Syst. **7**, 3971–3983 (2016)
5. Luo, F.L., Ye, H.: Small signal analysis of energy factor and mathematical modeling for power DC-DC converters. IEEE Trans. Power Electron. **22**(1), 69–79 (2007)
6. Inoue, S., Akagi, H.: A bi-directional dc–dc converter for an energy storage system with galvanic isolation. IEEE Trans. Power Electron. **22**(6) (2007)
7. Lai, C.-M., Li, Y.-H., Cheng, Y.-H., Teh, J.: A high-gain reflex-based bidirectional DC charger with efficient energy recycling for low-voltage battery charging-discharging power control. Energies **11**, 623 (2018)
8. Priyadarshi, N., Padmanaban, S., Bhaskar, M.S., Blaabjerg, F., Holm-Nielsen, J.B.: An improved hybrid PV-wind power system with MPPT for water pumping applications. Int Trans Electr Energy Syst. (2019)
9. Kumar, R., Singh, B.: BLDC motor driven solar PV array fed water pumping system employing zeta converter. In: 6th IEEE India International Conference on Power Electronics (IICPE), 8–10 Dec 2014
10. Parackal, R., Koshy, R.A.: PV powered zeta converter fed BLDC drive. In: Annual International Conference on Emerging Research Areas: Magnetics, Machines and Drives (AICERA/iCMMD), July 2014, vol. 24–26, pp. 1–5
11. Femi, R., Sree Renga Raja, T., Shenbagalakshmi, R.: A positive output-super lift Luo converter fed brushless DC motor drive using alternative energy sources. Int. Trans. Electr. Energ. Syst. 2020; e12740
12. Chandramouli, A., Sivachidambaranathan. V.: Extract maximum power from PV system employing MPPT with FLC controller. Power 1(4) (2019)
13. Riccardo, P., Kennedy, P., Blackwell, T.: Particle swarm optimization an overview. Swarm Intell. **1**, 33–57 (2007)
14. Gaing, Z.-L.: A particle swarm optimization approach for optimum design of pid controller in avr system. IEEE Trans. Energy Convers. **19**(2), 384–391 (2004)
15. Guman, S.K., Giri, V.K.: Speed control of DC motor using optimization techniques based PID controller. In: 2nd IEEE International Conference On Engineering and Technology (ICETECH), Coimbatore India, March 2016

Internet of Things-Based Design of Smart Speed Control System for Highway Transportation Vehicles

R. Senthil Ganesh, S. A. Sivakumar, V. Nagaraj, G. K. Jakir Hussain, M. Ashok, and G. Thamarai Selvi

Abstract The infrastructure of developing countries and developed countries mainly focuses on roadways that connect cities and towns. The roadways in India are broadly classified as Express highways, National highways, state highways and other roadways. The number of accidents in roadways are increasing eventhough the infrastructure of roadways are improved. This is due to the lack of control in speed by the drivers. The drivers are not following the speed limits of the respective highways which is determined by the authorities. This paper presents a smart speed control system for vehicles based on internet of things. The system is to be installed in all the vehicles. Based on the global positioning system coordinates, the speed of the vehicles is automatically optimized gradually so that the drivers follow the speed limit of the respective highways. In case of emergency, this system also sends request to the authorities for increasing the vehicle speed.

R. S. Ganesh (✉)
Department of ECE, Sri Krishna College of Engineering and Technology, Coimbatore, Tamilnadu, India
e-mail: drsenthilganesh@gmail.com

S. A. Sivakumar
Department of ECE, Ashoka Women's Engineering College, Kurnool, Andhra Pradesh, India
e-mail: drsasivakumar@gmail.com

V. Nagaraj
Department of ECE, Knowledge Institute of Technology, Salem, Tamilnadu, India
e-mail: nagarajphd16@gmail.com

G. K. J. Hussain
Department of ECE, KPR Institute of Engineering and Technology, Coimbatore, Tamilnadu, India
e-mail: gkjakirhussain@gmail.com

M. Ashok
Department of CSE, Rajalakshmi Institute of Technology, Chennai, Tamilnadu, India
e-mail: ashok.m@ritchennai.edu.in

G. T. Selvi
Department of ECE, Sri Sai Ram Institute of Technology, Chennai, Tamilnadu, India
e-mail: hodece@sairamit.edu.in

J. S. Raj et al. (eds.), *Intelligent Sustainable Systems*, Lecture Notes in Networks and Systems 213, https://doi.org/10.1007/978-981-16-2422-3_22

Keywords Smart system · Transportation · Speed control · Internet of things · Highway vehicles

1 Introduction

The speed controlling system was initially invented in the early years of 1900. The concept of the system is to maintain the constant speed of the vehicle by enabling the angular position of the throttle with respect to the load of the engine. Later by 1948, the motor speed control was invented which was commericalized in the year 1958. İn early 1970s, the growth of automobile industries and their marketing strategies were resulted in enormous increase in the usage of vehicles [1, 2]. This made the increase in demand of speed control system for maintaining proper speed and engine performance [3]. İn the year 1992, the Mitsubishi motors adopted an advanced speed control system than others by including detection of distance between the vehicles and by that maintaining the speed of its vehicles. İn the mid of 2000s, the engine control unit (ECU) was proposed along with the fixed electric air valves [4]. This ECU optimized the speed of the vehicles by measuring revoltiond per minute (RPM) of the respective motor [5]. Then, in 2013, adaptive cruise controller (ACC) was implemented in the vehicles to monitor the surrounding transportation environment by using smart sensors [6]. This ACC system was built with a technology similar to that of radar system. Recently, the speed control systems are developed by fuzzy logic based proportional integral dervative (PID) techniques [7]. This fuzzy PID technology [8, 9] regulates the speed of the vehicle by controlling the throttle valves. Figure 1 shows the timeline of the speed control system.

İn india, the roadways are broadly classified as express highways, national highways, state highways and other roadways. The express highways are very high speed roads whereas national highways are the network of trunk roads and state highways are provincial routes. The express highways are in top of the indian highways hirearchy, in which vehicles are allowed to travel at high speed than that of other

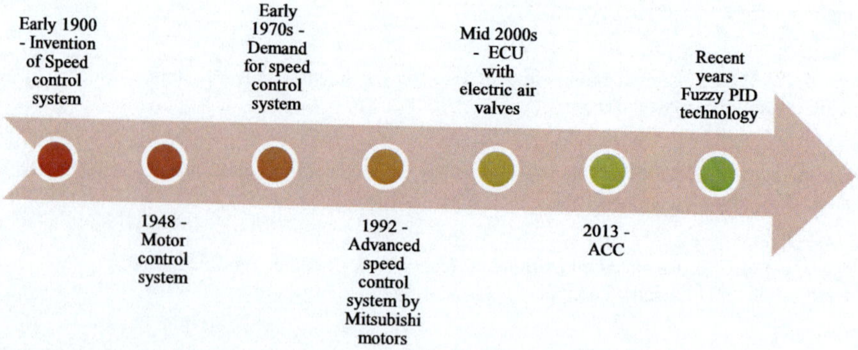

Fig. 1 Timeline of speed control system

roadways. The national highways come second in hirearchy, in which vehicles are allowed to travel at moderate speed and the state highways come last in hirearchy, in which vehicles are allowed to travel at limit speed. If these speed limits are not properly followed, then it could result in major accidents which will be loss of life, vehicles and also loss for the government [10, 11]. To avoid the accidents in the highways due to improper speeding, a smart speed control system is proposed.

In this proposed paper, an internet of things (IoT) [12] -based smart speed control system is designed for the transportation of vehicles specifically that are travelling on highways. The global positioning system (GPS) [13] and cloud server [14] are the two important blocks which supports the design of this smart speed control system. Based on the coordinates provided by the GPS module, the speed limit of the highways is detected and adjusted accordingly by referring the database in the cloud server. The maximum speed limit of the vehicle is either increased or decreased for the respective locations. The above stated procedure is a continuous process which will be carried out at regular intervals. This alloted maximum speed could not overcome by the driver even more acceleration for provided to the engine. If the driver attempts to damage or bypass the system, then the vehicle engine is turned off by limiting the engine RPM through ECU and throttle position [15]. Then an alert message with vehicle and driver details is sent to the authorities for taking legal actions against the driver. Also, if maximum speed limit of the vehicle is to be increased for any emergency situations, a request message with proper proofs is generated by using the user interface of the smart system and the message is communicated to the authorities. The authorities could verify the proofs and increase the speed limit for the appropriate vehicle through an acknowledgment or approval message in real time. The alert messages and request messages are sent to the authorities as well as approval message from the corresponding authorities is received through the GSM modules.

2 Proposed System Description

The proposed speed control system consists of two controller modules and dedicated cloud server as shown in Fig. 2. Module 1 is placed in the vehicles and the module 2 is with the authorities. The system consists of a cloud server which is loaded with the GPS coordinates and their respective speed limits. The controller modules communicate with the cloud server using WiFi module.

2.1 Cloud Server

The cloud server is limited access for the module 1 in vehicles and completely accessed by the module 2 in authorities' end. The limited access for the vehicles' module is that it can only read the speed limit of the respective locations, whereas

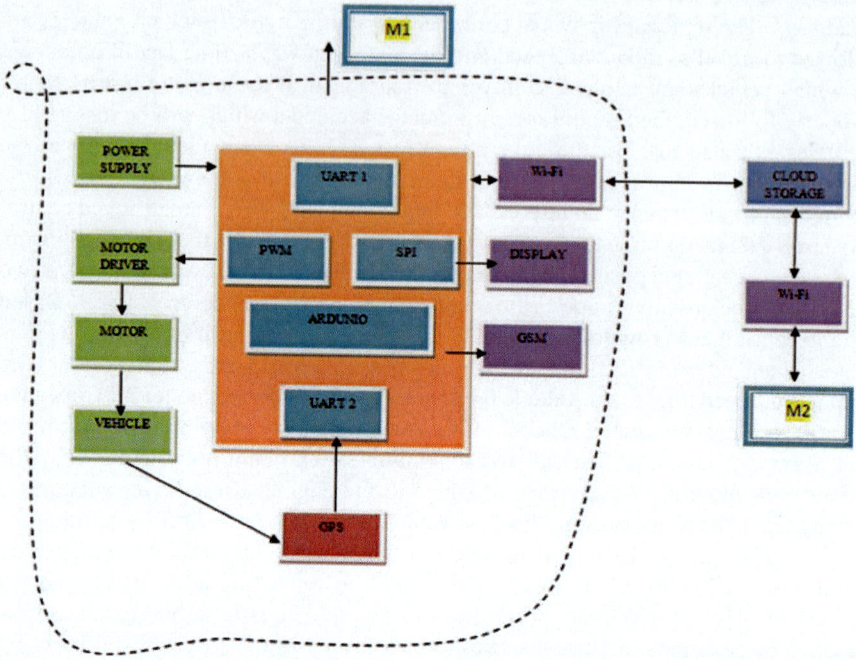

Fig. 2 Block diagram of the proposed speed control system

the module 2 is provided with complete access which can read as well as write the information in the cloud database. This implies that the cloud database is open access for the module 1 in the vehicles and it is secured access for the authorities.

2.2 *Arduino Uno*

Module 1 and module 2 of the proposed system are controlled by cost efficient microcontroller called Arduino Uno. The ardunio uno in module 1 is used to control the devices such as WiFi module, GSM module, GPS module, LCD display and motor driver to access throttle valve of the vehicle. Similarly, the arduino uno in module 2 is used to control GSM module and WiFi module.

2.3 *WiFi Module*

In this system, esp8266 WiFi module is accommodated in both module 1 and module 2.

The esp8266 is chosen because of its low cost, small size, secured, effective capability with complete TCP/IP stack and a powerful IoT device. This module is easily programmable and here it is efficiently used to communicate with the database in the dedicated coud server.

2.4 GSM Module

The GSM module, which acts as a transceiver, is used to send and receive messages in both module 1 (M1) and module 2 (M2) of the proposed work. In module 1, GSM module is used to send the request message by the driver for increasing speed limit, due to some emergency situations, to the authorities. The request message is received by the GSM module at the authorities' end. Once the verification is over, the acknowledgement message is sent by the authorities to the respective vehicle which initiated the request process. This acknowledgement message with the approval of change in speed limit is received by the GSM module in vehicle end.

2.5 Motor Driver

The L298N motor driver is a high power and cost efficient two channel H-bridge type motor driver which is highly suitable for driving dc motors. The motor driver is deployed in this system to adjust the throttle valve of the vehicle. The motor driver is activated and made it to drive the dc motor based on speed limit that attained from the cloud server through arduino uno. The motor driver also gets activated based on the acknowledgement message communicated from the authorities. So, either through GPS coordinates or acknowledgement message, the motor is enabled by the motor driver to adjust the throttle valve of the vehicle. This change in throttle valve is reflected in the ECU that controls the RPM of the engine accordingly. The Eq. (1) is used to convert the respective speed limit to RPM of the engine [16].

$$\text{Vehicle speed (miles/hr)} = \frac{5.95}{1000} * \frac{RPM \cdot R_{lt}}{R_{tg} \cdot R_{ra}} \tag{1}$$

where RPM refers engine speed, R_{lt} is of the loaded tyre, transmission gear ratio and rear axle ratio is given as R_{tg} and R_{ra} respectively. Also, the motor driver is made to drive the DC motors to shutdown the throttle valve if the driver attempts to damage the system or bypass the system.

2.6 GPS Module

In the vehicles, GPS module is placed to track the location of the vehicle. The GPS module is switched on along with the ignition of the vehicle. The microcontroller is programmed to receive the signals from GPS module in predefined time intervals. The GPS module provides the location of the vehicle in the form of coordinates. These coordinates are submitted to the arduino uno. The microcontroller accesses the cloud server with the help of WiFi module and compares the attained coordinates with the coordinates in the database. From the comparison, the microcontroller able to identify the highway category on which the vehicle is travelling and also the maximum allowable speed limit of that particular location.

2.7 LCD Display

The liquid crystal display (LCD) is used in module 1 to notify the information like type of highway and the alloted maximum speed limit for the highway. The display unit is also used to view the request message that has to be sent to authorities in emergency situations and acknowledgement message from the autorities with increase in speed limit.

3 Proposed System Implementation

The implementation of proposed speed control system is classified into hardware implementation and software implementation.

3.1 Hardware Implementation

Figure 3 shows the circuit diagram of module 1. The circuit diagram clearly illustrates the connections among Arduino Uno, motor driver, GPS module, GSM module and LCD display. The components used in the circuit are less cost and more efficient. The entire module 1 costs not more than thirty dollars.

There are fourteen digital pins and six analog pins in Arduino Uno. The digital and analog pins can be configured as both input and output pins. Among the fourteen digital pins, six pins can be deployed as pulse width modulation (PWM) output pins. Both the GPS module and the GSM module consists of four pins each which are transmitter pin (Tx), receiver pin (Rx), 5v power supply pin (vcc) and ground (GND). The L298N motor driver consists of six input pins, four output pins, 12v power supply pin and ground pin (GND). The six input pins of motor driver are classified as four control input pins as IN1, IN2, IN3 and IN4; and two enable pins as ENA and ENB. The output pins of motor driver are OUT1, OUT2, OUT3 and OUT4, which are used to interface two dc motors. The LCD display with I2C module

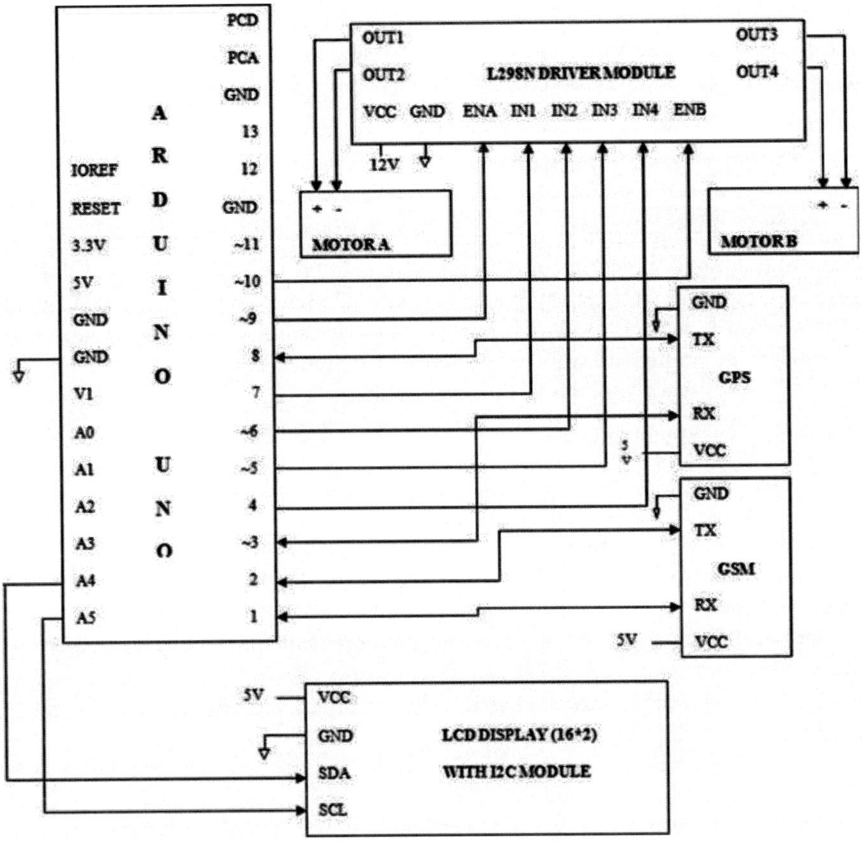

Fig. 3 Circuit Diagram of Module 1 implemented in the Vehicle

consists of four pins as serial data pin (SDA) and serial clock pin (SCL), 5v power supply pin (vcc) and ground (GND).

The Receiver and Transmitter pins of GPS module are interfaced to third and eighth pins of Arduino Uno. Similarly, Receiver and Transmitter pins of GSM module are interfaced to first and second pins of Arduino Uno. The fourth, fifth, sixth, seventh, nine and tenth pins of Arduino Uno are connected to IN1, IN2, IN3, IN4, ENA and ENB of L298N motor driver respectively. The OUT1 and OUT2 pins of motor driver are used to connect dc motor A where OUT3 and OUT4 pins are connected to motor B. The analog pins A4 and A5 are connected to SDA and SCL pins of LCD display with I2C module. The module 2 of the proposed system consists of Arduino Uno microcontroller, GSM module and WiFi module. The composition of the components not cost more than 15 dollars. The connections of microcontroller with GSM and WiFi modules are similar to that of module 1.

In this proposed system, Google cloud storage is used as cloud server. A unique Google cloud platform is created with required number of volumes and target buckets.

An authentication token is also generated and used as authorities' login and their accesses. The service account and service account Identification numbers are created for vehicles to access the cloud storage through WiFi module.

3.2 *Software Implementation*

The software implementation of the proposed speed control system is classified as main programming, functional programming and interface programming. All these programming are performed using integrated development environment of Arduino platform. Embedded C is used for programming in Arduino IDE. The programs, that is, sketches, are compiled and fetched into the Arduino Uno microcontroller. The algorithm of software implementation is as follows:

Algorithm for vehicle module

Step 1: Start the speed control sytem

Step 2: Read GPS coordinates periodically

Step 3: Access cloud storage

Step 4: Compare received GPS coordinates with the database and identify the respective highways.

Step 5: If the vehicle is in same highway then maintain the maximum speed limit as it is.

Step 6: If the vehicle switched to different highway then update the maximum speed limit accordingly.

Step 7: Based on database information, drive the motor to control the throttle valve.

Step 8: If vehicle moves from higher category of highway type to lower category, the maximum speed limit of the vehicle is limited and also when the vehicle moves from lower category of highway type to higher category, the maximum speed limit of the vehicle is increased.

Step 9: Else if the vehicle travels in same highway then no change is carried out on the throttle valve.

Step 10: In case of emergency, a request message is sent to the authorities.

Step 11: If acknowledge message came from authorities, read it and update the maximum speed limit accordingly else continue with same speed limit.

Step 12: If the system is either damaged or by-passed the speed limit, a warning message is displayed and also, an alert message is communicated to the authorities.

Step 13: If the alert message is communicated, decrease the RPM of the engine and then turn off the engine.

Step 14: Repeat from step 2

Algorithm for authorities' module

Step 1: Start the monitoring system

Step 2: Access the cloud storage through secured token if any updates in the cloud server.

Step 3: Wait for request message

Step 4: If any request message is received, verify the nature of vehicle's emergency and make decision to acknowledge the request.

Step 5: If the verification is done and positive, an acknowledge message is generated and communicated with new maximum speed limit for that respective vehicle.

Step 6: If the verification is negative, no acknowledge message is generated.

Step 7: If any alert message is received, issue charge sheet to the vehicle and update the same in the cloud database.

Step 8: Repeat from step 2.

4 Results and Discussions

The prototype of the proposed speed control system for vehicle is shown in Fig. 4. The prototype consists of quad wheeled car with breadboard mounted on it. The Arduino Uno, GPS module, GSM module, motor driver and LCD display are connected using the breadboard. A 12v battery is used s power source for the system. For explanation purposes, some locations are predefined as state highway, national and express highway as shown in Table 1.

When the vehicle is turned on, the GPS module also turned on. Then the GPS module starts to track the location of the vehicle as shown in Fig. 5. As shown in Fig. 6, the maximum speed limit of the vehicle is set to 100 km/h since it is identified as state highway with the help of cloud database. As the vehicle moves, the GPS coordinates are updated. Based on the new updated coordinates, the system identifies the national highway and express highway along with its maximum speed limits as shown in Fig. 7 and Fig. 8, respectively. If the driver attempts to bypass the system and tries to tamper the system, a warning message is displayed as shown in Fig. 9.

Fig. 4 Prototype of the proposed system

Table 1 Predefined GPS locations for highways and their respective maximum speed limit

S.No.	GPS coordinates	Highway type	Maximum speed limit
1	10.9375°N & 76.9558°E	State highway	100 km/h
2	10.9364°N & 76.9549°E	National highway	120 km/h
3	10.9359°N & 76.9537°E	Express highway	140 km/h

Fig. 5 GPS tracking of vehicle location

If any emergency situation, like accidents or medical issues, occurs, then the driver can load the proofs of the emergency situations to the cloud server through the WiFi module. The driver could use vehicle's service account identification number which was already created to the vehicles by the cloud server. A registration number is generated after the successful submission of the proofs. Then the driver can raise a request message along with the registration number to the authorities as shown in Fig. 10.

The authorities can verify the proofs of the emergency situations by referring the registration number that sent by the driver. Meanwhile, the LCD display in the vehicle shows a message as request is in check as shown in Fig. 11. When the verification is successful, the authorities can update those information to the cloud server and also send an acknowledgement message to the respective vehicle which raised the request as shown in Fig. 12. The acknowledgement message also contains updated maximum speed limit that sanctioned by the authorities. The microcontroller receives the message through GSM module and drives the motor driver accordingly to increase the RPM of the engine. Here the vehicle is considered to be travelled in express highway, due to emergency request, the maximum speed limit is raised from default limit of 140 km/h to 160 km/h as shown in the Fig. 13.

Fig. 6 Identification of state highway and its maximum speed limit as 100 km/h

Fig. 7 Identification of national highway and its maximum speed limit 120 km/h

Fig. 8 Identification of national highway and its maximum speed limit 140 km/h

Fig. 9 Display of Warning message

Fig. 10 Sending request to the authorities for emergency situations

Fig. 11 Display of Checking Request Message

Fig. 12 Display of request acceptance as acknowledgement message

Fig. 13 Display of updated maximum speed limit as 160 km/h for express highway

5 Conclusion

The paper proposes an internet of things-based speed control system for vehicles which are travelling on the highways. The GPS coordinates are considered as the fundamental key of this proposed system. Based on the GPS coordinates, the type of highway on which the vehicles is identified and by referring the cloud server, the maximum speed limit of the vehicle is tuned accordingly. The Arduino Uno is used to access the cloud server through the WiFi module. The maximum speed limit of the vehicle is adjusted by controlling throttle valve using dc motors and its drivers. The proposed system also paves way to request the authorities to increase the maximum speed limit in case of emergency situations using GSM module. The authorities verify the request and may grant the respective vehicle to increase the speed limit. The prototype of the proposed system is constructed and verified for its functionalities. The total cost of the prototype did not exceed fifty dollars. This shows that the proposed speed control system is cost-effective and highly efficient in avoiding accidents on the highways. In future work, the proposed approach will be adopted in semi autonomous vehicles and in driverless vehicles.

Acknowledgements The prototype is developed, programmed and verified in the research laboratory of ECE department, Sri Krishna College of Engineering and Technology, Coimbatore, Tamilnadu, India.

References

1. Afify, M., Abuabed, A.S.A.,Alsbou, N.: Smart engine speed control system with ECU system interface. In: 2018 IEEE International Instrumentation and Measurement Technology Conference (I2MTC), pp. 1–6. (2018)
2. Senthil Ganesh, R., Vinoth Kumar, K., Vijayagopal, N., Prithvi Raj, K., Mohanasundar, R.: Automated accidental precautions in public transportation management system. Int. J. Fut. Revol. Comput. Sci. Commun. Eng. **4**(3), 99–105 (2018)
3. Senthil Ganesh, R., Mohanasundar, R., Prithvi Raj, K., Vijayagopal, N., Vinoth Kumar, K.: A survey on issues in public transport management system. Int. J. Res. Appl. Sci. Eng. Technol. **6**(III), 866–870 (2018)
4. Saeed, M.A., Ahmed, N., Hussain, M., Jafar, A.: A comparative study of controllers for optimal speed control of hybrid electric vehicle. In:2016 International Conference on Intelligent Systems Engineering (ICISE), pp. 1–4. IEEE (2016)
5. Xu, S., Peng, H., Song, Z., Chen, K., Tang, Y.: Accurate and smooth speed control for an autonomous vehicle. In: 2018 IEEE Intelligent Vehicles Symposium (IV), pp. 1976–1982. IEEE (2018)
6. Muthu Brindha, G., Karishma, K.K., Mukhesh, A., Nilesh Veeramani, K.K., Mohamed Faizal, M., Maruthi Shankar, B., Vidhya, B.: Accident prevention by automatic braking system and multisensors. Int. J. Adv. Sci. Technol. **29**(8 s), 873–879 (2020)
7. Mahmud, K., Tao, L.: Vehicle speed control through fuzzy logic. In: 2013 IEEE Global High Tech Congress on Electronics, pp. 30–35. IEEE (2013)
8. Arif, S., Iqbal, J., Munawar, S.: Design of embedded motion control system based on modified fuzzy logic controller for intelligent cruise controlled vehicles. In: 2012 International Conference of Robotics and Artificial Intelligence, pp. 19–25. IEEE (2012)

9. Wang, J.F.,, Zhao, H.Q..: Speed control of tracked vehicle autonomous driving system using fuzzy self-tuning PID. In: 2019 4th International Conference on Mechanical, Control and Computer Engineering (ICMCCE), pp. 305–3053. IEEE (2019)
10. Senthil Ganesh, R., Vinothkumar, K., Vijayagopal, N., Prithviraj, K.R., Mohanasundar, R: Smart public transport management system via IOT based automation. Int. J. Sci. Res. Sci. Eng. Technol. **4**(7), 43–47 (2018)
11. Shankar, B.M.: Neural network based hurdle avoidance system for smart vehicles. Int. J. New Pract. Manag. Eng. **8**(4), 01–07 (2019)
12. Senthil Ganesh, R., Mohanasundar, P., Vijayagopal, R., Kumar, V.: Passenger identification system using iot based transportation smart card. Int. J. Sci. Res. Sci. Technol. **4**(5), 401–407 (2018)
13. Jahan, A.S., Hoq, I., Westerfeld, D.: GPS enabled speed control embedded system speed limiting device with display and engine control interface. In: 2013 IEEE Long Island Systems, Applications and Technology Conference (LISAT), pp. 1–7. IEEE, (2013)
14. Lalithavani, J., Senthil Ganesh, R., Sivakumar, S.A., Maruthi Shankar, B.: Cloud server based smart home automation and power management. Materials Today: Proceedings (2020)
15. Hunt, K.J., Johansen, T.A., Kalkkuhl, J., Fritz, H., Gottsche, T.: Speed control design for an experimental vehicle using a generalized gain scheduling approach. IEEE Trans. Control Syst. Technol. **8**(3), 381–395 (2000)
16. Hoess, A.: Realisation of an intelligent cruise control system utilizing classification of distance, relative speed and vehicle speed information. In: Proceedings of the Intelligent Vehicles' 94 Symposium, pp. 7–12. IEEE (1994)

Effective Ensemble Dimensionality Reduction Approach Using Denoising Autoencoder for Intrusion Detection System

Saranya Prabu, Jayashree Padmanabhan, and Geetha Bala

Abstract Intrusion detection system monitors network traffic and issues alert when it spots suspicious activities. Dimensionality reduction is the initial indispensable task while mining a large dataset for improving overall system performance. This preprocessing step is found to be more complex when it needs to analyze the heterogeneous data. In this paper, an ensemble approach of dimensionality reduction of features is presented. In the first phase, the features that have a high correlation score are filtered by evaluating the statistically significant relationship with each other. During second phase, the predictive denoising autoencoder model with bottleneck layer is used for extracting final feature subset along with attack classification. For performance evaluation of the proposed work, DARPA KDDCUP 99 large attack dataset is used and results reveal that ensemble dimensionality reduction approach provides better detection rate on attack dataset compared to many existing approaches.

Keywords Ensemble classifier · Dimensionality reduction · Deep learning · Autoencoder · Intrusion detection systems

1 Introduction

In recent years, billions of devices are connected to the internet, and there is a dramatic increase in data transmission and storage. These leads to a drastic burst in network traffic and equally expose to attackers for data intrusions. There are few traditional methods to safeguard the network including data encryption and firewalls. However, these methods may not be completely preventing intrusions due to the rapid development of intrusion techniques. Thus, there is a vital need for an Intrusion detection system (IDS) to monitor network traffic and alert users while it detects malicious activities [1].

S. Prabu (✉) · J. Padmanabhan
Department of Computer Technology, MIT Campus, Anna University, Chennai, India

G. Bala
Department of Electronics and Communication Engineering, Prince Dr. K. Vasudevan College of Engineering & Technology, Chennai, India

J. S. Raj et al. (eds.), *Intelligent Sustainable Systems*, Lecture Notes in Networks and Systems 213, https://doi.org/10.1007/978-981-16-2422-3_23

Depending on detection-based methodology, IDS can be classified into two types: misuse or signature based detection and anomaly detection [2]. In misuse detection technique, the key idea is to represent attack behaviors as signatures. It needs to maintain a massive database to store signatures and has a highly missed alarm rate. The idea behind anomaly detection system is to define the degree of deviation of abnormal behavior from the normal profile. Since this method has strong generalizability, it is preferred in this paper.

In real world, anomaly IDS needs to handle huge data and it is usually overwhelmed with superfluous and irrelevant features which reduce its detection speed. The big data have significant challenge in discovering the knowledge. Dimensionality reduction is an important preprocessing step to improve the performance of IDS [3]. It includes feature selection and feature extraction as the elementary steps in IDS construction. Feature selection holds the subset of original features but feature extraction generates a new set of features. It is a great tool for eliminating the curse of dimensionality as well as reducing the training time.

Feature selection is used to select the optimal subset of features that contribute useful information to describe the target [4]. It is categorized into filter and wrapper methods. In comparison to wrapper methods, filter methods are faster and also have less computational complexity since it depend only on correlation scores between features. However, wrappers use cross-validation to find the best subset of features by training the machine learning model. Since wrapper methods induce high computational cost, filter method is preferred in this work. Filter is used to pick uncorrelated features followed by feature extraction.

Feature extraction (Feature projection) is used to transform given set of features to new reduced set of features [5]. This transformation may be linear or nonlinear and leads to the removal of irrelevant information. Like PCA, there are many linear feature extraction techniques. But denoising autoencoder is utilized to promote nonlinear transformations in the given set of features. It is availed to efficiently classify network flow information as well as feature extraction synchronously. This work attempts to integrate the good qualities of both feature selection and feature extraction methods.

An autoencoder is a type of unsupervised artificial neural network to learn portrayal of given data. It can be utilized for both dimensionality reduction and classification. Denoising autoencoder is a version of autoencoder to suppress the network to learn the identity function. For this purpose, noise is introduced by corrupting the given data and that should be reconstructed (de-noise) by the autoencoder.

This paper is organized as follows. The related research works are described in Sect. 2. Section 3 introduces the methodologies adopted in this work. Section 4 presents the formulated proposed method. The assessment of the proposed method on the dataset and its findings are discussed in Sect. 5. Section 6 concludes the paper.

2 Related Works

At present, there are numerous studies for intrusion detection based on simple machine learning techniques that have limitations in handling high dimensional feature space. Thereby artificial neural network architectures, especially deep learning have abundantly been used for feature engineering due to its implicit fast learning characteristics [6]. Some works related to dimensionality reduction on network traffic data are analyzed and presented here.

Meng Wang et al. [7] suggested dynamic MLP (Multiayer Perceptrons) by amalgamating sequential feature selection along with the feedback approach. This feedback approach renovates the model while detecting significant detection errors. Samira Sarvariet et al. [8] configured MCF (Mutation Cuckoo Fuzzy) for selecting the appropriate features to enhance execution time and ENN (Evolutionary Neural Network) as a classifier to upgrade its performance by assessing optimal values of model parameters.

Manoharan et al. [9] analyzed the time and energy efficiency by reducing feature space using graph wavelet along with Hermitian wavelet transform. The authors of paper [10] applied packet scoring to detect attacks where leaky bucket flow mechanism has been adopted to identify features contributing towards computing packet legitimacy scores. Recent studies mostly insisted autoencoder for both classification and feature extraction on account of its unsupervised learning approach along with the nonlinear transformation as follows.

SharanRamjee and Aly El Gamal [11] proposed AMBER (Autoencoder and Model-Based Elimination of features using Relevance and Redundancy scores) that employed two models like ranker model and autoencoder. The former model focused on ejection of features that are not pivotal for classification tasks and the latter intents in the ejection of correlated features. It facilitates the advantages of both filter and wrapper methods and hence it is efficient in computation as well as retains high performance. [12] supported automatic feature extraction with help of a simple autoencoder. It concerns dimensionality reduction as well as preserves the relation between features. Then attacks are classified using SVM in this intrusion detection system.

Fangyi Wan et al. [13] used SAE (Stacked Autoencoder) that has a strong ability to extract features and also conserves the primary information of data. They proposed two criteria like Grubbs and PauTa to obtain the reconstruction error for outliers and so that this model promptly finds out the outliers from the normal data [14]. The powered model with SAE obsessed to detect anomalies in web application firewalls by extracting features from HTTP request logs. Isolation Forest is used as a classifier as well as the one class learner to differentiate abnormal from normal data.

Binghao Yan AndGuodong Han [15] exploited SSAE (Stacked Sparse Autoencoder) to learn high dimensional sparse features in a discriminative way. It appreciated the nonlinear mapping to reduce the dimensionality of features. The optimized lower-dimensional features are used to train a classifier and also to test its performance. [16] suggested stacked NDAE (Nonsymmetric Deep Autoencoder) to learn

Table 1 Summary of feature learning on attack detection

Reference paper	Dataset	Proposed feature learning techniques
[7]	ISOT, ISCX and Customized network data	Sequential backward selection – Multilayer perceptron
[8]	NSL-KDD	Mutation Cuckoo Fuzzy
[10]	Customized network data	Double-filtering mechanism using leaky buckets
[11]	MNIST, Wisconsin Breast Cancer and RadioML2016.10b	Autoencoder and Model Based Elimination of features using Relevance and Redundancy scores
[12]	KDD 99 and NSL-KDD	Autoencoder
[13]	ADIAC, FordA, and Mallat	Stacked Autoencoder
[14]	CSIC 2010	Stacked Autoencoder
[15]	KDD 99, NSL-KDD and Kyoto2006	Stacked Sparse Autoencoder
[16]	KDD 99 and NSL-KDD	Nonsymmetric Deep Autoencoder
[17]	KDD 99, UNSW-NB15 and CICIDS2017	Multichannel Autoencoders

features in an unsupervised manner and RF classifier is used to evaluate its performance. By blending both deep and shallow learning, it utilizes their strengths and alleviates their overheads.

GiuseppinaAndresiniet. al. [17] presented MINDFUL (MultI-chanNel Deep Feature Learning for intrusion detection) that unites an unsupervised model of two autoencoders to learn normal and attack flows with a supervised one to exploit dependency of features among channels. The autoencoders inferred the class wise specific features using a multi-channel depiction of network flows along with convolutions to ascertain the impact of one channel on the other.

Table 1 provides summary of various feature learning techniques applied over various network traffic data:

Based on the literature survey, it is inferred that complicated models are used for intrusion detection. Thus, there is a need to create a simple model that holds less computational cost without compromising its performance. This paper is employed with denoising autoencoder which corrupt the given data purposefully and train the model to learn a noisy data that increases its robustness.

3 Methodologies Adopted

3.1 Spearman's Cross-Correlation

Among the various statistical feature selection methods, Spearman's cross-correlation is a widely used method, on account of its ability to describe nonlinear relationship and stability even for outliers established by traffic bursts [18]. The Spearman's cross-correlation rank is calculated by finding the Pearson correlation coefficient between rank variables ρr_x, r_y as follows:

$$Rs = \rho r_x, r_y = \frac{Cov(r_x, r_y)}{\sigma r_x \sigma r_y} \tag{1}$$

where r_x, r_y are ranks of variables x and y; $Cov(r_x, r_y)$ is the covariance of rank variables; $Cov(r_x, r_y)$ are standard deviations of rank variables.

3.2 Autoencoder

Autoencoders are unsupervised neural network that adopt representation learning. It comprises of an encoder and a decoder network as illustrated in Fig. 1. Even though it acts as a classifier, this technique is especially used for dimensionality reduction by enforcing a bottleneck middle layer in the network [19]. This layer provides a compressed representation of the given original data.

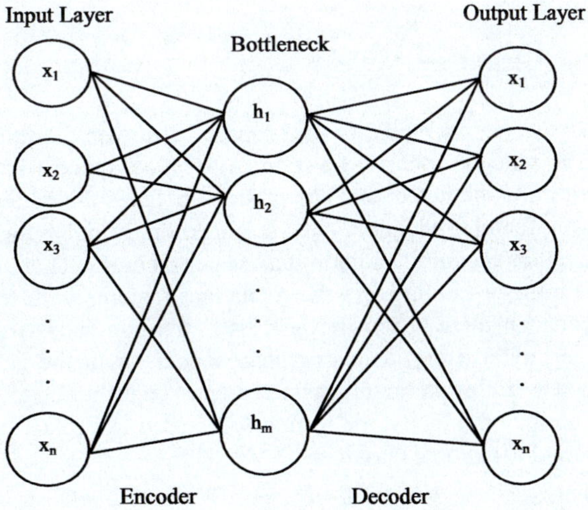

Fig. 1 Simple Architecture of Autoencoder

The model contains training set X = {x x x} with n features. The encoder maps the input training vector x to hidden vector h_i with the help of following mapping function

$$H = h_i = f_\theta(X) = r(WX + b) \tag{2}$$

The decoder maps the hidden vector h_i to the output or reconstruction vector y_i as defined in Eq. 3

$$Y = y_i = f'_\theta(H) = r\left(W'H + b'\right) \tag{3}$$

f_θ and f_θ denotes the mapping function from input to the hidden layer and hidden layer to reconstruction layer, respectively. W and b are the weight connection matrix and bias values from input to the hidden layer. The same values from hidden layer to reconstruction layer are denoted by W′ and b′. The ReLU activation function is denoted by "r" as in Eq. 4

$$r(z) = max(0, z) \tag{4}$$

The goal of autoencoder is to minimize the reconstruction error between input and output vector. Mean squared error metric L(X,Y) is used for optimizing the model parameters as given below

$$L(X, Y) = \frac{1}{m}\sum_{i=1}^{m} ||x_i - y_i||^2 \tag{5}$$

4 Proposed Work

The workflow model of the proposed intrusion detection system is shown in Fig. 2. The work is focused on reducing feature space without compromising accuracy on intrusion detection. The process includes the data preprocessing, feature selection, feature reduction, and classification. Initially, the data are preprocessed using one-hot encoding and min-max scaler methods. Then, Spearman's cross-correlation is utilized to eliminate the redundant features with the help of correlation scores between them. Then the processed training set is used to pretrain and fine-tune the model using denoising autoencoder. With the help of this denoising autoencoder, irrelevant features which are not contributing towards better classification are eliminated. Finally, the testing set is used to evaluate the performance of resultant low-dimensional classifier.

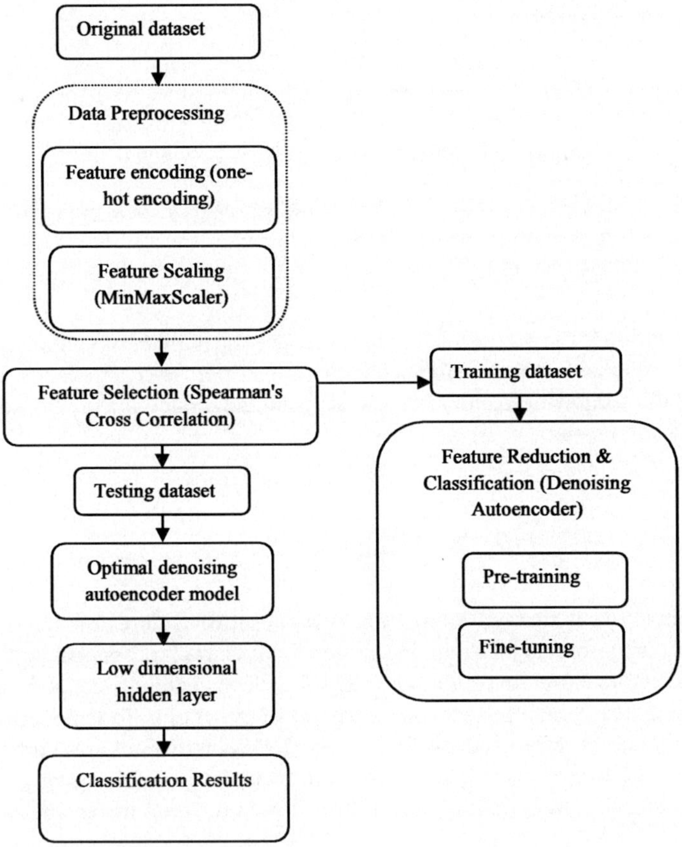

Fig. 2 Proposed Workflow Model

4.1 Feature Encoding

At first step, one-hot encoding is needed to transform symbolic features to numerical features which support machine learning techniques to enhance its prediction process. The symbolic feature values could be label encoded and then columnized. Then it is replaced by binary values, i.e., it assigned 1 for the existing feature values and 0 for leftover feature values.

4.2 *Feature Scaling*

Normalization is used to rescale the original feature values to a specific range so that its values are maintained in the same order of magnitude. For this, min-max normalization is used to transform the feature values between 0 and 1.

4.3 *Feature Selection*

Feature selection is the process that helps to reduce the number of features by identifying the correlated features. In this work, Spearman's cross-correlation is suggested to identify the redundant features because it promoted monotonic relationship rather than linear relationship.

4.4 *Feature Projection and Classification*

Feature extraction is utilized to create a new subset of features by combining the existing feature values. The denoising autoencoder is used to perform both classification and feature extraction simultaneously. Denoising autoencoder is an alteration on the autoencoder which helps to reduce the risk of overfitting. To achieve this goal, it intentionally makes certain input values to zero. Relevant features are extracted using the network bottleneck layer. The size of bottleneck layer is fixed after performing detailed sensitivity analysis on various feature sets towards increasing prediction accuracy.

5 Experiments and Results

5.1 *Dataset*

KDD Cup'99 dataset [20], typical dataset used for anomaly detection study, was developed by MIT Lincoln Labs used for The Third International Knowledge Discovery and Data Mining Tools Competition. Many studies on intrusion detection systems (IDS) rely on this dataset as a benchmark. It comprises a huge range of intruder flows in the military network atmosphere and documented it using 42 attributes with 24 varieties of attack [12]. This dataset comprises 4,898,431 samples for training and 311,027 samples for testing.

5.2 Data Preprocessing

The KDD Cup'99 dataset contains both symbolic and numeric features. The following preprocessing procedures are employed to map the symbolic to numerical features and to maintain the feature values in the same order of magnitude.

Feature and Class Numeralization. The symbolic features of KDD Cup'99 dataset include protocol_type, flag, and service. The protocol_type encompasses 3 feature values, flag has 11 different feature values and service has 70 different feature values. These values are mapped into numerical features with the help of one-hot encoding. As a result of this, 41 features are expanded into 122 features. Likewise, the non-numerical attack classification feature is converted into numerical category by binary coding. The normal and attack classes are assigned as 0 and 1 in binary classification.

Feature Normalization. The Min-max normalization method is used to normalize the feature values between 0 and 1 as given in Eq. 6.

$$f_{norm} = \frac{f - f_{min}}{f_{max} - f_{min}} \tag{6}$$

where f_{norm} denotes the normalized feature value for the original feature value f. Within the original feature values, f_{min} and f_{max} represent the minimum and maximum values.

5.3 Test Bed

GPU enabled Tensorflow is used to assess the proposed work. An Intel Core I7 processor, 32 GB RAM, and Nvidia GTX 1060 6 GB system was used as the test bed for evaluating the proposed work.

Out of 41 features, initially 36 features are selected by Spearman's cross-correlation filter. To further reduce the features, an experiment is conducted to study the optimal number of features that give high classification accuracy. For this purpose, denoising autoencoder with bottleneck layer is adopted. The size of bottleneck layer is varied from 25 to 6 indicating reduced number of features that can be selected. The back propagation approach is applied to adjust the model parameters with Adam optimizer, since the experimental studies indicate Adam optimizer surpass other stochastic optimizers [21]. Mean squared loss function is used to minimize the reconstruction error. ReLU activation function is used to overcome the vanishing gradient problem and allow models to perform better.

The size of bottleneck layer is fixed based on number of features for which accuracy is maximum and stable based on the results obtained with simple autoencoder and ensemble autoencoder as given in Table 2.

Table 2 Impact of bottleneck layer on the classification performance

No. of neurons in hidden layer	Metrics (%) (mean value)	Architecture	
		Simple Autoencoder	Ensemble Autoencoder
25 features	Accuracy	91.50	80.89
	F1-Score	94.98	89.39
	Precision	90.47	80.82
	Recall	99.97	99.99
20 features	Accuracy	90.26	87.40
	F1-Score	91.61	92.74
	Precision	84.52	86.48
	Recall	99.98	99.98
15 features	Accuracy	92.61	91.47
	F1-Score	95.60	94.96
	Precision	91.83	90.49
	Recall	99.69	99.90
10 features	Accuracy	92.92	93.90
	F1-Score	95.72	96.33
	Precision	93.33	93.31
	Recall	99.22	99.55
8 features	Accuracy	90.02	93.99
	F1-Score	96.41	96.39
	Precision	93.39	93.45
	Recall	98.62	99.59
6 features	Accuracy	89.11	90.46
	F1-Score	95.24	96.08
	Precision	92.70	92.81
	Recall	97.91	95.59

5.4 Evaluation Metrics and Result Analysis

The performance of proposed intrusion detection system is evaluated on four metrics namely accuracy, detection rate, precision, and F1 score. The metrics are based on the parameters True Positive (TP), True Negative (TN), False Negative (FN), and False Positive (FP), which are defined as in Table 3.

The level of accuracy or classification rate is the ratio of the number of correct decisions that are made to the total number of test samples.

$$Accuracy = \frac{TP + TN}{TP + FP + FN + TN} \tag{7}$$

Table 3 Confusion matrix parameters

TP	Number of anomaly links correctly classified as attackers
TN	Number of normal links correctly classified as normal
FP	Number of normal links that are misclassified as attackers
FN	Number of anomaly links that are misclassified as normal

Precision is the ratio of the number of correctly identified attackers to the total predicted values.

$$Precision = \frac{TP}{TP + FP} \tag{8}$$

Recall or detection rate is the ratio of the number of correctly identified attackers to the total actual values.

$$Detectionrate = \frac{TP}{TP + FN} \tag{9}$$

F1 score is calculated by finding the harmonic mean of precision and recall.

$$F1score = 2 * \frac{Precision * Recall}{Precision + Recall} \tag{10}$$

Srv_serror_rate, same_srv_rate, diff_srv_rate, dst_host_diff_srv_rate, and dst_host_serror_rate features are eliminated in phase I using Spearman's cross-correlation. Using denoising autoencoder 8, most relevant features are selected to identify attacks efficiently.

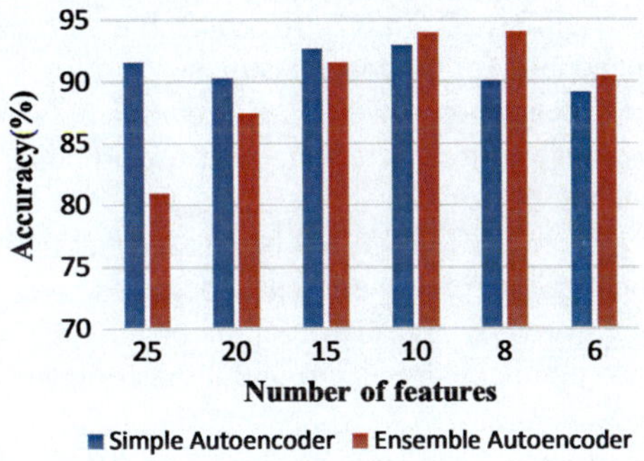

Fig. 3 Size of bottleneck layer versus accuracy

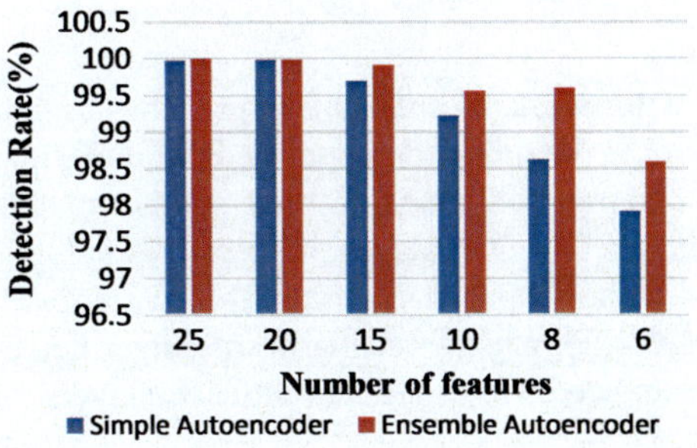

Fig. 4 Size of bottleneck layer versus detection rate

Table 4 Anomaly detection performance on KDDCUP dataset

Model	Accuracy (%)	Precision (%)	Recall (%)	F1-Score (%)
Stacked non-symmetric deep autoencoder [16]	97.85	99.99	97.85	98.15
Mirrored adversarial autoencoder [22]	–	95.27	96.77	96.01
Multi-channel deep feature learning [17]	92.49	–	–	95.13
Proposed ensemble autoencoder	93.99	93.45	**99.59**	96.39

Figures 3 and 4 show the effect of feature size variation on prediction accuracy and detection rate for simple autoencoder and ensemble autoencoder.

From the sensitivity analysis graphs the following are inferred. Initially, the accuracy of simple autoencoder is good when compared to ensemble autoencoder. But it gradually decreases, when the number of features in hidden layer is decreased. From 10 features onward, ensemble autoencoder started giving better accuracy than simple autoencoder.

It is also inferred that the detection rate of simple autoencoder is good when there are more number of features, while the detection rate of ensemble autoencoder is becoming better with reduction in features. Since the detection rate and accuracy are better after 10 features for ensemble autoencoder, the optimal number of features is fixed as 8 for proposed encoder

The proposed work is also compared with the existing works in the literature for comparing the performance of the proposed system and the results are tabulated as shown in Table 4.

From the Table 4, it is concluded that the proposed autoencoder-based dimensionality reduction system with 8 features gives good recall indicating that the attackers are correctly classified compared to other systems. The other metrics of accuracy and precision are also found to be comparable with existing systems.

6 Conclusion

In this work, an ensemble approach is used for dimensionality reduction for network traffic data towards better detection of attack data. By utilizing statistical machine learning techniques, a good number of features have been selected and validated on KDDCUP dataset. Comparative analysis on the performance of system indicates that the proposed system is able to compete well with other existing systems, while providing a better recall towards attack detection. The proposed system can be further enhanced with multiple bottleneck layers to improve accuracy of detection.

References

1. Mohammadi, S., Mirvaziri, H., Ghazizadeh-Ahsaee, M., Karimipour, Hadis: Cyber intrusion detection by combined feature selection algorithm. Journal of information security and applications **44**, 80–88 (2019)
2. Liu, H., Lang, B.: Machine learning and deep learning methods for intrusion detection systems: a survey. Appl. Sci. **9**(20), 4396 (2019)
3. Sisiaridis, D., Markowitch,O.: Feature extraction and feature selection: Reducing data complexity with apache spark. arXiv preprint arXiv:1712.08618 (2017)
4. Singh, S., Silakari, S.: An ensemble approach for feature selection of Cyber Attack Dataset. arXiv preprint arXiv:0912.1014 (2009)
5. Moore, K.L., Bihl, T.J., Bauer Jr, K.W., Dube, T.E.: Feature extraction and feature selection for classifying cyber traffic threats. J. Defense Model. Simul. **14**(3), 217–231 (2017)
6. Smys, S., Zong Chen, J.I., Shakya, S.: Survey on neural network architectures with deep learning. J. Soft Comput. Paradigm (JSCP) **2**(3), 186–194 (2020)
7. Wang, M., Yiqin, L., Qin, J.: A dynamic MLP-based DDoS attack detection method using feature selection and feedback. Comput. Secur. **88**, (2020)
8. Sarvari, S., Fazlida Mohd Sani, N., Mohd Hanapi, Z., Taufik Abdullah, M.: An efficient anomaly intrusion detection method with feature selection and evolutionary neural network. IEEE Access **8**, 70651–70663 (2020)
9. Manoharan, S.: Study on hermitian graph wavelets in feature detection. J. Soft Comput. Paradigm (JSCP) **1**(01), 24–32 (2019)
10. Jayashree, P., Easwarakumar, K.S., Anandharaman, V., Aswin, K.: A proactive statistical defense solution for DDoS attacks in active networks. In: 2008 First International Conference on Emerging Trends in Engineering and Technology, pp. 878–881. IEEE, 2008
11. Ramjee, S., El Gamal, A.: Efficient wrapper feature selection using autoencoder and model based elimination. arXiv preprint arXiv:1905.11592 (2019)
12. Kunang, Y.N., Nurmaini, S., Stiawan, D., Zarkasi, A., Jasmir, F.: Automatic features extraction using autoencoder in intrusion detection system. In: 2018 International Conference on Electrical Engineering and Computer Science (ICECOS), pp. 219–224. IEEE, 2018
13. Wan, Fangyi., Guo, Gaodeng., Zhang, Chunlin., Guo, Qing, Liu, Jie: outlier detection for monitoring data using stacked autoencoder. IEEE Access **7**, 173827–173837 (2019)
14. Vartouni, A.M., Sedighian Kashi, S., Teshnehlab, M.: An anomaly detection method to detect web attacks using stacked auto-encoder. In: 2018 6th Iranian Joint Congress on Fuzzy and Intelligent Systems (CFIS), pp. 131–134. IEEE, 2018
15. Yan, B., Han, G.: Effective feature extraction via stacked sparse autoencoder to improve intrusion detection system. IEEE Access **6**, 41238–41248 (2018)
16. Shone, N., Nguyen Ngoc, T., Dinh Phai, V., Shi, Q.: A deep learning approach to network intrusion detection. IEEE Trans. Emerg. Top. Comput. Intell. **2**(1), 41–50 (2018)
17. Andresini, G., Appice, A., Di Mauro, N., Loglisci, C., Malerba, D.: Multi-channel deep feature learning for intrusion detection. IEEE Access **8**, 53346–53359 (2020)
18. Wei, W., Chen, F., Xia, Y., Jin, G.: A rank correlation based detection against distributed reflection DoS attacks. IEEE Commun. Lett. **17**(1), 173–175 (2013)
19. Farahnakian, F., Heikkonen, J.: A deep auto-encoder based approach for intrusion detection system. In: 2018 20th International Conference on Advanced Communication Technology (ICACT), pp. 178–183. IEEE, 2018
20. KDD Cup 1999 Data. [online] Available: http://kdd.ics.uci.edu/databases/kddcup99/kddcup99.html
21. Kingma, D.P., Ba, J.: Adam: A method for stochastic optimization. arXiv preprint arXiv:1412.6980 (2014)
22. Wu, Y., Balaji, Y., Vinzamuri, B., Feizi, S.: Mirrored autoencoders with simplex interpolation for unsupervised anomaly detection. arXiv preprint arXiv:2003.10713 (2020)

Computer-Aided Detection for Early Detection of Lung Cancer Using CT Images

Usha Desai, Sowmya Kamath, Akshaya D. Shetty, and M. Sandeep Prabhu

Abstract Doctors face difficulty in the diagnosis of lung cancer due to the complex nature and clinical interrelations of computer-diagnosed scan images. Hence, the visual inspection and subjective evaluation methods are time consuming and tedious, which leads to inter and intra observer inconsistency or imprecise classification. The Computer-Aided Detection (CAD) can help the clinicians for objective decision-making, early diagnosis, and classification of cancerous abnormalities. In this work, CAD has been employed to enhance the accuracy, sensitivity, and specificity of automated detection in which, the phases of lung cancer are discriminated using image processing tools. Cancer is the second leading cause of death in non-communicable diseases worldwide. Lung cancer is, in fact, the most dangerous form of cancer that affects both the genders. Either or both sides of the lung begin to expand during the uncontrolled growth of extraordinary cells. The most widely used imaging technique for lung cancer diagnosis is Computerised Tomography (CT) scanning. In this work, the CAD method is used to differentiate between the phases of pictures of lung cancer stages. Abnormality detection consists of 4 steps: pre-processing, segmentation, extraction of features, and classification of input CT images. For the segmentation process, Marker-controlled watershed segmentation and the K-means algorithm are used. From CT images, normal and abnormal information is extracted and its characteristics are determined. Stages 1–4 of cancerous imaging were discriminated and graded with approximately 80% efficiency using neural network feedforward backpropagation algorithm. Input data is collected from the Lung Image Database Consortium (LIDC), which out of 1018 dataset cases uses 100 cases. For the output display, a graphical user interface (GUI) is developed. This automated and robust CAD system is necessary for accurate and quick screening of the mass population.

U. Desai (✉) · S. Kamath · A. D. Shetty
Department of Electronics and Communication Engineering, NMAM Institute of Technology, Nitte, Udupi 574110, Karnataka, India
e-mail: usha.nmamit@nitte.edu.in

M. S. Prabhu
Department of Electronics and Communication Engineering, Canara Engineering College, Benjanapadavu, Mangaluru 574219, Karnataka, India

J. S. Raj et al. (eds.), *Intelligent Sustainable Systems*, Lecture Notes in Networks and Systems 213, https://doi.org/10.1007/978-981-16-2422-3_24

Keywords Computer-Aided detection (CAD) · Lung cancer detection · CT scan images · Watershed algorithm · PCA · Neural network classifier · GUI

1 Introduction

Cancer is the world's second biggest disease, for cause of death, causing an estimated 9.6 million fatalities in 2018. Around 1 in 6 individuals worldwide die from cancer [1]. Lung cancer is a major cause of death in both men and women [2]. Cancer mortality rates are higher in men than in women. The probability of a man getting lung cancer in his lifespan is 1 in 15, however in a woman the risk is about 1 in 17. Smoking is the major risk factor that leads to cancer which includes cigarettes, cigars, and pipes. Cigarette smokers are 15–30 times more likely to get lung cancer than non-smokers, according to the Centers for Disease Control and Prevention (CDC) report. Due to cigarette smoking, around 85% of males and 75% of females suffer from lung cancer [2]. Remaining 10–15% occurrences are due to air pollution, second hand smoking, exposure to radon gases, asbestos, and other carcinogens. Cancer starts when cells begin to expand uncontrollably in the lungs. There are mainly two types of lung cancer, depending on the microscopic presence of tumour cells, i.e. "Small Cell Lung Cancer (SCLC)" and "Nonsmall Cell Lung Cancer (NSCLC)". The most well-known amoung the two is NSCLC, which makes up about 80–85% of all cases. Around 15–20% of lung tumours are alluded to by SCLC. SCLC's development speed is higher than NSCLC and much faster than NSCLC. Breathing in hazardous chemicals, especially over a long period of time, can also cause lung cancer, according to the American Lung Association. It can spread anywhere in the body once cancer has reached the lymph nodes and bloodstream. Before cancer spreads beyond the lungs, care needs to be taken. As a result of early cancer diagnosis, the probability of patient survival rates improves. The most widely used imaging technique used for diagnosing lung cancer is CT scanning [3, 4] (Fig. 1). The lung nodule is an abnormality that contributes to lung cancer, marked by a small circular or oval lung

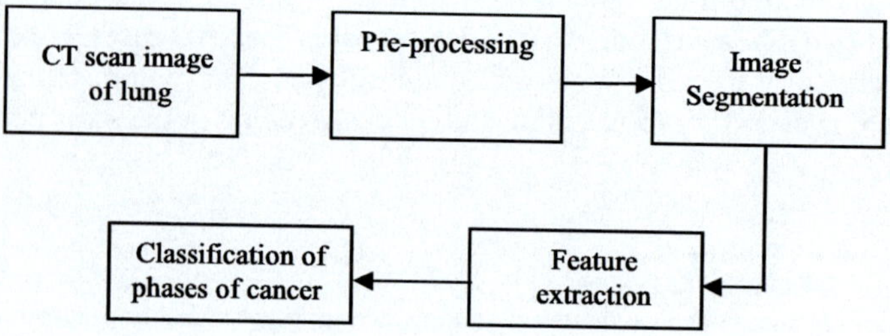

Fig. 1 Block diagram for lung cancer detection using CT scan images

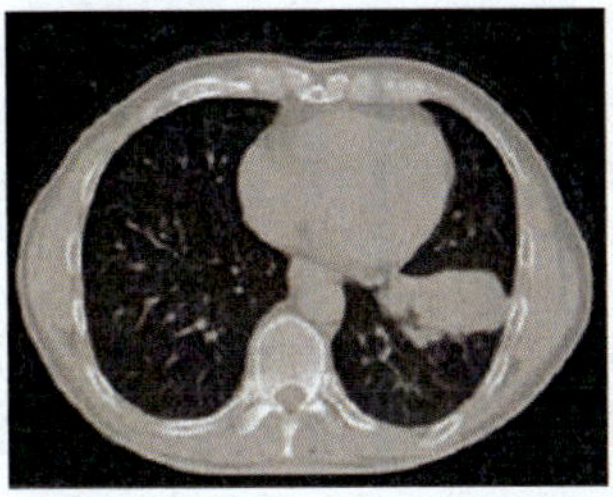

Fig. 2 A typical CT scan image of a male patient of age 57 years [5]

growth that appears in the CT scan picture as a white shadow (Fig. 2). The growth of healthy lung tissues is limited by these uncontrollable cells. In addition, this abnormality is not treated at the right time; this development in the nearby tissue, called metastasis, will spread beyond the lung. Early detection of cancer is important in order to save the life of a person suffering from the abnormality [6–13]. Using CAD techniques [14–24], this can be done.

In this research, abnormal (Stages I-IV) classes of lung cancer using CT scan images are discriminated with an efficiency measure of 80% using the feedforward backpropagation neural network. This paper is further organised as follows: Sect. 2 addresses the materials and methods applied. Section 3 describes the findings and Sect. 4 concludes the paper.

2 Materials and Methodology

In this paper, for segmentation and classification of normal and abnormal types, CT images are used. As compared with X-ray and Magnetic Resonance Imaging (MRI) methods, CT images have low noise. The key benefit of using CT images is that with less distortion, it gives clarity. For the retrieval of a range of patient CT scan images, the Lung Image Database Consortium Image Collection (LIDC-IDRI) [5] is used. This database consists of CT images of lung cancer screening for the creation, training, and evaluation of Computer-Assisted Diagnostic (CAD) methods for the detection and diagnosis of lung cancer, initiated by the National Cancer Institute. In this database, 1018 dataset cases contributed by seven academic centres and eight medical imaging companies are present in which the images are in DICOM format, 512 × 512 pixels in size. It is difficult to process the DICOM format; these files are therefore converted to the JPEG format. DICOM format CT scan images are converted using Micro DICOM software to a suitable JPEG format. A set of Digital Imaging and Communications in Medicine (DICOM) was collected [5]. As shown in Fig. 1, in the proposed system's flow diagram, Input CT images taken from the database [5] are pre-processed by means of a median filter and applied for segmentation. Using PCA, the dimensionality of segmented data is decreased and then supplied for function extraction and classification processes. The median filter is used in image pre-processing on the CT scan image greyscale image. In addition, the processed picture is segmented to section the lung tumours using K-means and watershed segmentation. Using Discrete Wavelet Transform (DWT) and Principal Component Analysis (PCA) algorithm, functions are extracted. Finally, using the neural network classifier, the segmented nodule was discriminated into stages and classified.

A. *Pre-processing:* Image pre-processing is done to improve the image characteristics that are necessary for further processing and to eliminate the noise factors present in the CT image. The initial stage in the diagnosis of lung cancer is pre-processing. Conversion and filtering of the greyscale are critical steps in image pre-processing. The greyscale picture typically consists of sources of noise such as Gaussian noise and salt-and-pepper noise [25], which can be filtered by means of a median filter method.
B. *Greyscale Conversion:* A greyscale image is one that depicts an image's individual pixel value. The greyscale image emits only a single light source and only contains luminance information, but it does not contain any colour information. For greyscale pictures, the sensitivity level is black at the lowest level and white colour at the maximum intensity level. The grey image consists of 8 bits per pixel and the picture representation is from 0 to 255, reflecting different brightness levels.
C. *Filtering:* It is a tool used to alter or improve an image. Median filter is a nonlinear approach with a very useful and effective technique that removes the noise in the image. The edges are identified by this filter; first noise in the image should be eliminated, and then edge removal should be carried out. Without eliminating edges, the primary use of the median filter is to eliminate the noise. A basic working procedure is followed by the Median filter. The neighbouring pixels are considered for any pixel in an image. First the neighbouring pixels are sorted out and then the median value of all the sorted pixels is substituted for the pixel, and if the neighbouring pixels are even in number, the median value of the middle two pixels is used for the substitution. Median filter function can be represented for a 2-dimensional image as

$$M = medfilt2(S[u, v]) \tag{1}$$

which performs median filtering of the matrix S in two dimensions [26].
D. *Segmentation using Watershed Algorithm*
By considering the plane in which the light elements are elevated and dark elements are low, the Watershed transform discovers the image's watershed ridge lines [27]. Compared to K-means segmentation, Marker-controlled watershed correctly transforms segments of the specified regions; thus the proposed

model uses watershed segmentation. Its key feature is that the touching objects in the image can be isolated and marked [4]. This characteristic assists in the sufficient segmentation of cancer images if other incorrect clusters are affected [28].

Steps involved in segmentation using marker-controlled watershed are

i. Colour image data is converted into Greyscale value.
ii. The image gradient value, which is high at the borders and low within the objects, is evaluated.
iii. Markers in the image are calculated and thresholding operation is done on the dark pixels that belong to the background.
iv. Watershed transform is applied to segment the image and results are visualised.

E. *K-means Clustering:* The algorithm finds a number of centres and then allocates each data point to the nearest centre, keeping the area of the centres small. K-means is used in this work to segment the field of analysis of the image. Based on the K-centroids given, it clusters the given data into K-clusters or sections. The steps in K-means algorithm are

i. Select a number of clusters.
ii. Select the centroids at random K points.
iii. Between each cluster centre and each pixel of the given image, compute the Euclidean distance.
iv. Depending upon the Euclidean distance, each pixel is assigned to a cluster with the closest centre.
v. There is a re-calculation of the location of the cluster centre.
vi. Until it converges, the process is repeated.
vii. After that, cluster pixels are reshaped into an image.

F. *Principal Component Analysis* (*PCA*): PCA is a strategy for minimising linear dimensionality, which by emphasising their similarities and differences projects the data into the directions of highest variability. This method calculates the Principal Components (PCs) in the direction of decreasing order of variability, which are diagonal vectors, and allows the main directions in which the data differs to be defined. The first PC provides the fundamental vector for the direction of maximum variability or maximum discrimination. For the next orthogonal direction to the first PC, the second PC provides the basis vector, and so on. The first few PCs reflect the maximum prejudice that occurs in the dataset. In order to choose the number of PCs, a percentage of total variability of the data is set as the threshold. The PCA computation for the image includes following steps:

(i) Covariance Matrix Computation,
(ii) Eigenvalue and estimation of own vector,
(iii) Structure of the eigenvectors in the downward direction of the eigenvalues, and
(iv) Projecting the original ECG beats by taking the internal product of the original signals and the sorted eigenvectors in the direction of the arranged eigenvectors.

G. *Discrete Wavelet Transform* (*DWT*): DWT is used to decompose data into various components of frequency, and then each component is considered for analysis. The DWT technique successively decomposes the signal into several levels of high-frequency and low-frequency sub-band components. The coefficients of the high-frequency signal are called the information and the equivalents of the low frequency are called the approximation. The shorter windows (fine resolution) are used to cover minute shifts at high frequencies and the long windows are used at low frequencies (coarse resolution). The DWT, scaling, and wavelet functions are expressed as

$$\psi_{high}\left(2^{j}t\right) = \sum_{k=-\infty}^{\infty} h_1(k)\sqrt{2}\varphi\left(2^{j+1}t - k\right) \tag{2}$$

$$\varphi_{low}\left(2^{j}t\right) = \sum_{k=-\infty}^{\infty} h_0(k)\sqrt{2}\varphi\left(2^{j+1}t - k\right) \tag{3}$$

In Eq. (2), the high-pass filter $h_1(k)$ corresponding to the wavelet function $\psi_{j,k}(t)$ extracts the image details, and in Eq. (3) the low-pass filter $h_0(k)$ corresponding to the scaling function $\varphi_{j,k}(t)$ extracts the approximation information in the image. There is a finite length of the DWT mother wavelet, and the frequency information derived from the transformation of the wavelet is localised in time. This results in excellent extraction of features solely from non-stationary data such as in pictures.

H. *Grey-Level Co-occurrence Matrix* (*GLCM*): It gives the distribution of co-occurring pixel coefficient at a particular threshold. It is used as a technique to texture study with different applications predominantly in medical image investigation. GLCM is expressed as

$$P_{i,j} = \frac{V_{i,j}}{\sum_{i,j=0}^{N-1} V_{i,j}} \tag{4}$$

where i is the row number and j is the column number.
Features that are extracted in this work are listed below.
Contrast: Contrast measures local variations in GLCM. It calculates intensity contrast between a pixel and its neighbour pixel for the whole image. Contrast is zero for a constant image. It is given by the equation:

$$\text{Contrast} = \Sigma\Sigma\,(i-j)^2 P(i,j) \tag{5}$$

where P(i,j) is the pixel at location (i,j).
Energy: The sum of square values in the GLCM is defined by this function.

$$\text{Energy} = \Sigma\Sigma(P(i,j))^2 \tag{6}$$

Correlation: Correlation tests the occurrence of the joint likelihood of the pixel pairs defined. It is given by the equation:

$$\text{Correlation} = \frac{\sum_{i=0}^{G-1}\sum_{j=0}^{G-1}(i-\mu_i)(j-\mu_j)P(i,j)}{\sigma_i\sigma_j} \tag{7}$$

Homogeneity: Homogeneity tests the closeness to the GLCM diagonal of the distribution of the elements in the GLCM.

$$\text{Homogenicity} = \frac{\sum_{i,j} P(i,j)}{1+|i-j|} \tag{8}$$

I. *Neural Network Classifier:* For the classification of cancer stages, this classifier is used. To distinguish stages of lung cancer, a distinctive feedforward back-propagation neural network is implemented in this paper. Neural networks act for the purpose of nonlinear information, simulating the reflection of biological neurons. Features derived from the previous step are added as an input to the neural network. With 12 and 8 neurons, respectively, the proposed neural network model has two hidden layers in which input information is distributed into samples of training and tests. The training process continues if the network on the validation set continues to develop. After preparation, experiments with new datasets are carried out. The output indicates whether or not the individual input sample selected has an abnormality and demonstrates the stage of cancer. Training the network with a wide range of input datasets will increase detection accuracy (Fig. 3).

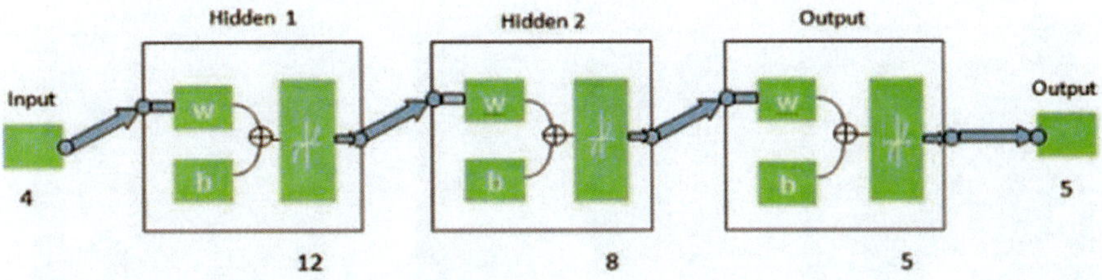

Fig. 3 The proposed neural network model

3 Results and Discussion

Figure 4 shows the output of the pre-processing stage. Here, the input image is converted into greyscale and denoised using the median filter. The proposed system uses watershed segmentation for the segmentation of lung nodules. The output of each step of the segmentation process is shown in Figs. 5, 6, 7, 8, and 9, respectively.

The proposed system uses feedforward neural network. Figures 9 and 10 shows the GUI to start training of neural network and to check the input image, respectively. Figure 11 shows the training of the neural network model.

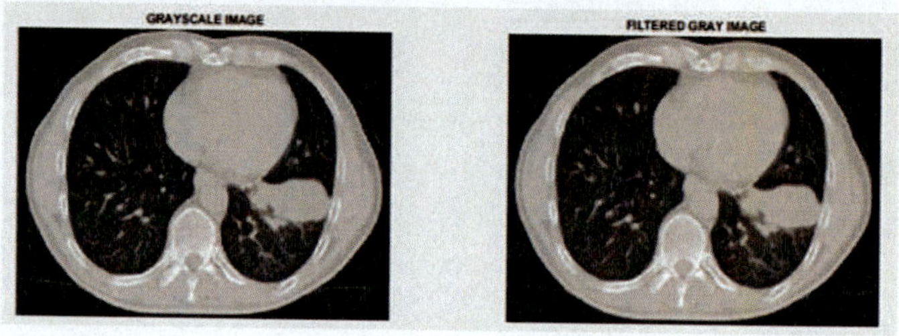

Fig. 4 Pre-processed output

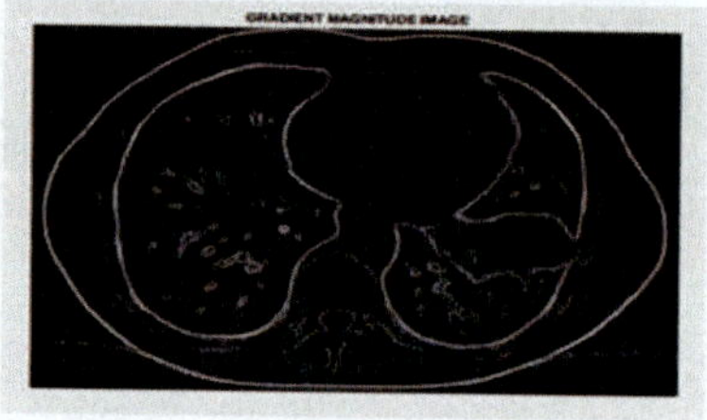

Fig. 5 Gradient magnitude image

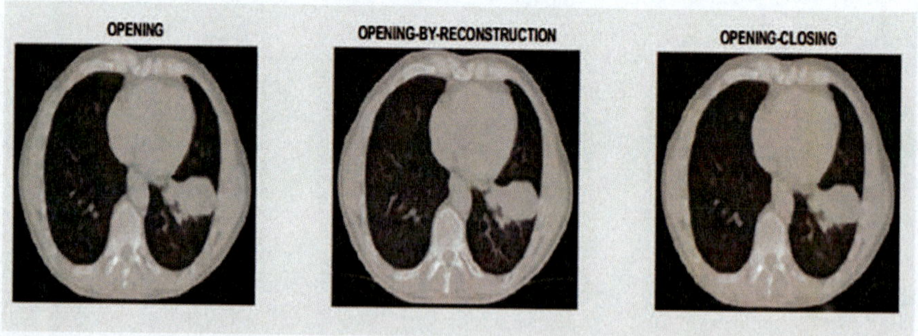

Fig. 6 Representation of foreground markers

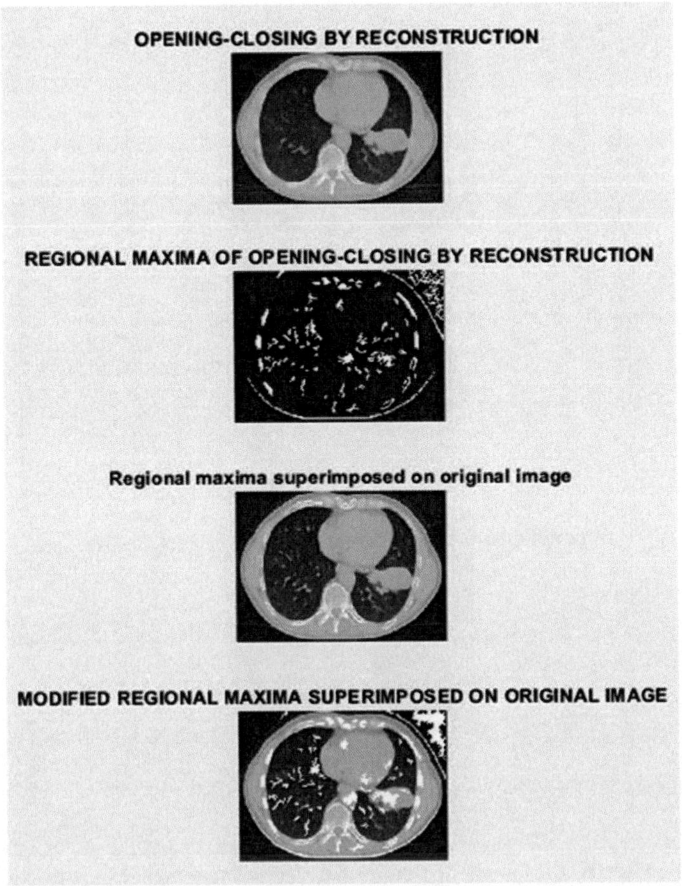

Fig. 7 Representation of foreground markers after processing

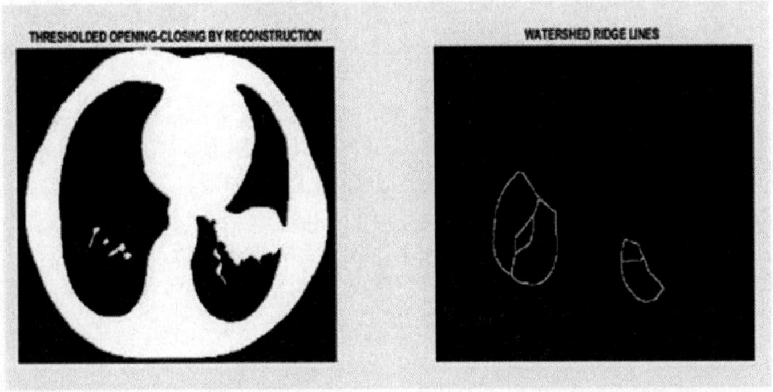

Fig. 8 Representation of background markers

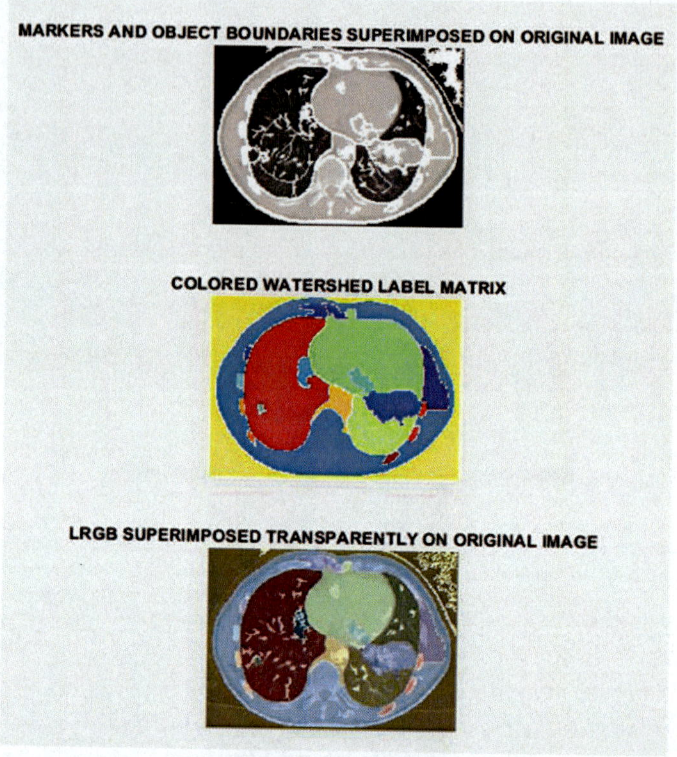

Fig. 9 Result of superimposing marked image on original image

Table 1 indicates the range of feature extracted values used to discriminate the stages. Figure 12 shows non-cancerous stage (normal) of cancer and Figs. 13 and 14 show discrimination of cancer in Stages I–IV, respectively. Stage I and Stage II of cancer are shown in Fig. 15. The tumour is present in Stage I cancer in the initial stage but has not spread beyond the lung. Cancer is limited to the lung and lymph nodes during Stage II.

Stage III and Stage IV of cancer are shown in Fig. 14. Cancer is located in the lungs and lymph nodes in the middle of the chest in Stage III. In Stage IV, cancer has spread to the lungs as well as to other areas of the body.

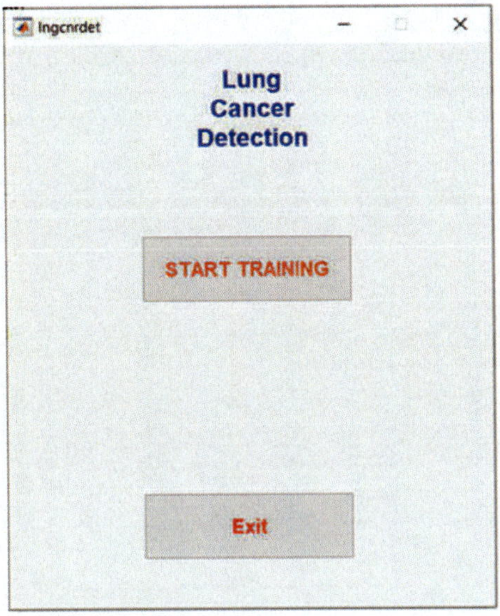

Fig. 10 GUI window to start the training of network

4 Conclusion

The proposed device detects lung cancer with an efficiency of 80% using the feed-forward neural network technique. Feature extraction methods are used to differentiate between the phases of lung cancer using an acceptable set of feature values extracted. The proposed method lacks the clarification of the distinct cancer phases IIIA and IIIB, respectively. For classification, it utilises only four characteristics of features. By selecting more features for classification, the efficiency of the system can be improved. Therefore, using nonlinear transformation techniques to distinguish subtle changes present in the images, cancerous stages can be clearly identified as a potential scope of improvement. This computer-aided system will therefore be a vital support system for medical diagnosis.

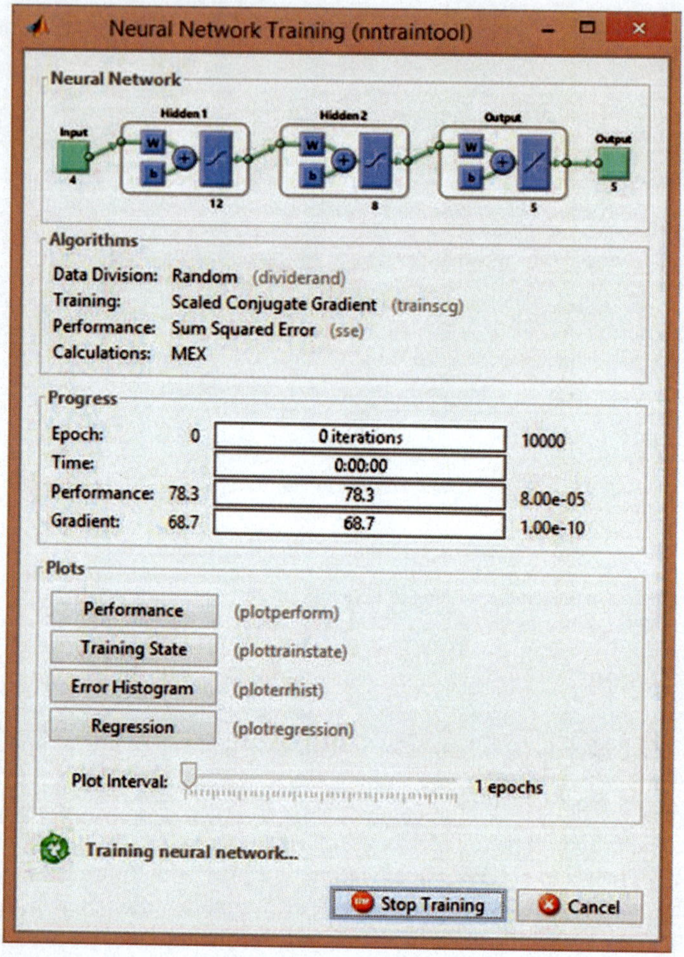

Fig. 11 Training of the neural network classifier

Table 1 Discrimination of stages based on feature value ranges

Correlation	Energy	Stage detected
0.1937–0.1659	0.9303–09275	Stage1
0.1584–0.1552	0.9238–0.9152	Stage2
0.1537–0.1409	0.9125–0.9103	Stage3
0.1126–0.1072	0.9102–0.8950	Stage4

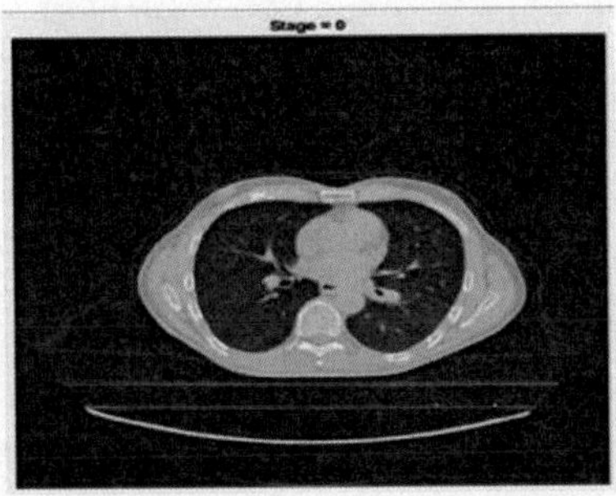

Fig. 12 Non-cancerous stage

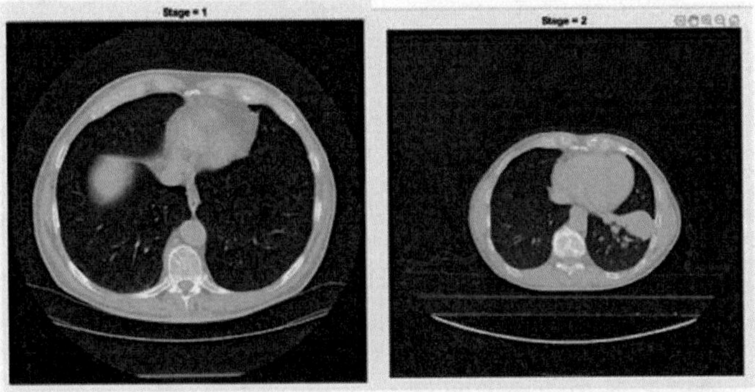

Fig. 13 Stage I and Stage II of lung cancer

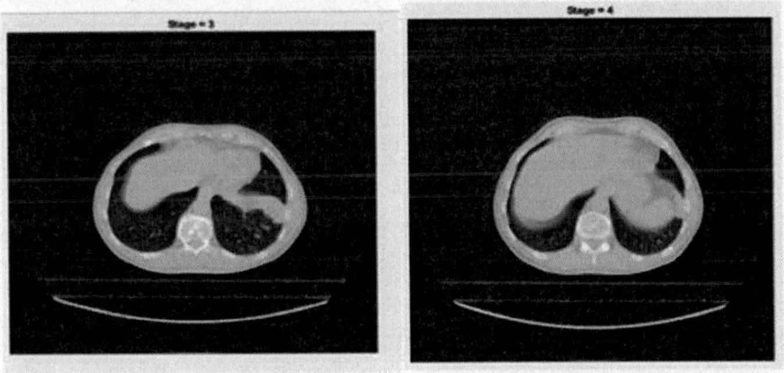

Fig. 14 Lung Cancer Stage I and Stage II

References

1. https://www.who.int/news-room/factsheets/detail/cancer
2. https://www.cancer.org/research/cancer-facts-statistics/all-cancer-facts-figures/cancer-facts-figures-2020.html
3. Torre, L.A., Siegel, R.L. Jemal, A.: Lung cancer statistics. In: Lung Cancer and Personalized Medicine, pp. 1–19. Springer, Heidelberg (2016)
4. Hancock, M.C., Magnan, J.F.: Lung nodule malignancy classification using only radiologist quantified image features as inputs to statistical learning algorithms: probing the lung image database consortium dataset with two statistical learning methods. SPIE J. Med. Imaging (2016)
5. https://wiki.cancerimagingarchive.net/display/Public/Lung+CT+Segmentation+Challenge+2017
6. Li, Y., Zhang, L., Chen, H., Yang, N.: Lung nodule detection with deep learning in 3D thoracic MR images. IEEE Access **7**, 37 822–37 832 (2019)
7. Sharma, D., Jindal, G.: Identifying lung cancer using image processing techniques. International Conference on Computational Techniques and Artificial Intelligence (ICCTAI) **17**, 872–880 (2011)
8. Abdillah, B., Bustamam, A., Sarwinda, D.: Image processing based detection of lung cancer on CT scan images. J. Phys.: Conf. Ser. **893**, 012063 (2017)
9. Makaju, S., Prasad, P., Alsadoon, A., Singh, A., Elchouemi, A.: Lung cancer detection using CT scan images. In: 6th International Conference on Smart Computing and Communications, Procedia Computer Science, vol. 125, pp. 107–114 (2018)
10. Talebpour, A.R., Hemmati, H.R., Hosseinian, M.Z.: Automatic lung nodules detection in computed tomography images using nodule filtering and neural networks. In: 22nd Iranian Conference on Electrical Engineering (ICEE), pp. 1883–1887 (2014)
11. Rahane, W., Dalvi, H., Magar, Y., et al.: Lung cancer detection using image processing and machine learning healthcare. In: International Conference on Current Trends towards Converging Technologies (ICCTCT), pp. 1–5 (2018)
12. Tekade, R., Rajeswari, K.: Lung cancer detection and classification using deep learning. In: Fourth International Conference on Computing Communication Control and Automation (ICCUBEA), pp. 1–5 (2018)
13. Senthil Kumar, K., Venkatalakshmi, K., Karthikeyan, K.: Lung cancer detection using image segmentation by means of various evolutionary algorithms. Computational and Mathematical Methods in Medicine (2019)
14. Desai, U.: Cardiac monitoring & resuscitation system for ventricular fibrillation detection. In: International Conference on Circuits and Systems in Digital Enterprise Technology (ICCSDET), pp. 1–3. IEEE, (2018)
15. Gurupur, V.P., Kulkarni, S.A., Liu, X., Desai, U., Nasir, A.: Analysing the power of deep learning techniques over the traditional methods using medicare utilisation and provider data. J. Exp. Theoret. Artif. Intell. **31**(1), 99–115 (2019)
16. Desai, U., et al.: Discrete cosine transform features in automated classification of cardiac arrhythmia beats. Emerging Research in Computing, Information. Communication and Applications. Springer, New Delhi (2015)
17. Desai, U., Martis, R.J., Acharya, U.R., et al.: Diagnosis of multiclass tachycardia beats using recurrence quantification analysis and ensemble classifiers. J. Mech. Med. Biol. **16**, 1640005 (2016)
18. Desai, U., Martis, R.J., Nayak, C.G., Sarika, K., Seshikala, G.: Machine intelligent diagnosis of ECG for arrhythmia classification using DWT, ICA and SVM techniques. India Conference (INDICON), Proceedings of the Annual IEEE India Conference (2015). https://doi.org/10.1109/INDICON.2015.7443220
19. Desai, U., Martis, R.J., Nayak, C.G., et al.: Decision support system for arrhythmia beats using ECG signals with DCT, DWT and EMD methods: a comparative study. J. Mech. Med. Biol. **16**, 1640012 (2016)

20. Desai, U., Nayak, C.G., Seshikala, G.: An application of EMD technique in detection of tachycardia beats. In: International Conference on Communication and Signal Processing (ICCSP) 2016 Apr 6, pp. 1420–1424. IEEE (2016)
21. Desai, U., Nayak, C.G., Seshikala, G.: An efficient technique for automated diagnosis of cardiac rhythms using electrocardiogram. In: Recent Trends in Electronics, Information & Communication Technology (RTEICT), IEEE International Conference. https://doi.org/10.1109/rteict.2016.7807770, pp. 5–8, (2016)
22. Desai, U., Nayak, C.G., Seshikala, G.: Application of ensemble classifiers in accurate diagnosis of myocardial ischemia conditions. Progr. Artif. Intell. **6**(3), 245–253 (2017)
23. Desai, U., Nayak, C.G., Seshikala, G., Martis, R.J.: Automated diagnosis of coronary artery disease using pattern recognition approach. In: Proceedings of the 39th Annual International Conference of the IEEE Engineering in Medicine and Biology Society (EMBC), pp. 434–437 (2017)
24. Desai, U., Nayak, C.G., Seshikala, G., Martis, R.J., Fernandes, S.L.: Automated diagnosis of Tachycardia beats. In: Smart Computing and Informatics. Smart Innovation, Systems and Technologies, vol. 77, Springer, Singapore. doi:https://doi.org/10.1007/978-981-10-5544-7-41, (2018)
25. Zayed, N., Elnemr, H.A.: Statistical analysis of Haralick texture features to discriminate lung abnormalities. International Journal of Biomedical Imaging (2015)
26. http://www.ece.northwestern.edu/local-apps/matlabhelp/toolbox/images/medfilt2.html
27. https://in.mathworks.com/help/images/marker-controlled-watershed-segmentation.html
28. Wason, J.V., Nagarajan, A.: Image processing techniques for analyzing CT scan images towards the early detection of lung cancer. Bioinformation **15**(8), 596 (2019)

Quote Prediction with LSTM & Greedy Search Decoder

Amarjit Malhotra, Megha Gupta, Kartik Vashisth, Naman Kathuria, and Sunny Kumar

Abstract Today, human beings want to express themselves smarter, better, faster and without mistakes. Text predictions increase the speed of typing significantly and also decrease the chance of typos. Auto-text completion property becomes a smart companion for users that helps them to respond to emails quickly, to think ahead for expressing thoughts and to compile documents faster. In this paper, the aim is to generate quotes using mood personalization which involves generating positive, negative or neutral codes on the basis of the input given by the user. The proposed approach deals with the generation of Quotes based on some training on Quotes given by some famous personalities using Long Short-Term Memories and the greedy search decoder approach. A spell checker library is used to replace meaningless words and to find closest in-vocabulary meaningful words depending on Levenshtein distance.

Keywords Quote generation · RNN · LSTM · Neural networks

1 Introduction

Sequence generation is used in various applications in fields of machine learning, artificial intelligence such as music generation, and image captioning project which

A. Malhotra · K. Vashisth · N. Kathuria · S. Kumar
Department of Information Technology, Netaji Subhas University of Technology, Delhi, India
e-mail: uppalz_amar@yahoo.com

K. Vashisth
e-mail: kartikvashisth25@gmail.com

N. Kathuria
e-mail: namankathu16@gmail.com

S. Kumar
e-mail: sunnyxyz184@gmail.com

M. Gupta (✉)
Department of Computer Science, MSCW, University of Delhi, Delhi, India
e-mail: meghabis@gmail.com

J. S. Raj et al. (eds.), *Intelligent Sustainable Systems*, Lecture Notes in Networks and Systems 213, https://doi.org/10.1007/978-981-16-2422-3_25

includes generation of sentences with respect to a given image. All these applications require to maintain some relationship between the type of words that occur after a particular word under some given circumstances.

On the similar lines, new motivational, sad or happy quotes can be generated using similar models and technologies. Neural networks may be used for such problem solution. Recurrent Neural Networks such as long short-term memory and bidirectional long short-term memory have properties of memorizing that can be used to predict the text. RNN helps us to generate variable-sized input and the corresponding variable-sized output, for example, if input is a sequence of characters, output could also be a sequence of the same length, or for one character input, output is a sequence of characters or if input is a sequence of characters, output generated may be a single character [1].

The quote generator has applications in various areas. For example, in Google search boxon typing some text, Google auto-complete feature automatically tries to complete it by suggesting the most probable word. It is done by analyzing the text that has been written and the characters/words that are currently typed by the user. This issue can also be defined as the quote generation problem.

Recurrent Neural Networks (RNNs) [1] are used for the generation of text. RNNs were designed to overcome the shortcomings of the traditional neural networks like Multilayer Perceptron and Convolutional Neural Networks (CNN). Neural Networks in their simple state don't maintain any kind of state or memory. For example, if an image of a dog is fed into a CNN model, it might be able to give correct prediction. But if the same image is again fed into the same model, it does not remember that it has already classified this image previously. RNNs, on the other hand, maintain a state. They are stateful neural networks. They remember what they have processed till now. This type of behavior is useful in many cases like Sentence generation, Music generation, etc., where the prediction of the next word or tone (in the case of music generation) is based on what has been already generated so far. Recurrent Neural Networks come under the category of Sequential Models as they deal with Sequential data wherein the order of training of data matters.

Recurrent neural networks not only have found applications in many areas but also have some shortcomings in case of sequential data. It has been observed that RNN is not able to remember the already processed information for a longer time. It only remembers the info of the recent past. But in some cases, in order to narrow down the possibilities of the next word (or tone) in the sequence, it is required to have info of the past. LSTMs (Long Short-Term Memories) [2] overcome this shortcoming of RNNs. LSTMs can remember the information for a longer period of time which helps us to give more accurate predictions. LSTM layers learn very quickly and if with this the Dropout layers are used, then every neuron will be having the probability of surviving as well as of getting dropped from the model.

The architecture of the neural network depends upon the application; therefore, generally multiple architectures are created having different number of hidden layers and varying number of neurons in these hidden layers. The performance of the network is tested for all of these architectures. The dropout layer helps in doing this task. Each neuron is removed at a time (with a certain probability) which essentially creates a different architecture. This helps in reducing the dependency of the model on any particular neuron or layer. It helps in generalizing the model rather than overfitting for a particular dataset [3, 4].

Section 2 discusses some already published work similar to the concept. In Sect. 3, the proposed approach has been discussed. Section 4 presents the results.

2 Literature Survey

In work [5], the author used a method called CGMH which uses Metropolis-Hastings sampling for constrained sentence generation and allows constraints such as the presence of multiple keywords in target sentences. This method can be evaluated on a variety of tasks, e.g., on unsupervised sentence error correction, unsupervised sentence paraphrasing and keyword-to-sentence generation.

The technique used in the research work [6] for sentence generation is Knowledge Bases (KBs). Knowledge base consist of a large repository of facts which mainly comprises three elements, namely a subject, a predicate (relationship) and an object. Knowledge bases are designed on the basis of W3C standards called the Resource Description Framework (RDF) consisting of three entities—subject, predicate and object—which describes the relationship between two entities. Various models are generated which returns RDF triples which need to be converted into natural sentences to make the sentences more readable. GTR-LSTM is used to achieve the sentence from these RDF triples. Three sets in RDF triples are used to generate a single sentence. Some easy methods to achieve target sentences is by concatenating the elements of RDF triplets in linear sequence and then training the model according to the sequence input to learn corresponding natural sentence, but this might lead to losing the relationship between entities that affect the semantics of resulting sentences, that's why GTR-LSTM has been introduced to overcome this inefficiency, which maintains the structure of triplets as a small knowledge graph known as the GTR-LSTM model. It preserves the relationship between elements of the triple by computing the hidden state of each entity in the knowledge graph, thus helping us achieve more accurate sentences.

Recurrent neural generative models are discussed in the work [7] related to variational auto-encoders. In prior models, little attention has been paid to reconstruction error. A new evaluation standard has been discussed which uses novel automatic and human evaluation metrics. In this work, the author introduces the technique for measuring the quality of generated texts which is 'FrechetInferSent Distance' which involves measuring the distance between the generative distribution and data distribution which is superior to Inception score and is motivated by the study of various Generative Adversarial Network models.

Out-of-domain images are not been generalized very well by existing image captioning models, which contain novel scenes or objects. In work [8], the flexible approach is used with the help of deep captioning architectures which enables us to take advantage of image taggers at the test time, without retraining. Constrained beam search is used which forces the inclusion of selected tag words in the target. Fixed pre-trained word embeddings are used to provide vocabulary expansion to previously unused and unseen tag words. State-of-the-art results are achieved for out of domain captioning in the MSCOCO dataset. The quality of generated ImageNet captions significantly improved by leveraging ground-truth labels.

The author in work [9] addresses the difficulties being faced above using a statistical model for natural language generation using neural networks. The model consists of a feedforward neural network which encodes each triple from input in a vector of fixed dimensionality in a continuous semantic space, and a decoder which is based on RNN is used to generate textual summary, one word at a time.

The author in work [10] suggests conversational service for psychiatric counseling. This is to understand the contents using high-level Natural Language Understanding (NLU) and emotion recognition using a multi-modal approach, which helps to observe continuous emotional changes. A lot of emotional labeled data is used to train various emotion classification models. Models are generated using deep learning methods such as convolutional neural network, attention network and recurrent neural networks. For making the training process more convenient, data is diversified into image, video, audio and text. Some of the intelligent assistant services are Google now, Apple siri, etc., which respond to user inputs such as queries , voice and suggest useful information too.

In work [11], a sequence-to-sequence recurrent neural model with attention is used to generate textual summaries and selected facts. It has been ensured that generated sentences should contain input facts. Various collaborative knowledge bases and Wikipedia lack coverage and quality as a long tail of special topics are not present in sufficient details. The resulting model helps to initialize sub-articles with structured data, mapping of value facts to one-sentence encyclopedic biographies, and also improving consistency and accuracy of existing articles. An attention mechanism is incorporated for focusing on the generation of specific facts, for complementary extraction task and to generate accurate words.

3 Proposed Work

The type of model used in the proposed approach is functional. In the Functional model, part or all of the inputs are directly connected to the output layer. This architecture makes it possible for the neural network to learn both deep patterns and simple rules. In the Sequential model, it is not possible to share layers, have branches, or multiple inputs or multiple outputs. Hence, the Functional Model can be extended to generate 3 or 4 characters at a time which can help in faster computation.

Types of layers used in the proposed approach as shown in Fig. 1 are input, Long Short-Term Memory (64 units), Bidirectional Long Short-Term Memory (256 units) and activation layer using softmax (73 units). Quotes are read in csv format, extra white spaces are removed, each input has no more than 15 characters and output has only one character (many-to-one relationship). Data is prepared according to the

Fig. 1 Functional Model used for Quote Generation

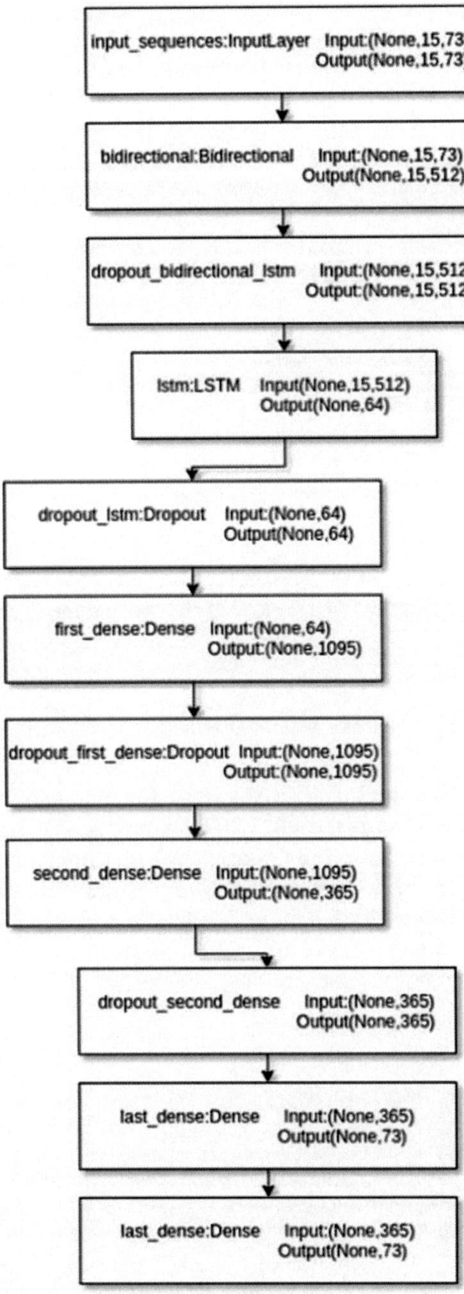

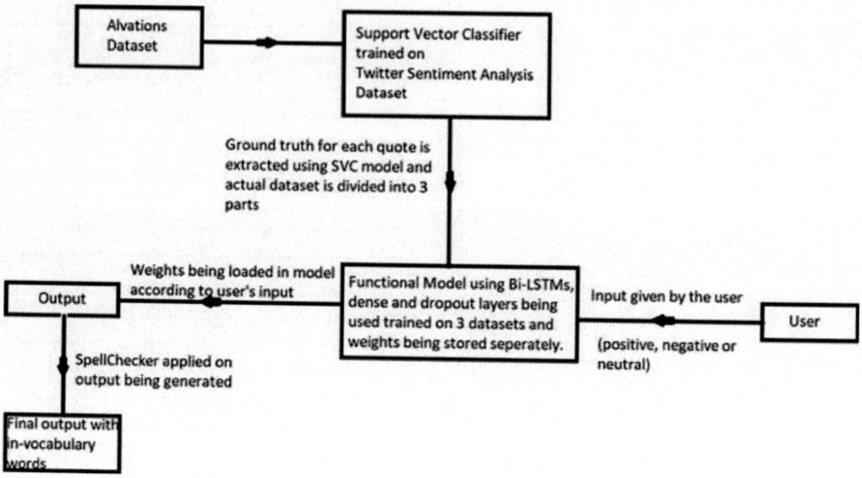

Fig. 2 Flow Chart for Quote Generation using Mood Personalization

type of input model (dimensions are 753142*15*73, where 753142 are the number of sentences used). Flow of the work done is shown in Fig. 2 shown below.

3.1 Dataset Description

Dataset used is alvations from Kaggle [12] which consists of 36,165 quotes with 878,450 words from 2,297 famous people (author, singer, politician, sportsman, scientist, etc.). Data consists of the names of the famous people who wrote those quotes in one column and their quotes in the other column. A column containing the length of each quote is added and quotes are cleaned by removing unused characters and extra blank spaces. Stopwords are not removed from the dataset so that meaningful sentences are generated.

3.2 Methodology

Values are mapped in the dictionary according to the sorted order, for the preparation of data needed to be fed to the proposed model. Input of size 15 is fed to the model, hence maximum length of each sentence has to be 15 and each input unit can take 73 values. Step size of 6 is used for generating sentences and can be changed. In order to keep the dimensionality the same for all the three set of weights, the same mapping is used for all the three datasets. The output unit with the highest probability

```
print(df['text'][1])
print("Output: " + df['airline_sentiment'][1])
```

```
@SouthwestAir seeing your workers time in and time out going above and beyond is why I love flying with you guys. Thank you!
Output: positive
```

Fig. 3 Twitter sentiment analysis dataset

would map with the correct alphabet in the dictionary, in order to generate the output sentence accurately.

The dataset used for training the quote generation model does not provide us with knowledge whether the quote given in the dataset is positive, negative or neutral. In order to determine the intent of a particular quote (positive, negative or neutral), a model based on multiclass classification is trained on the Twitter sentiment analysis dataset which consists of lot of features. Data being used in the model is shown below in Fig. 3.

The features used for training are 'airline_sentiment' and another feature required is 'text' which contains all the tweets. The data is organized in x_train and y_train formats. Training data is broken into individual words for each tweet using word tokenizer and stopwords are removed from the training set along with changing actual words to their root words using Lemmatizer, which takes the word and pos tags for that particular word which should be in simple words with their actual meaning such as 'wordnet.VERB' for verb and 'wordnet.NOUN' for noun because the function pos_tag provided by NLTK returns JJD for adjective, NNJ for noun, etc.

Using Count_Vectorizer, top 3000 words occurring maximum number of times are extracted which acts as our vocabulary and creates a dataset which produces a count of each word (feature) in each tweet (2-dimensional array). Support Vector Classifier is used on Twitter sentiment analysis data for training and GridSearchCV is used to find the best hyper-parameters to give better results on SVM. Preprocessing is done on the Twitter dataset as well as on the quotes dataset. Support Vector Classifier is used for extracting the output corresponding to each quote. The whole dataset is divided into three parts to extract three sets of weights being trained on the same model under three different classes, namely positive, negative and neutral.

3.3 Input

The input consists of a 2-D matrix of shape (Batch Size x Max length of a quote) where each element of the matrix is a word and Batch Size specifies the number of quotes trained in 1 epoch. Since the machine cannot understand English words, the words have been transformed to their corresponding word vectors using Transfer Learning. Also, Glove Embedding was used which is a pre-trained model developed by Stanford. It converts each word into a 50-dimensional vector. For doing this, 2-D Tensor is passed through the embedding layer consisting of an embedding matrix which converts the 2-D tensor into a 3-D tensor of shape (Batch Size x Max length $\times$ 50) which is then fed to the LSTM layer. So the input to the LSTM layer is a 3-D tensor. The max length of a quote taken is 15. So if any quote's length is less than 15, then post-padding of zeros will be performed and if its length is greater than 15, then the quote will be squeezed down to length 15 by removing extra words.

3.4 *Long Short-Term Memory*

In the proposed model, stacked LSTM is used in which the context or the activation vector of the first Bidirectional LSTM layer is of the shape (256,) and the context of second LSTM layer is of the shape (64,1). This helps to gradually reduce the number of features of the data. Since stacked LSTM is used, instead of using the final context vector of the first Bidirectional LSTM layer, the predictions (Yhat vectors) is used that are coming out of the LSTM layer as an input to the next LSTM layer. The first Bidirectional LSTM layer helps to maintain two states—one for the past and the other for the future. Both of these states are combined to produce much better predictions.

Dropout layer with a probability of 0.5 is used. Value of 0.5 is taken for probability so that every neuron will be having the same probability of surviving as well as of getting dropped from the model.

4 Results and Discussion

During output generation using greedy search decoder at character level, one or two random words are picked up, the function starts generating quotes from those random words with length equal to 15 because the predict function takes input of size 15*73. A prediction array of size 73 is obtained based on the input , using which the next possible alphabet is predicted.

The coded function calculates the probability of occurrence of each alphabet and returns the index of character having the best probability and that character is then appended to the sentence. Some of the quotes generated can be seen in Fig. 4.

```
In [63]: generate_quote()
            I couldn't walk in the church.

In [64]: generate_quote()
            You don't have to take on my family.
```

Fig. 4 Output of Character Level

```
embedding["play"]

array([-0.73571 ,  0.19937 , -0.89408 ,  0.36406 , -0.20246 , -0.034324,
       -0.63138 ,  0.76669 , -0.94343 ,  0.65883 ,  0.049478,  0.55608 ,
       -1.2809  ,  0.44575 ,  0.73791 ,  0.014728,  0.80956 , -0.35516 ,
       -1.0248  , -0.13845 , -0.47632 ,  0.32001 ,  0.35023 ,  0.77794 ,
        0.60233 , -1.2321  ,  0.043144, -0.41347 ,  0.34533 , -1.3093  ,
        3.4681  ,  0.9882  ,  0.038253, -0.33672 ,  0.30999 ,  0.6331  ,
        0.30798 ,  0.68528 , -0.21989 , -0.77505 , -0.036884,  0.051738,
       -0.25442 , -0.063405, -0.20665 ,  0.91281 ,  0.80133 , -0.075279,
       -0.44448 ,  0.47437 ])
```

Fig. 5 Embedding Matrix of 'play'

For output generation using greedy search decoder at word level, embeddings corresponding to each word available in the vocabulary are used to train the model. An embedding corresponding to word play is shown in Fig. 5.

Strategy of using 15 words in input and 1 word in the output is used to train the same model using embeddings. Output obtained using word level is shown in Fig. 6.

Since output being generated using word level is not as per the user's requirement of quality in sentences due to lack of data, character level is out-performing word level for quote generation. Therefore, the character-level model without an embedding layer is used for mood personalization.

For mood personalization quote generation, the dataset is divided into 3 different classes (positive, negative and neutral) using the results of Support Vector Classifier being trained on the Twitter sentiment analysis dataset. The dataset after concatenating ground truths of each quote with them looks as shown in Fig. 7.

Positive, negative and neutral quotes are separated, and are trained separately on the same model and their weights are stored in three different .h5 files. For neutral quotes, the length of dataset generated is 1,44,777 sentences of size 15 which is trained using the model and training looks like the image as shown in Fig. 8.

```
for i in range(10):
    print(predict_quote(model1))

think we are the world of the world
do fat the world is the world of the world
we re the best of the world
just consider the world is the world of the world
threatened nation is the world of the world
the pursuit of the world of the world is the world of the world
what we are the world is the world of the world
the beginning of the world is the world of the world
anticipate the world is the world of the world
we have to be the world
```

Fig. 6 Output of Word Level

	author	quote	Ground Truth
6199	Charles Spurgeon	Saving faith is an immediate relation to Chris...	negative
32369	Steve Jobs	Design is not just what it looks like and feel...	negative
6634	Chris Hemsworth	As kids, our experiences shape our opinions of...	negative
26383	Nick Cannon	I am a hopeless romantic, and so is my wife.	neutral
22441	Lisa Jakub	My memoir is being published by Beaufort Books...	negative
34088	Tryon Edwards	He that never changes his opinion never correc...	negative
33130	The Edge	Bono is chairman and founding member of Over-A...	negative

Fig. 7 Dataset after concatenating Ground Truths

```
In [49]: model.fit([x], y, batch_size=128, epochs= 15)

Epoch 1/15
144777/144777 [==============================] - 203s 1ms/step - loss: 2.2787
Epoch 2/15
144777/144777 [==============================] - 192s 1ms/step - loss: 1.9539
Epoch 3/15
144777/144777 [==============================] - 192s 1ms/step - loss: 1.8078
Epoch 4/15
144777/144777 [==============================] - 191s 1ms/step - loss: 1.7064
Epoch 5/15
144777/144777 [==============================] - 194s 1ms/step - loss: 1.6301
Epoch 6/15
144777/144777 [==============================] - 194s 1ms/step - loss: 1.5681
Epoch 7/15
144777/144777 [==============================] - 194s 1ms/step - loss: 1.5115
Epoch 8/15
```

Fig. 8 Training of Neutral Quotes

For each of these three datasets, first, the model is trained for 15 epochs with a batch size of 128. Then the model is trained for 2 epochs for batch size of 1024 and 2048, after which loss became stagnant.

Some of the positive quotes being generated consist of some positive intent in the sentence with some positive words as shown in Fig. 9.

Negative quotes being generated consist of some negative intent and words which show some kind of rejection or sadness as shown in Fig. 10.

The proposed technique is able to generate new quotes and also auto-completion of quotes by giving 1 or 2 words. There are 73 characters in total being used in quotes for generation. Therefore, the output layer gives an output vector of 73 float integers.

```
: quote_type = input("Enter the type of quote you want to generate - positive, negative or neutral : ")
generate_quote(quote_type)

Enter the type of quote you want to generate - positive, negative or neutral : positive
      No matter, and people and really didn't love in the person of country important to the heart, and an end in forath usings
between instakes.
```

Fig. 9 Output based on Positive Input

```
quote_type = input("Enter the type of quote you want to generate - positive, negative or neutral : ")
generate_quote(quote_type)
```

```
Enter the type of quote you want to generate - positive, negative or neutral : negative
        When you do demons, the other mistakes us ic someone is easys in work about director it is hard to the need hair with me
to stay his mmoties from the Rcsoma power to degering a black.
```

```
quote_type = input("Enter the type of quote you want to generate - positive, negative or neutral : ")
generate_quote(quote_type)
```

```
Enter the type of quote you want to generate - positive, negative or neutral : negative
        I myself sweal to, you can do almoked such bad works a lot easier is not importantly at least research of empathirab.
```

Fig. 10 Output based on Negative Input

```
Enter the type of quote you want to generate - positive, negative or neutral : positive
        I would love being any productive very entercing a solegring, and do just get not any hutfess, but I sain the heart, bu
t he's a quice mymors.

The new sentence with rectified words is :
I would love being any productive very entering a sobering , and do just get not any gutless , but I sain the heart , but he is
a quite memory .
```

Fig. 11 Output after applying Spellchecker on Quote

Also, quotes are generated according to user's current mood, positive, negative or neutral.

Since quotes being generated using character level might contain meaningless or incorrect words, it is important to replace those words with meaningful in-vocabulary words for which a spellchecker module available in Python is used which makes use of Levenshtein Distance. Levenshtein Distance is about how many operations (insert, delete or replace a word) need to be performed to make an incorrect word meaningful and the word with minimum number of operations is chosen and if multiple words having the same edit distance exist, then the word is chosen according to the words in the vicinity of that particular word using sequence-to-sequence model. An example of spellchecker being applied on generated quote is shown in Fig. 11.

Some experimental results are generated based on Support Vector Classifier, which is trained using the Twitter sentiment analysis dataset. Classification report for the same is shown in Fig. 12.

Decent accuracy is achieved with the available amount of data. In the worst-case scenario, a positive quote could be predicted as neutral quote and vice versa. Accuracy might be improved if Constrained Beam Search Decoder for sentence generation is

	precision	recall	f1-score	support
negative	0.96	1.00	0.98	24
neutral	0.80	1.00	0.89	35
positive	1.00	0.76	0.86	42
micro avg	0.90	0.90	0.90	101
macro avg	0.92	0.92	0.91	101
weighted avg	0.92	0.90	0.90	101

Fig. 12 Classification Report on Quotes generated

used. Such an approach can be used if there is a small amount of data, and some new data needs to be generated for different projects.

5 Conclusion and Future Work

The proposed technique defines a model that can generate quotes given some condition. This work is concentrated on generating new quotes or auto-completion of quotes by giving 1 or 2 words before generation based on the probability of occurrence of each word or special character after the set of words being passed on to the pre-trained model. Also, the proposed work is able to personalize quotes according to user's positive, negative or neutral mood. The work is done on character level instead of word level, hence the number of trainable parameters are reduced by 1.7 million because word level uses embedding layers in the proposed model which leads to very less training and processing time and also helps in faster generation of desired outputs. Instead of preparing three different models to generate three different types of quotes, three sets of weights with the same dimensionality has been prepared and loaded in the model as per the input given by the user which extensively reduces the space complexity.

In future, the work will be extended to personalize quotes depending on other different moods of the user.

References

1. Karpathy, A.: The Unreasonable Effectiveness Of Recurrent Neural Networks. [online] Karpathy.github.io. (2015). Accessed 1 Oct 2020
2. Graves, A.: Long short-term memory. in: supervised sequence labelling with recurrent neural networks. Studies in Computational Intelligence, vol. 385. (2012). Springer, Berlin, Heidelberg. https://doi.org/10.1007/978-3-642-24797-2_4
3. Sungheetha, A., Sharma, R.: Transcapsule model for sentiment classification. J. Artif. Intell. **2**(3), 163–169 (2020)
4. Mitra, A.: Sentiment analysis using machine learning approaches (Lexicon based on movie review dataset). J. Ubiquit. Comput. Commun. Technol. (UCCT) **2**(3), 145–152 (2020)
5. Miao, N., Zhou, H., Mou, L., Yan, R., Li, L.: CGMH: constrained sentence generation by metropolis-hastings sampling. Proceedings of the AAAI Conference on Artificial Intelligence **33**, 6834–6842 (2019)
6. Trisedya, B.D., Qi, J., Zhang, R., Wang, W.: GTR-LSTM: a triple encoder for sentence generation from RDF data. In: Proceedings of the 56th Annual Meeting of the Association for Computational Linguistics, vol. 1, pp. 1627–1637 (2018)
7. Cıfka, O., Severyn, A., Alfonseca, E., Filippova, K.: Eval all, trust a few, do wrong to none: comparing sentence generation models, Cornell University, pp. 1–12 (2018)
8. Anderson, P., Fernando, B., Johnson, M., Gould, S.: Guided open vocabulary image captioning with constrained beam search. In: Proceedings of the 2017 Conference on Empirical Methods in Natural Language Processing, pp. 936–945(2017)

9. Vougiouklisa, P., Elsaharb, H., Kaffeea, L.A., Gravierb, C., Laforestb, F., Harea, J., Simperla, E.: Neural Wikipedian: Generating Textual Summaries from Knowledge Base Triples, Cornell University, pp 1–16 (2017)
10. Oh, K., Lee, D.K., Ko, B., Choi, H.: A Chatbot for psychiatric counseling in mental healthcare service based on emotional dialogue analysis and sentence generation, In: 2017 IEEE 18th International Conference on Mobile Data Management, pp. 371–376 (2017)
11. Chisholm, A., Radford, W., Hachey, B.: Learning to generate one-sentence biographies from Wikidata. In: 15th Conference of the European Chapter of the Association for Computational Linguistics, vol. 1, pp. 633–642 (2017)
12. Tan, L.: Quotables. [online] Kaggle.com (2017). Accessed 1 Oct 2020

Skin Cancer Detection from Low-Resolution Images Using Transfer Learning

M. D. Reyad Hossain Khan, Abdul Hasib Uddin, Abdullah-Al Nahid, and Anupam Kumar Bairagi

Abstract Skin cancer is one of the worst diseases noticed in humankind. It beholds some types, which even experts find challenging to categorize. In recent times, neural network-based automated systems have been entitled to perform this difficult task for their amazing ability of pattern recognition. However, the challenge remains due to the requirement for high-quality images and thus the necessity of highly configured resources. In this research manuscript, the authors have addressed these issues. They pushed the boundary of neural networks by utilizing a low-resolution (80 × 80, 64 × 64, and 32 × 32 pixels), highly imbalanced, grayscale HAM10000 skin cancer dataset into several pre-trained network architectures (VGG16, DenseNet169, DenseNet161, and ResNet50) that have been successfully used for a similar purpose with a high-resolution, augmented RGB HAM10000 skin cancer image dataset. The image resolution of the original HAM10000 dataset is 800 × 600 pixels. The highest achieved performance for 80 × 80, 64 × 64, and 32 × 32 pixel images were 80.46%, 78.56%, and 74.15%, respectively. All of these results were accomplished from the ImageNet pre-trained VGG16 model. The second-best model in terms of transfer learning was DenseNet169. The performances demonstrate that even within these severe circumstances, neural network-based transfer learning holds promising possibilities.

Keywords Skin cancer · HAM10000 · Low-resolution image · Transfer learning · Neural networks · ResNet50 · DenseNet169 · DenseNet121 · VGG16

M. D. R. H. Khan · A. H. Uddin · A.-A. Nahid · A. K. Bairagi (✉)
Khulna University, Khulna 9208, Bangladesh
e-mail: anupam@ku.ac.bd

A.-A. Nahid
e-mail: nahid.ece.ku@ku.ac.bd

J. S. Raj et al. (eds.), *Intelligent Sustainable Systems*, Lecture Notes in Networks and Systems 213, https://doi.org/10.1007/978-981-16-2422-3_26

1 Introduction

The modern world deals with human problems from modern technological perspectives. However, situations may demand human expertise to solve classical difficult human complications, such as cancer. Among them, skin cancer, perhaps, can be considered as one of the striking models, which requires express visual aptitude. In ongoing technological developments, neural networks are considered as truly outstanding, if not the most ideal alternative for pattern recognition from images and texts. Numerous research methodologies have been evolved and deployed for skin cancer detection from image datasets.

Dorin Moldovan proposed a strategy that comprises two stages dependent on deep learning and transfer learning for the classification of skin cancer by utilizing the HAM10000 dataset [1]. The accuracy for testing of the first and second steps are 85% and 75%, respectively. They used a DenseNet121 PyTorch pre-trained model. In Fig. 1, their proposed model is illustrated briefly. Ardan Adi Nugroho et al. utilized CNN to detect skin cancer by using RGB images as input data [2]. The accuracies of training and validation models are 80% and 78%, respectively. At the first step, they extracted features from the data, then they used it to train their model [3]. Using transfer learning on the ResNet pre-trained model trained

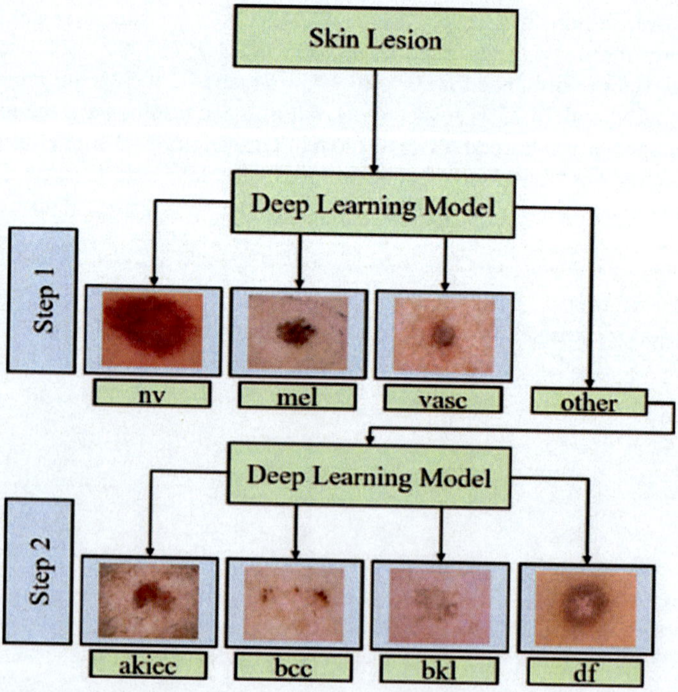

Fig. 1 Two-step method for skin cancer image classification [1]

with ImageNet dataset, they achieved ideal result. The accuracy was 90.51%. They also applied XGBoost, Random Forest, and SVM algorithms for comparison. They also achieved 78% accuracy by using VGG16. Muhammad Attique Khan et al. came up with an automated strategy for skin lesion classification-based deep convolutional neural network (DCNN) feature extraction through transfer learning and kurtosis controlled principle component (KcPCA)-based optimal feature selection [4]. Deep CNNs (DCNNs) are neural network architectures that consist of a significantly larger number of layers than typical CNN architectures, where filters become deeper rightward. The pre-trained deep neural network (DNN) such as RESNET-101 and RESNET-50 are utilized for feature extraction. At that point, they intertwined their data and chose the best features which were later used to supervise learning strategy, for example, SVM of radial basis function (RBF) for the classification task. HAM10000, ISBI 2017, and ISBI 2016 are the three datasets utilized for experimental results and achieved an accuracy of 89.8%, 95.60%, and 90.20%, respectively. By using RBF, they acquired the ideal result. ND Reddy et al. published a paper where they trained a CNN using ResNet-50 architecture and generated a balanced 91% of validation accuracy and 73.5% of testing accuracy [5]. Manu Goyal et al. proposed the utilization of two object-locating meta-architectures for end-to-end ROI (Region of Interest) skin lesion detection with dermoscopic images [6]. They trained the SSD-InceptionV2 and Faster-RCNN-InceptionV2 on the training dataset of ISBI-2017 and evaluated their model performance on the ISBI-2017 testing set, PH2, and HAM10000 datasets. The Faster-RCNN-InceptionV2 method outperformed the state-of-the-art segmentation model for ROI detection on all testing sets. Wannipa Sae-Lim et al. brought forward a skin lesion ordering approach based on the lightweight deep CNNs, called MobileNet [7]. They achieved improved accuracy by their modified MobileNet. Emrah ÇEVİK et al. came up with a powerful CNN model with VGGNET-16 architecture to classify seven different categories of skin cancer on dermoscopic images [8]. This paper used HAM10000 with VGGNET-16 architecture. Philipp Tschandlthey et al. used the ImageNet pre-trained model with ResNet34 as encoding layers of U-Net style design [9]. They also applied the encoding layers with the He uniform initialization. They achieved better results from ResNet34 architecture with the ImageNet pre-trained model and fine-tuning the model through the random initialization. This paper used SVM in the last step to categorize the seven skin lesions. They also applied image augmentation to balance the dataset. Abdullah-Al Nahid et al. proposed a method to identify breast cancer by using the BreakHis dataset with the DNN technique [10]. This paper proposed a CNN, a Long Short-Term Memory (LSTM), and a combination of CNN and LSTM techniques for the classification. After extracting features, they utilized Support Vector Machine (SVM) layers and Softmax classifier for the decision-making stage using the novel DNN models. They achieved the best accuracy estimation of 91%. Dr. Samuel Manoharan et al. came up with an algorithm to improve the typical graph cut methods for the edge segmentation [11]. Dr. T. Vijayakumar analyzed and showed the CNN performs better than Feed Forward Neural Networks (FNN) and Recurrent Neural Networks (RNN) for the investigation and detection of Tumor and Cancer diseases [12]. Moreover in this paper, the authors classified breast images into Malignant and

Benign classes by utilizing an advanced technique of CNN [13]. They normalized the images by applying the Multiscale Retinex algorithm. They achieved the best result with Convolutional Residual and MaxMin CNNs.

In this manuscript, Deniz et al., utilized transfer learning to classify breast cancer from the histopathology images of the their dataset [14]. They resized the input image size to 227 × 227 pixels and 224 × 224 pixels from the original image size of 700 × 460 pixels, and they used different magnifying factors like 40X, 100X, 200X, and 400X. They used VGG16 and AlexNet models and also their experimented platform was configured with an Intel Core i7-4810 CPU with 32 GB of memory. In our manuscript, we showed how we can solve the memory issue and addressed the utilization of transfer learning with low-resolution images. We used the HAM10000 dataset to classify skin cancer disease with downscaled images such as 80 × 80, 64 × 64, and 32 × 32 pixels. We used various architectures such as ResNet50, VGG16, DenseNet121, and DenseNet169 to train the CNN model with transfer learning. We have demonstrated that low-resolution grayscale images can be effectively classified using proper transfer learning. Even with 32 × 32 pixel images, the ImageNet VGG16 pre-trained model was able to identify more than 74% of skin cancer lesion images.

The rest of the manuscript is structured as follows. Section 2 describes the data collection and processing steps. Section 3 narrates the Transfer Learning processes, while Sect. 4 catalogs the corresponding results and discussions. Finally, Sect. 5 concludes the work and its limitations and describes some of our future works.

2 Data Description

2.1 Data Collection

The HAM10000 (Human Against Machine with 10000 training images) dataset was published by Philipp Tschandl et al. in 2018 for the purpose of using machine learning in dermatology [15]. It comprises 10015 dermatoscopic images with a resolution of 800 × 600 which were published as a training image dataset for educational machine learning purposes and is available through the ISIC archive publicly. More than 50% of lesions of this dataset have been confirmed by pathology. Some samples of this dataset are shown in Fig. 2.

The images of this dataset are collected from different populations acquired and gathered by various modalities. This dataset was provided to the contestants of the ISIC 2018 classification challenge hosted by the annual MICCAI conference in Granada, Spain. The benchmark can be utilized for machine learning and comparisons of human experts.

The 10015 dermatoscopic images of this dataset were gathered throughout a time span of a long time from two different locales, the skin cancer practice of Cliff Rosendahl in Queensland, Australia, and the Department of Dermatology at the Medical University of Vienna, Austria. The Australian site put away pictures

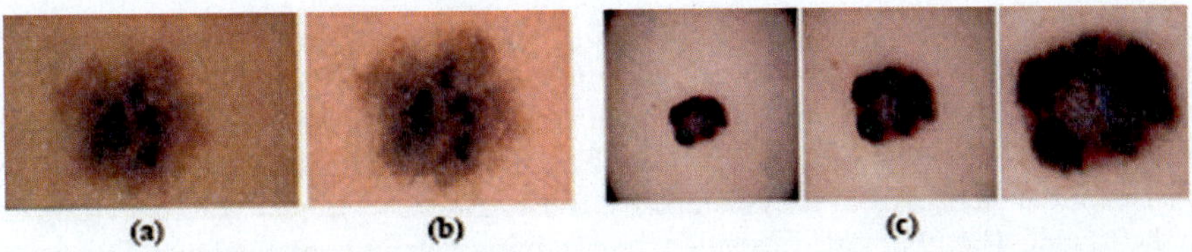

Fig. 2 Original scanned image (**a**) without quality review and (**b**) with a manual quality review and (**c**) taken from different angles and applying magnification [15]

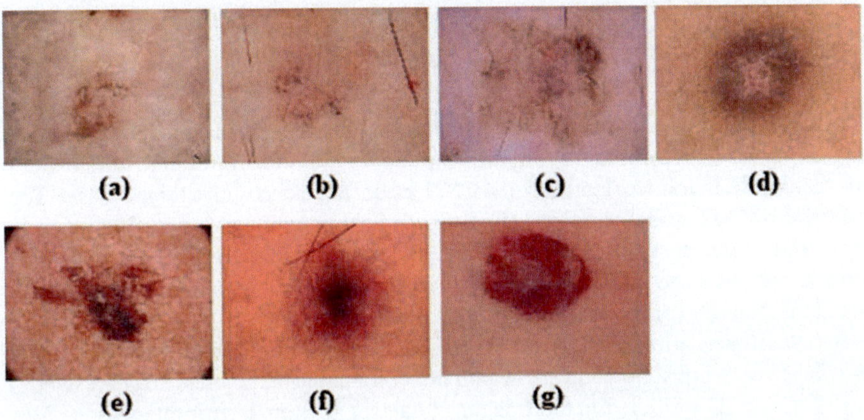

Fig. 3 Seven types of skin lesions images. **a** akiec, **b** bcc, **c** bkl, **d** df, **e** mel, **f** nv, **g** vasc

and meta-information in PowerPoint documents and Excel datasets. However, the Austrian site began to gather pictures before the time of computerized cameras and put away pictures and metadata in various configurations during various time frames.

This dataset consists of seven categories of skin cancer images such as Actinic keratoses (akiec), Basal cell carcinoma (bcc), Benign keratosis (bkl), Dermatofibroma (df), Melanocytic nevi (nv), Melanoma (mel), and Vascular skin lesions (vasc). These seven classes of skin disease are visually represented in Fig. 3. The dataset range of vasc includes cherry angiomas to angiokeratomas [16] and pyogenic granulomas [17]. This category also includes Hemorrhage.

2.2 *Data Processing*

In the first step of data processing, from each of the seven categories of lesion images we split 10% of images as the test dataset, another 10% of the images as validation dataset, and the rest 80% of images were used for training purposes. Hence, the final test image set contained 10% images from each category. Processing RGB images requires a substantial amount of memory. Nonetheless, we wanted to address this issue and therefore we converted all RGB images into grayscale before applying any processing steps to the images. On the next step after grayscale conversion, we downsampled each image and created three different instances of the dataset containing 80 × 80, 64 × 64, and 32 × 32 pixel images. Finally, we converted the images of the dataset from grayscale to RGB using the CV2 library before feeding it to the model. The final test, validation, and training dataset contained 998, 998, and 37555 images accordingly. In Fig. 4, there is a simple flow diagram where the first four steps are represented as the data processing steps.

3 Transfer Learning for Skin Cancer Detection

Transfer learning is a technique in deep learning where we can transfer knowledge learned earlier in one domain to another domain for classification and feature extraction tasks [18]. In transfer learning, the CNN model is trained with a large dataset. The pre-trained CNN model is trained (fine-tuned) further with a new and small dataset compared to the previous dataset. Our model was pre-trained with the ImageNet dataset, and we used various architectures like ResNet50, VGG16, DenseNet121, and DenseNet169 for the fine-tuning of the model.

ResNet50 was introduced by He et al. in 2015 [19]. This structure was influenced by VGG16 network architecture. The number of trainable parameters in the ResNet50 model is over 23 million. On the ImageNet dataset, the top-1 error rate was 20.74% and the top-5 error rate was only 5.25%.

VGG16 was introduced by K Symonyan in 2015 [20]. This network architecture comprises 16 Convolutional layers along with 3 fully connected layers. It uses the ReLU activation function and on the final layers, it applies softmax as the classifier. The accuracy of this network on the ImageNet dataset was 92.70% (top-5 test accuracy).

DenseNet121 and DenseNet169 were introduced by Gao Huang in 2017 [21]. The top-1 error rate on the ImageNet validation dataset was 23.80%. On the other hand, the top-5 error rate was 6.85%. Additionally, the corresponding top-1 and top-5 error rates in the case of DenseNet121 on the ImageNet validate set were 25.02% and 7.71%, respectively.

Mobiny et al. showed the procedure of improvement of performance by several deep learning models such as VGG16, DenseNet169, and ResNet50 [22]. They used the HAM10000 dataset to classify skin cancer lesions. They trained the model to decrease the weighted cross-entropy loss using the ADAM optimizer. In the beginning, they started training with an initial learning rate of 0.001; after that, they reduced it with a factor of 0.2 after every 10 epochs following a step-wise approach. They did set the batch size to 128 and performed the training with a maximum range of 100 epochs. The input image size was (224 × 224) and the training and testing data ratio was 80% and 20%, respectively. They evaluated the validation accuracy after every epoch and saved the model with the best prediction accuracy on the validation set. For VGG16, they achieved 79.63% accuracy on the HAM10000 dataset. As for

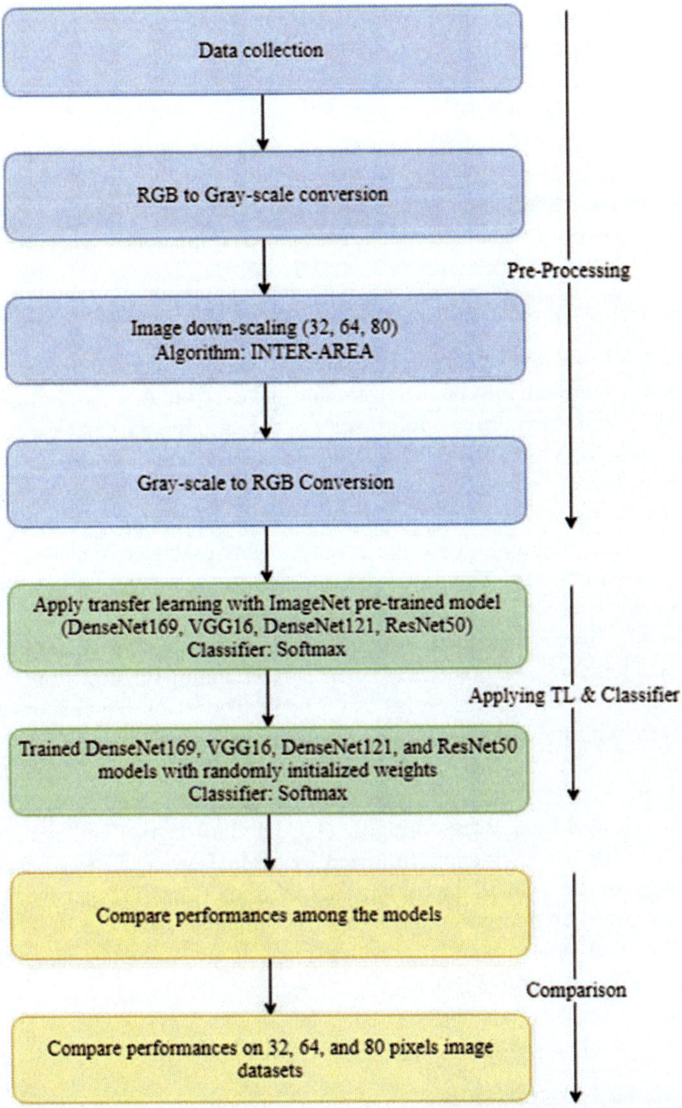

Fig. 4 Flow diagram of skin cancer detection from low-resolution images using transfer learning

ResNet50, their achieved performance was 80.45%, while 81.35% accuracy was due to DenseNet 169.

Nils et al. deployed DenseNet121, DenseNet169, and ResNet50 on the HAM10000 dataset [23]. They applied image augmentation and utilized the Adam optimizer. The initial learning rate for their implementation was 0.0005 which they reduced step by step with a factor of 0.2 over 50 epochs. After that, they decreased the value of the learning rate after every 25 epochs. They trained the models for 125 epochs. They applied fivefold cross-validation. For DenseNet121, the mean accuracy was 82.30% and for DenseNet169 it was 85.20%. Furthermore, they achieved 86.20% accuracy for ResNet50.

For our VGG16 model, batch size was 128, the learning rate was 0.001, and the decay rate was 0.02. For DenseNet121, the respective batch size, learning rate, and decay rate were 40, 0.005, and 0.02. As for DenseNet169, the batch size was 128, the learning rate was 0.001, and the decay rate was 0.02. Similarly, the batch size, the learning rate, and the decay rate were 128, 0.001, and 0.02, respectively. We generally trained each model until 10 epochs with no improvement in terms of validation loss were encountered.

In each model, we used categorical cross-entropy as the loss function and softmax as the classifier. We set all the layers as trainable and used the Adam optimizer, except in DensNet121, where the Nadam optimizer was applied. In each case, we considered two scenarios on each of the three image sets. In the first scenario, we initialized the weights of the models with corresponding ImageNet pre-trained values and in the second scenario instead of using ImageNet pre-trained models, we initialized the weights randomly. The final layers of each of the networks were densely connected layers with seven neurons. The output of the models was flattened before feeding into the final dense layer.

Since the highest performance was encountered from the ImageNet pre-trained VGG16 model with 80×80 pixel images, the corresponding illustration is visualized in Fig. 5. The size of the input layer is $80 \times 80 \times 3$. The VGG16 part starts with two Conv2D layers with the size (80, 80, 64) followed by a MaxPooling2D (40, 40, 64) layer. Next, it has also two Conv2D layers with the size (40, 40, 128) and followed by a MaxPooling2D (20, 20, 128) layer. Afterward, it has three Conv2D (20, 20, 256) layers followed by a MaxPooling2D (10, 10, 256) layer. Furthermore, it has three Conv2D (10, 10, 512) layers followed by a MaxPooling2D (5, 5, 512) layer. In the last steps of the VGG16 part, it has another three Conv2D layers with the size (5, 5, 512) followed by a MaxPooling2D (2, 2, 512) layer. After the VGG16 part, it has one Flatten and one Dropout layer with the size 2048. At the last step, the output layer has a Dense layer with size 07.

4 Results and Discussion

In Table 1, we tabulated the best cases of test accuracy and loss for various architectures with and without the pre-trained models. We used ImageNet for the pre-trained model. We have 3 types of image sizes such as 80×80, 64×64, and 32×32 pixels. From this table, we can describe that the image size, number of epochs, test accuracy, and the test loss of VGG16 architecture are (80, 54, 0.8046, 2.8469), (64, 12, 0.7856, 2.4131), (32, 13, 0.7415, 1.8319) with the pre-trained model and (80, 15, 0.6743, 2.9122), (64, 15, 0.6994, 3.5306), (32, 22, 0.6723, 2.3453) without the pre-trained model. Likewise, for the DenseNet121 the image size, number of epochs,

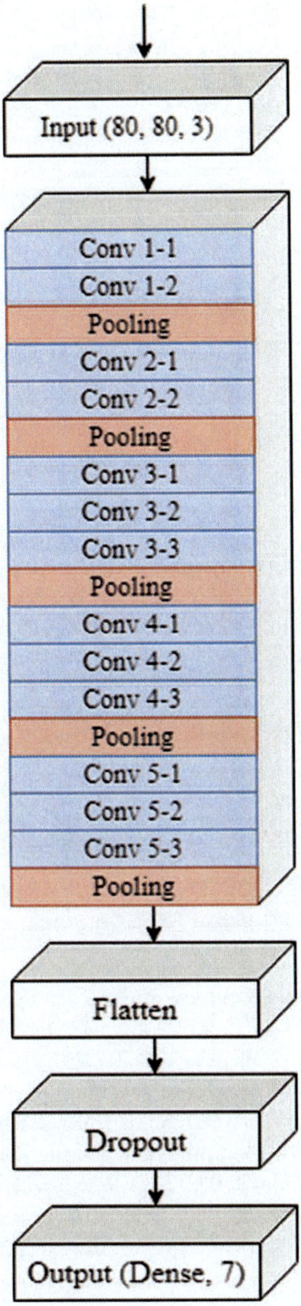

Fig. 5 Transfer learning for skin cancer detection using ImageNet pre-trained VGG16 model with 80 × 80 pixel lesion images

Table 1 Test accuracy and loss of VGG16, DenseNet121, DenseNet169, and ResNet50

#	Model Name	Pre-trained Weights	Image Size (px)	Number of Epochs	Test Accuracy	Test Loss
01	VGG16	ImageNet	80 × 80	54	0.8046	2.8469
02			64 × 64	12	0.7856	2.4131
03			32 × 32	13	0.7415	1.8319
04		None	80 × 80	15	0.6743	2.9122
05			64 × 64	15	0.6994	3.5306
06			32 × 32	22	0.6723	2.3453
07	DenseNet121	İmageNet	80 × 80	53	0.7786	2.2325
08			64 × 64	11	0.7224	1.1259
09			32 × 32	12	0.6994	2.5825
10		None	80 × 80	13	0.6804	2.1629
11			64 × 64	11	0.6934	2.0901
12			32 × 32	12	0.6904	2.4415
13	DenseNet169	İmageNet	80 × 80	83	0.7896	1.5961
14			64 × 64	11	0.7605	1.8405
15			32 × 32	11	0.7104	2.4648
16		None	80 × 80	12	0.6052	2.3836
17			64 × 64	18	0.6954	2.0169
18			32 × 32	12	0.6323	1.9187
19	ResNet50	İmageNet	80 × 80	45	0.7665	1.3087
20			64 × 64	14	0.7605	1.7091
21			32 × 32	14	0.7074	2.8826
22		None	80 × 80	15	0.7104	2.8504
23			64 × 64	13	0.6162	2.5876
24			32 × 32	13	0.6253	1.9332

test accuracy, and the test loss are (80, 53, 0.7786, 2.2325), (64, 11, 0.7224, 1.1259), (32, 12, 0.6994, 2.5825) with the pre-trained model and (80, 13, 0.6804, 2.1629), (64, 11, 0.7224, 1.1259), (32, 12, 0.6904, 2.4415) without the pre-trained model. For the DenseNet169, the image size, number of epochs, test accuracy, and the test loss are (80, 83, 0.7896, 1.5961), (64, 11, 0.7605, 1.8405), (32, 11, 0.7104, 2.4648) with the pre-trained model and (80, 12, 0.6052, 2.3836), (64, 18, 0.6954, 2.0169), (32, 12, 0.6323, 1.9187) without the pre-trained model. For the ResNet50, the image size, number of epochs, test accuracy, and the test loss are (80, 45, 0.7665, 1.3087), (64, 14, 0.7605, 1.7091), (32, 14, 0.7074, 2.8826) with the pre-trained model and (80, 15, 0.7104, 2.8504), (64, 13, 0.6162, 2.5876), (32, 13, 0.6253, 1.9332) without the pre-trained model. It is clear that the performance of the VGG16 architecture with pre-trained weights and with the image size 80px is the best among all of the deployed models.

Afterwards in Table 2, we recorded Precision, Recall, and F1-Score of seven classes of skin lesions for all architectures according to the serial number of Table 1. From this table, we can demonstrate that for the VGG16 model, the Precision, Recall and F1-Score of the akiec class with image size 80px, 64px, and 32px are (0.67, 0.50, 0.57), (0.59, 0.41, 0.48), and (0.50, 0.31, 0.38), respectively, with the pre-trained model and (0.18, 0.25, 0.21), (0.45, 0.28, 0.35), and (0.46, 0.19, 0.27), respectively, without the pre-trained model. Likewise, the Precision, Recall, and F1-Score of the bcc class with image size 80px, 64px, and 32px are (0.78, 0.49, 0.60), (0.69, 0.57, 0.62), and (0.58, 0.37, 0.45), respectively, with the pre-trained model and (0.29, 0.35, 0.32), (0.39, 0.33, 0.36), and (0.30, 0.59, 0.39), respectively, without the pre-trained model. For the bkl class, the Precision, Recall, and F1-Score with image size 80px, 64px, and 32px are (0.61, 0.61, 0.61), (0.61, 0.45, 0.52), and (0.43, 0.50, 0.46), correspondingly, with the pre-trained model and (0.32, 0.19, 0.24), (0.40, 0.19, 0.26), and (0.31, 0.27, 0.29), respectively, without the pre-trained model. For the df class, the Precision, Recall, and F1-Score with image size 80px, 64px, and 32px are (1.00, 0.36, 0.53), (0.80, 0.36, 0.50), and (0.57, 0.36, 0.44), correspondingly, with the pre-trained model and (0.20, 0.09, 0.13), (0.43, 0.27, 0.33), and (0.57, 0.36, 0.44), respectively, without the pre-trained model. For the mel class, the Precision, Recall, and F1-Score with image size 80px, 64px, and 32px are (0.68, 0.38, 0.49), (0.67, 0.32, 0.44), and (0.56, 0.34, 0.42), respectively, with the pre-trained model and (0.38, 0.16, 0.23), (0.34, 0.15, 0.23), and (0.39, 0.24, 0.30), respectively, without the pre-trained model. For the nv class, the Precision, Recall, and F1-Score with image size 80px, 64px, and 32px are (0.85, 0.97, 0.90), (0.83, 0.97, 0.89), and (0.84, 0.91, 0.87), correspondingly, with the pre-trained model and (0.79, 0.91, 0.85), (0.76, 0.94, 0.84), and (0.81, 0.86, 0.83), correspondingly, without the pre-trained model. For the vasc class, the Precision, Recall, and F1-Score with image size 80px, 64px, and 32px are (0.33, 0.14, 0.20), (0.33, 0.14, 0.20), and (0.22, 0.14, 0.17), correspondingly, with the pre-trained model and (0.00, 0.00, 0.00), (0.00, 0.00, 0.00), and (0.00, 0.00, 0.00), correspondingly, without the pre-trained model. Likewise, the Precision, Recall, and F1-Score are also recorded for the models DenseNet121, DenseNet169, and ResNet50 in Table 2. By analyzing the results, it is clear that the generated Precision, Recall, and F1-Score for the VGG16 model with image size 80 pixel is the best among all the deployed models.

In Table 3, we tabulated the Weighted average of Precision, Recall, and F1-Score of seven classes of skin lesions for VGG16 (serial 01 to 06), DenseNet121 (serial 07 to 12), DenseNet1169 (serial 13 to 18), and ResNet50 (serial 19 to 24) architectures according to the serial number of Table 1. From this table, we can see that the weighted average of Precision, Recall, and F1-Score of VGG16 with image size 80px, 64px, and 32px are (0.79, 0.80, 0.79), (0.76, 0.79, 0.76), and (0.72, 0.74, 0.73), respectively, with the pre-trained model and (0.63, 0.67, 0.64), (0.64, 0.70, 0.65), and (0.66, 0.67, 0.66), respectively, without the pre-trained model.

Likewise, for DenseNet121, the weighted average of Precision, Recall, and F1-Score for image size 80px, 64px, and 32px are (0.76, 0.78, 0.76), (0.70, 0.72, 0.69), and (0.68, 0.70, 0.64) accordingly with the pre-trained model and (0.69, 0.68, 0.68), (0.67, 0.69, 0.67), and (0.63, 0.69, 0.62) accordingly without the pre-trained model.

Table 2 Precision, Recall, and F1-Score

# from Table 1	Model Name	Akiec			bcc			bkl			df			mel			nv			vasc		
		P	R	F1	P	R	F1	P	R	F1	P	R	F1	P	R	F1	P	R	F1	P	R	F1
01	VGG16	0.67	0.50	0.57	0.78	0.49	0.60	0.61	0.61	0.61	1.00	0.36	0.53	0.68	0.38	0.49	0.85	0.97	0.90	0.33	0.14	0.20
02		0.59	0.41	0.48	0.69	0.57	0.62	0.61	0.45	0.52	0.80	0.36	0.50	0.67	0.32	0.44	0.83	0.97	0.89	0.33	0.14	0.20
03		0.50	0.31	0.38	0.58	0.37	0.45	0.43	0.50	0.46	0.57	0.36	0.44	0.56	0.34	0.42	0.84	0.91	0.87	0.22	0.14	0.17
04		0.18	0.25	0.21	0.29	0.35	0.32	0.32	0.19	0.24	0.20	0.09	0.13	0.38	0.16	0.23	0.79	0.91	0.85	0.00	0.00	0.00
05		0.45	0.28	0.35	0.39	0.33	0.36	0.40	0.19	0.26	0.43	0.27	0.33	0.34	0.15	0.23	0.76	0.94	0.84	0.00	0.00	0.00
06		0.46	0.19	0.27	0.30	0.59	0.39	0.31	0.27	0.29	0.57	0.36	0.44	0.39	0.24	0.30	0.81	0.86	0.83	0.00	0.00	0.00
07	DenseNet121	0.65	0.41	0.50	0.61	0.37	0.46	0.61	0.52	0.56	0.56	0.45	0.50	0.63	0.32	0.43	0.82	0.96	0.89	1.00	0.21	0.35
08		0.60	0.09	0.16	0.45	0.67	0.54	0.40	0.36	0.38	1.00	0.18	0.31	0.49	0.26	0.34	0.81	0.92	0.86	0.00	0.00	0.00
09		0.20	0.09	0.13	0.27	0.69	0.39	0.50	0.06	0.10	0.60	0.27	0.37	0.63	0.15	0.25	0.78	0.95	0.86	0.00	0.00	0.00
10		0.21	0.44	0.28	0.27	0.35	0.31	0.39	0.44	0.41	0.00	0.00	0.00	0.49	0.32	0.39	0.85	0.84	0.84	0.00	0.00	0.00
11		0.29	0.38	0.32	0.27	0.29	0.28	0.43	0.29	0.35	0.23	0.27	0.25	0.48	0.25	0.33	0.80	0.90	0.85	0.12	0.07	0.09
12		0.21	0.19	0.20	0.00	0.00	0.00	0.44	0.28	0.34	0.12	0.27	0.17	0.62	0.07	0.13	0.75	0.96	0.84	0.00	0.00	0.00
13	DenseNet169	0.65	0.41	0.50	0.71	0.53	0.61	0.60	0.59	0.59	0.50	0.36	0.42	0.64	0.31	0.41	0.84	0.96	0.90	0.67	0.14	0.24
14		0.46	0.50	0.48	0.52	0.25	0.34	0.58	0.42	0.49	0.67	0.36	0.47	0.58	0.28	0.38	0.81	0.97	0.88	1.00	0.07	0.13
15		0.20	0.09	0.13	0.38	0.22	0.28	0.46	0.15	0.22	0.75	0.27	0.40	0.62	0.23	0.34	0.75	0.97	0.85	0.00	0.00	0.00
16		0.19	0.56	0.29	0.25	0.31	0.28	0.29	0.41	0.34	0.00	0.00	0.00	0.34	0.31	0.32	0.87	0.73	0.79	0.05	0.07	0.06
17		0.27	0.34	0.30	0.34	0.33	0.34	0.38	0.39	0.39	0.30	0.09	0.13	0.42	0.25	0.31	0.82	0.89	0.85	0.00	0.00	0.00
18		0.19	0.16	0.17	0.27	0.35	0.31	0.35	0.28	0.31	0.33	0.09	0.14	0.27	0.21	0.23	0.80	0.82	0.81	0.05	0.14	0.08
19	ResNet50	0.50	0.44	0.47	0.59	0.47	0.52	0.54	0.45	0.49	0.83	0.45	0.59	0.54	0.32	0.40	0.83	0.95	0.89	0.25	0.07	0.11
20		0.47	0.28	0.35	0.59	0.51	0.55	0.46	0.42	0.44	0.57	0.36	0.44	0.54	0.34	0.42	0.84	0.95	0.89	1.00	0.07	0.13

(continued)

Table 2 (continued)

# from Table 1	Model Name	Akiec			bcc			bkl			df			mel			nv			vasc		
		P	R	F1	P	R	F1	P	R	F1	P	R	F1	P	R	F1	P	R	F1	P	R	F1
21		0.45	0.16	0.23	0.00	0.00	0.00	0.42	0.21	0.28	0.00	0.00	0.00	0.65	0.15	0.25	0.23	0.99	0.84	0.00	0.00	0.00
22		0.33	0.22	0.26	0.52	0.24	0.32	0.44	0.26	0.32	0.33	0.09	0.14	0.43	0.18	0.25	0.76	0.96	0.85	0.00	0.00	0.00
23		0.28	0.38	0.32	0.41	0.47	0.44	0.39	0.32	0.35	0.00	0.00	0.00	0.42	0.24	0.31	0.81	0.90	0.86	0.00	0.00	0.00
24		0.22	0.19	0.20	0.26	0.37	0.30	0.28	0.38	0.32	0.00	0.00	0.00	0.30	0.28	0.29	0.83	0.79	0.81	0.00	0.00	0.00

Table 3 Weighted average

# from Table 1	Weighted Avg. of precision	Weighted Avg. of recall	Weighted Avg. of F1-score
01	0.79	0.80	0.79
02	0.76	0.79	0.76
03	0.72	0.74	0.73
04	0.63	0.67	0.64
05	0.64	0.70	0.65
06	0.66	0.67	0.66
07	0.76	0.78	0.76
08	0.70	0.72	0.69
09	0.68	0.70	0.64
10	0.69	0.68	0.68
11	0.67	0.69	0.67
12	0.63	0.69	0.62
13	0.77	0.79	0.77
14	0.74	0.76	0.73
15	0.66	0.71	0.65
16	0.67	0.61	0.63
17	0.67	0.70	0.68
18	0.63	0.63	0.63
19	0.74	0.77	0.74
20	0.74	0.76	0.74
21	0.62	0.71	0.63
22	0.65	0.71	0.66
23	0.67	0.70	0.68
24	0.64	0.63	0.63

For DenseNet169, the weighted average of Precision, Recall, and F1-Score for image size 80px, 64px, and 32px are (0.77, 0.79, 0.77), (0.74, 0.76, 0.73), and (0.66, 0.71, 0.65), respectively, with the pre-trained model and (0.67, 0.61, 0.63), (0.67, 0.70, 0.68), amd (0.63, 0.63, 0.63), respectively, without the pre-trained model. For ResNet50, the weighted average of Precision, Recall, and F1-Score for image size 80px, 64px, and 32px are (0.74, 0.77, 0.74), (0.74, 0.76, 0.74), and (0.62, 0.71, 0.63), correspondingly, with the pre-trained model and (0.65, 0.71, 0.66), (0.67, 0.70, 0.68), and (0.64, 0.63, 0.63), correspondingly, without the pre-trained model.

From this table, we can say that VGG16 with image size 80px has acquired greater values of Precision, Recall, and F1-Score, which signifies that transfer learning on the VGG16 model with moderately high-resolution images can achieve optimal performances for skin cancer detection.

Table 4 Confusion matrix

# from Table 1	Model name	Image size (px)	Pre-trained weight	Correctly classified						
				akiec	bcc	bkl	df	mel	nv	vasc
01	VGG16	80 × 80	ImageNet	16	25	67	4	42	647	2
02		64 × 64		13	29	49	4	36	651	2
03		32 × 32		10	19	55	4	38	612	2
04		80 × 80	None	8	18	21	1	18	607	0
05		64 × 64		9	17	21	3	17	631	0
06		32 × 32		6	30	29	4	27	575	0
07	DenseNet121	80 × 80	ImageNet	13	19	57	5	36	646	3
08		64 × 64		3	34	39	2	29	614	0
09		32 × 32		3	35	6	3	17	634	0
10		80 × 80	None	14	18	48	0	36	563	0
11		64 × 64		12	15	32	3	28	601	1
12		32 × 32		6	0	30	3	8	642	0
13	DenseNet169	80 × 80	ImageNet	13	27	64	4	34	645	2
14		64 × 64		16	13	46	4	31	648	1
15		32 × 32		3	11	16	3	26	650	0
16		80 × 80	None	18	16	45	0	34	490	1
17		64 × 64		11	17	43	2	28	594	0
18		32 × 32		5	18	30	1	23	552	2
19	ResNet50	80 × 80	ImageNet	14	24	49	5	35	637	1
20		64 × 64		9	26	46	4	38	635	1

(continued)

Table 4 (continued)

# from Table 1	Model name	Image size (px)	Pre-trained weight	Correctly classified						
				akiec	bcc	bkl	df	mel	nv	vasc
21		32 × 32		5	0	23	0	17	661	0
22		80 × 80	None	7	12	28	1	20	641	0
23		64 × 64		12	24	35	0	27	603	0
24		32 × 32		6	19	41	0	31	527	0

In Table 4, we recorded the confusion matrix. From this table, we can describe that for the VGG16 architecture the correctly classified images from the seven classes of our dataset with various image sizes are 80px(16, 25, 67, 4, 42, 647, 2), 64px(13, 29, 49, 4, 36, 651, 2), 32px(10, 19, 55, 4, 38, 612, 2) with the pre-trained model and 80px(8, 18, 21, 1, 18, 607, 0), 64px(9, 17, 21, 3, 17, 631, 0), 32px(6, 30, 29, 4, 27, 575, 0) without the pre-trained model.

For DenseNet121 architecture, the correctly classified images from the seven classes of our dataset with various image sizes are 80px(13, 19, 57, 5, 36, 646, 3), 64px(3, 34, 39, 2, 29, 614, 0), 32px(3, 35, 6, 3, 17, 634, 0) with the pre-trained model and 80px(14, 18, 48, 0, 36, 563, 0), 64px(12, 15, 32, 3, 28, 601, 1), 32px(6, 0, 30, 3, 8, 642, 0) without the pre-trained model. For the DenseNet169 and ResNet50 models, the confusion matrix is also recorded in Table 4, where the highest performance for both of DenseNet169 and ResNet50 are due to 80-pixel images with the ImageNet pre-trained model in six classes (except nv). For the nv class, the highest performance is for 32-pixel images.

5 Conclusion, Limitations, and Possible Future Contributions

In this work, we have thoroughly investigated the commitments of neural networks and transfer learning for low-resource and low-quality dataset. Undoubtedly, the challenge of dealing with downsampled or partial images becomes more intrigued with the decrease in the image dimension. Nonetheless, we have shown that even in the worst scenarios, transfer learning can provide trustful outcomes. The VGG16 with ImageNet weights achieved the best performance for all of the 80 × 80, 64 × 64, and 32 × 32 pixel images. Pre-trained DenseNet169 performs as the second-best model for all of the previously mentioned images.

Regardless of the benefits of this work, there are certain limitations. This manuscript only deals with the HAM10000 dataset. We need to examine more datasets on similar or difficult types of cancer and examine the effects. Additionally, the lowest resolution of the utilized images was 32 × 32 pixels. However, experimentations for images with lower resolutions, such as 16 × 16 and 8 × 8 pixels should also be conducted. Images with lower resolutions require less memory space, which could be feasible for developing portable skin cancer detection devices for medical practitioners.

6 Author Contribution

M.R.H.K. and A.H.U. proposed the topic, conducted the research, and assembled the results. M.R.H.K. prepared the manuscript. M.R.H.K and A.H.U. revised it. A.N. reviewed the research and A.K.B. supervised the process.

References

1. Moldovan, D.: Transfer learning based method for two-step skin cancer images classification. In: 2019 E-Health and Bioengineering Conference (EHB), pp. 1–4. IEEE (2019)
2. Nugroho, A.A., Slamet, I., Sugiyanto: Skins cancer identification system of HAMl0000 skin cancer dataset using convolutional neural network. In: AIP Conference Proceedings, vol. 2202, no. 1, p. 020039. AIP Publishing LLC (2019)
3. Garg, R., Maheshwari, S., Shukla, A.: Decision support system for detection and classification of skin cancer using CNN. arXiv preprint arXiv:1912.03798 (2019)
4. Khan, M.A., Younus Javed, M., Sharif, M., Saba, T., Rehman, A.: Multi-model deep neural network based features extraction and optimal selection approach for skin lesion classification. In: 2019 International Conference on Computer and Information Sciences (ICCIS), pp. 1–7. IEEE (2019)
5. Reddy, N.D.: Classification of Dermoscopy Images using Deep Learning. arXiv preprint arXiv: 1808.01607 (2018)
6. Goyal, M., Hassanpour, S., Hoon Yap, M.: Region of interest detection in dermoscopic images for natural data-augmentation. arXiv preprint arXiv:1807.10711 (2018)
7. Sae-Lim, W., Wettayaprasit, W., Aiyarak, P.: Convolutional neural networks using MobileNet for skin lesion classification. In: 2019 16th International Joint Conference on Computer Science and Software Engineering (JCSSE), pp. 242–247. IEEE (2019)
8. Çevik, E., Zengin, K.: Classification of skin lesions in Dermatoscopic images with deep convolution network. Avrupa Bilim ve Teknoloji Dergisi, pp. 309–318 (2019)
9. Tschandl, P., Sinz, C., Kittler, H.: Domain-specific classification-pretrained fully convolutional network encoders for skin lesion segmentation. Comput. Biol. Med. **104**, 111–116 (2019)
10. Nahid, A.-Al, Ali Mehrabi, M., Kong, Y.: Histopathological breast cancer image classification by deep neural network techniques guided by local clustering. BioMed Res. Int. 2018 (2018)
11. Manoharan, S.: Improved version of graph-cut algorithm for CT images of lung cancer with clinical property condition. J. Artif. Intell. **2**(04), 201–206 (2020)
12. Vijayakumar, T.: Neural network analysis for tumor investigation and cancer prediction. J. Electron. **1**(02), 89–98 (2019)
13. Nahid, A.-Al, Ali, F.B., Kong, Y.: Histopathological breast-image classification with image enhancement by convolutional neural network. In: 2017 20th International Conference of Computer and Information Technology (ICCIT), pp. 1–6. IEEE (2017)

14. Deniz, E., Şengür, A., Kadiroğlu, Z., Guo, Y., Bajaj, V., Budak, Ü.: Transfer learning based histopathologic image classification for breast cancer detection. Health Inf. Sci. Syst. **6**(1), 18 (2018)
15. Tschandl, P., Rosendahl, C., Kittler, H.: The HAM10000 dataset, a large collection of multi-source dermatoscopic images of common pigmented skin lesions. Sci. Data **5**, (2018)
16. Zaballos, P., Daufí, C., Puig, S., Argenziano, G., Moreno-Ramírez, D., Cabo, H., Marghoob, A.A., Llambrich, A., Zalaudek, I., Malvehy, J.: Dermoscopy of solitary angiokeratomas: a morphological study. Arch. Dermatol. **143**(3), 318–325 (2007)
17. Zaballos, P., Carulla, M., Ozdemir, F., Zalaudek, I., Bañuls, J., Llambrich, A., Puig, S., Argenziano, G., Malvehy, J.: Dermoscopy of pyogenic granuloma: a morphological study. Br. J. Dermatol. **163**(6), 1229–1237 (2010)
18. Orenstein, E.C., Beijbom, O.: Transfer learning and deep feature extraction for planktonic image data sets. In: 2017 IEEE Winter Conference on Applications of Computer Vision (WACV), pp. 1082–1088. IEEE (2017)
19. He, K., Zhang, X., Ren, S., Sun, J.: Deep residual learning for image recognition. In: Proceedings of the IEEE Conference on Computer Vision and Pattern Recognition, pp. 770–778 (2016)
20. Simonyan, K., Zisserman, A.: Very deep convolutional networks for large-scale image recognition. arXiv preprint arXiv:1409.1556 (2014)
21. Huang, G., Liu, Z., Van Der Maaten, L., Weinberger, K.Q.: Densely connected convolutional networks. In: Proceedings of the IEEE Conference on Computer Vision and Pattern Recognition, pp. 4700–4708, 2017
22. Mobiny, A., Singh, A., Van Nguyen, H.: Risk-aware machine learning classifier for skin lesion diagnosis. J. Clin. Med. **8**(8), 1241 (2019)
23. Gessert, N., Sentker, T., Madesta, F., Schmitz, R., Kniep, H., Baltruschat, I., Werner, R., Schlaefer, A.: Skin lesion diagnosis using ensembles, unscaled multi-crop evaluation and loss weighting. arXiv preprint arXiv:1808.01694 (2018)

CNN-Based Vehicle Classification Using Transfer Learning

G. M. Rajathi, J. Judeson Antony Kovilpillai, Harini Sankar, and S. Divya

Abstract This paper focuses on the classification of vehicles using Convolutional Neural Network (CNN) which is a class of deep learning neural network. This work makes use of transfer learning using the pre-trained networks to extract powerful and informative features and apply that to the classification task. In the proposed method, the pre-trained networks are trained on two vehicle datasets consisting of real-time images. The classifier performance along with the performance metrics such as accuracy, precision, false discovery rate, recall rate, and false negative rate is estimated for the following pre-trained networks: AlexNet, GoogLeNet, SqueezeNet, and ResNet18. The classification model is implemented on the standard vehicle dataset and also on a created dataset. The model is further used for the detection of the different vehicles using Regions with a Convolutional Neural Networks (RCNN) object detector on a smaller dataset. This paper focuses on finding the perfect network suitable for the classification problems which have only a limited amount of non-labeled data. The model makes use of limited pre-processing and achieves greater accuracy on continuous training of the networks on the vehicle images.

Keywords Convolutional neural network (CNN) · Pre-trained networks · AlexNet · GoogLeNet · SqueezeNet and ResNet18 · Accuracy · Precision · False discovery rate · Recall rate · False negative rate · Vehicle classification

G. M. Rajathi · J. J. A. Kovilpillai (✉) · H. Sankar · S. Divya
Department of ECE, Sri Ramakrishna Engineering College, Coimbatore, India
e-mail: judeson@srec.ac.in

G. M. Rajathi
e-mail: rajathi.gm@srec.ac.in

H. Sankar
e-mail: harinisankar.1602048@srec.ac.in

S. Divya
e-mail: divya.1602035@srec.ac.in

J. S. Raj et al. (eds.), *Intelligent Sustainable Systems*, Lecture Notes in Networks and Systems 213, https://doi.org/10.1007/978-981-16-2422-3_27

1 Introduction

An Artificial Intelligence-based Intelligent transportation system includes various different functionalities and can provide greater leeway for managing traffic. The possibility of an AI-based classification and detection of vehicles provides better accuracy and ease in handling operations. AI-enabled intelligent systems and the key tech components are the basic building blocks of smart city-based traffic control systems. Intelligent automated systems are developed using proficient deep neural networks which adopts various advanced network architecture with several layers of computations for evaluation and optimization. The neural network imports a huge amount of real-time data and extracts different features from it continuously and makes the prediction or classification based on the dataset and learning rate. Transfer learning is a much efficient method of solving the problem in question with the limited amount of non-labeled data available. This research focuses on developing a Convolutional Neural Network (CNN)-based vehicle classification where the methodology is tested with a pre-existing dataset and a created dataset in which the different accuracy and performance metrics are calculated. The model adopts transfer learning where the model is verified with different CNN-based architectures like AlexNet, GoogLeNet, SqueezeNet, and ResNet18 and the accuracy of detection is analyzed.

2 Related Works

Patil et al. [1] proposed deep learning-based car damage classification. A new dataset is created by collecting images from the Web and manually annotating them. Multiple deep learning-based techniques such as training CNNs from random initialization, Convolution Auto-encoder-based pre-training followed by supervised fine-tuning, and transfer learning methods are implemented. Transfer learning has a better performance.

Han et al. [2] proposed a pre-trained convolutional neural network for image-based vehicle classification. A classification method based on CNNs has been detailed in this paper. To improve the accuracy in the vehicle detection, the unrelated background was removed for facilitating the feature extraction and vehicle classification. Then, an auto-encoder-based layer-wise unsupervised pre-training is introduced to improve the CNN model by enhancing the classification performance. Experimental results have demonstrated that the pre-trained CNN method based on vehicle detection is the most effective for vehicle classification.

Krizhevsky et al. [3] demonstrated impressive classification performance using a large ConvNet. The paper shows that a large, deep convolutional neural network is capable of a highly challenging dataset using purely supervised learning.

Xing et al. [4] achieved a binary detection rate of 91.4% accuracy with their driver activity recognition system designed based on the deep convolutional neural networks.

3 Methodology

The proposed system makes use of a transfer learning approach [5] using the pre-trained convolutional neural networks such as AlexNet, GoogLeNet, ResNet, and SqueezeNet for the classification and detection task. The transfer learning approach helps to reduce the execution time, and the networks using the approach have the capacity to learn the weights faster. The performance and accuracy of the networks are compared and analyzed. Pre-trained networks are convolutional neural networks that use the learned features and apply those to the destined task. This eliminates the tedious process of rebuilding a network and deciding on the various processes related to it like training and configuring the network The pre-trained networks have the capacity to classify different object categories which have been learned previously from the source task. The pre-trained networks are then fine-tuned to suit the classification [6] problem. There are many pre-trained networks which are trained on the ImageNet [7] database. The model is implemented using five of such pre-trained networks which are, namely: AlexNet, GoogLeNet, ResNet18, and SqueezeNet. The model makes use of limited pre-processing and achieves greater accuracy on continuous training of the networks on the vehicle images. In pre-processing, the images are resized into the images of the same size as they are collected from a different source, and grayscale images are converted into color images (Fig. 1).

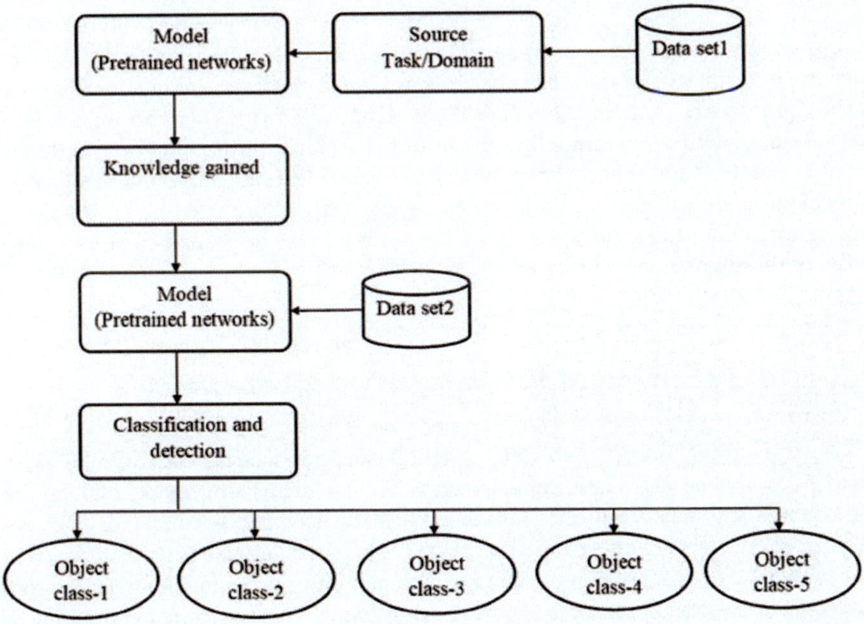

Fig. 1 Block diagram

This work comprises of two dataset labels (dataset1 and dataset2) to classify the vehicles using CNN. The dataset1 denotes the ImageNet dataset and dataset2 denotes standard (origin Nepal created in the year 2015) and created datasets (images collected from flickr, pixabay, shutterstock, and alamystock). The pre-trained networks are trained on the created and standard datasets making use of the weights previously learned from the source task. The inputs are the images of the five different object classes and the two datasets comprise the five object classes car, truck, bus, auto, and two-wheeler. The pre-trained networks extract the features which are the weights which help in defining and recognizing the objects. The process gives results with higher speed and accuracy. The feature extraction is a process in which the original weights are subjected to the CNN functions and the new convolved weights are passed on from one layer to another layer. The classified object classes and detected results are obtained after the successful completion of training and testing.

3.1 Transfer Learning

Transfer learning is an AI-based deep learning technique that adopts the knowledge obtained by the model learned from a vast set of labeled data and applies it in domain-specific areas where the amount of labeled data available is considerably low. It uses different pre-trained neural network models on a newly designed model created using a dataset with a lesser amount of labeled data for efficient problem-solving. They use problem-specific and model-specific models in which the time to train is low and the performance of the neural network is high. Transfer learning adopts several techniques to pre-train and initialize the weights of a neural network-based classification model by reconstructing the different features of the images at different epochs and iterations. The techniques used in transfer learning improve convergence, model reusability, faster data training rate, and also the feature extraction followed by reconstruction of the extracted information optimizes the performance of the model in a significant manner.

3.2 Using Pre-Trained Networks for Transfer Learning

Transfer learning imports pre-trained neural network models for different task-specific problems [8]. These models can make use of existing models which are pre-trained and they are further fine-tuned and the accuracy is validated. The pre-trained parameters are not changed at first and a low learning rate is set to ensure that the network does not unlearn the already obtained information and knowledge from the source dataset in the process of continuous training. This can avoid overtraining of data and yield better accuracy. Transfer learning can easily optimize feature extraction with learned and updated representation and model-specific modeling of data. Pre-trained networks use the learned features and apply those to the task. This eliminates the tedious process of rebuilding a network and deciding on the various processes related to it like training and configuring the network. Training the network is an essential process in classification problems. The training involves several changes made to the input images like geometric transformations, resizing, rotations, translations, etc. The pre-trained networks extract the features from the new images based on the features learned from the source task and classify the images.

3.3 *Implementation of Transfer Learning*

- **Loading the pre-trained network**: The lower layers of the Convolutional Neural Network (CNN) capture features like edges which are considered as low level, while the higher layers of the CNN capture intricate and elaborate details of the image and other attributes of composition (Fig. 2).
- **Replace the final layers**: The CNN's lower layers that constitute the generic features are regained and the final layers are reconstructed with the updated layers extracted from the dataset.
- **Train the network**: The neural network model is then trained using the new dataset. The model uses a data augmentation feature to avoid overfitting the CNN and storing the equivalent details of the training images and dataset. The training options are declared and the extracted features from the primary layers of the pre-trained neural network are obtained (the transferred layer weights). The learning rate of transferred layers is slowed down and the initial learning rate is set very low. Then the network is verified and evaluated for every 'Validation Frequency' iteration at different epochs during the training process.
- **Predict and assess network accuracy**: The updated neural network can now classify different images and the predicted results can also be obtained along with the accuracy evaluation.
- **Results**: The results are acquired and the accuracy metrics are analyzed.

4 Experimental Results

The datasets consist of five object classes in total comprising of vehicles, namely truck, bus, two-wheeler, car, and auto-rickshaw.

Fig. 2 Workflow of transfer learning

The vehicles are classified using the pre-trained CNN networks through transfer learning. The classification algorithm is implemented on the vehicle dataset and the created dataset. The model has been trained using all five object classes.

4.1 Dataset Description

Standard vehicle dataset: 2138 images origin Nepal created in the year 2015.
Standard vehicle dataset total number of images: 2138. Images for training: 1497. Images for testing: 641.
Created dataset: 180 images collected from flickr, pixabay, shutterstock, alamystock.
Created dataset total number of images: 180 Images for training: 126 Images for testing: 54.

4.2 Training the Pre-Trained Networks on the Dataset

The proposed system makes use of the knowledge already learned by the pre-trained networks on the desired task, and it is implemented by loading the input images and by replacing the final layers of the network which are the fully connected layer, softmax layer, and output layer to match the task at hand with trained weights.

4.3 Training Details

The training is done several times and the best classification accuracy is achieved by modifying the minibatch size and epochs. The training of the network is done on a single CPU and each time the confusion chart and matrix are plotted (Tables 1, 2, 3, 4, 5 and 6).

All the data in the above tables represent the detection accuracy and other metrics with respect to different vehicles, and set1, set2 represent the created and standard dataset.

4.4 Training the Pre-Trained Network Alexnet for Detection

The pre-trained network is trained on the created dataset using Alexnet [9] and the weights are used to detect the vehicles in the test images, and it has better detection accuracy comparatively (Fig. 3).

Table 1 Parameters involved in training (created dataset)

Created dataset: 180, training: 126, test: 54

AlexNet	Minibatch size	Epoch	Iterations	Validation frequency	Iterations per epoch	Accuracy	Time elapsed
	20	6	36	6	6	100	3 min 57 s
		9	54	6	6	96.36	6 min 8 s
	30	8	32	4	4	100	5 min 41 s
		9	36	4	4	98.18	6 min 51 s
		11	44	4	4	98.18	4 min 34 s
		12	48	4	4	100	5 min 41 s

Created dataset: 180, training: 126, test: 54

GoogLeNet	Minibatch size	Epoch	Iteration	Validation frequency	Iterations per epoch	Accuracy	Time elapsed
	20	6	36	6	6	100	3 min 26 s
		9	54	6	6	100	4 min 19 s
	30	8	32	4	4	100	8 min 7 s
		9	36	4	4	100	7 min 19 s
		11	44	4	4	100	7 min 10 s
		12	48	4	4	100	6 min 34 s

Created dataset: 180, training: 126, test: 54

ResNet18	Minibatch size	Epoch	Iteration	Validation frequency	Iterations per epoch	Accuracy	Time elapsed
	20	6	36	6	6	96.36	2 min 45 s
		9	54	6	6	96.36	4 min 10 s
	30	8	32	4	4	92.73	3 min 39 s
		9	36	4	4	96.36	4 min 1 s

(continued)

Table 1 (continued)

Created dataset: 180, training: 126, test: 54							
		11	44	4	4	94.55	5 min 10 s
		12	48	4	4	98.18	5 min 53 s
Created dataset: 180, training: 126, test: 54							
SqueezeNet	Minibatch size	Epoch	Iteration	Validation frequency	Iterations per epoch	Accuracy	Time elapsed
	20	6	36	6	6	94.55	1 min 30 s
		9	54	6	6	87.27	2 min 30 s
	30	8	32	4	4	92.73	2 min 40 s
		9	36	4	4	98.18	2 min 28 s
		11	44	4	4	98.18	5 min 59 s
		12	48	4	4	100	2 min 59 s

Table 2 Parameters involved in training (standard dataset)

Standard vehicle dataset: 2138; training: 1497; test: 641							
AlexNet	Minibatch size	Epoch	Iterations	Validation frequency	Iterations per epoch	Accuracy	Time elapsed
	20	10	740	74	74	90.80	68 min 14 s
		15	1110	74	74	87.27	72 min 20 s
	30	8	392	49	49	92.98	42 min 40 s
		9	441	49	49	88.92	42 min 52 s
		11	539	49	49	91.11	168 min 52 s
		15	735	49	49	90.17	112 min 48 s
Standard vehicle dataset: 2138; training: 1497; test: 641							

(continued)

Table 2 (continued)

Standard vehicle dataset: 2138; training: 1497; test: 641

GoogLeNet	Minibatch size	Epoch	Iteration	Validation frequency	Iterations per epoch	Accuracy	Time elapsed
	20	10	740	74	74	92.82	6 min 11 s
		15	1110	74	74	94.54	109 min 29 s
	30	8	392	49	49	95.79	39 min 47 s
		9	441	49	49	94.38	47 min 42 s
		11	539	49	49	93.76	92 min 52 s
		15	735	49	49	94.85	214 min 16 s

Standard vehicle dataset: 2138; training: 1497; test: 641

ResNet18	Minibatch size	Epoch	Iteration	Validation frequency	Iterations per epoch	Accuracy	Time elapsed
	20	10	740	74	74	92.98	52 min 11 s
		15	1110	74	74	92.82	81 min 30 s
	30	8	392	49	49	92.36	46 min 52 s
		9	441	49	49	92.82	47 min 25 s
		11	539	49	49	93.29	106 min 52 s
		15	735	49	49	92.82	69 min 21 s

Standard vehicle dataset: 2138; training: 1497; test: 641

SqueezeNet	Minibatch size	Epoch	Iteration	Validation frequency	Iterations per epoch	Accuracy	Time elapsed
	20	10	740	74	74	90.48	24 min 55 s
		15	1110	74	74	90.95	69 min 18 s
	30	8	392	49	49	90.80	19 min 43 s
		9	441	49	49	91.58	22 min 27 s
		11	539	49	49	89.70	26 min 4 s
		15	735	49	49	92.51	35 min 3 s

Table 3 Performance metrics (AlexNet)

AlexNet	Batchsize: 30	Object classes					
		Data	Auto	Bus	Car	Truck	Two-wheeler
	Precision/positive predictive rate (%)	Set1 98.2	90.9	100	100	100	100
		Set2 91.1	95.5	87.5	96.5	62.2	100
	False discovery rate (%)	Set1 98.2	9.1	0.0	0.0	0.0	0.0
		Set2 91.1	4.5	12.5	3.5	37.8	0.0
	True positive/recall rate (%)	Set1 98.2	100	100	00	100	93.8
		Set2 91.1	77.8	89.1	82.8	93.2	100
	False negative rate (%)	Set1 98.2	0.0	0.0	0.0	0.0	6.3
		Set2 91.1	22.2	10.9	17.2	6.8	0.0

Table 4 Performance metrics (GoogLeNet)

GoogLeNet	Batchsize: 30	Object Classes					
		Data	Auto	Bus	Car	Truck	Two-wheeler
	Precision/ positive predictive rate (%)	Set1 100	100	100	100	100	100
		Set2 94.9	100	92.7	98.1	74	100
	False discovery rate (%)	Set1 100	0.0	0.0	0.0	0.0	0.0
		Set2 94.9	0.0	7.3	1.9	26	0.0
	True positive/ Recall rate (%)	Set1 100	100	100	100	100	100
		Set2 94.9	96.3	92.7	89.3	95.9	100
	False negative rate (%)	Set1 100	0.0	0.0	0.0	0.0	0.0
		Set2 94.9	3.7	7.3	10.7	4.1	0.0

Table 5 Performance metrics (ResNet18)

ResNet18	Batchsize: 30	Object Classes					
		Data	Auto	Bus	Car	Truck	Two-wheeler
	Precision/positive predictive rate (%)	Set1 98.2	100	100	100	100	94.1
		Set2 93.33	96.7	86.5	96.8	71.4	100
	False discovery rate (%)	Set1 98.2	0.0	0.0	0.0	0.0	5.9
		Set2 93.33	4.3	13.5	3.2	28.6	0.0
	True positive/ Recall rate (%)	Set1 98.2	10	0.0	0.0	0.0	0.0
		Set2 93.33	18.5	18.2	10.3	5.4	0.0
	False negative rate (%)	Set1 98.2	0.0	0.0	0.0	0.0	0.0
		Set2 93.33	3.7	7.3	10.7	4.1	0.0

Table 6 Performance metrics (SqueezeNet)

SqueezeNet	Batchsize: 30	Object Classes					
		Data	Auto	Bus	Car	Truck	Two-wheeler
	Precision/Positive predictive rate (%)	Set1 92.7	88.9	90.9	100	100	88.9
		Set2 92.5	93.1	100	98	62.8	100
	False discovery rate (%)	Set1 92.7	11.1	9.1	0.0	0.0	11.1
		Set2 92.5	6.9	0.0	2.0	37.2	0.0
	True positive/ Recall (%)	Set1 92.7	80	100	90.9	87.5	100
		Set2 92.5	100	80	85.8	95.9	99.6
	False negative rate (%)	Set1 92.7	20	0.0	9.1	12.5	0.0
		Set2 92.5	0.0	20	14.2	4.1	0.4

The training of the model on different minibatch sizes shows that accuracy can be achieved effectively by training the model on a minibatch size of 30, which has been found to produce better results in all cases. The iteration value depends on the minibatch size and epoch. The minibatch size and epoch values are chosen so as to keep the iteration value at an optimum level. The minibatch loss and validation loss parameters are initially high; by continuous training, it reaches a lower value at the end. The validation accuracy is initially low, which improves on training and reaches a higher value. The model was implemented in MATLAB using deep learning and computer vision toolbox [10] and has been able to produce accurate and effective results by making use of the minimal data available and also using only simple methods for training and execution. Intel® Core™ i5-8250U Processor@1.60–1.80 GHz with 4 GB RAM is used to implement this work.

5 Conclusion

Thus, the pre-trained neural networks were used for the classification and detection of the vehicles using transfer learning. The overall validation accuracy of GoogLeNet is higher when compared to the other networks. The accuracy of GoogLeNet is about 100% on an average in the created dataset and 94.3566% on an average in the standard vehicle dataset. The performance of AlexNet and ResNet18 are at par and they take up optimal execution time. SqueezeNet has an efficient execution time but it has an effect on the accuracy. ResNet18 works fine even on the standard vehicle dataset

```
****************************************************************
Training an R-CNN Object Detector for the following object classes:

* auto

--> Extracting region proposals from 24 training images...done.

--> Training a neural network to classify objects in training data...

Training on single CPU.
Initializing input data normalization.
|=======================================================================================|
|  Epoch  |  Iteration  |  Time Elapsed  |  Mini-batch  |  Mini-batch  |  Base Learning  |
|         |             |   (hh:mm:ss)   |   Accuracy   |     Loss     |      Rate       |
|=======================================================================================|
|       1 |           1 |       00:00:06 |       53.33% |       1.1473 |      5.0000e-06 |
|       4 |          50 |       00:02:56 |       90.00% |       0.1880 |      5.0000e-06 |
|       7 |         100 |       00:05:02 |       90.00% |       0.2499 |      5.0000e-06 |
|      10 |         150 |       00:07:06 |       93.33% |       0.1319 |      5.0000e-06 |
|=======================================================================================|

Network training complete.

--> Training bounding box regression models for each object class...100.00%
```

Fig. 3 Object detection using AlexNet

consisting of a large number of images and achieves good accuracy on an average of about 92.84% taking up only optimal time for execution. AlexNet and GoogLeNet work very efficiently on the smaller dataset achieving an accuracy of more than 98% overall. The detection using Alexnet has an overall accuracy of about 80% approximately which can be improved by the inclusion of more training images The model was implemented on two datasets consisting of real-time images. GoogLeNet was found to have greater accuracy of about 90% in both cases. The vehicles were also detected using AlexNet and the accuracy also improved on training. GoogLeNet works fast with greater accuracy in the new task but takes up a longer execution time. ResNet18 works and achieves near-perfect accuracy taking up optimal execution time. AlexNet takes up more execution time and achieves comparatively lesser accuracy.

SqueezeNet takes up a lesser execution time but this affects the accuracy.

References

1. Patil, K., Kulkarni, M., Sriraman, A., Karande, S.: Deep Learning Based Car Damage Classification (2017)
2. Han, Y. et al.: Pretraining convolutional neural networks for image-based vehicle classification. Hindawi, Adv. Multimedia (2018)
3. Krizhevsky, I.S., Hinton, G.E.: Imagenet classification with deep convolutional neural networks. In: NIPS 2012: Neural Information Processing Systems
4. Xing, Y., Lv, C., Wang, H., Cao, D., Velenis, E., Wang, F.-Y.: Driver activity recognition for intelligent vehicles: a deep learning approach. IEEE Trans. Vehicul. Technol. (2019)
5. Almisreb, A.A., Jamil, N., Md Din, N.: Utilising alexnet deep transfer learning for ear recognition. In: Fourth International Conference on Information Retrieval and Knowledge Management (2018)
6. Shakya, S.: Analysis of artificial intelligence based image classification techniques. J. Innov. Image Process. (JIIP) **2**(01), 44–54 (2020)
7. Russakovsky, O., Deng, J., Su, H., Krause, J., Satheesh, S., Ma, S., Huang, Z., Karpathy, A., Khosla, A., Bernstein, M., Berg, A.C., Fei-Fei, L.: ImageNet large scale visual recognition challenge. Int. J. Comput. Vis. **115**(3), 211–252 (2015)
8. Pan, S.J., Yang, Q.: A survey on transfer learning. IEEE Trans. Knowl. Data Eng. **22**(10), 1345–1359 (2010)
9. Felzenszwalb, P.F. et al.: Object detection with discriminatively trained part-based models. IEEE Trans. Pattern Anal. Mach. Intell. **32**(9) 1627–1645 (2010)
10. https://in.mathworks.com/help/deeplearning/ug/pretrained-convolutional-neural-networks.html

IoT: A Revolutionizing Step for Pharmaceutical Supply Chain Management During and After Covid19

Amrita Verma Pargaien, Tushar Kumar, Monika Maan, Suman Sharma, Himanshu Joshi, and Saurabh Pargaien

Abstract Internet of Things (IoT) is an interrelated system of computing devices, machines, sensors, and objects. It can transfer data over a network without the Human-to-computer or Human-to human interaction. This article is focused on the progression in IoT. It was analyzed that how IoT can truly revolutionize the oppositions faced by Manufacturing, Logistics, and Supply chain logistics in the pharmaceutical industry during the Covid-19 pandemic and **following**. In the present scenario of this pandemic and annihilation, IoT has emerged as the need of the time for the pharmaceutical industries. In order to develop a well-coordinated logistic and supply chain network, it is required to make their **prominent occupancy** in the market. Patients are unavailable for the essential medicines and medications that are critical and strong to protect. With the current and recent developments in automation technology, the need for consistency of supply chain visibility in the pharmaceutical industries has increased enormously. The implementation of these automation technologies will certainly result in getting serious confidence in understanding the statistics of the supply chain. This technology is also built to invalidate the risk factors that accompany the pharmaceutical supply chain network. Generally, these systems are available and operate as a decentralized system. The information or data transmitted via these systems is isolated and the protection of vulnerable data is therefore not compromised.

Keywords Internet of things · Supply chain · Pharmaceutical manufacturing · Management · Covid-19

A. V. Pargaien · T. Kumar · H. Joshi · S. Pargaien (✉)
Graphic Era Hill University, Bhimtal, India

M. Maan · S. Sharma
Department of Pharmacy, Banasthali University, Vanasthali, Rajasthan, India

J. S. Raj et al. (eds.), *Intelligent Sustainable Systems*, Lecture Notes in Networks and Systems 213, https://doi.org/10.1007/978-981-16-2422-3_28

Abbreviations

IoT	Internet of Things
RFID	Radio Frequency Identification
EPC	Electronic Product Code
ALE	Analytics and Location Engine
ONS	Object Naming Services
AIDC	Automatic Information Data Collection
IoT-PM	Internet of things-Pharmaceutical Manufacturing
SCM	Supply Chain Management

1 Introduction

IoT has emerged in the past as an influential and motivating technology that has the potential to accelerate the coming era and make it more sustainable and help humanity survive in the difficult times as people are going through in this pandemic [1]. It's a network of varying components which are inter-connected and even has a network connectivity to perform a wide spectrum of tasks without much interference of human experts. At present the pharmaceutical industry represents a significant role in the healthcare sector optimistically [2, 3]. The idea behind IoT is that the conventional benefits are originated with sensing, identifying, notifying, rectifying, monitoring, and networking potentials permitting each of them to connect and share the data and services having a common objective to build up an IoT network in order to perform their assisted task at their excellent possibilities [4, 5]. Primarily for its application in the industries IoT's can be equipped for the commercial as well as the financial transactions to deal with the affairs of production, banking, logistics, service sector, financial and governmental authorities. And this leads to the removal without human involvement of most of the insignificant networks [6].

In the environment, it can be designed to inspect, protect, and enhance the flora and fauna present in a specific region and to deal with its recycling, management, breeding, energy harvesting and management, etc. It can even be used in the society for its development, inclusion of cities, people, societies, etc. to deal with the e-participation of public, e-inclusion, society structure services, etc. [7, 8]. IoT is used in a very unique manner in the field of medical and health care. Initially the RFID tags and the wireless sensors can be implanted in the medical equipments or the clinical devices which are connected to the hospital's network as well as the patients. The use of these RFID tags prevents the theft and piracy of the medical devices, medical equipments, medicines, etc. The use of these even helps us to maintain a well tabulated and organized record of these basic utilities in the hospital [5, 9]. The use of these also includes maintaining the hospital a well tabulated and coordinated record of these essential utilities [5, 9]. By utilizing the IoT in resource transfer results in autonomous tracing and sorting, which can accelerate this process insidiously and

Table 1 IoT implementation in pharma industries

Health monitoring	Manufacturing and supply chain	Sales and marketing	Patient Access
Real-time diagnostics	Automated Information Data Collection (AIDC) for smart serialization	Drug interaction checker—interactive ecosystem with HCPS using database encompassing	Wearable devices Mobile-Based App
Wearable devices to monitor health	Real-time logistics visibility using RFID	Use of sensors and devices to monitor clinical sites	Drug usage tracking and medication compliance

even increase security. This also prevents the misplacement of products, both globally and domestically, from contributing to needless dumping [10]. Implementation of IoT in the pharmaceutical industry is summarized in Table 1.

2 Literature Review

The application of the IoT in the manufacturing and the supply chain network tends to result in productive in the near-coming future. On condition that the money, facilities, protection, alliances, public confidence consumer demands, etc. in position can also revolutionize the whole way in the pharmaceutical companies currently operate. Table 2 summarizes recent advances in the application of IOT in the area of the pharmaceutical supply chain.

3 Pharmaceutical Logistics Issues

The supply chain of pharmaceuticals is connected to the medical and healthcare sectors. This is why a very high production standard, a strong supply chain infrastructure and logistics are needed. IoT can be used in this genre for real-time tracking and remote monitoring of both warehouses and operations. And such an order with this kind of network that helps businesses to produce widely recognized standard quality medicines and also helps businesses perform their operations efficiently and accurately concerning state, government regulations. It also needs plenty of technical improvements to function such an operational infrastructure at a well-defined level of protection that can only be accomplished by digitizing and improving the critical standards of the manufacturing process and supply chain management [12–14]. Pharmaceutical logistics failures are illustrated in Fig. 1.

Table 2 Current developments in the application of IOT in the field of pharmaceutical supply chain

S. no.	Author	Problem specification	Proposed approach	Advantage
1	Ting et al. [9]	Supply chain visibility and transparency in pharmaceutical industry, counterfeiting issue, product lifecycle, product authentication	EPC tag (RFID coded with a unique EPC)	Higher information visualization, medicine recall, correct inventory control, stock management, replenishment
2	Marathe et al. [1]	Collusion deficits, Handling of Data, Inventory control	EPC-IS, Local ONS, Root ONS, ERP	Lowers the risk of drug trafficking, controls temperature nicely, reduces the collusion deficits
3	Ravi Pratap Singh et al. [3]	Problems due to Covid-19 Pandemic in the classical method of operation in healthcare system	Managing the process virtually, Visualization of data in remote areas	Increased control, lowering expenses, quality diagnosis
4	Catarinucci et al. [5]	The ineffective RFID-based tractability	Using the High-Performance UHF RFID Tags, Item-Level Tracing	Increased spectrum of the tracing area, multilevel tracing, better tracking of the inventory goods
5	Singh et al. [11]	Remote tracking, real-time tracing, shop floor perceptibility, remote distribution	Introduction of the IoT-based network architecture in the current running pharma technologies	Enhances the goods tracing ability, processes and develops new business models, Reinforcements in packaging
6	Turcua et al. [12]	Lack of the important medical information of the patient, poor quality of information sharing in the healthcare sector	The collusion of RFID technologies and multiagent technologies for a better approach in healthcare via IoT	RFID-based identification of the patients, medical staff and medical equipments, tracking RFID-tagged things
7	Alagarsamy et al. [13]	Huge data generation, harvesting the huge data, analysis of the huge data	Individualized Medicines, logistic traceability using the RFID tags, machine and equipment maintenance forecasting	Can act as a catalyst for paperless data production, enhanced safety, protection of the brands market value

(continued)

Table 2 (continued)

S. no.	Author	Problem specification	Proposed approach	Advantage
8	Yan et al. [14]	Lag in data transmission in traditional method of supply chain management	Chain information transmission model, RFID tracking, IoT-based network structure for supply chain	Solves the information asymmetry, helps to come up with a global supply chain management strategy
9	Pachayappan et al. [6]	Risk factors in the overall supply chain network, ineffective information control, unintegrated data, and information	RFID- tagging, GPS-tagging, introducing decentralized systems in the supply chain network	Real-time goods tracing from the supplier to the customer, lower cost investments, rapid reaction to the changes in shipments by the huge pharmacy
10	De Blasi et al. [15]	The use of the near field passive UHF-tags	Implementation of the far field UHF-tags,	Enhances the performance at the item level tagging and tracing
11	Wang [16]	High infection risk due to patient movement, risk of patient's clinical history theft	Ethereum hybrid network certification system; collusion of the Block-chain technologies, cloud technologies and IoT	Cost effectiveness, minimized response time, ease of scalability, minimized risk of data theft

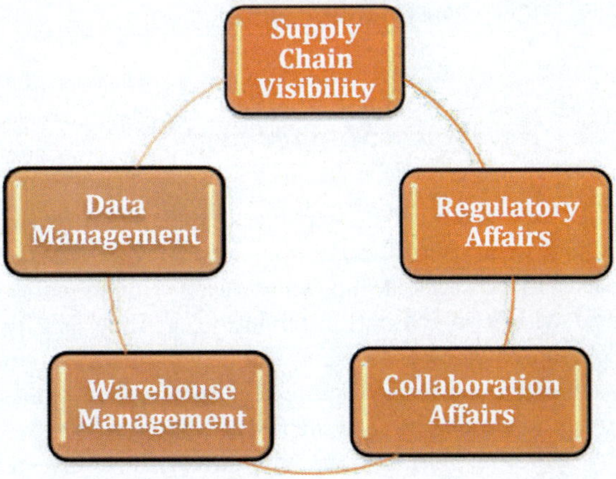

Fig. 1 Hindrances in the pathway of pharmaceutical logistics

3.1 Supply Chain Visibility

The thieving and the piracy of the drugs during their transportation is a very serious issue. If a pharmaceutical company is not capable to reconstruct such disruptions or mishaps then they may face serious problems regarding the regulations. And all of this calls for the product or the product to be remotely traced.

3.2 Regulatory Affairs

It is a serious requirement for the pharma industries to manufacture the drugs as per the norms and regulations of the state across the world, creating more seriousness in supply chain network. And any disruption or mishap in these may lead to the serious issues of the company with the state.

3.3 Collaboration Affairs

The lack of accountability and severe trust problems in the exchange of data in the pharma supply chains are attributable to the above. Significant multinationals are participating in a very extreme and dynamic data flow in the partnership of this large supply chain. In this case, the exchange of data by such a large multi-strata organization is a very significant obstacle to the cooperation of the supply chain network, which allows for technical developments to upgrade this supply chain network and address trust issues.

3.4 Warehouse Management

The pharmaceutical industry has a rapid pace of growth and it's a tremendous growth in industrial sector and simply implies that it also has a very large scale which needs to be stocked and warehoused. But logistics are not so well developed to deal with it in the current situation, as they lack the infrastructure to satisfy the demand of these pharmaceutical sectors. But if anyone urges the construction of a well-organized infrastructure equipped with the latest technological developments to introduce this kind of stability, then they must be prepared to face the wrath of the enormous corporations who already have invested and are equipped with such a good infrastructure. Even after all this, the confidentiality of the information and the information exchanged from the businesses to the stockers or the keeper of the warehouse is still a matter of very serious concern.

3.5 *Data Management*

The pharmaceutical industries will need to efficiently and effectively resolve to cope with this development in this period of massive digitization and when all start as online. And to do so, industries need to reliably monitor their information and prevent it from compromising the data in the hands of hackers, ransom-wares, etc.

4 Supply Chain Difficulties Faced by the Pharmaceutical Companies During the Covid-19 Situation

China is the world's largest producer of pharmaceutical ingredients and India is the wholesome importer of about 70% of global pharmaceutical APIs for the manufacture of drugs and drug products. In the early stages of the Covid-19 situation, the Indian government limited imports from China to about 26 separate APIs, making up a total of about 10% drug exports from India, really caused a situation of scarcity and shook the global pharmaceutical market [17].

Despite the fact that stock levels and buffer inventory across the supply chain of biopharmaceuticals were equipped, and there exist a serious evidence of supply chain delays that may significantly hinder the ability of the industry to treat patients. A total of around 15 new drugs were added to the FDA's drug shortage list as the pandemic began. India removed export restrictions on various drugs and drug products APIs, but their obtainability still remains as a problem [18].

With the U.S. being the largest consumer of prescription medicines and drug products, the demand is projected to be approximately 45–50%. As the use of antivirals is seriously growing, due to their concern for exposure to Covid-19, this may lead to a situation where chronic diseases remain untreated in various patients and individuals. This has contributed to a shift in demand for medications from various manufacturers of branded pharmaceuticals. As the predominant way of producing pharmaceutical drugs and drug products, the pandemic situation has become severely complicated and tampered upon. An additional issue emerged with regard to the disturbance caused in the delivery of the goods, as enormous constraints were imposed on massive population movements creating a serious problem of distribution and shipping [19].

5 Improvements

As already discussed with the problems that are faced in the pharmaceutical industry in manufacturing, supply chain, and logistics were solved. There are several approaches that can be used to enhance production, supply chain, and logistics problems, as shown in Fig. 2. These methods need to be implied along with the classical or the traditional ways of operations or functioning.

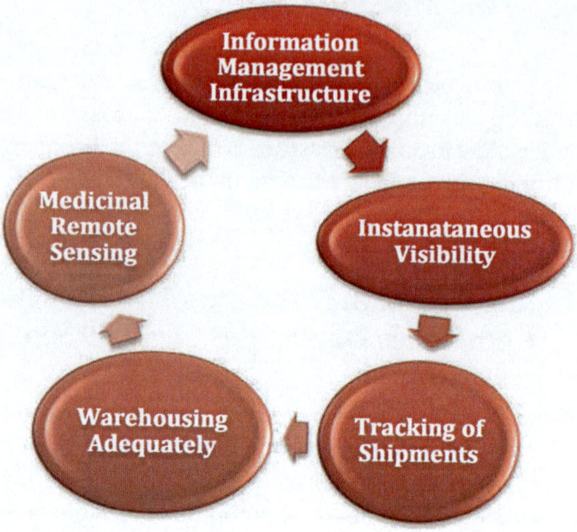

Fig. 2 Methods for improving pharmaceutical supply chain and logistics issues

5.1 *Information Management Infrastructure*

Varying RFID tags along with the EPC's can be stacked, installed, or pasted on the medicinal coverings, medicinal equipment, clinical tools, etc. Then the necessary data can be published on those RFID tags or the EPC's, such as the date of manufacturing, date of expiring, batch codes, and chemical compositions. This can be used for easy information sharing and transmission. With the help of these, the information regarding the utilities can be transferred or shared among the bodies engaged in the operations. With this, the ALE and the RFID tag readers are installed in the concerned premises or areas for their eased traceability [20]. As the medicines and the equipment enter the premises, they are tracked and their information is shared with the concerned parties. Then the information is transferred to the required ONS, helping the buyers to know that the utility they have purchased is a genuine product or a product of theft and counterfeit. In this way, the consumers can protect themselves from the counterfeits available in the market. This is what how easily IoT can help us with the information management.

5.2 *Instantaneous Visibility*

Instantaneous visibility can help in the real-time tracking and viewing the current operations in the concerned premises, enable to track the crucial areas or the areas of serious concern. It allows monitoring the ongoing activities in the warehouses and to regulate or adjust them on the demands and needs in the markets. It can also control the yield of output according to the demands of the market.

5.3 *Tracking of Shipments*

The pharma companies face a serious loss due to the cargo thieving every year. To reduce this mishap the RFID tags along with the 2-D bar codes and NFC sensors can be used to pack the utilities for shipping. The use of these can continuously track the products, and also helps to track down their thefts during shipping. Moreover, the AIDC along with the electronic chips can also be used for the packaging, so that the entry of these drugs and the other utilities in varying areas of need and implementation can be easily tracked down for their controlled and safe delivery.

5.4 *Warehousing Adequately*

The management of the warehouse plays a key role in their functioning, development, and operations in any pharmaceutical industry. It can be used to store the utilities even at their unrealized storage capacities with instantaneous notifications about the warehouses [12]. Here, the NFC sensors can provide real-time warehouse processing information that can be arranged as data for further warehouse processing. Such data can also be used to plan the yield rates of the goods and meet the demands of the industry.

5.5 *Medicinal Remote-Sensing*

Medicinal Remote-Sensing helps to alter and enhance the medication and the diagnosis of a patient by accumulating the data from his/her medical records, medical programs, lifestyle habits, etc. To create and develop a patient-committed system helps to minimize the healthcare expenses. Here the sensors and the RFID tags help us to accumulate the data and to work, process it to come up with the latest Drugs and Drug delivery systems, within a short period, by connecting up to the regulations of the state and even at a global level. It may even help the pharmaceutical companies to reduce the costs of the drugs and to gear up their production at a rampant pace [11].

6 Effect of IOT in Pharma Industries

Several firms have embraced IoT solutions in their manufacturing and supply chains. Some of the applications are listed in Table 1 [21]. IoT-PM is currently being used by **Apotex,** to upgrade its manufacturing process via IoT automation by utilizing automated forklifts and RFID tracking, even the sorting process is tracking enabled helping them with uninterrupted batch production. IoT was effectively put into practice in the company **Johnson & Johnson**, enabling them to obtain the FDA clearance regarding an HIV medication Prezista, helping them to move from the batch production form to continuous manufacturing form. Hence eradicating the need for separate testing, sampling tests by using the sensors in the process of manufacturing. Here IoT secured the data integrity by interlinking those separate processes. **GlaxoSmithKline** (GSK) is also implementing the IoT-equipped and IoT-based systems in major manufacturing plants so that it becomes easy to switch from traditional batch manufacturing to continuous flow processing. Even **Pfizer** is readily implementing IoT across its vital manufacturing plants to build up a unified manufacturing facility. **Merckis** also implemented IoT to analyze the biologic system data, enabling them to discover a serious flaw in their manufacturing plant.

7 Discussion

It would be apparent from this article that supporting the adequate application of the methods and the mainstream IoT, as discussed in the scenario of challenging times such as the current pandemic, would be evident. These are to be introduced only after the requisite facilities, production quality, operational capacity, protection, etc. have been provided by the companies. This also results in the generation of vulnerable data which is required to maintain the secrecy. And therefore, necessary preventive measures must be taken to prevent the companies as well as the patient's data, from any sort of data breach and ransomware attacks. In the Pharmaceutical industry, the problems regarding the supply chain visibility and the counterfeiting issue during the transfer of logistics from one geolocation to the other can be treated via the use of EPC-IS tags. But if the EPC tag is somehow tampered with/harmed during the transportation of logistics then it may compromise the integrity of the system. According to literature, one of the major benefits of the involvement of IoT in pharmaceutical manufacturing is that it would help to create individualized medicines. But if this is to be done on a macro scale then its feasibility may get compromised pharmacologically. It was obviously undeniable that all the gadgets involved in framing IoT-based network architecture for the smooth operation of the pharmaceutical manufacturing and supply chain operations on a decentralized basis that ultimately renders the process of data transmission, data transfer, and data collection exhaustible and eventually leads to the formation of enormous data. The major concern while operating on a network that is architectured on IoT technology, RFID technology, and multiagent technology is that in some cases it may emerge with a sense of hardware incompatibility which leads to disruption in the network architecture. Even after a complete IoT-based network infrastructure is introduced, there may be a risk of drug embezzlement during the transport of pharmaceutical products from a distributor to a stockist and then to a retailer through the supply chain.

8 Conclusion

With the involvement of the various stakeholders, the pharmaceutical industries operate at a broad-spectrum level any decision concerning the presence of IoT in the industry's manufacturing and supply chain network must be addressed with the company so that no further commitments would occur, as well as whether their current operating methods will be compatible with the application of IoT. It is evident that, as an IoT, complex network architecture works persuasively for the pharmaceutical production and supply chain. It must then function competently as a centralized system and ultimately streamline the process of data transmission, data generation, and data harvesting. As it has become apparent that an overwhelming number of the technologies used by the pharmaceutical industries to forge their IoT-based network are prominently outsourced, they should use their indigenous systems and software to the fullest extent according to their needs to prevent these industries from serious complications. In order to remove the risk of drug embezzlement during transport, it is required to create such a supply chain network that is exclusively dedicated to the supply of prescription drug products from their producers to their distributors.

References

1. Marathe, A., Enterprise, D., New, C.: Internet Of Things : Opportunities And Applications (2018)
2. Nazir, S., Ali, Y., Ullah, N.: Internet of Things for Healthcare Using Effects of Mobile Computing : A Systematic Literature Review, Vol. 2019 (2019)
3. Pratap, R., Javaid, M., Haleem, A., Suman, R.: Internet of things (IoT) applications to fight against COVID-19 pandemic. Diabetes Metab. Syndr. Clin. Res. Rev. **14**(4), 521–524 (2020). https://doi.org/10.1016/j.dsx.2020.04.041
4. Usak, M., Kubiatko, M., Salman, M.: Health Care Service Delivery Based on the Internet of Things : A Systematic and Comprehensive Study, pp. 1–17 (2019). https://doi.org/10.1002/dac.4179
5. Catarinucci, L., Colella, R., De Blasi, M.: High Performance UHF RFID Tags for Item-Level Tracing Systems in Critical Supply Chains (2011). https://doi.org/10.5772/19164.
6. Pachayappan, M., Rajesh, N., Saravanan, G.: Smart logistics for Pharmaceutical Industry Based on Internet of Things (IoT), Vol. 14, pp. 31–36 (2016)
7. Liao, B.I.N., Ali, Y., Nazir, S.: Security analysis of IoT Devices by Using Mobile Computing : A Systematic Literature Review, Vol. 8, pp. 120331–120350 (2020). https://doi.org/10.1109/ACCESS.2020.3006358.
8. Islam, S.M.R., Kwak, D., Kabir, H.: The internet of things for health care : a comprehensive survey. IEEE Access **3**, 678–708 (2015). https://doi.org/10.1109/ACCESS.2015.2437951
9. Ting, S.L., Kwok, S.K., Tsang, A.H.C., Lee, W.B.: Enhancing the Information Transmission for Pharmaceutical Supply Chain Based on Radio Frequency Identification (RFID) and Internet of Things
10. Da Xu, L., Member, S., He, W., Li, S.: Internet of Things in Industries : A Survey (2015). https://doi.org/10.1109/TII.2014.2300753.
11. Singh, M., Sachan, S., Singh, A. Internet of Things in Pharma Industry : Possibilities and Challenges. Elsevier (2020)
12. Elena, C., Octavian, C.: Internet of things as key enabler for sustainable healthcare delivery. Proc. Soc. Behav. Sci. **73**, 251–256 (2013). https://doi.org/10.1016/j.sbspro.2013.02.049

13. Alagarsamy, S., Kandasamy, R., Latha, S.: Applications of Internet of Things in Pharmaceutical Industry (2019). https://doi.org/10.2139/ssrn.3441059
14. Yan, B.: Supply Chain Information Transmission based on RFID and Internet of Things (2009)
15. De Blasi, M., Mighali, V., Patrono, L., Stefanizzi, M.L.: Performance evaluation of UHF RFID tags in the pharmaceutical supply chain. In: Giusto, D., Iera, A., Morabito, G., Atzori, L. (eds.) The Internet of Things. Springer, New York (2010). https://doi.org/10.1007/978-1-4419-1674-7_27
16. Wang, H.: IoT based Clinical Sensor Data Management and Transfer using Blockchain Technology, Vol. 02, no. 03, pp. 154–159 (2020)
17. Armstrong,R.: How Covid-19 is affecting the pharma supply chain. https://www.epmmagazine.com/opinion/how-covid-19-is-affecting-the-pharma-supply-chain/. Accessed 20 Dec 2020
18. Henderson, S.: Addressing the pharma supply chain problem during Covid-19. https://medcitynews.com/2020/04/addressing-the-pharma-supply-chain-problem-during-covid-19/. Accessed 20 Dec 2020
19. Balfour, H.: Has COVID-19 changed public perception of pharmaceutical companies. https://www.europeanpharmaceuticalreview.com/article/132222/has-covid-19-changed-public-perception-of-pharmaceutical-companies/. Accessed 20 Dec 2020
20. Summary, E.: The Internet of Things : The New Rx for Pharmaceuticals Manufacturing & Supply Chains (2017)
21. Shrivastava, A.: Nextgen-pharma-takes-smart-strides-with-internet-of-things. http://www.wipro.com/businessprocess/what-can-iot-do-for-healthcare-/. Accessed 25 Dec 2020

Social Network Mining for Predicting Users' Credibility with Optimal Feature Selection

P. Jayashree, K. Laila, K. Santhosh Kumar, and A. Udayavannan

Abstract The distribution of information in social media is very ad hoc in nature because of the increasing social media and the sophistication of social networks. Identifying fraudulent accounts is one of the key security issues. The task of identifying the reputation of social media users is difficult. In order to prevent malicious users from developing fake accounts, a social network mining model is suggested to evaluate and forecast user's trust. The proposed work mines text data and metadata on Twitter accounts to achieve the spammicity classification of the user. To evaluate the optimal characteristics that lead to optimizing the classification accuracy, the recursive Feature Elimination technique is used. The proposed work is evaluated on a real-time Twitter dataset and the empirical findings indicate that the proposed method performs better.

Keywords User credibility · Spam tweet detection · Machine learning · Recursive feature elimination · Feature scaling

1 Introduction

Despite the controversy around fake news, privacy, hacking, and other disadvantages of online communication, people persist in embracing the platform of social media. Twitter, one of the best platforms that would enable users to get knowledge associated with their topic of interest [1]. In November 2017, the character limit was increased from 140 characters to 280. Each tweet can have markup symbols that can be explored by the researchers toward determining spammicity. For example, (1) reply/mention "@" followed by another user's username, i.e., @username + message, this means that the message is a response to the user who likes to even get attention to the tweet subject from the message receiver. Users can post responses to anyone on Twitter, regardless of their friends or followers. Another user can also be listed by users

P. Jayashree (✉) · K. Laila · K. Santhosh Kumar · A. Udayavannan
Department of Computer Technology, MIT Campus, Anna University, Chennai, India

K. Laila
e-mail: laila@mitindia.edu

J. S. Raj et al. (eds.), *Intelligent Sustainable Systems*, Lecture Notes in Networks and Systems 213, https://doi.org/10.1007/978-981-16-2422-3_29

anywhere in the tweet other than at the start of the tweet. The hashtag '#' (2) defines trending topics that allow users to track recent developments and events in real-time. (3) 'RT' means that tweets created by other users are re-posted. For researchers to use them and extract detailed information, these markup symbols add a virtual space to tweets. Perhaps not all information produced from Twitter is correct and reliable. Verifying the user's and content's usability is challenging as this platform is not only available to follow the activities of other users through Tweets, account profile pages, but also to track their favorite celebrities in real time. As tweets are in the public domain and feasible to bring out the mining Twitter data. Therefore, this turned out to be a promising field of research.

Social networking sites deal with the challenges of spam content as it is cost-effective to connect and exchange information across these platforms [2]. Analyzing the trustworthiness of users has become a modern research area. In order to detect micro-blogging service (Twitter) consumer reputation, several articles have been adopted. However, because of the unpredictable existence of Twitter content, the implementation of machine learning methods to differentiate against credible and non-credible tweets proves to be an extreme challenge. To recognize credibility tweets by minimizing the effects of posting false information, this task involves automated learning methods [3, 4]. Furthermore, automated spam filtering is a required social data operation [5, 6]. To recognize malicious, ingenuous accounts, and spam posts, Twitter uses automated detection software. There are several signs to recognize the user's reputation according to the Twitter privacy policy [7], including whether the Twitter account sends massive duplicate mentions, hashtags, and the post with URL Links. Despite the widespread dissemination of erroneous information, such as rumors and untrustworthy information, Twitter has made strenuous efforts to ensure that people have access to important, reliable, and high-quality information. Measures to reduce the ill-effects of unworthy posts were discussed in [8]. In order to create an appropriate model that can exclude hypocritical tweets, the veracity of a tweet must be clearly defined and discriminated toward as trustworthy or not. A reliable tweet is generally defined, as stated in [9], as a discerned quality consisting of multiple dimensions, such as authenticity, relevance, and quantitative details. A non-credible tweet, on the other hand, perceives the purpose of misleading individuals who transmit fabricated information.

The aim of this article was to classify User Credibility (UC) by developing machine learning techniques based on spam features from a large-scale Twitter dataset. This contribution mainly highlighted feature engineering, which is the extraction of features with a function selection process that plays a major role in deciding the user's reliability in terms of spam detection. For spam classification, a binary classification model is used.

2 Related Work

Researchers have gained attention by evaluating the veracity of the tweets intended to classify unrealistic content along with spammers. There are several research on reliable social media user identification that have been conducted primarily to detect spammicity in online tweets. By using four interconnected components to prevent the dissemination of untrustworthy information on Twitter, the authors of the article [10, 11] introduced a new information reputation analysis framework. The author of the article [12] also addressed the problem of spreading false information by analyzing the reputation of a user on a subject and measuring the sentiment of the user within social networks, integrating the reputation of a user on a topic, and measuring the user's sentiment within the social networks, and integrating features in a reputation-based credibility assessment method. The authors specifically evaluated statistical features in the article [6, 13], taking into account the complex essence of misleading tweets. The approach to "learning from unlabeled tweets" (Lfun) addressed the point of recognizing Twitter Spam Drift through unlabeled tweets. Opinion mining was used in article [14, 15] to forecast fake Arabic news through classification models that recognize credible and non-credible users with an accuracy of 76%. In order to discover the similarities between the Twitter user name and the display name, the reputation identification also centered on combining social media mining and NLP techniques. Machine learning models such as the Decision Tree (DT) and the Naïve Bayes (NB) algorithms are used in the article [16] to prevent the spread of inaccurate social media information.

The reliability of the Twitter services is assessed in the article [17] using time-sensitiveness on online content by network structure of the users. Heuristic cues such as authorization, identification, and bandwagon are used to evaluate the source credibility of tweets and retweets for disseminating public health on social platforms [18]. For social media mining, Wu et al.[19] suggested a new "TraceMiner" method that pays more attention to modeling the spread of knowledge in a social network, and this approach uses network communication to solve the problem of sparsity,—for example, friendship and membership of the social community to obtain user embedding. In article [20] the author designed a capsule network to tackle high dimensional features in image classification. The author of the article [21] concentrated on reducing complexity using the Hermitian graph wavelet transform approach.

In classification, extracting relevant features plays a crucial role. The author of the article [22] suggested an Adversarial Network to discover differential features with the help of reinforcement learning for knowledge credibility assessment on the Internet. By reducing the irrelevant and generic characteristics of the extracted characteristics, shared private models which improve efficiency. The authors of the article [23] emphasized the significance of the credibility of content on social media and the identification of trustworthy users using the proposed multilayer perceptron method to compute the degree of credibility of the post with restricted content-based features such as extreme content-based features such as several unique-hashtags, mentions, and URLs.

From the literature survey, the following limitations are identified. User's reputation is identified based on tweet and retweet data or Twitter account data. Many reputation models are not supporting data scalability. Some research used information on network access and authorization of network users to assess the credibility of information, which leads to increased complexity. There are several conventional methods for selecting characteristics, such as Principal Component Analysis (PCA), in which identical attributes are combined to create new attributes. Through robust Recursive Feature Elimination (RFE), the proposed work scalability is addressed and computational complexity is reduced with simpler statistical and textual analysis.

3 System Methodology

The spam classification model has the following sub-modules to analyze the discriminating features toward a better classification of tweets (Fig. 1). The novelty of the proposed work occupies in extricating two types of data to analyze spammicity without increasing the complexity. The two types of data used are (1) Data characterizing user behavior in tweets, (2) Data related to actual tweets content. The reduction in the complexity feature reduction was performed using the Recursive Feature Elimination (RFE) method.

3.1 Data Acquisition

Natural Language Processing (NLP) approaches are used to process Twitter data obtained using the Twitter Streaming API. Meta-features such as friends_count/following_count, account age, status_count, average tweets per day and week, followers_count, tweet date, and time, retweet_count, mentions,

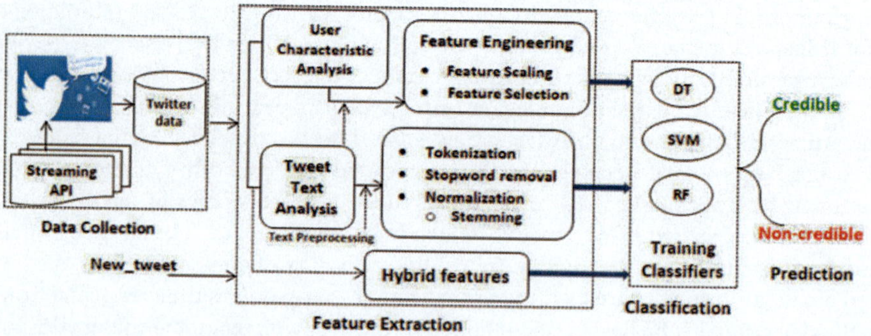

Fig. 1 Twitter spam filtering methodology

unique_mentions, URLs; and most used hashtags are extracted and subjected to feature engineering.

3.2 *Feature Extraction*

The discriminating features from the meta-tweet data and tweets collected are extracted based on their characteristic to identify fake tweets.

(1) **User Characteristic features**: reflecting characteristic properties of the user into account which mainly indicates the users' reputation and their experience.
(2) **Tweet Text features**: reflecting the tweet content such as tweet size, number of retweets, user mentions, or hashtags presence in a tweet.
(3) **Hybrid features**: combining both user characteristics and tweet text features.

3.3 *Feature Engineering*

(a) **Feature Scaling**

For many machine learning algorithms, standardization is a general prerequisite to prevent overfitting and bias. For the feature set, performing feature rescaling is essential as it includes heterogeneous characteristics with a wide range of values. For instance, tweets-length, retweet-count, hashtag-count, mean of hashtags, followers count, friends' ratio, etc. have large values while some other features have 0 or 1. For StandardScaler, Sklearn, its key scaler, is used for this task to downscale the characteristics based on the standard normal distribution The mathematical formula of feature scaling is mentioned in the algorithm where Z_{score} is the standardized score.

(b) **Recursive Feature Elimination**

A variant of a multivariate feature selection method termed *Recursive Feature Elimination (RFE)* is applied to enhance the model performance and diminish the model complexity by removing less important features, and these are not contributing toward spam prediction. Moreover, with fewer features, model implementation is simple and more straightforward. Recursive feature elimination (RFE) is a feature selection method that fits a model and removes the weakest feature till the specified number of features is reached. Ranking of features is done by model's coef_ or feature_importances_ attributes. RFE attempts to eliminate dependencies and colinearity that exists in the model by recursively eliminating a small number of features per loop.

(c) **Tweet Text Preprocessing**

Text preprocessing is to mine relevant information from tweet data. Transforming text into a canonical form reduces noise, computational cost, and clusters words with similar semantic meanings. The preprocessing of tweets involves tokenization, stopword removal, normalization, and stemming.

(1) **Tokenization**: It helps to decide the granularity of the text by using a tokenizer that splits documents into tokens based upon special characters and white space. This tokenizer can extract more information than commonly used word analyzers.
(2) **Stopword Removal**: This process removes the sparse terms from tweets that do not add predictive value in text classification such as "a", "the", "wasn't", "doing", "that'll", and punctuation and special symbols.
(3) **Text Normalization**: The normalization or text canonicalization process on tweets includes, converting all tweets to upper or lower case, transforming numbers into words, removing leading or ending space (white spaces), and abbreviation expansion, etc.
(4) **Stemming**: This process exploits a *snowball stemmer* (stemming algorithm) in the text to stem each word when computing n-gram frequency revealing the root word.

3.4 Classification

(a) Supervised learning approaches are used to assess if the target variable is trustworthy or not with Tweets in order to test the efficiency of the collected features in defining user credibility. **Decision Tree**: It is a commonly used non-parametric tree-structured supervised method used for classification and regression tasks. Some of the attribute selection methods, namely, Entropy, Information Gain, and Gini Index are used to select the best optimal features for the parent, and the sub-nodes. To create a decision tree at each node, it considers the records of datasets recursively. To outliers, it is comprehensible and solid. On the other hand, due to high variation in the research dataset, it is responsible for overfitting.
(b) **SVM Classifier**: This prominent classification method is aimed at the best possible hyperplane, which divides the data points linearly between two groups by maximizing the n-dimensional feature space margin. The optimization strategies enhance the prediction accuracy.
(c) **Random forest**: This classifier uses multiple decision trees to make decisions using the voting or average method.

The proposed user credibility process is defined below.

Algorithm

Input: Twitter data T_d, Tweets T= {t_1, t2,...,t_d},
User characteristic features F= {f_1, f2,....fn},
Text Token= {tk_1, tk2,....tk_n }
Output: credible, non-credible users
//selecting User characteristic features
Procedure
for each feature f in T_d do
%feature scaling: standardization

$$Z_{score} = \frac{(\text{initial value}_x - \mu)}{\sigma} \quad \%\mu = 0, \sigma = 1$$

%feature selection
Select relevant features F'= {fl', f2',....fn'} using RFE
Training SVM, RF using F'
//selecting Tweet text features
% preprocessing
for each tweet t in T do
tk_i←Tokenizer()
for each tk_i in T do
T' : remove stop-words, normalization, & stemming
Training DT, SVM using t ϵ T'
//selecting hybrid features
Training SVM, RF using F'& t ϵT'
end Procedure
new_instance (*x*):
return credible or non-credible users

4 Implementation

4.1 Dataset and Preprocessing

Twitter data was collected using a Python library called "*Tweepy*". User characteristic data were collected for 1331 users through Twitter streaming API. Extracted features are listed in Table 1. Tweet datasets of 5695 rows containing 4327 credible and 1368 non-credible are collected from the Kaggle benchmark dataset. In the statistical dataset of Twitter users, the correlation between features is very less, but the number of features is high. Since dimensionality reduction techniques like PCA which attempts to reduce dimension by combining attributes are not best suited for this particular dataset, RFE is used. Through RFE the top-10 features characterizing the users are selected and listed as in Fig. 2.

4.2 Classification and Evaluation

The reduced feature's ability to classify spam users is evaluated using simple machine learning algorithms, namely, Decision Tree (DT), Support Vector Machine (SVM), and Random Forest (RF). The parameters of these models are tabulated in Table 2.

Table 1 Extracted features for trust evaluation

User characteristic features	Tweet text features
Followings count	Retweet_count
Followers count	Tweet length in characters
Listed count	Tweet length in words
Description length	Average tweet-length
URL ratio	Has URLs?
Created at(Account age)	Has user mentions?
Hold profile picture?	Has hashtags?
Followings/followers-ratio	URL count
Average hashtag figures	User mentions count
Average URL-ratio	Hashtags count
Average mention-ratio	Average tweets per day
Retweet_ratio	Average tweets per week
Reply ratio	Tweet-count
Unique-mentions	Most used hashtags
Mean of inter-tweet delay	
Standard deviation of inter-tweet delay	

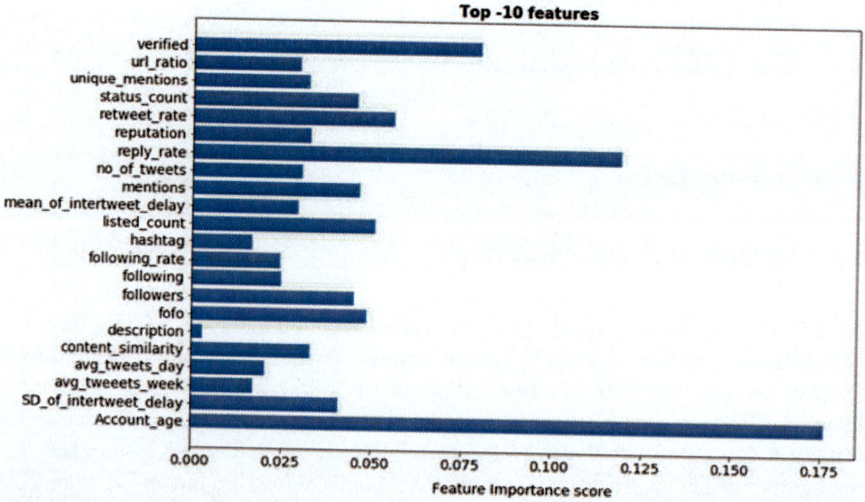

Fig. 2 Reduced statistical features

The classification performance is evaluated using the following metrics where the true and false alarm factors are defined in Fig. 3.

(a) ***Precision***: It measures the amount of credible tweets that were classified as credible.

$$\frac{TP}{TP + FP} \tag{1}$$

Table 2 Parameters of the classifiers used in the model

Model	Parameter	Value
Decision tree	Criterion	Entropy
	Min samples_split	10
	Min samples_leaf	50
SVM	Kernel	RBF
	C: Regularization	1.0
Random forest	Bootstrap	True
	Criterion	Gini
	n_estimaters	100
	Max_features	Auto
	Min samples_split	10
	oob_score	False

		Actual class	
		Positive :**Credible**	Negative: **Non-credible**
Predicted	Positive: **Credible**	True Positive (TP)	False Positive (FP)
	Negative: **Non-credible**	False Negative (FN)	True Negative (TN)

Fig. 3 Confusion matrix parameters

(b) ***Recall***: It measures the amount of actual credible tweets that were correctly classified.

$$\frac{TP}{TP + FN} \tag{2}$$

(c) ***F1-score***: It measures the *harmonic mean* of both recall and precision into a single score that is higher score as a better model.

$$\frac{2 * (Precision * Recall)}{Precision + Recall} \tag{3}$$

(d) ***Accuracy***: It measures the proportion of accurately classified tweets out of all instances.

$$\frac{TP + TN}{TP + TN + FP + FN} \tag{4}$$

Table 3 Classification performance with feature reduction on SVM and RF classifiers

Before RFE and feature scaling			After RFE and feature scaling	
Evaluation parameters	SVM	RF	SVM	RF
Accuracy	0.60	0.92	0.92	0.925
Precision	0.98	0.921	0.924	0.926
Recall	0.271	0.956	0.934	0.956
F1-score	0.432	0.938	0.929	0.941

5 Result Analysis

5.1 The Performance of the Proposed Method with Statistical Features

User characteristics, tweet text features, and hybrid features were evaluated separately for the performance of the proposed system. The accuracy of spammer detection is tabulated in Table 3. It is found that the accuracy of SVM is increased from 60 to 92% after standardization and feature selection as a matter of fact that SVM is a distance-based algorithm that is sensitive to unscaled data. On the contrary, Random forest does not require feature scaling. Since the tree split at each node is to improve the homogeneity of that node. Other features in the dataset will not be affected by the split.

5.2 The Performance of the Proposed Method with Tweet Text Features.

For extracting textual features, the tweet corpus was analyzed using PythonNLTK. Fourteen textual features extracted are fed to classifiers for spammicity detection using Decision Tree and SVM with 80:20 as train to test split. The results are tabulated in Table 4. This experimental result indicates that the textual features have a less significant impact on evaluating users' credibility compared to statistical features.

Table 4 Textual features-based classification performance

Model	Accuracy	AUC	Precision	Recall	F1-score
Decision tree	0.887	0.868	0.883	0.883	0.882
SVM	0.824	0.908	0.811	0.899	0.811

Table 5 Comparative performance analysis of the proposed work on hybrid features

Model	Accuracy	AUC	Recall	Precision	F1-score
Random forest	0.930	0.958	0.959	0.934	0.934
SVM	0.904	0.947	0.899	0. 925	0.912
MLP [24]	0.824	0.704	0.548	0.665	0.603

5.3 *The Performance of the Proposed Method with Hybrid Features.*

Using Random Forest and SVM, the output of the proposed framework with combined user features and tweet text functionality is evaluated. Cross-validation of tenfold has been used. The data was split and replicated ten times in order to use each fold exactly once and to measure the mean accuracy. The proposed model is compared with an existing MLP-based spam classification model [24] on the same dataset as tabulated in Table 5. It is observed that the proposed system can find spammers with increased accuracy. The ROC curve plotted as shown in Fig. 4a, b indicates the supremacy of the proposed system in evaluating user credibility.

6 Conclusion

As Twitter is a popular social media site, identifying the user's reputation has developed to be a non-trivial job, and users have a tendency to post false information intended for political or financial benefits. To classify spammers in social media, the machine learning system is proposed with social data mining. Using the tweet data and Twitter metadata, spammers are detected. Recursive Features Elimination (RFE) is used to identify the most promising features in social media to determine user credibility. In contrast to current works, the proposed system is found to have a better 93% accuracy in detecting spammers. More variables from sentiment analysis and popularity metrics will be included as a potential task to carry out context-based reputation analysis. To maximize the optimal feature collection, feature selection using evolutionary approaches will be extended.

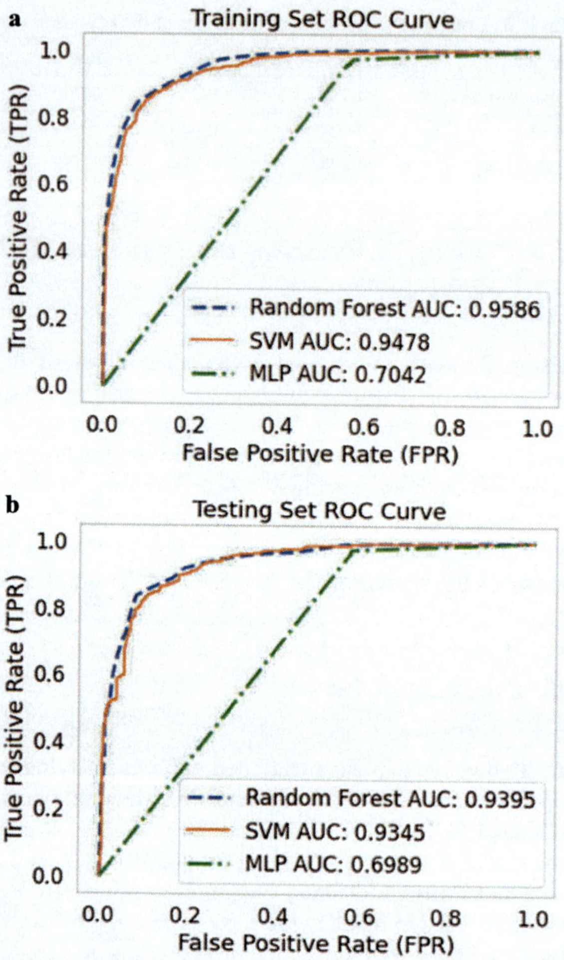

Fig. 4 Training features ROC curve for Random forest and SVM classifiers for hybrid features. b Testing features ROC curve for Random forest and SVM classifiers for hybrid features

Acknowledgements This work is financially supported under grants provided by Anna Centenary Research Fellowship (ACRF), Anna University, Chennai.

References

1. https://iag.me/socialmedia/guides/do-you-know-the-twitter-limits/
2. Adewole, K.S., Anuar, N.B., Kamsin, A., Sangaiah, A.K.: SMSAD: a framework for spam message and spam account detection. Multimedia Tools Appl. **78**(4), 3925–3960 (2019)
3. Torabi, Z.S., Nadimi-Shahraki, M.H., Nabiollahi, A.: Efficient support vector machines for spam detection: a survey. Int. J. Comput. Sci. Inf. Secur. **13**(1), 11 (2015)

4. Akinyelu, A.A., Adewumi, A.O.: Classification of phishing email using random forest machine learning technique. J. Appl. Math. **2014** (2014)
5. Ning, H., Dhelim, S., Aung, N.: PersoNet: friend recommendation system based on big-five personality traits and hybrid filtering. IEEE Trans. Comput. Soc. Syst. **6**(3), 394–402 (2019)
6. Chen, C., Wang, Y., Zhang, J., Xiang, Y., Zhou, W., Min, G.: Statistical features-based real-time detection of drifted twitter spam. IEEE Trans. Inf. Forens. Secur. **12**(4), 914–925 (2016)
7. Twitter Privacy Rules. https://help.twitter.com/en/rules-and-policies/twitter-rules, http://snap.stanford.edu/class/cs224w-2011projividya_Finalwriteup_v1.pdf. Accessed 20 Apr 2018
8. Vaidya, T., Votipka, D., Mazurek, M.L., Sherr, M.: Does being verified make you more credible? Account verification's effect on tweet credibility. In: 2019 Proceedings of the CHI Conference on Human Factors in Computing Systems, pp. 1–13 (2019)
9. Setiawan, E.B., Widyantoro, D.H., Surendro, K.: Measuring information credibility in social media using combination of user profile and message content dimensions. Int. J. Electr. Comput. Eng. (IJECE) **10**(4), 3537–3549 (2020)
10. Alrubaian, M., Al-Qurishi, M., Hassan, M.M., Alamri, A.: A credibility analysis system for assessing information on twitter. IEEE Trans. Depend. Sec. Comput. **15**(4), 661–674 (2016)
11. Alrubaian, M., Al-Qurishi, M., Alamri, A., Al-Rakhami, M., Hassan, M.M., Fortino, G.: Credibility in online social networks: a survey. IEEE Access **7**, 2828–2855 (2018)
12. Alrubaian, M., Al-Qurishi, M., Al-Rakhami, M., Hassan, M.M., Alamri, A.: Reputation-based credibility analysis of Twitter social network users. Concurr. Comput. Pract. Exp. **29**(7), e3873 (2017)
13. Chen, C., Zhang, J., Xie, Y., Xiang, Y., Zhou, W., Hassan, M.M., AlElaiwi, A., Alrubaian, M.: A performance evaluation of machine learning-based streaming spam tweets detection. IEEE Trans. Comput. Soc. Syst. **2**(3), 65–76 (2015)
14. Jardaneh, G., Abdelhaq, H., Buzz, M., Johnson, D.: Classifying Arabic tweets based on credibility using content and user features. In: 2019 IEEE Jordan International Joint Conference on Electrical Engineering and Information Technology (JEEIT), pp. 596–601 (2019). IEEE
15. Mouty, R., Gazdar, A.:The effect of the similarity between the two names of twitter users on the credibility of their publications. In: 2019 Joint 8th International Conference on Informatics, Electronics & Vision (ICIEV) and 2019 3rd International Conference on Imaging, Vision & Pattern Recognition (icIVPR), pp. 196–201 (2019). IEEE
16. Tyagi, S., Pai, A., Pegado, J., Kamath, A.: A proposed Model for Preventing the spread of misinformation on online social media using machine learning. In: 2019 Amity International Conference on Artificial Intelligence (AICAI), pp. 678–683 (2019). IEEE
17. Afify, E.A., Eldin, A.S., Khedr, A.E., Alsheref, F.K.: User-generated content (UGC) credibility on social media using sentiment classification. FCI-H Inform. Bull. **1**(1), 1–19 (2019)
18. Tajalizadeh, H., Boostani, R.: A novel stream clustering framework for spam detection in twitter. IEEE Trans. Comput. Soc. Syst. **6**(3), 525–534 (2019)
19. Wu, L., Liu, H.: Tracing fake-news footprints: characterizing social media messages by how they propagate. In: 11th ACM international conference on Web Search and Data Mining, pp. 637–645 (2018)
20. Sungheetha, A., Sharma, R.: A novel CapsNet based image reconstruction and regression analysis. J. Innov. Image Process. (JIIP) **2**(03), 156–164 (2020)
21. Manoharan, S.: Study on Hermitian graph wavelets in feature detection. J. Soft Comput. Paradigm (JSCP) **1**(01), 24–32 (2019)
22. Wu, L., Rao, Y., Nazir, A., Jin, H.: Discovering differential features: adversarial learning for information credibility evaluation. Inf. Sci. **516**, 453–473 (2020)
23. Dhingra, A., Mittal, S.: Content based spam classification in twitter using multi-layer perceptron learning. Int. J. Latest Trend. Eng. Technol.
24. Herzallah, W., Faris, H., Adwan, O.: Feature engineering for detecting spammers on Twitter: modelling and analysis. J. Inf. Sci. **44**(2), 230–247 (2018)

Real-Time Security Monitoring System Using Applications Log Data

S. Pratap Singh, A. Nageswara Rao, and T. Raghavendra Gupta

Abstract Nowadays, the data are generally stored in digital form and utilized for further analysis. Digital data can be very crucial for further analysis. The data generated by all these electronic devices can be made informative depending on the purpose. Every device makes use of some applications that may be related to us or others to serve the intended purpose. During the application or website or system software access, there is a chance that every activity can be recorded in the system either as a client system or server system. The data recording can be in any form or format as log files. These log files can be used to even track the intruder's or hacker's interest or any anonymous user, who attempts for gaining unauthorized access to confidential information and so on. The information that is related to such a user is also recorded in the system as log files. The main aim of the paper is to collect the log file data from the systems' security logs or website of different organizations and preprocess the data to make it ready for analysis and apply the metrics to send the alerts to the administrative team or quality assurance team regarding the crucial data loss or their assets and make them alerted by sending Email alert message and get precautions before they may get heavy loss.

Keywords Log monitoring · Security attacks · Log file · Log analysis · System log

S. Pratap Singh (✉)
Marri Laxman Reddy Institute of Technology and Management, Dundigal, Hyderabad, India

A. Nageswara Rao
CSE Department, CMR Institute of Technology, Medchal, Hyderabad, India
e-mail: nageswararao.a@cmritonline.ac.in

T. Raghavendra Gupta
CSE Department, Hyderabad Institute of Technology and Management, Medchal, Hyderabad, India

J. S. Raj et al. (eds.), *Intelligent Sustainable Systems*, Lecture Notes in Networks and Systems 213, https://doi.org/10.1007/978-981-16-2422-3_30

1 Introduction

The use of electronic devices in society is increasing at an unprecedented rate. The amount of data generated from all the resources may suffer cyberattacks or threats. The attacks can be from different resources that may be related to access of information, damage to assets, administration, etc. The attacks are also getting sophisticated due to the increase in the new evolution of technologies and tools. Because of these sophisticated threats, it is becoming difficult to maintain organization data or assets and the threats that may exploit vulnerabilities. The vulnerabilities may be related to access control, non-repudiation, integrity, confidentiality, authentication, and denial of service (DoS). The attacks may be either active attacks or passive attacks that may happen during active or inactive states. To detect the attack, most of the organizations are dealing with logs generated in the system. Logs are nothing but the messages generated during the communication between the user and system. These logs are stored in a file called log files. Log files automatically record and store information like login time, logout time, authorization, content accessed, etc. The software applications are developed and the errors, faults, bugs, etc. generated are recorded in the log file. While developing the software application, the developers are asked to maintain a log, where the events of the logs can be stored. Image logs generated by the different applications and devices are vast. The logs can be generated from software application, devices, software system, network devices, and so on. Figure 1 gives details of how logs are generated from various sources and stored, and Fig. 2 is an example of a log summary of different applications or hardware in the system.

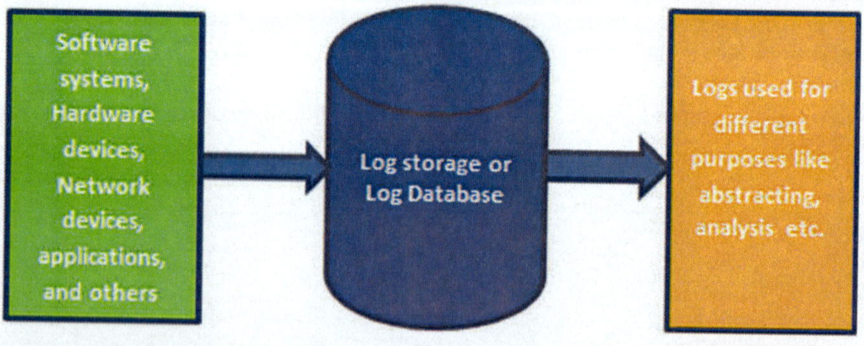

Fig. 1 Different applications log files

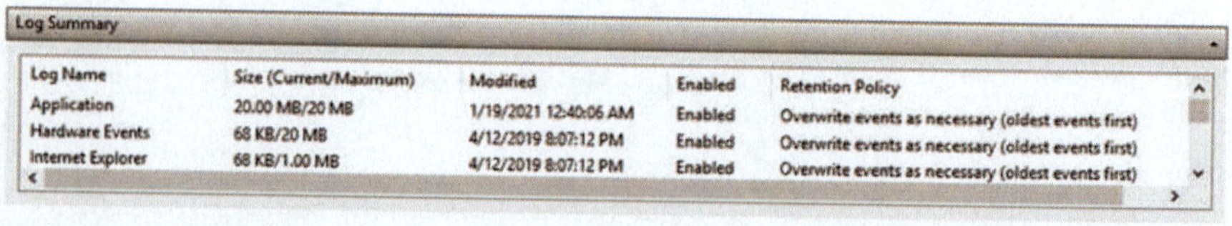

Log Summary

Log Name	Size (Current/Maximum)	Modified	Enabled	Retention Policy
Application	20.00 MB/20 MB	1/19/2021 12:40:06 AM	Enabled	Overwrite events as necessary (oldest events first)
Hardware Events	68 KB/20 MB	4/12/2019 8:07:12 PM	Enabled	Overwrite events as necessary (oldest events first)
Internet Explorer	68 KB/1.00 MB	4/12/2019 8:07:12 PM	Enabled	Overwrite events as necessary (oldest events first)

Fig. 2 Example of log summary of different applications or hardware in the system

Level	Date and Time	Source	Event ID	Task Category
Information	11/21/2020 8:15:02 PM	Kernel-General	1	(5)
Information	11/21/2020 5:15:10 PM	Kernel-Power	107	(102)
Information	11/21/2020 5:15:01 PM	Kernel-Power	42	(64)
Error	11/21/2020 4:32:29 PM	DistributedCOM	10016	None
Error	11/21/2020 4:32:27 PM	DistributedCOM	10016	None
Information	11/21/2020 4:31:48 PM	Service Control Manager	7045	None
Warning	11/21/2020 4:22:21 PM	i8042prt	18	None
Information	11/21/2020 4:22:19 PM	Power-Troubleshooter	1	None
Warning	11/21/2020 4:22:14 PM	WHEA-Logger	17	None
Warning	11/21/2020 4:22:14 PM	WHEA-Logger	17	None
Warning	11/21/2020 4:22:14 PM	rt640x64	1	None
Information	11/21/2020 4:22:10 PM	Kernel-Power	131	(33)
Information	11/21/2020 4:22:10 PM	Kernel-General	1	(5)
Information	11/21/2020 3:41:54 PM	Kernel-Power	107	(102)
Information	11/21/2020 3:41:51 PM	Kernel-Power	42	(64)
Information	11/21/2020 2:56:08 PM	BROWSER	8032	None
Warning	11/21/2020 2:55:05 PM	BROWSER	8021	None

Fig. 3 Example of the severity level of logs

There are different types of log files depending on the source like event log (system log, computer logs, application logs, and execution log), audit log, transaction log, query logs, error logs, and webserver logs. The common terms that are used in computer science for all these logs or logging are faults, errors, and failure. A fault is an incorrect step or process or data definition in a program, an error is the difference between the observed and expected outcome, and a failure is the inability of a system or component to perform its required function. All these are interrelated from fault to error and error to failure. A fault is the root cause which can produce an error. Errors are the root cause for failures and failures are the incorrect functionality for any bad results.

Apart from maintaining the logs, the application contains the log messages and these log messages vary depending upon the fault, error, or failures. These messages contain the severity level like error, warn, info, and debug as shown in Fig. 3.

2 Literature Survey

This section of the paper presents a brief literature review of previous work done by various authors. Aue in "Log Analysis from A to Z: A Literature Survey" presents a literature survey of log file, error, faults, and failure;also presents the log severity and priorities; concludes that the maintaining of logs will help for log analysis and detection of anomalies; and allso suggests that even proper abstraction and detection techniques do not guarantee effectiveness. Where, what, and how to log are of key importance to enable effective further analysis [1]. DeLaRosa presents in his paper how the log is maintained and monitored. He also suggests that there is no standard of log and no proper format of logs, but suggests the log file of any system, application, or device must include: timestamp, category, and description. He suggests that log monitoring tools should be evaluated with their objectives and goals [2].

Awontis Word Press defines the life cycle of log management as 10 steps as follows: policy definition, configuration, collection, normalizing, indexing, storing, correlationing, baselining, alerting, and reporting. Almgren and Lindqvist focused on the collection of data produced by the applications and referred to it as application-based Intrusion Detection System. They presented an application-integrated approach so that they can have access to more information than external monitors so that they can detect more attacks [3]. Cao et al. focused on anomaly detection system based on machine learning for weblog files. They collected the weblog files data, preprocessed the data by applying a decision tree algorithm, modeled the data, and got the accuracy of the system [4]. Latib et al. [5] focused on large log files in a Big Data environment using the Hadoop framework. The results generated by this framework have increased performance in response and operation time.

Shan et al. [6] carried the analysis and monitoring based on DNS log files and provide a visual detection of abnormal queries in log files and identify the anomalies. The authors in this paper conclude from the review of the above papers that there is no single way or common way how this analysis and monitoring of the security system is done. It all depends on the developer and organization's need to utilize the security of the application using logs. Logs analysis can be done with respect to clients or user or application specific. But monitoring and analysis of log data help to secure and protect from heavy losses. Yang et al. present an IP address monitoring system in the cloud. The main focus is to build a trustworthy and secure model for cloud tenants, where the quality of service can be provided and operational efficiency is maintained. They could accomplish this task by regularly monitoring the IP address, collecting all the illegal IP addresses, and storing them into the database. Later, the analysis of IP address is done and the alert messages about the security status and security risks are sent to cloud tenants [7].

Yadav et al. [8] present a review on log anomaly detection using deep neural networks. Their survey presents a brief log analysis based on the types of datasets used for log analysis, how this log is helpful for understanding the behavior of the system, failure cause detection, performing security scanning, and failure prediction. Ogie presents a report on how the cyberattacks had happened on the key national critical infrastructure and industrial processes and presents the key points where we can plan toward protecting critical industrial networks. The author uses a classification scheme to analyze cyberattacks [9].

Li et al. [10] proposed a log monitoring architecture for cyber-physical power systems (CPPS) to detect the log analysis. The proposed mechanism first trains the network protocol feature database and improves the efficiency of log analysis. The proposed method is an ensemble prediction algorithm based on time series (EPABT) which is used to predict abnormal features during network traffic flow. Later presents a new method of evaluation criteria to meet the characteristic of CPPS. The new method for evaluation considered is asymmetric error cost (AEC).

Janos and Dai expressed some security threats and their solution to fight against security operation center (SOC) in order to enhance the ability of organizations to mitigate the security issues[11].

Onwubiko proposed a framework called cybersecurity operations center (CSOC) in which it collects the logs, analyzes the logs, incidence response and forensic investigation are the phases of the CSOC. This framework is applied to ICT monitoring [12].

Ahmad et al. proposed the information security assurance behavior framework, which is developed by using an empirical study. This framework is built by conducting a survey on the internal and external experts both from industry and academics, and their study helps the organizations to assess their current practices that may be used to nurture information security [13]. Adithya et al. [14] used log detection in the cloud and Shakya [15] proposed an error detection mechanism for securing the IoT devices.

3 Methodology

This section of the paper deals with the methodology adopted to secure the organizational valuable information by monitoring and analysis of log files data. Figure 4 shows the log data that is stored in the system that is application log data which is related to the application that is installed and running in the system and their logs.

A sample methodology is adopted, where a number of applications are installed and running in the systems or webserver. All these applications, system software,

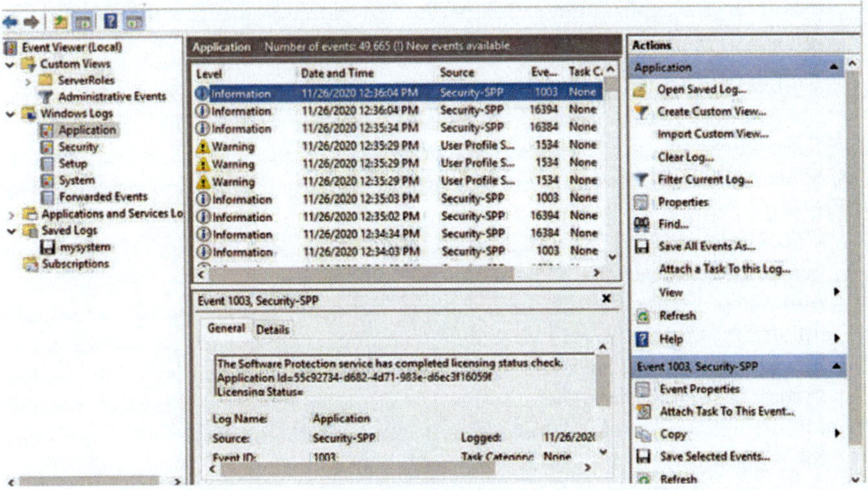

Fig. 4 Shows the log data that is stored in the system

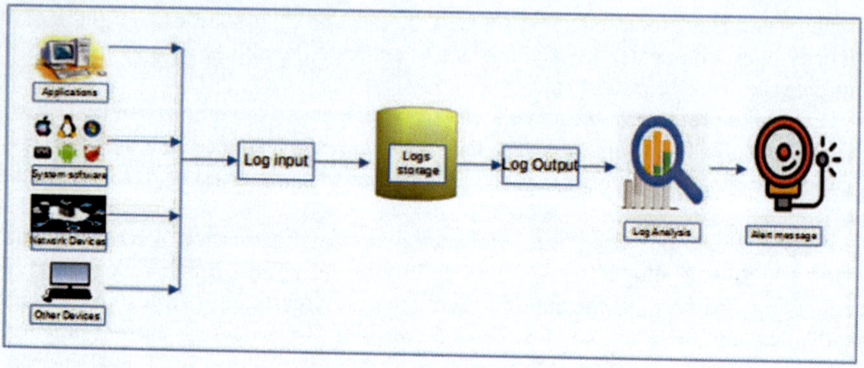

Fig. 5 Block diagram of the security monitoring system

networking devices, etc. will generate the events during their installation, their working, any user login to that application, timestamp of the user that how long was the application active(session of the user), and logout. All this information is stored in the log storage. We can see the example of window system which generated the logs for all the events on the system and is able to view in event viewer in windows as shown in Fig. 4. The main ideation of this paper is to analyze the log files that are stored in the system by monitoring the events that are generated and sending the alert message to the user of the system if there is important information lost or to know any unauthorized access. This type of log analysis system is mainly used for security monitoring purpose and security testing. Figure 5 shows the method of how the system is implemented. The steps involved in this implementation of the security monitoring system are as follows.

Steps (here for explanation and experimental purpose apache server log files is considered).

1. Collect the events or logs of the different applications (either software or hardware), webservers, network devices, and other applications.
2. Store the log data in log storage for analysis.
3. If there is any update to the log file (such as error.log file), then identify the type of log.
4. Check for the log types of data and check the level of log data. The log level can be critical, warning, information, debug, and error.
5. In the initial stage, the threshold value is user-defined when the file is read, the no. of times the log level like critical, warning, error, etc. are raised, the count value of the type error is send, if the threshold value which is defined by the system or user, then it generates the alert message. In later stages as soon as the log file updates or a log is recorded, it should send a message with any one of the log levels.
6. If the type of log data is warning, information, or error, then the system should generate an alarm or send the alert message to the user through email or mobile number.

4 Implementation

The implementation of the log analysis and monitoring system is implemented in Python programming. The implementation of this system requires two important things like the installation of the SMTP library in Python (smtplib) and the configuration of Gmail to send and receive the Gmail from the application. The implementation flowchart of the system is given as in Fig. 6. Firstly, the user collects the events logs that have been stored in the log files in the log storage which are generated by different applications, devices, webservers, databases, etc. Here event log file of the window system is taken to demonstrate. The log file is analyzed and identified parameter values (like the number of times the warning, errors, and information and critical errors occurred). Then the parameter or security level is compared with the

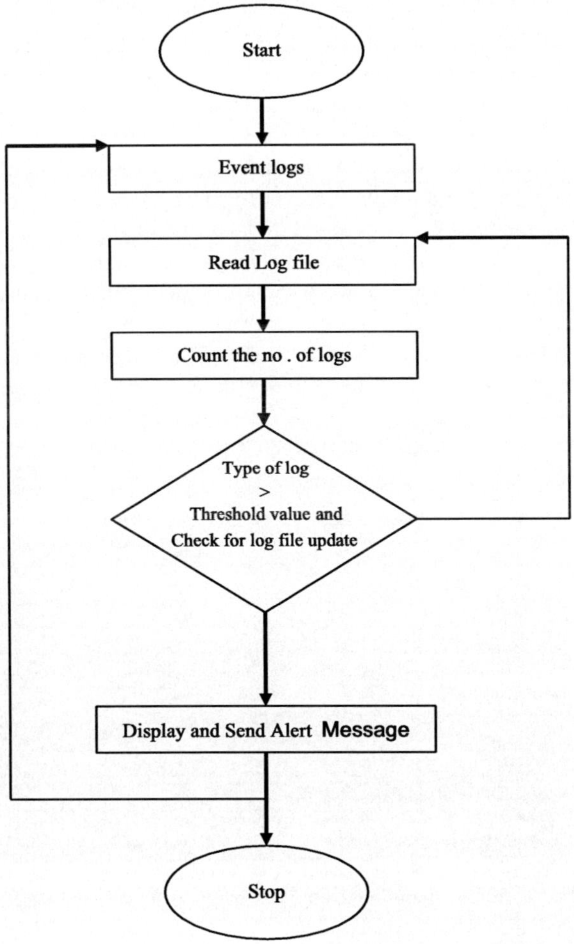

Fig. 6 Flowchart of the security monitoring system

threshold value, if the security level is greater than the threshold value, then the system should send an alert message. The implemented model sends a Gmail, which is configured with the smtp.gmail.com in the system.

5 Result

The results generated by the proposed method of the security monitoring system by collecting and analyzing the log data of the organization are to protect information or assets by monitoring them from unauthorized access. The results which are generated are based on log file data collected from systems (that is event viewer of Windows 10 system). The security monitoring system collects the security level value or parameter values as the number of times a particular parameter is generated. From Fig 7, screenshot of the monitoring system, the parameter values for warnings, error,s and information are seen as warning are 16, errors are 14, and the other parameter information (Info) is 0. The system sends the alert message when the threshold value of the parameter is greater. If the threshold value is 5 and the parameter or security level value of warning or errors is more than that, the system sends the alert message as "Alert message from the mail server regarding critical error is send" or "Alert message from the mail server regarding warning is send". This message is nothing but the way we want to get the alert message.

From Fig. 8, screenshot of the security monitoring system, which is showing that the alert message has been sent through Gmail. The Gmail details that are configured in the implementation are not shown as they reflect the privacy of my Gmail and Gmail credentials; therefore, the Gmail account details are not shared. The other way of setting the Gmail credential in the code itself so that the application does not ask for Gmail credentials is shown in Fig. 8 screenshot of security parameter values and alert message sent by the Gmail account that is configured in the application.

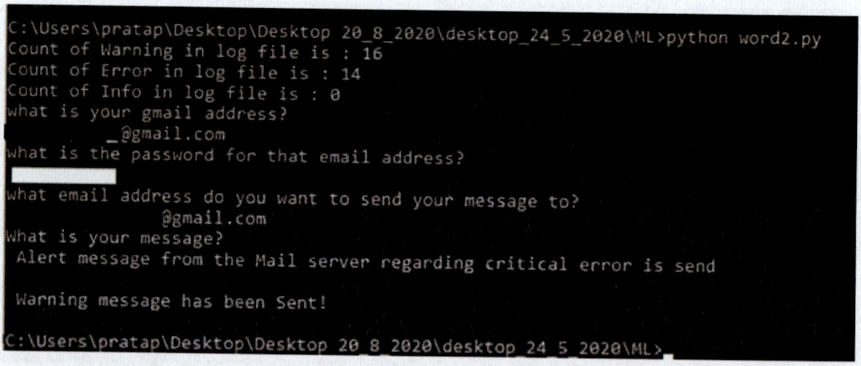

Fig. 7 Screenshot of a number of security levels or parameters with their values and sending an alert message

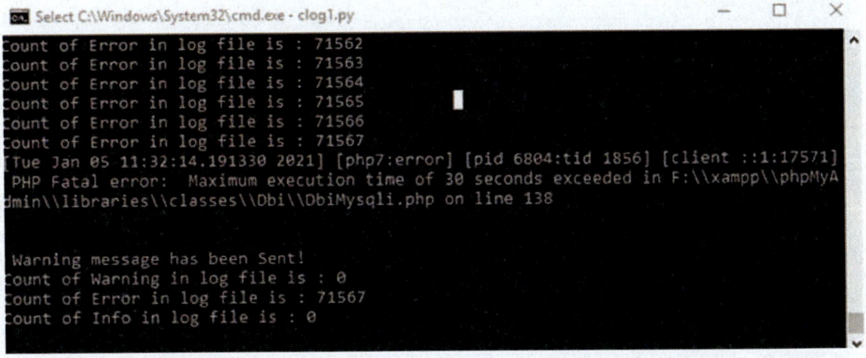

Fig. 8 Showing that the alert message has been sent through Gmail

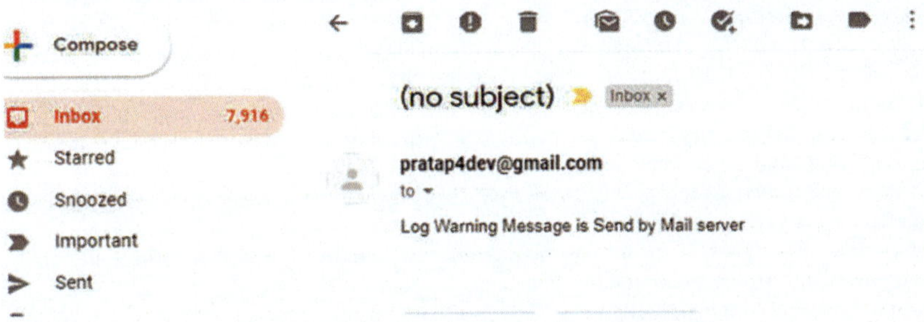

Fig. 9 Screenshot of receiving an alert message in Gmail account

Figure 9 shows the alert message received by the security monitoring system when the system security parameter exceeds the threshold value or as decided by the developer in the system.

6 Conclusion

This research work concludes that the implementation of a security monitoring system by collecting and analysis of applications log data is conducted, and based on log analysis, the alert message is generated. The alert message is sent to the application user to protect and secure the assets or information of the organization. The author also concludes that there is no single way or common way how this analysis and monitoring of the security system is done. It depends on the developer and organization's need how to use the security of application using logs. Logs analysis can be done with respect to clients or user or application specific. But monitoring and analysis of log data help to secure and protect the user from heavy losses.

References

1. DeLaRosa, A.: Log monitoring: not the ugly sister. Pandora FMS (2018). Retrieved 14 Feb 2018
2. Awontis Word Press. https://awontis.com/2018/12/12/the-10-steps-of-the-log-management-lifecycle-infographic/
3. Almgren, M., Lindqvist, U.: Application-integrated data collection for security monitoring. Lect. Notes Comput. Sci. (including Subser. Lect. Notes Artif. Intell. Lect. Notes Bioinformatics) (2015)
4. Cao, Q., Qiao, Y., Lyu, Z.: Machine learning to detect anomalies in web log analysis. In: 2017 3rd IEEE International Conference on Computer and Communications, ICCC 2017 (2018)
5. Latib, M.A., Ismail, S.A., Sarkan, S.Md., Yusoff, R.C.M.: Analyzing log in big data environment: a review. In: ARPN J. Eng. Appl. Sci. (2015)
6. Shan, G., Wang, Y., Xie, M., Lv, H., Chi, X.: Visual detection of anomalies in DNS query log data. In: IEEE Pacific Visualization Symposium (2014)
7. Yang, S.J., Cheng, I.C.: Design issues of trustworthy cloud platform based on IP monitoring and file risk. In: Proceedings—2015 IEEE 5th International Conference on Big Data and Cloud Computing, BDCloud 2015 (2015)
8. Yadav, R.B., Kumar, P.S., Dhavale, S.V.: A survey on log anomaly detection using deep learning. In: ICRITO 2020—IEEE 8th International Conference on Reliability, Infocom Technologies and Optimization (Trends and Future Directions) (2020)
9. Ogie, R.I.: Cyber security incidents on critical infrastructure and industrial networks. In: ACM International Conference Proceeding Series (2017)
10. Li, Q. et al.: Safety risk monitoring of cyber-physical power systems based on ensemble learning algorithm. IEEE Access (2019)
11. Janos, F.D., Dai, N.H.P.: Security concerns towards security operations centers. In: SACI 2018—IEEE 12th International Symposium on Applied Computational Intelligence and Informatics, Proceedings (2018)
12. Onwubiko, C.: Cyber security operations centre: Security monitoring for protecting business and supporting cyber defense strategy. In: 2015 International Conference on Cyber Situational Awareness, Data Analytics and Assessment, CyberSA 2015 (2015)
13. Ahmad, Z., Ong, T.S., Liew, T.H., Norhashim, M.: Security monitoring and information security assurance behaviour among employees: an empirical analysis. Inf. Comput. Secur. (2019)
14. Adithya, M., Scholar, P.G., Shanthini, B.: Security analysis and preserving block-level data deduplication in cloud storage services. J. Trend. Comput. Sci. Smart Technol. (TCSST) **2**(02), 120–126 (2020)
15. Shakya, S.: Process mining error detection for securing the IoT system. J. ISMAC **2**(03), 147–153 (2020)

Comparison of Machine Learning Methods for Tamil Morphological Analyzer

M. Rajasekar and Angelina Geetha

Abstract Morphological Analysis is the study of word formation which explains how a word is evolved from smaller pieces called root word. Morphological analysis is an important task in natural language processing applications, namely, POS Tagging, Named Entity Recognition, Sentiment Analysis, and Information Extraction. The heart of the morphological analysis process is to find out the root words from the given documents that is exactly matched with the corpus list. There are many research works that have been done in this area of research however not much contribution has been made in domain specific in the area of domain specific analysis regional languages. Morphological analysis for regional languages is complex and demands extensive analysis of natural language rules and syntax pertaining to specific regional language of focus. In order to improve the quality of natural language processing, generally research works are restricted to domain specific analysis. Morphological analysis in Tamil language documents is quite complex and valuable for Tamil NLP process. Our work focuses on a comparative study of three different approaches in performing morphological analysis on the regional language called Tamil. The scope our work is restricted to Gynecology domain text in represented in Tamil language. The analysis of morphological process is done in three different machine learning methods for the Gynecological documents. The performance analysis is carried out on the three morphological analysis models, namely, Rules-based lemmatizer (IndicNLP), Paradigm-based Tamil Morphological Analyzer (Tacola), and N-gram-based lemmatizer (UDPipe), and our experimental results proved that paradigm-based finite state model gives optimal results (0.96).

Keywords Morphological analysis · Finite state automation model · Rules-based approach · Morphology · Gynecological domain

M. Rajasekar (✉)
Department of Computer Applications, Hindustan Institute of Technology and Science, Padur, Chennai, India

A. Geetha
Department of Computer Science and Engineering, Hindustan Institute of Technology and Science, Padur, Chennai, India
e-mail: angelinag@hindustanuniv.ac.in

J. S. Raj et al. (eds.), *Intelligent Sustainable Systems*, Lecture Notes in Networks and Systems 213, https://doi.org/10.1007/978-981-16-2422-3_31

1 Introduction

To make a machine to understand the human language is a big challenge. The basic process to make the machine to understand the human language is Natural Language Processing. Natural Language Processing is a tract of Artificial Intelligence and Linguistics [1]. This growing area has spread its applications in various recent fields such as Machine translation, information extraction, text summarization, question answering system, spam mail detection, medical record extraction, and virtual medical consultation. In the field of information extraction, the language components identification and extraction are the major tasks.

Natural Language Processing in morphologically rich language is very hard task. The morphologically rich language, Tamil, has complex language components. Many reserachers have contributed in Language processing task in Tamil language. The language processing task comprises Preprocessing, Word count, Morphological analysis, Parts of Speech Tagging, Named Entity recognition, Language Identification, and Sentiment analysis.

Morphological analysis is the basic and important component of Natural Language Processing [2]. Morphological analysis is defined as a grammatical analysis of words, how they are formed by morphemes. A morpheme is the minimum unit in a language model [3]. It is used in speech recognizer, speech synthesizer, lemmatization, spell and grammar checker, noun decompounding, and machine translation [4]. The language Tamil has the very strong grammatical rule set. Hence designing a morphological analyzer for Tamil language is the challenging task. Many research contributions in the recent past as addressed issues in the process. The morphological analysis is the basic preprocessing task for Natural Language Processing applications, such as POS Tagging, NER Tagging, and Information retrieval. In this work we have attempted to compare three Tamil language morphological analysis models, namely, Rules-based lemmatizer (IndicNLP), Paradigm-based Tamil Morphological Analyzer (Tacola), and N-gram-based lemmatizer (UDPipe).

2 Background

The text processing task in medical field is growing recently. Based on the medical data, many fields are evolved. Such as drug suggestion, medical records automation, medical record classification, decease prediction, and medical image processing. The medical and healthcare information are very costly in current world. Many research have been done in the medical data processing field. As a multi-lingual country, such as India, it is very complex to implent in our local language. In this research, the genealogical information extraction has been done in Tamil Language. The genealogical data in Tamil language has been processing via Natural Language Processing task. The basic task is to identify the root word of the each and every word in given sentence. For that the morphological analysis has been done.

2.1 Morphology

The structural grammar of any language is divided into two parts, namely, Morphology and Syntax. The term morphology was minted by August Schleicher in 1859. The term morphology deals with the words and their structure. Syntax is called as how to arrange words to given meaningful sentences [5]. Computational morphology is the development of computational tools and framework to implement the morphological process for such language structure. The morphological analysis task is very common for English. But the classical language Tamil morphologically affluent. Tamil is one of the ancient, morphologically rich, and agglutinative.

In general, every root word is affixed with one or more morphemes to form a word. In Tamil language the root words are postpositionally inflected with morphemes. Tamil language takes lexical morphology and inflectional morphology. The term morphological analyzer is the tool to find the inflected morphemes and root word of a particular word. For example, in Tamil language:

Noun $\rightarrow$ marangaLai – maram + kk + aL + ai.
Verb $\rightarrow$ OdukiRaan – Odu + kiRu + An.

The morphological analysis plays a major role in most of the applications in Natural Language Processing such as Information Extraction and Retrieval, Grammar Checker, Spell Checker, Machine Translation, and Search Engines. Our work focuses on domain specific Tamil language documents. The domain of concern here is gynecology. In this paper, we have considered different machine learning tools for morphological analysis and applied it on Tamil documents pertaining to gynecology. We have compared the tools based on the accuracy of results produced.

3 Related Work

There are many research works that have been carried out in the field of morphological analysis task in Tamil language context. They have employed different approaches, namely, rules based, machine learning based, finite state transducer method, and such others.

The two run rules-based morphological analyzer and POS Tagger by Anthony et al. [5]. Their approach used grammar rules to extract the root words. It follows five steps. Untagged words extraction, Identification of the root words from the Wikipedia corpus. Matching the region from word list for this process it uses the tamil-wikipedia-word-list corpus. It has the list of 1,198,774 words. Analyzing the captured word with morphological rules. This system was evaluated by Precision Recall evaluation. The Overall F-Score measure was 92.1%.

The morphological analyzer for classical Tamil text using rules-based approach was developed by Agilan et al. [6]. This approach used rules to classify the post-positions of the word. Functional rules were used to stripping of post-positions, adding affixes, doubling post-positions. Finally it processes to add the post-positions for suitable and correct pronunciation is called sandhi process. This rules-based approach produced 72% of accuracy for classical Tamil text.

The Piripori morphological analyzer was based on word level analysis. It was based on the morphological rules based on the word level [7]. The root word and inflection details were in a corpus. The computational time was analyzed with Tacola analyzer based on input of 137,144 words. Tacolaa analyzer took four minutes for the word list. But Piripori approach proved that their work took twenty-four seconds in word level.

The morphological generator is designed using word formation rules and morphophonemics with the Finite state automata mechanism by Menaka et al. [8]. The paradigmatic classification method for verbs and nouns was used. The source words were taken from Cre-A dictionary of Tamil. The system was evaluated based on Precision and Recall values. Totally 2556 noun root words and 19,152 verb root word were used. For the noun and verb category it achieved 99% of F-Score accuracy.

The Finite state morphological analyzer and generator for Tamil have been developed in the concept of computational grammar [9]. The meta-morphological rules and scripts are developed to translate these rules into finite state grammar processable concepts. This tool could extract 330 verb roots and their inflected forms. It has 100,000 noun lexicons. From AUKBC 500K corpus used for evaluation. 19,250 verbs terms with their inflected forms and 26,000 noun tokens were found as correct.

Finite State Automation (FSA) approach was used to analyze the morphemes and the root word for the given word by Ram et al. [10]. Using FSA, all the possible morphemes are built for the given root word. They categorized the root word based on paradigm method to improve the orthographical rules. Morphosyntax rules are used to analyze correct morpheme. They have categorized 36 noun paradigms and 34 verb paradigms. Their work processed totally 44,055 root words and it has 286 morphosyntax rules. It has general domain words and tourism domain words. The accuracy of this approach was 93.24% in general domain and 90.17% in tourism domain.

The FSA method used to analyze the morphemes using finite state transducer approach was also employed by Umamaheswari et al. [11]. This approach concentrates on compound, numeral, and colloquial words with general words. The spelling variation rules were applied to transform informal written words into formal words. The morphological analysis for compound, numeral, and colloquial words was performed in the documents related to blogs, lyrics, newspapers, and tourism pages. Totally 100,596 words were taken from the above different domains. They have divided the words into three categories, namely, Compound words, Numeral, and Colloquial. Precision and recall evaluation method was used for evaluation. The Precision score is high in Tourism for compound words (0.84), Blog words for Numerals (0.92), and Tourism for Colloquial (0.84). The recall score is high in News for Compound words (0.9), Blogs for numerals (0.95), and News for Colloquial (0.98).

A sequence labeling method is implemented to develop the morphological analyzer for Tamil language content by Anandkumar et al. [12]. In this approach the morphological analysis problem is redefined as classification problem. The approach is based on sequential labeling. The kernel method of labeling and training captures the nonlinear relationships of the morphological terms from training sample data. The Recall–Precision and F-Score evaluation method is used to evaluate the accuracy. In this method untagged words and tagged words are used. For verb category it achieved 93.3% of accuracy. In Noun category it achieved 94.03% of accuracy. Word level accuracy is 90% in verb and 91.5% in noun. Finally the accuracy of this model is compared with Memory based tagger and Conditional random fields methods. This method reached high accuracy based on the experimental results carried out 22222.

The unsupervised lexical normalization process for Hindi and Urdu sentiment analysis was done by Mehmood et al. [13]. It is a transliteration-based encoding for Roman Hindi and Urdu normalization. They used linguistic aspects of Roman Hindi/Urdu to transform the variant lexical forms into their canonical form. They have classified the process into three modules as transliteration-based encoder, filter module, and hash code ranker. The encoder generates all the possible hash-codes for a single word from Roman Hindi/Urdu. The filter filters the irrelevant hash-codes. Finally hash code rankers rank filtered hash-codes based on their relevance. They have used 11,000 sentiment analysis reviews of Roman Hindi\Urdu language. The data set with the transliteration model they performed using phonetics algorithms. They proved that the proposed model significantly reduced the error rate from the baseline.

A proposal of Morpho-semantic knowledge Graph for Arabic information retrieval was proposed by Bounhas et al. [14]. They proposed to build a morpho-semantic knowledge graph from Arabic corpora. They used a tool combination of Ghwanmeh stemmer and MADAMIRA tools for Arabic to analyze and disambiguate Arabic texts with reduced ambiguities. This proposed model extracted multi-level lexicon from Arabic vocalized corpora. Both morphological and semantic links were represented by compressed graphs. The BM25 ranking algorithm was used for retrieving related morphological description for a given query. They used two datasets as Tashkeela and ZAD. The proposed model evaluated by query expansion and experimented on 25 queries from ZAD dataset. They proved that the proposed model gives optimal results for the given input datasets.

An analysis of movie reviews based on sentiments using machine learning approach is proposed by Mitra et al. [15]. The dataset is the collection of movie reviews to analyze based on the sentiments. They used lexicon-based hybrid approach to analyze the text. The pre-defined rules are used to analyze the reviews. There are three categories of results, positive, negative, and neutral. They used F-score evaluation method to get the optimal results. Finally they achieved 0.80 accuracy level in the selected domain.

An analysis of emotions of theme park visitors using giospatial and social media analytics has been done by Samuel et al. [16]. They have selected the Circumplex model of affect for emotional words classification. There are four categories of emotions with the combination of pleasure, displeasure, low arousal, and high arousal. They have processed 20,400 tweets and compared with existing techniques. They found that the quality of the analysis result increased.

The gyneacological text in Tamil used to analyze the risk factors of PCOD problem in girl students in Bishop Heber college by Nivetha et al. [17]. They have collected the risk factor details by simple questionnaires from the students. By using these data they categorized the high, average and low risk students from polycystic ovarian disease.

4 Proposed Work

The Tamil text processing is done by various machine learning methods. Most of the text processing methods focused on general Tamil words. This research work focus on the processing of text in gynecological documents in Tamil language. The words and terms in gynecological documents mostly different from general text. Some text processing methods proved as best in general Tamil text. In this, the most optimal methods are taken to process the specified domain text. Gynecological Tamil text to be analyzed morphologically and find list of root words of the given text. The result of analysis of various methods to be compared and find out the suitable method for further process. There are many morphological analysis methods available. For this proposed work three methods are chosen. Paradigm-based finite state machine for morphological analyzer, unsupervised IndicNLP morphological analysis, rules-based UDpipe model for lemmatization.

4.1 Paradigm-Based Finite State Machine

The Ffinite state machine is the powerful method for text processing. The process of the FSM model is based on N-gram comparison with the corpus data. In the initial state the text has been selected as per the N-gram methods and does the iterations in processing state till the text is to be identified from corpus list (Fig. 1).

The paradigm-based approach for Tamil morphological analyzer is implemented in finite state machine. This approach has 95% of accuracy when test with millions of words in CIIL corpus [18]. It will analyze 3.5 million words forms in Tamil corpus. This approach gives high accuracy in general domain. This approach is implemented in Tacola Tamil morphological analyzer. The paradigm is used to find out the root word and the finite state transducer is built to find out the best morpheme affixes for the given word. This tool is well developed to analyze nouns, verbs, compound nouns, numerals, sandhi, finite and infinite verbs, and it gave the best result in large size of corpus.

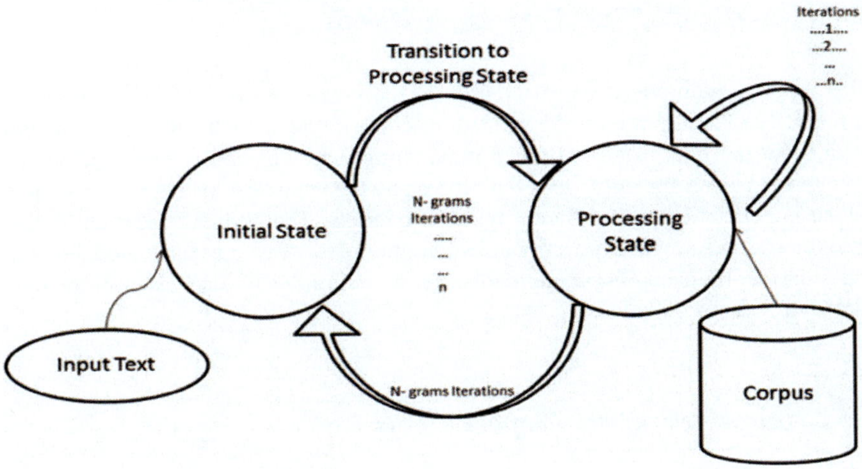

Fig. 1 Finite State Machine model

4.2 *Unsupervised Approach*

The Indic NLP has model to process the text documents in Indian region. It has general Indian language corpus and models to process such data. Indic NLP library is used to support most of the text processing and NLP applications for Indian Languages [19]. The unsupervised model uses statistical approach to find the optimum match for the root word. The Indic NLP master libraries have all the language processing tools such that Text normalization, tokenization, word segmentation, lemmatization, Indication, Transliteration. This tool uses the rules to select the inflected morphemes from the corpus. It uses the large corpus with root words, affixes, and tagged information. This method gives 92% of accuracy in general domain in Tamil.

4.3 *UD2.0 Lemmatization*

UDPipe is a trainable pipeline that performs tokenization, sentence segmentation, POS Tagging, dependency parsing, and lemmatization [20]. UDPipe baseline system automatically generates suffix rules and a token with distinct suffix is to be split from actual word. The rules are automatically generated by attaching all rules in the training data, which cannot trigger incorrectly usually. The UDPipe lemmatizer is comparable. The lemmatization guesser gives pairs (lemma rule, UPOS). The lemma rule generates a lemma by stripping prefixes or suffixes from the input word. The guesser generates outputs with affixes and word prefixes. The disambiguation is attained by averaged perceptron tagger. This model gives 89% of accuracy in Tamil general domain.

4.4 Variations of Text Processing Models

The above three models are selected to do the processing of text of this selected domain. The Paradigm-based finite state machine model is the mathematical model to find out the final results of given input using iterations and N-gram technique. The unsupervised approach does the process by the statistical history of the model, which is performed with different types of datasets. The optimal statistical values give the best result. UDPipe lemmatization, the rules-based approach works on the input dataset by using its pre-defined rules. It checks the similarities with the data already it has via the rules.

4.5 Uniqueness of the Proposed Work

The above three models are used to process the specified domain text in an effective manner. The source document text is the actual input of the above three models. These models are performed well in the general Tamil text. But here we have compared the performance of the models by the quality and accuracy of the output text. Based on the number of identified unknown text the model is evaluated. The higher number of unknown text is affecting the model evaluation. The output of the process is the root word list of gynecological data in Tamil. From this root word, we can do more further text process. We can identify the suitable and optimal performed model for text processing in gynecological data.

5 System Design

5.1 Preprocessing

The source documents are the data collected from various Internet sources based on women health issues and tips in Tamil language. The text documents are cleaned by preprocessing techniques. In Python the text processing tool kit NLTK is used to do the preprocessing task. Sentence truncation, removal of punctuation marks, unwanted dotted letters removal in each word, word wise truncation have to be done. We have implemented Tamil NLTK libraries to do the preprocessing of the specified domain text.

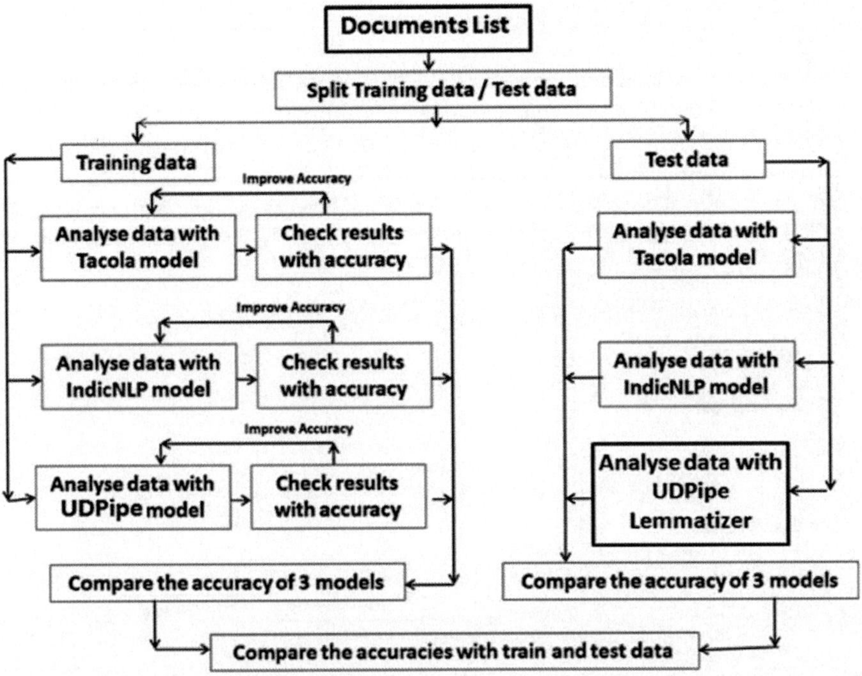

Fig. 2 Morphological analysis by various models

5.2 *Implementation*

The proposed morphological analysis is designed as in Fig. 2. From the total source data can be spitted as training data and test data 70 and 30% accordingly. The training data is checked with the entire three models to analyze the morphological terms and find the accuracy. To improve the accuracy of a particular model, the unstrapped words can be processed and added in the model corpus. The combination of three model accuracies is compared. Then the same process can be done with test data but without improving accuracy. Finally the training data and test data accuracy can be compared and find out the best model for this selected domain.

5.3 *Implementation*

There are totally 470 data documents which are used to analyze. The data documents are classified as 330 training documents and 140 testing documents.

5.3.1 Use of Tacola Morphological Analyzer

The words in the sentence are normalized by removing punctuation marks. The sentence is tokenized as a separate word. The morpheme and the root word are separated from the actual word using the corpus. The stripped morphemes can be checked from right to left of the actual word. It generates the affix from the root word using finite state automation. The tacola analyzer is designed in java code. In the GUI the file can be uploaded to analyzer. The process is done by checking all the affix files and POS tag corpus (Figs. 3 and 4).

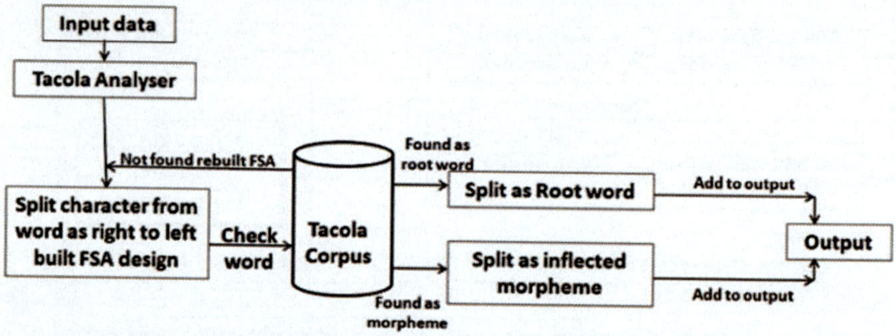

Fig. 3 Process flow of tacola analyzer

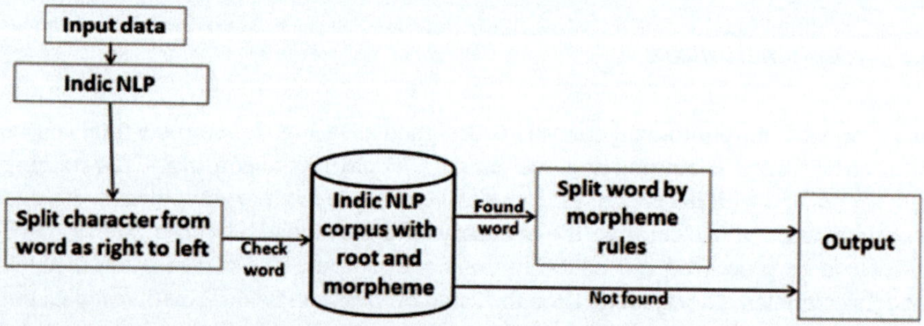

Fig. 4 Process flow of IndicNLP analyzer

Input:கர்ப்பபப்பையை வலுவாக்குவதில் முதல் மருந்து கற்றாழை. வெள்ளைப்படுதல், ரத்தப்போக்கு, பிசிஓடி, சிஸ்ட் கர்ப்பபப்பை புற்றுநோய் என் எந்த கர்ப்பபப்பை தொடர்பான பிரச்சனைக்கும் கற்றாழை ஒரு அருமருந்து.

Output: <கர்ப்பபப்பையை>:கர்ப்பபப்பை < Noun >ய் < Sandhi >ஐ < Accusative Case ><வலுவாக்குவதில்>:வலு < Noun >வ் < Sandhi >ஆக்கு < Auxiliary Verb >வ் < Future Tense Marker >அது < Neuter OR Pronominal >இல் < Locative Case ><வலு+வ்+ஆக்கு+வ்+அது+இல்><முதல்>:முதல் < Particle ><முதல்>முதல் < Postposition ><முதல்>முதல் < Noun ><முதல்><மருந்து>:மருந்து < Noun ><கற்றாழை.>:<unknown>
<வெள்ளைப்படுதல்,>:<unknown><ரத்தப்போக்கு,>:<unknown><பிசிஓடி,>:<unknown>< சிஸ்ட்>:ட் < Past Tense Marker ><ட்><unknown><கர்ப்பபப்பை>:கர்ப்பபப்பை < Noun ><புற்றுநோய்>:புற்றுநோய் < Noun ><என>:என < Particle ><எந்த>:எந்த < Interrogative Adjective ><கர்ப்பபப்பை>:கர்ப்பபப்பை < Noun ><தொடர்பான>:தொடர்பு < Noun >ஆன < Adjectival Suffix ><பிரச்சனைக்கும்>:பிரச்சனை < Noun >க்கு < Dative Case >உம் < Clitic ><கற்றாழை>:கற்றாழை < Noun ><ஒரு>:ஒரு <charNumbers>அருமருந்து>:
ந்த் < Past Tense Marker >உ < Verbal Participle Suffix ><ந்த்+உ><unknown>

5.3.2 Use of IndicNLP Analyzer

Indic NLP morphological analyzer uses the morphemic rules to split root word from actual word then check with the corpus entry to get the optimal result of the given word. This tool is implemented in Python code.

Output:['கர்ப்பபப்பை', 'யை', 'வலுவாக்க', 'ுவதில்', 'முதல்', 'மருந்து', 'கற்றாழை.', 'வெள்ளைப்படுதல்,', 'ரத்தப்போக்கு,', 'பிசிஓடி,', 'சி', 'ஸ்ட்', 'கர்ப்பபப்பை', 'புற்றுநோய்', 'என்', 'எந்த', 'கர்ப்பபப்பை', 'தொடர்பான', 'பிரச்சனை', 'க்கும்', 'கற்றாழை', 'ஒரு', 'அருமருந்', 'து']

5.3.3 Use of UDPipe Lemmatizer

The UDPipe model for lemmatization is trained with Tamil words. It loads the UDPipe lemmatizer model for the Tamil language. The text file is accepted as frame file to process as table. At first the model split each and every sentence from the given text and indexed with word count. Then tokenize the sentence into words with its repetition index. It lists out the words in ascending order. The tokenized index words can be checked with the model corpus then finds the root word. Finally it lists the root words only.

Output:,. அருமரு எந்த என் ஒரு கர்ப்பப்பு கர்ப்பப்பை கற்றாழை சிஸ்ட் தொடர்பு பிசிஓடி பிரச்சனை புற்றுநோ மரு முதல் ரத்தப்போ வலுவாக்குவது வெள்ளைப்படுதல்

Figure 5. explains the comparison results of three models for the morphological analysis. For the example totally 19 words are taken as input. The Tacola model gives high number of correct word roots. Others give moderate number of root words and their analysis (Fig. 6).

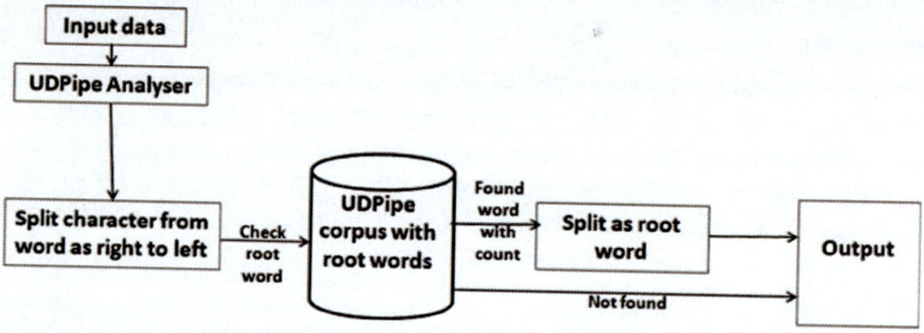

Fig. 5 Process flow of UDPipe Model

List of Words	Correct Roots	Results		
		Tacola	IndicNLP	UDPipe
கர்ப்பப்பையை	கர்ப்பப்பை	கர்ப்பப்பை +ய் +ஐ	கர்ப்பப்பை +யை	கர்ப்பப்பை
வலுவாக்குவதில்	வலுவாக்கு	வலு +வ் +ஆக்கு +வ் <+அது +இல்	வலுவாக்க + ுவதில்	வலுவாக்குவது
முதல்	முதல்	முதல்	முதல்	முதல்
மருந்து	மருந்து	மருந்து	மருந்து	மருந்து
கற்றாழை	கற்றாழை	கற்றாழை	கற்றாழை	கற்றாழை
வெள்ளைப்படுதல்	வெள்ளப்படு	வெள்ளை +ப் +படு +த் +அல்	வெள்ளைப்படுதல்	வெள்ளை
ரத்தப்போக்கு	ரத்தப்போக்கு	ரத்த +ப் +போ +க்கு	ரத்தப்போக்கு	ரத்தப்போக்கு
பிசிஓடி	பிசிஓடி	பிசிஓடி	பிசிஓடி	பிசிஓடி
சிஸ்ட்	சிஸ்ட்	சிஸ்ட்	சிஸ்ட்	சிஸ்ட்
கர்ப்பப்பை	கர்ப்பப்பை	கர்ப்பப்பை	கர்ப்பப்பை	கர்ப்பப்பை
புற்றுநோய்	புற்றுநோய்	புற்றுநோய்	புற்றுநோய்	புற்றுநோய்
என	என	என	என	என
எந்த	எந்த	எந்த	எந்த	எந்த
கர்ப்பப்பை	கர்ப்பப்பை	கர்ப்பப்பை	கர்ப்பப்பை	கர்ப்பப்பை
தொடர்பான	தொடர்பு	தொடர்பு +ஆன	தொடர்பான	தொடர்பு
பிரச்சனைக்கும்	பிரச்சனை	பிரச்சனை +க்கு +உம்	பிரச்சனை +க்கும்	பிரச்சனை
கற்றாழை	கற்றாழை	கற்றாழை	கற்றாழை	கற்றாழை
ஒரு	ஓர்	ஒரு	ஒரு	ஒரு
அருமருந்து	அருமருந்து	அருமருந்து	அருமருந் +து	அருமரு
Total	19	16	12	10

Fig. 6 Result Comparison of Morphological Analysis

Table 1 Training model results

Training module						
Models	Total no. of words	Analyzed words	Unknown error	Input error	Correctness	Correctness (%)
Tacola	7260	6978	148	27	6803	97.5
IndicNLP	7260	6723	245	18	6460	96.2
UDPipe	7260	6501	221	35	6245	96.6

Table 2 Testing model results

Testing module						
Models	Total no. of words	Analyzed words	Unknown error	Input error	Correctness	Correctness (%)
Tacola	3080	3011	47	8	2956	98.2
IndicNLP	3080	2947	68	12	2867	97.3
UDPipe	3080	2912	72	11	2829	97.1

5.3.4 Findings and Discussions

The results of training and testing module can be evaluated using actual input words, analyzed words, Unknown error produced as result, and Input error also calculated. The evaluation method is shown in Tables 1 and 2.

Based on the above results from training module and testing module, the Tacola Tamil morphological analyzer model gives good results for the problem. Tacola model results and evaluation method are as follows:

$$Precision = \frac{True\ Positive}{True\ Positive + False\ Positive} = \frac{True\ Positive}{Total\ Predicted\ Positive}$$

$$Recall = \frac{True\ Positive}{True\ Positive + False\ Negative} = \frac{True\ Positive}{Total\ Actual\ Positive}$$

$$F1 = 2 \times \frac{Precision * Recall}{Precision + Recall}$$

For the evaluation calculations, the Precision, Recall, and F1 Score method are used to find the accuracy of the three models. Based on the calcuation the Tacola model is performed well for our domain text (Fig. 7 and Table 3).

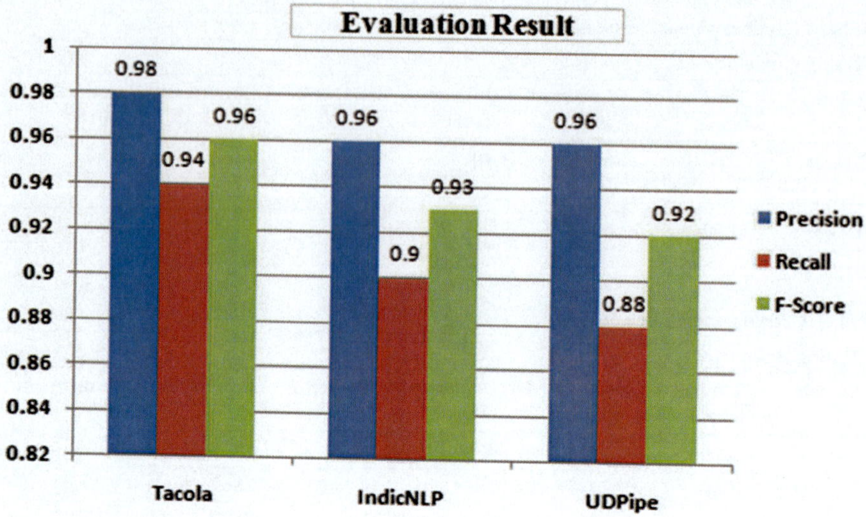

Fig. 7 Evaluation result

Table 3 Final results

Models	Tacola	IndicNLP	UDPipe
Precision	0.98	0.96	0.96
Recall	0.94	0.9	0.88
F-score	0.96	0.93	0.92

6 Conclusion

We have taken as input of our work, the morphological analysis of gynecology domain text documents in Tamil language. Based on the literature study on previous works, three real-time open source tools were selected to do analysis task. The analysis was done on three standard models and evaluated the outcomes. Tacola analyzer has proved to give good outputs with detailed morphemes and root words. IndicNLP morphological analyzer has found the list of root words and it truncates the affixes from the root word. UDPipe Lemmatizer stripes the list of root word from the given sentences in alphabetical order with occurrence count and list of affixes separately. It is also found that Tacola Tamil morphological analyzer produced good results in this domain. For the Tacola model the Precision, Recall, and F-score evaluations have been done. It gave good accuracy value of 96% for gynecology domain documents in Tamil Language. Our future work is focused in implementing Tacola model and proceeding with POS Tagging. Our work can be extended to domain specific Tamil documents question answering system with enhanced accuracy levels and processing speed.

References

1. Khurana, D., Koli, A., Khatter, K., Singh, S.: Natural language processing: state of the art, current trends and challenges. Comput. Sci. Comput. Lang. (2017). arXiv:1708.05148
2. Lushanthan, S., Weerasinghe, A.R., Herath, D.L.: Morphological analyzer and generator for Tamil Language. In: IEEE 14th International Conference on Advances in ICT for Emerging Regions (ICTer), 10–13 December 2014, IEEE Xplore: 16 April 2015. Electronic ISBN: 978-1-4799-7732-1, Print ISBN: 978-1-4799-7731-4
3. Thanaki, J.: Python Natural Language Processing, Book of Packt (2017). ISBN 9781787121423
4. Betina Antony, J., Mahalakshmi, G.S.: Two run morphological analysis for POS tagging of untagged words. Int. J. Comput. Sci. Inf. Secur. (IJCSIS) **14** CIC (2016). ISSN 1947-5500
5. Dayanand, A.: Morphological analyzer for Tamil. Computational Analysis of Tamil at Morphological level Thesis Report, October 2017
6. Akilan, R., Naganathan, E.R.: Morphological analyzer for classical tamil texts: a rulebased approach. Int. J. Innov. Sci. Eng. Technol. **1**(5) (2014)
7. Suriyah, M., Anandan, A., Narasimhan, A., Karky, M.: Piripori: morphological analyser for Tamil. In: Springer Nature Switzerland 2020: Proceedings of International Conference on Artificial Intelligence, Smart Grid and Smart City Applications (2020)
8. Ram, V.S., Menaka, S., Devi, S.L.: Morphological generator for Tamil. In: Conference: Knowledge Sharing Event—Mysore (2016)
9. Sarveswaran, K., Dias, G., Butt, M.: Using meta-morph rules to develop morphological analysers: a case study concerning Tamil. In: Proceedings of the 14th International Conference on Finite-State Methods and Natural Language Processing, pp. 76–86, Dresden, Germany, September 23–25, 2019
10. Menaka, S., Ram, V.S., Devi, S.L.: Morphological generator for Tamil. In: Parakh, M. (ed.) Morphological Analysers and Generators, pp. 82–96. LDC-IL, Mysore (2010)
11. Umamaheswari, E., Ranganathan, K., Geetha, T.V., Parthasarathi, R., Karky, M.: Enhancement of morphological analyzer with compound, numeral and colloquial word handler. In: Proceedings of ICON-2011: 9th International Conference on Natural Language Processing, Macmillan Publishers, India. http://ltrc.iiit.ac.in/proceedings/ICON-2011
12. Anand Kumar, M., Dhanalakshmi, V., Soman, K.P., Sankaravelayuthan, R.: Sequence labeling approach to morphological analyzer for tamil language. Int. J. Comput. Sci. Eng. **02**(06) 1944–1951 (2010)
13. Mehmood, K., Essam, D., Shafi, K., Malik, M.K.: An unsupervised lexical normalization for Roman Hindi and Urdu sentiment analysis. ELSEVIER—Inf. Process. Manage. 102368 (2020)
14. Bounhas, I., Soudani, N., Slimani, Y.: Building a morpho-semantic knowledge graph for Arabic information retrieval. ELSEVIER—Inf. Process. Manage. **25**, 102124 (2019)
15. Mitra, A.: Sentiment analysis using machine learning approaches (Lexicon based on movie review dataset). J. Ubiquit. Comput. Commun Technol. (UCCT) **2**(03), 145–152 (2020)
16. Manoharan, S.: Geospatial and social media analytics for emotion analysis of theme park visitors using text mining and gis. J. Inf. Technol. **2**(02), 100–107 (2020)
17. Nivetha, M., Suganya, S.G.: Survey of poly cystic ovarian disease (PCOD) among the girl students of bishop Heber college. IOSR J. Nurs. Health Sci. (IOSR-JNHS) e-ISSN: 2320-1959. p- ISSN: 2320-1940, Vol. 5, Issue 4 Ver. I (Jul.–Aug. 2016), pp. 44–52
18. Ramesh Kumar, S., Viswanathan, S.: Tamil Morphological Analyser. Documentation of Funded Project on AU-KBC Research Centre (2014)
19. Rizvi, M.S.Z.: 3 Important NLP Libraries for Indian Languages You Should Try Out Today. Analytics Vidhyalaya Online Lessions, January 2020
20. Straka, M., Strakova, J.: Tokenizing, POS tagging, lemmatizing and parsing UD 2.0 with UDPipe. In: Proceedings of the CoNLL 2017 Shared Task: Multilingual Parsing from Raw Text to Universal Dependencies, August 2017. https://doi.org/10.18653/v1/K17-3009

Regression-Based Optimization and Control in Autonomous Systems

J. Judeson Antony Kovilpillai, S. Jayanthy, and T. Thamaraikannan

Abstract Machine learning techniques are currently used in many industries, smart factories for predictive maintenance, fault tolerance, and optimization. One of the biggest advancements in the field of automation is the ability of the machines to adopt machine to machine (M2M) or edge computing technologies to provide a secure layer of communication and superior control over different units. Regression is a commonly used machine learning technique for analytical insights, estimating performance metrics, prediction, etc. and most of the industrial embedded appliances have automated machines driven by controllers which can be tuned using machine learning algorithms to have an effective control over its functional units. This paper focuses on developing an autonomous module which integrates M2M and machine learning techniques between two units (MSP430—a TI-based microcontroller unit and a PC) with the ability to predict, analyze the data, optimize different functional metrics using a neural network model, and establish proficient communication and control. The prediction model is designed using Spyder IDE which imports an open source dataset to predict the CO_2 emission values using multiple linear regression and evaluate the accuracy of the predicted output. The predicted output is sent to the TI-MSP430 MCU via serial communication which uses the predicted values to control the hardware peripherals that cool down the source of emission. The mean absolute error, residual sum of squares, r-2 score, and loss values are calculated and it is observed that the model with ReLu activation function predicts the CO_2 emission level accurately.

Keywords Regression · M2M · Machine learning · Automation · TI-MSP430

J. Judeson Antony Kovilpillai (✉) · S. Jayanthy · T. Thamaraikannan
Department of ECE, Sri Ramakrishna Engineering College, Coimbatore, India
e-mail: judeson@srec.ac.in

S. Jayanthy
e-mail: jayanthy.s@srec.ac.in

T. Thamaraikannan
e-mail: thamraikannan.1957002@srec.ac.in

J. S. Raj et al. (eds.), *Intelligent Sustainable Systems*, Lecture Notes in Networks and Systems 213, https://doi.org/10.1007/978-981-16-2422-3_32

1 Introduction

Machine to machine (M2M) communication plays a key role in industrial telematics, automation, and control. Industries are stepping into another era of Industrial revolution termed as Industry 4.0 where various technologies like M2M, IoT, robotics, and artificial intelligence are integrated together to hasten the working process, get deeper analysis on key performance indicators, and acquire profitable business insights. One of the biggest advancements in the field of automation is the ability of the machines to adopt machine to machine (M2M) or edge computing technologies to provide a secure layer of communication Industry 4.0 lead to the development of smart machines that operate in a framework that can have a huge repercussion on manufacturing, delivery, reducing human labor, and upgrading the industrial work environment. Most of the companies use machine learning or deep learning techniques just as an analysis or estimating tool but its area of application can be extended to controlling various drives and processes involved in an industry. Many small scale industries can also adopt these techniques to transform their working methodologies very effectively thereby increasing productivity and profitability. TI-MSP430 is a 16-bit Texas-based microcontroller with ultra-low-power RISC mixed-signal processor with integrated analog and peripherals for sensing and measurement used to detect different events and set triggers to different conditions. This paper focuses on developing a module which has two units (i) Prediction Unit: this unit has the role of predicting the output by developing a neural network using multiple linear regression algorithm with an open source dataset from IBM Watson. When an input data is given, this unit predicts an output and sends it to the control unit through serial communication. (ii) Control unit: the control unit is developed using TI-MSP430g2553 which receives the signal via the serial port from the prediction unit and controls different hardware interfaces. Both the prediction and control unit are integrated together using serial communication and they establish a M2M-based communication and control between each other.

2 Related Works

Subramanya [1] proposed a branch prediction technique for embedded systems to improve the performance of a processor with high prediction rate. Conventional techniques have a major disadvantage of miss prediction, but it consumes low power with high speed. A neural network-based prediction can improve the accuracy of a system but there are many challenges in its architectural design. This paper implemented a branch prediction algorithm particularly for superscalar and pipelined processor where the instruction is decoded from the opcode for branching at its fetching stage and it notifies the hardware that there is a branch instruction if the condition is true. The process continues until the condition is true, and the data pipeline has continuous stream of incoming data. If the CPU finds the branching condition is false, then it will flush the data in pipe and the execution continuous. This method yields us almost 99 percent accuracy so it can be deployed in high performance system where the cost of hardware system is manageable making it easy to implement in industrial embedded appliances. The work can be extended further with branch detection in the same decoding stage and thus the latency can be further reduced.

Muninathan et al. [2] proposed an experimental and numerical studies on catalytic converter to reduce CO_2 emissions in IC Engines. In atmosphere 0.04% of carbon dioxide present in air which contributes a major role for increasing photosynthesis and maintaining ideal condition of the earth. But in recent days the level of carbon dioxide gets increased and created a drastic change in environment and a major reason for increased greenhouse gas level. Automobiles are the major emitter of this carbon dioxide about 85 percent which causes severe effects in our environment. The main objective is to implement a numerical analysis of CFD to reduce and control the carbon dioxide emission from automobiles by absorbing it using activated charcoal and based absorbents in engines that are operated using petrol and comparing it with others. Charcoal has a high-level porosity, so it absorbs carbon dioxide at its best when compared with other absorbents, in addition to that charcoal's adsorption capacity increases when ordinary charcoal reacted with other reagents. Similarly, alumina is introduced to a phase transformation to obtain γ alumina to increase its porosity. The authors reduced the carbon dioxide emission level by using the charcoal and alumina to a greater extent.

Wenjun and Marek [3] implemented a ML algorithm to reduce the error in prediction and to accelerate learning curve for large datasets. In previous works Ir-partitions algorithm was designed in a way to improve C4.5. However, there is learning error and high percentage of undefined attribute values in final occurs in it. The paper proposed a new type of clustering method to overcome those errors and the original datasets are divided for learning, training, and testing then the divided learning and training dataset are applied on the core algorithm (Seleukos algorithm) which contains the Ir-partition. Then the process is followed by clustering in which the data is merged to make one function, along with testing sets taken for cross validation. This seleukos algorithm accelerates the learning curve, yields 4% better points when compared C4.5 at earlier stage, and has a high accuracy. The algorithm fully automates every single process and only requires users to upload a dataset and to feed the input parameters.

Kirci [4] suggests a novel model for routing vehicles using techniques to minimize CO_2 emissions with a time window factor. It also proposes a design google maps are used to find real routes along with routes that are presented over graphs denoting areas with high CO_2 emission. The main parameter considered is the emitted CO_2 with spent time rates during the delivery of a vehicle. In addition to that the basic parameters like speed of a vehicle, time windows, customers demand, and vehicle capacity also taken for consideration. The paper also focuses on the dataset in turkey as a case study analysis and a relationship between the path followed and CO_2 emitted is derived. The limitation of this paper is that GPS-based analysis may be used as an avoidance strategy and not as a proper methodology to control the emission or analyze the source of emission.

Wang and Yang [5] proposes a machine learning technique to predict CO_2 Emissions from manufacturing industrial dataset based on GM(1, N) Model and SVM. This paper suggests a proficient prediction algorithm for predicting carbon dioxide emission from a manufacturing industry in Chongqing. The variables are selected using 'Pearson Correlation Coefficient' method and the selected variables are divided for training and testing samples. At the GM (1, N) model, development factor coordination coefficient is calculated, and at SVM model RBF function parameter and penalty factor are calculated for the comparative model accuracy. SVM has better accuracy metrics and performs better than GM (1, N) model. The main drawback of their project is this technique only applicable for small samples and high-dimensional pattern recognition.

Pooja and Suhasini [6] proposed a prediction model for CO_2 emission using machine learning and deep learning techniques. The paper proposes an iterative and continuous prediction methodology using a design regression model to predict CO_2 emission value from a historical data. After data cleaning and cross validation 70% of data is taken for training and 30% is taken for testing. The collected data is imported to design regression model for accuracy evaluation and RMSE calculation. Finally, the CO_2 emission value is predicted but for better accuracy we can choose trial and error method so that we can achieve low RMSE value. This paper has a limited scope as it calculates only RMSE values as an accuracy metric and doesn't cover any other metrics like R2 score, log loss, etc.

3 Methodology

The proposed methodology constitutes a prototype (Fig. 2) which mimics a machine to machine communication designed using a PC and TI-MSP430 with essential software tool-chains. As discussed earlier the prototype has two units (i) Prediction unit (ii) Control unit. They communicate with each other using serial communication by configuring their individual COM ports and equivalent baud rates. The deep learning techniques are used in the prediction unit. They are developed in the PC using a software tool-chain known as Spyder IDE which uses Python as a programming language. The designed neural network imports the pre-existing CO_2 emission data, extracts features, analyzes the accuracy across multiple epochs, and then sends the predicted output to the control unit. The control unit is designed using MSP430 microcontroller unit, motor to control the coolant fan, buzzer to alert a manual response, and HC-05 Bluetooth module to turn on the coolant manually. The control unit receives the predicted output and uses it to control the coolant and alerting mechanisms. The control unit also waits for a manual response using HC-05 Bluetooth module to turn on the coolant by sending an alert through buzzer and if there is no response it switches on the coolant fan by itself to cool down the source of CO_2 Emission. The layout of the proposed methodology (Fig. 1), block diagram (Fig. 2) and the schematics (Fig. 3) of the workflow are shown in the diagrams below.

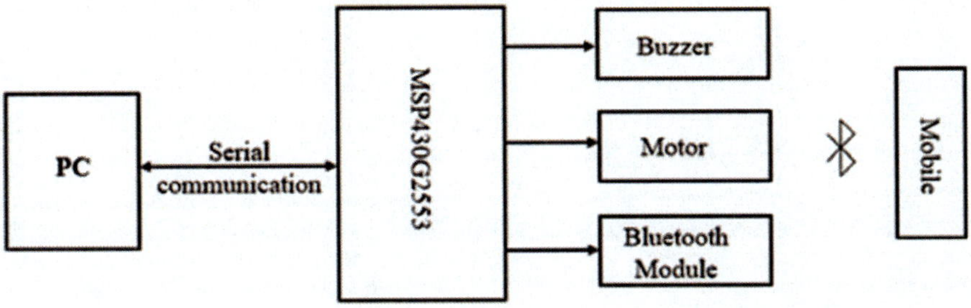

Fig. 1 Layout of the proposed methodology

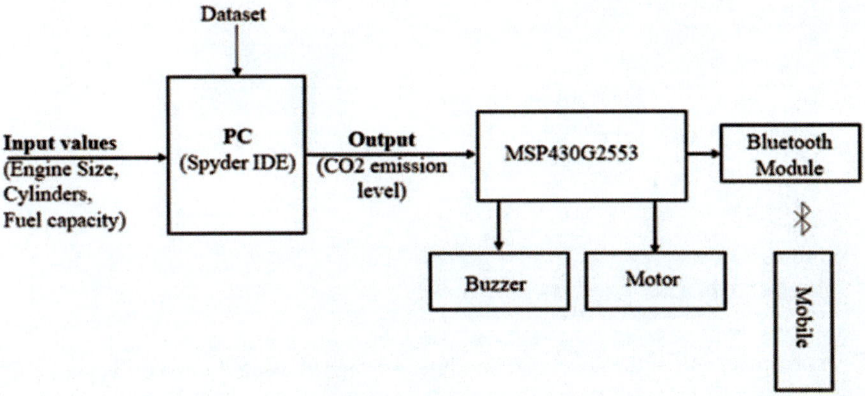

Fig. 2 Block diagram of the proposed methodology

The prediction unit uses multiple regression algorithm implemented using a Python package known as Keras and the accuracy metrics are also calculated using sci-kit learn Python package. The input parameters extracted from the dataset are engine size, cylinders, and fuel consumption, and the output values are respective CO_2 Emission value. By using multiple linear regression model the coefficient and intercept values are derived. Then the activation functions like ReLu, sigmoid functions are applied and accuracy metrics are evaluated. Then the prediction unit uses the user's input parameter to estimate the CO_2 emission level and if the emission level is high it signals the control unit to turn on the coolant mechanism [7]. The communication is done by configuring the serial port of the control unit and setting a common baud rate using a Python package called pyserial. The control unit is designed using TI-MSP430 interfaced with a Bluetooth module, buzzer, and a motor. They communicate with the prediction unit via the same COM port and equivalent baud rates. By using the predicted results, the control unit signals HIGH or LOW to the buzzer and motor to alert and control the coolant fan. A diagrammatic representation of the workflow of prediction unit (Fig. 5) and control unit (Fig. 4) is shown below.

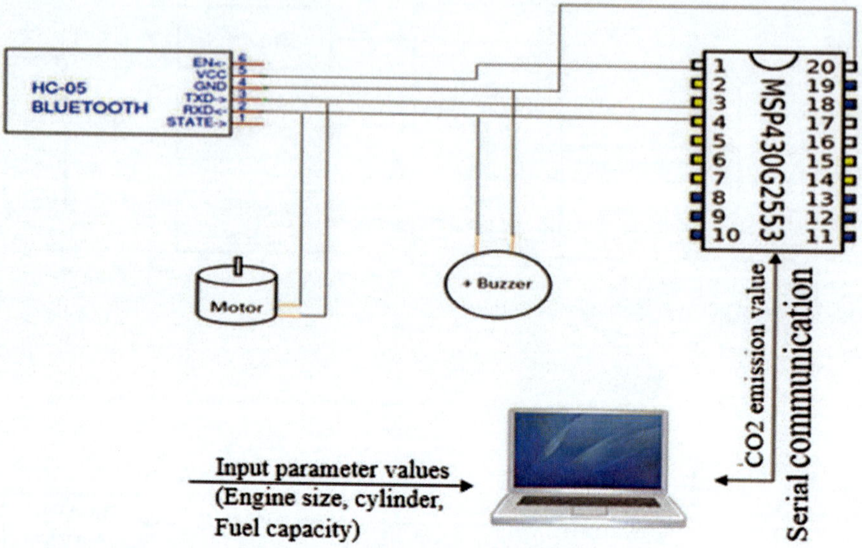

Fig. 3 Schematic diagram of the proposed methodology

4 Experimental Results

The prediction unit of the developed module is designed using Python on Spyder IDE (Intel i5—64bit processor with 8 GB RAM) in which the multiple linear regression algorithm is applied and the accuracy metrics are also evaluated. Then the results are optimized by designing a neural network with various activation functions where the loss and accuracy metrics are improved across different sets of epochs. The obtained results are optimized using the RMS (Root Mean Square) Prop technique where exponential moving average and uses that average to calculate the variance is evaluated at the set epoch value. The different phases involved in designing the neural network for prediction unit, calculating the accuracy, and optimization techniques used are explained below.

Phase 1: Importing Dataset, Extracting Features, and Applying Multiple Linear Regression

The dataset is obtained from IBM Watson and its imported as a.csv file in the Python program using the functions in the pandas package. The null values, garbage values, and the columns with unprocessed and incomplete data are removed by importing them as a dataframe into the Python script. The unnamed columns are then renamed using the df.rename() function so that rows and columns can be easily imported to perform mathematical and numerical functions. Since the dataset obtained from IBM Watson is a large dataset, it could be split and used as training and testing data using df.mask() function. The regression model is imported from the sklearn library where

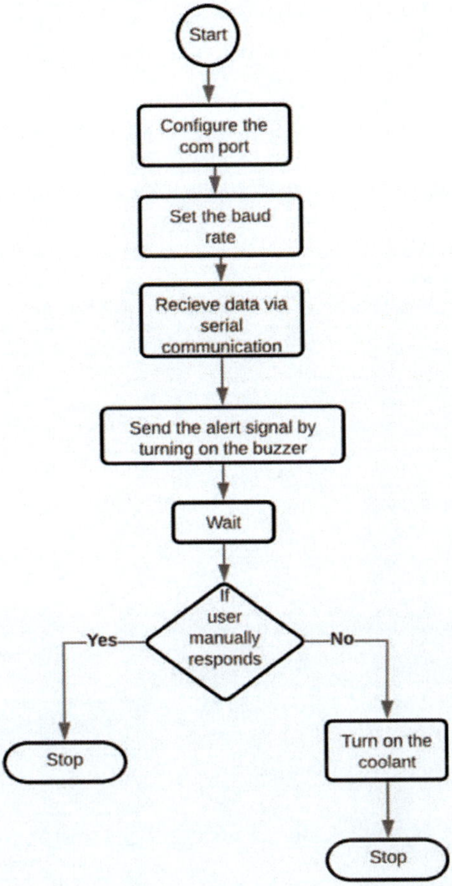

Fig. 4 Flow chart of the control unit

the input parameters extracted from the labeled dataset are engine size, cylinders, and fuel consumption, and the corresponding CO_2 Emission value is set as at the output parameter. **Yi =** $\boldsymbol{\beta 0 + \beta 1.Xi1 + \beta 2.Xi2 + \ldots \beta p.Xip}$ where Yi, Xi is the dependent (CO_2 Emission value) and independent variable (engine size, cylinders, and fuel consumption) [8] and $\beta 0$, βp is intercept and slope values. The coefficient and slope values obtained from the dataset can be used to predict the output (i.e.,) CO_2 emission when a new set of inputs are given in the Python script. The coefficient and intercept values are calculated by using the parameters in the dataset where input is set as Xi and output label is set as Yi and slope values are calculated using the traditional polynomial slope equation. The accuracy metrics are Mean Absolute error, R2 score, and Residual sum of squares are also imported using the sklearn [9] library where the prediction is done and both the variables are tested for the accuracy. MAE is used to measure the mean of magnitude in the errors during predictions of results by taking the average over the test input's absolute differences between prediction and

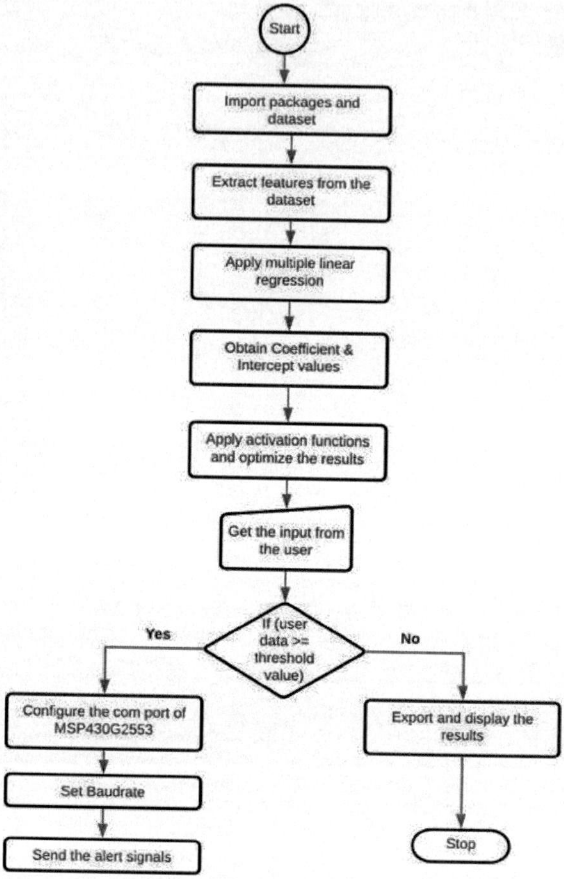

Fig. 5 Flow chart of the prediction unit

actual observation with equivalent weight values without considering their direction. Residual Sum of squares checks for variance by changing the independent variables and R2 score is the coefficient that measure the fit of the regression line. These metrics are commonly used to evaluate the performance and accuracy of machine learning or deep learning-based models. The steps involved in this phase are given below.

1. Import necessary packages like numpy, pandas, keras, sklearn.
2. Preprocess the data.
3. Load the dataset as a dataframe by using pd.read_csv function.
4. Initialize the input parameters (engine size, cylinders and fuel consumption) as float.
5. Split the dataset for training and testing.
6. Apply multiple linear regression by importing linear model from sklearn.
7. Initializing inputs to x.

8. Initializing output to y.
9. Fitting x and y.
10. Printing coefficients.
11. Printing intercepts.
12. Importing mean square error and r2-score from sklearn.
13. Predicting y using x.
14. Testing x.
15. Testing y.
16. Printing Mean absolute error.
17. Printing Residual sum of squares.
18. Printing R2-score.

The following diagram displays the coefficient and intercept values obtained by implementing multiple linear regression in the Spyder IDE and the resultant accuracy metrics are also obtained (Fig. 6).

Phase 2: Designing Neural Network model and Applying Activation Functions

The results and accuracy metrics can be improved and optimized by constructing neural network module with different activation functions like that determines the output of a node in a network. The neural network is designed by importing keras.layers(), keras.models(), and different activation functions like ReLu, Linear, Sigmoid, softmax, TanH from the keras library. ReLu function is an activation function which will return the direct output if it is positive and returns zero otherwise. Sigmoid functions return a real-valued output in the range of 0–1 but take input at any range. Softmax function also returns real values from 0 to 1 with sum of probability equivalent to one. Linear function returns the weighted sum of the neuron which is proportional to the input. TanH is a nonlinear zero centered activation function which returns the output in the range of –1 to 1. The optimizer used is RMSprop

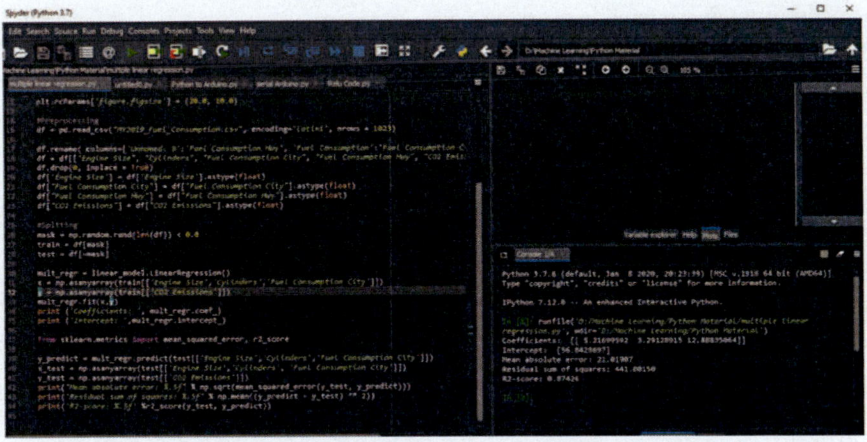

Fig. 6 Implementation of multiple regression using Spyder IDE

optimizer [10] which converges faster and the learning rate is also high. The optimizer uses squares of the gradient $\mathbf{g}t$ into a state vector in which loss is at the rate 0.245, 0.050 s/epoch with $\mathbf{s}t = (1 - \gamma)\mathbf{g}2t + \gamma\mathbf{s}t - 1 = (1 - \gamma)(\mathbf{g}2t + \gamma\mathbf{g}2t - 1 + \gamma 2\mathbf{g}t - 2 + \ldots,)$ where st denotes the state vector, γ represents the learning rate and gt represents the gamma function. The steps to design the neural network and apply the activation functions are given below.

1. Import necessary packages like numpy, pands, keras, sklearn.
2. Preprocess and load the dataset.
3. Initialize the input parameters(engine size, cylinders and fuel consumption) as float.
4. Split the dataset for training and testing.
5. Apply multiple linear regression by importing linear model from sklearn.
6. Design a neural network by importing keras.model() and keras.layer() function.
7. In the model.add() function set the activation function as ReLu.
8. In the model.compile() function set the optimizer as RMSprop.
9. Initializing inputs to x.
10. Initializing output to y.
11. Fitting x and y.
12. Printing coefficients and intercept.
13. Importing mean square error and r2-score from sklearn.
14. Predicting y using x.
15. Testing x and y.
16. Printing Mean absolute error, Residual sum of squares and R2-score.
17. Set another activation function at Step 7 (Linear, sigmoid, softmax, and TanH) and repeat the process.

The diagram below displays the Mean absolute error, Residual sum of squares, and R2-score obtained while using the designed neural network with ReLu activation function (Fig. 7).

Similarly the accuracy metrics are obtained for other activation functions of the neural network and it is shown in the Table 1.

Phase 3: Loss calculation and analysis

Mean square loss is the basic regression-based loss function which calculates the distance between the target values and predicted values. It is implemented in our paper by importing the model.compile() function from the keras library. The loss is evaluated by iterating the process into a different set of epochs using various activation functions (Fig. 8).

Similarly the loss is calculated at different sets of epochs for other activation functions of the neural network and it is shown in the Table 2.

The neural network created using ReLu active function has better accuracy metrics than other models. It has lesser Mean squared Error, Residual sum of squares, and significantly low loss compared to the other models. The Linear model has the best R2 score and it also has other metrics better than TanH, sigmoid and softmax models.

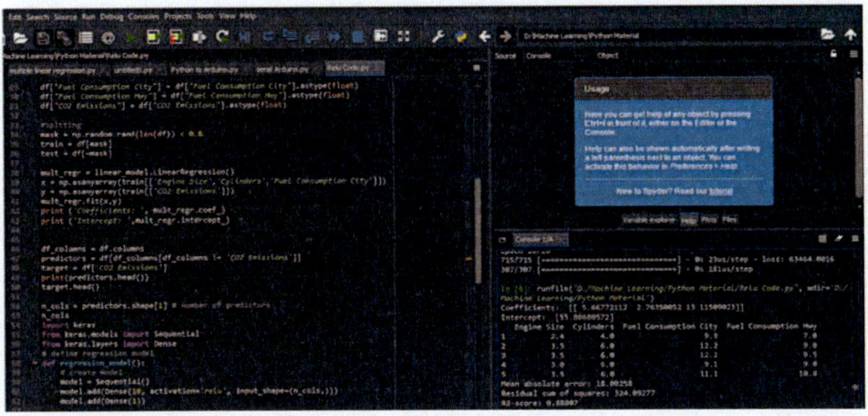

Fig. 7 Design of neural network with ReLu activation function

Table 1 Accuracy Metrics of the designed neural network

	Mean absolute error	Residual sum of squares	R2-score
Multiple linear regression	21.01907	441.80150	0.87426
MLR with ReLu	18.00258	324.09277	0.88007
MLR with Linear	19.49881	380.20362	0.86467
MLR with Sigmoid	18.53642	343.59902	0.90260
MLR with Softmax	18.91685	357.85573	0.89754
MLR with TanH	20.27753	411.17819	0.86083

The optimization and performance improvement of the model at different epochs can be seen in Figs. 9, 10, 11, 12, 13 and 14.

Phase 4: Configuration, Communication, and Control

The prediction unit gets the input from the user and calculates the output by using the predicted coefficient and intercept values from the pre-trained dataset. Then the predicted output is sent to the control unit by serial communication where both the units use the same baud rate and com port. This configuration is done by using pyserial library which is installed using the command Conda install pyserial. If the predicted output denotes a higher level of CO_2 emission, it sends the binary value '1' or it sends a null value. The control unit (TI-MSP430) has motor and buzzer interfaced at its GPIO ports and once it receives the value '1' from the prediction unit, it sends

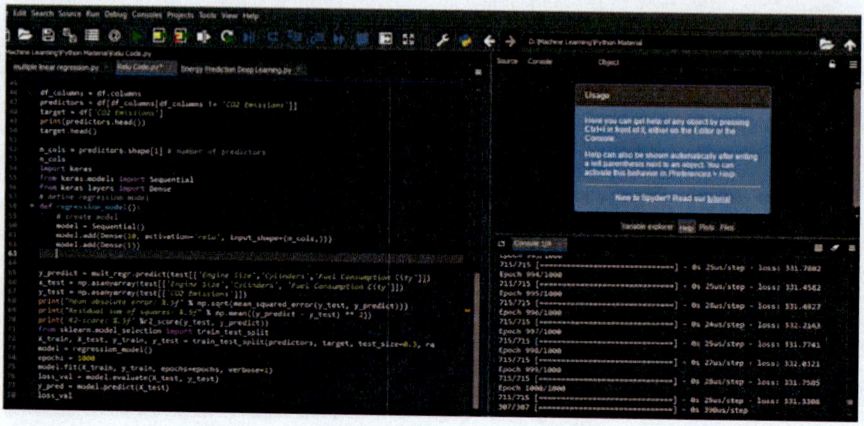

Fig. 8 Performance of the model at 1000 epochs

Table 2 Loss calculation

Activation functions	Epochs	Loss
ReLu	50	3174.9403
	100	463.1492
	500	338.9436
	1000	331.1390
Linear	50	426.4615
	100	398.3633
	500	341.3162
	1000	331.3306
Sigmoid	50	59,724.1829
	100	53,470.9398
	500	26,013.4024
	1000	4070.5647
Softmax	50	63,238.8383
	100	61,901.7859
	500	53,444.4546
	1000	43,748.6531
TanH	50	57,333.2663
	100	51,475.8836
	500	18,644.6101
	1000	2529.3380

Fig. 9 Linear function

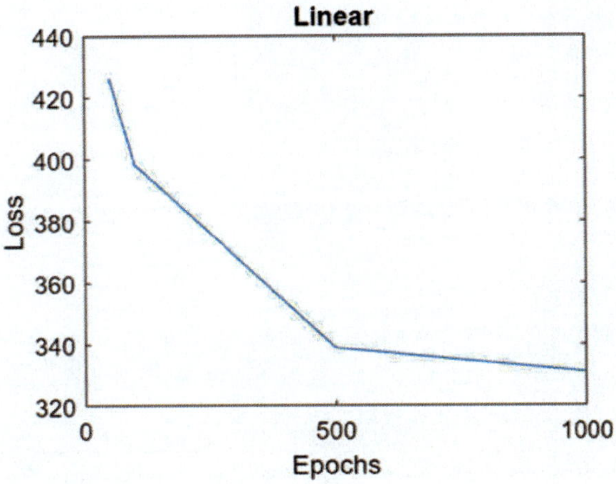

Fig. 10 ReLu function

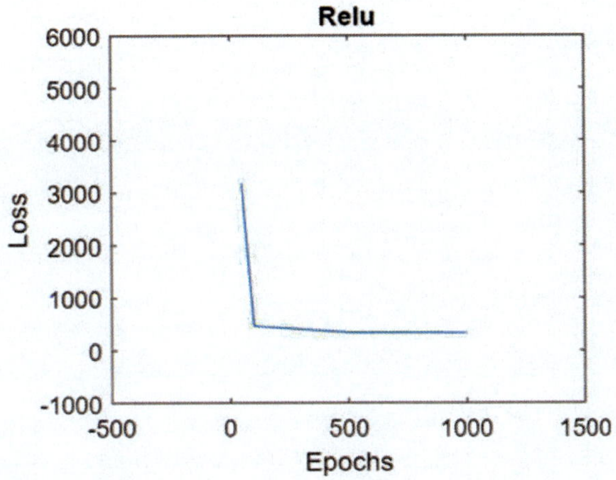

an alert by switching on a buzzer and waits for a manual response. The manual response can be given by the user to switch on the coolant by using the Bluetooth module(HC-05). If there is no manual response, the control unit automatically sends an "HIGH" signal to the motor of the coolant fan to control the source of the CO_2 emission. The result of the prediction unit and the hardware setup of the control unit is shown in the Figs. 15 and 16.

Fig. 11 Softmax function

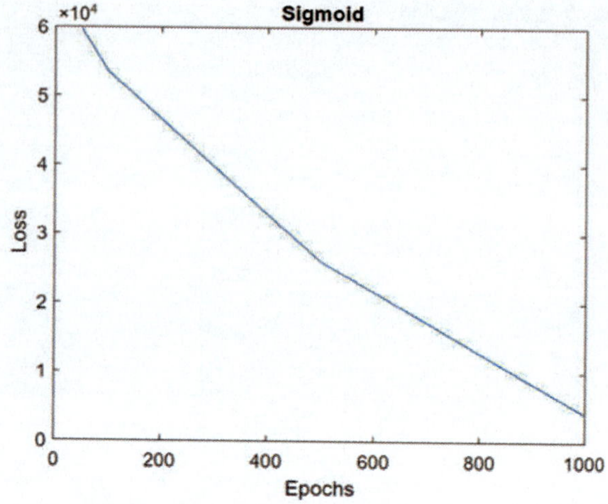

Fig. 12 Sigmoid function

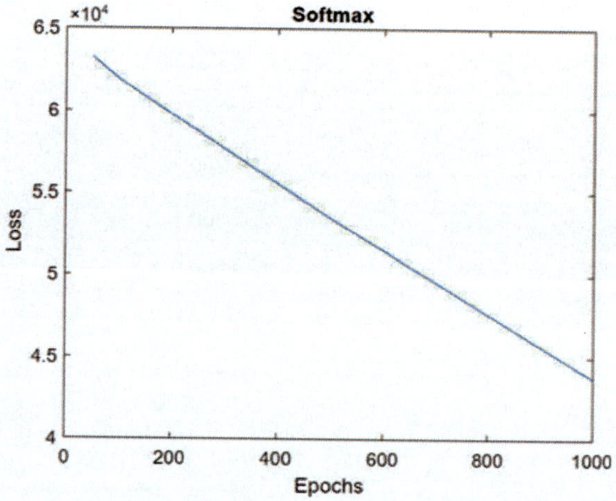

5 Conclusion

In this paper, an autonomous module was developed to determine output values using multiple linear regression, optimize the results, improve the accuracy metrics and use the predicted results to control the functional units. The accuracy of prediction was evaluated for the neural network model with different activation functions. The model with the ReLu activation function has better accuracy compared to other models with the mean square loss of 331.13. The ReLu model also has the best residual sum of squares value of 18.002 and mean absolute error value of 324.092. The Linear model also has better performance with the R2 score of 0.86467 which is comparatively greater than the ReLu Model. The TanH model has lesser accuracy of prediction

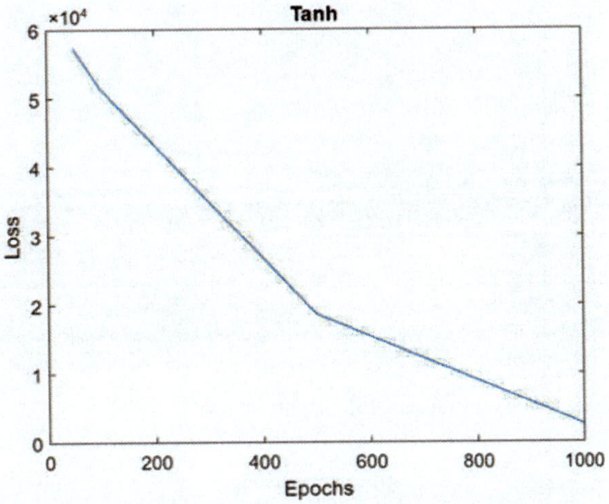

Fig. 13 Tanh function

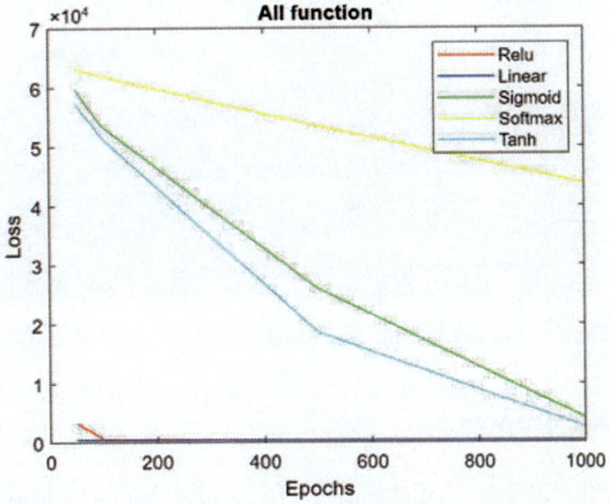

Fig. 14 Performance of all the functions

compared to the other models. In this paper the linear functions (i.e.,) ReLu and Linear model has significantly better performance metrics than non-linear functions like TanH. This paper can be further extended by using other embedded network protocols or different types of optimizers to improve the results.

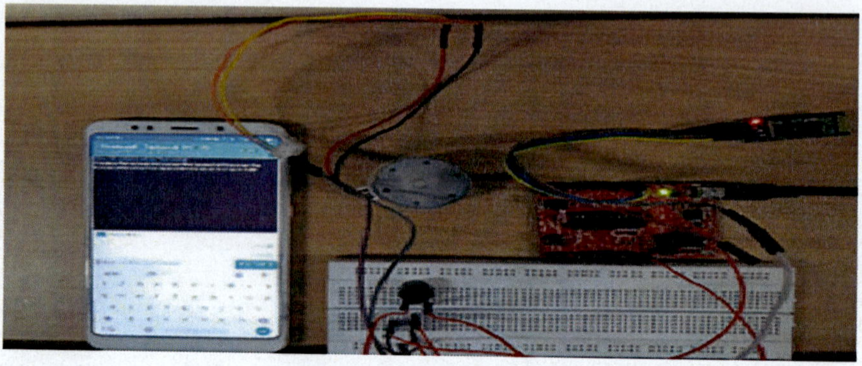

Fig. 15 Hardware setup of the control unit

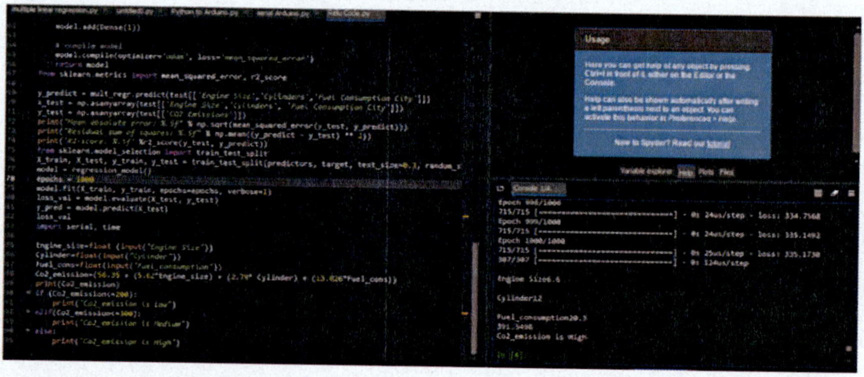

Fig. 16 Prediction using user's input

References

1. Subramanya, G.: Dynamic branch prediction for embedded system applications. In: International Conference on Communication and Electronics Systems (ICCES), pp. 966–969 (2019)
2. Muninathan, K., Vishnu Prasanna, D., Gowtham, N., Jerickson Paul, B., Vikneshpriya, K.: Experimental and numerical studies on catalytic converter to reduce CO_2 emissions in IC engines. In: International Conference on System, Computation, Automation and Networking (ICSCAN) (2020)
3. Wenjun, H., Marek, P.: A novel machine learning algorithm to reduce prediction error and accelerate learning curve for very large datasets. In: IEEE 49th International Symposium on Multiple-Valued Logic (ISMVL) (2019)
4. Kirci, P.: A novel model for vehicle routing problem with minimizing CO_2 emissions. In: 3rd International Conference on Advanced Information and Communications Technologies (AICT) (2019)
5. Wang, Y., Yang, S.: The prediction of CO_2 emissions from manufacturing industry based on GM(1, N) model and SVM in chongqing, In: International Conference on Sensing, Diagnostics, Prognostics, and Control (SDPC) (2018)

6. Kadam, P., Vijayumar, S.: Prediction model: CO_2 emission using machine learning. In: 3rd International Conference for Convergence in Technology (I2CT) (2018)
7. Salehl, C., Dzakiyullah, N.R., Bayu Nugroho, J.: Carbon dioxide emission prediction using support vector machine. In: IOP Conference Series: Materials Science and Engineeri, pp. 114 (2016)
8. https://towardsdatascience.com/a-look-at-gradient-descent-and-rmsprop-optimizers-f77d483ef08b
9. https://scikit-learn.org/stable/modules/classes.html
10. https://keras.io/api/optimizers/rmsprop/

Intelligent Transportation System Applications: A Traffic Management Perspective

Varsha Bhatia, Vivek Jaglan, Sunita Kumawat, Vikas Siwach, and Harkesh Sehrawat

Abstract The rapid increase in population has increased the traffic density in urban areas. In current scenario, the major challenges faced by transportation systems are congestion, accidents and pollution control, etc. In recent decade's development of Internet of Things (IoT) and communication technologies and connected devices have paved way for the involvement of information and communication in all aspects of smart city. IoT has transformed the transportation system by enhancing customer experience, safety, and operational performance. The smart or intelligent transportation system is possible that can provide in vehicle transportation and navigation system and assistance. Various new techniques are required to further enhance the intelligence and capabilities of applications. Cyber-physical system is an integral part of intelligent transportation system. CPS can help in gathering information in distributed manner over various diverse networks and analysing useful information to control operations of Intelligent Transportation system. This paper provides the description of the main elements of traffic management system in intelligent transportation systems. The focus is on the algorithms used to increase the efficiency of transport management in Intelligent Transportation System.

Keywords Intelligent transportation system (ITS) · Connected vehicles · Intelligent transportation system applications · Traffic management system · Traffic data sources · Deep learning algorithms

---

V. Bhatia (✉) · S. Kumawat
Amity University, Gurugram, Haryana, India

S. Kumawat
e-mail: skumavat@ggn.amity.edu

V. Jaglan
Graphic Era Hill University, Dehradun, India

V. Siwach · H. Sehrawat
University Institute of Engineering and Technology, MD University, Rohtak, India

J. S. Raj et al. (eds.), *Intelligent Sustainable Systems*, Lecture Notes in Networks and Systems 213, https://doi.org/10.1007/978-981-16-2422-3_33

1 Introduction

An intelligent transportation system can be a technology application or a platform that integrates information, compution, and advanced communication technologies to solve various transport-related issues. The aim of ITS is to improve the quality of transportation by utilizing applications capable of monitoring, managing, and enhancing transportation system. ITS is crucial in reducing traffic congestion, accidents, fuel consumption, and air pollution.

ITS system uses diverse data sources such as smart card, RFID tags, sensors, video cameras, GPS and social media, etc. to collect required information. The data collected is large in volumes and generally disorganized in nature, in current scenario the data collected from ITS can be in trillion byte to Peta byte. In order to deal with vast amount of complex data new and efficient data processing systems are required.

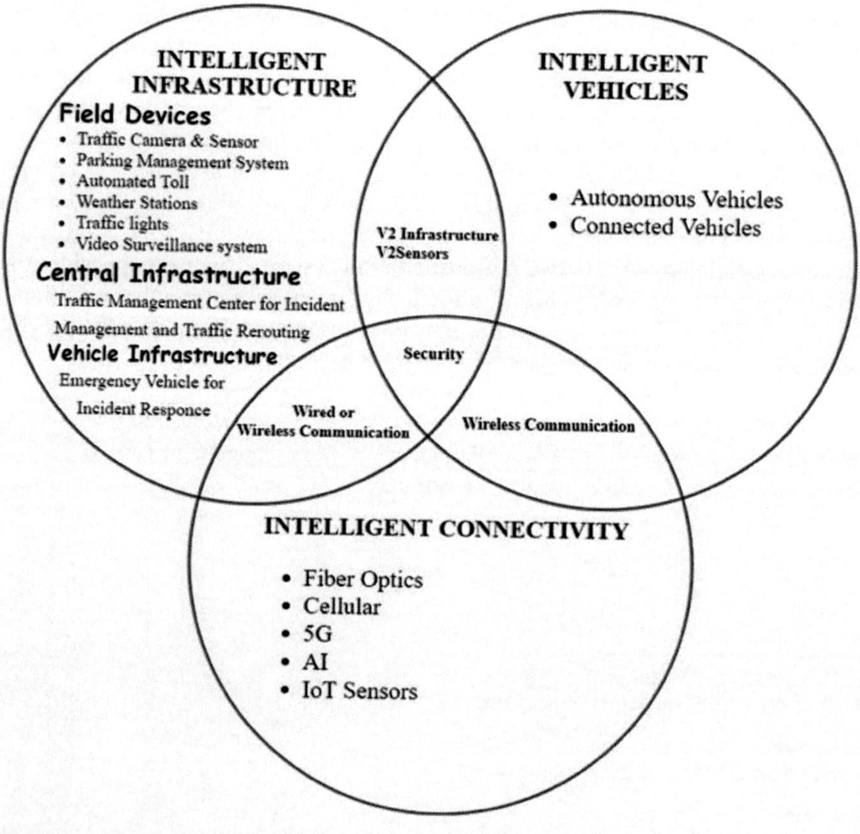

Fig. 1 Interactions among different constituent of ITS

Intelligent Transportation System is based on some fundamental components Intelligent Infrastructure, Intelligent Vehicles, and Intelligent Connectivity [1]. Figure 1 depicts each constituent in detail.

Intelligent Connectivity is combination of fibre optics, cellular, 5G technologies, Artificial Intelligence, and IoT devices. Intelligent Infrastructure can be broadly classified into three types field infrastructure, central infrastructure, and vehicle infrastructure. Intelligent Vehicles are connected vehicles or autonomous vehicles. ITS applications include Traveller information management, Public vehicle management, transportation management, vehicle control system, and logistic vehicle management [2]. ITS collects large amount of data from varied sources and volume data generated is still expanding. Many sub-systems in ITS operate in real time and they have to collect, analyse, and provide decisions for real-time traffic management.

Some of the benefits of ITS are

- Selecting shortest route from source to destination.
- ITS can be used to detect road anomalies by analysing data from sensors attached to car or mobile phone to prevent road accidents.
- Slow moving vehicle warning, Speed Control, and Collision warning.
- Smart Parking, Driver assistance.
- Trip time prediction.
- Traffic congestion information.

The benefits are vast and still expanding. The most integral part is the inference drawn from traffic information and its implementation into physical system to obtain the required results. In this work the key elements and techniques of ITS and algorithms used in road transportation system are discussed.

1.1 Traffic Data Sources

In order to plan and implement intelligent transportation system, different data sources are used to collect traffic information Some of these data sources [2, 3] for ITS are:

Smart Card: They are used to gather and analyse individual traffic pattern. The smart card data can be useful in predicting the traffic flow based on travel direction, start and end time, frequency of visit to destination, etc.

GPS and GIS: These are both sources of collecting travel data. It helps in collecting real-time data of traffic distance, time, and speed of the vehicle. This method of collecting data is expensive and can be incomplete as it requires all subjects to have GPS.

Mobile Phone: It can provide individual level data which can be more accurate. It will help creating user personal records. The major drawback of mobile phone-based data collection is the cleaning of the large volume of data generated, and its dependency on power and charging of devices.

Call Detail Record: It is a more regular and economical way of collecting travel information. CDR is the normal way of billing by mobile phone carrier. CDR store time-stamped coordinates of phone user hence are helpful in capturing users individual trip routes.

Traffic Flow Sensors: With the advancement in low cost technologies, it is now possible to instal sensors and detectors on important locations to collect traffic information. Traffic flow sensors are the device capable of capturing presence or passage of vehicles at a location and determining traffic states and parameters. The Traffic sensors can be installed either in embedded or attached to road surfaces or can be mounted above the surfaces like tall poles or traffic lights etc. Some of the sensors used are microwave radar sensors, Video Image processors (VIP), ultrasonic and acoustic infrared sensors, etc. Detection system can be built on combination of traffic flow sensors. These devices can help in collecting information about the volume of traffic per hour and average speed of vehicles, etc.

Data From Passive Collection: Social media data is an example of passive data collection, which includes applications and websites where user interacts with each other to share views and information. Some of the examples of these applications include twitter, Facebook, Linkedin, etc.

Other Sources: Other possible sources include Floating Car sensors, Airborne sensors, connected autonomous vehicles, etc. The sensors embedded in intelligent vehicle can be used for traffic signs detection and recognition to help drivers in weather conditions [4].

1.2 Communication Technologies

The vehicular communication in ITS can be by the standalone sensors and sensor systems or through vehicles sharing real-time information and messages [5]. The standalone sensors can help in alerting about the rear end or speed of driver, etc. The vehicle sharing information in real time will be helpful in sharing traffic conditions and road environment. Vehicles can exchange information using peer to peer, Adhoc networks, or road side infrastructure [6]. The next generation smart transportation comprises connected vehicles (CV). This technology will pave way for intelligent and green transportation in smart transportation. Connected vehicles are vehicles capable of connecting and communicating wirelessly with their external and internal environments [7]. The wireless is the most preferred communications' technologies for ITS which includes cellular network, Bluetooth, Wi-Max, Wi-Fi, infrared, etc. The communications in ITS can be Vehicle to sensors (V2S), Vehicle to Road infrastructure (V2R), vehicle to vehicle (V2V), and vehicle to internet (V2I).

1.3 Security

Information, Intelligence, and communication are three pillars of ITS ecosystem. Security in ITS refers to securing the transportation infrastructure and information to ensure reliable and efficient transportation. Security becomes imperative as the ITS is evolving with new technologies that are generally prone to attacks. These attacks can cause major disruption in ITS. ITS applications share a lot of information about traffic conditions flow of traffic and management. The security of this information is also essential as cyber attacks and breaches are inevitable. The major concern should be confidentiality and authentication [8] of data. Efficient strategies must be used to prevent and recover from cyber attacks and security breaches with minimum loss.

2 ITS Applications

The goal of ITS to improve traffic management and effectively utilize the existing transportation infrastructure. ITS services can be divided into different groups based on their basic technical purpose. ITS services can be classified into four groups that are Efficiency applications, Autonomous driving, Safety / Security, and Comfort applications. These groups are further sub-divided into various groups comprising of ITS user services. Efficiency applications comprises traffic management and road monitoring applications. Safety and security applications include vehicle identification, Incident management, and collision avoidance. Infotainment and comfort applications provide internet connectivity and information about weather, restaurants parking, etc. Figure 2 depicts ITS applications' groups and sub-groups.

ITS applications intend to minimize accidents and reduce pollution. This paper intends to cover algorithms that models, prediction, and plan traffic management system to facilitate intelligent transportation systems (ITS).

3 Algorithms Used for Traffic Management Systems (TMS)

Some of the algorithms which are more prominently used in Intelligent transportation management system are described in following paragraphs.

3.1 Artificial Neural Network (ANN)

Neural network are nonlinear in nature and are generally used to determine the approximate functioning of the system. ANN observes nodes as artificial neuron similar to neurons present in our brain. The artificial neurons consist of inputs which

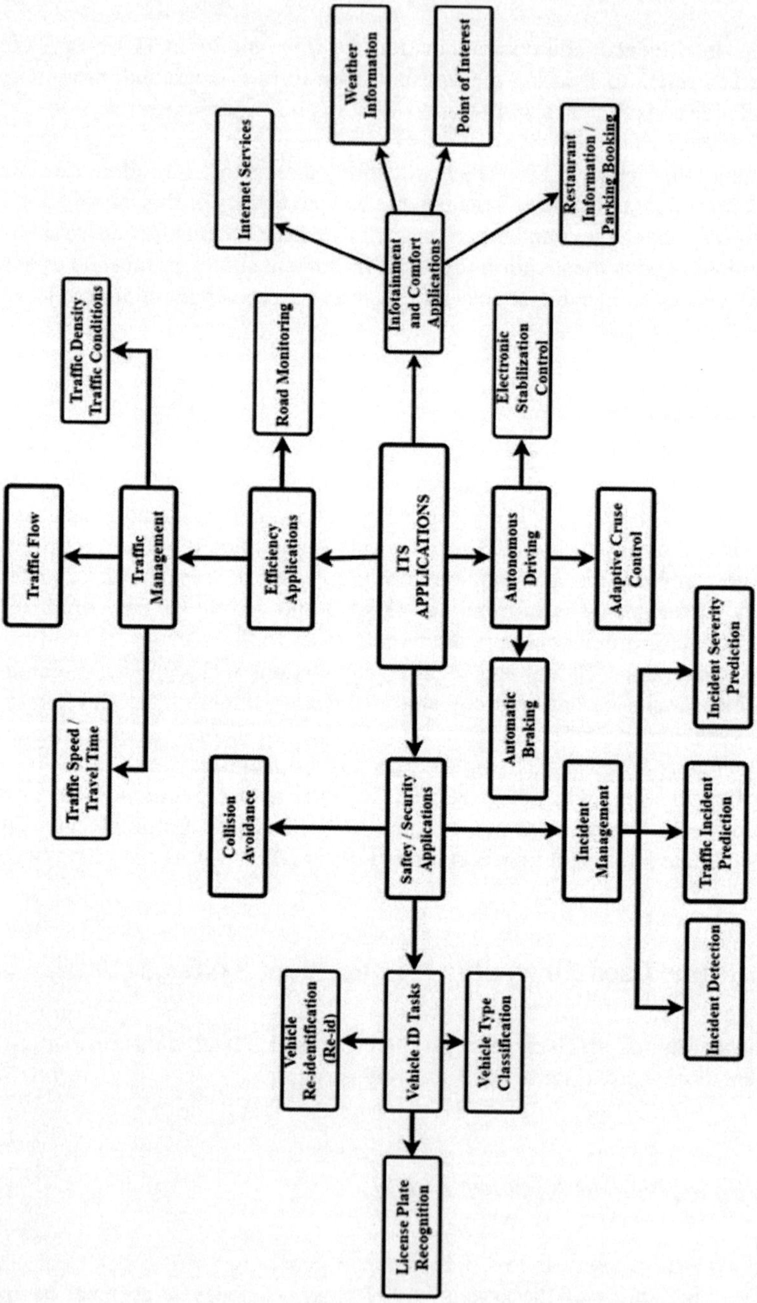

Fig. 2 Intelligent transportation system (ITS) applications

are multiplied by weights and computed using activation function. ANN forms a layered network of neurons to process information. ANN is effectively used to detect and localize damage in transportation infrastructure. ANN can be helpful for optimization and periodic maintenance of transportation infrastructure [9], road planning, public transport, prediction traffic condition, and traffic incident detection [10].

3.2 Convolution Neural Network (CNN)

CNN is a type of deep neural network preferably used to process multiple array data like video, image text, or audio. The CNN comprises three main sub-structures convolution layer, pooling layer, and fully connected layer. The convolution layer consists of various feature maps wherein each feature map is made by convoluting a small region of input data. The pooling layer also known as sub-sampling layer is used for dimensionality reduction of feature maps obtained from convolution layer and selecting most significant feature. Fully connected layer is the final layer that composites and aggregates the information obtained from previous layers. CNN is mainly used for object detection and image classification.

3.3 Recurrent Neural Network (RNN)

RNN is supervised learning algorithm to capture dynamic sequence of data. A single layer RNN can be represented as input layer, hidden layer, and an output layer. RNN stores recent input and provides the same through a feedback in order to predict the output of the layer at next time step.

Each state depends on the current input and state of network in previous time step. Two major problems faced while training RNN are exploding and vanishing gradient problem. In order to deal with these problems long short-term memory (LSTM) architecture is used. LSTM contains memory cell to store the state for a definite time, the role of this cell can be regulated to control its effect on network. RNN is used for natural language processing, time series prediction, and speech recognition. Various variants of LSTM architecture are possible and they can perform better than basic RNN model.

3.4 Deep Belief Network (DBN)

DBN is an unsupervised learning algorithm based on restricted boltzman machines. It is a combination of probability and statistics with machine learning. The nodes in each layer are connected to all nodes in previous or subsequent layer. A network must have minimum two hidden layers to be considered as DBN. The hidden layer performs dual role it acts as hidden for the layer precedent to it and acts as visible layer to the succeeding layer [11]. The goal is to help the system to classify data into different categories. DBN are used for motion capture, video recognition, and image recognition.

3.5 Auto Encoder (AE)

AE is an unsupervised training model similar to principal component analysis model. The model works on the principal of compression of input data and its subsequent reconstruction. AE is a form of neural network that uses backpropagation technique, which tries to learn an approximation of identity function so as output values are similar to input values. AE is of two kinds, i.e., under complete and over complete autoencoders. The aim is to learn salient features of input. Under complete AE has hidden layer dimensions less than that of input parameters and learning is achieved by minimizing the loss function. Over complete AE has a high dimension hidden layer than input layer. It aims of learning in overcomplete, AE is to copy input to output [12]. AE is used for traffic flow prediction, object detection, etc.

4 Algorithms in ITS Based on Traffic State Variables

The traffic management group of services share real-time information to improves efficiency and safety of transportation system. Some of the user services offered in this group are traffic control, incident management, emission control, route guidance, ride sharing, travel time prediction, parking management, etc.

The traffic management systems are based on estimation of traffic state variables. The most important traffic state variable is traffic flow, travel time, and traffic speed. The traffic state variables are dependent on each other; hence it is possible that same method can be used to predict more than one state variable. The traffic state variables are predicted considering various parameters obtained from data sources installed in the region of interest, through GPS or social media, etc. The prediction results can be classified as short-term, medium-term, or long-term based on the duration of prediction. In the following section, a brief review on techniques used for estimation of traffic state variables is covered.

4.1 Travel Time Prediction

Travel time is average time taken to complete the journey. Travel time prediction is a challenging problem of advanced traveller information system components of ITS. The work [13] implements a deep learning-based Auto Encoders (AE) model for continuous travel time prediction problem. The prediction model considers signal time as factor and results are satisfactory with mean absolute error less than 4 s. Siripanpornchana et al. [14] implemented travel time prediction method based on deep belief network (DBN) where generic features of the system are learned in unsupervised learning and predictions are done using sigmoid regression in a supervised way. Short-term travel time prediction is useful for accuracy and reliability of route selection in ITS applications. The work [15] predicts short-term travel time using different settings for 16 hyper parameters using long short-term memory neural network with deep neural layer (LSTN-DBN). The LSTN-DNN is compared with other linear models such as ARIMA, linear regression, Ridge and Lasso regression, and DNN models. Hou and Edara [16] proposed two deep learning model one using CNN and other based on Long Short-Term Memory (LSTM) to predict travel time in a road network.

4.2 *Traffic Flow*

Traffic flow or volume can be defined as a number of vehicles crossing lane/highway in specified time interval. MF-CNN [17] is CNN-based multi-feature prediction model. The model considers multiple spatiotemporal features, holidays, and weather conditions as parameters. The MF-CNN model performs better as compared to five base-line models considered in the work. A Stack Auto Encoder Neural Network (SAE-DNN) is implemented for predicting short-term traffic flow [18]. The SAE model is used to mine useful information from historical data and the result from SAE model is used by DNN model to predict traffic flow for next period. The work in [19] uses long short-term memory Recurrent Neural Network (LSTM RNN) to predict traffic flow by determining optimal time lag dynamically. A machine learning-based short-term traffic flow prediction model is proposed [20]. The model considers speed, occupancy, traffic flow, and time of day as parameters for prediction. A model for short-term traffic prediction for long road was proposed in [21]. The model uses irreducible and a periodic Markov chain to model dynamics of traffic flow on road network. In [22] a traffic forecasting algorithm using Graphical CNN (GCNN) is proposed for urban areas.

4.3 *Traffic Speed*

Traffic speed prediction is crucial for incidence response management. Nguyen et al. [23] present a deep learning-based decision support system for traffic speed prediction for multiple road segments, the model is capable of predicting traffic speed in arterial road network from 5 to 30 min in advance. Convolution Neural network (CNN) based architecture is used to learn the sequence and capture dependencies in dataset. The long short-term memory units (LSTM) are used to learn temporal correlations with long-term dependencies. The work in [24] uses deep long short-term memory neural network (DLSTM-NN) to estimate traffic speed using cellular data under extreme road conditions.

4.4 Traffic Conditions

Traffic conditions include congestion and jams which are often seen in urban scenarios. Guo et al. [25] proposed a hybrid deep learning architecture using combined three-dimensional CNN (3DCNN) with normal CNN and Recurrent Neural Network (RNN). The model uses CNN-RNN for temporal and spatial features and three-dimensional CNN for traffic state prediction for large-scale road network. The work [26] uses camera images to detect traffic congestion. The work proposes two approaches based on deep learning one using you only look once (YOLO) based on simple CNN and other one using Deep CNN(DCNN). The result of both these algorithms was compared with Support Vector Machine (SVM) based algorithms. The PCNN [27] algorithm is proposed to detect short-term traffic prediction using both real time and historic data. The algorithm uses deep convolution network.

4.5 Traffic Density Estimation

Traffic density estimation is the key factor in determining traffic flow. Traffic density is defined as a number of vehicle on highway or lane and expressed as vehicles/km. Traffic density estimation can help in reducing environmental pollution by minimizing traffic jams at junctions. The reduction in time and cost for users and companies. The work in [28] proposes algorithm based on CNN for traffic density estimation. The result of CNN-based algorithm is used by traffic control algorithms at the junction for better traffic control. Chen et al. [29] propose a Hidden Markov Model(HMM) for real-time traffic estimation over various weather and illumination situations. The HMM model is initialized and constructed using auto-class clustering technique.

4.6 Vehicle Classification

Vehicle classification refers to identifying colour and type of vehicles. Vehicle classification using CNN-based method is implemented [30], the method uses type and colour as attributes to classify vehicles. The attribute type consists of four classes and colour consists of seven classes. An automated video-based vehicle detection and identification system are implemented in conjunction with an electronic toll plaza (ETP) [31]. The identification system is based on CNN for classifying trucks, cars, and buses. Automated vehicle identification is performed using K nearest neighbour and decision tree-based classifier [32]. The image data is obtained using light curtain installed in each lane.

4.7 Traffic Signal Timing

Traditional traffic signals' timers have fixed time period to switch traffic in different directions, but this technique does not consider traffic density at real time into consideration. The users have to wait even if the traffic density is low at the intersection. With the help of ITS traffic signal timings can be adaptively controlled according to traffic density. The work in [33] proposes a neural network-based approach to optimize the traffic signal timings to maintain a normal traffic flow and avoid congestion. The method uses historical data at intersection,

time series, and environment variables to optimize traffic signal timings. The work in [34] implements traffic signal timing optimization using reinforcement learning. A multi-agent reinforcement learning-based framework [35] for automatic optimization of traffic light controller. The framework is able to discover control strategies on the basis of local traffic information, a probabilistic model of car behaviour, and a learned value function.

4.8 Traffic Incident Risk and Detection

Traffic incident risk for a particular location can be calculated using accident records and surveillance camera data. Traffic incident risk variable will help in reducing incident in dangerous locations by alerting the drivers. Traffic incident detection is also possible using social media data and surveillance cameras. The work [36] implements two algorithms using Deep belief network (DBN) and long short-term memory (LSTM) to predict traffic accidents from social media data. The algorithms use social media platform Twitter to detect traffic accidents.

4.9 Summary of Algorithms

Table 1 summarizes algorithms used based for state variable prediction and applications of ITS and possible data sources to collect the required information from the respective environment.

Table 1 Algorithms used in intelligent transportation system

State variables/applications	Possible sources of data	Algorithm used	References	Year
Travel time prediction	GPS, cell phone tracking Roadside sensor	Deep belief network (DBN) Long short-term Network (LSTM) + DBN DBN	Xiong et al. [13] Siripanpornchana et al. [14] Liu et al. [15] Hou and Edara [16]	2015 2016 2017 2018
Traffic flow	Smart card Floating car sensors Wide area sensors	Stack auto encoder deep neural network (SAE-DNN) LSTM recurrent neural network (LSTM RNN) Deep neural network Graph convolution network (GCN) Graphical CNN (GCNN)	Yang et al. [17] Zhao et al. [18] Tian, and Pan [19] Mohammed, and Kianfar [20] Zhang et al. [21] Kumar [22]	2019 2019 2015 2018 2019 2020
Traffic speed	Connected vehicles GPS Roadside sensor and video	DLSTM-NN	Nguyen et al. [23] Ding et al. [24]	2019 2019
Traffic conditions	Roadside sensor Wide area sensor CCTV	YOLO + DCNN CNN	Guo et al. [25] Chakraborty et al. [26] Chen et al. [27]	2019 2018 2018
Traffic density estimation	GPS, road side sensor, surveillance camera		Nubert et al. [28] Chen et al. [29]	2018 2009
Vehicle Classification	Floating car sensor, surveillance camera, automated video from electronic toll plaza	CNN K nearest neighbour	Maungmai and Nuthong [30] Wong et al. [31] Sarikan et al. [32]	2019 2020 2017
Traffic signal Timing	Video Road side sensor	RL MRL	Lawe et al. [33] Bakker et al. [35]	2016 2020 2010
Traffic incident risk	Social media data Accident data	RF	Zhang et al. [36] Zuccarelli [37]	2018 2020

5 Conclusion

The work depicts the recent algorithm used for traffic management and vehicle identification. The algorithms were grouped based on traffic state variables required for classification and prediction. The most popular among them are Convolution Neural Network (CNN), Recurrent Neural Networks(RNN), and hybrid model based on CNN + RNN. Two most common data sources are cameras and Loop detectors. The cameras are installed at traffic intersection, in-vehicle, and autonomous vehicles. Cameras are cheap and easy to instal.

However, security is a major concern in IoT-enabled transportation solutions as they are prone to cyberattacks. The major hurdle lies in dealing with large amount of data generated by sensors and video cameras, etc. The vast amount of data generated requires robust data storage, analytics, and data collection strategies. Another important aspect is development of few more benchmark dataset to facilitate comparison of various models and algorithms.

Hence, for designing ITS models efficient data storage and management strategies are required. The machine learning and deep learning techniques are the future of intelligent transportation system. Future work involves predicting traffic characteristic to predict shortest path and travel time in smart transportation system.

References

1. ITS Infrastructure|RNO/ITS—PIARC (World Road Association). https://rno-its.piarc.org/en/intelligent-transport-systems-what-its2100-website-basic-its-concepts/its-infrastructure. Accessed 20 Jan 2021
2. Zhu, L., Yu, F.R., Wang, Y., Ning, B., Tang, T.: Big data analytics in intelligent transportation systems: a survey. IEEE Trans. Intell. Transport. Syst. **20**, 383–398 (2019). https://doi.org/10.1109/TITS.2018.2815678
3. Dabiri, S., Heaslip, K.: Transport-domain applications of widely used data sources in the smart transportation: a survey (2018). arXiv:1803.10902
4. James, H.: Computer vision based traffic sign sensing for smart transpor. J. Innov. Image Process. 1, 11–19 (2019). https://doi.org/10.36548/jiip.2019.1.002
5. Lytrivis, P., Amditis, A.: Intelligent transport systems: co-operative systems (vehicular communications). Wirel. Commun. Netw. Recent Adv. (2012). https://doi.org/10.5772/34970
6. Rawat, D.B., Bajracharya, C., Yan, G.: Towards intelligent transportation cyber-physical systems: real-time computing and communications perspectives. SoutheastCon **2015**, 1–6 (2015). https://doi.org/10.1109/SECON.2015.7132923
7. Lu, N., Cheng, N., Zhang, N., Shen, X., Mark, J.W.: Connected vehicles: solutions and challenges. IEEE Internet Things J. **1**, 289–299 (2014). https://doi.org/10.1109/JIOT.2014.2327587
8. Harvey, J., Kumar, S.: A survey of intelligent transportation systems security: challenges and solutions. In: 2020 IEEE 6th Intl Conference on Big Data Security on Cloud (BigDataSecurity), IEEE International Conference on High Performance and Smart Computing, (HPSC) and IEEE International Conference on Intelligent Data and Security (IDS), pp. 263–268 (2020). https://doi.org/10.1109/BigDataSecurity-HPSC-IDS49724.2020.00055
9. Gharehbaghi, K.: Artificial neural network for transportation infrastructure systems. MATEC Web Conf. **81**, 05001 (2016). https://doi.org/10.1051/matecconf/20168105001

10. Abduljabbar, R., Dia, H., Liyanage, S., Bagloee, S.A.: Applications of artificial intelligence in transport: an overview. Sustainability **11**, 189 (2019). https://doi.org/10.3390/su11010189
11. Haghighat, A.K., Ravichandra-Mouli, V., Chakraborty, P., Esfandiari, Y., Arabi, S., Sharma, A.: Applications of deep learning in intelligent transportation systems. J. Big Data Anal. Transp. **2**, 115–145 (2020). https://doi.org/10.1007/s42421-020-00020-1
12. Types of Autoencoders—CellStrat. https://www.cellstrat.com/2017/11/01/types-of-autoencoders/. Accessed 20 Jan 2021
13. Xiong, G., Kang, W., Wang, F., Zhu, F., Yisheng, L., Dong, X., Riekki, J., Pirttikangas, S.: Continuous Travel Time Prediction for Transit Signal Priority Based on a Deep Network (2015). https://doi.org/10.1109/ITSC.2015.92.
14. Siripanpornchana, C., Panichpapiboon, S., Chaovalit, P.: Travel-time prediction with deep learning. In: 2016 IEEE Region 10 Conference (TENCON) (2016). https://doi.org/10.1109/TENCON.2016.7848343
15. Liu, Y., Wang, Y., Yang, X., Zhang, L.: Short-term travel time prediction by deep learning: a comparison of different LSTM-DNN models. In: 2017 IEEE 20th International Conference on Intelligent Transportation Systems (ITSC), pp. 1–8 (2017). https://doi.org/10.1109/ITSC.2017.8317886
16. Hou, Y., Edara, P.: Network scale travel time prediction using deep learning. Transp. Res. Rec. **2672**, 115–123 (2018). https://doi.org/10.1177/0361198118776139
17. Yang, D., LI, S., Peng, Z., Wang, P., Wang, J., Yang, H.: MF-CNN: Traffic flow prediction using convolutional neural network and multi-features fusion. IEICE Trans. Inf. Syst. **E102**(D), 1526–1536 (2019). https://doi.org/10.1587/transinf.2018EDP7330
18. Zhao, X., Gu, Y., Chen, L., Shao, Z.: Urban short-term traffic flow prediction based on stacked autoencoder. 5178–5188 (2019). https://doi.org/10.1061/9780784482292.446
19. Tian, Y., Pan, L.: Predicting short-term traffic flow by long short-term memory recurrent neural network. In: 2015 IEEE International Conference on Smart City/SocialCom/SustainCom (SmartCity) (2015). https://doi.org/10.1109/SmartCity.2015.63
20. Mohammed, O., Kianfar, J.: A machine learning approach to short-term traffic flow prediction: a case study of interstate 64 in Missouri. In: 2018 IEEE International Smart Cities Conference (ISC2), pp. 1–7 (2018). https://doi.org/10.1109/ISC2.2018.8656924
21. Zhang, Y., Cheng, T., Ren, Y.: A graph deep learning method for short-term traffic forecasting on large road networks. Comput.-Aided Civ. Infrastruct. Eng. **34**, 877–896 (2019). https://doi.org/10.1111/mice.12450
22. Kumar, D.: Video based traffic forecasting using convolution neural network model and transfer learning techniques. J. Image Process. **2**, 128–134 (2020). https://doi.org/10.36548/jiip.2020.3.002
23. Nguyen, H., Bentley, C., Kieu, L.M., Fu, Y., Cai, C.: Deep learning system for travel speed predictions on multiple arterial road segments. Transp. Res. Rec. **2673**, 145–157 (2019). https://doi.org/10.1177/0361198119838508
24. Ding, F., Zhang, Z., Zhou, Y., Chen, X., Ran, B.: Large-scale full-coverage traffic speed estimation under extreme traffic conditions using a big data and deep learning approach: case study in China. J. Transp. Eng. Part A: Syst. 145, 05019001 (2019). https://doi.org/10.1061/JTEPBS.0000230
25. Guo, J., Liu, Y., Wang, Y., Yang, K.: Deep learning based congestion prediction using PROBE trajectory data (2019). https://doi.org/10.1061/9780784482292.271
26. Chakraborty, P., Adu-Gyamfi, Y.O., Poddar, S., Ahsani, V., Sharma, A., Sarkar, S.: Traffic congestion detection from camera images using deep convolution neural networks. Transp. Res. Rec. **2672**, 222–231 (2018). https://doi.org/10.1177/0361198118777631
27. Chen, M., Yu, X., Liu, Y.: PCNN: Deep convolutional networks for short-term traffic congestion prediction. IEEE Trans. Intell. Transp. Syst. 1–10 (2018). https://doi.org/10.1109/TITS.2018.2835523
28. Nubert, J., Truong, N.G., Lim, A., Tanujaya, H.I., Lim, L., Vu, M.A.: Traffic density estimation using a convolutional neural network (2018). arXiv:1809.01564

29. Chen, J., Tan, E., Li, Z.: A machine learning framework for real-time traffic density detection. IJPRAI **23**, 1265–1284 (2009). https://doi.org/10.1142/S0218001409007673
30. Maungmai, W., Nuthong, C.: Vehicle classification with deep learning. In: 2019 IEEE 4th International Conference on Computer and Communication Systems (ICCCS), pp. 294–298 (2019). https://doi.org/10.1109/CCOMS.2019.8821689
31. Wong, Z.J., Goh, V.T., Yap, T.T.V., Ng, H.: Vehicle classification using convolutional neural network for electronic toll collection. In: Alfred, R., Lim, Y., Haviluddin, H., and On, C.K. (eds.) Computational Science and Technology, pp. 169–177. Springer, Singapore (2020). https://doi.org/10.1007/978-981-15-0058-9_17
32. Sarikan, S.S., Ozbayoglu, A.M., Zilci, O.: Automated vehicle classification with image processing and computational intelligence. Proc. Comput. Sci. **114**, 515–522 (2017). https://doi.org/10.1016/j.procs.2017.09.022
33. Lawe, S., Wang, R.: Optimization of traffic signals using deep learning neural networks (2016). https://doi.org/10.1007/978-3-319-50127-7_35
34. Joo, H., Lim, Y.: Reinforcement learning for traffic signal timing optimization. In: 2020 International Conference on Information Networking (ICOIN), pp. 738–742 (2020). https://doi.org/10.1109/ICOIN48656.2020.9016568
35. Bakker, B., Whiteson, S., Kester, L., Groen, F.C.A.: Traffic light control by multiagent reinforcement learning systems. In: Babuška, R. and Groen, F.C.A. (eds.) Interactive Collaborative Information Systems, pp. 475–510 (2010). Springer, Berlin, Heidelberg. https://doi.org/10.1007/978-3-642-11688-9_18
36. Zhang, Z., Heb, Q., Gao, J., Ni, M.: A deep learning approach for detecting traffic accidents from social media data (2018). arXiv:1801.01528
37. Zuccarelli, E.: Using Machine Learning to Predict Car Accidents. Towards Data Science. https://towardsdatascience.com/using-machine-learning-to-predict-car-accidents-44664c79c942. Accessed 03 Jan 2021

Bottleneck Features for Enhancing the Synchronous Generator Fault Diagnosis System Performance

C. Santhosh Kumar, A. Bramendran, and K. T. Sreekumar

Abstract In this work, we show how bottleneck features (BNF) with backend support vector machine (SVM) classifier can enhance the machine fault diagnosis system performance. Bottleneck features are extracted from convolutional neural networks (CNN) as they are good in extracting the local patterns from the raw data. As part of this work, we first developed a baseline system for 3kVA synchronous generator fault diagnosis system with backend-SVM classifier. At the point when we use backend-SVM classifier, the extracted features should match with the SVM kernels. This may affect the classifier performance. Next, we developed a CNN-based fault diagnosis system with the raw current signal as its input. Further, we extracted BNF from CNN and classified with backend-SVM classifier. It is noted that the BNF-SVM system outperformed by 23.16%, 18.57%, and 17.47 % for R, Y, B phases, respectively, compared with the SVM-based baseline fault diagnosis system.

Keywords Fault diagnosis · Support vector machine (SVM) · Convolutional neural network (CNN) · Bottleneck features (BNF)

1 Introduction

Rotating machines are extensively used in industries, space crafts, automobiles, etc. Continuous operation with various load conditions can lead to component level failures which can further lead to catastrophic failure of the machine. Condition-based monitoring (CBM) [1] methods are used to continuously monitor the machine by analyzing various internal parameters and maintenance decision are taken accordingly. CBM involves in fault diagnosis, fault isolation, and fault identification. CBM uses one or more of the following three approaches:

C. Santhosh Kumar (✉) · A. Bramendran · K. T. Sreekumar
Machine Intelligence Research Laboratory, Department of Electronics and Communication Engineering, Amrita School of Engineering, Coimbatore, Amrita Vishwa Vidyapeetham, India
e-mail: cs_kumar@cb.amrita.edu

K. T. Sreekumar
e-mail: kt_sreekumar@cb.amrita.edu

J. S. Raj et al. (eds.), *Intelligent Sustainable Systems*, Lecture Notes in Networks and Systems 213, https://doi.org/10.1007/978-981-16-2422-3_34

1. Model-based approach
2. Data-driven approach
3. Hybrid Approach

The mathematical modeling of physical system is used in the model-based approach. The mathematical modeling includes the entire knowledge about the control system design. For a complex system it is difficult to develop a mathematical model and therefore the data-driven or hybrid approaches need to be used. In data-driven approach [2], health status of machines is analyzed using historical data. Machine learning (ML) algorithms are widely used to handle this historical data through which health condition of the machine can be captured. Decision tree algorithms, Fuzzy methods, support vector machine classifier (SVM), K-nearest neighbor algorithms, artificial neural networks (ANN) are the mainly used ML algorithms in the area of machine fault diagnosis. R.Perez et al.[3] used fuzzy logic to perform fault diagnosis in electrical distribution systems. Samantha [4] has applied SVM and ANN with generic algorithm to detect the fault of gears. A.S. Raghunath et al. [5] applied feature mapping and feature normalization techniques to improve the performance of SVM-based multicomponent fault diagnosis system.

Synchronous generator is a key element in the electrical energy generation process and its application varies from power houses to complex space and aeronautics. Many researchers have been investigated fault diagnosis of synchronous generators to date. Nanidi et al. [6] explained various types of electrical machines faults, out of which most commonly occurring faults are stator winding related. Inter-turn faults are the key cause of stator winding faults, which may mainly happen due to the problems in insulation. When short circuit occurs, heavy current flows through the windings thus increasing the temperature and causes more damage to the insulation of the other windings. Hence, early detection of inter-turn faults can avoid catastrophic failures of electrical machines. Siddique et al. [7] described various methods for identifying faults in the stator windings of induction machines. Motor current is a parameter which is commonly used in monitoring methods because most of the machines have inbuilt current sensors and can measure signal non-invasively.

Various signal processing and artificial intelligence techniques are used for analyzing the recorded stator winding current signals for fault signs in any of the phases of the rotating machines with less human intervention. Since the signals are dynamic, time domain signal processing is incompatible.

In this work, we first implemented a backend-SVM classifier-based baseline system that uses frequency domain statistical features. The SVM-based system performance relies on the extracted features and SVM kernel. If the features are not matching with the kernel, the fault diagnosis system performance may reduce. Further, we explored how the performance of a data-driven machine fault diagnosis system can be improved using convolutional neural networks (CNN). K. K. Agrawal et al. [8] reviewed the neural network-based monitoring of the machine states. Further we explored bottleneck features (BNF) [9] derived from CNN and it was seen that the BNF with backend-SVM helps in further improvement of fault diagnosis system performance.

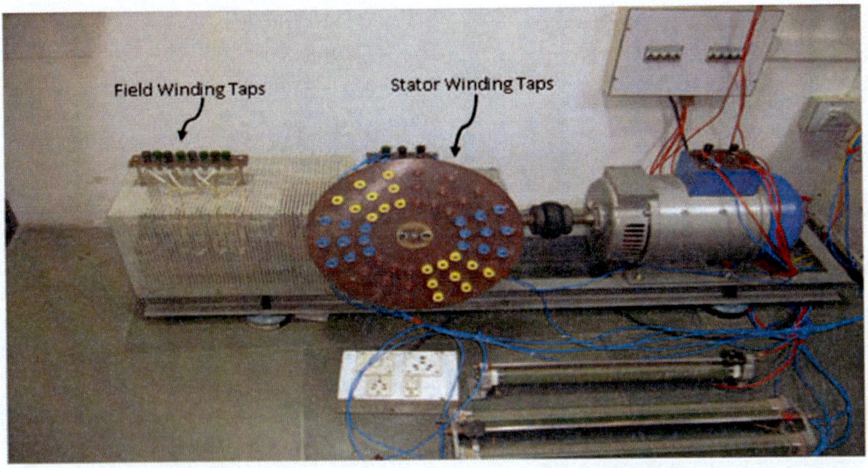

Fig. 1 Synchronous generator test setup

2 Experimental Setup

In synchronous generator, short circuit fault might happen in the field winding and stator winding coils. In this work, we used 3 phase 3kVA synchronous generator with fault injection capability. In each phase, the stator windings of synchronous generator have 18 taps. The stator winding has six coils in each phase and each coil has 28 turns. In this experiment, we shorted only 6, 8, and 14 turns of 168 turns in each phase respectively.

Various combinations of the winding taps are connected for data collection. The winding setup and methodologies for data collection from 3kVA synchronous generator are explained by Gopinath et al. [10] in detail. Fault injection capable 3kVA synchronous generator and winding diagram with coil taps are shown in Figs. 1 and 2.

3 Data Collection

The data is collected from the machine with the stator winding tapped for injecting inter-turn fault manually. Since the CBM is data-driven model, recordings of the stator current signal during healthy and faulty conditions are captured. Inter-turn fault with various severity levels is injected with taps provided in the all the phases of stator windings using the terminal board. To have industrial type environment we emulate by adding resistive load banks as the loading in the synchronous machine is changing.

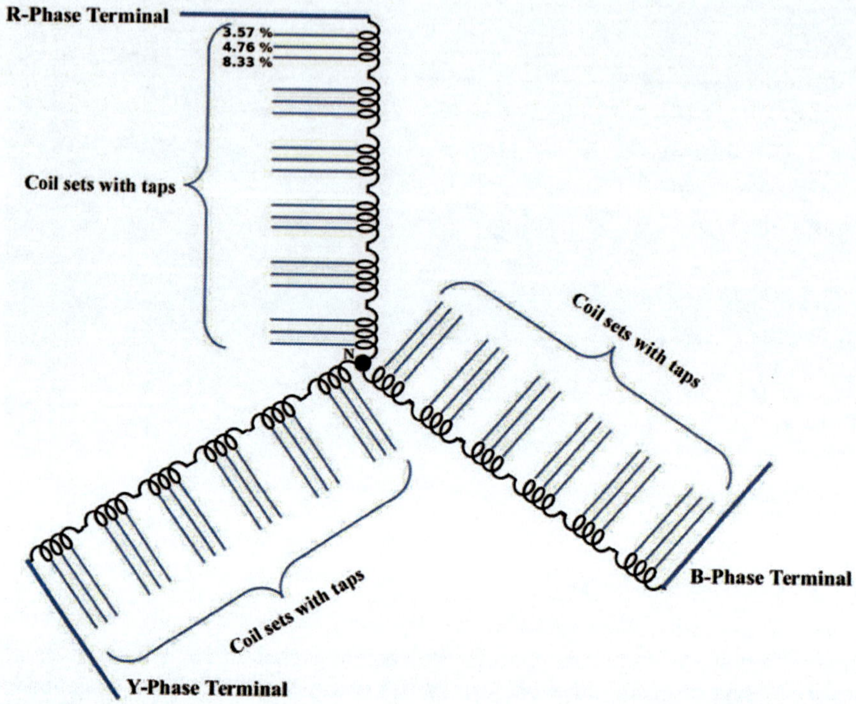

Fig. 2 winding diagram with coil taps

A three-phase resistive load is connected and experiments are carried out at various loads, i.e., 0.5A, 1A, 1.5A, 2A, 2.5A, 3A, 3.5A. To record the stator current signals, National Instruments data acquisition system (NI-PXI6221) with hall effect current sensors and LabVIEW signal express software are used. In each experiment the machine is made to run for 10 s at a sampling frequency of 1 kHz. Therefore 10000 samples are collected in each trial. The stator winding current signals during healthy and faulty condition is in the Fig. 3.

4 System Description

The acquired current signatures are first divided into training data and test dataset. Feature extraction technique is popularly used to reduce the undesired information from the acquired [11]. In this work, we explored frequency domain features to study the robustness of the fault models. Table 1 describes the list of frequency domain features used in this work. The time domain to frequency domain conversion is carried out using fast Fourier transform (FFT). From the transformed data we

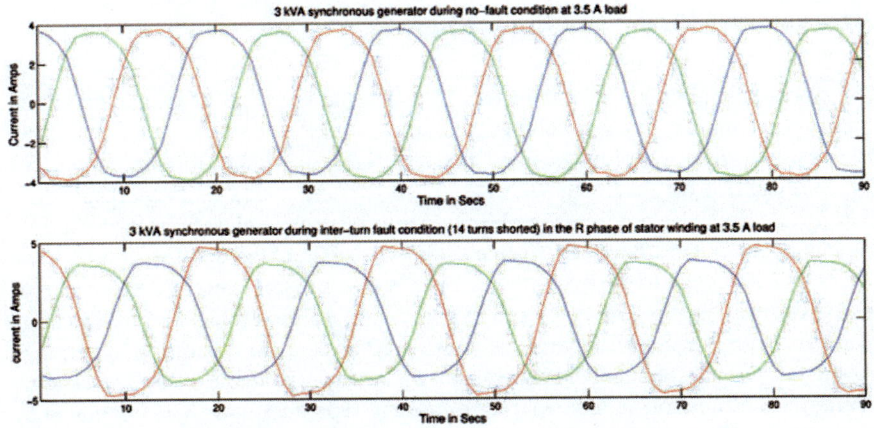

Fig. 3 Current signature of no fault and fault data

Table 1 Frequency domain features

Description	Features
Mean	$Feat_1 = \frac{1}{K}\sum_{k=1}^{K} g(k)$
Variance	$Feat_2 = \frac{1}{K-1}\sum_{k=1}^{K}(g(k) - Feat_1)^2$
Skewness	$Feat_3 = \frac{\sum_{k=1}^{K}(g(k)-Feat_1)^3}{K(\sqrt{Feat_2})^3}$
Kurtosis	$Feat_4 = \frac{\sum_{k=1}^{K}(g(k)-Feat_1)^4}{K(Feat_2)^2}$
Frequency center	$Feat_5 = \frac{\sum_{k=1}^{K} f_k g(k)}{\sum_{k=1}^{K} g(k)}$
Standard deviation	$Feat_6 = \sqrt{\frac{1}{K}\sum_{k=1}^{K}(f_k - Feat_5)^2 g(k)}$
Coefficient variability	$Feat_7 = \frac{Feat_6}{Feat_5}$
Stability factor	$Feat_8 = \frac{\sum_{k=1}^{K} f_k^2 g(k)}{\sqrt{\sum_{k=1}^{K} g(k) \sum_{k=1}^{K} f_k^4 g(k)}}$
Root Mean Square frequency	$Feat_9 = \sqrt{\frac{\sum_{k=1}^{K} f_k^2 g(k)}{\sum_{k=1}^{K} g(k)}}$
Skewness frequency	$Feat_{10} = \frac{\sum_{k=1}^{K}(f_k-Feat_5)^3 g(k)}{K(Feat_6)^3}$
Kurtosis frequency	$Feat_{11} = \frac{\sum_{k=1}^{K}(f_k-Feat_5)^4 g(k)}{K(Feat_6)^4}$
Spectrum power convergence	$Feat_{12} = \sqrt{\frac{\sum_{k=1}^{K} f_k^4 g(k)}{\sum_{k=1}^{K} f_k^2 g(k)}}$
Spectrum power positional factor	$Feat_{13} = \frac{\sum_{k=1}^{K}(f_k-Feat_5)^{\frac{1}{2}} g(k)}{K\sqrt{Feat_6}}$

extracted 13 different statistical features and that are used to train the SVM model. Evaluated this model using statistical features extracted from the test data.
[1]

[1] where $g(k)$ is the spectrum for $k = 1, 2 \ldots K,$. f_k is the frequency value of kth spectrum line.

4.1 Support Vector Machine (SVM)

Binary classification is a ML algorithm in which the given set of data is classified in to two categories, as per predefined classification rules. SVM [12] is a popularly used classification algorithm, and works effectively for binary classification. An SVM model can be generated with the aid of training data and its corresponding class labels. Input data points are classified into respective classes by generating a hypothetical decision boundary (separating hyperplane) between the classes. The points that are nearer to the hyperplane are known as support vectors, and the sum of distances from the support vectors to hyperplane is referred to as margin. Decision boundary with a maximum margin is considered as the best separating hyperplane for classification. Mathematical expression of a hyperplane is $x^T w + b = 0$. w being the normal to hyperplane, considering a linearly separable data. SVM can also be used for classification of non-linear data. Classification is performed by mapping the non-linear data to a higher dimensional space through kernel methods. Further explanation on binary class SVM is given in [13, 14].

4.2 Convolutional Neural Network (CNN)

CNN is a deep learning neural network algorithm which is commonly used for image classification [15]. Based on its translational invariant features and shared weight architecture CNN has an application on various fields, e.g., natural language processing (NLP), recommended system, medical image analysis, analyzing time series data in finance for future predictions, etc. CNN is a regularized version of multilayer perceptron where neuron is not fully connected and hence avoiding overfitting of the data. The feature extraction and classification are the two main sections of CNN. The input, convoultional, and pooling layers belong to feature extraction section and fully connected layers and the output layer are performing classification.

As the name suggests it uses convolutional multiplication of the data points with weights assigned. It uses filters to obtain feature map from the input image. Convolutional layer and pooling layer are stacked together for getting best features out from the input data. Pooling is of 2 types average pooling and maxpooling where maxpooling is most commonly used. Pooling layer selects appropriate features from the feature map and reduces the computational complexity. ReLU activation function provides better performance and fast learning in convolutional. Fully connected layer is the array of all the features in one dimension after the pooling layer. As there are many redundant features in the fully connected layer, dropout layer drops some units to reduce computational complexity and make performance better. Hence dropout layer avoids overfitting in data. CNN is used in many fields because of its automatic feature extraction capability. Here the CNN is used for performing fault classification. The block diagram of CNN is shown Fig. 4.

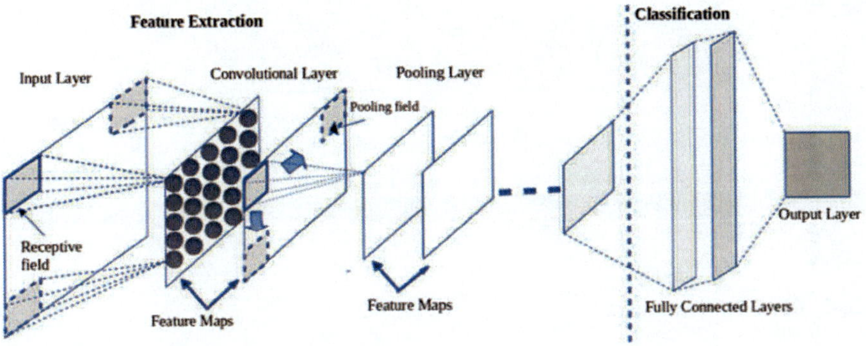

Fig. 4 Block Diagram of CNN

The ReLU activation is used in the convolutional layer for fast learning of the features of the input layer with filters and better performance. Since we are classifying fault and no fault data, sigmoid activation function is used in the fully connected layer. The CNN model designed has a loss function as Mean square error, optimizer as Adam (Adaptive Moment Estimation), and accuracy as its metrics to find the model accuracy and test accuracy. Adam is a combination of RMS prop and Adagrad which performs better optimization.

4.3 Bottleneck Features (BNF)

The bottleneck features from a CNN architecture are generally extracted from hidden layer of a neural network. Compared to other hidden layers, the bottleneck layer has smaller number of hidden units. This bottleneck layer produces a compression in the network that forces the relevant information required for classification into a low-dimensional representation.

5 Experiments and Results

5.1 Baseline SVM System

The experiments are carried out using a synchronous generator with a power rating of 3kVA. The stator current was collected from R,Y, B phases in a faulty and no-faulty conditions. The faulty and no-faulty signals at different loading conditions are also explored in this experiment. The acquired signal was splitted into frames with the size of 512. From the collected data, we used 70% for training the baseline SVM model and 30% for evaluating the model. Figure 5 shows the experimental procedure

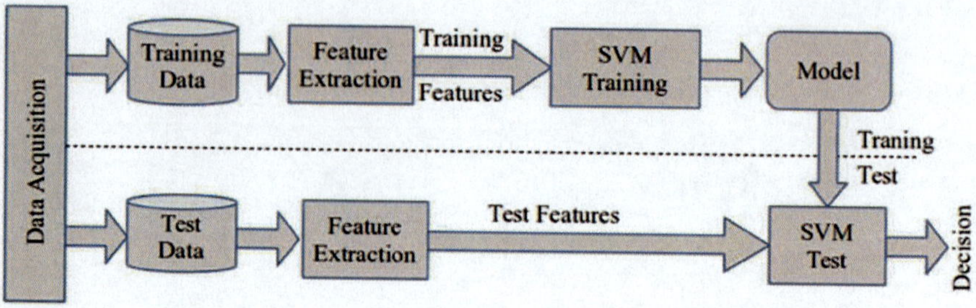

Fig. 5 Baseline SVM system

Table 2 Baseline fault diagnosis system performance of 3 kVA generator

SVM Kernel	Accuracy (in %)		
	R Phase	Y Phase	B Phase
Linear	67.45	61.46	66.68
Polynomial	76.66	81.35	82.31
RBF	65.33	67.94	72.86

to develop SVM-based baseline system. Linear, polynomial and radial basis function (RBF) kernels are also used to evaluate performance and the results are listed in the Table 2.

5.2 *CNN-Based Fault Diagnosis System*

The CNN architecture [16] consists of 3 channel input layer, three pairs of convolutional-maxpooling layer, a fully connected layer and an output layer. The fault signal collected from R, Y, B phases are considered as 3 channel in the input layer The shape of input layer is $512 \times 1 \times 3$. The one-dimensional convolution was carried out in the first convolutional layer. Here, we used 64 3×1 filters and ReLU activation function. The output of first convolutional layer was fed into maxpooling layer with the window size of 2×1. The same operation was carried out in the second and third conv-maxpooling layer pairs. The filter size used in the second convolutional layer was 32 and third layer was 16. The last maxpooling layer output was flattened and fed into the fully connected layer with 256 neurons. The activation function used in the fully connected layer was also ReLU. Finally, the classification was obtained from the softmax output layer. Since our experiments are based on two classes, we used two neurons in the output softmax layer. The output of softmax algorithm is interpreted as the probability estimate of no fault and fault class. The CNN architecture which was used in this experiment is shown in Fig. 6. The CNN-based fault diagnosis system performance is shown in Table 3.

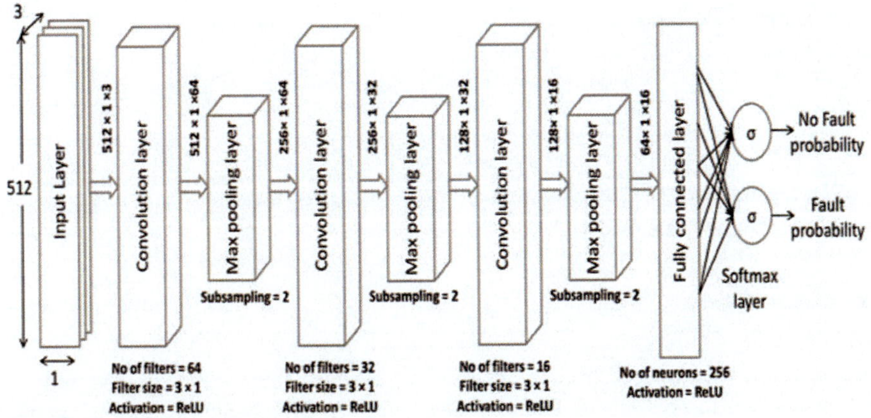

Fig. 6 Block Diagram of CNN

Table 3 Performance of CNN-based system

System	Accuracy (in %)		
	R Phase	Y Phase	B Phase
CNN	96.66	98.22	97.73

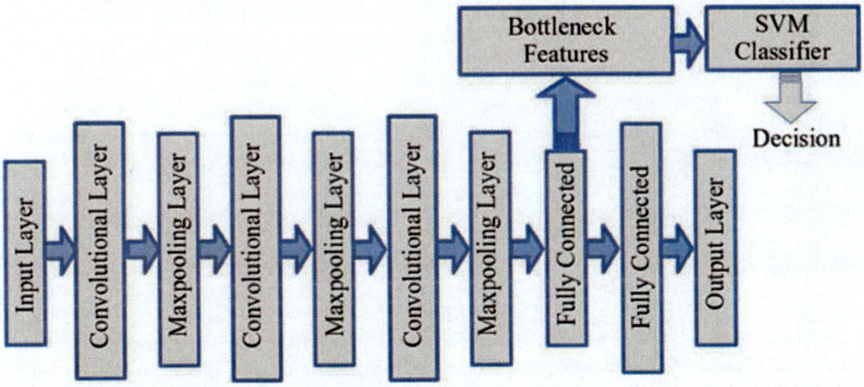

Fig. 7 Bottleneck features based SVM system

5.3 *Bottleneck Features (BNF) Based SVM System*

In this work we extracted the BNF, with the dimension of 20, directly from the current signal. The SVM model was trained from this BNFs. Figure 7 shows the procedures to extract BNF and classify using SVM classifier. The BNF-SVM system performance is illustrated in Table 4.

Table 4 BNF-SVM Performance

SVM Kernel	Accuracy (in %)		
	R Phase	Y Phase	B Phase
Linear	99.22	99.55	99.62
Polynomial	99.59	99.68	99.67
RBF	**99.82**	**99.92**	**99.78**

6 Conclusion

In this work, we explored how to enhance the machine fault diagnosis system performance by using bottleneck features (BNF) derived from the convolutional neural network (CNN). First, we developed support vector machine (SVM) classifier-based baseline system with frequency domain statistical features. We obtained a classification accuracy of 76.66%, 81.55%, and 82.31% for R phase, Y Phase, and B phase faults, respectively, for the baseline system. If the extracted features are not matching with the SVM kernel, it may affect the performance of machine fault diagnosis system. In CNN, features are automatically extracted by convolutional method using filters. We achieved a performance improvement 20.00%, 16.67%, and 15.42% classification accuracy for R,Y, B phases respectively, over the SVM-based baseline system. Further, we derived BNF from CNN and trained the SVM model. Using BNF-SVM system, we achieved fault classification accuracy of 99.82 %, 99.92%, and 99.78% for R, Y, and B phases, respectively.

Acknowledgements Authors would really like to acknowledge the assistance and support of Ms. Haritha H., Ms. Mrudula G. B., and Ms. Pooja Muralidharan toward the work.

References

1. Jardine, A.S., Lin, D., Banjevic, D.: A review on machinery diagnostics and prognostics implementing condition-based maintenance. Mechanical systems and signal processing **20**(7), 1483–1510 (2006)
2. Neethu Mohan, K.P. Soman, S. Sachin Kumar,A data-driven strategy for short-term electric load forecasting using dynamic mode decomposition model, Applied Energy, Volume 232,Pages 229-244, (2018)
3. Perez, Ramn, Esteban Inga, Alexander Aguila, Carmen Vsquez, Liliana Lima, Amelec Viloria, and Maury-Ardila Henry. Fault diagnosis on electrical distribution systems based on fuzzy logic. In International Conference on Sensing and Imaging, pp. 174-185. Springer, Cham, (2018)
4. Samanta, B.: Gear fault detection using artificial neural networks and support vector machines with genetic algorithms. Mechanical systems and signal processing **18**(3), 625–644 (2004)
5. A. S. Raghunath, K. T. Sreekumar, C. S. Kumar and K. I. Ramachandran, "Improving Speed Independent Performance of Fault Diagnosis Systems through Feature Mapping and Normalization," 2016 15th IEEE International Conference on Machine Learning and Applications (ICMLA), Anaheim, CA, 2016, pp. 764-767, https://doi.org/10.1109/ICMLA.2016.0136.

6. Nandi, S., Toliyat, H.A., Li, X.: Condition Monitoring and Fault Diagnosis of Electrical Motors A Review. IEEE Transactions on Energy Conversion **20**(4), 719–729 (2005)
7. Siddique, Arfat, G. S. Yadava, and Bhim Singh. A review of stator fault monitoring techniques of induction motors. IEEE transactions on energy conversion 20, no. 1 : 106-114 (2005)
8. K. K. Agrawal, G. N. Pandey and K. Chandrasekaran, Analysis of the condition based monitoring system for heavy industrial machineries, 2013 IEEE International Conference on Computational Intelligence and Computing Research, Enathi, pp. 1-4 (2013)
9. Yu, Dong, Seltzer, Michael L.: Improved bottleneck features using pretrained deep neural networks. Twelfth annual conference of the international speech communication association (INTERSPEECH) **237–240**, (2011)
10. Gopinath, R., T. Nambiar, S. Abhishek, S. Pramodh, M. Pushparajan, K.I. Ramachandran, C. S. Kumar and R. Thirugnanam. Fault injection capable synchronous generator for condition based maintenance. 2013 7th International Conference on Intelligent Systems and Control (ISCO), pp: 60-64, (2013)
11. Sreekumar, K. T., Kuruvachan K. George, C. Santhosh Kumar, and K. I. Ramachandran. Performance enhancement of the machine-fault diagnosis system using feature mapping, normalisation and decision fusion. IET Science, Measurement & Technology 13, no. 9 : 1287-1298 (2019)
12. V. N. Vapnik and V. Vapnik, Statistical learning theory. Wiley New York, vol. 1 (1998)
13. Chang, C.-C., Lin, C.-J.: LIBSVM: a library for support vector machines. ACM Transactions on Intelligent Systems and Technology, pp. 2:27:127:27 (2011)
14. Sreekumar, K. T., R. Gopinath, M. Pushparajan, Aparna S. Raghunath, C. Santhosh Kumar, K. I. Ramachandran, and M. Saimurugan. Locality constrained linear coding for fault diagnosis of rotating machines using vibration analysis. In 2015 Annual IEEE India Conference (INDICON), pp. 1-6. IEEE, (2015)
15. A. Jayanth Balaji, D.S. Harish Ram, Binoy B. Nair, Applicability of Deep Learning Models for Stock Price Forecasting An Empirical Study on BANKEX Data, Procedia Computer Science, Volume 143,Pages 947-953, (2018)
16. Sreekumar, K. T., C. Santhosh Kumar, and K. I. Ramachandran. System Independent Machine Fault Diagnosis Using Convolutional Neural Networks. In 2018 15th IEEE India Council International Conference (INDICON), pp. 1-6. IEEE, (2018)

Feasibility Study of Combined Cycle Power Plant in Context of Bangladesh

Md. Sazal Miah, Shishir Kumar Bhowmick, Md. Rezaul Karim Sohel, Md. Abdul Momen Swazal, Sazib Mittro, and M. S. Hossain Lipu

Abstract The government of Bangladesh set the target to alleviate poverty in the shortest possible time by achieving high economic growth. Currently 98% of our population have access to electricity. The Government has committed to provide electricity to 100% of the population by 2021. Sustainable Development Goals (SDG) also focus on energy sector as priority. The SDG target has been set to guarantee universal access to reasonable, dependable, and smart power service by 2030. Stable and secured power supply is one of the preconditions for faster economic growth. Bangladesh is already going through more than 8.1% GDP growth. The expected electricity growth per year would be around 10–12%. Bangladesh needs large power plants to meet up these targets. A 400–500 MW dual fuel Combined Cycle Power Plant (CCPP) will contribute to meet the growing demand and achieve grid system stability. Here we Study the Feasibility study of combined cycle power plant in context of Bangladesh (Case Study on Dual Fuel CCPP at Sonagazi, Feni).

Keywords Combined cycle power plant · Power generation · Feasibility study

1 Introduction

The site is located at Purbo Borodholi Mouza of Sonagazi Upazilla in Feni District and Musapur Union of Companiganj Upazilla in Noakhali District (22°46′45" North Latitude and 91°91′22" East Longitude). The Musapur closer dam and Choto Feni river in the region are adjacent to the site. Total area is 498.973 acres.

Md. Sazal Miah (✉)
School of Engineering and Technology, Asian Institute of Technology, Pathumthani 12120, Thailand

S. K. Bhowmick · Md. Rezaul Karim Sohel · Md. Abdul Momen Swazal · S. Mittro
Department of Electrical and Electronic Engineering, University of Asia Pacific, Dhaka 1205, Bangladesh

M. S. Hossain Lipu
Department of Electrical, Electronic and Systems Engineering, Universiti Kebangsaan Malaysia, 43600 Bangi, Malaysia
e-mail: lipu@ukm.edu.my

J. S. Raj et al. (eds.), *Intelligent Sustainable Systems*, Lecture Notes in Networks and Systems 213, https://doi.org/10.1007/978-981-16-2422-3_35

A gas fired CCPP with an oil backup facility will be arranged. Currently about 54% of our electricity is generated from domestic natural gas due to cheap primary fuel cost. In the meantime, the gas pressure and reserve is gradually decreasing [1–3]. Bangladesh government has planned to import Liquefied Natural Gas (LNG) from abroad to offset the shortage of domestic gas availabilty. An alternative fuel is to be considered in case shortage of gas supply occurs. High speed diesel and heavy fuel oil have been studied comparatively as alternative fuel in this report.

The proposed power plant would require 65 million cubic ft of natural gas per day. The proposed plant site is close to existiareng Chittagong to Bakhrabad gas pipeline. The pipeline is 24 inch diameter and distance to site is 15–20 km. A new 30 inch pipeline near the existing one has been planned by GTCL (Gas Transmission Company Limited).

The plant will be capable of running on liquid fuel so that shortage or crisis of natural gas supply does not hamper power generation [4]. The liquid fuel can be high speed diesel or heavy fuel oil. Liquid fuel will be chosen on the basis of price, availability, etc. High speed diesel appears more suitable for this project. Liquid fuel can be transported to the site from Chittagong port via Sandwip Channel by lighter ship.

The site is suitable for waterways transportation. Navigation route is shown at Fig. 1. Distance from Chittagong port to the site is about 100 km by Sandwip channel. The road transportation is not very good at this moment. The existing roads are required to be widen and paved for plying heavy vehicle (Figs. 2, 3, 4, 5 and 6).

The project construction work is expected to start on 2020 and subsequent completion on 2024–25. Presently the site has no occupants, some area is used for grazing of bufallos & cattle, so there would minimum environmental and social effect.

The estimated cost of the project is about 642 Million US$. Average prevailing tariff of 3.22 bdt/kWh was used for financial analysis which produces Financial Internal Rate of return (FIRR) as 5.88%. To calculate Economic Rate of Return (EIRR), Willingness-to-Pay (WTP) method was applied for tariff. WTP represents the economic value that consumers assess in their consumption pattern. Using this method, tariff for gas power generation was found 6.06 bdt/kWh and tariff for diesel power generation was found 11.06 bdt/kWh. EIRR was calculated as 20.64% [5].

2 Current Scenarıo of Power Sector in Bangladesh

Power sector in Bangladesh is growing rapidly to accommodate the rising demand. In the last decade the generation, transmission, and distribution systems went through major development [6–8].

The achievements in last 11 years are shown below (Table 1).

Natural gas has always been the main resource of fuel for power generation in our country. Diesel, Furnace oil, and coal are also used in some power plants. Power generation by type of fuels as on March 2020 are shown below.

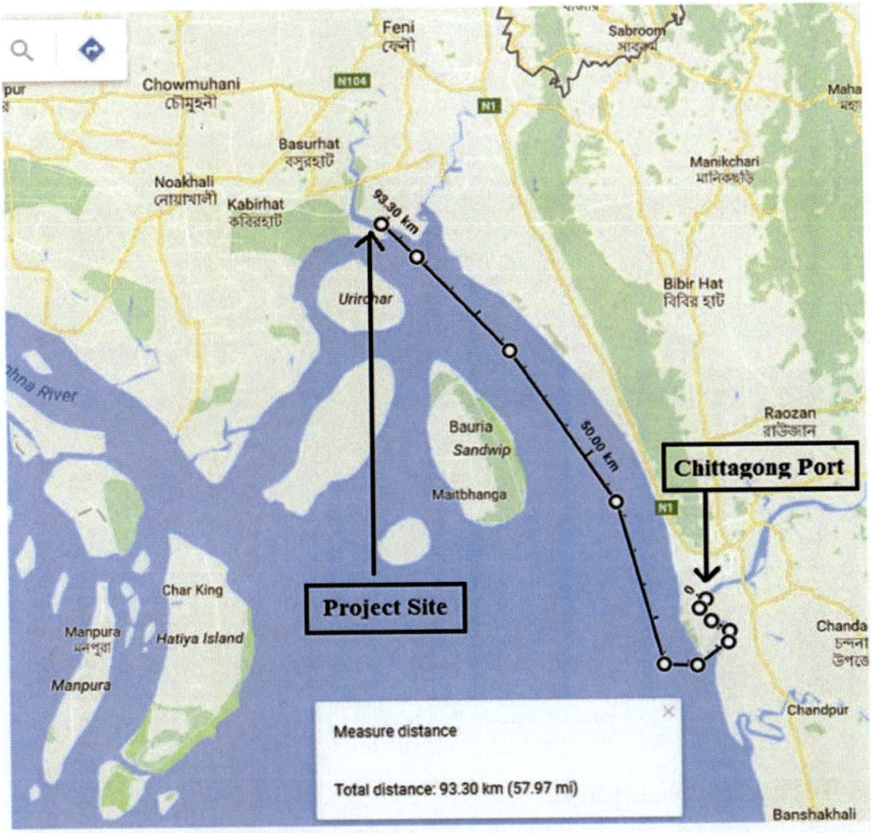

Fig. 1 : Transportation route from Chittagong port to proposed site

Though power sector has expanded a lot, still the whole country does not have access to electricity yet. As of now 98% people of the country consume electricity while 2% people cannot. Sometimes industrial production gets hampered due to shortage of electricity in the peak hours. Bangladesh is now going through a high GDP growth rate of 8.1%. The high GDP growth rate will cause the electricity demand to grow by 10–12%. Bangladesh government has taken firm initiatives to meet the growing demand [9].

An ambitious target has been set for growth of power generation. The long-term goal for energy production is shown below (Table 2).

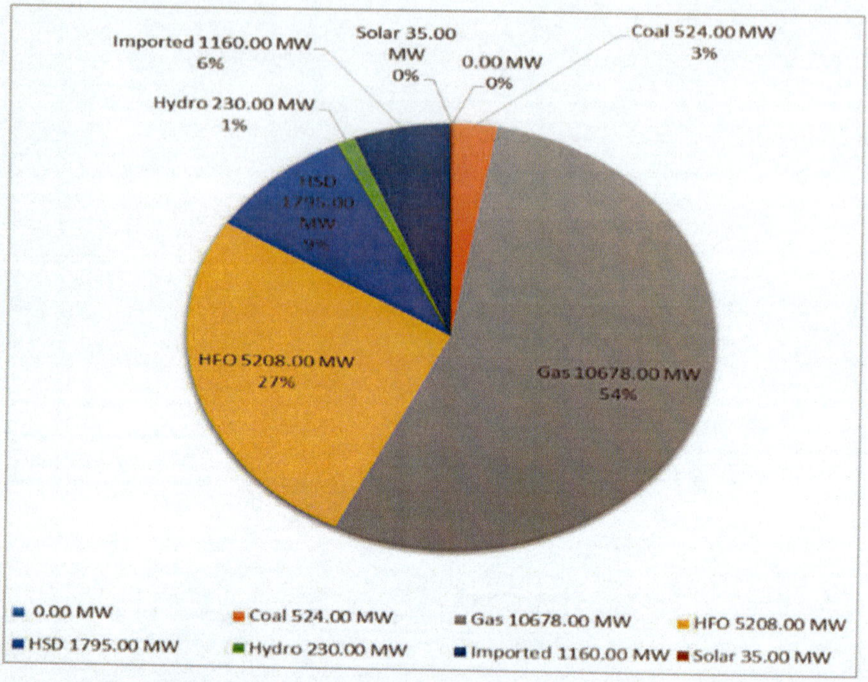

Fig. 2 Installed Capacity by Fuel Type-as on March, 2020

3 Technical Feasibility

Natural gas and liquid oil will be the fuels in the proposed power plant. Their current usage and supply situation are presented below.

3.1 *Natural Gas*

Gas generation from the current local gas field in 2015 was 2,500 MMCFD and has arrived at a pinnacle generation of 2,700 MMSCFD in 2017, at that point begin to decay. Moreover, gas requirement in Bangladesh predicts a huge increment later on. The requirement and flexibly gap must be filled by gas (LNG) imports. The initial LNG was presented in 2019 at the percentage of 500 MMSCFD, which relates to 17% of gas request. From the prediction analysis, the requirement for natural gas will be increased by 40%, 50%, and 70% within 2023, 2028, and 2041, respectively [10, 11].

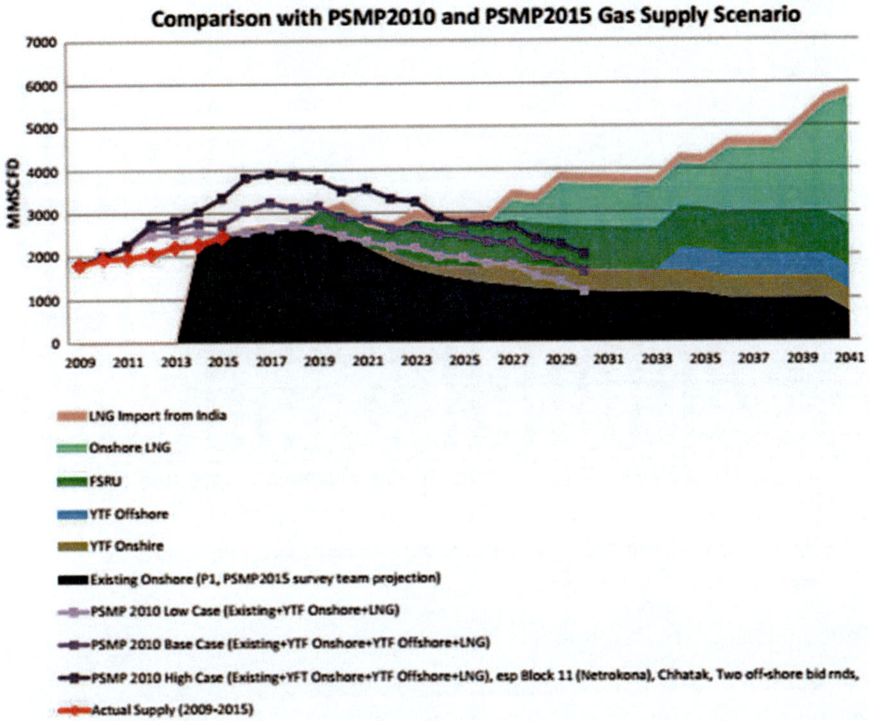

Fig. 3 : Gas Supply Forecast 2016 ~ 2041

3.2 Oil

Bangladesh's present oil yearly requirement is approximately 5 million tons, and the independence percentage is just 5%. Moreover, Bangladesh anticipates continuous economic growth, and the business segment and transportation segment requirements will prompt drastic oil requirement development: six times greater in 2041 than in 2016 (normal development percentage 7.4% p.a.), significantly under the "Power Proficiency and Conservation Scenario". Bangladesh has a few goals to expand or recently create petroleum treatment facilities; moreover, if the oil requirement develops as predicted, oil imports will be required to satisfy the need and continue expanding [10].

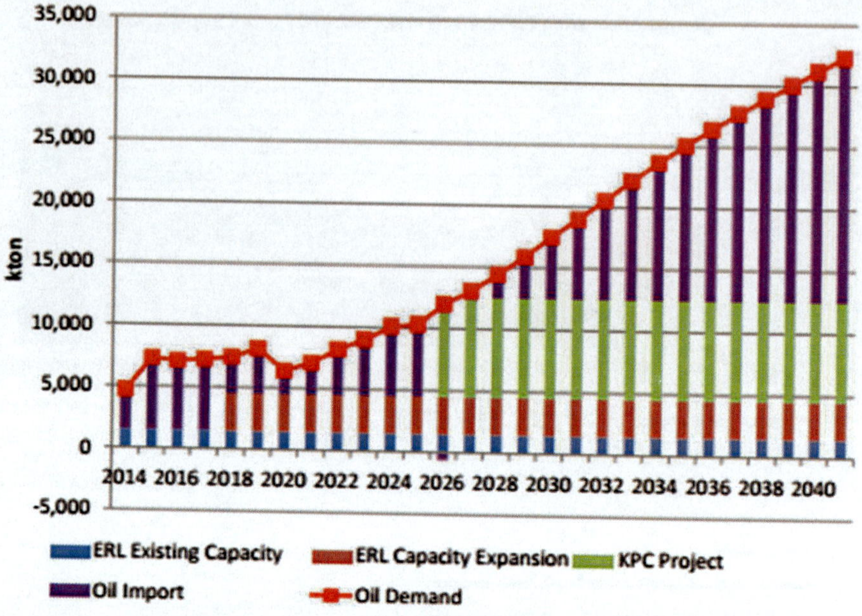

Fig. 4 : Oil Demand and Supply Balance, 2014 to 2041

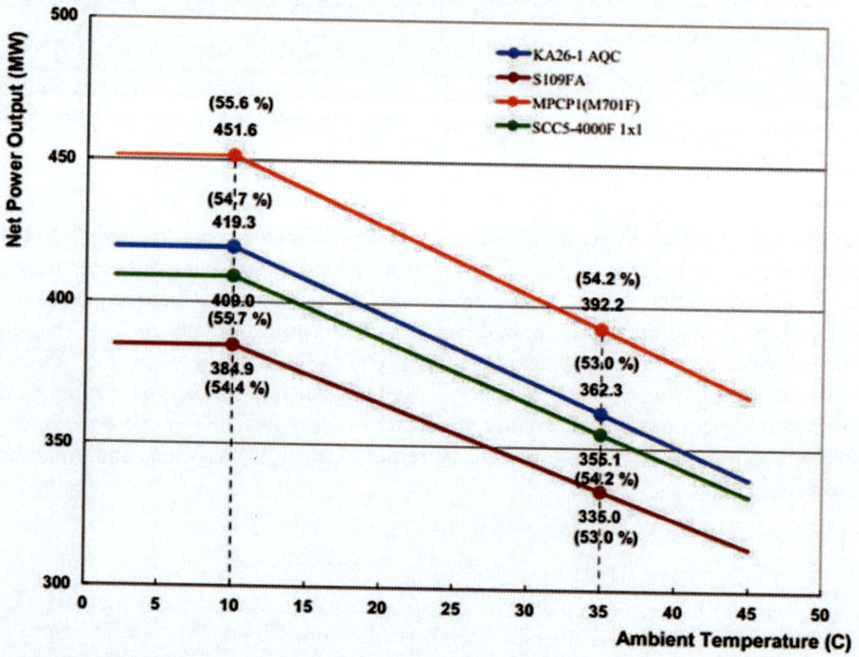

Fig. 5 Ambient Temperature Performance Characteristics Curve

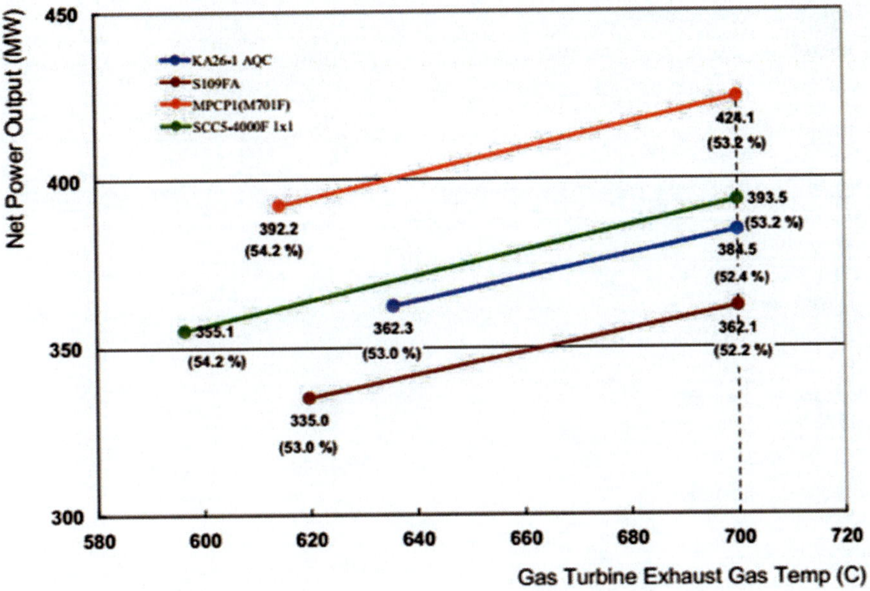

Fig. 6 Relationship between Power Output and Duct Firing Temperature

4 Fundamental Design

4.1 Candidate Models of CCPP

Four models are available in the international market as the combined cycle power plant (CCPP) which is comprised of a 50 Hz use largest capacity Gas Turbine model of which turbine inlet temperature is of F-class level. The F-class models of Gas Turbines of four original equipment manufacturers (OEMs) are matured with much working experience and are regarded to be most appropriate for the Project from operating experience points of view. As per the Gas Turbine World 2007 to 2008 GTW Handbook, the four techniques of consolidated cycle energy plants are as organized below:

Name of OEM of GT	Model of CCPP
Alstom	KA26-1 (with air quench cooler)
General Electric	S109FA, S109FB
Mitsubishi	MPCP1(M701F)
Siemens	SCC5-4000F 1 × 1

In choice of the candidate models, it will be viewed as that the utilized Gas Turbine can be worked in a basic cycle mode considering that it might be placed

Table 1 Current Scenario

Subject	in 2009	in 2020	Achievement in 11 years (2009–2020)
(i)	*(ii)*	*(iii)*	*(iv= iii – ii)*
Number of power plants	27	137	(+) 110
Installed Capacity (MW)	4,942	22,787	(+) 17,845
Highest Power Generation (MW)	3,268 (6th January, 2009)	12,893	(+) 9,625
Total Transmission Line (Circuit km)	8,000	12,119	(+) 4,119
Grid Substation Capacity (MW)	15,870	44,340	(+) 28,470
Purchase of electricity (MW)	-	1160	(+) 1160
Distribution Line (km)	2,60,000	5,60,000	(+) 3,00,000
Population with access to electricity	47	96	(+) 49
Per head electricity production (kwh)	220	510	(+) 290
Number of electricity consumers	1,08,000,000	3,64,00,000	2,56,00,000
Number of irrigation connections	2,34,000	4,24,000	(+) 1,90,000
Budget in Annual Development Plan (crore taka)	2,677	28,862	(+) 26,185
System loss (%)	14.33	9.35	(-) 4.98

Table 2 Long-Term Goal for Energy Production in Bangladesh

Year	Target (MW) of electricity generation
2021	24,000
2030	40,000
2041	60,000

into a commercial activity ahead of time to resolve the looming deficiency of energy supply. For instance, Alstom can provide two sorts of GT26 Gas Turbine. First one is GT26 with an air extinguish cooler technology, second one is GT26 with a single through cooler technique which utilizes the steam to cool the air extricated from the air blower for inside cooling of hot pieces of the Gas Turbine. Thusly, the last kind of GT26 Gas Turbine cannot be worked without a cooling mechanism of steam. For such an explanation, the GT26 Gas Turbine where encompassing air is utilized as a cooling channel picked as an applicant CCPP for the Plant. Similarly, out of the two (2) types of S109 CCPP models that GE has, the model of S109FB is specified as a model for only combined cycle mode use in the said Handbook, therefore, this model is excluded from the study. For the heat balance calculation of each model of CCPP, a bypass stack and a damper are considered [12, 13].

4.2 CCPP Efficiency Data on ISO Conditions

In the mentioned GTW Handbook, execution information of the above models of CCPPs are depicted at ISO terms (101.33 kPa, 15OC, 60% RH) on ordinary gas through some other fundamental conditions than encompassing temperature and weight are not generally determined. The presentation information of the four CCPP models are as portrayed below:

Model of CCPP	Net Plant Output (kW)	Net Plant Efficiency (%)
KA26-1 with AQC	Not specified	Not specified
S109FA	390,800	56.7
MPCP1 (M701F)	464,500	59.5
SCC5-4000F 1 × 1	416,000	58.2

4.3 Calculation Outcomes of CCPP Heat Balance on Unfired Modes

The plant net thermal proficiencies are estimated to rate from 53.0% to 54.2% under the same terms. Thus, the demand of the plant's total thermal proficiency under unfired conditions to be prescribed in the tender documents should be "not less than 52.0%".

The maximum net power outputs are estimated to be ranging from 396.3 MW to 465.0 MW. Any issues in electrical network system in Bangladesh shall be analyzed against the power output of 500 MW in consideration of a certain margin.

The ambient temperature performance characteristics of four (4) models of CCPPs are shown in the next page. From this figure, it is found that every model of CCPP has the same power outcome features against the ambient temperature and that its plant net thermal efficiency is so close that it is within the range of ± 0.7%.

4.4 Heat Balance Calculation Results Under Duct-Fired Conditions

The duct-fired CCPP is a commonly employed system to augment the power output of the bottoming system of the CCPP [14, 15]. There are many experiences with this system which is a matured technology without any difficulties. The following table shows the sample of experiences of HRSGs with a duct firing system in Japan and outside countries of Japanese HRSG manufacturer.

Range of GT Power Output (MW)	Units of HRSG
MW ≤ 50	18
50 < MW ≤ 200	5
200 < MW	5

It is well known that the duct firing limit temperature without large design change of HRSG casing is generally said to be 750OC. The above performance calculation was carried out for the firing temperature of 700 OC in consideration of proper tolerance. The next figure shows the relationship between the net power output and the duct firing temperature of four CCPP models [16].

4.5 Heat Balance Calculation Results Under Duct-Fired Conditions

The primary model functions to be demanded to the Gas Turbine which will be utilized for this project are as illustrated here on.

The Gas Turbine shall be of an open cycle over duty single-shaft type of which turbine inlet temperature level is of F-class. The Gas Turbine shall be supplied by original equipment manufacturers. The Gas Turbine shall be capable of operating on a simple cycle mode because it is scheduled to put into commercial work in advance uniquely from the bottoming framework considering the present impending power supply shortage situation in Bangladesh. For the purpose, an exhaust gas bypass system shall be equipped. The following four categories of Gas Turbines could be identified with Gas Turbine World 2007–08 GTW Handbook (Volume 26) as F-class Gas Turbines.

Name of OEM	Type of Model
Alstom Power	GT26 with air quench cooler
GE Energy Gas Turbine	PG9351 (FA)
Mitsubishi Heavy Industry	M701F4
Siemens Power Generation	SGT5-4000F

The Gas Turbine power outcome will be determined on a premise of continuous base-load with the load weighting variable of 1.0 for computation of the proportionate working hours which will be a size of the investigation interval of warm gas way parts.

4.6 Water Treatment System

The process water for demineralized water, potable water and sanitary water, fire fighting water, and miscellaneous service water shall be produced through pretreatment system from underground water. The process water for cooling tower shall be produced from underground directly. The demineralized water will be utilized as HRSG make-up water, assistant cooling water, chemical dosing preparation, etc. The EPC Contractor will affirm the nature of the generated demineralized water whether it is worthy to the HRSG. The pretreatment framework consists of a coagulator and filter, etc. The demineralized framework comprises chemical storage and recovery gear, and so on. The need and detail of the pretreatment framework will be chosen dependent on the nature of groundwater. The EPC Contractor will take proper counter measures whenever mandatory.

4.7 Wastewater System

Wastewater comprises neutralized reproduced waste from HRSG blowdown, floor channels from the Gas Turbine and steam turbine structures, defiled yard channels from the transformer region. Sewage and clean wastewater will be treated in a decontamination of the office. Floor channels from the Gas Turbine and steam turbine building and polluted yard channel from the transformer territory will be treated in oil or water separators. Moreover, after treatment, these perfect wastewater streams will be released through the primary seepage funnel to the river. The cooling tower channel without treatment will be released through the primary waste funnel to the waterway.

4.8 Fire Fighting System

The CCPP will be planned and worked with the arrangement of a safe working condition and faculty. This will be accomplished by division and isolation of hardware with adequate separations and by a choice of reasonable gear and materials. Hazardous territories are assigned and appropriate component is chosen for utilizing in these. Diverse firefighting frameworks will be introduced relying upon the operational attributes of the gear, region, and building to be secured. The firefighting limit of the CCPP has to withstand a fire during two hours as per NFPA 850 will be at least 300 m3 and weight of around 10 bar. The CCPP will have its own firewater

Table 3 List of touchy zones and firefighting and recognition framework types

Zone	Firefighting technique
Gas Turbine	CO_2 combined framework
Steam turbine lube oil package, lube oil piping	Water spray system
Steam turbine bearings	Water spray system
Steam turbine zone indoor	Wet stand pump house framework
Generator unit, auxiliary and start-up transformer	Water spray system
Oil tankers	Build in Dike protection system
Controller building	sprinkler system (Cable basement) argonite or similar (Controller room)
Electrical or Switchgear	Sprinkler framework if needed and portable fire extinguishers
Yard	Hydrants

battling framework with pump house and the firewater will be given from the crude water tanks (Table 3).

4.9 Electrical Equipment

The electrical system will be structured based on the multi-shaft design of the having two generator step-up transformers, Steam Turbine Transformer (ST transformer), and Gas Turbine Transformer (GT transformer) and two generators, Steam Turbine Generator (STG) and Gas Turbine Generator (GTG). The voltage of the energy yield from the steam turbine and gas turbine generators will be ventured up to 230 kV by means of ST transformer and GT transformer. The outcome from these two GT transformers and ST transformer is transmitted to the 230 kV substation, respectively. The bus switching arrangement utilizes breaker and one half bus scheme.

During the unit tasks, the energy resource to the unit auxiliary load will be taken care of from the GTG by means of the unit transformer. During the unit shutdown and the unit start-up, the energy resource to the unit auxiliary load will be taken care of from 132 kV substation by means of the start-up transformer. The unit transformers will be associated with the 6.9 kV unit bus A by means of the circuit breakers. Then again, the start-up transformer will be associated with the 6.9 kV unit bus B by means of the circuit breakers. The energy will be appropriated to the helper loads from the unit bus.

GT Generator is synchronized at a 230 kV energy framework through a GT electrical switch when GTG is accomplished at evaluated speed and voltage. Then other

Table 4 Overview Specifications of the Generators

Generator	GT Generator	ST Generator
Type	3-Phase Synchronous	3-Phase Synchronous
Poles	2	2
Phases	3	3
Rated capacity	248MVA	131.6MVA
Frequency	50 Hz	50 Hz
Ordinary speed	3,000 rpm	3,000 rpm
End voltage	16 kV	11 kV
Power factor	Lagging 0.80	Lagging 0.80
Rotor cooler technique	Hydrogen or Water	Hydrogen or Water
Stator cooler	Hydrogen or Water	Hydrogen or Water

ST Generator is synchronized at 230 kV energy framework by means of ST electrical switch when STG is achieved at appraised speed and voltage.

GTG and STG can be synchronized at 230 kV power system breaker which is formed by one half bus scheme. For that reason there is no need to introduce GT and ST circuit breakers.

The design of the generator main circuit shall be based on the multi-shaft configuration of the having two (2) generators (GTG and STG) and two (2) generator step-up transformers (GT transformer and ST transformer). Each generator, transformer, PT is connected to the Isolated Phase Bus (IPB) and transmitted 230 kV substation via each generator circuit breaker and generator disconnecting switch (Table 4).

An air-cooled framework has few advances from hydrogen gas-based cooled framework such as ordinary framework, simple working principle and maintenance, saving cost. On the other hand adoption of air-cooled system makes the generator downsized so that air-cooled system has the advantage for transportation and construction stage [17].

Therefore, the generators for the steam turbine and gas turbine shall be of air-based cooled or H2 gas-cooled type. The Bidder shall have the application experience with similar capacity to the generator specified in his Bid.

The generator manufacturer should have the experience to have provided at least two air-cooled generators and/or two H2 gas-cooled generators. The capacity of the generators should not be less than 280MVA on the IEC conditions.

The generator maker will have AVR. AVR recognizes generator voltage and controls the receptive capacity to control the generator voltage.

There are two types of GT start-up techniques available, one is a motor-driven torque converter (MDTC) and another one is the thyristor start-up method (TSM). The utilization of the GT starts up technique depends on the requirement of the contractor. The components like a GT circuit breaker, disconnecting switch, and ST circuit breaker, disconnecting switch are placed at the second side of the GT and ST transformer, respectively for synchronization.

GT Generator is synchronized at a 230 kV energy framework by means of a GT electrical switch when GTG is achieved at appraised speed and voltage. Then, next ST Generator is synchronized at 230 kV force framework by means of ST electrical switch when STG is accomplished at evaluated speed and voltage. GT and STG can be synchronized at 230 kV power system breaker which is formed by one half bus scheme. For that reason there is no need to introduce GT and ST circuit breakers. GT and ST circuit breakers will adjust the load limit. The typical particulars of the GT and ST circuit breaker are demonstrated as follows.

- Rated Open Circuit Current: 800 – 1,250 A.
- Rated Short-Circuit Current: 25.0 – 31.5 kA.

The unit electric flexibly will be designed from a start-up transformer and unit transformer. The gear utilized for energy plant activity will be fueled from the unit transformer. The hardware utilized for normal gear (water taking care of, wastewater dealing with, and so on) will be controlled from the start-up transformer framework.

6.9 kV Unit Bus will flexibly fundamental auxiliary energy for plant activity. The plan of the generator main circuit shall be based on the two configurations of A and B. Unit Transformer will step down from GTG voltage (16 kV) to Unit Bus A (6.9 kV) and Unit Bus A shall supply necessary auxiliary power. Start-up Transformer will step down from transmission line voltage (230 kV) to Unit Bus B (6.9 kV) and Unit Bus B will flexibly vital helper power. Unit Bus An and B (6.9 kV) are associated by means of transport tie electrical switch and disengaging switch. Essentially the transport tie electrical switch and separating switch are opened. The transport tie electrical switch and detaching switch are shut toward the beginning down and shutdown stage. Unit Bus B evacuates Unit Bus A the electric power in that case. Likewise, Unit Bus B clears Unit Bus A the electric force when the plant coincidentally stumbled.

220 V DC Electric Supply System will have two battery gear and DC load will be provided the energy from DC conveyance board. The plant can stop securely by DC power from the battery under power outage conditions.

The plant will have one Emergency Diesel Generator Equipment. It will be fit for restart-up of the plant by power from Emergency Diesel Generator Equipment.

5 Cost, Economic, and Financial Analyses

The output and thermal efficiency of the power plant must be assumed for economic and financial analysis. There is a slight difference according to the supplier of the power plant. Purchase of the power plant is commonly carried out based on the EPC contract and the EPC contractor is selected through the international competitive

Table 5 Technical assumptions for dual fuel CCPP

Gross Power Output	451 MW
Net Power Output	450 MW
Net thermal efficiency	50%
Construction Period	42 months
Plant load factor	70%
Project period	30 years

bidding. In this case, the bid price is evaluated with consideration given to the difference in the proposed specifications, and the difference in performances including output and thermal efficiency. Accordingly, the bidder of the lowest price does not always win the contract. The technical assumptions for the plant are shown in the table below (Table 5).

The project cost includes the EPC cost, consultant fee, contingency, various taxes and duties, interest rate during construction, and direct administrative expenses incurred on the BPDB. The cost analysis is shown in Tables 6–2 under financial cost.

5.1 Scope of the Project

The project is to construct a mixed cycle Gas Turbine power generation plant in Feni. The plant is planned to operate as the base-load supplier. The fuel gas is to be supplied by Petrobangla possibly from the nearby gas pipeline of GTCL or Feni gas field. A branch pipeline is to be constructed to receive the gas from the trunk pipeline. For transmission of power, the project is to utilize the existing transmission line which will deliver the power generated by the project to the power grid for sale to the single buyer BPDB.

5.2 Fuel Price

In addition to the tariffs of power and gas, attention should be given to the tariff of the petroleum products, in particular, High Speed Diesel Oil (HSDO). Bangladesh produces no crude oil but imports all of its petroleum consumption. The import and marketing of petroleum products in the country is handled by Bangladesh Petroleum Corporation (BPC). The tariff matter is understood to be within the jurisdiction of BERC. Nevertheless, the government has been by-passing BERC in determining and announcing the tariff of petroleum products. The prices are kept under frequent revision, once every year lately, and the level of the tariff is treading the international prices closely. As of now (September, 2017), the local selling price of HSD is 65 tk/liter.

5.3 Operation and Maintenance Cost

Operation and maintenance costs include pay for officers, various inspection tests, repair and purchase of plant equipment whenever necessary, etc.

5.4 Profit and Loss Analysis

The financial analysis model enables us to review the profit and loss conditions of the project. To begin with, we have to make one reservation that the analyses conducted here in this item through item (5) ratio analysis are carried out by utilization of the model developed for FIRR and therefore does not take into the account of the price contingency while the tariff of the electricity generated is assumed to be sold to BPDB at the bulk selling rate of BPDB. The fuel price is considered the prevailing gas tariff of Petrobangla. The gas tariff of Petrobangla is under the appraisal of BERC for its upward revision. The potential increase of the tariff will invoke the subsequent revision of power tariff that might maintain at least the present cost benefit relationship. The model analysis indicates that the earnings before interest and tax (EBIT) are shown to be positive from the very first year of operation through the end of the project. For the net profit after tax, the project will take the deficit on the first year but will turn to be positive on the second year (FY 2026) and will keep going through the final end of the project. To be also learnt from the model, the project will have profits from the second year of operation and thus will be subject to the income tax.

5.5 Cash Flow Analysis

The major stream of the cash flow is seen to be

(1) Cash is constantly generated through the daily operation;
(2) Expenditures for the operation and maintenance are paid out of the cash flow;
(3) The cash flow covers payment of principal and interest of the loan;
(4) Cash flow remaining after 1) through 3) will be declared and paid as the dividend.

The prominent feature of the cash flow is that the free cash flow generated through the operation sufficiently covers the operational expenses, debt service, and the return on the equity. The profits after tax are added by the depreciation to produce the significant amount of the free cash flow. The annual cash flow appears to be negative, though in small amounts, in the first year and 6 consecutive years starting from 2030 the repayment of loans commence. The setbacks in the cash flow will be in minimal amount and will be accommodated by the accumulation of cash from during prior years. The free cash flow which is described as the cash flow from operating activities less the cash flow from investment activities maintains the comfortably positive level supported by the constant cash flow generated from operating activities, as the project anticipates no investment during its life with exception of maintenance. There will be neither shortfall nor insufficiency of funds for the financial operation of the project including the payment of principals and interest of the loans, payment of fiscal levies as well as the declaration of dividend.

5.6 *Debt Service*

The net cash flow from operating activities starts the first year with a negative amount, but generates surplus of Taka 2,000 million or above in the second year and steadily increases to stay at Tk 2,000–3,000 million till the end of the project. The repayment of the principal of the loan will commence in the 6th year (FY 2030) and will be over in FY 2049.

6 Conclusion

The project will greatly contribute to increase generation capacity and in turn provide access of power to more people and development of the country. Dual fuel arrangement will make the power plant operable even in case of gas crisis. Modern fuel efficient techniques will ensure lower Green House Gas production while emission of other gases will be controlled with advanced filters and safety procedures.

The ecological investigation has uncovered that the venture can be set up as indicated by the proposed structure and design in the proposed site and area. The ecological effects are of restricted nature, though the advantages of the task are many. The assessment process included scoping, site visits, site surveys for impact assessment based on project level information provided by the project developer, primary baseline studies and monitoring, and extensive stakeholder consultations along with reviewing of Site and Configuration Selection Report, Reconnaissance Survey Report and studying satellite imageries. Through this procedure, an appraisal has been attempted of the likely natural and social dangers and effects that might be ascribed to the improvement of the task in its pre-development, construction, and activity stages. Appraisals of the effects have been given an effective rating of every likely effect. Options in contrast to the Project and key plan perspectives were additionally taken into consideration. The Project Proponent should ensure that construction activities be limited in the project site only. The project proponent should also consult with Forest Department and NGOs involved in conservation of species for in situ conservation measures of species of significance, if any.

Combine Cycle frameworks are proficient minimal cost frameworks that give affirmations of execution and working destinations. Combine Cycle frameworks can be modified to the utility requirements and inclinations. They offer lucrative conservative flexible energy production. Working adaptability of Combine Cycle power producing warrant their thought for most energy production applications. Overall the proposed power plant can contribute to the betterment of the socio-economic condition of the country.

References

1. Hossain, Mohammad Imtiaz, Ishtiaque Abedin Zissan, Md Shariare M. Khan, Yasin Rashiq Tushar, and Taskin Jamal. "Prospect of combined cycle power plant over conventional single cycle power plants in Bangladesh: A case study." In 2014 International Conference on Electrical Engineering and Information & Communication Technology, pp. 1–4. IEEE, 2014.
2. Miah, Md Sazal, Md Abdul Momen Swazal, Sazib Mittro, and Md Morshedul Islam. "Design of A Grid-Tied Solar Plant Using Homer Pro and an Optimal Home Energy Management System." In 2020 IEEE International Conference for Innovation in Technology (INOCON), pp. 1–7. IEEE, 2020.
3. Alam, Mahmudul, Md Ashraful Dewan, Shikder Shafiul Bashar, Md Sazal Miah, and Anupom Ghosh. "A Microcontroller Based Dual Axis Tracking System for Solar Panel." In 2019 3rd International Conference on Electrical, Computer & Telecommunication Engineering (ICECTE), pp. 25–28. IEEE, 2019.
4. Shetol, M. Hassan, M. Moklesur Rahman, Ratneshwar Sarder, M. Ismail Hossain, and F. Kabir Riday. "Present status of Bangladesh gas fields and future development: A review." Journal of Natural Gas Geoscience 4, no. 6 (2019): 347–354.
5. http://egcb.gov.bd/
6. Lipu, Molla Shahadat Hossain, Md Shazib Uddin, and Muhammad Ahad Rahman Miah. "A feasibility study of solar-wind-diesel hybrid system in rural and remote areas of Bangladesh." International Journal of Renewable Energy Research (IJRER) 3, no. 4 (2013): 892–900.
7. Bhuiyan, N., Ullah, W., Islam, R., Ahmed, T., Mohammad, N.: Performance optimisation of parabolic trough solar thermal power plants–a case study in Bangladesh. Int. J. Sustain. Energ. **39**(2), 113–131 (2020)
8. Chowdhury, Nusrat, Chowdhury Akram Hossain, Michela Longo, and Wahiba Yaïci. "Feasibility and Cost Analysis of Photovoltaic-Biomass Hybrid Energy System in Off-Grid Areas of Bangladesh." Sustainability 12, no. 4 (2020): 1568.
9. https://www.bpdb.gov.bd/bpdb_new/index.php/site/page/5a3f-2fdb-e75f-3cab-e66b-f70d-5408-cbc9-f489-c31c
10. https://documents.worldbank.org/en/publication/documents-reports/ documentdetail /649111468740689490/bangladesh-energy-assessment-status
11. Das, A., McFarlane, A.A., Chowdhury, M.: The dynamics of natural gas consumption and GDP in Bangladesh. Renew. Sustain. Energy Rev. **22**, 269–274 (2013)
12. https://new.siemens.com/global/en.html
13. https://www.mhi.com/mitsubishi-heavy-industries-ltd-global-website
14. Kehmna, Mark David, and Seyfettin Can Gulen. "Duct fired combined cycle system." U.S. Patent 9,500,103, issued November 22, 2016.
15. Copen, John. "Exhaust heat augmentation in a combined cycle power plant." U.S. Patent Application 11/297,063, filed June 14, 2007.
16. William A Vopat , Power station economy and engineering , 2010–2011
17. Gadhamshetty, V., Nirmalakhandan, N., Myint, M., Ricketts, C.: Improving air-cooled condenser performance in combined cycle power plants. Journal of energy engineering **132**(2), 81–88 (2006)

A Brief Study on Applications of Random Matrix Theory

N. Siva Priya and N. Shenbagavadivu

Abstract Random matrix theory proves itself to be an efficient tool in recent technologies and many application areas. The main scope of this paper is to do a brief study on various application fields of Random Matrix Theory (RMT), for future analyses purposes. It is used as an analytical tool in many developing applications and analytical fields like optical physics, cellular networks, wireless communication, cell biology and even in big data analytics, to find out the hidden correlation between variables, hypotheses and to provide knowledgeable information to end users. The random matrix theory is believed to have many applications from quantum physics to biological networks. The analysis is made simple by using many laws like Semi-circle law, Full circle law, Marcenko-Pastur law, Stieltjes's transform, eigenvalue distribution and so on.

Keywords Random matrix theory · Random variables · Eigenvalues · Correlation · Probability density function

1 Introduction

Random matrix is a row vector matrix containing random variables. Random matrix theory is the mathematical science that deals with laws and theories used for analysing data in the form of a matrix, whose entries are random. The name random itself implies that this theory can be used to analyse the data whose occurrence is unpredictable. The eigenvalue determination and study of eigenvalue distribution is also an important task while employing Random Matrix Theory (RMT) tools for studying or analysing any user defined data. In accordance to linear algebra, as per the definition cited from Wikipedia [24] "An eigenvector or characteristic vector of a linear transformation is a non-zero vector that only changes by a scalar factor when that linear transformation is applied to it; the roots to this eigenvector are called as eigenvalues" [1]. The entries of the random matrix are converted to the corresponding eigenvectors

N. Siva Priya (✉) · N. Shenbagavadivu
Department of Computer Applications, University College of Engineering, Anna University, Bharathidasan Institute of Technology Campus, Trichirapalli-24, India

J. S. Raj et al. (eds.), *Intelligent Sustainable Systems*, Lecture Notes in Networks and Systems 213, https://doi.org/10.1007/978-981-16-2422-3_36

and eigenvalues for ease of analysis. Since their occurrence is random the eigenvalues and eigenvectors are also random, and the main objective of this theory is to understand the distribution of eigenvalues. This theory is not only applicable to small pairs of data but also can detect groups of many correlated units and even groups that change over time, and consistently add and lose members. Random matrix theory also acts as a statistical tool with universality in its eigenvalues and eigenvectors, which are used to find hidden correlations within masses of random data and to gain knowledgeable information. Various laws involved in RMT has already been discussed elaborately in the references [1–4], and in the below subsection we have a brief view on the history of RMT.

The main goal of the Random Matrix Theory is to offer in depth knowledge on the varied properties, derived mostly from the statistics of matrix eigenvalues of matrices with entries drawn randomly from various probability distributions traditionally referred to as the random matrix ensembles. Three classical random matrix ensembles are the Gaussian Orthogonal Ensemble (GOE), the Gaussian Unitary Ensemble (GUE) and the Gaussian Symplectic Ensemble (GSE). They are composed, respectively, of real symmetric, complex Hermitian and complex self-ad joint quaternion matrices with independent, normally distributed mean-zero entries whose variances are adjusted to ensure the invariance of their joint probability density with respect to Orthogonal (respectively, Unitary or Symplectic) similarity transformations. Such invariance is also shared by the corresponding Lebesgue measures. If one keeps the requirement of invariance of the joint probability density of all entries but relaxes the property of entries being independent, one arrives at broader classes of invariant non-Gaussian Ensembles (Orthogonal, Unitary or Symplectic). Note that being self-ad joint the matrices of those types have all their eigenvalues real.

Equally important are the Circular Ensembles composed of complex unitary matrices sharing the same invariance properties of the measure as their Gaussian counterparts, but with eigenvalues confined to the unit circle in the complex plane rather than to the real line. Those ensembles are known as the Circular Orthogonal Ensemble (COE), the Circular Unitary Ensemble (CUE) and the Circular Symplectic Ensemble (CSE). Finally deserves mentioning the so-called Ginibre Ensemble of matrices with independent, identically and normally distributed real, complex or quaternion real entries, and no further constraints imposed. The corresponding eigenvalues are scattered in the complex plane.

1.1 Brief History of Random Matrix

Random matrix theories have been a great tool for both mathematicians and physicists. Wishart in 1928 discovered Random Matrix Theory for learning mathematical statistics. Despite a slow start, random matrix theory became more prominent after the introduction of the concept of statistical distribution of nuclear energy levels by Wigner in 1950. But only after 1955 when Wigner introduced ensembles of random matrices it came to popularity.

Random matrices have been gaining popularity year by year resulting in the application of RMT extending in various fields.

The above table discusses a brief history of random matrix theory in various field. There has been various developments and discoveries in RMT consistently the table above just discusses few improvements and discoveries in this field, the authors in [2–4] has discussed the history of RMT in detailed manner (Table 1).

Table 1 Historical review of random matrix theory [2, 3, 5]

S.No	Year	Author	Discovery	Area
1	1928	Wishart	Wishart distribution	Mathematical statistics
2	1939	Fisher, Hsu, Roy & Girshick	Fisher-Hsu-Roy distribution	Mathematics
3	1950	Eugene Wigner	Wigner matrix	Nuclear physics
4	1955	Potter & Thomas	Potter-Thomas distribution	Nuclear decays
5	1956	Wigner	Wigner surmise	Nuclear physics
6	1957 1980	Harish-Chandra Itzykson-Zubers	Harish-Chandra-Itzykson-Zuber integral	Unitary matrix integral in mathematics
7	1960 to 1962	Dyson	Classification of random matrix ensembles according to their invariance	Mathematical physics
8	1960	Potter & Rosenweig	Invariant random matrix ensembles	Physics- Spectra in complex atoms
9	1960	Mehta	Orthogonal ensemble Gou,Goe,Gos	Mathematical physics
10	1963	L.K.Hua	Integration measure of invariant random matrix ensembles	Mathematical science
11	1963	Dyson-Mehta	Dyson-Mehta statistics	Mathematical statistics
12	1964	James	Zonal polynomials	Mathematics
13	1964	Gaudin & Mehta	Mathematical analysis of spacing distribution	Mathematics
14	1965	Gorkov & Eliasberg	Theory of small metallic particles	Mesoscopic physics
15	1965	C.E.Potter	Statistical theories of spectra	Spectra in complex atoms
16	1965	Anderson	One dimensional disordered systems, eigenfunctions of disordered systems	Solid state physics
17	1965	Ginibre	Ginibre ensemble	Nuclear physics

(continued)

Table 1 (continued)

S.No	Year	Author	Discovery	Area
18	1967	Marchenko and Pastuer	Marchenko-Pasteur law	Physics
19	1967	M.L.Mehta	A study on the link between random matrices and the energy levels theory	Nuclear physics
20	1968	Balian	Gausssian random matrix for minimising the random matrix ensembles in information entrophy	Physics and nuclear physics
21	1982	Muirhead	Matrix integrals and zonal polynomials	Mathematical statistics
22	1983	Efetov	Supersymmetric functional integrals of disordered solids	Nuclear physics
23	1984	Efetov	A generic link between RMT and the spectral fluctuation properties of classically chaotic quantum systems with few degrees of freedom	Bohigas conjecture
24	1985	Silverstein	Asymptotic expression for multivariate F-matrix	Statistics
25	1990	Korepin, Bogoliubov and Izergin	Review on one-dimensional integral systems	Quantum physics
26	1993	Forrester	Relation between random matrix theory and information systems	Monographs
27	1997	Aggassi	Theory of S-matrix fluctuations	Optical physics

2 Brief Account on Applications of Random Matrix Theory

Though the random matrix theory was introduced to address the problem in Mathematical statistics, it became popular only after the introduction of Wigner ensembles and Wigner Surmise by Eugene Wigner to address the problems in nuclear physics. The random matrix theory has so many application areas from physics to recent trends like big data analytics, data mining and so on. The main reason behind this extensive use of Random matrix theory in many application areas is that any physical system can be represented and modelled in the form of a matrix for its ease of analysis. Not only for theoretical analysis can real-time analysis of any physical system be accomplished using this vast theory. The theory does not bind itself to modelling and analysis of linear systems; it has also paved way for the analysis of any nonlinear system with random variables as its outcome. Recent research has also showed that random matrix theory can also be applied to the big data analytics [10, 20, 22]. Random matrix theory has also set path to the analysis of biological data like EEG data, feature extraction of brain images and so on [11–13]. The major application area of theory is wireless communication and information theory. In wireless communication random matrix theory is used extensively in channel modelling, calculation of Signal to Noise Ratio (SNR), information entropy and so on.

Random matrix theory turns out to be an irreplaceable tool for analysis and modelling of versatile data types. The basic criterion to be satisfied by the data for analysis is that the data set should be random. Irrespective of type of data, predictability, volume and so on all kinds of data can be modelled/ analysed. The area of application is also not a constraint for this theory, as discussed in the previous paragraph this theory has been used almost in all the fields. In the forthcoming sections the application areas of random matrix theory is discussed in brief.

3 Random Matrix Theory in Physics

As mentioned in the previous section random matrix theory became popular in the mid 1950s after it has been deployed to nuclear physics. Gradually this theory has been deployed to other branches of physics like optical physics, quantum chaos, condensed matter physics, disorder and localization, mesoscopic transport, optic lasers, quantum entanglement, neural networks, gauge theory, QCD, matrix models, cosmology, string theory, statistical physics (growth models, interface, directed polymer,etc.,), cold atoms and so on.

3.1 RMT in Nuclear Physics

In 1950 random matrix theory was used by Wigner and Dyson for the analysis of nuclear spectra. Not only in the analysis of nuclear spectra but it also plays a major role in the theory of nuclear reactions. It was also used in the study of compound nucleus for CN reactions and CN scattering. Various experimental reviews conducted by Wigner revealed that there was a strong relationship between RMT and CN scattering. Later along with CN scattering various researches revealed a strong relationship between RMT and chaotic quantum scattering. This further progressed by combining scattering theory based on shell model with novel techniques using super symmetric generating function.

3.2 RMT in Condensed Matter Physics

Random matrix theory was deployed in condensed matter physics in two different categories. The first one to study the thermodynamic properties of closed systems like metal grains or semiconductor dots and the second category is to study the transport properties of open systems like metal wires or quantum dots with point contact. The applications of random matrix theory in these two broad categories have flourished the development in nanotechnology.

3.3 RMT in Optical Physics and Disordered Systems

The random matrix theory finds its application in optical physics in both classical optical systems and quantum optical systems. In classical optical systems the interference patterns of optical speckle and coherent backscattering was observed using random matrix theory. Apart from this the Random matrix theory was also used to study reflection from an absorbing random medium, long-range wave function correlation and open transmission channels. In quantum optical systems random matrix theory has been deployed in the study of grey body radiation and chaotic laser cavity. Random matrix theory has been used to determine the statistical energy levels and Eigen functions in disordered systems [6, 7].

The following table summarises the application of Random matrix theory in physics including nuclear physics, condensed matter physics, optical physics and disordered systems (Table 2).

4 Random Matrix Theory in Wireless Communication

Random matrix theory is being extensively used in the field of wireless communication. It performs the important task of separating the required signal from noise. The theory gained its popularity after it has been involved in wireless communication. It has been deployed in this area during 1960s, as mentioned in Table 1, it was used for the very first time in information entropy by Balian in 1968. This theory is used to study various characteristics of wireless communication channels like channel capacity, noise filtration, signal to noise ratio and so on.

This theory is used widely to study the characteristics and spectral analysis of both wired and wireless communication channels. It is also used to filter the noise from the signal. The applications also include the determination of Channel Capacity and SNR in Direct Sequence CDMA, Multiple Carrier CDMA, Racian Channel, etc. In DS CDMA channels the channel capacity is determined in three cases, viz., channel capacity without fading, channel capacity with fading and Channel capacity with frequency selective fading.

Table 2 Applications of Random Matrix Theory to Physics

S.No	Research Area	Application	Parameters	Analysis result
1.	Nuclear physics	• Analysis of nuclear spectra. • Statistical theory of nuclear reactions. • Study and analysis Compound Nuclear (CN) scattering.	• Resonances • Dynamic origin resonances • Resonance reactions in chaotic nuclear system • N resonances • Scattering matrices (S-matrix). • Stochastic S matrices. • Average S matrices. • Violation of symmetry or invariance. • Ericson fluctuations • Isospin	• Spectrum of neutrons taken from G.E.Mitchell.et.al [8].  • Autocorrelation function of compound nuclear scattering taken from G.E.Mitchell.et.al [8]. 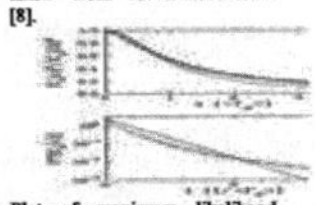 • Plot of maximum likelihood function versus root mean square of values obtained from U_{238} taken from G.E.Mitchell.et.al [8]. 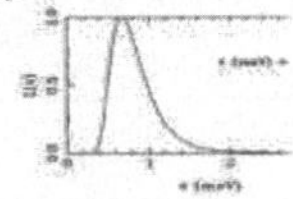
2.	Condensed matter physics	• Searching for universal features in the electronic pro-perties of metals, originating from the universality of eigenvalue repulsion. • Study of thermodynamic properties of closed systems, such as metal grains or semiconductor quantum dots. • Study of transport properties of open systems, such as metal wires or quantum dots with point contacts	• Eigenvalue repulsion. • Quantum wires • Conductance fluctuations • Non lograthimic eigen value repulsion • Sub poissonian shot noise. • Quantum dots • Level and wave function statistics • Scaterring matrix ensembles • Conductance distribution • Sub poission shot noise • Thermopower distribution.	• Fluctuations as a function of perpendicular magnetic field taken from Beenakker.et.al[6]. 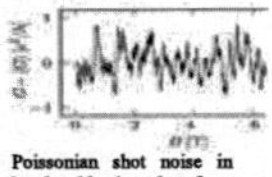 • Sub Poissonian shot noise in disordered gold wire taken from Beenakker.et.al [6]. 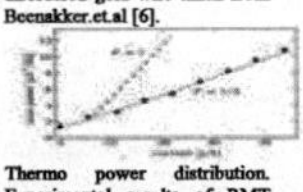 • Thermo power distribution. Experimental results of RMT taken from Beenakker.et.al [6]. 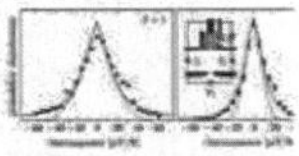
3.	Optical Physics	• Analysis of optical speckle and coherent backscaterring. • Analysis of reflection from an absorbing random medium. • Correlation Long range wave function. • Distribution of open transmission channel. • Distribution of grey body radiation in quantum optics. • Random matrix theory analysis of a chaotic laser cavity.	• Classical optics • Optical speckle and coherent backscattering. • Reflection from an absorbing random medium. • Long range wave function cor-relation.. • Open transmission channels. • Quantum optics • Grey body radiation. • Random matrix of chaotic laser cavity.	• Speckle pattern produced by laser beam taken from Beenakker.et.al [6]. 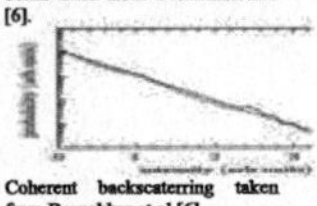 • Coherent backscaterring taken from Beenakker.et.al [6].  • Speckle pattern of open transmission channels taken from Beenakker.et.al [6]. 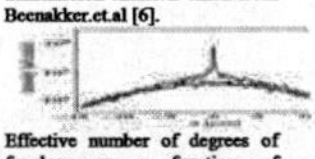 • Effective number of degrees of freedom as a function of normalized amplification rate in chaotic laser cavity taken from Beenakker.et.al [6]. 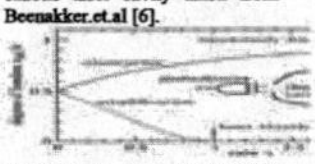
4.	Disordered systems	• Description of statistical properties of energy level in disordered systems. • Generation of disorder-averaged correlation function. • Statistical characteristics of disordered systems. • Theoretical study of eigenfunction statistics.	• Eigenfunction fluctuations. • Energy levels. • Deviations from universality. • Quasi one dimensional geometry. • Average of eigenvalues.	• Distribution of eigenfunction taken from Efetov.et.al[7]. 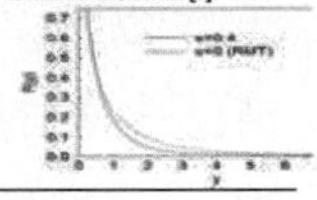

4.1 *Random Matrix Theory in Multiple Antenna and Massive MIMO Wireless Channels*

Random matrix theory was also used to study the characteristics of multiple antenna wireless communication channels. As per the work done in [4] the theory was used in characterising the fundamental channel capacity limits in Multiple Input Multiple Output antenna paradigm, performance analysis of practical MIMO transmission Strategies of real-time propagation areas and so on. The characteristics of microarray antenna were also studied using random matrix theory.

In a recent study [23] random matrix theory is also used to model and characterise the massive data MIMO wireless communication channels. This is yet another milestone of random matrix theory to the emerging field of big data analytics.

The above table summarises some of the applications of random matrix theory to wireless communication channels and networks (Table 3).

Table 3 Summarising RMT applications to wireless communication

S.No	Research Area	Application	Parameters	Analysis result
1.	Wireless communication	• Analysis of Direct Sequence CDMA wireless channel. • Determination of Signal to Noise Ratio (SNR) in wireless networks. • Modelling and analysis of MIMO systems. • Modelling and analysis of multiple antenna wireless communication systems. • Big data analytics of MIMO wireless communication System.	•Capacity of Direct Sequence CDMA. •Signal to Noise Ratio of Direct Sequence CDMA. •Channel capacity of Multi Carrier CDMA. •Signal to Noise Ratio of Multi Carrier CDMA. •Ergodic capacity and average symbol rate in multiple antenna communication systems.	•Capacity of CDMA taken from Tulino et.al.[4]. 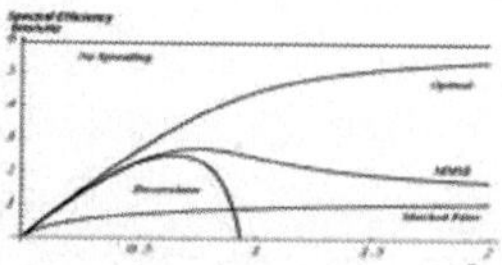 • Probability density function of massive MIMO system taken from Tulino et.al. [4]. 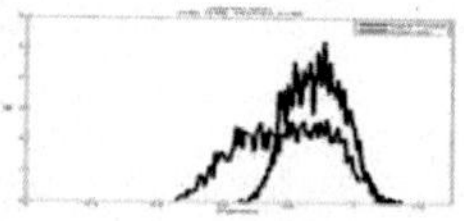 • Analytical result for multiple antenna wireless communication system taken from Tulino et.al. [4]. 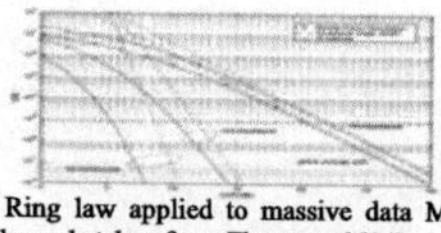 • Ring law applied to massive data MIMO channels taken from Zhang.et.al.[24] 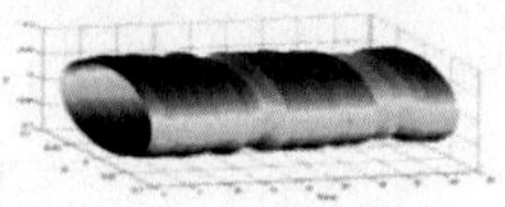

5 Random Matrix Theory in Cell Biology

In biological networks Random matrix theory is deployed to model the biological cellular structure. The yeast protein–protein interaction networks and yeast metabolic network is found to follow the Gaussian Orthogonal Ensembles of Random Matrix Theory. The protein cell without interaction is found to follow Poisson's distribution. Random matrix theory was also used to identify the functional modules within global biological networks. Biological networks have modular structure with stronger interaction hence to distinguish between these components random matrix theory was used. Apart from the biological cellular networks in the field of biological sciences it is also used in the AIDS study. From the study of random matrix theory application to stock market and analysis of biological enzymes, Dr. Arup Chakraborthy, Chemical Engineering professor of MIT thought that this theory could be adopted in the study of sector of HIV that rarely undergoes multiple mutations.

5.1 *Analysis of Cancer Cell*

Cancer is found to be a dreadful disease arising due to topological change in gene interaction. A perfect tool is required to analyse this change for the early prediction or prevention of this disease. In [11] the author has preferred using random matrix theory for such analysis because of its universality and symmetry property of some parameters of random matrix theory like Gaussian Orthogonal ensembles, Wigner matrix distribution, etc. The cancel cell gene interaction data collected from TNCG data repository from many number of patients has been modelled in the form of matrix for analysis. The gene interaction is modelled in the form of cumulative distribution function of Wigner matrices. In [11] the author has presented the analysis result of cancer cell gene interaction in a detailed manner. Thus random matrix theory has become a powerful tool surpassing decades in modelling and analysis of biological network.

5.2 *Denoising and Feature Extraction using RMT*

The other fields of biology where random matrix theory is applicable include denoising and feature extraction of brain images and analysis of EEG data. In [13] the author has deployed random matrix theory for denoising the brain MRI images. This is done by exploiting the redundancy in MRI images. The author has made use of the universality property of the Eigen spectrum of random covariance matrices thereby reducing the noise without compromising the image accuracy. The author has also obtained reasonable increase in signal to noise ratio. By some statistical methods the author was able to obtain improvement in the image precision. This not only comes under biological application of RMT but also an image processing application of this theory, where the precision of the image is improved by reducing noise and intrinsic redundancy.

In the analysis of EEG data [11] the author has aimed to analyse the correlation of the EEG data by using the tools in random matrix theory. For this application the EEG data is modelled as a random matrix, as the human brain signal is not always constant and unpredictable. In this work the spectral density of the EEG signal is found to be a best fit in Wishart ensemble rather than Marchenko-Pasteur distribution. As a result the modelled EEG data of about 90 people bounds under the universality property of Random Matrix Theory.

Mathematical modelling of biological data has always been a challenge. But by the use of Random matrix theory the mathematical modelling of data like EEG, Cancer cell data, etc. is becoming feasible. Not only mathematical modelling but also by the knowledge of application of RMT in image processing many biological images have been processed for viewing the image in a better way so that prognosis and diagnosis are made in a proper manner. Biology being a most prominent field, accuracy is always more important, which is acquired by the use of RMT.

The applications of Random Matrix Theory to biology covering the areas like data analysis, feature extraction, study of spectral density, etc. has been discussed in brief and the below table summarises the discussion with some results obtained (Table 4).

6 Random Matrix Theory in Data Mining and Data Analysis

Recent areas of research focus highly on data mining and data analysis. Data mining involves extraction of useful information from a given database, similar to denoising of signals or images. As discussed in the previous sections RMT has proved to be a suitable tool for denoising purposes. RMT has also been used in feature extraction, which according to data science is termed as data mining [16, 21]. The Principal Component Analysis (PCA) of RMT is used to reduce the size (dimensionality reduction) of the data without missing out the important values (Table 5).

The above table also discussed some of the emerging trends which use RMT for analysis, data cleaning, data modelling, etc. But RMT is not restricted for all the above-mentioned applications it has many more fields like statistics, communication, etc. to mention some.

Table 4 RMT application to biological networks

S.No	Research Area	Application	Parameters	Analysis result
1.	Biology	• Modelling of yeast biological network. • Analysis of cancer cell gene interaction. • Denoising of brain images. • Analysis of EEG data. • Analysis of biological network, yeast protein-protein interaction. • Random matrix theory approach in AIDS studies (Theoretical study).	•Universality property of eigenvalues. • Signal to Noise Ratio. • Covariance spectrum. • Wigner matrices. • Cumulative distribution function.	Cancer cell gene interaction taken from Luo.et.al.[13]. 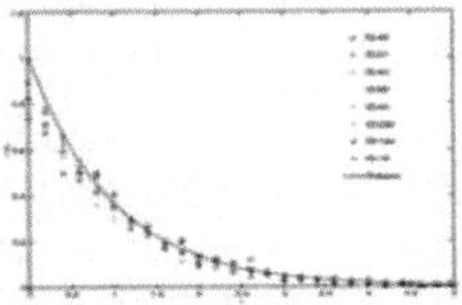 • Spatial distribution of human EEG data taken from Seba.et.al.[12]. 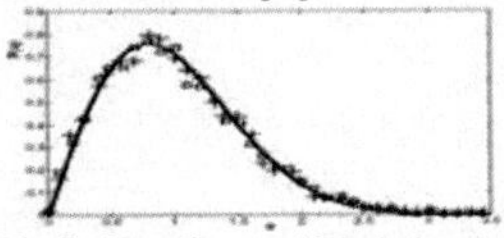 • Distribution of yeast biological network taken from Luo.et.al. [13]. 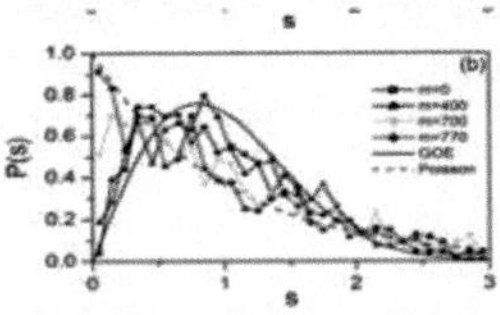 • Effect of denoising brain images using random matrix theory taken from Veeratta.et.al.[14]. 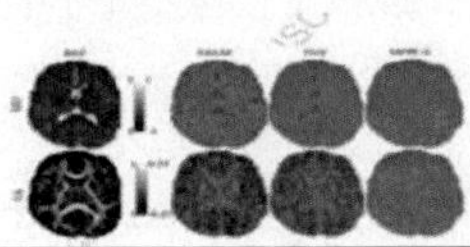 •

7 Conclusion

In this paper, we have discussed various application areas of random matrix theory in a brief way irrespective of limitations in the methods used, we find that RMT has had been a very expedient tool in various research areas. But, the above-mentioned areas are to cite some, random matrix theory has broader application areas than this. Random matrix theory can be applied to any application area whose output is unpredictable and random by viewing the data in terms of spectrum researchers find it valuable to gather handy information from any random data. A study made by Arup Chakraborthy, chemist and chemical engineering professor, MIT based on random matrix theory applications to stock market, revealed the application of random matrix theory to AIDS study. Dr.Chakraborty thought that random matrix theory could help find sectors of HIV that rarely undergo multiple mutations. The research based on his study ended positively. Likewise random matrix theory had always been a better tool not only for statistical study and spectrum analysis but also for emerging trends like

Table 5 RMT applications to emerging technologies

S.No	Research Area	Application	Parameters	Analysis result
1..	Data mining and data analysis	• Data mining of power system in non Gaussian environment using random matrix theory. • Dimensionality reduction for analysis of huge volume of random data.	• Non gaussian data • Time series data. • Power equipment conditioning • Voltage stability • Low frequency oscillation • Signal detection.	• Distributon of various phase lines in power system taken from Han.et.al. [22]. 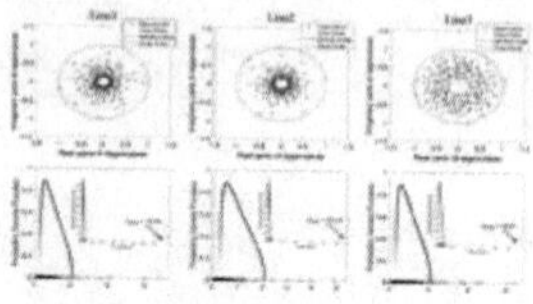
2.	Financial applications	• Correlation cleaning recipes of financial data inspired from random matrix theory. • Risk and portfolio analysis of stock market using statistical tools of random matrix theory.	• Correlation • Pearson estimator • Probability density function • Multivariate distribution of returns. • Risk analysis.	• Emprical average eigen value distribution of financial data taken from Philippe.et.al.[17]. 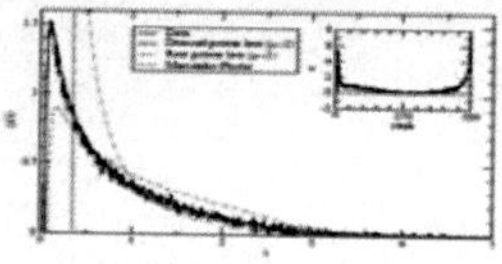
3.	Other emerging application areas.	• Big data analytics of mobile cellular networks. • Big data analytics of smart grid. • Random matrix theory approach for machine learning and deep learning of neural networks.	• Big location data, big signalling data, big heterogeneous data, etc. • Mean spectral radius • Emprical spectral density • Kernel density estimator.	• A framework for big data analytics in mobile cellular networks taken from He.et.al.[21]. 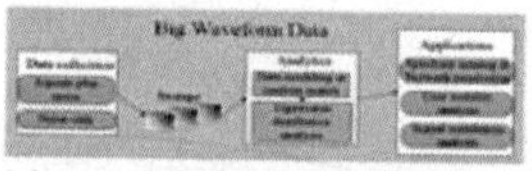 • A framework for big data analytics in smart grid taken from He.et.al.[23]. 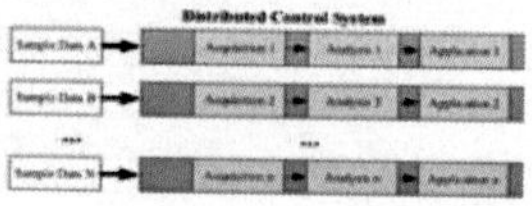

feature extraction, data mining, data analysis, big data analytics and so on especially for dimensionality reduction and smoothening of voluminous data.

References

1. Krishnapur, M.: "Random Matrix Theory," Lecture notes on course on random matrix theory held at IISC, Bangalore (2011)
2. Forrester, P.J., Snaith, N.C., Verberscoth, J.J.M.: Developments in random matrix theory. J. Phys. A: Math. Gen, 36, pp R1- R10 (2003) stacks.iop.org/JPhysA/36/R1
3. Edelman, A., Wang, Y.: Random matrix theory and its innovative applications
4. Tulino, A.M.: Random matrix theory and wireless communication, Foundation and Trends in Communication and Information Theory. (2004)
5. Mehta, M.L.: Random matrices, vol. 142. Academic press (2004)
6. Beenakker, C.W.J.: Applications of random matrix theory condensed matrix and optical physics, The oxford handbook of random matrix theory, oxford publications (2008) arXiv: 0904.1432V2
7. Efetov, K.B.: Random matrices and supersymmetry in disordered systems (2005) arXiv.cond-mat/0502322v1

8. Mitchell, G.E., Richter, J., Weidenmuller, H.A.: Random matrices and Chaos in Nuclear Physics: Nuclear reactions (2010) arXiv.1001.2422v1
9. Khorunzhy, A.M. Khoruzhenko, B.A., Pastur, L.A.: Asymptotic properties of large random matrices with independent entries. J. Math. Phys. **37**(10), 5033–5060 (1996)
10. Dieng, M., Tracy, C.A.: Application of random matrix theory to multivariate statistics, (2006) arXiv:math/0603543v1
11. Qiu, R.C.: Large random matrices and big data analytics. In: Big Data of Complex Networks. CRC Press, Boca Raton, FL, USA (2016)
12. Seba, P.: Random matrix analysis of human EEG data, Phys. Rev. Lett. **91**(19), (2013)
13. Feng, L., Jianxin, Z., Yunfeng, Y., Scheuermann1, R.H. Jizhong, Z.: Application of random matrix theory to biological networks, research supported by United State Department of Energy under Genomics
14. Veraarta, J., Novikov, D.S., Christiaens, D., Ades-aronb, B., Sijbersa, J., Fieremansb, E.: Denoising of diffusion MRI using random matrix theory. Neuroimage (2016). https://doi.org/10.1016/j.neuroimage.2016.08.016
15. Rojkova, V.B.: Feature extraction using random matrix theory, Electronic theses and dissertations. Paper 1228.https://doi.org/10.18297/etd/1228.
16. Tao, T., Vu, V., Krishnapur, M.: Random matrices: universality of ESDs and the circular law, Ann. Probab. **38**(5), 2023–2065 (2010)
17. Bouchaud, J.P., Potters, M.: Financial applications of random matrix theory: a short review. (2009) arXiv:0910.1205v1
18. McKay, M.R.: Random matrix theory analysis of multiple antenna communication systems. A dessertion submitted for the degree of Doctor of Philosophy at the University of Sydney, Australia (2006)
19. Louart, C., Liao, Z., Couillet, R.: A random matrix approach to neural networks, Ann. Appl. Probab. (2017) arXiv:1702.05419v2
20. Achlioptas, D.: Random matrices in data analysis. In: Boulicau,t J.F. et al. (eds.)ECML 2004, LNAI 3201, pp. 1–7, Springer-Verlag Berlin Heidelberg (2004)
21. He, Y., Yu, F.R., Zhao, N., Yin, H., Yao, H., Qiu, R.C.: Big data analytics in mobile cellular networks. In: Special section on theoretical foundations for big data applications: challenges and opportunities. IEEE Access (2016) https://doi.org/10.1109/ACCESS.2016.2540520
22. Han, B., Luo, L., Sheng, G., Li, G., Jiang, X.: Framework of random matrix theory for power system data mining in a Non-Gaussian environment. IEEE Access (2017). https://doi.org/10.1109/ACCESS.2017.2649841,Jan
23. He, X., Ai, Q., Qiu, R.C., Huang, W., Piao, L., Liu, H.: A big data architecture design for smart grids based on random matrix theory. IEEE Trans. Smart Grid (2015) arXiv 1501.07329v3 [Stat.ME]
24. Zhang, C., Qiu, R.C.: Massive MIMO as a big data system: random matrix models and testbeds. (2015) arXiv:1503.06782v1
25. Wikipedia: The free encyclopedia, Wikimedia Found., Inc., (2016)

Development of Feedback-Based Trust Evaluation Scheme to Ensure the Quality of Cloud Computing Services

Sabout Nagaraju and C. Swetha Priya

Abstract The expeditious rise in cloud computing facilitates economical data storage, expertise software, and high-speed and scalable computing resources through infrastructures, platforms, and software (as a service). However, due to the openness and highly non-transparent nature of cloud computing, trust is the pressing issue that may hamper the adoption and growth of cloud services. Moreover, conventional trust management solutions are inadequate and cannot be directly adopted in cloud computing environments. Many existing techniques have addressed the above issue, and a comprehensive study of the literature helped in formulating the current research objectives. This research intends to develop an effective feedback-based trust evaluation scheme to ensure the quality of cloud computing services. The proposed scheme finds an accurate global trust value of the cloud services based on the aggregation of the genuine feedbacks trust, reputation trust, and Quality of Service (QoS) trust. Suitable performance analysis is done using Google cloud trace logs. The results show that the feedback-based trustworthy monitoring system is very much desirable and solely needed for reliable cloud computing service provisioning.

Keywords Cloud computing · Feedbacks trust · Reputation trust · Quality of service trust · Witness reputation and neighborhood reputation

---

S. Nagaraju (✉)
Department of Computer Science, PUCC, Pondicherry University, Lawspet 605008, India
e-mail: saboutnagaraju1983@gmail.com

C. S. Priya
S-201, Shri Annai Apartment, 12th Cross, Bharathi Nagar, Lawspet, Pondicherry 605008, India

J. S. Raj et al. (eds.), *Intelligent Sustainable Systems*, Lecture Notes in Networks and Systems 213, https://doi.org/10.1007/978-981-16-2422-3_37

1 Introduction

Due to recent advancements in cloud computing architectures, stakeholders are migrating and relying on different third-party service providers. So, the user applications, data, and processes are running on some third-party platforms. Thus, wherever user applications run on the thirty parties' environment, trust becomes an issue. Especially, sensitive services like banking, defense, academic, and health care need to be looked into the trust aspects. In the near future, consumers who have a service in one cloud may access various services from the other cloud service providers. In this collaborative scenario also, the trust will be playing a major role to protect customer's sensitive data and operations from malicious insiders and outsiders. Even though cloud service providers enforce appropriate necessary countermeasures, malicious attackers try to exploit or damage outsourced assets [2]. Moreover, consumers' feedback is a good source to assess the overall trustworthiness of a cloud service provider. However, malicious CSP or users may have multiple accounts (is called Sybil Attack) to give numerous fake feedbacks for self-promoting or hampering cloud services (is called Collusion Attack) [18, 1, 8]. To address these untrusted issues, an accurate feedback-based trust evaluation process plays a critical role in establishing a trustworthy relationship between cloud service consumers and providers. In fact, traditional trust management schemes cannot be directly adopted in collaborative cloud environments due to the openness and highly non-transparent nature of the cloud services [23].

In general, trust functions have been broadly categorized into two comparative dimensions [15] such as subjective and objective trust. In objective trust, the dependability of a specific entity is measured or quantified by keeping the outcome or result of an event as a benchmark. But in subjective trust, the results are dependent on individual attention. Wang et al. [20] have concentrated on the web service environment and determined the factors affecting the trust management in it. A trust model has been proposed by the authors who talk about the subjective uncertainty of trust factors. By using an algorithm and trust model, the service provider's degree of dynamic trust is calculated. However, Wang et al. proposed approach fails to identify malicious users who will be submitting fake feedback and reputation ratings. Liu Xin et al. [11] have designed an algorithm for deducting dynamic trust depending on communication among users and the trust chain. Based on communication among the user and his friends in the social network trust value is updated dynamically. However, Xin et al. approach cannot apply practically for cloud computing environments because of unrealistic theoretical assumptions. In recent years, some worth considering trust management solutions have been developed. A literature study indicated that the existing schemes fail to enforce the genuineness of historical trust feedbacks and other multifaceted trust factors accountability.

A. Research Contributions

The following is the summary of main research contributions.

(1) To evaluate feedback trust by ensuring the consumers feedback genuineness.
(2) To compute an accurate global trust value by aggregating genuine feedbacks trust, reputation trust, and SLA quality of service attributes trust.
(3) To compare the performance of the proposed scheme with the existing schemes.

B. Organization of the Paper

This paper is further organized as follows. Section 2 summarizes the literature study. Section 3 describes the Proposed Trustworthy Monitoring Scheme. Section 4 deals with the graphical interpretations of the overall performance analysis of the proposed scheme. Section 5 summarizes the research and future scope.

2 Literature Study

The related works carried out by different researchers in the cloud computing context are discussed in this section. To calculate competence and trustworthiness, Ghosh et al. [16] have constructed a trust evaluation framework. By using feedback about the cloud vendors' reputations and direct trust methods evaluation of trustworthiness is done. Calculation of competency is done based on provider's SLA assurances transparency. Manuel (2015) has proposed a different trust model organized on previous authorizations and present competencies of a cloud service provider. Trust value has been determined by the author with the help of parameters such as availability, reliability, data integrity, and turn around efficiency. To put the trust model into action a trust management system is also offered by the author. However, Ghosh et al. and Manuel models lack in the detection of the genuineness of the users' feedbacks and not focused on context-aware multifaceted global trust.

Talal H. Noor et al. [18] designed an adaptive and robust credibility model for reputation-based trust management. The scheme predicts appropriate reputation trust by using detective controls. A crowdsourcing model is developed by Wang et al. (2016) by making use of the social cloud. Social cloud connects sensing objects and users and also, it behaves like a cloud service provider by providing storage and compute services. Sensing objects provide feedback for the jobs proffered by the cloud provider. By measuring the trustworthiness of the crowdsourcing participants' expense and champion determination are done. This is accomplished by combining a reputation-based algorithm with crowdsourcing. However, Talal H. Noor et al. and Wang et al. schemes fail to combine the reputation trust of multiple sources like the business, the group derived, witness, and technology partnership reputations. Moreover, these schemes also lack in providing aggregated context-aware multifaceted global trust.

On the basis of feedback from various sources, Varalakshmi et al. (2017) have proposed a framework for recognizing a trustworthy service provider. The authors have confirmed that the proposed model works efficiently in collaborative cloud environments. Tang et al. (2017) have presented a framework to select a trustworthy service in the cloud environment. Subjective and objective trust measures have been considered by the authors. Based on cloud client feedback and QoS monitoring both the trust measures have been combined. Authors have proved that the model's performance can be improved by integrating both objective and subjective measures. To calculate absolute trust in the cloud computing environment, Mohannad et al. [13] presented a context-aware multifaceted trust evaluation model. This model overcomes the limitations of the Xiaoyong Li et al. trust model. In this model, user feedbacks, expert opinion, application context, and SLA evidence will be extracted in the preparation phase. The global trust value is computed in the evaluation phase by assigning the user preferred weights to the evidence collected from the preparation phase. Mohannad et al. model allow the consumers to customize their preferences in selecting desired reliable cloud services. However, Varalakshmi et al., Tang and et al., and Mohannad et al. models lack in the detection of the genuineness of the users' feedbacks and are not focused on multiple perspectives like competence, business, membership, and technology partnership reputations.

Xiaoyong et al. [23] presented an adaptive and attribute-based trust model to evaluate the global trustworthiness of CSPs based on security, reliability, and availability attributes. The Talal H. Noor et al. and Xiaoyong Li et al. schemes fail to consider security standardization parameters, the genuineness of user feedback, and other multifaceted activity log accountability. Waqas Ahmad et al. [21] presented a privacy preserved reputation-aware approach based on anonymous secure-shell cipher-text policy. The scheme resists collude and malicious insider attacks but fails to focus on multiple perspectives like competence, genuine feedbacks, and business and technology partnership reputations. For the cloud-based mobile IoT smart service networks, scalable trust management variant protocols are designed and analyzed by Hamid et al. [6] and Ing-Ray et al. [7] based on the user application-specific context-aware criteria such as credibility rating and service ratings. Their performance analysis shows that the proposed protocols outperform in resilience, convergence, and accuracy. However, Hamid et al. and Ing-Ray et al. protocols fail to mitigate the effect of identity-based attacks such as spoofing, collusion, and Sybil attacks. Hala Hassan et al. [5] evaluated the trustworthiness based on its provider dynamic reputation history. Yakun Li et al. [24] presented an implicit trust recommendation approach for rating prediction. However, Hala Hassan et al. and Yakun Li et al. solutions lack the detection of the genuineness of the users' trust feedbacks and not focused on multiple perspectives attributes. The disadvantages of the existing works in this literature are resolved in the proposed scheme.

3 Proposed Trustworthy Monitoring Scheme

Basically, there are two types of trust factors one is SLA QoS trust factors, and the other is Non-SLA trust factors [13]. SLA QoS trust factors provide first handed evidence for overall trust calculation. This first handed evidence can be collected by monitoring and analyzing the service log information. The SLA trust factors considered in the proposed scheme are data processing, data privacy, and data transmission factors. Non-SLA trust factors provide second-handed evidence from various sources such as feedbacks, direct reputation, and other sources. To calculate absolute trust in the cloud computing environment, Mohannad et al. [13] presented a context-aware multiple trust factors evaluation models. Figure 1 illustrates the context-aware multifaceted trust model used in our proposed scheme. First, feedback extractor computes historical feedback trust by ensuring the genuineness of trust feedbacks. Next, a Non-SLA extractor determines reputation trust by aggregating direct and indirect reputation. Next, QoS Trust Extractor computes SLA trust by aggregating the QoS attributes. Finally, the Global Trust Evaluator determines the global trust value and

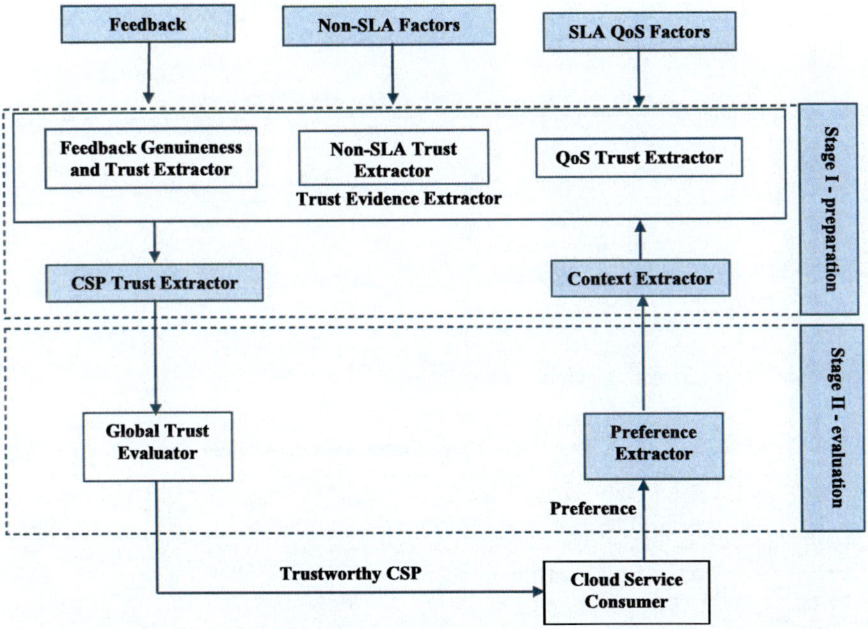

Fig. 1 Context-aware multifaceted trust model

selects an appropriate and trustworthy cloud service provider by aggregating SLA and Non-SLA trust values based on consumer preferences.

3.1 Genuine Feedback Trust Evaluation

Genuine feedback trust is the amount of feedback trust that one genuine user has in *CSP* for a specific cloud service usage in a time window θ to t. Let $F = \{F_1, F_2, F_3, ..., F_n\}$ be the set of trust feedbacks received from an identity record I'_u for a cloud service s and SF_c be the similarity frequency count of I'_u. SF_c is set to 1 if I'_u matches with only one identity record I_u in *IdP* database. In each feedback F_i, let $R_P = \{R_{P1i}, R_{P2i}... R_{Pni}\}$ be the ratings given for each QoS parameter and Let $W = \{W_{1i}, W_{2i}, \ldots W_{ni}\}$ be the set of weights assigned for each QoS parameter, where $\sum W_j \leq 1$. Let Δ_f be the feedbacks threshold which indicates the maximum number of feedbacks that can be considered in a time window $[\theta, t]$, where θ indicates the past time and$t$ indicates the current time. Let $T_f(u, s)^t$ be the user feedback trust in s with respect to time window bound $[\theta, t]$ and is calculated as follow:

$$T_f(u,s)^t = \frac{\sum_{i=1, i\in|F|, if\ I'_{ui}==I_{ui}\ I_{ui}\in IdP \&\& SF_c==1 \&\& |F|\leq\Delta_f \in[0,1]}^{|F|} T_{fi}(u,s)}{|F|} \quad (1)$$

$$|pi| \quad //i\text{th feedback trust with respect to } R_P \text{ and } W \quad (2)$$

where $T_{fi}(u, s) = \sum f_i(R_{Pj}, s) R_{Pj\epsilon QoS} * W_j$.
$j = 1, j\epsilon |p_i|, \sum W_j \leq 1.$

Let $T_{fu}(s)$ be the overall feedback trust of a user in cloud service s with respect to time window $[t_0, t]$ and is calculated as follow:

$$T_{fu}(s) = (T_f(u,s)^{[t_0, \theta-1]} + T_f(u,s)^t)/2, \text{ where } t_0 \text{ indicates service registration time} \quad (3)$$

Algorithm 1 summarizes the feedback trust calculation of a genuine user.

Algorithm 1 Feedback trust calculation by ensuring user genuineness

---

Input: I'_u, Rp, W, Δ_f and t_0.

Assumption: The user needs to register their identity record I_u with IdP for a cloud service usage as specified in the proposed authentication scheme before sending any feedback.

Begin

1. If $(I_u == I'_u \;\&\&\; SF_c == 1)$
2. The total number of feedbacks $|F|$ in a time window $[\theta, t]$ will be counted.
3. If $(|F| \leq \Delta_f \in [\theta, t])$

   User feedback trust will be calculated from genuine and valid user feedbacks.

   $T_{tf} \leftarrow 0$

   $T_f \leftarrow 0$

   For $(i=1 \text{ to } |F|)$

   For $(j=1 \text{ to } |P_j|)$ //Sum of each parameter of a feedback trust calculation

   $T_f \leftarrow T_f + f_u(p_{rj}, s) p_{rj \in QoS} * W_{fj}$

   End For

   $T_f \leftarrow T_f / |p_i|$ //Aggregation of feedback parameters' trust

   $T_{tf} \leftarrow T_{tf} + T_f$ //Sum of feedback trusts' in a time window $[\theta, t]$

   End For

   $T_{tf} \leftarrow T_{tf} / |F_j|$ // Aggregation of feedback trusts' in a time window $[\theta, t]$

   $T_{tf} \leftarrow T_{tf} + T_{pf}(u, s)^{[t0,\, \theta-1]}/2$ //Aggregation of each user feedback trusts' in a time window $[t_0, t]$
4. End If
5. End If

End

Output: Genuine User feedback trust.

3.2 *Global Trust Calculation*

To calculate the accurate global trustworthiness of a cloud service provider for a particular service provisioning, the desired SLA and Non-SLA trust factors are described in this section.

3.2.1 Service Level Agreement Trust

The service level agreement (SLA) trust is the aggregation of direct interaction quality of service (QoS) attributes trust with respect to user preferences. In SLA trust evaluation process mainly focused on the following three kinds of trust attributes.

Data Processing Trust is calculated based on whether the authentication process and user actions on service data are taken place successfully or not. Trustee stores the number of non-error requests (R_{ne}) and error requests (Re). Data Processing Trust (T_{dp}) can be calculated through Eq. (4).

$$T_{dp} = (R_{ne}/(R_{ne} + R_e)) \tag{4}$$

Data Privacy Trust is computed based on the unauthorized accesses and governed privacy compliance. Trustee stores M_u= [1 or 0] that indicate misuse or unauthorized access and the rating given to the governed privacy compliance laws (CL_r) [0–9]. The Data Privacy Trust (T_{priv}) value can be computed using Eq. (5).

$$\text{Pr} = a * Mu + (1 - a) * CL_r \tag{5}$$

$$T_{\text{priv}} = \begin{cases} 0 \text{ If Pr} = 0 \\ 1 \text{ If Pr} > 0 \end{cases}$$

Data Transmission Trust is calculated based on whether the communication data is transmitted successfully or not. Trustee stores the number of *successful data transmission requests* (DT_s) and *failure data transmission requests* (DT_f) and the data transmission trust is calculated via Eq. (6).

$$T_{dt} = (DT_s/(DT_s + DT_f)) \tag{6}$$

Thus, the SLA trust value (SLA_{tval}) is computed by aggregating the Eqs. (4) to (5) and formed an Eq. (7) as given below.

$$\text{SLA}_{\text{tval}} = \sum (I_1 * T_{dp} + I_2 * T_{\text{priv}} + I_3 * T_{dt})/3 \tag{7}$$

where I_1, I_2, and I_3 are the preferences of processing, privacy, and transmission parameters and $\Sigma\, I_i,\ _{1 \le i \le 3} = 1$.

3.2.2 Non-SLA Trust

In this subsection, a mathematical equation is formulated for evaluating service Non-SLA trust in terms of feedback trust, direct, and indirect reputation trust values.

Overall Feedbacks Trust (FT) is the amount of feedback trust that one user has in *CSP* for specific service usage, and it is calculated via Eq. (3) and is given in (8).

$$T_{fb}(s) = T_{f_u}(s) \quad (8)$$

Direct Reputation (D_r) can be calculated as the number of users currently using the service (indicated as N_u') is divided by the number of users who have chosen service of the *CSP* (denoted as N_u), where $N_u \geq N_u'$. The direct reputation value is computed by using Eq. (9).

$$Dr = Nu'/Nu \quad (9)$$

Indirect Reputation is a reputation based on second-hand evidence. It deals with the three kinds of reputation attributes such as witness reputation, neighborhood reputation, and the group-derived reputation.

(i) *Witness Reputation* is the evidence information gathered from the business and technology partners about the service of a *CSP*.
(ii) *Neighborhood Reputation* is the reputation based on social prejudice
(iii) *Group-derived reputation* is computed based on membership to certain groups.

For each kind of indirect reputation, the rating range considered from 0 to 10 and the aggregation value of these three evidences is the indirect reputation. Thus the reputation trust value is obtained by using Eq. (10).

$$\text{Reputation } (R_{\text{tval}}) = a * Dr + (1 - a) * \text{Indirect Reputation} \quad (10)$$

where α is the importance of service reputation.

Thus, the Non-SLA trust is the aggregation of overall feedback trust and reputation trust with respect to user preferences is calculated by using Eq. (11).

$$\text{NSLA}_{\text{tval}} = I_1 * R_{\text{tval}} + (1 - I_1) * T_f(s) \quad (11)$$

Therefore, a formal equation to measure the accurate global trustworthiness of a cloud service provider is formulated based on SLA_{tval} and $\text{NSLA}_{\text{tval}}$ trust Eqs. (7) and (11). Thus, the global trust value is calculated as per Eq. (12).

$$\text{Global Trust value } (\text{GT}_{\text{val}}) = \alpha * \text{SLA}_{\text{tval}} + \beta * \text{NSLA}_{\text{tval}} \quad (12)$$

where α and β are the user chosen weights for SLA trust and Non-SLA trust and $\alpha + \beta = 1$.

3.2.3 Selection of the Trustworthy Cloud Service Provider

This section mainly concentrates on selecting appropriate and highly trustworthy cloud service providers using a case study with two weight sets. The selection of an appropriate and highly trustworthy cloud service provider is illustrated in Algorithm 2. In the case study, cloud users ($U_i's$) can find trustworthy cloud service providers ($CSP_i's$) by calculating global trust value from Eq. (12) and by following Algorithm 2. In Eq. (12), the values of SLA trust and Non-SLA trust of CSP_j can be computed from Eq. (7) and Eq. (11).

Case Study: Table 2 represents three cloud service users and their qualified cloud service providers based on the service provider SLA and Non-SLA trust values, and minimum acceptable SLA and Non-SLA trust values. By using Table 1 and Algorithm 2, a user U_i can find the highly trustworthy service provider CSP_j based on his/her preferred weights for the SLA and Non-SLA trust values as provided in Tables 2 and 3. *Weight-set-1* represents three users and their choice of trustworthy *CSPs* as illustrated in Table 2, where user U_1 selects CSP_3 by assigning weight 2/3 to SLA trust and 1/3 to Non-SLA trust, respectively. Similarly, U_2 and U_3 can select CSP_2 and CSP_1, respectively by assigning their desired weights. *Weight-set-2* represents three users and their chosen trustworthy *CSPs* as given in Table 3, where user U_1 can select CSP_1 because CSP_1 has the highest global trust value than CSP_2 and CSP_3. Similarly, U_2 and U_3 can select CSP_3 has the highest global trust value than CSP_1 and CSP_2 based on the weights assigned to cloud service SLA and Non-SLA trusts.

Algorithm 2 Selection of Cloud Service Provider based on the Global Trust Value

Input: User preferences *(α, β)*, SLA_{tval}, $NSLA_{tval}$ Trust values and user acceptable SLA_{tval} and $NSLA_{tval}$

Begin

1. If ($SLA_{tval} \geq$ min user acceptable SLA_{tval} and $NSLA_{tval} \geq$ min user acceptable $NSLA_{tval}$) then

    Finds the global trust value (Gt_{val}) based on user preferences.

    Global Trust value $(GT_{val}) = \alpha * (SLA_{tval}) + \beta * NSLA_{tval}$, where $0 < \sum(\alpha, \beta) \leq 1$

    If (user chooses a CSP based on maximum GT_{val}) then

        A user is directed to the identity provider for cloud service registration.

    End If

   End If
2. Repeat

   STM monitors the quality of service parameters and records the evidence for each legitimated user sessions.
3. Until (the service session is terminated).
4. STM also stores genuine user feedback details in the highly secured and distributed databases.

End

Output: Returns highly trustworthy CSP.

Table 1 Trustworthy CSP selection based on the user preferences

Choice	SLA_{tval}	$NSLA_{tval}$	$SLA_{tminaccval}$	$NSLA_{tminaccval}$	GT_{val}
$U_1 \leftrightarrow CSP_1$	0.9	0.8	0.8	0.6	**0.866**
$U_1 \leftrightarrow CSP_2$	0.8	0.9	0.8	0.6	0.833
$U_1 \leftrightarrow CSP_3$	0.8	0.7	0.8	0.6	0.766
$U_2 \leftrightarrow CSP_1$	0.8	0.8	0.8	0.6	0.800
$U_2 \leftrightarrow CSP_2$	0.9	0.8	0.8	0.6	**0.875**
$U_2 \leftrightarrow CSP_3$	0.8	0.9	0.8	0.6	0.825
$U_3 \leftrightarrow CSP_1$	0.8	0.8	0.8	0.6	0.800
$U_3 \leftrightarrow CSP_2$	0.8	0.7	0.8	0.6	0.760
$U_3 \leftrightarrow CSP_3$	0.9	0.7	0.8	0.6	**0.820**

Table 2 User weight set-1 and selected CSP

	SLA_{tval}	$NSLA_{tval}$	α	β	GT_{val}	Choice
U_1	0.9	0.8	2/3	1/3	**0.866**	$\boldsymbol{CSP_1}$
U_2	0.9	0.8	3/4	1/4	**0.875**	$\boldsymbol{CSP_2}$
U_3	0.9	0.7	3/5	2/5	**0.820**	$\boldsymbol{CSP_3}$

Table 3 User weight set-2 and selected CSP

	SLA_{tval}	$NSLA_{tval}$	α	β	GT_{val}	Choice
U_1	0.9	0.8	1	0	**0.90**	$\boldsymbol{CSP_1}$
U_2	0.9	0.8	0	1	**0.90**	$\boldsymbol{CSP_3}$
U_3	0.9	0.7	1	0	**0.90**	$\boldsymbol{CSP_3}$

4 Performance Analysis

In this section, we describe the selection of desired and trustworthy cloud service providers using two case studies. In these two case studies, cloud service consumer (U_i) can find a trustworthy cloud service provider (CSP_j) by calculating the global trust value of a cloud service using Algorithm 2. To analyze the selection of trustworthy cloud service providers, more than a thousand sample trust rating records are used. In this case study, the trust weights α, and β are assigned with the random values for the selection of trustworthy cloud service providers. Initially, three hundred genuine users trust rating records are considered for the selection. Random trust importance weights are set for both the SLA and Non-SLA trusts. Figure 2 illustrates the users and their choice of trustworthy cloud service providers based on both SLA trust and Non-SLA trust preferences. Next, one thousand genuine users trust rating records are considered for the selection. Random trust importance weights are set for only SLA trust. Figure 3 depicts the users and their choice of trustworthy

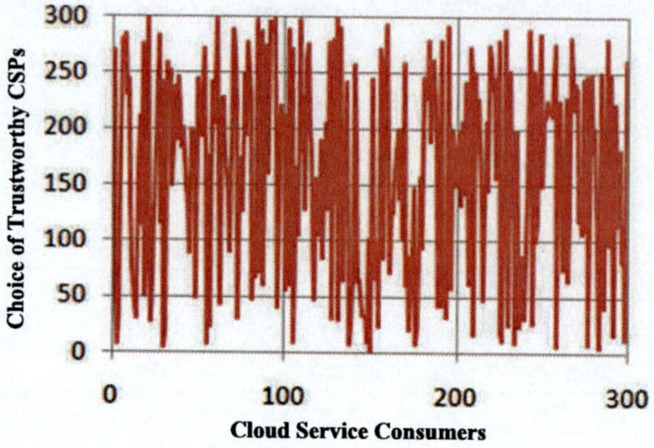

Fig. 2 Trustworthy CSPs selection based on SLA and Non-SLA trusts preference

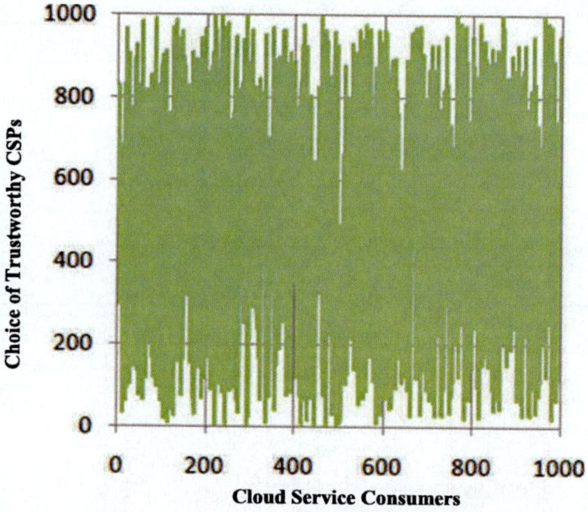

Fig. 3 Trustworthy CSPs selection based on SLA trust preference

cloud service providers based on SLA trust preference. Finally, one thousand genuine users trust rating records are considered with random trust importance weights for only Non-SLA trust. Figure 4 depicts the users and their choice of trustworthy cloud service providers based on Non-SLA trust preference.

The trust characteristics comparison is summarized in Table 4. Talal et al. scheme fail to support context awareness into their trust evaluation process. The existing schemes partially support to evaluate the multifaceted trust in the cloud computing environments (for instance, few of the CSA Cloud Controls Matrix attributes taken into consideration).

None of the existing schemes fully support to verify the genuineness of the consumers' trust feedbacks. Due to this reason, the existing schemes are not suitable to strengthen the consumers and cloud service providers trust relationships. Table 4

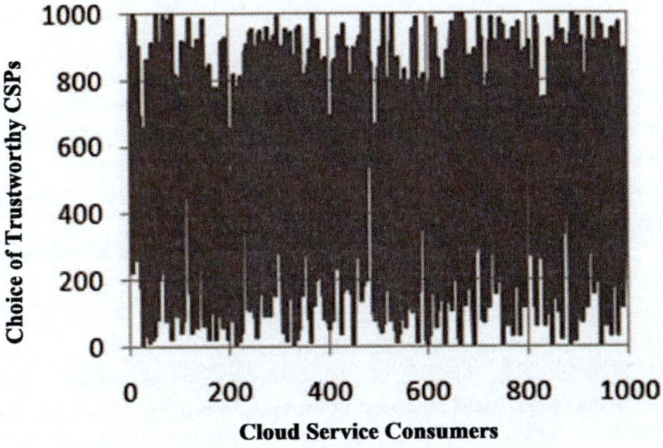

Fig. 4 Trustworthy CSPs selection based on Non-SLA trust preference

Table 4 Trust characteristics comparison

Schemes	Standardization Efforts	Multifaceted	Genuine Feedbacks	Expandable	Context awareness
Talal H. N. et al., 2016	◐	◐	◐	◐	○
Mohannad et al., 2017	◐	◐	○	◐	◐
Waqas A. et al., 2018	◐	◐	○	◐	○
Hamid et al., 2019	◐	◐	○	◐	◐
Ing-Ray et al., 2019	◐	◐	○	◐	●
H. Hassan et al., 2020	◐	◐	○	◐	◐
Yakun Li et al., 2020	◐	◐	○	◐	◐
Proposed Scheme	●	◐	●	●	●

● Fully Achieved ◐ Partially Achieved ○ Not Achieved

illustrates that the proposed scheme not only supports to calculate security standardization trust, context-aware trust, expandable to aggregate the user preferred trust factors but also ensures the genuineness of the consumers' trust feedbacks.

5 Conclusion and Future Scope

A trustworthy monitoring scheme has been designed, developed, implemented, and compared to its intended functionalities. This scheme helps cloud service consumers in selecting appropriate trustworthy cloud service providers and to ensure the quality of cloud services. The proposed scheme withstands against Sybil and collusion attacks. The research can be further extended with machine learning analytics to detect identity attacks, and calculate feedback and SLA QoS-based trusts in a faster way.

References

1. Zhu, C., Nicanfar, H., Victor, C.M., Yang, L.T.: An authenticated trust and reputation calculation and management system for cloud and sensor networks integration. IEEE Trans. Inf. Forensics Secur. **10**(1), 118–131 (2015)
2. Cloud Security Alliance: Top threats working group (2018) Available from: https://cloudsecurityalliance.org/working-groups/top-threats/#_overview
3. Cloudwards: "Best Cloud Storage Reviews 2019," in Cloudwards (2019). Available from: https://www.cloudwards.net/cloud-storage-reviews/
4. Garg, S.K., Versteeg, S., Buyya, R.: A framework for ranking of cloud computing services. Future Gener. Comput. Syst. **29**(4), 1012–1023 (2013)
5. Hassan, H., El-Desouky, A.I., Ibrahim, A., El-Kenawy, E.-S.M., Arnous, R.: Enhanced QoS-based model for trust assessment in cloud computing environment. IEEE Access. **8**, 43752–43763 (2020)
6. Al-Hamadi, H., Chen, I.-R., Cho, J.-H.: Trust management of smart service communities. IEEE Access. **7**, 26362–26378 (2019)
7. Chen, I.-R., Guo, J., Wang, D.-C., Tsai, J.J.P., Al-Hamadi, H., You, I.: Trust-based service management for mobile cloud IoT systems. IEEE Trans. Network Serv. Manage. **16**(1), 246–263 (2019)
8. John (JD) Douceur: "The Sybil Attack," in Microsoft Research (2002). Available from: https://www.microsoft.com/en-us/research/publication/the-sybil-attack/
9. Tung, L.: "Mozilla to China's WoSign: We'll kill Firefox trust in you after mis-issued GitHub certs," in ZDNet (2016). Available from: https://www.zdnet.com/article/mozilla-to-chinas-wosign-well-kill-firefox-trust-in-you-after-mis-issued-github-certs/
10. Qu, L., Wang, Y., Orgun, M.A.: Cloud service selection based on the aggregation of user feedback and quantitative performance assessment. In: 2013 IEEE International Conference on Services Computing, vol. 5, pp. 2187–2199 (2013)
11. Xin, L., Leyi, S., Yao, W., Zhaojun, X., Wenjing, F.: A dynamic trust conference algorithm for social network. In: 8th International Conference on P2P, Parallel, Grid, Cloud and Internet Computing (3PGCIC), IEEE, pp. 340–346 (2013)
12. Manuel, P.D., Abd-El Barr, M.I., Thamarai Selvi, S.: A novel trust management system for cloud computing IaaS providers. J. Comb. Math. Comb. Comput. **79**(3), 1–20 (2011)
13. Mohannad, A., Bertok, P., Tari, Z.: Trusting cloud service providers: trust phases and a taxonomy of trust factors. IEEE Cloud Comput. **4**(1), 44–54 (2017)
14. Prosanta, G., Hwang, T.: Lightweight and energy-efficient mutual authentication and key agreement scheme with user anonymity for secure communication in global mobility networks. IEEE Syst. J. **10**(4), 1370–1379 (2016)
15. Zhang, Q., Yu, T., Irwin, K.: A classification scheme for trust functions in reputation based trust management. In: International Conference on Trust, Security, and Reputation on the Semantic Web, vol. 127, pp. 52–61 (2004)
16. Ghosh, S.K.: "Cloud Computing," in *YouTube* (2015). Available from: https://www.youtube.com/watch?v=ZHCtVZ6cjdg
17. Habib, S.M., Ries, S., Mühlhäuser, M., Varikkattu, P.: Towards a trust management system for cloud computing marketplaces: using CAIQ as a trust information source. Secur. Commun. Networks **2**, 71–81 (2013)
18. Noor, T.H., Sheng, Q.Z., Yao, L., Dustdar, S., Anne, H.H.: CloudArmor: supporting reputation-based trust management for cloud services. IEEE Trans. Parallel Distrib. Syst. **27**(2), 367–380 (2016)
19. Tamram, sptramer, and ppluijten: "Use the Azure storage emulator for development and testing," in *Microsoft Azure* (2018). Available from: https://docs.microsoft.com/en-us/azure/storage/common/storage-use-emulator
20. Wang, S.-X., Zhang, L., Wang, S., Qiu, X.: A cloud-based model for evaluating quality of web services. J. Comput. Sci. Technol. **25**(6), 1130–1142 (2010)

21. Ahmad, W., Wang, S., Ullah, A., Sheharyar, Z.M.: Reputation-aware trust and privacy-preservation for mobile cloud computing. IEEE Access **6**, 46363–46381 (2018)
22. Li, X., Du, J.: Adaptive and attribute-based trust model for service level agreement guarantee in cloud computing. IET Inf. Secur. **7**(1), 39–50 (2013)
23. Li, X., Yuan, J., Ma, H., Yao, W.: Fast and parallel trust computing scheme based on big data analysis for collaboration cloud service. IEEE Trans. Inf. Forensics Secur. **13**(8), pp. 1917–1931 (2018)
24. Li, Y., Liu, J., Ren, J., Chang, Y.: A novel implicit trust recommendation approach for rating prediction. IEEE Access **8**, 98305–98315 (2020)

Analysis of Hypertensive Disorder on High-Risk Pregnancy for Rapid Late Trimester Prediction Using Data Mining Classifiers

Durga Karthik, K. Vijayarekha, B. Sreedevi, and R. Bhavani

Abstract Pregnancy is one of the important stages for women. As several complications arise during pregnancy, it may lead to high risk for both mother and foetus. One such is the incidence of hypertension associated with pregnancy which leads to increased maternal and foetal deaths during pregnancy and childbirth. Due to its high incidence rate and associated complications, the study of this disorder has been a subject of numerous investigations in an attempt to determine its prevention and improve the treatment. In this context, this work uses the data mining (DM) technique to predict rules for high-risk pregnancy due to hypertension. The JRip classification rule and EM clustering are applied on various parameters from the collected data for predicting hypertensive disorders during III trimester of pregnancy. The Jrip classification rules generated for the data have 97% true positive rate and EM clustering generated similar clusters enabling a good proposed model.

Keywords Preeclampsia · Eclampsia · Data mining · Jrip · EM cluster

1 Introduction

World Health Organization (WHO) reports that in the world, yearly around 6 million women are pregnant. Nearly 1000 women die during pregnancy and childbirth every day [1]. Systolic blood pressure greater than 140 mm Hg or with diastolic blood pressure greater than 90 mm Hg is considered hypertensive disorder. Hypertensive

D. Karthik (✉) · B. Sreedevi · R. Bhavani
Department of CSE/SRC, SASTRA Deemed University, Kumbakonam, India
e-mail: durgakarthik@src.sastra.edu

B. Sreedevi
e-mail: sreedevi@src.sastra.edu

R. Bhavani
e-mail: bhavani@src.sastra.edu

K. Vijayarekha
School of EEE, SASTRA Deemed University, Tirumalaisamudram, India
e-mail: vijayarekha@eee.sastra.edu

J. S. Raj et al. (eds.), *Intelligent Sustainable Systems*, Lecture Notes in Networks and Systems 213, https://doi.org/10.1007/978-981-16-2422-3_38

disorder is one of the common causes for complications that may lead to the death of both mother and baby during pregnancy. It is observed that maternal death due to hypertensive disorders is nearly 15% of the total for the past 5 years. Also, every seven minutes a woman is reported to die due to hypertension.

Hypertension is due to high blood pressure that generally leads to heart attacks, vision problem, kidney problem, etc. High blood pressure is systolic pressure more than 160 mmHg and/or diastolic pressure more than 110 mmHg [2]. The National High Blood Pressure Education program classifies hypertension during pregnancy into preeclampsia or gestational. Preeclampsia is added to both the categories based on maternal or foetal symptoms [3, 4].

2 Literature Review

Obese woman under age 20 or above 40 with renal problems and diabetes have higher risk during pregnancy [5, 6]. One such is the incidence of hypertension associated with pregnancy which leads to increased maternal and foetal deaths during pregnancy and childbirth. Due to its high incidence rate and associated complications, the study of this disorder has been a subject of numerous investigations in an attempt to determine its prevention and improve the treatment.

Data mining provides various algorithms for classification, clustering or prediction rules for high-risk pregnancy due to hypertension. Decision Tree (DT), Generalized Linear Models (GLM), Support Vector Machine (SVM) and Naïve Bayes (NB) are used for risk prediction for pregnant women [7]. Various risk factors and complications on hypertensive disorder duing pregnancy are identified for prediction [8]. Prediction using cluster analysis on various parameters for pregnant women has evolved with 3 classified groups [9]. A boosted Random Forest Approach for predicting complication for maternal hypertensive disorder using ID3 and C4. Five algorithms have given better results [10]. Data mining [11–13] can be employed for identifying prediction rules on varying age groups of women. Big data analytics [14, 15] provides valuable tools that are useful for prediction on healthcare data.

3 Materıals and Methods

3.1 Data Collection

The data was collected from Government Hospital (GH), Kumbakonam, and Primary Health Centre (PHC), Swamimalai, Tamilnadu. For a period of 4 years, 2014–2018, nearly 617 samples were used for analysis. Data cleaning followed by data selection and transformation was carried out before generating the prediction rule. The model included data from a semi-urban region that can be used to determine the underlying problems of the location. The rules generated using this region can be used for other similar regions.

3.2 Data Pre-Processing

A persistent format for the data model was developed which handles missing data or noisy data, finding duplicate data and removal of invalid data. A few women had not mentioned data clearly on certain variables, and they were treated as missing data. Those missing data, if found to be required variables, were selected and based on age they was appropriately filled with mean data for that group. Finally, cleaned data was stored for further processing. Followed by data selection, data related to analysis was decided to retrieve from the dataset. The collected data was used in a 60:40 ratio for training and testing. The Hypertensive dataset had eight (8) attributes; their type and description is given in Table 1.

Followed by data transformation is data consolidation. The raw data file was saved in Comma Separated Value (CSV) file format. Finally, Rule generation was carried out using JRip rule miner followed by verifying the rules using the EM Clustering method. The Jrip initially splits the data into growing and pruning sets. An initial rule set is formed using heuristic methods in the growing phase. The large growing set is pruned using operators until the rules are formed. The EM algorithm iteratively estimates parameters that depend on unobserved latent variables.

4 Model Formulatıon and Sımulatıon Results

In this paper, we have proposed a technique that uses the JRip classifier and EM clustering for the formulation of the predictive model using the simulation environment. Novelty in this work includes collecting data on pregnant women job status and

Table 1 Attributes of Hypertensive Dataset

Attributes	Type	Description
Year	Numerical	Year considered
Month	Alphanumerical	Month considered (Trimester month)
City	String	City around the Dataset where collected
Gender	String	Gender considered (Female only)
Age	Numerical	Age Consideration (18–40)
Diet	String	Food habit either Veg. or Non-Veg
Job Status	Boolean	The employment status either employee or home maker
Blood Pressure	Boolean	BP raw data

Fig. 1 Flow diagram for prediction

BMI along with other factors for identifying the influence of those factors in causing hypertension during pregnancy. The results of the formulation of the predictive model for the risk prediction showed that four variables were the most important risk factors of hypertension. The variables identified in the order of their importance were Age, Diet, Job Status and BMI. The four variables identified by the JRip rule miner which were used for the risk identification due to hypertension during pregnancy were formulated using the EM clustering on the WEKA simulation environment.

Figure 1 shows the steps involved in the prediction model generation using data mining.

4.1 JRip Classification Rule

This rule implements a propositional rule learner, Repeated Incremental Pruning to Produce Error Reduction (RIPPER). The JRip classifier uses association rule mining to select the most relevant features from the given dataset. It is applied on a tenfold cross-validation testing mode which extracts the rules from a dataset and then the association rules mining technique is applied to rank the features. The JRip classifier applied to Patients' Age, BMI, Diet conditions and Job status extracted the following rules.

The rules extracted by the JRip rule miner are as follows:

1. (BMI = Obesity) and (Age >= 23) and (Job = Yes) => BP=Yes (36.0/0.0).
2. (Age >= 31) and (BMI = Overweight) and (Job = Yes) => BP=Yes (20.0/0.0).
3. (BMI = Obesity) and (Diet = Non Veg) and (Age >= 23) => BP=Yes (18.0/0.0).
4. (Job = Yes) and (Diet = Non Veg) and (Age >= 26) and (BMI = Overweight) => BP=Yes (5.0/0.0).
5. (Age >= 36) and (BMI = Normal) and (Diet = Non Veg) and (Job = Yes) => BP=Yes (5.0/0.0).
6. BP=No (284.0/0.0).

Rule 1 states that working obese women and age greater than or equal to 23 is predicted as BP during their III trimester.

Rule 2 states that working overweighted women and age greater than or equal to 31 is predicted ass BP during their III trimester.

Rule 3 states that obese and non veg. women and age greater than or equal to 23 is predicted as BP during their III trimester.

Table 2 Average standard metric values of JRip rule miner classifier

Weighted Avg:	TP Rate	FP Rate	Precision	Recall	F-Measure	ROC area	Class	Kappa stats
	0.989	0.083	0.976	0.989	0.983	0.772	No	0.9216
	0.917	0.011	0.963	0.917	0.939	0.957	Yes	
	0.973	0.067	0.973	0.973	0.973	0.814		

Rule 4 states that overweight and working women along with non veg. food habit and age greater than or equal to 26 is predicted as BP during their III trimester.

Rule 5 states that working normal BMI women along with non veg. food habit and age greater than or equal to 36 is predicted as BP during their III trimester.

Rule 6 states that the women have no BP during their III trimester.

The algorithm given below describes the generated rules for prediction.

Step 1: Start.

Step 2: Get the age, BMI, Job status and BP.

Step 3: Check the above rules for various criteria depending on age, obesity, working status and present BP.

Step 4: If the obtained data correlates with rules from 1 to 5, then predict as high-risk pregnancy with BP in third trimester, else no high risk.

Step 5: Stop.

Table 2 shows the stastistics for the data that was used in the Jrip rule miner for classification. The true positive rate is nearly 98% and the false positive is 0.067%. Similarly, the above rules have a 97% of precision and recall, hence it suggests the above rules are significant.

4.2 EM Clustering

In the statistics approach, an expectation–maximization (EM) algorithm is a cyclic method to find the maximum likelihood or maximum a posteriori (MAP) estimates of parameters, where the model depends on unnoticed latent variables. EM assigns a probability distribution to each instance which indicates the probability of it belonging to each of the clusters. EM can decide how many clusters to create by cross validation, which may specify a priori to generate cluster. EM generated 5 clusters as shown in Fig. 2, and the cluster instances are shown in Table 3.

The model does not reveal the class labels hence based on JRip rules, a GUI was developed and was tested on the data. The results of classification are shown in Table 4.

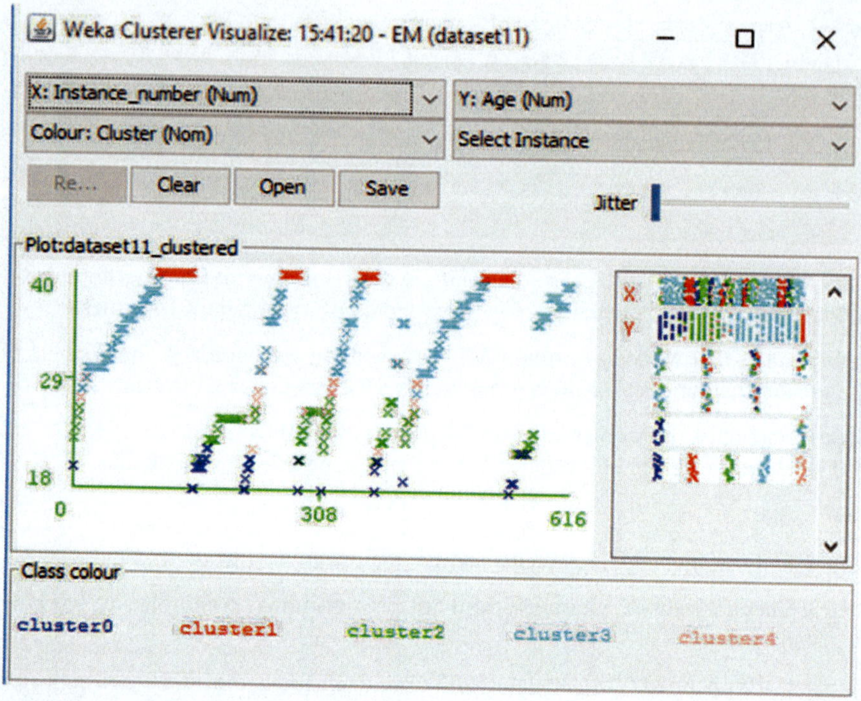

Fig. 2 Visualizing cluster assignments

Table 3 Model and evaluation on training set and clustered instances

Cluster No.	Cluster instances	Percentage (%)
0	53	9
1	290	47
2	125	20
3	114	18
4	35	6

Table 4 Results of BP prediction

Age group	Number of occurrences of last trimester BP	Total no. of women	Percentage (%)
18 to 22	6	61	10
23 to 25	132	312	42
26 to 30	64	202	31
31 to 35	27	32	84
36 to 40	9	10	90

Table 4 reveals that women in the age group 18 to 22 did not have BP during pregnancy rather than other groups. It entirely differs for the age group 23 to 25, because this group has many women being employed. The number of predicted high-risk people in age group 26 to 30 was less as the total number of pregnant women declined. In the age groups 31 to 35 and 36 to 40, the percentage of high risk is nearly 90% that would account to the age factor.

5 Conclusion

The JRip rule miner generated five (5) rules for predicting the last trimester hypertension for pregnant women. The rules had very high (0.973) as true positive (TP) and false positive (FP) is very less (0.067). Also, kappa statistics for both the class labels is near 1 (0.9216). The results show a high correlation between age and risk (age between 31 and 35 84% risk, and age > 36 risk is 90%). EM clustering was used for testing on the collected data that revealed a similar region on it. Hence, the rules can be used for modelling the hypertensive disorder on high-risk pregnancy for rapid late trimester. In future, the work can be extended by collecting data from other urban and rural areas and a few more other parameters can be included.

References

1. Sharmilan, S., Chaminda, H.T.: Pregnancy complications diagnosis using predictive data mining. www.researchgate.net/publication/308415497 (2016)
2. Moreira, M.W., Rodrigues, J.J., Oliveira, A.M., Saleem, K., Neto, A.J.V.: Performance evaluation of the tree augmented naïve bayes classifier for knowledge discovery in healthcare databases, 17° WIM—Workshop de Informática Médica (2017)
3. Moreira, M.W., Rodrigues, J.J., Oliveira, A.M., Saleem, K., Neto, A.J.V.: Predicting hypertensive disorders in high-risk Pregnancy using the random forest approach. In: IEEE ICC 2017 SAC Symposium E-Health Track, 978–1–4673–8999–0/17 (2017)
4. Adebayo, I.P.: Predictive model for the classification of hypertension risk using decision trees algorithm. Am. J. Math. Comput. Model. **2**(2), 48–59 (2017)
5. Mathew, N.: (2017) A study on hypertensive disorders in pregnancy using data mining techniques. Int. J. Adv. Res. Sci. Eng. **12**(06–12)
6. Patel, R., Barial, H., Patel, H.R., Nayak, S.: A study on pregnancy induced hypertension and foetal outcome among patient with PIH at tertiary care hospital, Valsad. In: Patel, R., et al. (eds.) Int. J. Community Med. Public Health **4**(11), 4277–4281 (2017)
7. Brandao, A., Pereira, E., Portela, F., Santos, M., Abelha A., Machado, J.: Predicting the risk associated to pregnancy using data mining. ALGORITMI Research Centre, Universidade do Minho, Guimarães, Portugal (2015)
8. Ye, C., Ruan, Y., Zou, L., Li, G., Li, C., et al.: The 2011 survey on hypertensive disorders of pregnancy (HDP) in China: Prevalence, risk factors, complications, pregnancy and perinatal outcomes, 9–6 (2014)
9. Kenjeric, I.B.D., Solic, K., Mandic, M.L.: Cluster analysis as a prediction tool for pregnancy cluster analysis as a prediction tool for pregnancy outcomes. Coll. Antropol. **39**(1), 247–252 (2015)

10. Mathew, N.: A boosting approach for maternal hypertensive disorder detection. In: Proceedings of the 2nd International Conference on Inventive Communication and Computational Technologies, IEEE Xplore Compliant—Part Number: CFP18BAC-ART (2018). ISBN: 978–1–5386–1974–2
11. Senthil, K.T.: Data mining based marketing decision support system using hybrid machine learning algorithm. J. Artif. Intell. **2**(03), (2020)
12. Karthik, D., Vijayarekha, K.: Clustering regions based on ground water levels using machine learning algorithms. Int. J. Pharm. Technol. **16** (2016)
13. Tekieh, M.H., Raahemi, B.: Importance of data mining in healthcare: a survey. In: Proceedings of the 2015 IEEE/ACM International Conference on Advances in Social Networks Analysis and Mining 2015, pp. 1057–1062 (2015)
14. Raghupathi, W., Raghupathi, V.: Big data analytics in healthcare: promise and potential. Health Inf. Sci. Syst. **2**(1), 3 (2014)
15. Smys, S.: Survey on accuracy of predictive big data analytics in healthcare. J. Inf. Technol. **1**(02), 77–86 (2019)

Condition Monitoring of a Sprinkler System Using Feedback Mechanism

S. Mahalakshmi, N. Veena, S. Guruprasad, and Pallavi

Abstract In India, traditional irrigation techniques, those that are operated by hand, are used widely because of their low price. There are different kinds of ways of irrigating fields. An automatic irrigation system isn't thus rife here. The proposed system permits versatile management of mechanical device valves for releasing water to fields, and data concerning the current standing and quantity of water discharged is distributed to the user by an app through the online server which acts as an information collector of this method. A mobile application is employed to indicate the action performed within the garden. The proposed solution provides simple and versatile management of irrigation methods and provides optimum water consumption. The shrewd sprinkler water system innovation depends on real water needs. This framework can be modified to naturally begin a set time and day consistently. Little versatile sprinklers can be incidentally positioned on gardens if extra watering is required or if no perpetual framework is set up. The framework has been effectively picked, actualized, and worked in the field. The framework setup has been changed to meet the water necessities as per the yield development stages. Using IoT for water management systems can be extended out to different techniques in irrigation. It helps to reduce human involvement in irrigation.

Keywords IoT · Sensors · Mobile app · Irrigation

S. Mahalakshmi (✉) · N. Veena · S. Guruprasad · Pallavi
Department of ISE, BMS Institute of Technology and Management, Bengaluru, India
e-mail: maha.shanmugam@bmsit.in

N. Veena
e-mail: veena@bmsit.in

S. Guruprasad
e-mail: guruprasad@bmsit.in

Pallavi
e-mail: pallavimahesh98@gmail.com

J. S. Raj et al. (eds.), *Intelligent Sustainable Systems*, Lecture Notes in Networks and Systems 213, https://doi.org/10.1007/978-981-16-2422-3_39

1 Introduction

In Asian nations, most of the irrigation systems are opened manually. Mechanical device irrigation may be a technique of applying irrigation water that is comparable to natural precipitation. One of the most benefits of the mechanical device irrigation technology is a lot of economical use of water for irrigation in agriculture. Water is additionally distributed a lot equally across crops serving to avoid wastage. A mechanical device is employed to irrigate agricultural crops, golf courses, landscapes, lawns, and different areas. Sprinklers are often used for agricultural, industrial, and residential usage. It's helpful on uneven land wherever water availability isn't possible. A sprinkler system uses valves to manage water flow to every selected space. This magnetically controlled coil element uses electrical signals from the irrigation controller to physically open and shut every valve to permit water to the maneuverer through the system. An unsuccessful magnet, however, causes system leaks and improper watering amounts throughout the sprinkler system.

We want to ascertain the health risks of those elements. An appropriate level of water in the soil may be a necessary prerequisite for the optimal growth of any plant. Also, water being an important component for all time nutrition, there is a requirement to evade its excessive usage. Irrigation may be a dominant shopper of water. This infers the necessity to manage water for agriculture. Fields should neither be over-sprinkled nor under-sprinkled. The target of this project is to style an easy, straightforward process to observe and designate the extent of soil wetness that's incessantly controlled so as to realize growths of plants and at the same time optimize the irrigation resources on an observation hardware Wireless Relay Module, and therefore the device knowledge is often seen on the Web. The info transmission on the net is predicated on wireless sensor network (WSN), general packet radio sensor (GPRS), and global position system (GPS). The use of simple elements reduces the production and maintenance prices and makes a cost-effective, suitable, and a less maintenance application, for small agriculture and farmers. This analysis work assists the trivial farmers and can increase the profit.

Over time, systems are enforced toward understanding the goal that machine-driven processes are foremost well-liked as they permit information to be acquired at a higher rate with less worker necessities. Majority of the prevailing architecture uses circuitry-based systems. IoT is modifying the way of agriculture and reducing the problems of agriculturists. Automatic systemizing of water sprinkling for the farm or nursery allows farmers to use the right amount of water at the correct time, which still needs the availability of labor to open and close the valves. This can finish with the assistance of an associate automatic valve mechanism and system. Victimizing these techniques farmers would be ready to utilize the proper quantity of water at the right time by automating at the right time. Avoiding irrigation at the incorrect time of the day, scale back runoff from over-watering saturated soil can improve crop performance.

2 Related Work

Much analysis has been done to improve the performance in irrigation. In [1], a mixed transmission theme of both uplink and downlink has been projected for Wireless Sensor Networks in agriculture, where that ancient Orthogonal Multiple access is applied in downlink transmission and relay-aided Multiple OMA theme is applied in uplink transmission from the device nodes to the sink node [2]. Owing to large connectivity and excessive data rate to the sink node which improves the statistics rate to the sink node, thus lessen the energy consumption of device nodes, and extend the lifetime . The paper [3] shows that Modernization of IoT helps in assembly data on environments like level of water, humidity, temperature, and prosperity of soil, climate, crop net based totally inspection allows discovery of wild flower, locating infections, gardening, creature interruption in to the arena, slender development to examine remotely the conditions as image and video where distant cameras are used. IoT development can lessen the worth and improve the yield of the traditional method [4]. The paper [5] provides a WSN microcomputer appliance preciseness irrigation, especially things within that device nodes are additionally placed in a way in which within which from the arranger.

The planned version relies on a tailored hardware and code package program attending to every device node's life. The arranger is alert concerning device nodes to be had strength which feature permits estimating last period of every device node for info acquisition of the abstract version for a sensing part node hardware and code package program system vogue, about electricity management and messages grouping, rising each device nodes and community life. The paper [6] explores driving McKibben actuators with high stress mechanics and displays some disputes as related to modeling the actuators at hydraulic pressures. Four modeling methods [7] were evaluated toward experimental records in conjunction with a current version that eliminates kind of the restrictions of existing models with the facilitate of predicting output pressure as a operate of initial mechanism mathematics handiest and shooting the nonlinear variation in elastomeric wall thickness as a results of the mechanism is strained, absolutely the fee of the error from experimental values of every of the four fashions ranged from 9.1% for the re- creation developed throughout this paper to 9.9%, 10.0%, and 10.5% for the three totally different current approaches, but the new model has distinctive blessings in modeling hydraulic McKibben actuators that require thicker tube walls. In the paper [8], a bearing irrigation system is incontestable that absolutely automates the delivery of irrigation water and calculates in real time the water demand from satellite photos. Irrigation [9] is managed mentally by a central computer that issues commands to 693 management nodes to start irrigation-supported analytics extracted from satellite photos. The management nodes are sorted to create 15 m by 15 m cells and each cell is typically addressed severally and may irrigate differentially.

3 Proposed System

It permits versatile management of mechanical device valves for releasing water to fields and data concerning current standing and quantity of water discharged that is distributed to the user by an app by the online server which acts as an information collector of this method. A mobile application is employed to indicate the action performed within the garden. The proposed solution provides simple and versatile management of the irrigation method and provides optimum water consumption. The smart sprinkler irrigation technology is predicated on how much actual water needs for irrigation. During this, the system irrigation happens once the water is needed by the soil. It provides solely that quantity of water. The proposed system is easy to deploy and controlled mechanically. The sensor, actuators, and every hardware device are checked mechanically for their regular operation, and feedback is distributed to the user concerning the harm to the element. The projected system is reliable and simply deployable so as to figure underneath harsh outside conditions while not the requirement for direction or regular watching (Fig. 3 and Tables 1, 2).

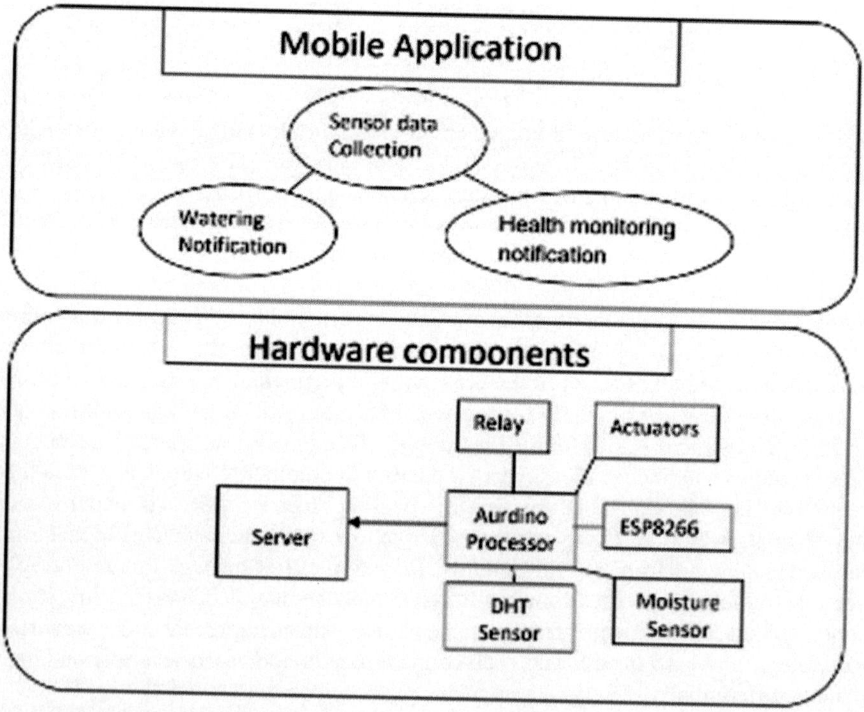

Fig. 1 Overall system architecture

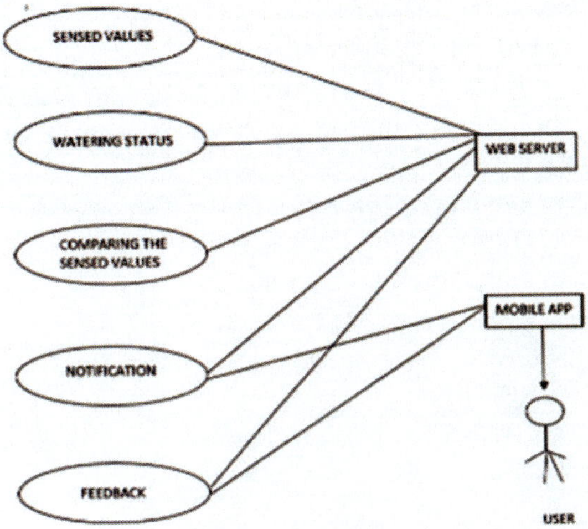

Fig. 2 Use case diagram

Fig. 3 Proposed System

Table 1 Arduino connection

Ardiuno pins	Connected to
GNDPin	GND(T.CD display, WiFi module, Soil moisture sensor, DHT1 I
5 V	I,CD display, soil moisture sensor,dHTI 1
Pin 13(PB5)	BCD display
Pm 12(PB4)	LCD display
Pin 11(PB3)	LCD display
Pin 10(PB2)	LCD display
Pin 9(PR1)	LCD displav
Pin S(PBO)	LCD display
Pin 7(PD7)	dill 11
Pin 6 (PL > 6)	Re lay (Input)
Pin 5(PD5)	Soil moisture sensor

Table 2 Components and specification

Material	Specifications
Arduino UNO	Microcontroller AT mega328 Voltage of operating 5 V Recommended input voltage 7–9 V Range of input voltage 6–20 V Pins of digital I/O 14 (of which 6 provide PWM output) Pins of Analog Input 6 Per I/O Pin IX' Current 40 mA Pin Current DC for 3.3 V 50 mA Memory of Flash 32 K1Ì (aTmega328) (0.5 KB used by boot loader)
Temperature Sensor (DIIT11)	Operating Voltage: 3.5 V to 5.5 V Operating current: 0.3 mA (measuring) 60uA (standby) Output: Serial data Temperature Range: () °C–50 °C Humidity Range: 20–90% Resolution: Temperature and Humidity both are 16-bit
Soil Moisture Sensor (FC-28)	Input voltage: 3.3–5 v Output voltage: 0–4.2 v Input current: 35 mA Output Signal: Both Analog (A0) and Digital (DO)
Humidity Sensor (SYH-1 Specs)	Manufacturer SYHITECH Measuring range 20%. 0.95% RII Tolerance ± 5% RH Output configuration analog voltage Operating temperature (I.., 60 °C Resistance 23kQ

4 System Implementation and Results

4.1 Sample Source Code

```
void setup()
{
Serial.begin(9600);
lcd.begin(16, 2);
dht.begin();
pinMode(6,OUTPUT);}
void loop()
{
lcd.clear();
lcd.setCursor(0, 0); // set the cursor to column 0, line1
lcd.print(" health monitoring "); //print name
lcd.setCursor(0, 1); // set the cursor to column 0, line
lcd.print("sprinkler system "); //print name
delay(1000);
//delay(2000);
//Read data and store it to variables hum and temp
hum = dht.readHumidity();
temp= dht.readTemperature();
//Print temp and humidity values to serial monitor
Serial.print("Humidity: ");
Serial.print(hum);
Serial.print(" %, Temp: ");
Serial.print(temp);
Serial.println(" Celsius");
lcd.clear();
lcd.setCursor(0, 0); // set the cursor to column 0, line1
lcd.print(" TEMP:"); //print name
lcd.print(temp); //print name
lcd.print(" 'C"); //print name
lcd.setCursor(0, 1); // set the cursor to column 0, line
lcd.print("HUMIDITY:"); //print name
lcd.print(hum); //print name
lcd.print(" %"); //print name
delay(10000); //Delay 2 sec.
if(temp>33)
{
digitalWrite(6,HIGH);
lcd.clear();
lcd.setCursor(0, 0); // set the cursor to column 0, line1
lcd.print(" HIGH "); //print name
lcd.setCursor(0, 1); // set the cursor to column 0, line
lcd.print("TEMPERATURE"); //print name
Serial.println(" HIGH TEMPERATURE");
delay(10000);
```

Closed-loop irrigation systems [10] require feedback from sensors. The system continuously checks the status of the sensors. A closed-loop control system is a set of electronic or mechanical devices which regulates a process variable automatically to a set point or the desired state without human interaction. First, the temperature and humidity values are sensed by the DHT sensor, and based on the sensed data the comparison is done with the threshold values in the Web server. If the sensed values are lesser than the threshold values, then the relay is turned on which in turn leads to watering [11]. A moisture sensor is used to check the moisture level of the field in order to ensure that proper watering is done [12]. All the sensed data including the condition of sensors and the watering status is sent to the user app. The functions of the components used are as follows:

Temperature Sensor As temperature builds the dampness, content in the dirt declines which makes the soil dry [13]. At the point when the temperature surpasses as far as possible, a notice will be sent to the Android application naturally.

Humidity Sensor At the point when the sensor esteem goes below the edge esteem, it educates the client with a warning with each refreshed information. At the point when the sensor esteem arrives at the expected edge esteem, the water system will begin consequently.

Notification Part At the point when the sensor esteem goes below the edge esteem, it educates the client with a warning with each refreshed information.

1. The Need for Automated Irrigation: A keen water system is a key part of accurate farming [14]. It encourages ranchers to stay away from water wastage and improve the nature of harvest development in their fields by a) inundating at the right occasions, b) limiting overflows and different wastages, and c) deciding the dirt dampness levels precisely, in this way, finding the water system prerequisites at wherever.

2. The IoT-based Irrigation System: Vian et al. [15] wireless Architecture includes soil dampness sensor, stickiness sensor, and temperature sensor, which are set on the fields, send ongoing information to the microcontroller. The microcontroller additionally has servo engines to ensure that the funnels are really watering the fields consistently so no region gets stopped or is left excessively dry. The whole framework can be overseen by the end client through a devoted versatile application. A brilliant water system makes it workable for cultivators to screen and flood their fields remotely.

3. Soil Moisture Sensor: Senseing the soil moisture level is a part in the irrigation procedure. Temperature Sensor: To check the temperature threshold value before starting irrigation. Humidity Sensor: To check the humidity level in the soil.

4. The Role of LED Lights: A shrewd water system unit, with a microcontroller at its center, likewise has pre-tried Drove bulbs. When the on-field sensors report that the dampness level has fallen below the suggested or edge level, the bulb sparkles, showing that sprinkler system must be started.

5. The Placement of Sensors: It's all about setting up portals and siphons and different devices; in any case, except if the sensors are set effectively in the fields, the choices taken by the brilliant water system can possibly be wrong. Specialists prescribe clients to ensure that the sensors stay in contact with the dirt surface consistently, precluding the nearness of any air holes, and are put at least 5 feet from water system heads, property lines, homes, and high-traffic zones [16]. For best outcomes, the sensors ought to be deliberately positioned in the regions that get the most extreme daylight and inside the root zones of the plants. A dirt dampness sensor must be secured with soil, yet the encompassing weight ought not to be excessively high (Fig. 4).

6. Fault Detection: When the components fail to work as programmed, notifications will be sent to the user's phone through an android application. This makes the system more useful, efficient, and reliable (Table 3).

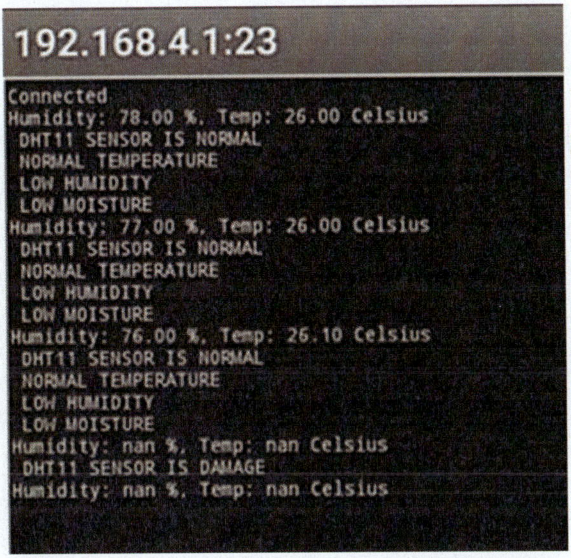

Fig. 4 Output of the application

Table 3 Component testing

Test Unit	Testcase	Result
Temperature and humidity sensor	Check serial monitor for the values 0–100	Invalid values result in sending feedback to user
Soil moisture sensor	Check serial monitor for the values 0–100	The sensor gives us the value 0–1023.and these values are mapped to 0–100. If the values are beyond the range, then sensor is damaged

4.2 Component Testing

4.3 System Testing

UNIT	TEST CASE	RESULT
Mobile application, firebase	Sensed values are less than threshold value	Sensed values and watering status are displayed in the message sent to the user
Mobile application, firebase	Sensed values are greater than threshold value	Sensed values and watering status are displayed in the message sent to the user
Mobile application, sensors	No value sent from the sensor	Feedback message is sent to user about component damage
Mobile application, sensors	The values from sensors are beyond the range of sensor	Feedback message is sent to user about component damage

5 Conclusion

The shrewd sprinkler water system innovation depends on real water need. This framework can be modified to naturally begin at a set time and day consistently. Little versatile sprinklers can be incidentally positioned in gardens if extra watering is required or if no perpetual framework is set up. The framework has been effectively picked, actualized, and worked in the field. The framework setup has been changed to meet the water necessities as per the yield development stages. Water is provided by the requirements of plants. The keen water system framework alongside the controllers works satisfactorily and in an exact way the input instrument sends a fitting message to the client about the state of the segments and the watering status of the field. The proposed framework is solid, effectively deployable, and demonstrated its capacity to provide the necessary amount of water. Generally speaking, the best possible establishment and setup of every one of the advancements tried here were a significant factor in deciding the viability to which every framework could decrease water wastage.

6 Future Enhancement

This task can be made further progressively creative by making it into an astute framework, wherein the framework predicts client activities, precipitation design, time to reap, and a lot more highlights which will make the framework autonomous of human activity. We can likewise show the graphical portrayal of the dampness content levels in the field.

To improve the proficiency and viability of the framework, the accompanying suggestions can be placed into thought. An alternative to controlling the water tap can be given to the farmer. The farmer may decide to stop the development of harvests, or the yields may get harmed because of unfriendly climate conditions. In such cases, farmers may need to stop the framework remotely. Using IoT for water management systems can be extended out to different techniques in irrigation. It helps to reduce human involvement in irrigation.

References

1. Wang, L., et al.: Application of non-orthogonal multiple access for iot in food traceability system. In: 2019 IEEE 4th Advanced Information Technology, Electronic and Automation Control Conference (IAEAC), vol. 1. IEEE (2019)
2. Nalajala, P.: P Sambasiva Rao, Y Sangeetha, Ootla Balaji, K Navya", Design of a Smart Mobile Case Framework Based on the Internet of Things". Adv. Intell. Syst. Comput. **815**, 657–666 (2019)
3. Nalajala, P. et al.: Design of a smart mobile case framework based on the internet of things. In: First International Conference on Artificial Intelligence and Cognitive Computing. Springer, Singapore (2019)
4. Seo, J.-B., Kwon, T., Choi, J.: Evolutionary game approach to uplink NOMA random access systems. IEEE Commun. Lett. **23**(5), 930–933 (2019)
5. Vieira, R.G. et al.: On the design of a long range wsn for precision irrigation. IEEE Sensors J **18**(2), 773–780 (2017); Thomalla, Steven D., James D. Van de Ven.: Modeling and implementation of the McKibben actuator in hydraulic systems. IEEE Trans Robot **34**(6), 1593–1602 (2018)
6. Anwar, A., Seet, B.-C., Ding, Z.: Non-orthogonal multiple access for ubiquitous wireless sensor networks. Sensors **18**(2), 516 (2018)
7. Su, X, et al.: Power domain NOMA to support group communication in public safety networks. Future Gener Comput Syst. **84**, 228–238 (2018)
8. Moltafet, M. et al.: A new multiple access technique for 5G: Power domain sparse code multiple access (PSMA). IEEE Access **6**, 747–759 (2017)
9. Zhang, J. et al.: Performance analysis of user ordering schemes in cooperative power-domain non-orthogonal multiple access network. IEEE Access **6**, 47319–47331 (2018)
10. Kader, M.F., Shin, S.Y.: Coordinated direct and relay transmission using uplink NOMA. IEEE Wirel. Commun. Lett. **7**(3), 400–403 (2017)
11. Sedaghat, M.A., Müller, R.R.: On user pairing in uplink NOMA. IEEE Trans. Wirel. Commun. **17**(5), 3474–3486 (2018)
12. Wan, D., et al.: Cooperative NOMA systems with partial channel state information over Nakagami-$ m $ fading channels. IEEE Trans. Commun. **66**(3), 947–958 (2017)
13. Ferentinos, K.P. et al.: Wireless sensor networks for greenhouse climate and plant condition assessment. Biosyst. Eng. **153**, 70–81 (2017)
14. Daskalakis, S.N. et al.: Backscatter morse leaf sensor for agricultural wireless sensor networks. In: 2017 IEEE Sensors. IEEE (2017)

15. Viani, F., Bertolli, M., Polo, A.: Low-cost wireless system for agrochemical dosage reduction in precision farming. IEEE Sens. J. **17**(1), 5–6 (2016)
16. Khan, R. et al.: Smart sensing-enabled decision support system for water scheduling in orange orchard.IEEE Sens. J. (2020)

Cognitive Radio Networks for Internet of Things

K. Leena and S. G. Hiremath

Abstract The Internet of Things is a promising subject, both strategically and socially, of increasing technical and economic significance. The key feature of IoT is that Internet connectivity and powerful data collection capabilities are integrated with separate computers. The IoT meaning is about several devices and sensors, i.e. connections to the Internet. The IoT is a worldwide network of linked processors. Objects, however, are not always palpable for interconnection. According to many estimates, the effect of IoT on the Internet and on the economy will be really inspiring, with a large global economic impact in the coming years. IoT may establish interconnectivity for organisations by specific wireless networking technologies such as cost-effectiveness problems and remote interface accessibility, rendering wireless communications a feasible option. However, the IoT model argues that communication technology poses new obstacles as a number of heterogeneous systems can be interconnected, and one of the main chaos Cognitive Radio (CR) networks and the incorporation of CR into IoT will increase spectrum precision. This paper explores the choices for applying the cognitive radio network to IoT.

Keywords Cooperative communication · Pervasive computing · Internet of things · Cognitive radio

1 Introduction

The wireless spectrum that is used by a lot of radio equipment in the world is getting more and more congested. As the demand for data traffic increases, the competition between the different wireless carriers will become fierce. Since authorities under some circumstances should supply spectrum lightly to satisfy the need, but not sufficiently to fulfil the need, one option may be to use the spectrum more dynamically, like in spectrum management by authorities, whether the spectrum has been allocated to using more flexibly. Through the capacities of computational wireless

K. Leena (✉) · S. G. Hiremath
Department of Electronics and Communication Engineering, East West Institute of Technology, Bangalore, Karnataka, India

J. S. Raj et al. (eds.), *Intelligent Sustainable Systems*, Lecture Notes in Networks and Systems 213, https://doi.org/10.1007/978-981-16-2422-3_40

technologies, radio systems have the potential to detect the world surrounding them and opportunistically absorb the usable RF spectrum.

Software Defined Radios that manage frequency band exchange and spectrum sharing, spectrum guard, spectrum usage, and noise reduction may become important in future Networked Application Styles and Networked Device Architectures, for example, in the Complex Spectrum Access method (focusing on high-frequency spectrum), in the Wireless Phone (focusing on the cellular spectrum), and in the Intelligent Radio Network (focusing on sparse frequency bands). Because SDR (Software Defined Radio) technology is able to execute various communication functions, e.g. basic communication functions, reliable distribution function of broadcast information, and other functions for D2D nodes of an edge computing infrastructure or for secondary users of CR-IoT, it can be used for several other related applications.

In order to resolve spectrum access for potential IoT technologies focused on cognitive networks, cognitive radio networks should be proposed. Cognitive IoT (CR-IoT) technologies, especially for WSN, have improved intelligence to achieve successful data sensing and interpretation. Smart and enhanced medium access control (MAC) architecture to completely integrate CR-IoT into WSN, enabling the presence of the sensor network with the current wireless infrastructure. In addition, in order to completely integrate CR-IoT, perception in spectrum access needs to be discussed. That is, much of the spectrum is used on and off, and some parts of the spectrum are also congested. There are cases of artificial and genuine spectrum scarcity that make it important for the spectrum to be exploited efficiently.

By implementing the opportunistic spectrum access (OSA) mechanism, which is often referred to as the mechanism of dynamic spectrum access (DSA), spectrum inefficiency can be tackled. OSA is known to be a big CR function where wireless devices can hear, understand, and adjust. While CRSN helps to improve the efficiency and user specifications of the IoT network. Sensor systems, however, need energy overhead to conduct spectrum sensing and spectrum sharing in order to meet QoS and high application throughput specifications. These applications include distributed access to the channel, taking into account undisclosed network knowledge. This architecture was introduced in [1] for single cognitive users and in [2] for multiple users. In view of the multi-user situation, though, it causes network collision. Any of the users on CR-IoT are greedy because of intrusion. Priority, random access, and an equal resource sharing mechanism have been presented to minimise collisions. These models are, however, restricted to supplying a cognitive user with just one platform at a time. As a consequence, if a chosen channel is busy, the customer has to wait for another slot and the bandwidth is lost if another channel is open at the moment.

Various existing research, [3, 4] and [4], have been adopted to achieve a fair resource allocation evolutionary computing model such as particle swarm optimisation, reinforcement learning, game theory model, and so on. These models, however, are built to take into account static secondary consumers, single channels, and homogeneous worlds, and [5] have implemented an evolutionary computing paradigm that utilises multi-channel systems and mobile secondary computer game theory. However, because of intervention when mobility speed is raised, the presence of

Nash equilibrium is not presented and incurs channel switching overhead. In addition, the new paradigm has tackled problems of interference in the same area, and very little analysis is carried out while communication systems run within an overlapping segment of two neighbouring primary consumer cells. In addition, user mobility is very constrained in CR-IoT applications such as CR-WSN but needs less latency with good energy consumption; user mobility is very quick in CR-VANET, and user mobility is complex in nature in CR-D2D. Such a heterogeneous criterion in the modelling spectrum allocation process for CR-IoT is not considered in the current model proposed so far.

This work provides a reliable estimate of the possibility of channel accessibility using an evolutionary computation approach to solve overhead attributable to versatility and model an effective spectrum control model for multi-channel CR-IoT.

2 Literature Survey

In order to increase spectrum access and usage of CR-IoT, this segment carries out a detailed survey relevant to the different current methods offered. In addition, the study voids for achieving an appropriate spectrum resource distribution model for CR-IoT describe state-of-the-art work.

Opportunistic spectrum access model for CR-IoT: In [6], the reciprocal regulation of sample rates and channel connectivity of sensor nodes under energy usage, channel bandwidth, and interference constraints were considered as an optimiser network utility. With fluctuating rates of energy harvesting and channel transition costs, we formulate network efficiency maximisation as a non-linear programming problem with a mixer and cope with it effectively, coupled with double decline. A joint channel access and sampling rate control framework called JASC would then be presented, taking account of the results of real-time channel sensing and energy recovery rates. If nodes have data to relay, spectrum sensing on demand is initiated and a subset of spectrum sensing nodes are often used to gather information on spectrum availability across all nodes. In addition, CRSN nodes establish an adaptive duty period that enables them to sleep consistently and stay awake if transmission data is necessary. CAMAC is thus a sophisticated approach that uses a limited number of spectrum sensing nodes with an adaptive sensing period to achieve low energy consumption. [7] The complex channel access model was introduced in light of the energy usage of channel sensing and exchange; the state of the sensor nodes and the transfer to a licenced channel in order to increase energy efficiency are calculated by the packet loss rate on the license-free channel. In order to define channel sensing and interchange sequences for intracluster and intercluster data transfer, two complex channel control schemes are also proposed. In [8], a branch-and-bound solution focused on algorithms was proposed, with a reduction in search space. Their approach constructs the best routes from each source to the sink node, including the best set of hops in each direction, the best variety of relays, and the maximum allotment of power for cooperative transmission connections. In [9], a QoS-conscious packet scheduling approach is implemented to boost secondary consumer transmission efficiency. A QoS-based priority model is specifically structured to resolve data classification within this framework. And then channel efficiency and channel transmission impact are implemented in the packet scheduling process on the basis of priority.

Opportunistic spectrum access employing evolutionary computing model for CR-IoT: Increased channel sensing learning was carried out in the cluster-based wireless sensor network [10]. The efficiency technique is built to reduce sensing by adding the problem of the Markov Decision. The result indicates a rise in the identification of primary customers and the cost of energy over conventional greedy searches. The spectrum sensing time period model has been developed [10] to provide estimated data for the secondary consumer channel sensing sources. They also introduced a complex sensing paradigm for the Markov chain, which facilitates listening before talking. The simulation involves various time scales. The findings indicate that the model for the Markov chain enhances cognitive radio WSN energy efficiency, but the network density is known to differ. The dilemma of sequential multi-user channel sensing and links to complex neural networks, in which the active user collection is discussed [11]. In addition, each user has just his or her particular details without information sharing between users. The goal of users is to decide their channel sensing order. In order to resolve the Channel Sensing Order overlap, they first set a standard interference metric and set two maximising objectives: to reduce aggregate interference for any active user set and to minimise anticipated aggregate interference for all prospective users. However, two-problem optimisation cannot be addressed centrally since the active user collection varies arbitrarily and the distributions of the active user sets are not specified for the user. Two non-cooperative game models were proposed in order to address ptimization problems: a state-dependent single-shot game and a robust game. They have shown themselves to be viable games and that the strongest Nash balance of the two games leads, respectively, to optimum strategies for all optimisation problems. They proposed a stochastic learning algorithm for coping with unknown, complex, and incomplete knowledge constraints in dynamic networks, which is analytically seen to converge on the equilibria in the presence of an evolving community of active participants in the two formulated games. In [12], we set up a delayed market mechanism for coping with the inefficiency of Nash's balance and are suggesting a different replicator dynamic for this price mechanism. Channel regulations and pricing can be modified asynchronously, so that vehicle buyers may improve their channel rate awareness before making decisions about an entry. In [13], an online learning system for improving the energy efficiency of cognitive radio sensor networks (CRSN) was developed [14], such as Particle Swarm Optimising (PSO) by controlling packet lengths. Fairness between different consumers and the usage of energy [15] are seen as critical problems for the growth of potential contact networks.

3 Research Challenges

The overall survey indicates that cognitive radio plays an important role in enhancing the overall efficiency of the Internet of Things application network linked to the heterogeneous wireless networking world of the future century. In order to increase the prediction of the precision of channel state knowledge for successful spectrum sensing, the usage of evolutionary computing techniques helped. The ultimate result obtained demonstrates that dispersed spectrum sensing for CR-IoT is carried out with very minimal work considering complex mobility. In order to increase precision for channel state prediction for effective channel entry, there is a need to build a better optimisation technique. In addition, given the multi-channel CR-IoT setting, the current model suffers from the provision of an equal and effective spectrum access model. It may help to solve certain spectrum allocation issues by utilising evolutionary computation techniques. However, the presence of Nash equilibrium (i.e. optimum distribution of resources) is not provided by the current paradigm of evolutionary computing-based spectrum allocation and suffered overhead channel swapping due to interruption as the speed of mobility is increased. Designing an effective dispersed opportunistic spectrum access model for CR-IoT utilising spatial distribution and temporal channel knowledge patterns is critical for overcoming research challenges. In addition, to better use spectrum more efficiency under the multi-channel heterogeneous mobile CR-IoT environment, use a successful evolutionary computing technique. In addition, the new effective spectrum sharing model solves the issue of competition between the contending devices in the conflicting area of the neighbouring cellular network.

4 CR-IoT Architecture

The simple IoT architecture is presented in Fig. 1. The topology illustrates a network necessity for facilitating data flow with worldwide interoperability, i.e. Internet access. This interconnectivity has historically been established by GSM or some other conventional technologies.

However, IoT artefacts should provide a cognitive facility in the cognitive IoT network to render wise spectrum judgments and execute intelligent operations by observing network circumstances. Figure 2 illustrates the cognitive wireless system for interconnectivity provision.

The smooth cognitive functions of these CR artefacts must be tested. Figure 2 demonstrates the topology of a data flow network with global bandwide network and portability, an Internet Protocol (IP) network, and a global network-based wireless system/network. Wireless networking is focused on CR, where the entire device should be clever enough to reach the usable spectrum holes. Since large IoT systems may use large quantities of data, this information is worthless unless it is handled carefully and translated into useful information. A cloud-based infrastructure may

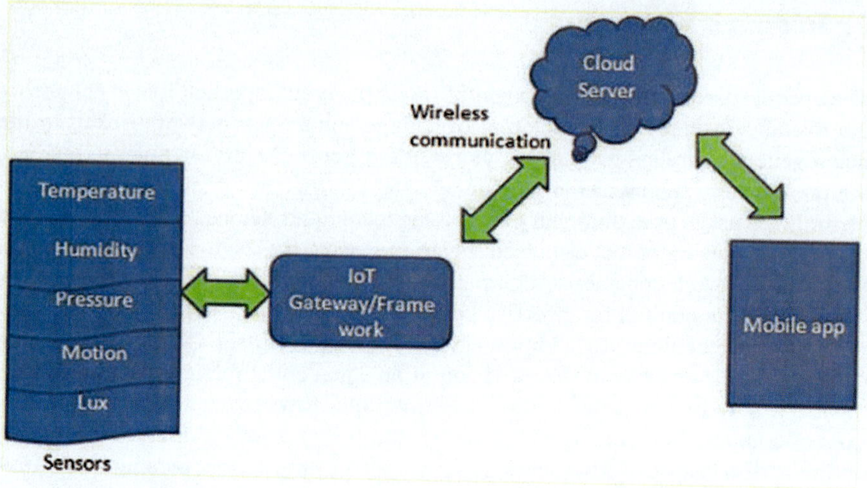

Fig.1 Basic IoT network architecture

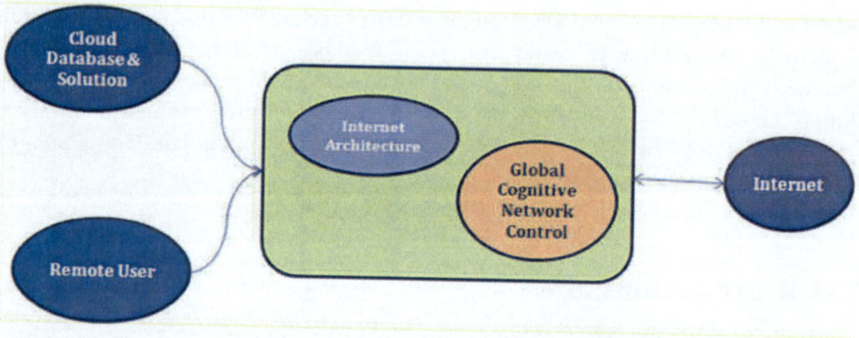

Fig.2 Cognitive arrangement for IoT network

also be a proposal, as mentioned earlier, to mitigate these problems. For data analysis and knowledge decision-making, a data-oriented CR-based IoT architecture is feasible. Computational capability of the network allows good choices for automated service management. Integrating IoT sensors increase the energy-saving problem. CR may be seen to be a promising enabler for IoT.

5 Issues Associated with CR-Based IoT

In order to build an alliance (Fig. 3) between both CR and IoT innovations, there are several tough questions that need to be tackled. There are some examples that require careful attention for all technologies to play a leading role in the foreseeable future, such as central concerns such as interoperability of interconnected systems, offering a higher degree of smartness, improving adaptability depending on the current climate, industrialization, and standardisation and, most notably, guaranteeing protection and belief in the ubiquitous environment. We address the CR network implementations for IoT in this paper and offer answers to the problems that both systems are currently experiencing.

The IoT is a mixture of heterogeneous items. An optimal approach for heterogeneous networks is the CR infrastructure that usually involves node-based design with proper control strategies. However, since uniform security requirements are not available on all heterogeneous networks, this results in some security concerns. This is a key issue for the heterogeneous network focused on CR [16]. Regularisation and standardisation have become a critical source of tension that needs to be resolved urgently. The regulatory issues of the use of CRNs in the approved range ought to

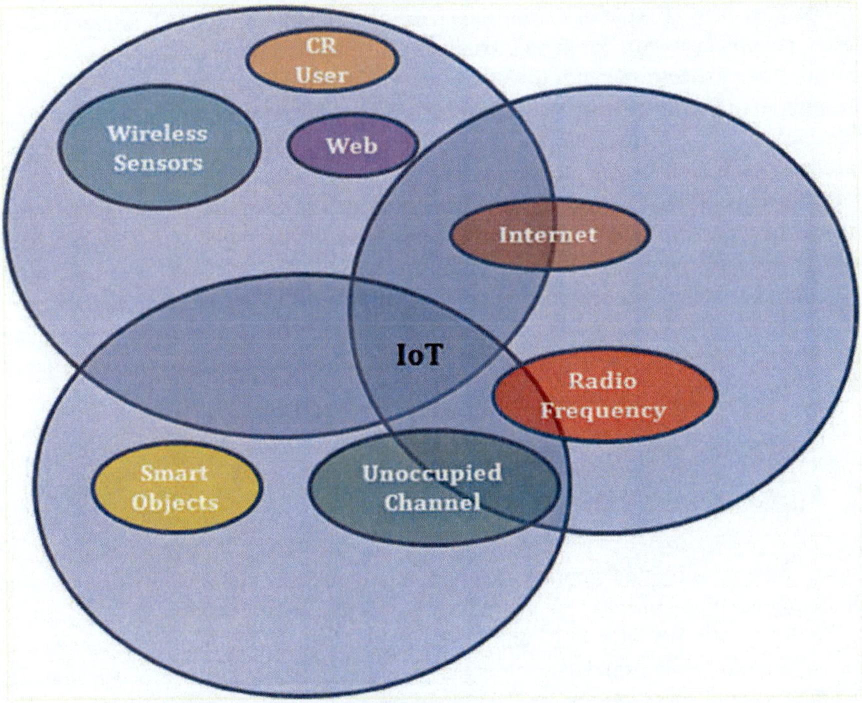

Fig.3 Cognitive Radio for Internet of Things Paradigm

be discussed by the authorities involved, as without appropriate authorisation, no unlicensed wireless device will be permitted to enter the ownership spectrum. This may generate nuisance, protection and surveillance risks and also conflict with the PU services [17].

The right identification of PU existence is most important, i.e. a difficult job is to categorise between the Pus signal and the SUs signal. In addition, there would be a number of signals in the same band for the involvement of several approved consumers, which is another key challenge [18].

6 Motivations for Using CR in IoT

It is envisaged that CR-based IoT structures will become essential requirements in the future with the continuing growth of CRNs and IoT. In order to read, think, and make choices through knowing both the social and physical realms, the IoT artefacts must be fitted with cognition [3, 6, 8]. Smart decision-making, the interpretation response loop, large data analytics, on-demand service provisioning, semantic derivation, and the exploration of information are additional criteria. Therefore, in the future, a CR-based IoT is a predictable requirement; this could be for the following purposes.

The key inspiration arises from the sharing of bandwidth for IoT artefacts. It is anticipated that the number of IoT artefacts will rise in significant amounts, and the distribution of spectrum bands to these objects will be very complicated. In addition, the amount of Pus would also rise, allowing unlicensed consumers to face difficulties. The strategy of fixed spectrum assignment involves spectrum purchasing costs; thus, spectrum assignment will produce needless expenses for such a large number of items. In any of these circumstances, CRNs can aid. The spectrum exchange between different users is not enabled by conventional contact strategies. CRN would be a help in the future for its spectrum sharing benefit as artefacts will increase, each searching for access to the spectrum [8, 9]. CRNs will then interpret the spectrum landscape and, by informed decision-making, provide on-demand services among consumers. Cellular connectivity entails prices, although there is a small choice of Bluetooth and Zigbee.

7 Applications of CR-Based IoT

Some of the possible IoT applications that may benefit from CRNs are provided in this segment (Fig. 4).

Health Care: We have IoT healthcare implementations in the functional realm now. In order to track vital data such as temperature, blood pressure, glucose content, and others, intelligent sensors are deployed on and around a patient. With remote control, the criteria are constantly observed by medical personnel. There are already wireless solutions; nevertheless, maintaining seamless management is desperately

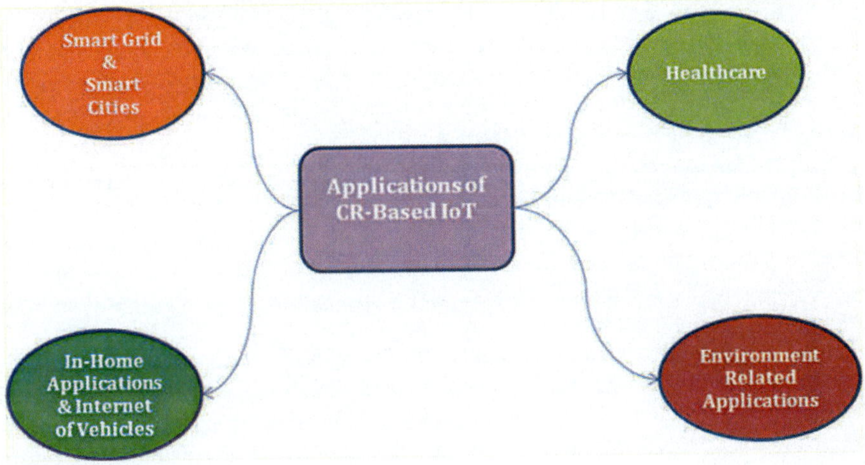

Fig.4 Applications of CR-Based IoT

important. For this reason, health details can be distributed to medical professionals without the requirement for bandwidth allocation. Without any questions about bandwidth accessibility, CR-based IoT architectures will accomplish this across long ranges [19]. In the social realm, a traffic control scheme in the form of intelligent traffic lights may also be used as an illustration. Another indication of a social practise is the usage of CR-based IoT artefacts for commercial uses, such as consumer wireless broadband [5, 10].

Environment-Related Applications: Environmental IoT-focused technologies are currently developing with great interest. Waste management, noise and temperature measurement, humidity estimation, and CO2 emission monitoring are some of the major areas. Besides utilising sensors, dedicated cars are also possible for these purposes. However, efficient and successful monitoring requires installing and utilising large quantities of heterogeneous devices at sensitive places. Under the current fixed assignment scheme, this large amount of consumers cannot share the spectrum. IoT functionality could be possible for network-congested/scarce locations with cognitive capabilities [20].

In-Home Applications: IoT-based technologies are already within the working realm of home/building, and IoT is predicted to be required in the future through evolving technology. Homes are integrated with sensors/devices to conduct regular tasks for improved quality of life. Some instances, including smart fridges and smart lamps, are already present in home appliances and home energy management. Wi-Fi access points are typically installed for these purposes, but this can significantly conflict with the commercial, science, and medical (ISM) band. Equipping sensors with cognitive abilities to stop ISM bands [2] is a better option.

Smart Grid: Smart grid is becoming an evolving model. Consumers like to hear about their energy use anywhere, wherever. This demonstrates the need for IoT in the smart grid. Here, transmitting vast amounts of data from many meters/devices to long distances in a small spectrum bandwidth without interruption is a significant drawback. Current wireless methods involve difficulties. Wired techniques such as DSL and optical fibre and wireless techniques such as cellular can solve these issues, but large costs are needed for cable/fibre construction or spectrum acquisition. As a feasible option, we're left with CRNs [2].

Smart Cities: Smart cities are a sustainable planning model combining ICT structures and IoT systems. The impetus behind the smart city is to include e-services in an eco-friendly fashion with an improved lifestyle. Continuous networking is important to allow this, because the backbone is a knowledge and communication infrastructure. Also essential is data collection and user engagement. CRNs [12] will endorse continuous communication.

Internet of Vehicles (IoV): Recently, developments have changed to less human dependency, contributing to the IoV model where vehicle regulation is accomplished by Integrated communications, controls, and embedded devices. IoV is supposed to be an automated travel decision-maker. Safe navigation may be feasible in the future by sharing details from vehicles to vehicles, from sensors mounted to vehicles, and users' intentions. The problem for IoVs is spectrum availability for mobile vehicles, and because of their long range and interference-free spectrum detection, CRNs may be a good option [4, 10].

8 Conclusion

The convergence of CRNs and IoT threatens to transform the direction of advanced wireless networks, but they are also at a transitional stage and a significant field of study. While researchers are focused on the usage of cognitive IoT radios, the reasons behind this approach are not detailed. In this field, we aim to provide greater support for evaluating the capacity of cognitive radios in IoT. Millions to trillions of IoT artefacts would require smooth access to the spectrum in the future. Orthodox networking systems cannot combat these circumstances. Thus, a transformation from ordinary IoT to cognitively competent things is required that allows IoT artefacts to manage spectrum congestion circumstances. IoT is supposed to be nothing more than a drain on current network networks without cognitive functionality. IoT artefacts may easily leverage a current radio continuum with spectrum capacity that is limited and underused through the acquisition of cognitive skills. Studies and studies should be solid and relevant enough to address IoT and CR convergence concerns and challenges.

References

1. Kakkavas, K. Tsitseklis, V. Karyotis and S. Papavassiliou, "A Software Defined Radio Cross-layer Resource Allocation Approach for Cognitive Radio Networks: From Theory to Practice," in IEEE Transactions on Cognitive Communications and Networking. doi: https://doi.org/10.1109/TCCN.2019.2963869
2. J. Ren; Y. Zhang; R. Deng; N. Zhang; D. Zhang; X. Shen, "Joint Channel Access and Sampling Rate Control in Energy Harvesting Cognitive Radio Sensor Networks," in IEEE Transactions on Emerging Topics in Computing , vol.PP, no.99, pp.1–1,21 April, 2016.
3. Shah, G., Akan, O.: Cognitive adaptive medium access control in cognitive radio sensor networks. IEEE Trans. Veh. Tech. **64**(2), 757–767 (2015)
4. N. Li, M. Xiao and L. K. Rasmussen, "Spectrum Sharing with Network Coding for Multiple Cognitive Users," in IEEE Internet of Things Journal. doi: https://doi.org/10.1109/JIOT.2017.2728626.
5. Aijaz, A., Aghvami, A.H.: Cognitive machine-to-machine communications for Internet-of-Things: A protocol stack perspective. IEEE Internet Things J. **2**(2), 103–112 (Apr. 2015)
6. T. M. Chiwewe and G. P. Hancke, "Fast Convergence Cooperative Dynamic Spectrum Access for Cognitive Radio Networks," in IEEE Transactions on Industrial Informatics. doi: https://doi.org/10.1109/TII.2017.2783973.
7. S. Aslam, W. Ejaz and M. Ibnkahla, "Energy and Spectral Efficient Cognitive Radio Sensor Networks for Internet of Things," in IEEE Internet of Things Journal. doi: https://doi.org/10.1109/JIOT.2018.2837354.
8. Ejaz, W., Shah, G.A., Kim, H.S., et al.: Energy and throughput efficient cooperative spectrum sensing in cognitive radio sensor networks. Transactions on Emerging Telecommunications Technologies **26**(7), 1019–1030 (Jul. 2015)
9. Maghsudi S, Stanczak S. Hybrid Centralized-Distributed Resource Allocation for Device-to-Device Communication Underlaying Cellular Networks[J]. IEEE Transactions on Vehicular Technology, 2016.
10. Yuhua Xu; Jinlong Wang; Qihui Wu; Anpalagan, A.; Yu-Dong Yao. "Opportunistic Spectrum Access in Unknown Dynamic Environment: A Game-Theoretic Stochastic Learning Solution," IEEE Trans. Wireless Communications, pp. 1380–1391, 2012.
11. A. Anandkumar, N. Michael, A. Tang, and A. Swami, "Distributed algorithms for learning and cognitive medium access with logarithmic regret," IEEE J. Sel. Areas Commun. on Advances in Cognitive Radio Networking for Communications, vol. 29, pp. 731–745, Mar. 2011.
12. Vakili, S., Liu, K., Zhao, Q.: Deterministic sequencing of exploration and exploitation for multi-armed bandit problems. IEEE J. Sel. Topics Signal Process. **59**(3), 1902–1916 (Oct. 2013)
13. Y. Gai and B. Krishnamachari, "Decentralized online learning algorithms for opportunistic spectrum access," in Proc. IEEE Global Commun. Conf. (GLOBECOM), Dec. 2011, pp. 1–6.
14. Xu, Y., Wu, Q., Wang, J., Shen, L., Anpalagan, A.: Robust Multiuser Sequential Channel Sensing and Access in Dynamic Cognitive Radio Networks: Potential Games and Stochastic Learning. IEEE Trans. Veh. Technol. **64**(8), 3594–3607 (Aug. 2015)
15. F. Zhou, N. C. Beaulieu, J. Cheng, Z. Chu and Y. Wang, "Robust Max–Min Fairness Resource Allocation in Sensing-Based Wideband Cognitive Radio With SWIPT: Imperfect Channel Sensing," in IEEE Systems Journal. doi: https://doi.org/10.1109/JSYST.2017.2698502.
16. Jang, H., Yun, S.Y., Shin, J., Yi, Y.: Game Theoretic Perspective of Optimal CSMA. IEEE Trans. Wireless Commun. **17**(1), 194–209 (Jan. 2018). https://doi.org/10.1109/TWC.2017.2764081
17. Lu, Y., Duel-Hallen, A.: A Sensing Contribution-Based Two-Layer Game for Channel Selection and Spectrum Access in Cognitive Radio Ad-hoc Networks. IEEE Trans. Wireless Commun. **17**(6), 3631–3640 (June 2018)
18. Gowda, V., Sridhara, S.B., Naveen, K.B., Ramesha, M., Pai, G.N.: Internet of things: Internet revolution, impact, technology road map and features. Adv. Math. Sci. J. **9**(7), 4405–4414 (2020). https://doi.org/10.37418/amsj.9.7.11

19. Ren, J., Zhang, Y., Zhang, N., Zhang, D., Shen, X.: Dynamic Channel Access to Improve Energy Efficiency in Cognitive Radio Sensor Networks. IEEE Trans. Wireless Commun. **15**(5), 3143–3156 (May 2016)
20. F. Mansourkiaie and M. H. Ahmed, "Optimal and Near-Optimal Cooperative Routing and Power Allocation for Collision Minimization in Wireless Sensor Networks," in IEEE Sensors Journal, vol. 16, no. 5, pp. 1398–1411, March1, 2016.

Comprehensive View of Low Light Image/Video Enhancement Centred on Deep Learning

C. Anitha and R. Mathusoothana S. Kumar

Abstract In computer vision, enhancing the image/video quality is strongly influenced by darkness and the noise is considered as an important challenge. Currently, different methods are used for performing image/video enhancement. However, deep learning-based methodologies have introduced huge changes in image/video quality improvement. This paper introduces more classic image/video enhancement network in recent years. Additionally, the elementary understanding of CNN components, current challenges, comparison of traditional enhancement with deep learning methods along with their benefits and drawbacks are analysed.

Keywords Computer vision · Enhancement · Deep learning · Video

1 Introduction

The significance of image enhancement is very broad. Image enhancement is adopted to highlight the general or partial characteristics of the images such as colour enhancement, brightness or contrast of the image, and helps to obtain a clear image. Purpose of the exercise is to expand the difference between the features of different objects in the image, suppress features that are not of interest, and improve the visual effect of the image.

Capturing videos under extreme low light conditions is fraught with many challenges. One such challenge is the low level of photons in the background. To overcome this, we can make use of the High-ISO System. The problem is using such a system is that along with substantial increase in the brightness of the image, the associated noise levels also get implied to the same extent. When images are captured by using consumer grade cameras or mobile phones, the aperture size of the cameras limit the quality of images captured. If on the other hand flash is used to improve the ambient lighting, then such an attempt totally alters the nature of the scene and

C. Anitha (✉)
Department of CSE, NICHE, Kumaracoil, Tamil Nadu, India

R. M. S. Kumar
Department of IT, NICHE, Kumaracoil, Tamil Nadu, India

J. S. Raj et al. (eds.), *Intelligent Sustainable Systems*, Lecture Notes in Networks and Systems 213, https://doi.org/10.1007/978-981-16-2422-3_41

affects the video. Another problem encountered is that we cannot have the luxury of long exposure extending to tens of seconds while shooting dynamic scenes. Due to the above-stated reasons, we are left with the only option of adopting computational techniques for low light video processing.

Figure 1 displays the block illustration of traditional and existing categories of image/video enhancement. Traditional image enhancement [1] has been considered forever. The main methods for processing are self-enhancement and Context-based fusion enhancement. The surviving approaches can be roughly divided into three classifications. The spatial method [2] is to directly process the pixel value, such as histogram equalization and gamma transformation; the frequency domain method is based on a certain transformation. Operations in the domain, such as wavelet transform, and hybrid domain methods are some methods that combine the spatial and frequency domains. Traditional methods are generally simpler and faster, but they do not take into account the contextual information in the image, so the effect is not very good. Over the last few years, convolutional neural networks (CNN) have achieved boundless developments in numerous low-level computer vision (CV) tasks including image super-resolution, deblurring, dehazing, denoising, image enhancement, video enhancement, etc. Associated with traditional approaches, some methods built on CNN using Generative Adversarial Nets (GAN) have greatly improved the quality of image enhancement. Most of the existing methods are supervised learning [15, 16]. For an original image and a target image, the mapping relationship between them is analysed to obtain an enhanced image. However, such datasets are relatively small and many of them are artificially adjusted. Therefore, self-supervised or weakly supervised methods are required to solve this problem.

The paper is structured as follows (shown in Fig. 2): Sect. 1 describes the purpose of low light contrast improvement, challenges in extremely low light conditions, and video enhancement categories. Section 2 explains the comparison between deep

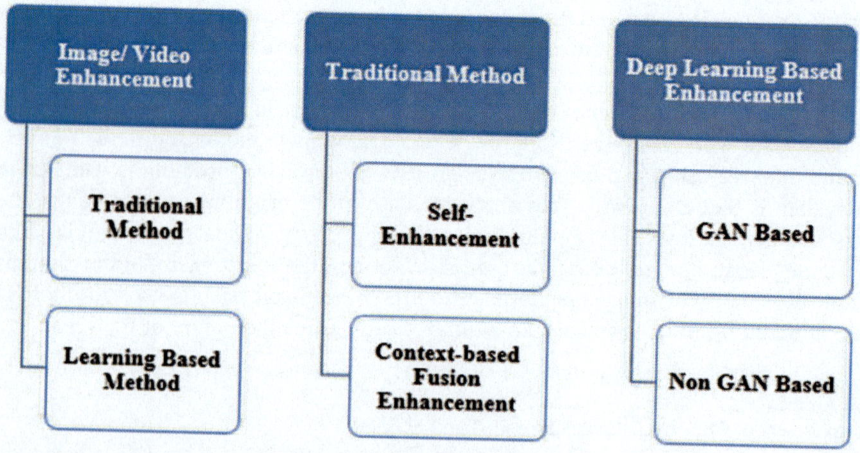

Fig. 1 Block Illustration of Video Enhancement Categories

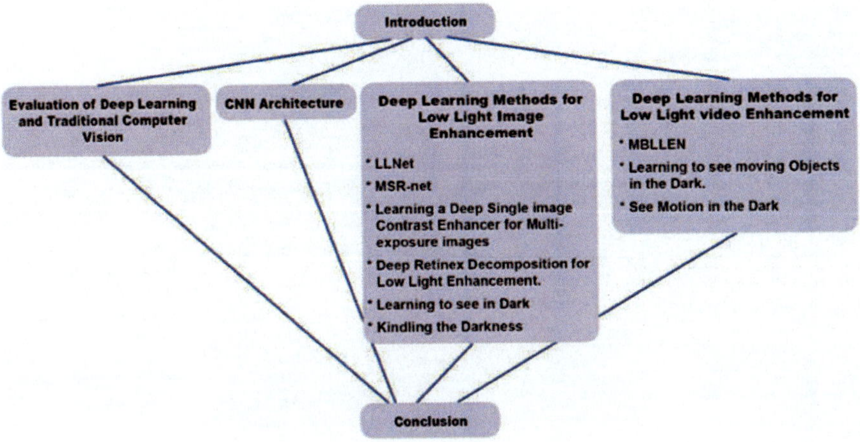

Fig. 2 Body of the survey paper with different sections

learning methods with traditional method for low light enhancement. Section 3 discusses the basic CNN architecture whereas, Sect. 4 discusses the existing six deep learning methods for low light image enhancement. Section 5 discusses the existing three deep learning methods for low light video enhancement. Finally, the last section has the conclusion.

2 Evaluation of Deep Learning (DL) and Traditional Computer Vision

Figure 3 shows the traditional and DL approach block diagram for enhancing low light image/video. As per the traditional method in Fig. 3(a), the input of the image/video frame of low light is pre-processed and then the pre-processed image is forwarded to any one of the enhancement techniques. The enhancement transformation can be spatial or frequency domain. The output from the enhancement techniques is the enhanced image/video frame. Figure 3 (b) displays the block diagram of the DL approach for enhancing the image/video frame affected by low light conditions. The input image/video frame affected by low light is forwarded to the trained deep neural network model to produce an enhanced result. The DL models are trained on the assigned input image dataset. The CNNs learn the fundamental features of images and robotically effuse the best expressive besides relevant features. It has been firm that Deep Convolutional Neural Network [DCNN] performs far superior than the traditional approach. Rapid advancement in DL and enhancements in device abilities such as memory capacity, computing power, power consumption, optics, and image sensor resolution have accelerated the blowout of CV-centred applications along with improved performance and cost-effectiveness. As DL models are trained more than

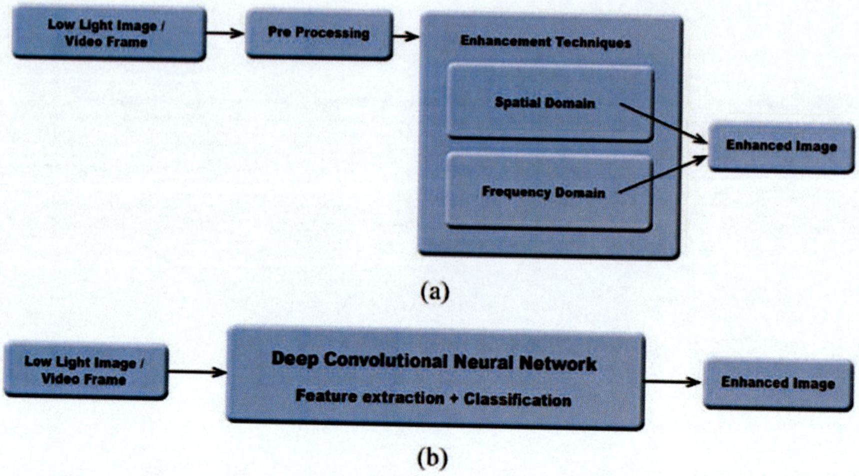

Fig. 3 **a** Traditional approach. **b** Deep learning approach

programmed, models following this method frequently need lesser fine-tuning plus skilful examination. The availability of a humongous quantity of video data in today's system supports this cause. While CV algorithms are likely to be extra domain-specific, DL [3], on the other hand, offers more flexibility since CNN models and frameworks can be retrained by means of a convention dataset. Table 1 summarizes the comparison between deep learning and traditional image processing [4, 5]. Table 2 shows some typical applications of DL and traditional image processing.

Table 1 Comparison between deep learning and traditional image processing

Selection criteria	Traditional Image Processing	Deep Learning
Computing Power	Low	High
Training dataset	Small	Large
Training Time	Less	More
Organization Flexibility	More	Less
Outflow	Low	High
Algorithm Transparency	More	Less
Annotation Time	Small	Long

Table 2 Some typical applications of DL and traditional image processing

Traditional Image Processing	Deep Learning
Image Transformation (Lens distortion correction, view changes)	Image Classification (OCR and Handwritten character recognition)
Image Signal Processing	Object detection/identification
Camera calibration	Semantic segmentation
Stereo image processing	Image synthesis
3D data processing	Image Super-resolution
Calculating Geometries	Scene understanding

3 CNN Architecture

The structure of CNN is layers. These layers are divided into three and are broadly called convolutional, pooling, and fully connected layers.

A layer consisting of a sequence of filters forms the convolution layer. The neural network trains the filters to do their functions. Each of these filters plays a role in convolving the image and the output is kept in the n x n dimensional portion of the next layer. The convolution process is aimed at capturing various features of the input image. Function of the first layer is to segregate characteristics like lines, corners, and edges from the low light images. The other layers capture a well-defined characteristic of the image. Figure 4 shows the basic components of CNN. It's the convolutional layers which process the image, resulting in a filtered output from the network.

The role of pooling/subsampling layers is to curtail the dimension of the earlier layer to a size appropriate for easy accommodation to the following layer of the network. The pooling is carried out in two ways. They are called max- and average-pooling. In either case, the inputs are arranged in a 2-dimensional matrix without any overlapping. The limitation in CNN is not on the number order of pooling or

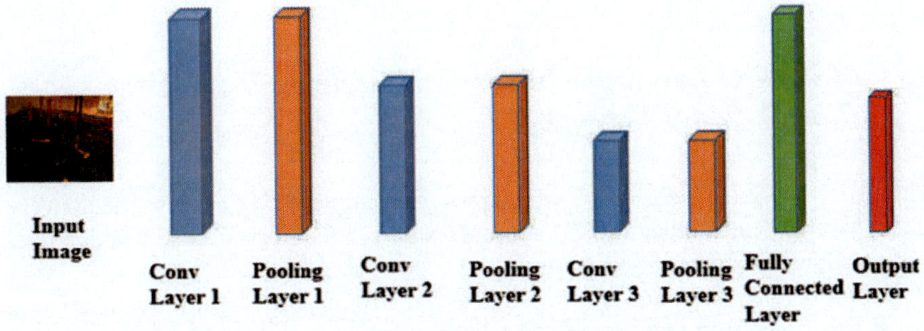

Fig. 4 Basic CNN components

convolutional layers, but the limitation is on the computational capability and the time which may end up in over fitting.

Fully Connected Layer: The final step of regular neural network is the fully connected layer. Fully connected layer learns nonlinear groupings of the high-level features generated by the convolutional layers. It is simply a feedforward neural network. It contains the probability of images in each of the categories.

4 Deep Learning Methods for Low Light Image Enhancement

4.1 LLNet: A Deep Autoencoder Approach to Natural Low Light Image Enhancement [6]

This article [6] is grounded on deep lighting enhancement tasks using deep learning methods. It demonstrates stack sparse denoising centred on synthetic data training. Self-encoder ability to increase and denoise low light, noisy images. Training of the model is centred on image patches, by means of sparsity standardized reconstruction loss as a loss function.

Various key aids are as follows:

- LLnet is a method of data generation, for activating a low light environment (adding Gaussian noise combines with gamma correction).
- A couple of structures considered are (a) LLNet, simultaneously learned contrast improvement and denoising; and (b) S-LLNet, employs a couple of modules to improve contrast and reducing noises in stages.
- Appraisals were carried out on images taken on low light conditions, validity of the model created with synthetic data.
- Visualize the layer weights besides provide understanding about the features learned by this process.

4.2 MSR-Net: Low Light Image Enhancement Using Deep Convolutional Network [7]

This article [7] introduces CNN, which provides an interesting aspect. The conventional Multi-Scale Retinex (MSR) approach stands a feedforward CNN which uses various Gaussian convolutional filters.

By employing MSR-net, it is possible to directly map low light images captured under low light from end to end and turn into bright images. MSR-net is divided into modules called (i) multi-scale logarithmic transformation, (ii) convolution differential, and (iii) colour recovery.

The purpose of training data is to adjust the high-quality images and compare with corresponding synthetic low light images. The loss function defines the sum of squares of error matrix at regular intervals.

4.3 *Learning a Deep Single Image Contrast Enhancer from Multi-Exposure Images [8]*

This article [8] is to highlight single image contrast enhancement (SICE) for low contrast images in underexposure and overexposure situations. Its key aids are

- A multi-exposure image dataset is constructed, comprising low contrast images of unlike exposures and equivalent high contrast reference images.
- The proposed enhancement model is of two steps. In the initial step, original image is separated as low-frequency modules and high-frequency modules by weighted least squares filtering method, afterwards the two modules are separately boosted; the next phase is fused to the enhanced low-frequency and high-frequency modules, and later enhanced again to output the result.

Two-stage structure is designed, because the enhancement result of the single-stage CNN is not reasonable, and there is colour shift phenomenon, which perhaps is because single-stage CNN is hard to balance the smooth and texture components of the image.

It is worth mentioning that the decomposition step of the initial phase of the model uses the traditional method, while the Retinex-Net describes the CNN structure.

4.4 *Deep Retinex Decomposition for Low Light Enhancement [9]*

This article [9] applies Retinex's theory. This model is fully implemented using the CNN structure and the model is divided into two steps, decomposition and adjustment steps. For the training of Decom-Net, the stability of reflectance and the smoothness of illumination are presented, which is very simple to reproduce and the observational effect is good.

The main contributions are as follows:

- The low light/normal illumination LOL dataset is constructed, which is a paired dataset gathered in the actual scene. Division of dataset is done as following: data of the image of the actual scene is changed by modifying the sensitivity of the camera plus the exposure time. Data of synthesized image is modified by the Adobe light room interface, and the adjusted Y-channel histogram of the image should resemble as closely as possible to the real low light scene.

- Retinex-Net is separated into Decom-Net which decouples the image and decomposes as a light map and a reflection map; Enhance-Net enhances the previously obtained light map, and the enhanced light map multiplying the original reflection image yields an enhanced result. In addition, seeing the noise difficulty, a joint denoising and enhancement strategy is accepted, and the denoising technique adopts BM3D.
- Proposed structure aware total variation constraint makes use of information from gradient of reflection graph as the weight to weight the total variation minimization (TV) loss, such that the smoothing of image constraint is maintained without abolishing the texture details and boundary information.

4.5 Learning to See in the Dark [10]

This article [10] focusing on images taken under extremely low light situations and short time exposure provides excellent results. The model customs a CNN to perform raw image-to-RGB image processing.

Fully convolutional network (FCN) is the basic structure of CNN that is directly trained in the form of an end-to-end fashion, with the loss function. Furthermore, the article recommends a See-in-the-Dark (SID) dataset consisting of a short exposure image and a matching long exposure reference image.

4.6 Kindling the Darkness: A Practical Low Light Image Enhancer [11]

This article [11] presents three difficulties in the existence of enhancement on low light conditions:

- Way of accurate discerning illumination using a mono image and adjusting.
- After attaining image brightness to the expected level, remove noise and colour distortion.
- Training of a model without a ground truth.

The network is separated as Decomposition-Net, Restoration-Net, and Adjustment-Net, which accomplish image decomposition and reflection recovery, respectively, lighting map tuning. The following innovations may be seen:

(i) For Decomposition-Net, in addition to the consistent loss of reconstruction loss and reflection maps with Retinex-Net, two new loss functions are added in order to get regional smoothness and mutual consistency of the illumination map.

(ii) For Restoration-Net, it is considered that the reflection pattern often has a degrading effect in the case of low light, so the reflection map in noble illumination is applied as a reference. The dispersal of degradation properties in the reflection map is complicated and extremely dependent on the illumination, thus introducing illumination map information.

(iii) For Adjustment-Net, a tool for constantly adjusting the intensity of light is implemented (the enhancement ratio is combined as a feature map and a light map as inputs). By comparing with gamma correction, it is proved that their adjustment method is matched with the real condition.

5 Deep Learning Methods for Low Light Video Enhancement

5.1 MBLLEN: Low Light Image/Video Enhancement Using CNNs [12]

The core idea of this article [12] is 3D convolution in place of 2D convolution to develop the network and increase the video enhancement enforcement.

In addition, there is a negative situation in the low light enhancement of video, flickering, that is, there may be unexpected brightness jumps between frames. AB (var) metric is used to measure this problem.

Network structure is separated as Feature Extraction Module (FEM), Enhancement Module (EM), and Fusion Module (FM). FEM is considered as a single-flow neural network using 10 convolution layers. The result of individual layer is inputted into each EM sub-module to extract hierarchical features. Eventually, these hierarchical features are stitched together and the ultimate outcome is achieved by 1X1 convolution fusion.

Loss function: The conventional MSE or MAE loss function is not used in this paper, but three new loss functions are proposed: structural loss, content loss, and region loss. Structural loss is a combination of SSIM and MS-SSIM metrics; content loss is based on VGG-19 network; regional loss makes the network more focused on low light areas of the image.

5.2 Learning to See Moving Objects in the Dark [13]

The main contributions of this article [13] are

- Establish a new coaxial optical system to capture time synchronized plus spatially aligned low light video for training as well as bright videos as ground truth.
- Provide the low light street view with dynamic vehicles and pedestrians/Normal light video data pair.
- The improved U-net structure combined with 3Dconv and 2D pooling/deconv is used as the framework for low light video enhancement, showing excellent presentation in colour fidelity and time consistency.

5.3 *Seeing Motion in the Dark [14]*

In the paper [14], SID does low light image enhancement, and this article does video enhancement. The main motivation is extremely low light video enhancement is a thorny issue, because it is hard to collect video dataset corresponding to ground truth. Its main aids are as follows:

- Recommended a static video under extreme low light and the corresponding ground truth data to the Dark Raw Video (DRV) dataset, and also shot a real dynamic low light video (without ground truth) for testing;
- A deep Siamese network and a specific loss function are designed to ensure time and space stability and consistency. The network is trained on static datasets and can be extended to dynamic video processing, and obtain state-of-the-art results.

6 Conclusion

This paper presents an extensive review of different deep learning-based approaches for enhancing image and video affected by environmental conditions especially in dark or night time. This review not only affords a better understanding of image/video enhancement models but also helps upcoming research activities and application developments in this field. Deep learning-based method overpower the performance of the traditional methods because it robotically states and extracts different features of the input. The obtainability of datasets at the training makes the testing more accurate. In spite of great accomplishment, there are quiet various unresolved problems.

References

1. Anitha, C., & Kumar, R. M. S.(2019). Extremely Low Light Video Enhancement along with No-Reference Video Quality Measurements, International Journal of Advanced Trends in Computer Science and Engineering, Trends in Computer Science and Engineering, Vol. 8 , No.5, September - October 2019
2. Rao, Y., Chen, L.: A survey of video enhancement techniques. Journal of Information Hiding and Multimedia Signal Processing. **3**(1), 71–99 (2012 Jan)
3. Hemanth DJ, Estrela VV, editors. Deep learning for image processing applications. IOS Press; 2017.
4. Anitha, C., & Kumar, R. M. S. (2019). Naturalness Preserved Extremely Low Light Video Frame Enhancement by Correcting Illumination and Reducing Fixed Pattern Noise, Jour of Adv Research in Dynamical & Control Systems, Vol. 11, 07-Special Issue, 2019
5. Anitha, C., & Kumar, R. M. S. State of the Art Analysis of Low Light Video Noise and Various Environmental Conditions for Enhancement, International Journal of Engineering and Advanced Technology (IJEAT) ISSN: 2249–8958, Volume-8, Issue-6, August 2019
6. Lore, K.G., Akintayo, A., Sarkar, S.: LLNet: A deep autoencoder approach to natural low-light image enhancement. Pattern Recogn. **1**(61), 650–662 (2017 Jan)

7. Shen L, Yue Z, Feng F, Chen Q, Liu S, Ma J. Msr-net: Low-light image enhancement using deep convolutional network. arXiv preprint arXiv:1711.02488. 2017 Nov 7.
8. Cai, J., Gu, S., Zhang, L.: Learning a deep single image contrast enhancer from multi-exposure images. IEEE Trans. Image Process. **27**(4), 2049–2062 (2018 Jan 15)
9. Wei C, Wang W, Yang W, Liu J. Deep retinex decomposition for low-light enhancement. arXiv preprint arXiv:1808.04560. 2018 Aug 14.
10. Chen C, Chen Q, Xu J, Koltun V. Learning to see in the dark. InProceedings of the IEEE Conference on Computer Vision and Pattern Recognition 2018 (pp. 3291–3300).
11. Zhang Y, Zhang J, Guo X. Kindling the darkness: A practical low-light image enhancer. InProceedings of the 27th ACM International Conference on Multimedia 2019 Oct 15 (pp. 1632–1640).
12. Lv F, Lu F, Wu J, Lim C. MBLLEN: Low-Light Image/Video Enhancement Using CNNs. InBMVC 2018 Sep 3 (p. 220).
13. Jiang H, Zheng Y. Learning to see moving objects in the dark. InProceedings of the IEEE International Conference on Computer Vision 2019 (pp. 7324–7333).
14. Chen C, Chen Q, Do MN, Koltun V. Seeing motion in the dark. InProceedings of the IEEE International Conference on Computer Vision 2019 (pp. 3185–3194).
15. Pandian, A. Pasumpon. "Identification and classification of cancer cells using capsule network with pathological images." Journal of Artificial Intelligence 1, no. 01 (2019): 37–44.
16. Manoharan, S.: Performance analysis of clustering based image segmentation techniques. Journal of Innovative Image Processing (JIIP) **2**(01), 14–24 (2020)

Internet of Things in Precision Agriculture: A Survey on Sensing Mechanisms, Potential Applications, and Challenges

R. Madhumathi, T. Arumuganathan, and R. Shruthi

Abstract Precision agriculture is one of the modern farming practices that gathers, processes, and analyzes data with the goal of increasing the agricultural production and reducing the usage of resources. Agriculture contributes an important share in the Gross Domestic Product (GDP) of our country. Precision agriculture involves the usage of Internet of Things (IoT) along with Wireless Sensor Networks (WSN) which provides an intelligent farm management system. IoT in agriculture provides decision support systems that help farmers to know the real-time information of their field. The applications of IoT becomes a game changer in agriculture as it monitors and transfers information without human intervention. Agricultural IoT systems are implemented with the help of sensors and actuators that senses and responds to different inputs and provides instant feedbacks. The key sensors involved in precision agriculture are used in both small and large-scale farmlands for effective production. The main objective of this study is to review the potential application of sensors used in agriculture, to describe the layers of IoT in agriculture, discuss the existing sensing approaches for monitoring the agricultural parameters effectively, and deliver the general challenges encountered while implementing IoT systems.

Keywords Internet of Things · Sensors · Wireless sensor networks · Agricultural applications · IoT challenges · Smart agriculture · Automation in agriculture

R. Madhumathi (✉) · R. Shruthi
Department of Computer Science and Engineering, Sri Ramakrishna Engineering College, Coimbatore, India
e-mail: madhumathi.r@srec.ac.in

R. Shruthi
e-mail: shruthi.1801217@srec.ac.in

T. Arumuganathan
ICAR-Sugarcane Breeding Institute, Coimbatore, India

J. S. Raj et al. (eds.), *Intelligent Sustainable Systems*, Lecture Notes in Networks and Systems 213, https://doi.org/10.1007/978-981-16-2422-3_42

1 Introduction

Agriculture is considered as the primary source of food production and it is the basic foundation in the developing countries which creates opportunities to raise the country's economy. Agriculture along with its associated sectors is unquestionably the largest livelihood provider in India. Around 70 percent of its rural households in India still depend primarily on agriculture for their livelihood, with 82 percent of farmers being small and marginal. The total food grain production was estimated at 275 million tons (MT) in 2017–18 [1]. In the current era, agricultural sector is experiencing a transition from the current industrial practices to modern practices like precision farming, by the convergence of numerous information technologies. Precision agriculture encompasses a set of leading-edge advancements which combines sensors, information systems, and information management to enhance the crop cultivation. It provides innovative solutions to numerous unsolved problems as it optimizes field-level management and matches farming practices according to the crop needs [2]. Precision agriculture incorporates sensors, microcontrollers, and their networks deploying Information and Communication Technologies (ICT) to obtain real-time information by intensive monitoring of crops. Several terminologies like Smart Farming, Smart Agriculture, Variable Rate Technology (VRT), Site-Specific Management of crops are put forward which all convey the same underlying concept as sensor-based precision agriculture [3].

IoT is employed in agriculture to increase efficiency, maintain security, and to meet the ongoing demands of food production [4]. The capability of IoT systems to configure, operate, and restore information makes it a good choice for precision agriculture. The global market size of IoT in agriculture is growing rapidly due to its cost efficiency and improved productivity and these technological advancements have allowed farmers to plant more acres of land in a short span of time. Along with IoT, few other emerging technologies like cloud computing, big data analytics, and Machine Learning (ML) also play a major role in the development of agricultural systems. Many Artificial Intelligence (AI) based models involving various algorithms are developed to collect farm-related data which facilitates crop price and effective yield prediction. Additionally, various cloud computing and big data techniques help in collecting, storing, and processing large volumes of essential agricultural data. These technological developments provide more opportunities for numerous innovations and are a boon to agriculturists.

In this paper, the recent ideas for IoT in agriculture and description of various sensing methods are reviewed and its potential advantages and applications are discussed along with the following points.

- Provides an outline on the types of sensors deployed in precision agriculture for effective monitoring of the field.
- Gives a brief idea about the sensors and actuators available for determining the characteristics of soil, plant, leaf, and weather factors.
- Explains the applications of IoT in various agricultural monitoring systems.

- Describes the challenges encountered and explains the architecture of IoT in implementing precision farming.

2 Architectural Overview of IoT

IoT automates our work by facilitating connection between the devices and allows in exchanging data between them. The architectural framework of IoT for precision agriculture is shown in Fig. 1 where the gateways provide connectivity as the data moves from things to the cloud and user applications. The data is effectively transmitted from the device layer and stored in the data warehouse. The stored data would be useful to perform data analytics, generate machine learning models, or develop user applications. The application layer provides the user interface for viewing the processed data.

Few applications of IoT include smart homes, industrial automation, precision agriculture, wearable devices, and smart cities. Several IoT frameworks including Microsoft Azure IoT, Amazon Web Services (AWS) IoT, Google Cloud Platform-IoT, IBM Watson- IoT, Cisco IoT cloud connect, ARM mbed IoT, Calvin, and SmartThings [5] provide cloud support in order to satisfy the needs to implement IoT technology in precision agriculture. The success of these frameworks depends on its characteristics and the way of emphasizing the security mechanisms in it. IoT allows agriculture to be data-driven and improves agricultural processes by ensuring reliable, cost and power efficient, scalable, and cleaner agricultural processes.

The architecture of IoT has three layers, namely, device layer (perception layer), network layer, and application layer as shown in Fig. 2. Data production occurs in the device layer and the network layer ensures data transfer followed by data manipulation in the application layer [6]. The layers consisting of different devices along with the required protocols are described and its application in precision agriculture is reviewed in this survey.

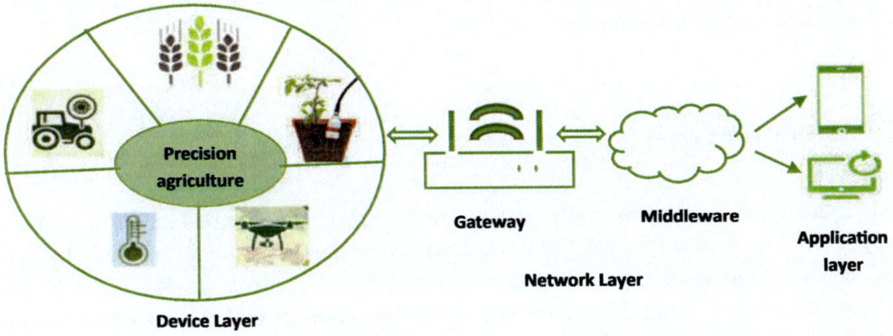

Fig. 1 Agricultural IoT framework architecture

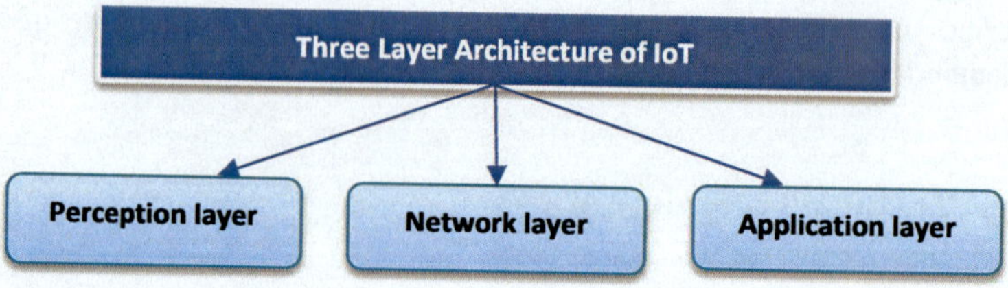

Fig. 2 Three-layer architecture of IoT

2.1 Perception Layer

The perception layer also called physical or device layer recognizes the physical objects and consists of sensors, actuators, and other smart devices where the production of data occurs. Data security is an important feature that has to be maintained in this device layer as many common security threats namely capturing of node, malicious attack, replay attack, and timing attack takes place [7].

Wireless Sensor Network

One of the precision agricultural techniques that monitors and collects data efficiently is WSN. It is a wireless communication-based environment that consists of sensor nodes used for real-time detection of several physical quantities with minimal consumption of time and with a greater accuracy. The main components of the WSN nodes include sensing unit, processing unit, and communication connected to a power unit for active transmission. WSN monitor parameters even in hazardous and remote areas and it provides flexibility for installation, increases robustness, portable, and reduces fitting costs [8]. Figure 3 describes the architectural diagram of the WSN with several sensor nodes connected to the Internet.

The challenges in WSN may emerge due to the deployment strategies, measurement interval, routing protocols involved, and also due to the region of deployment selected. Agriculture focuses mainly on WSN technology for effective transmission of field data using sensor nodes.

2.2 Network Layer

In IoT, the devices are connected to various networking technologies which are wired or wireless enabling a bi-directional communication between the real-world objects and service applications. Gateway provides additional security for IoT network as it protects data from malicious attacks and enables connection between devices and servers. IoT networking technologies facilitate communication between the sensor nodes and gateways. A good overview of various wireless technologies is listed in Table 1. Zigbee, Bluetooth Low Energy (BLE), Lora-WAN, Wi-Fi, Radio Frequency

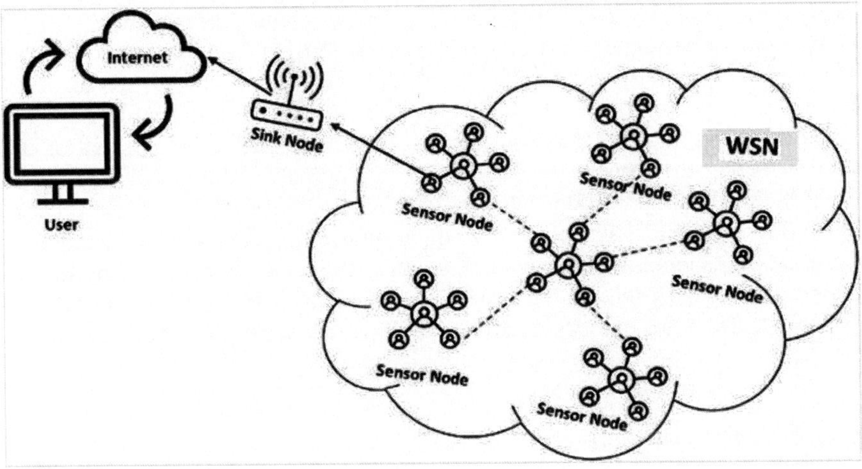

Fig. 3 WSN nodes integration

Table 1 Summary table of IoT communication technologies

Wireless technology	Standard	Frequency	Transmission range	Energy consumption
Wi-Fi	IEEE 802.11a, 11b, 11 g, 11n	2.4 GHz, 3,6 GHz, 5–60 GHz	20–100 m	High
Classic Bluetooth	IEEE 802.15.1	2.4 GHz	10–100 m	Medium
BLE	IEEE 802.15.1	2.4 GHz	10 m	Ultra-low
LoRa-WAN	LoRa-WAN	868–915 MHz	2–15 km	Very low
Zigbee	IEEE 802.15.4	868/915 MHz and 2.4 GHz	100 m	Low
RFID	Many standards	13.56 MHz	1 m	Low
NFC	ISO/IEC 13,157	13.56 MHz	0.1–0.2 m	Low
6 LowPan	IEEE 802.15.4	908.42 MHz – 2.4 GHz	100 m	Low
Sigfox	IEEE 802.15.4 g	868/915 MHz	10 km	Low
NB-IoT	3GPP Rel.13	180 kHz	10 km	Low
Mobile networks	2G-GSM, CDMA2000, 4G-LTE	865 MHz – 2.4 GHz	Entire cellular area	Medium
Z-Wave	Z-Wave	908.42 MHz	30 m	Low

Identification (RFID), Near Field Communication (NFC), 6LowPan, Sigfox, Z-Wave, Long Term Evolution (LTE), Narrowband (NB) IoT, and Wireless Highway Addressable Remote Transducer Protocol (HART) are reviewed in [9] and according to the devices selected, the communication technologies are to be employed. These technologies are majorly deployed in agriculture introducing the concept of ICT after analyzing the advantages, challenges, and efficiency with WSN to improve yield. Mobile communication technologies such as LTE, 3G/4G, or General Packet Radio Service (GPRS) also play an action in transferring data to the cloud. Finally, the latest 5G technology has higher data rates which adopts many possibilities to overcome the challenges in IoT. These communication technologies support IoT devices for data transportation enabling real-time monitoring in agriculture.

2.3 *Middleware and Application Layer*

Middleware layers are software layers that hide the complexities of the lower device and network layers perform discovery of information, provide access to devices responsible for applications by interaction with them. Heterogeneity should be shielded and secure, scalable, reliable, service-oriented middleware platforms are to be developed Popular middleware platforms as mentioned in [10] includes ThingSpeak, Firebase, SensorCloud, AWS IoT, Amazon IoT, IBM IoT, Microsoft Azure IoT, Oracle IoT platform, Kaa, Carriots, Temboo, and many other platforms provide services which focuses on context aware functionality for IoT implementation. Figure 4 shows the various service approaches in middleware. As mentioned in [11], service-oriented middleware approach can support distributed sensor applications by adopting different performance needs. The IoT middleware technology provides features to build advanced application for product and device ecosystem management and end-to-end processing.

The application layer is the upper layer of IoT architecture as it acts as the interface between network and the end devices that deliver application specific services as it deploys IoT to the users. The services include data storage, accessing data through an Application Program Interface (API) in user-friendly application and data analytics [12]. A set of protocols are determined in the application layer for message passing between internet and the application. Transmission Control Protocol (TCP) and User

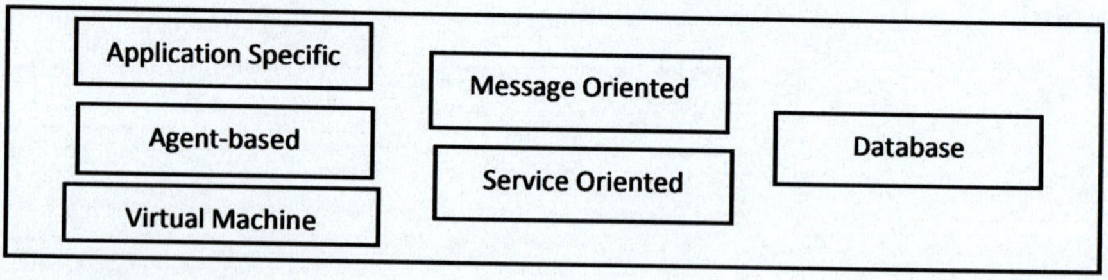

Fig. 4 Middle layer approaches

Table 2 Application layer protocols transport standards and architecture

Protocol	Transport protocol	Architecture
CoAP	UDP	Request/Response
MQTT	TCP	Publish/Subscribe
XMPP	TCP	Request/Response Publish/Subscribe
RESTful HTTP	HTTP	Request/Response
DDS	TCP/UDP	Publish/Subscribe
Web Socket	TCP	Client/Server Publish/Subscribe
AMQP	TCP	Publish/Subscribe
Light Weight M2M	TCP/UDP/LoRa-WAN	Request/Response
Simple Sensor Interface (SSI)	TCP/IP	Request/Response

Datagram Protocol (UDP) are used by the transport standards for forwarding information. These are listed in Table 2 which includes Constrained Application Protocol (CoAP), Message Queue Telemetry Transport (MQTT), Extensible Messaging and Presence Protocol (XMPP), Representational State Transfer (RESTful Services), Advanced Message Queuing Protocol (AMQP), Data Distribution Service (DDS), Web Socket and Secure MQTT [13]. Network capability, reliability, and security are the main challenges encountered while using these protocols.

3 Sensors and Actuators for Precision Agriculture

Sensors are a counterpart of IoT that monitors the physical characteristics effectively and detects external information which is useful for a specific purpose. In agriculture, a wide range of sensors are used to measure the field attributes, namely, soil, crop, and weather factors. The sensing mechanisms include electric and electromagnetic sensors, optical sensors, mechanical sensors, acoustic and airflow sensors, and dielectric sensors which work along with few types of actuators for continuous monitoring of agricultural systems. These mechanisms provide huge number of commercially available sensors that directly measure the field parameters. An overview of the application of different sensors and actuators is described in this review. The temperature is measured using thermistor or thermocouple which is an electrical sensing mechanism. Another electrical sensing mechanism involves fence sensors which are attached to the farm fence and detects faults through an electrical shock system or by an alarm notification system ensuring livestock management [14]. Various sensors are designed on different electromagnetic phenomena, namely, Electrical Resistivity (ER), Electromagnetic Induction (EMI), reflectometry approaches like Time Domain Reflectometry (TDR), Amplitude Domain Reflectometry (ADR), and Frequency Domain Reflectometry (FDR) to measure the soil Electrical Conductivity (EC). A research conducted showed that EMI technique has proved to give high accuracy in determining the EC and the soil types can also be differentiated easily along with the yield data of the crop [15].

The development of optical system helps in measuring nutrients, pH, moisture in the soil, and chlorophyll content of the crop. The efficiency of any nutrient will result in retarded crop growth and decreases the crop yield [16]. Hence, early detection of soil nutrients will have greater influence in the crop yield. The Leaf Area Index (LAI), crop nitrogen status, leaf chlorophyll, and yield are measured by the leaf sensor which senses the green light wavelengths. A system was developed based on the optical sensing technique to detect the parasite infestation based on the plant's reflection signal and so the crops are protected from such parasite infestations [17]. An important application of wireless acoustic sensor is pest detection. The sensor located in the field picks up the sound waves of insects and prevents the pests from damaging the crops [18]. The amount of moisture, nitrates, phosphates, potassium, and hydrogen for pH are detected with electrochemical sensors. Dielectric sensors measure moisture levels by determining the dielectric constant of the soil. Soil capacitive moisture sensors measure moisture using the dielectric permittivity of the soil. The sensor must be calibrated for different types of soil, as the measurement is soil dependent. A literature review shows that precision agricultural techniques can contribute to agricultural sustainability by optimizing the usage of pesticides and fertilizers thereby ensuring quality produce by retaining the fertility of the soil [19]. Although development of various sensors is in progress, majority of the agriculture technology use electrical, optical, and electrochemical systems for sensing and monitoring the field.

Table 3 describes the summary of sensors used in agriculture. Actuators are categorized based on the energy sources as pneumatic, hydraulic, thermal, and electrical. Actuator technology involving spraying, planting, irrigation, harvesting, and seed drilling also improves the capabilities in agriculture. Tractors in the field use electrical actuators for steering wheel adjustments, mirror adjustments, opening and closing of windows, weed detection, and for many more potential applications [20]. The actuators withstand rough weather conditions and exposure to all types of fertilizers. These sensors and actuators in the device layer gather data and transmit it to the second layer of IoT for further processing. Communication technologies in agriculture are deployed depending on the parameters measured and a review mentions that Zigbee, Wi-Fi, LoRa-WAN, Bluetooth, Sigfox, and RFID are relatively common for agricultural measurements [21]. The application layer services in agriculture include monitoring, event management, tracking, and logistics control of agricultural products and forecasting. Few applications of these communication technologies integrated with the sensing systems for effective farm monitoring are discussed in the next section.

Table 3 Summary of agricultural sensors

Sensing technology	Working principle	Examples of applications in agriculture
Electrical and electromagnetic	Electromagnetic induction	Determine soil EC, salinity and temperature, capability of soil particles to conduct charge, organic matter, and moisture content
Optical	Light reflection, absorption, transmission, and scattering	Soil moisture and nutrients, organic matter, parasite infestation, leaf turgor pressure, pH, Cation Exchange Capacity (CEC)
Mechanical	Use of probes in different equipments	Measures soil mechanical resistance, soil compaction, leaf turgor pressure, tractor control system, and irrigation investigations
Acoustic and airflow	Emission of sound signal with certain frequency and measures the air/water flow rate	Determine soil texture, depth variability, pest detection, and soil compaction
Electrochemical	Usage of electrodes and electrolytes	Measures soil pH, nutrients, and moisture content
Dielectric	Capacitance to measure dielectric constant	Temperature and soil moisture

4 Potential Applications of IoT in Agriculture

Numerous applications of IoT in precision agriculture are implemented using the three layers as described above. Using WSN, automation of agricultural process occurs by monitoring the soil, crop, and irrigation systems effectively and measuring the climatic changes.

Other applications of IoT in agriculture include farmland monitoring, greenhouse gases monitoring, agricultural production process management, pest and disease control, groundwater quality monitoring, cattle monitoring, asset tracking, and remote-control diagnosis [22]. The smart agricultural field system framework sequence consisting of sensing, data acquisition, communication, storage, and visualization is shown in Fig. 5. The sensors are connected to the cloud through various communication technologies that ensure effective monitoring of the field parameters. Automatic measurement of soil and plant characteristics is performed based on the strategically connected sensors in the farm. Sensors monitor soil parameters such as soil moisture, soil temperature and humidity [23], soil pH, salinity, EC, compaction, and crop parameters such as leaf area index, leaf color, and its texture.

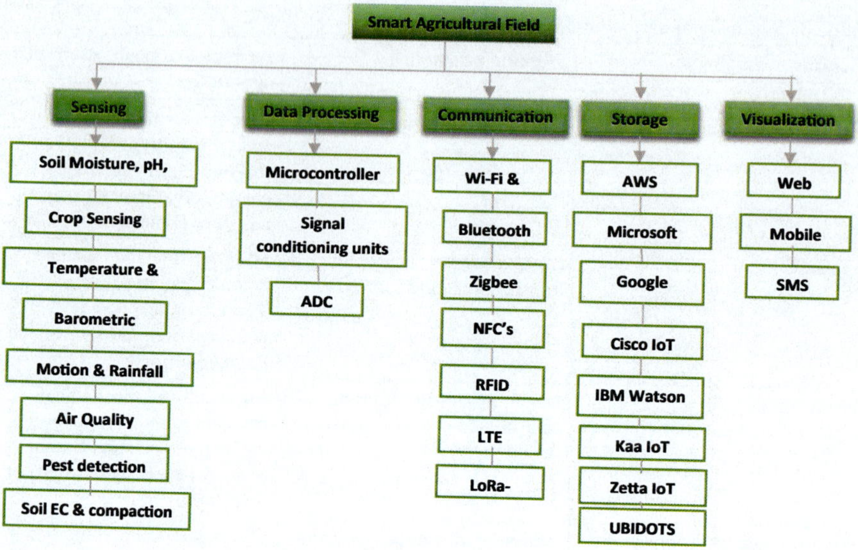

Fig. 5 Smart agricultural field system framework

Several sensor-based projects are being developed to control irrigation system. A distributed sensor-based irrigation system was developed by using Bluetooth to support site-specific irrigation management [24]. This Irrigation Management System (IMS) with Zigbee was implemented using WSN which monitors the water in the soil. BLE was used for air monitoring and actuating the irrigation system through WSN. These irrigation control systems store data for statistical analysis and improve the decision-making ability by receiving real-time system feedback. NFC's support distributed agriculture management service system and smartphone operations [25] by providing in-field farm information's which is widely used for agricultural-products tracking and agricultural product packages.

Horticulture deals with gardening, cultivation, and distribution of fruits, vegetables, and usage of greenhouse. WSN is deployed on greenhouse to monitor environmental parameters, namely, light, temperature, and humidity [26]. Zigbee also monitors the environmental parameters of field, prevents pest attacks, and supervises mobile plants in greenhouses. Greenhouse monitoring has also been deployed through LoRa-WAN as it has long coverage and low operation complexity [27]. This is a low-cost greenhouse system which allows users to access the agro data with a User Interface (UI). These Greenhouse environmental monitoring systems long coverage and low operation complexity.

RFID's were mainly used for tracking and identification of animals, for traceability control in the food chain, horticulture, precision viticulture, cold chain process, and farm machinery as well as for traceability of fruits or vegetables [28]. Animal monitoring systems are essential for resource production and are developed using wireless technologies. An IoT-based system was designed to supervise the health of the dairy cows using BLE which precisely differentiates the cow activities [29]. Another proposed method tracks the animal behavior based on IoT technologies which determines the posture of the animal in the vineyard areas [30]. These systems are designed to monitor the livestock continuously. Irrigation control, soil temperature and moisture detection, soil health monitoring, weather, and rainfall were monitored by a platform based on smart farming modular architecture [31]. NFC's support distributed agriculture management service system and smartphone operations by providing in-field farm information's which is widely used for agricultural-products tracking and agricultural product packages.

Few middleware platforms like HYDRA, FIWARE, SmartFarmNet, and Agri M2M are implemented widely in precision agriculture and the data transmission occurs through an API. A precision agriculture software architecture using FIWARE cloud was developed which reduced the usage of water and monitored environment effectively [32]. For efficient management of the smart field, the appropriate device management technology should be selected based on the system specifications, networking capabilities, deployment type, pricing models, security requirements, and communications standards. The application layer includes module interfaces for implementing irrigation and fertilizer control, crop monitoring, air quality monitoring, disease detection, weed mapping, weather predictions, and visualization of data [33]. The notification system enables farmers to inspect the data ranging from planting to harvesting and make decisions accordingly in order to manage the fields effectively and improve cultivation. These applications can be deployed in smartphones, tablets, or web applications improving precision farming. Smartphones for agriculture provide the ability to detect disease in plants, calculate fertilizer, perform soil, and water study, estimate crop water needs, and perform crop production analysis resulting in reduction of time and expenses. Farm Management Information Systems (FMIS) are increasing in many farms [34]. It supports automatic data acquisition, data processing, monitoring, and managing farm operations effectively through IoT. The involvement of autonomous vehicles and robotics in near future will completely change the agricultural practices enabling to fully adopt IoT.

5 Challenges in IoT

When implementing IoT in farming, diverse challenges are encountered which affects the overall operability of the system. With the aim of increasing the efficiency of the system, these challenges must be taken into consideration when designing an automation system.

- IoT still has competing standards, insufficient security, and complex communications which leads to chaos. To prevent chaos, efforts should be taken to reduce complexity, standardize applications, and guarantee the privacy and security ubiquitously.

- A major networking challenge is the unavailability of Internet connectivity at the same speed in all places. Networking functionalities comprising communication protocols and connectivity should be well defined to establish quick and quality communication between the devices.
- Interoperability and compatibility are major hurdles which are handled through different interoperability handling approaches [35] like Software Defined Networking (SDN), Service-Oriented Architecture (SOA), Open API's, Open Standard frameworks, and network function virtualization.
- The data heterogeneity affects the performance of a protocol employed in communicating the sensor data as the data differs in volume, velocity, and variety. Using standardized data formats and enabling the functionalities of Service-Controlled Networking (SCN) overcomes the data heterogeneity problems. Power consumption and device environments should be monitored while procuring sensor networks.
- Sensors developed should match the minimum requirements of the system so that the device functionality and durability is preserved. Fault-tolerant systems should automatically ensure operation without interruption when any component in WSN suffers from the harsh environmental effects.
- Common challenges include scalability, reliability, localization, complexity, high investment, lack of products, and robustness. Software as a Service (SaaS) and public infrastructure usage control the expenses and complexity that can be managed through protocol proliferation and integrating new technologies into existing system, identify communication requirements, and leverage cloud process and data integration.
- Lack of security in the IoT systems would involve in unauthorized activities leading to threats, software vulnerabilities, credential theft, and manipulation of data. By ensuring device authentication, securing the network, developing dynamic analysis and intrusion prevention systems, firewalls into products, the security of the IoT devices are ensured [36].

6 Conclusion and Future Scope

Precision agriculture aims to produce crops with good quality and has high potential to increase the income of farmers. Many technical solutions are required in agriculture to increase the harvest as the production is greatly affected by applying over dosage of fertilizers and other inorganic chemicals which harms the crops. Data-driven agriculture enables the agricultural process to carry on with less human intervention. IoT and its deployment offer numerous benefits to the agricultural industry due to the involvement of sensors and other integration systems. It is essential to know that each sensing system has its own advantages and limitations. Reliability, interoperability, and ease of use becomes a major concern when developing these commercial systems. Several researches are going on to reduce these concerns and improve the efficiency of the system by integrating different mechanisms. The site-specific crop management will be improved very soon by integrating various sources that predict better soil and crop properties. IoT solve different problems in the agricultural context and so this review would be a point of reference in identifying the major applications of agricultural IoT. With the future of highly precise farming methods and various AI-based analytical models, the entire agricultural sector would be transformed to digital agriculture and this would be a driving force for increasing the production in a rapid and cost-efficient manner.

Acknowledgements The authors sincerely thank the Science and Engineering Research Board (SERB) of DST for the financial support and the Director, ICAR-SBI, Coimbatore and the Management and Principal of Sri Ramakrishna Engineering College, Coimbatore for extending the required facilities.

References

1. Food and Agricultural Organization of the United States, http://www.fao.org/india/fao-in-india/india-at-a-glance/en/#:~:text=Agriculture%2C%20with%20its%20allied%20sectors,275%20million%20tonnes%20(MT).
2. Manishkumar Dholu, Ghodinde, K. A.: Internet of Things (IoT) for Precision agriculture Application. In: 2nd International Conference on Trends in Electronics and Informatics (ICOEI), pp.339–342, IEEE, Tirunelveli (2018)
3. Veena, S., Rajesh, M., Mahesh, K., Salmon, S.: Survey on Smart Agriculture Using IoT. International Journal of Innovative Research in Engineering & Management. **5**, 63–66 (2018)
4. Kunal Goe, Amit Kumar Bindal: Wireless Sensor Network in Precision Agriculture: A survey report. In: 5th IEEE International Conference on Parallel, Distributed and Grid Computing(PDGC), IEEE, Himachal Pradesh (2018)
5. Ammar, M., Russello, G., Crispoa, B.: Internet of Things: A survey on the security of IoT frameworks. Journal of Information Security and Applications. **38**, 8–27 (2018)
6. Tzounis, A., Katsoulas, N., Bartzanas, T., Kittas, C.: Internet of Things in agriculture, recent advances and future challenges. Biosys. Eng. **164**, 31–48 (2017)
7. Burhan, M.: Rana Asif Rehman, Bilal Khan, Byung-Seo Kim: IoT Elements, Layered Architectures and Security Issues: A Comprehensive Survey. Sensors. **18**, 2796–2832 (2018)
8. Kiani, F., Seyyedabbasi, A.: Wireless Sensor Networks and Internet of Things in Precision Agriculture. Int. J. Adv. Comput. Sci. Appl. **9**, 99–103 (2018)
9. Anna Triantafyllou, Panagiotis Sarigiannidis, Thomas, D., Lagkas: Network Protocols, Schemes, and Mechanisms for Internet of Things (IoT): Features, Open Challenges, and Trends. Wirel. Commun. Mob. Com. **2018**, 1–24 (2018)
10. Partha Prathim Ray: A Survey of IoT cloud platforms. Future Computing and Informatics Journal. **1**, 35–46 (2016)
11. Liang Zhao, Liyuan He, Xing Jin, Wenjun Yu: Design of Wireless Sensor Network Middleware for Agricultural Application. In: Computer and Computing Technologies in Agriculture. **6**, pp.270–279, Springer, China (2012)
12. Hernandez-Rojas, D.L., Fernandez Carames, T.M., Fraga-Lamas, P., Escudero, C.J.: Design and evaluation of a family of light weight protocols for heterogenous sensing through BLE beacons in IoT Telemetry applications. Sensors. **18**, 1–13 (2020)
13. Asim, M.: A Survey on Application Layer Protocols for Internet of Things(IoT). Int. J. Adv. Res. Comput. Sci. **8**, 996–1000 (2017)
14. Vikhram, B., Revathi, B., Shanmugapriya, R., Sowmiya, S., Pragadeeswaran, G.: Animal Detection System in Farm Areas. International Journal of Advanced Research in Computer and Communication Engineering. **6**, 587–591 (2017)

15. Anderson-Cook, C.M., Alley, M.M., Roygard, J.K.F., Khosla, R., Noble, R.B., Doolittle, J.A.: Differentiating Soil Types Using Electromagnetic Conductivity and Crop Yield Maps. Soil Sci. Soc. Am. J. **66**, 1562 (2002)
16. Gouri, V., Bharata Lakshmi, M., Seetaramalakshmi, C., Kumari, M.B.G.S., Ramanamurthy, Chitkala Devi, T., Enhancement of growth and yield of sugarcane through drip fertigation and soil application of micronutrients. J. Sugarcane Res. **7**, 60–63 (2017)
17. Uros Zibrat, Sasa Sirca, Nik Susic, Matej Knapic, Barbara Generic Stare, Gregor Urek: 18 - Noninvasive detection of plant parasitic nematodes using hyperspectral and other remote sensing systems. In: Hyperspectral Remote Sensing Theory and Applications. 1st Edition, pp.357–375 (2020)
18. Gavaskar, S., Sumitha, A.: Design and Development of Pest Monitoring System for Implementing Precision Agriculture using IoT. International Journal of Science, Technology and Engineering. **3**, 46–48 (2017)
19. Bongiovanni, R., Lowenberg-DeBower, J.: Precision Agriculture and Sustainability. Precision agriculture. **5**, 359–387 (2004)
20. Paster, F.J.F.: Juan Manuel Garcia-Chamizo, Mario Neito-Hidalgo, Jeronimo Mora Pascual, Jose Mora Martinez: Developing Ubiquitous Sensor Network Platform Using Internet of Things: Application in Precision Agriculture. Sensors. **16**, 1141 (2016)
21. Andres Villa Henriksen, Gareth T.C., Edwards, Liisa A., Pesonen, Ole Green, Claus Aage Gron Sorenson: Internet of Things in arable farming: Implementation, applications, challenges and potential. Biosyst. Eng. **191**, 60–84 (2020)
22. The state-of-the-art in practice and future challenges: Tamoghna Ojha, Sudip Misra, Narendra Singh Raghuwanshi. Wireless sensor networks for agriculture. Comput. Electron. Agric. **118**, 66–84 (2015)
23. Prahlad Bhadani, V., Vasudha Vashisht: Soil Moisture, Temperature and Humidity Measurement using Arduino. In: 9th International Conference on Cloud Computing, Data Science and Engineering", pp. 597–571, IEEE, Noida (2019)
24. Yunseop Kim, Robert G., Evans, William M., Iversen: Remote Sensing and Control of an Irrigation System Using a Distributed Wireless Sensor Network. IEEE Trans. Instrum. Meas. **57**, 1379–1387 (2008)
25. Wan, X.-F., Zheng, T., Cui, J., Zhang, F., Ma, Z.-Q., Yang, Yi.: Near Field Communication-based Agricultural Management Service Systems for Family Farms. Sensors. **19**, 4406 (2019)
26. Marwa Mekki, Osman Abdallah, Magdi B.M Amin, Moez Eltayeb, Tafaoul Abdalfatah, Amin Babiker: Greenhouse monitoring and control system based on wireless Sensor Network. In: International Conference on Computing, Control, Networking, Electronics and Embedded systems Engineering (ICCNEEE). pp.384–387, IEEE, Khartoum, Sudan (2015)
27. Ioannis Gravalos, Zisis Tsiropoulos, Dimitrios Moshou, Xyradakis P. A low-cost greenhouse monitoring system based on internet connectivity. Acta Hortic. **952**, 937–944 (2012)
28. Ruiz-Garcia, L., Lunadei, L.: The role of RFID in agriculture: Applications, limitations and challenges. Comput. Electron. Agric. **79**, 42–50 (2011)
29. Valeria V., Krzhizhanovskaya, Gábor Závodszky, Michael H., Lees, Jack J., Dongarra, Peter M.A., Sloot, Sérgio Brissos, Joao Teixeira, Olgierd Unold, Maciej Nikodem, Marek Piasecki, Kamil Szyc, Henryk Maciejewski, Marek Bawiec, Paweł Dobrowolski, Michał Zdunek: IoT-Based Cow Health Monitoring System. In: 20th International Conference on Computational Science ICCS pp. 344–356, Netherlands (2020)
30. Deepa, S., Vitur, H., Navaneeth, K., Vijayrathinam, S.: Animal monitoring based on IoT technologies. Waffen-und Kostumkunde Journal. **11**, 332–336 (2020)
31. Codeluppi, G., Cilfone, A., Davoli, L., Ferrari, GianLuigi: LoraFarM: A LoRa-WAN based Smart Farming Modular IoT Architecture. Sensors. **20**, 2028 (2020)
32. Lopez-Riquelme, Pavon Pulido, J.A., Navarro-Hellin, H., Soto Valles, F., Torres-Sanchez: A software architecture based on FIWARE cloud for Precision Agriculture. Agric. Water Manag. **183**, 123–135 (2017)
33. Triantafyllou, A., Sarigiannidis, P., Bibi, S.: Precision Agriculture: A Remote Sensing Monitoring System Architecture. Information **10**, 348 (2019)

34. Koksal, O., Tekinerdogan, B.: Architecture design approach for IoT-based farm management information systems. Precision Agric. **20**, 926–958 (2019)
35. Noura, M., Atiquzzaman, M., Gaedke, M.: Interoperability in Internet of Things: Taxonomies and open challenges. Mobile Networks and Applications. **24**, 796–809 (2019)
36. Zhi-Kai Zhang, Michael Cheng Yi Cho, Chia-Wei Wang, Chia-Wei Hsu, Chong-Kuan Chen, Shiuhpyng Shieh: IoT Security: Ongoing Challenges and Research Opportunities. In: 7th International Conference on Service-Oriented Computing and Applications. pp. 230–234, IEEE, Matsue, Japan (2014).

SURF Algorithm-Based Suspect Retrieval

Swati Srivastava, Mudit Mangal, and Mayank Garg

Abstract In this work, we have discussed the method to retrieve the suspect from the gallery of images with the help of a forensic sketch. A forensic sketch is defined as a container of a number of different features: image vector. Face outlines have the fundamental data about the spatial geography and mathematical subtleties of appearances while missing some significant facial credits, for example, identity, hair, eye, and skin color. On the basis of this number of features, the sketch is then iterated through a gallery of images to find the closest match. The proposed idea utilized the sketch and free facial characteristic data to prepare a machine learning model contrasted with the regular sketch-photograph acknowledgment techniques.

Keywords SURF algorithm · Key points · Descriptors · Sketch recognition · Suspect retrieval

1 Introduction

Humans have an inherent sense for justice, developed at infancy and nurtured through society. So, when one witnesses a crime happening, his/her instinct wants to stop it. And after the crime, one wishes to see the criminal punished. Individuals who have been at the scene of the crime can report to appropriate law enforcement agencies about the criminal. This process is often complicated by the fact that witnesses to crimes cannot identify the individual; except through his face. Sketch to image matching is a topic that has been in discussion for the past decade. In the present scenario, we are dealing with manual searching from thousands of criminal records present in the record room. Any automated system for the same purpose is not

S. Srivastava (✉) · M. Mangal · M. Garg
Department of Computer Engineering & Application, GLA University, Mathura, India
e-mail: swati.srivastava@gla.ac.in

M. Mangal
e-mail: mudit.mangal_cs17@gla.ac.in

M. Garg
e-mail: mayank.garg_cs17@gla.ac.in

J. S. Raj et al. (eds.), *Intelligent Sustainable Systems*, Lecture Notes in Networks and Systems 213, https://doi.org/10.1007/978-981-16-2422-3_43

in existence yet. However, most algorithms proposed have been used earlier in face recognition systems. The major challenge for this task is overcoming the abstractness or inaccuracy of a sketch. Like beauty being in the eye of the beholder, the description of and interpretation of a facial feature varies with the witness's memory and the artist's skill. Some sketches are purposely abstract so as to draw attention to their few concrete features. As this research is purely technical, these limitations can be ignored. The proposed thought can utilize the sketch and correlative facial attribute information to train a machine learning model with the help of the SURF algorithm as contrasted with the regular sketch-photograph acknowledgment strategies. Face outlines have the basic data about the mathematical subtleties of countenances while missing some significant facial credits, for example, facial cuts, eyes, nose, and lips. Therefore, a sketch drawn by a forensic artist is processed to define it as a combination of features like key points and descriptors of a sketch. Each image in the gallery is then iterated through and processed similarly to find the closest match for the sketch with the help of good points calculated out from key points and descriptors which best fit it. Though a challenge that still remains in this algorithm is when a person dresses in an unidentifiable manner, changes eye retina color, wear hats, fake beard, mustache, etc. remains non-identified when matched with the gallery of images stored. The paper is organized as follows: Sect. 2 includes related work. The proposed work is explained in Sect. 3. Section 4 represents the results. Finally, the work is concluded in Sect. 5.

2 Related Work

In the present scenario, we are dealing with annual searching from thousands of records. Many face identification systems concentrate continuously on image-based face identification, but most of the conditions need sketche-based identification. Pseudo-sketch amalgamation and sketch recognizable proof techniques were used to recognize face sketch [1]. Photo sketch origination procedure depends upon local linear by take care of geometry among image and sketches. Nonlinear distinguish research is for to allow analysis sketch which is coming out of blend pseudo-sketches. Face sketch synthesis procedure using residual training by the difference of standard perspective has been proposed in [2] where the goal was to rebuild the sketch image quickly (grasp of aligning connection with image and sketch). The target was to focus on the remaining image by studying the rhyming connection with the image and residual. That method provides progress in residual grading which is simple as compared to modify the exact scaling and obtain unique feature data. Further, joint word reference learning calculation is also proposed in [2] by taking care of the local pattern shape of information storage. From learned joint word reference, convert the sketch combination through the image area into the new and compressed area. The new and compressed area is passed over by learned dictionary molecules, whatever premise it should be considered. The outcome of the work was obtained by differentiating this method with several methods. Nowadays, some researchers work on deep learning-based Face Sketch Synthesis (FSS) with a focus on the application area with conventional face recognition. Optimal Deep Learning-based Convolutional Neural Network (ODL-CNN) for FSS to help in suspicious recognition system becomes more efficient with IOT setup [3]. The specification reform of the ODL-CNN model executes by the Improved Elephant Herd Optimization (IEHO) algorithm. Initially, the work trap observation video by IoT. Currently, researchers focus on the analysis of one depict area to another area. In [4], the authors concerned on the difficult task of face sketch recognition and suggested Deep Transformer which is a transformed learning-based approach that analysis of modification and depict function among characteristic of two areas. This method is not dependent on input data that can be put in any predefined learning characteristic. They suggested two models of Deep Transformer; both focus on eliminating the difference between two areas: 1. Semi-coupled; 2. Symmetrically coupled deep transform. They use an IIIT-D Composite Sketch with Age (CSA) unevenness information base that conveys pictures of sketches around 150 areas with age-distinct digital images. The execution of the model analyzes the novel steadiness of sketch – sketch, sketch-digital photo matching. The underlined part of the model is the adaptability of existing attribute eradication and separator in the framework. The result represents the stability of the model as differentiation with modern algorithms and commercial systems. Some researchers addressed the problem of image transformation. For image conversion, where the input image is converted in a sketch where the major application area is video observing-based law implementation for such type problem, convolutional neural network (CNN) or graphical probabilistic method is used [5]. The idea was influenced by the generative adversarial network (GAN) which creates images wherever they go with the structure created by GAN. They presented a bunch of models to rebuild the mash-up result. by performing this model on campuses database that performance is more efficient as compared to the modern method. A bunch of model strategies is further used to another GAN-based image-to-image convert problem. Many face identification systems emphasize image-based face identification. Face identification systems depend on face sketches where the project contains two modes [6]: 1. Pseudo-sketch combination and 2. Sketch acknowledgment, where the pseudo-sketch model is dependent on linearly take care of geometry between image and sketches that are influenced by the design of locally linear submerge. Nonlinear differentiate the survey that is used to allow the analysis of sketch through the synthesis sketches. By performing this model on 600 pairs of sketches, its result was effective. Computer vision-based traffic sign sensing emphasizes utilizing the capsules neural network that beats the convolutional neural network by escaping the necessities for the manual exertion [11]. The capsule network gives superior protection from a spatial change and the high unwavering quality in the detecting of the traffic sign is contrasted with the convolutional network. However, the enhancement of vision has been in continuation for the future. A psychological characterization-based visual saliency guided model for the productive recovery of data from media information stockpiling [12]. The Itti visual saliency model is portrayed here for the age of a general saliency map with the joining of shading saliency, power, and course maps. Multi-include combination ideal models are utilized for giving away from of the picture design.

However, it does not create a strong model that checks the closeness in the picture properties on numerous data set stages.

3 Proposed Method

In order to reduce the manual work of searching the suspect from thousands of records present in the record room as it takes a lot of time of around 10–15 days, this system tends to search the suspect with the help of a sketch with just a single click. The proposed idea utilizes the sketch and free facial characteristic data to prepare a machine learning model with the assistance of the SURF algorithm when contrasted with the regular sketch-photograph acknowledgment techniques. Face portraits have the basic data about the mathematical subtleties of appearances while missing some significant facial credits, for example, facial cuts, eye, nose, and lips. Therefore, a sketch drawn by a forensic artist is processed to define it as a combination of features like key points and descriptors of a sketch through the SURF algorithm. Each image in the gallery is then iterated through and processed similarly to find the closest match for the sketch with the help of good points calculated out from key points and descriptors through the SURF algorithm.

SURF Algorithm

SURF [7, 15] represents Speeded-Up Robust Features. It is a methodology that is for the most part used to develop a hearty picture that includes indicator and descriptor. It tends to be utilized in PC vision errands like article acknowledgment and 3D recreation. The primary interest of the SURF approach lies in its quick calculation of administrators utilizing box channels, subsequently empowering continuous applications, for example, following and item acknowledgment. SURF is composed of two steps.

- Feature Extraction;
- Feature Description.

Detection

SURF utilizes square-molded channels as an estimate of Gaussian smoothing. Sifting the picture with a square is a lot quicker if the necessary picture is utilized; the sum area equation S(x, y) at point x and y of any image is

$$S(x, y) = \sum_{i=0}^{x} \sum_{j=0}^{y} I(i, j) \tag{1}$$

where I(i, j) is the value of the pixel at point (x, y). In order to detect the interest points, SURF uses a blob detector [8] based on the Hessian [10] lattice. Mass-like

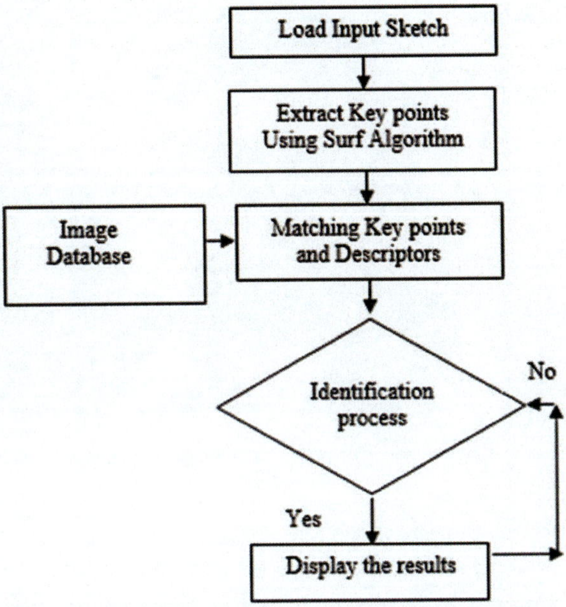

Fig. 1 Proposed Framework

structures are recognized in the picture, where the neighborhood determinant of the Hessian framework is greatest. SURF utilizes a mass finder dependent on the Hessian lattice to discover focal points. The determinant of the Hessian network is utilized as a proportion of nearby change around the point and focuses are picked where this determinant is maximal. As opposed to the Hessian-Laplacian identifier by Mikolajczyk and Schmid, SURF likewise utilizes the determinant of the Hessian for choosing the scale. The proposed framework is shown in Fig. 1.

Descriptors

The objective of a descriptor is to give a special and vigorous portrayal of a picture by depicting the power dispersion of the pixels inside the neighborhood of the focal point. Most descriptors are in this way figured in a neighborhood way; consequently, a depiction is gotten for each focal point recognized beforehand. To remove the descriptors, "amount of Haar wavelet reactions" [9] is utilized. We first build a square locale focused at the interest point and arranged along the direction chosen by an exceptional determination technique. We then use these descriptors to describe any given image and compare these descriptors to give us the similarity between sketch and image. The dimensionality of the descriptor has a direct effect on the two: its computational multifaceted nature and point-coordinating vigor/precision. Moreover, just 64 measurements are typically utilized in SURF [13, 14] to diminish the time cost for both element calculation and coordinating. Since every one of SURF highlights has just 64 measurements as a rule and an ordering plan is worked

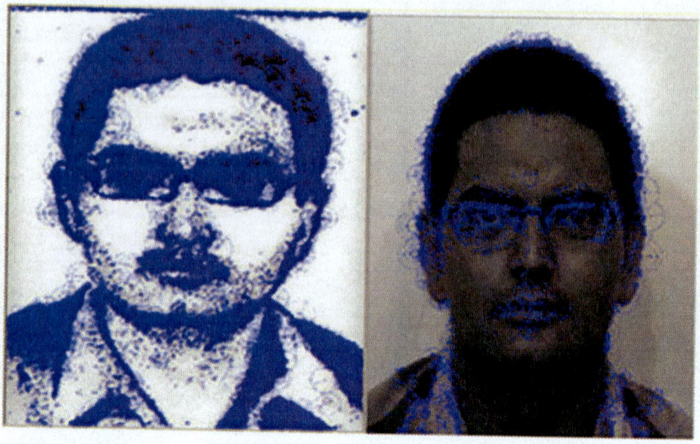

Fig. 2 Example of Key points and Descriptors in a sketch and image

Table 1 Matching Scores between sketch and closest matches

Matching Points	Sketch 1	Sketch 2	Sketch 3	Sketch 4	Sketch 5
Closest match	554	600	522	474	424
2nd Close match	432	521	402	462	381
3rd Close match	312	468	305	314	314
4th Close match	206	252	231	301	296
5th Close match	154	106	200	252	262

by utilizing the indication of the Laplacian, SURF is a lot quicker. The key points and descriptors are shown in Fig. 2.

4 Results

Table I shows the matching scores of the 5 closest matches of the sketch input calculated out with the help of the SURF algorithm. The results of four input sketches are shown in Table I. When we give a sketch as an input, it shows us 5 possible closest matches or photos from the gallery. The closest match and four other matches are shown in Fig. 3.

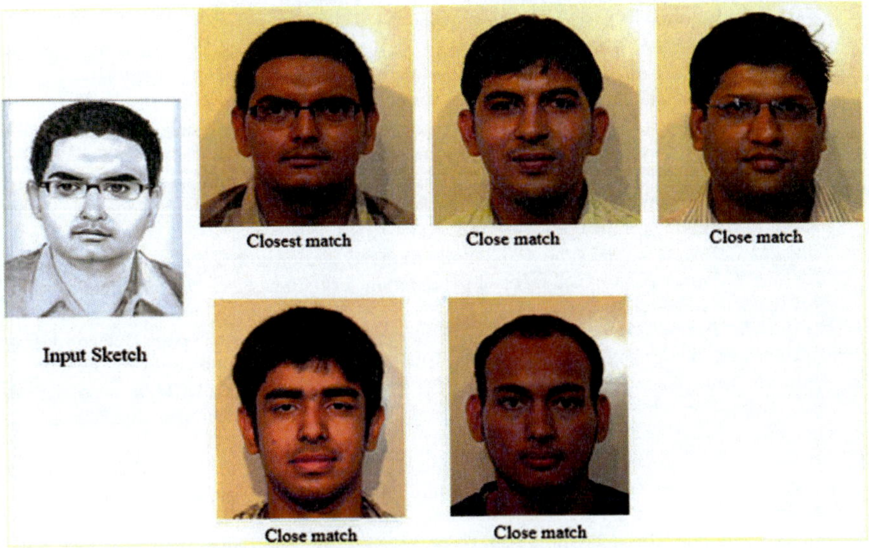

Fig. 3 Closest match and other close matches to Input Sketch

5 Conclusion

Sketch to face matching has become an important and challenging requirement. The requirement of forensic or spy agencies lacks in finding out the closest result quickly. In this paper, we have used the SURF algorithm which extracts out major key points and descriptors from the sketch as well as from the images. On the basis of these key points and descriptors input sketch, and images are matched with each other. If the closest match is found, it will get displayed on the screen otherwise it will search again for the 5 closest matches. To retrieve the image from the database with the help of voice as an input can be addressed as future work. This new model will help in saving a lot of time which could increase the efficiency of finding the suspect for the police. Secondly, there will be no need of hiring a forensic expert for making a sketch.

References

1. Tang, X., Wang, X.: Face sketch recognition. IEEE Trans. Circuits Syst. Video Technol. **14**(1), 50–57 (2004)
2. Jiang, J., Yu, Y., Wang, Z., Liu, X., Ma, J.: Graph-regularized locality-constrained joint dictionary and residual learning for face sketch synthesis. IEEE Trans. Image Process. **28**(2), 628–641 (2018)
3. Elhoseny, M., Selim, M. M., & Shankar, K. (2020). Optimal deep learning based convolution neural network for digital forensics face sketch synthesis in internet of things (IoT).

International Journal of Machine Learning and Cybernetics, 1–12.
4. Nagpal, S., Singh, M., Singh, R., Vatsa, M., Noore, A., & Majumdar, A. (2017). Face sketch matching via coupled deep transform learning. In Proceedings of the IEEE international conference on computer vision (pp. 5419–5428).
5. Wang, N., Zha, W., Li, J., Gao, X.: Back projection: An effective postprocessing method for GAN-based face sketch synthesis. Pattern Recogn. Lett. **107**, 59–65 (2018)
6. Liu, Q., Tang, X., Jin, H., Lu, H., & Ma, S. (2005, June). A nonlinear approach for face sketch synthesis and recognition. In 2005 IEEE Computer Society conference on computer vision and pattern recognition (CVPR'05) (Vol. 1, pp. 1005–1010). IEEE.
7. Aslan, M.F., Durdu, A., Sabanci, K.: Human action recognition with bag of visual words using different machine learning methods and hyperparameter optimization. Neural Comput. Appl. **32**(12), 8585–8597 (2020)
8. Xu, Y., Wu, T., Charlton, J. R., Gao, F., & Bennett, K. M. (2020). Small Blob Detector Using Bi-Threshold Constrained Adaptive Scales. IEEE Transactions on Biomedical Engineering.
9. Kumar, S., Kumar, R., Agarwal, R.P., Samet, B.: A study of fractional Lotka-Volterra population model using Haar wavelet and Adams-Bashforth-Moulton methods. Mathematical Methods in the Applied Sciences **43**(8), 5564–5578 (2020)
10. Xu, Y., Wu, T., Gao, F., Charlton, J.R., Bennett, K.M.: Improved small blob detection in 3D images using jointly constrained deep learning and Hessian analysis. Sci. Rep. **10**(1), 1–12 (2020)
11. Koresh, M.H.J.D., Deva, J.: Computer vision based traffic sign sensing for smart transport. Journal of Innovative Image Processing (JIIP) **1**(01), 11–19 (2019)
12. Vijayakumar, T., and Mr R. Vinothkanna. "Retrieval of Complex Images Using Visual Saliency Guided Cognitive Classification." Journal of Innovative Image Processing (JIIP) 2, no. 02 (2020): 102–109.
13. Gupta, S., Thakur, K., & Kumar, M. (2020). 2D-human face recognition using SIFT and SURF descriptors of face's feature regions. *The Visual Computer*, 1–10.
14. Karami, E., Prasad, S., & Shehata, M. (2017). Image matching using SIFT, SURF, BRIEF and ORB: performance comparison for distorted images. *arXiv preprint* arXiv:1710.02726.
15. Li, A., Jiang, W., Yuan, W., Dai, D., Zhang, S., Wei, Z.: An improved FAST+ SURF fast matching algorithm. Procedia Computer Science **107**, 306–312 (2017)

An Analysis of the Paradigm Shift from Real Classroom to Reel Classroom During Covid-19 Pandemic

Pawan Agarwal, Kavita A. Joshi, and Shweta Arora

Abstract Covid-19 pandemic brought turmoil to the entire world. It adversely affected diverse areas like economic, political, social, medical, education, etc. The outbreak of the peril stimulated the research fraternity in various areas chiefly in medical science, environment, and education. Primarily focusing on Education, the sudden lockdown brought a complete shut down to the long-standing physical mode of teaching/learning in the education system. To combat the menace of the Covid-19 pandemic, alternative teaching-learning mechanism through online mode was adopted. The only gateway for the teachers and students to remain connected was through a virtual platform. Though this counteractive phenomenon was an aid at this critical time, there were certain emerging anomalies like efficacy and acceptance of a virtual platform. Various apprehensions during this shift like psychological barriers, maintaining attendance, participation, etc. which seeped into the education system were to be studied and analyzed. In the present study, the concern of the researchers was to observe and examine the significant drift of teaching/learning platform from a real classroom to a reel classroom. It aims to study the girth of online teaching/learning in the present education system. The present study also focuses on the role of technology as an important tool in enhancing the Teaching/Learning mechanism during the Covid-19 pandemic. It was a big challenge as well as an opportunity for both the learners and teaching fraternity to cope with the stagnancy like carrying out interactive and team building activities, developing rapport and facilitating a learning process, etc. percolated during the outburst of the Covid-19 pandemic.

Keywords Covid-19 · Pandemic · Online · Teaching/learning · Virtual · Real · Classroom

P. Agarwal (✉) · K. A. Joshi · S. Arora
Graphic Era Hill University, Bhimtal, Uttarakhand, India

J. S. Raj et al. (eds.), *Intelligent Sustainable Systems*, Lecture Notes in Networks and Systems 213, https://doi.org/10.1007/978-981-16-2422-3_44

1 Introduction

Conventionally, education has always been assumed or portrayed as a teaching-learning process in a closed classroom with a blackboard and chalk method. With time, smart classes were introduced in the education system followed by a change in the modus operandi of teaching/learning as well as conducting exams—from pen-paper to online exams [1]. This incorporation of technology was a revolution in the history of the Indian education system. But nowhere it was thought to become an indispensable element for the functioning of the system [2]. We had never even realized that the advancement in technology would become the only means to carry out the teaching/learning and bridge the gap created in the education system during this crucial time [3]. The use of technology became an inevitable part to run the academic industry during the Covid-19 pandemic. The uncontrolled spread of the virus was likely to keep the student community away from the physical classroom until the rejuvenation of the situation [4]. To deal with the situation, we not only switched from the chalk-blackboard mode to a virtual online mode but also shifted from following a fixed schedule of classroom teaching/learning to a self-paced learning mode. Teaching/Learning could be carried out beyond the classroom during this pandemic, i.e. learning could take place anywhere and anytime irrespective of the place and time. A Strengths, Weaknesses, Opportunities, Challenges (SWOC) analysis of the current situation was done in order to get a clear picture and understanding to fight the exigency [5, 6]. It was revealed that it was not a complete dark situation, though there were weaknesses and challenges like psychological barriers for the acceptance of the new technology, lack of expertise to tackle the technological glitches, minimum possibility of offline clearance of doubts pertaining to this new learning yet to focus on the strengths and opportunities the advanced technology was already in existence [7]. As there was a need for the faculty to switch from the traditional teaching-learning methods to the new technology-centered teaching [8], this situation exposed the need to provide teachers and students with the skillsets to adapt in the new digital environment because there were many adaptable challenges to be faced during this period [9, 10]. The virtual Faculty Development Programmes, Webinars, and trending videos helped the teachers as well as the learners to work gradually and effectively on a virtual platform [11].

2 Background

The online or virtual mode became the only platform for imparting education in times of the Covid-19 emergency. The challenges encountered in online teaching/learning were many. Digital literacy turned out to be a boon for some and a bane for others [12]. It came up as an opportunity for some and a hurdle or challenge for others. This miscellaneous response from those involved in the academic community—students and teachers—invoked the researchers to conduct the present study. The present study involves an analysis of the academic cadre working in the virtual setup using various technological applications like Zoom, Google Classroom, WebEx, MS Teams, HackerEarth, Moodle, etc. [13]. It aims to study the change in the perspective of teaching/learning during COVID-19 which affected the entire education system [14]. It was important to analyze the efficacy of the newly incorporated teaching/learning mechanism [15].

3 Diagnosis and Objectives

1. Initially it was found that the sudden shift from real to reel classroom that online teaching/learning would affect the learning tendencies of the learners. It could be negative or positive.
2. This transition would also have an impact on the education system.

The diagnosis of the current situation led to the need for the present study.

It was found that the knack of digital literacy had an impact on online teaching/learning. Due to the Covid-19 pandemic, digital literacy became inevitable for the continuation of education in colleges as well as schools. The aim of this study was to analyze the paradigm shift from a real classroom to a reel classroom during this crucial period.

4 Sample Tools and Methods

It was an empirical study conducted on a sample size for the following:

Students: B.Tech CSE 5th Semester

Sample Size: 35

The following tools were employed:

1. Essay Writing submitted on Moodle platform.
2. Questionnaire set to the sample through Online Google Forms.

5 Activity Roadmap

With the announcement of the first lockdown by the end of March 2020, we had to abruptly switch from classroom teaching/learning to the online mode of education. There was a silent turmoil experienced in the academic community which demanded an intricate study of the situation. The researchers conducted the analysis for the aforesaid study in two phases with the following activities:

- Essay Writing submitted on Moodle platform.
- It was followed by a survey questionnaire on online education versus classroom learning.

Phase I was conducted after 1 month of opting for the online mode, i.e. end of April 2020, and Phase II was conducted after 5 months of online teaching/learning, i.e. end of August 2020.

5.1 *Activity: Essay Writing*

Essay writing on the topic "A Switch from Real Classroom to Reel Classroom" with a word limit of 250–300 words was given to the learners. The same topic was given in both the phases (Phase I and Phase II) followed by the assessment of the manuscripts.

5.2 *Online Survey: Questionnaire*

A set of 5 questions was prepared for the survey. A comprehensive comparative analysis of the data of Phase I and Phase II was done in which the same set of questions was used.

6 Data Analysis and Findings

6.1 *Activity: Essay Writing*

A comparative study of Phase 1 and Phase 2 was made. The following point cues were given for content development.

Virtual Teaching as a substitute to a physical setup, technological glitches, attendance, interaction, and psychological barriers (Tables 1 and 2).

Table 1 Phase 1 analysis

S.No.	Parameters	High	Medium	Low
1	Virtual teaching as a substitute	14	8	7
2	Technological glitches	23	9	3
3	Attendance	21	7	7
4	Interaction	8	12	15
5	Psychological barriers	28	3	9

Table 2 Phase 2 analysis

S.No.	Parameters	High	Medium	Low
1	Virtual teaching as a substitute	7	10	18
2	Technological glitches	11	7	17
3	Attendance	10	10	15
4	Interaction	20	11	4
5	Psychological barriers	13	10	12

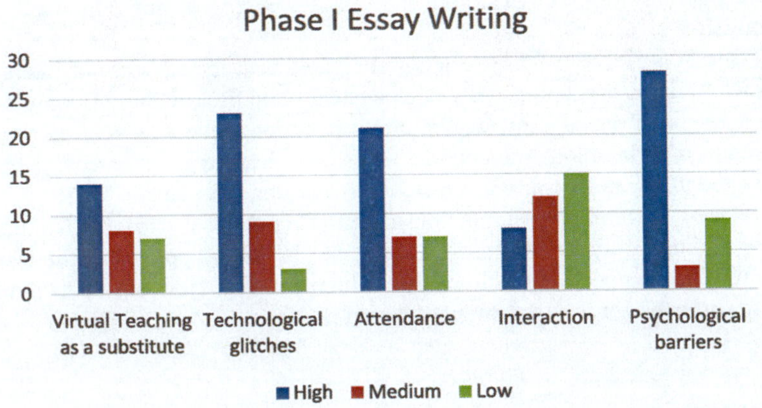

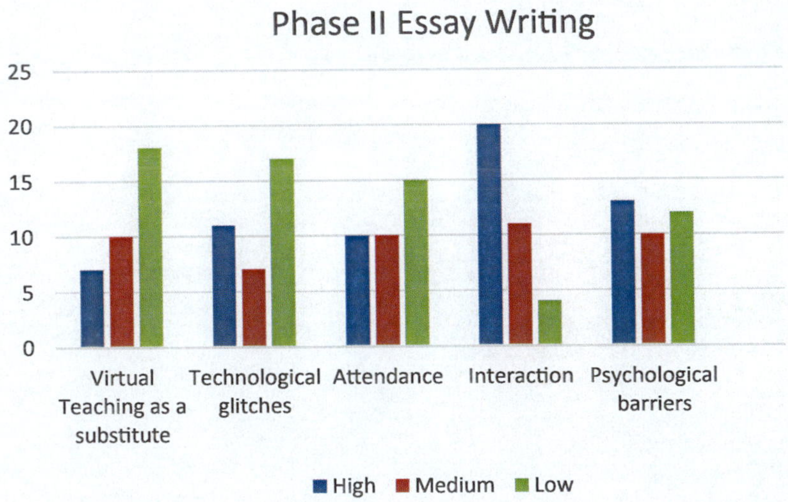

It was observed that there was 50% increase in the number of learners who considered virtual teaching as a substitute for physical classroom teaching. This indicates that there is a good adaptably of this contemporary mode of teaching/learning.

65.8% of the learners consider that technological glitches are not a hindrance to the online teaching/learning method. Surprisingly, this is approximately the same figure (65.7%) which was obtained in Phase I for learners considering technological glitches as a major hindrance in online teaching/learning method. This shows a positive drift toward transformation from real to reel classroom.

In Phase I, 60% of the learners highly believed that maintaining attendance would be a major issue. This figure rolled down to 28.5%. This depicts that when the platform was new to both the teachers and the learners, many important factors were unknown which resulted in the mushrooming myth of difficulty of maintaining attendance in a virtual platform. But with exploring and self-learning, these difficulties were seen to diminish.

In Phase I only 22.8% of the learners participated actively in classroom interaction which was quite low as compared to the physical setup. But in Phase II, only 11.4% of the learners exhibited a low degree of participation while 88.6% of the learners were engaged in classroom interactions. This shows that a large section of the sample has adapted to this newly introduced teaching/learning mechanism.

In Phase I, 80% of the sample considered psychological barriers affecting the learning through online platforms. But in Phase II, 43% (approx.) of the learners have shown a changed approach toward online learning. Now only 37.1% of the learners still consider it as a prominent factor.

6.2 *Activity: Survey Questionnaire*

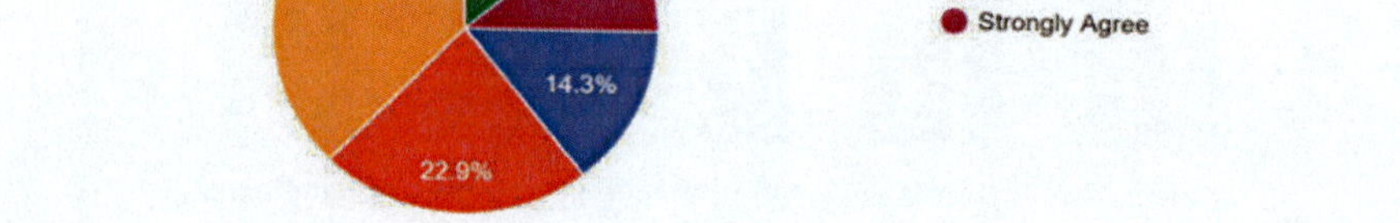

Q.1(ii)Online education can replace face to face learning in the next five years.(Your view in August,2020)

35 responses

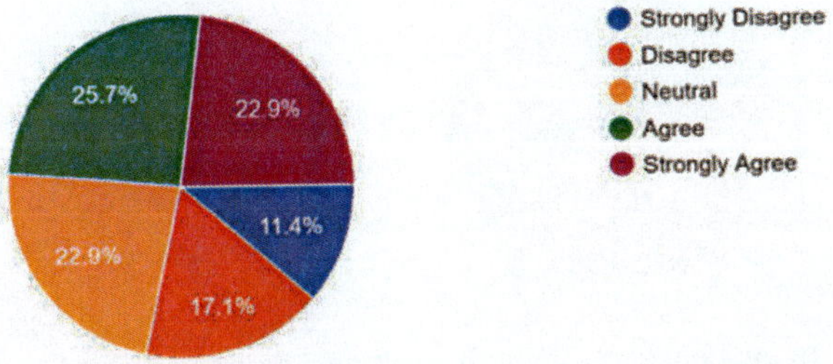

In Phase I, to the question of whether online education can replace face-to-face learning in the next 5 years, 11.4% of the study sample strongly agree to it. However, 14.3% of the selected sample strongly disagree with the possibility of the replacement of real classroom teaching/learning by online education.

In Phase II, 22.9% of the study sample strongly agree to the question. Only 11.4 of the selected sample strongly disagree with such a possibility. 37.1% of the sample gave a neutral response to the situation. This reflects that they are not in a state to form a clear-cut opinion regarding it. However, when compared with Phase I, there is a decline of 14.2% of this category in Phase II which shows a positive drift toward the clarity of thought regarding online teaching/learning.

Phase I reveals that 25.7% of the given study sample shows a positive inclination toward such change, however, 37.9% shows a negative scope for such a transition in the education system.

A total of 48.6% of the sample corresponds favorably to the drift in the online mode of education which is 22.9% more than that of Phase I.

28.5% of the sample still differs in such a change in the education system which is 8.7% less than that of Phase I. Thus, the analysis reflects a positive inclination toward online education in 5 years from now.

Q.2.(i) Online education is quite stressful for teachers and students.(Your view in April ,2020)

35 responses

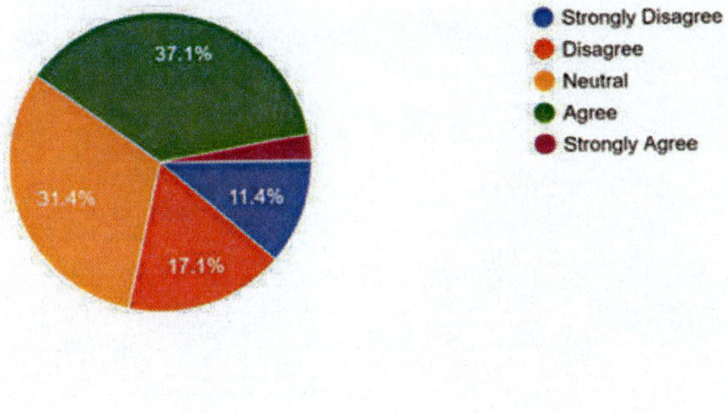

Q.2.(ii)Online education is quite stressful for teachers and students.(Your view in August ,2020)
35 responses

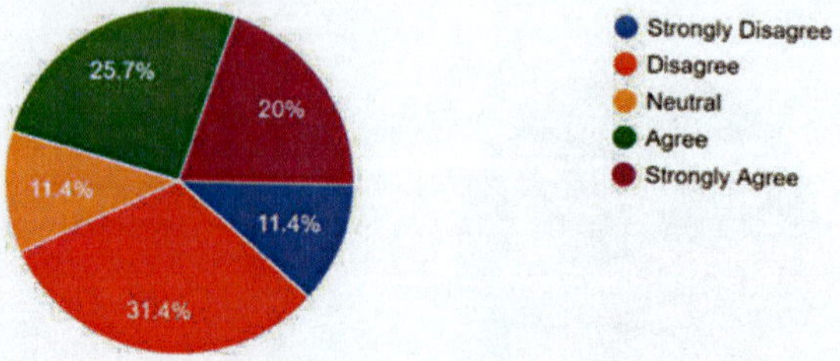

In Phase I, to the question of whether online education is stressful for the learners, 37.1% of the study sample approves it with 0% of the sample that strongly agrees that online education is stressful, i.e. the sample does not consider the online mode quite stressful. However, 28.5% of the selected sample disapproves of online education as a cause of stress. And 31.4% are in a state of indecision.

In Phase II, there is a noteworthy reflection that the percentage of the sample that strongly disagreed was nil in the first phase has now turned to 20%. And overall more than 50% of the sample considers online education quite stressful. This portrays a repellant behavior of the learners toward such a transition.

Q.3.(i) Online education relieves the students of peer pressure and students can learn at their own pace.(Your view in April,2020)
35 responses

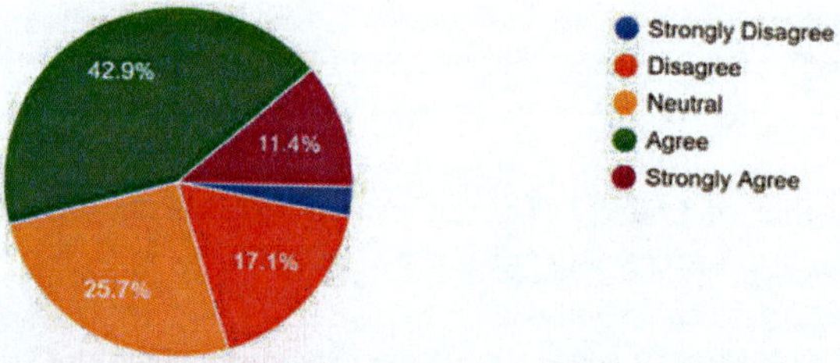

Q.3.(ii) Online education relieves the students of peer pressure and students can learn at their own pace.(Your view in August,2020)
35 responses

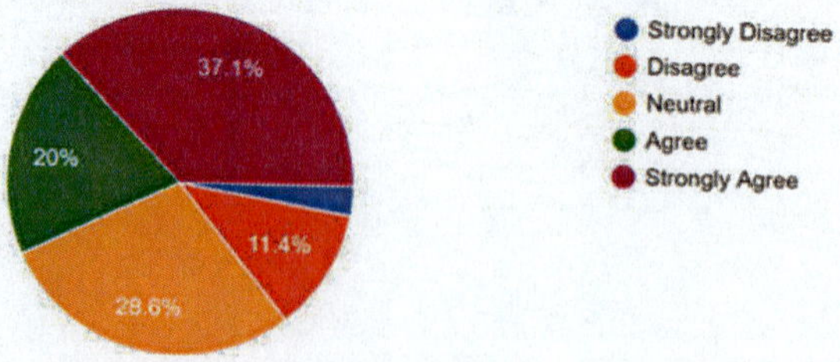

In Phase I, to the question of whether online education relieves the students of peer pressure and students can learn at their own pace, 42.9% of the study sample agreed to the fact. Interestingly, in Phase II this statistic decreased by nearly 50 percent which indicates that the students require teaching/learning in a physical setup in addition to online learning. Conversely, an interesting fact pertaining to peer pressure and pace of learning, crystalized in this analysis, 25.7% of the sample strongly agrees to the given question.

Q.4.(i) Attendance is a big issue in live online classes.(Your view in April,2020)
35 responses

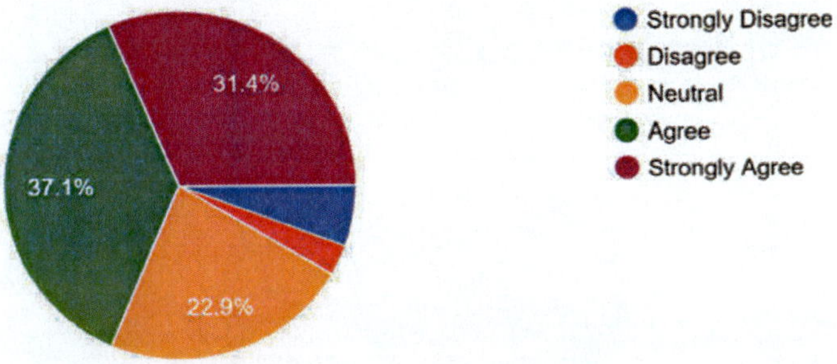

Q.4.(ii) Attendance is a big issue in live online classes.(Your view in August,2020)
35 responses

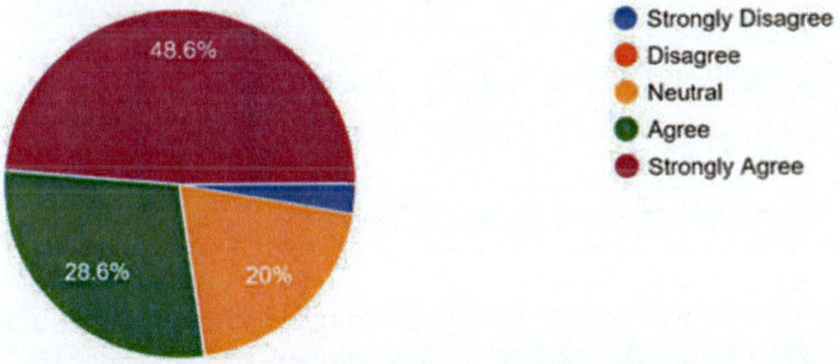

In the answer to the question of whether attendance is a big issue in online classes, 68.5% of the sample in Phase I was in favor whereas in Phase II the statistics increased to 77.2% which shows that despite various modes of marking/taking/downloading attendance on various platforms used, the students still found attendance as a big debatable issue.

Q.5.(i) It is difficult to reach all students due to technological limitations(Your view in April,2020)
35 responses

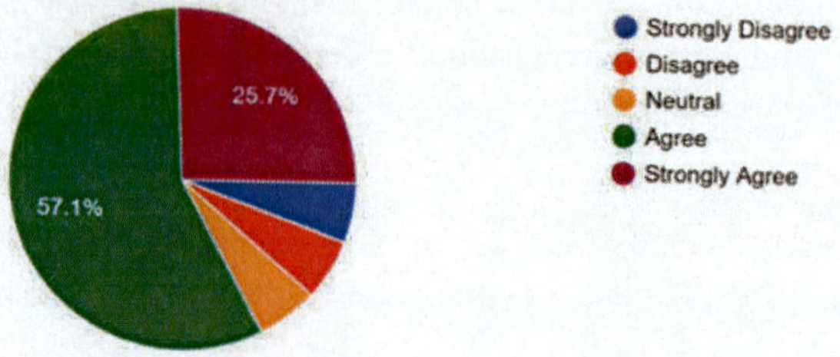

Q.5.(ii) It is difficult to reach all students due to technological limitations(Your view in August,2020)
35 responses

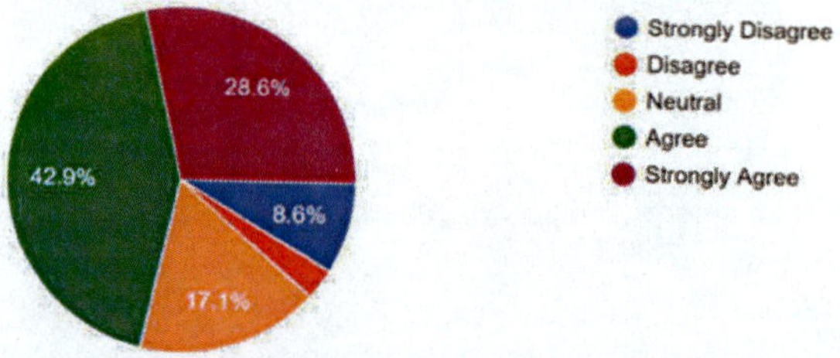

In Phase I surprisingly, 82.8% of the study sample opines that it is difficult to reach all the students due to technological limitations. In Phase II, these statistics fell to 71.5% which reveals that there is an insignificant positive drift. This shows that due to technological glitches, online education will take time to become a universal phenomenon specifically in developing countries like India.

7 Conclusion and Suggestions

- The new platforms of online teaching/learning although enhanced the quality of imparting education on virtual settings, brought intricacies in learning. This proves to be an underlying reason for stress and anxiety among learners.
- Also, in the present study, it has been deciphered that students require teaching/learning in a physical setup in addition to online learning.
- Though various manners of marking/taking attendance on various platforms like MS Teams, Moodle, etc. were available, the students still found attendance as a prevalent issue. This again highlighted the importance of a physical classroom setup.
- Due to technological glitches, online education will take time to become a universal phenomenon specifically in the developing countries like India.

- Students are in favor of interactive sessions, and they find live classes quite analogous to physical teaching/learning.
- Critical learning/thinking is an inevitable aspect of education; unfortunately, online education impedes it.
- According to some academicians, this method was affordable, self-paced, and adaptable whereas, for others it was secluded in terms of interaction and functional advantage. The technical glitches in the online teaching/learning has proven to be useful for conventional courses like humanities, etc. but professional courses with practical components required application-based learning in addition to online lectures and tutorials. In such cases, online teaching-learning has supplemented the education in the pandemic scenario; however, there is no substitute for classroom teaching and learning. Although it became an important tool for engaging the students in the pandemic, there were certain glitches as well which cannot be compromised.
- The current analysis also reflects a positive indication for the physical mode of teaching-learning, and it cannot be substituted by the online mode. However, online mode can be a supplement to the conventional education system.
- Moreover, the present analysis gives an insight that interactions in physical settings are inevitable.

References

1. Jena, P.K.: Impact of pandemic COVID-19 on education in India. Int. J. Curr. Res. **12**(7), 12582–12586 (2020)
2. Dziuban, C., Graham, C.R., Moskal, P.D., Norberg, A., Sicilia, N.: Blended learning: the new normal and emerging technologies. Int. J. Educ. Technol. High. Educ. **15**, 3 (2018)
3. Islam, N., Beer, M., Slack, F.: E-learning challenges faced by academics in higher education: a literature review. J. Educ. Train. Stud. **3**(5) (2015)
4. Dorn, E., Hancock. B., Sarakatsannis, J., Viruleg, E.: COVID-19 and learning loss—disparities grow and students need help (2020). https://www.mckinsey.com/industries/public-and-social-sector/our-insights/covid-19-and-learning-loss-disparities-grow-and-students-need-help
5. Arora, S., Joshi, K.A., Koshy, S., Tewari, D.: Application of effective techniques in teaching learning english. Engl. Lang. Teach. **10**(5), 193–203 (2017)
6. Joshi, K., Arora, S., Sabarwal, P., Agarwal, P., Gabbhir, A.: Contemporary personality litmus test through SWOC analysis, test engineering and management **83**, 659–666 (2020)
7. Bhardwaj, K.: Professional Communication. I.K. International Publishing House Pvt. Ltd., India (2012)
8. Chandrasekharam, D.: Post covid-19 education system, times of India, May 17 (2020). https://timesofindia.indiatimes.com/blogs/dornadula-c/post-covid-19-education-system/
9. Rioz, B.: Changes in education as a result of COVID-19 crisis are here to stay, experts say, June5 (2020). https://www.euractiv.com/section/economy-jobs/news/changes-in-education-as-a-result-of-covid-19-crisis-are-here-to-stay-experts-say/
10. Reimers, F.M.: What the Covid-19 Pandemic will change in education depends on the thoughtfulness of education responses today, April 9 (2020). https://www.worldsofeducation.org
11. Rawat, C.D.: Digital literacy in the midst of an outbreak, India Bioscience (2020). https://indiabioscience.org/columns/education/digital-literacy-in-midst-of-an-outbreak

12. Gomathy, C.K.: A study on the effect of digital literacy and information management. Int. J. Sci. Res. Rev. **7**(3), 51–57 (2018)
13. Rembousek, V., Stipek, J., Vankova, P.: Contents of digital literacy from the perspective of teachers and pupils. Proc.-Soc. Behav. Sci. **217**, 354–362 (2016)
14. Nicola, M., Alsafi, Z., Sohrabi, C., Kerwan, A., Jabir, A., Iosifidis, C., Agha, M., Agha, R.: The socio-economic implications of the coronavirus pandemic (COVID-19): A review. Int J Surg. **78**, 185–193 (2020)
15. Arora, S., Joshi, K.A., Koshy, S.: Professional Communication: Practical Workbook. Spire Publications, India (2018)

PoemAI: Text Generator Assistant for Writers

Yamini Ratawal, Vaibhav Singh Makhloga, Kartikay Raheja, Preksh Chadha, and Nikhil Bhatt

Abstract Text Generation has become one of the functional areas of Natural Language Processing (NLP) and is considered a major advancement in this generation of creative texts such as literature. Every year, the field is receiving increased future research. With the advent of Machine Learning and the Deep Neural Network, it has become possible to process and find hidden patterns from huge amounts of information (Big Data). Our research is focused on the use of the Neural Network Model to construct an AI assistant for writers, primarily poets, using Bi-Directional Long Short-Term Memory (LSTM), a variant of the Recurrent Neural Network (RNN), to generate poetry in the English language by providing it with a huge corpus of poems compiled specifically for research by renowned poets. With two unique themes, love and nature, the dataset is specifically chosen. Most of the poems are from the Renaissance and Modern period. This method will function in two ways, firstly by creating full-length poems and sonnets, and secondly as a provocative tool for the writer to make it a hybrid system in which a creative piece of work is produced by both the writer and AI. Some issues with previous versions and models have been improved, such as lack of coherence and rhyme, along with a newly designed use case introduced in our report, such as AI assistant. This model can produce unique poetry along with providing meaningful suggestions with a suggestion of rhyming with appropriate precision in some outputs.

Keywords Natural language processing · Recurrent neural network · Long short-term memory · Text generation · AI assistant · Poetry generation

1 Introduction

Two of the nine muses in ancient mythology known as Calliope and Erato represented the art of poetry as a means of knowledge and were considered to be a central part of life. All were written in poetic form from the Vedas to the Odyssey, and even earlier examples can be seen on monoliths, rhinestones, and steals. People have

---

Y. Ratawal · V. S. Makhloga (✉) · K. Raheja · P. Chadha · N. Bhatt
Akhilesh Das Gupta Institute of Technology and Management, Delhi 110053, India

J. S. Raj et al. (eds.), *Intelligent Sustainable Systems*, Lecture Notes in Networks and Systems 213, https://doi.org/10.1007/978-981-16-2422-3_45

indulged in poetry throughout human history as a means of inspiration, expression, and recreation. This makes people speculate about the reasons behind our long poetry obsession and separates it from other types of written works.

The development of neural networks in the human brain has assisted better than other species to identify auditory and visual patterns. It happens with the aid of short-term memory, which can allow assumptions and assumptions about the near future. Some of the patterns are the harmonic and periodic nature of music, rhyme, poetry meter, and so on.

With the advent of Machine learning and Artificial Neural network [1], there has been an exponential growth in pattern recognition capabilities. The progress in computer hardware and massive boom in data production (in the form of big data) has opened new boundaries in the field of ML. The ultimate aim is to find hidden patterns that humans do not observe themselves. This contributes to the greater concern that machines can be used as creative agents, thereby advancing the ML field and expanding its use case by replicating and studying poetic patterns and creating unique poetry.

In this article, an application is built in which it works in the following two ways.

- Firstly, Suggestive Text tool for writers, focusing primarily on poets and poetry formation. The collective expertise (trained on) of great poets would help the user use a deep learning model behind the scenes to create meaningful and distinctive phrases. Then it would function as a hybrid tool in which the author involved in the process of writing a series of poetry with active suggestions of the next few words from the trained Neural Network Model trained on the works of some of the century's greatest poets, including Shakespeare, Byron, T. S. Elliot, among others, from the Renaissance to the post-Modern period.
- Secondly, generating full-length poetry and sonnets using bi-directional Long Short-Term Memory (LSTM) [2], a variant of Recurrent Neural Network (RNN) [14], and investigating its ability to learn the poetic pattern and generating meaningful words, lines, and Rhyme. RNN is a sequence modeler designed to predict the likelihood of a token occurring next in the sequence (i.e., unit of languages such as a word, character, and phoneme) [3]. The next word in our rhyming couplet based on the training data is to be forecast by this model. The similarity between the poems produced and the data from the training was analyzed. As this is a search to produce more human-like text, one of the key attributes of the generated poem is its uniqueness.

Our bi-directional LSTM layer-trained model had results that suggest that the model was capable of producing unique lines of poetry effectively and add punctuations at the end of each line. Some rhyming has also been used in performances. With some informative specific lines, this model is used as a recommendation tool for writers. This research creates new use cases in the area of creative writing for cutting-edge AI technology. This has a potential scope to be trained with different styles of writing such as short stories and novels on more detail.

2 Background and Related Work

Hugo Goncalo Oliveira [4] (2009): the author suggested automatic generation of poetry, and focuses on the generation of last referred text genre involving several-level languages like phonetics, lexical choice, syntax, and semantics and usually demands a considerable amount of input knowledge. It includes techniques for the generation of poetry such as POEVOLVE, WASP, ASPERA, and COLIBRI.

Hisar Manurung, Graeme Ritchie, and Henry Thompson [5]: the authors used a stochastic hill-climbing model that proposes to model poetry generation as an explicit search, where a state in the search space is a possible text with all its underlying representation, and a 'move' in the space can occur at any level of representation, from semantics down to phonetics. The stochastic hill-climbing search model is an evolutionary algorithm, which is an iteration of two phases such as evaluation and evolution, applied to an ordered set (the population) of candidate solutions (the individuals).

Ruli Manurung, Graeme Ritchie, and Henry Thompson: the authors used Genetic Algorithms (GA) to create meaningful poetic text [6]. The later described McGonagall, an implemented system that adopts the model and uses genetic algorithms (Mitchell 1996) to produce syntactically well-formed texts, fulfill some pre-specified meter patterns (Attridge 1995), and express some basic meaning generally.

Zhe Wang, et al.: the authors recommended the Chinese poetry generation [7] with a planning-based Neural Network and used a recurrent neural network-based encoder-decoder model (RNN enc-dec) that generates poem line by line, and each line is generated according to the corresponding sub-topic and the preceding generated lines.

Jukka M. Toivanen, et al. [8] used a corpus-based approach to generate poetry. It used text mining and morphological analysis to generate poetry in Finnish.

Marek Korzeniowski and Jacek Mazurkiewicz used a data-driven system to generate Polish poetry [9] where the semantic and grammatical structures are derived from the input, followed by a poetic Turing test.

Some other work by Dr. I. Jeena Jacob on text classification [21] in which CapsNet Architecture is used to demonstrate that it is better than a Convolution Neural Network in the field of clustering and text classification was also studied for our research. Some other applications and usage of Natural Language processing were discussed by Ayushi Mitra [22] in the research of sentiment analysis based on text reviews where machine learning algorithms are used extensively to analyze and investigate this model.

3 Recurrent Neural Network

RNN works in a very peculiar way, one which can intrigue the most inquisitive of minds. It is generally used in use cases where to predict or generate the next word depending on the previous word or text. RNN is a layer of neural network cells where the input to each cell is of two kinds. The first is the previous token which could be anything ranging from a single character to a complete word. The second input given to the cell is generally the state which describes or contains information (encoded) about the previous text/words (tokens). Using this combination of input, the RNN layer predicts the next word based on the text data that was used in the training of the model.

This generated word will now be fed to the model again, and additional words will be generated in the same way. The hidden state will get updated at each step in the process. RNN uses the probability distribution model to select the words. These probabilities are updated, as explained, by the hidden state and word embedding. Every word in the language model is described by a vector (Word embedding). RNN is a cutting-edge technique to learn from the data and train the model to generate text retaining the context, as the input always contains the previous state.

4 Dataset

The dataset forms one of the most crucial parts of a machine learning or neural network model. The selection of corpus used to train a model is critical to the success of a poetry generation system [11]. A dataset needs to be wide enough without effective transfer learning to show not only the qualities that make the texts poetic in the corpus but also all the information about the world that is important in writing a semantically meaningful poem.

For an effective poem suggestion system, it is required to train the model on a wide variety of poets with different poetry styles throughout history. By merging two time periods such as Renaissance and Modern, the dataset is constructed. It was also necessary to maintain a common theme for initial training to investigate how well the model works in the direction given (topic). Around half of our dataset consists of ancient English poetry.

Table 1 gives an overview of the data used in this research: for instance, a sample poem by two pioneers Thomas Bastard and William Shakespeare from the Renaissance era. Apart from the poem, the Author, Poem name, Age, Type, and Length were also recorded so that the data is perfect in completeness. The training data consisted of about 500 Poem/Sonnets with Love, Nature, and Mythology being the predominant type averaging out at a length of about 300 words each.

Table 1 Overview of the Database

Author	Poem	Poem name	Age	Type	Length
Thomas Bastard	Fishing, if I a fisher may protest, of pleasures is the sweetest, of sports the best, of exercises the most excellent.	DePiscatione.	Renaissance	Nature	244
William Shakespreare	Where the bee sucks, there suck I: In a cowslips bell I lie; There I couch when owls do cry. On the bats back I do fly	Where the bee sucks, there suck I	Renaissance	Nature	226

5 Experiment

5.1 *Text Processing and Sequence Tokenization*

The text was tokenized to convert text/words into quantifiable sequences [13], indicating that the whole text was transformed into a series of integers. The objective was not only to produce sentences but also punctuations, so no punctuations were filtered and tokenized from the document.

The tokenized text was then converted into sequences of different lengths that are n-gram [16] sequences which were then padded such that each sequence length becomes equal as the model could not be trained with variable input sequences. N-gram sequences help train the model on the maximum possible number of sequences in a single sonnet as it creates contiguous sequences from the integral text array.

5.2 *LSTM Model*

LSTM is a variant of the artificial recurrent neural network used in the field of deep learning. LSTM's distinctive feature is that it can process not only single data points but also sequential data points, such as time series datasets, speech patterns, messages, and other sequential signals. People are also associated with its capacity to overfit [12]. LSTM is a variation of the Recurrent Neural Network as well. Some of the other features and use cases are identification of handwriting, voice, and text of an LSTM Model/Layer.

Table 2 Bi-Directional LSTM Model used with each of its layers including Parameters

Layer	Output shape	Param #
Embedding_9	(None, 345, 10)	121300
Bidirectional_12	(None, 345, 512)	546816
Droupout_12	(None, 345, 512)	0
Bidirectional_13	(None, 512)	1574912
Dropout_13	(None, 512)	0
Dense_9	(None, 12130)	6222690

5.3 *Embedding Layer*

The trained LSTM Model had an embedding layer that was fed with a dictionary/vocabulary size of 12131 words and converted each integer to a dense vector of fixed size that is 10. The input size of the model is fixed even with variable sequence lengths. The semantic similarity between each word in the vocabulary was defined by this embedding layer and the vector values were adapted accordingly, which passed into the LSTM layer as output.

5.4 *Bi-Directional LSTM*

Bi-directional LSTM is two independent RNNs placed together that allow the network to have both backward and forward sequence information at every step. This allows the network to determine the context of the sentence better than vanilla LSTM models.

Two bi-directional LSTM layers were added in the model with 256 units in each of them followed by a dropout layer [18] with 0.2 parts of the output being dropped out. The dropout layer acted as a regularization function for the output of the respective LSTM layer thus preventing the drawback of LSTMs, i.e., overfitting.

A final dense layer was added to the model, to combine the output of the LSTM layers using a softmax activation function [17] as it is used in the final output layers of models that work on multi-class classification; the softmax function in the dense layer returns the probability distribution over the target classes/words in our vocabulary. The model was compiled and the entire model summed to have 8465716 trainable parameters (Table 2).

Total Params: 8465718

Trainable Params: 8465718

Non-Trainable Params: 0

6 Result and Discussion

On 100 epochs, the model was trained and a final loss of 0.66 was addressed, which was considered acceptable as the model began to overfit if the loss was further reduced. The model created dense vectors that were then processed using the tokenizer to convert them back into words. Now for the text to be generated, the seed text had to be given a seed, translated into a sequence, padded, and then fed into the model to generate more text. One word/token was created at a time by the model, so that the generated token was applied to the seed to generate more text.

The first variant of the formation is based on the seed word provided by the user [19]. The user can also manipulate the length of the generated poem. An example output is given below.

Generative -

Seed - **"lovers are"**
Output Length - 40

Output Generated –

"Lovers Are Like A Painted Fair, Their Mouths Filled With Hopeful Expectation.

Make Lips Of Any Flint, Thou A Mine; For Fair Meritorious To Chase Design With Injustice Revengeful Arms:

By Their Knights, Should Right Poor Oaths, Ladies' To The Ground;"

The second Use case, as an AI assistant to writers where the user inputs the first few words and the trained model gives the suggestion to complete the line. After testing, One

Suggestive Seed - **"lovers are"**

Suggestions Generated –

(1) lovers are like a painted fair,
(2) lovers are like a tangled net,
(3) lovers are like a sad champion is not the gaining.

In the first output, the model can add punctuations such as ',' , '.' and ';' moderately effectively. It can break lines by itself and retain a length similar to that expected from a poem. The model produces words such as 'Thou' from the old English quite successfully. The second part of the system gives fairly comprehensible and meaningful suggestions like 'Lovers are like a tangled net', which can then be used to create a poem by the user selecting one of the following suggestions/directions (Fig. 1).

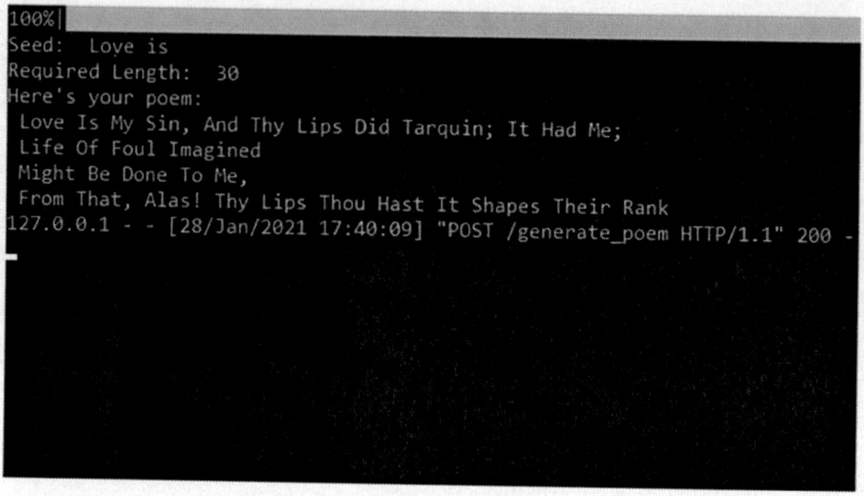

Fig. 1 Software Output of our Model with suggested word 'Love is' and Length 30

7 Conclusion

It is a popular notion that neural networks operate as a black box, and there is no simple way to understand how the forecast performance happened toward. Experiments with the number of layers and units in each cell and performance analysis will enable to better understand how the neural network functions and predict the outcome accordingly.

The Neural Network is useful in the field of creative writing through research and can be used to assist authors in writing poetry. In order to produce literary items such as short stories and novels, this can be extended further. There may be a risk of not getting enough data to train on even deeper and smaller sects of poetry than there may be (including other language poems). There is a need in such a case to find even more advanced technique algorithms that take less data to train and provide better results.

The innovative text generation in other languages such as French, Spanish, Russian, Japanese, and more is another domain that expands after this review. Some examples of such models were used for the Chinese language during our study. In this report, the primary emphasis and incorporation of the English language is to expand and strengthen. The dataset will have some drawbacks to such training, as generative models need a huge body of data to deliver satisfactory results. In order to improve the meaning of the created poems, further research is also needed. The randomness of the terms and phrases increases without the context, and the meaning is lost. The next advancement in the field of natural language processing and Artificial Intelligence will be the creation of such a model that retains the meaning, has coherence, and preserves Rhyme with great efficiency in several languages.

This study serves as a step in that direction and forms the basis for future work in the field of innovative text generation and extension of the use case.

References

1. Alpaydin, E.: Introduction to Machine Learning. MIT Press (2020)
2. Huang, Z., Xu, W., Yu, K.: Bidirectional LSTM- CRF models for sequence tagging. arXiv preprint arXiv:1508.01991 (2015)
3. Mikolov, T., et al.: Extensions of recurrent neural network language model. In: 2011 IEEE International Conference on Acoustics, Speech and Signal Processing (ICASSP). IEEE (2011)
4. Oliveira, H.G.: PoeTryMe: a versatile platform for poetry generation. Computational Creativity, Concept Invention, and General Intelligence, vol. 1 (2012), p. 21
5. Manurung, H.: An evolutionary algorithm approach to poetry generation (2004)
6. Manurung, R., Ritchie, G., Thompson, H.: Using genetic algorithms to create meaningful poetic text. J. Exp. Theor. Artif. Intell. **24**(1), 43–64 (2012)
7. Wang, Z., et al.: Chinese poetry generation with planning based neural network. arXiv preprint arXiv:1610.09889 (2016)
8. Toivanen, J., Toivonen, H., Valitutti, A., Gross, O.: Corpus-based generation of content and form in poetry. In: Maher, M.L., Hammond, K., Pease, A., Pérez y Pérez, R., Ventura, D., Wiggins, G. (eds.) Proceedings of the Third International Conference on Computational Creativity. University College Dublin, Dublin, pp. 175–179, International Conference on Computational Creativity (ICCC) (2012)
9. Korzeniowski, Marek, and Jacek Mazurkiewicz. "Data- Driven Polish Poetry Generator." *International Conference on* Artificial Intelligence and Soft Computing. Springer, Cham, 2017
10. Mikolov, T., Zweig, G.: Context dependent recurrent neural network language model. In: 2012 IEEE Spoken Language Technology Workshop (SLT). IEEE (2012)
11. Roh, Y., Heo, G., Euijong Whang, S.: A survey on data collection for machine learning: a big data-ai integration perspective. IEEE Trans. Knowl. Data Eng. (2019)
12. Lipton, Z.C., et al.: Learning to diagnose with LSTM recurrent neural networks. arXiv preprint arXiv:1511.03677 (2015)
13. Pentheroudakis, J.E., Bradlee, D.G., Knoll, S.K.: Tokenizer for a natural language processing system. U.S. Patent No. 7,092,871. 15 Aug 2006
14. Sherstinsky, A.: Fundamentals of Recurrent Neural Network (RNN) and Long Short-Term Memory (LSTM) Network. arXiv:1808.03314
15. Mandelbaum Adi Shalev, A.: Word Embeddings and Their Use In Sentence Classification Tasks. arXiv:1610.08229v1 [cs.LG] 26 Oct 2016
16. Van Gompel, M., Van den Bosch, A.: Efficient n-gram, Skipgram and Flexgram modelling with Colibri Core. J. Open Res. Softw. **4**. https://doi.org/10.5334/jors.105 (2016)
17. Enyinna Nwankpa, C., Ijomah, W., Gachagan, A., Marshall, S.: Activation Functions: Comparison of Trends in Practice and Research for Deep Learning. arXiv:1811.03378v1 [cs.LG] 8 Nov 2018
18. Pham∗†, V., Bluche´ ∗‡, T., Kermorvant∗, C., ome Louradour ˆ ∗ ∗ A2iA, J.: 39 rue de la Bienfaisance, 75008 - Paris - France † SUTD, 20 Dover Drive, Singapore ‡LIMSI CNRS, Spoken Language Processing Group, Orsay, France. Dropout improves Recurrent Neural Networks for Handwriting Recognition. arXiv:1312.4569v2 [cs.CV] 10 Mar 2014
19. Shwartz-Ziv, R., Tishby, N.: Opening the black box of deep neural networks via information. arXiv preprint arXiv:1703.0081 (2017)
20. Jain, P., et al.: Story generation from sequence of independent short descriptions. arXiv preprint arXiv:1707.05501 (2017). Wang, Z., et al.: Chinese poetry generation with planning based neural network. arXiv preprint arXiv:1610.09889 (2016)

21. Jacob, I.J.: Performance evaluation of caps-net based multitask learning architecture for text classification. J. Artif. Intell. **2**(1) (2020)
22. Mitra, A.: Sentiment analysis using machine learning approaches (Lexicon based on movie review dataset). J. Ubiquitous Comput. Commun. Technol. (UCCT) **2**(03), 145–152 (2020)

Design of Wearable Goniometer

A. Siva Sakthi, B. Mithra, M. Rakshana, S. Niveda, and K. Gayathri

Abstract Arthritis is one of the common rheumatoid or musculoskeletal disorders, and it is a significant contributor to the world's disability burden. It mainly affects knee joints, and the prevalence of disease increases with age. In several cases, people with arthritis suffer from the loss of mobility and hardening of joints. Common factors that result in decreased mobility include aging, obesity, poor physical activity, and an improper diet. To improve limited motion, a smart wearable monitoring device can be used to assist a person to maintain regular physical activity and to boost their mobility status. Our proposed idea is to develop a wearable goniometer with a mobile application to determine the active and passive range of motion, such as flexion and extension in the knee for senior citizens. This device can help the physiotherapists to virtually guide and supervise the patients in their rehabilitation phase.

Keywords Rheumatoid arthritis · Wearable goniometer · Senior citizens · Range of motion · Virtually supervise · Rehabilitation

---

A. Siva Sakthi (✉) · B. Mithra · M. Rakshana
Department of Biomedical Engineering, Sri Ramakrishna Engineering College, Coimbatore, India
e-mail: sivasakthi.a@srec.ac.in

B. Mithra
e-mail: mithra.1707036@srec.ac.in

M. Rakshana
e-mail: rakshana.1707048@srec.ac.in

S. Niveda · K. Gayathri
Department of Electronics and Communication Engineering, Sri Ramakrishna Engineering College, Coimbatore, India
e-mail: niveda.s@srec.ac.in

K. Gayathri
e-mail: gayathri.1707022@srec.ac.in

J. S. Raj et al. (eds.), *Intelligent Sustainable Systems*, Lecture Notes in Networks and Systems 213, https://doi.org/10.1007/978-981-16-2422-3_46

1 Introduction

Rheumatoid Arthritis is the musculoskeletal or an autoimmune disease [1, 2]. It is a progressive long-term disease characterized by inflammatory changes occuring in the connective tissues of the human body. It mainly affects the supporting structures of the body such as joints, tendons, ligaments, bones, muscles, and spine. In the initial stages, the disease causes pain and stiffness in the small joints. After several weeks, swelling and heat occur on the infected joints, and the pain gradually increases on the muscles. Moreover, it can affect the organs. Patients with active rheumatoid arthritis also experience a systemic inflammation associated with a variety of comorbidity, most importantly cardiovascular disease, which contributes to a major increase in mortality [1]. For rheumatoid arthritis, approximately 75% of patients are women [3] and it usually starts at the age between 30 and 50.

There are four stages of rheumatoid arthritis, as shown in Figs. 1 and 2. In the first stage, inflammation occurs in the joints [2, 3]. In addition, the tissue swells up and there is no damage to the bones. In the second stage, the inflammation in the synovial fluid leads to the damage of cartilage. Patients in this phase may experience pain and loss of mobility. In the third stage, the damage extends from cartilage to bones, and there is pain, swelling, muscle weakness, and more mobility loss. In the fourth stage, the bones fuse together [3].

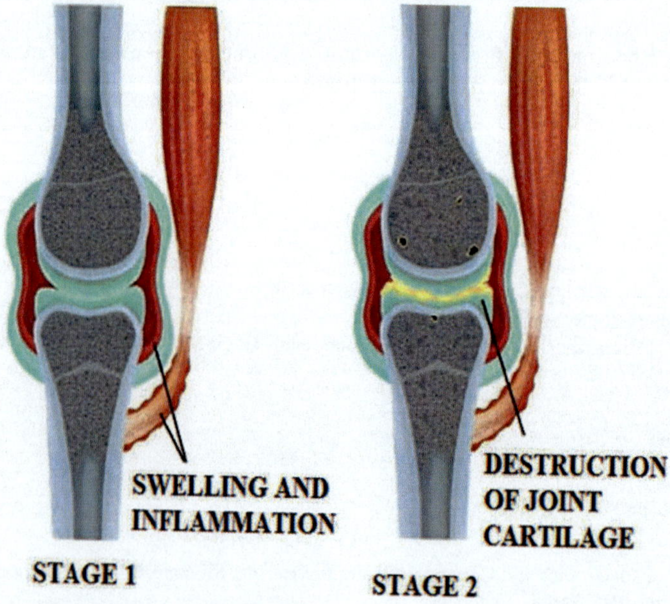

Fig. 1 Stages 1 and 2 of Rheumatoid Arthritis

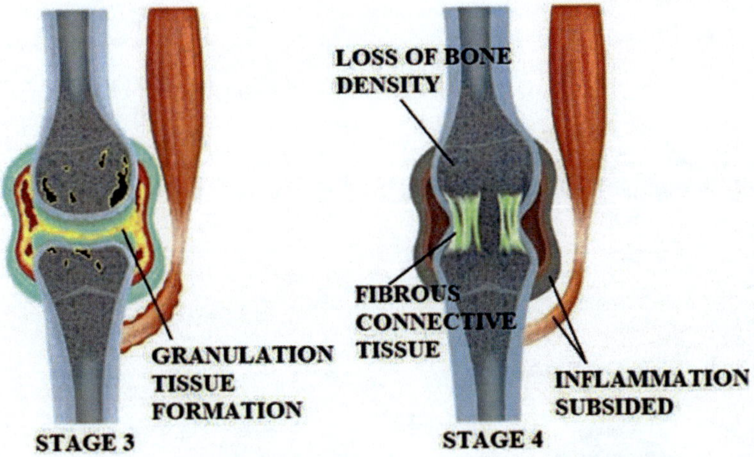

Fig. 2 Stages 3 and 4 of Rheumatoid Arthritis

To prevent joint damage, early diagnosis and treatments are recommended [2]. To treat the disease, medications such as nonsteroidal anti-inflammatory drugs and corticosteroids [3] are prescribed to the patients. Physical exercises such as mobilization and strengthening exercises are the other treatment options.

Physical exercises are recommended to the patients based on the assessment of joint range of motion. Existing methods to measure joint range of motion are manual goniometer, vision-based motion capture system, and a fiber optic goniometer. İn physiotherapy, a manual goniometer, which is manufactured by Kristeel, is used to measure a range of motion of joints. Stationary arm and movable arms are the two arms present in the manual goniometer. Movable arm is positioned with the moving limb and the stationary arm acts as a reference. In vision-based motion capture systems, multiple cameras are used to capture the human postures [4, 9], and it provides accurate measurements. But the major problem in this system includes placement of markers to the human body, increase in time consumption, high cost, and bulky and high space requirements [12]. İn fiber optic goniometer, an optical fibre-based sensor for angular measurements was developed [5, 9], but it has drawbacks such as inaccuracy and potential errors.

In the project, we designed a cost-effective wearable goniometer. It helps in monitoring a patient's mobility status in the field of rehabilitation [6, 11] and also used by physiotherapists to better assess and evaluate a patient's current condition. It is a low-cost, affordable, and effective system for personal use of the patients and therapists. İn the device, an accelerometer sensor is used to measure the range of motion in the knee joint. Then, for wireless communication, Bluetooth, is used to send the data from the device to the mobile app.

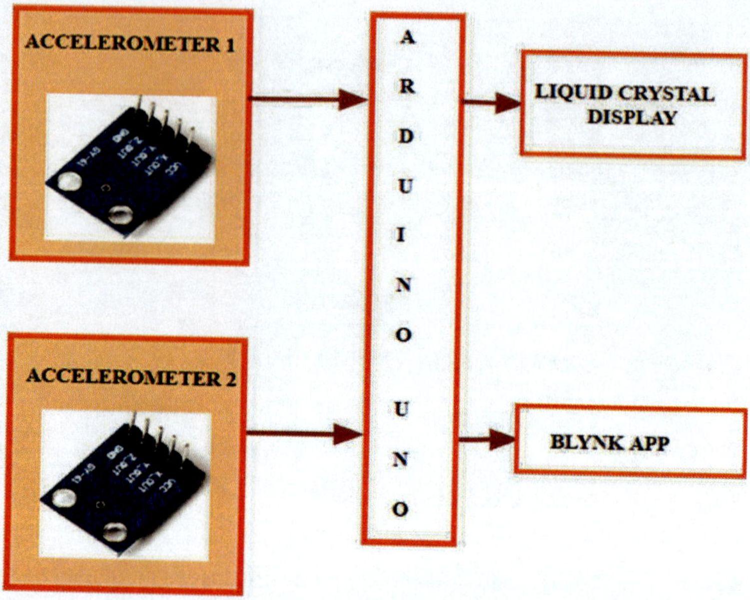

Fig. 3 Block Diagram of the Accelerometer sensor-based Wearable Goniometer

2 Materials and Methods

2.1 Hardware Platform

The wearable goniometer consists of two tri-axial accelerometer sensors, ADXL335, as shown in Fig. 3. It is a small and low-powered device [14] with signal conditioning output voltage [7]. It measures acceleration in the range of $\pm$ 3 g [10] and works on the principle of piezoelectric effects. The dimensions of ADXL335 used in the project were 4 mm $\times$ 4 mm $\times$ 1.45 mm, and the operating voltage is 0.8 V–3.6 V.

The sensor units were connected to the Arduino UNO Microcontroller, which is based on the ATMega328 microchip. The analog signals from the accelerometer sensors were transmitted to the Arduino board. Then, the controller processes the output voltages and digitizes the x, y, and z signals [15] from the accelerometer.

2.2 Software Platform

The software application used in the project to acquire and process the data is Arduino. It is an Integrated Development Environment tool. In this software, the program to control and configure the sensor was written.

To display the joint range of motion and to send notifications to the patients, the Blynk mobile application was used. It remotely controls and monitors the hardware. It is operated on both IOS and Android operating system devices. The data can be sent to the mobile application through Bluetooth communication.

2.3 Sensor Placement

The knee or tibiofemoral joint is a hinge joint, which exhibits mainly six degrees of freedom. It includes flexion and extension, varus and valgus angulation, internal and external rotation, medial and lateral shift, compression and distraction, and anterior and posterior glide.

The knee joint is present between the thigh and shank. One of the accelerometer sensors is placed on the thigh, [8, 9] which acts as the reference sensor, as shown in Fig. 4. Another sensor module is placed on the shank [8, 9], which is a movable part. During flexion and extension, the accelerometer senses the angle of movement [13]. The designed prototype of the device is shown in Fig. 5.

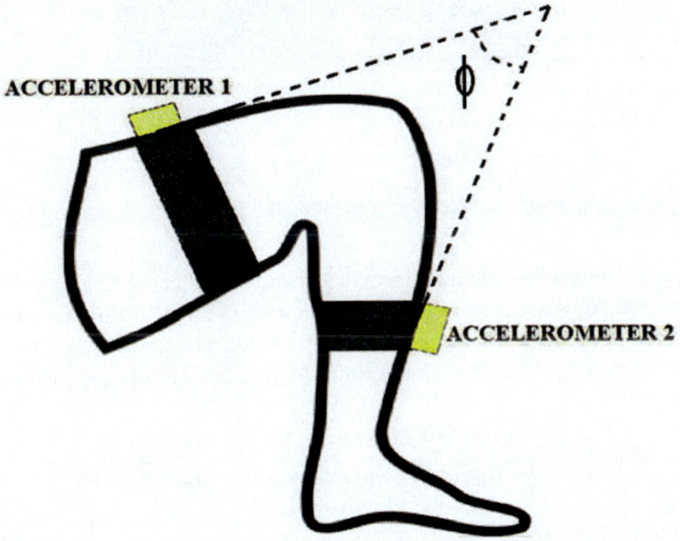

Fig. 4 Accelerometer sensor position on knee joint for angle measurement

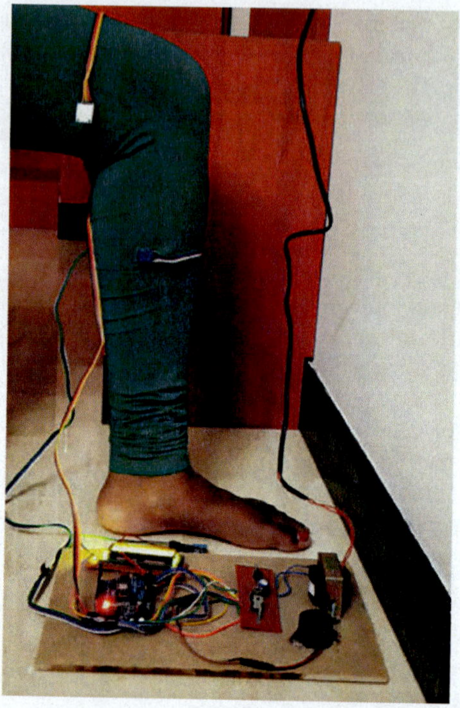

Fig. 5 Device Prototype

2.4 Joint Range of Motion Estimation

In the project, the accelerometer sensor was used to collect the x, y, and z signals. Then, the three signals such as x, y, and z were converted to Euler angles, mainly pitch and roll. Pitch is the rotation on the side-to-side axis and roll is the rotation on the front to back axis. To calculate the Euler angle, the formula is given below.

$$Ax = arctan\,(aX/sqrt((aY * aY) + (aZ * aZ)))$$
$$Ay = arctan\,(aY/sqrt((aX * aX) + (aZ * aZ)))$$

Ax = angle on x axis
Ay = angle on y axis
aX = acceleration on x axis
aY = acceleration on y axis
aZ = acceleration on z axis

For fusion of two accelerometer sensors, the Kalman filtering technique [16] was used. The filter can be used to determine the angular estimation in the presence of noise by combining results from the accelerometer sensors.

Table 1 Range of Motion in Knee

Knee activity	Normal range of motion
Full extension	0°
Safely climb stairs	83°
Safely descend stairs	90°
Get up from a chair	105°
Ride a bike	115°
Full flexion	135°

Then, the joint range of motion is estimated by the difference between an angle measured by the reference sensor and an angle measured by the movable sensor (i.e., on the shank). To calculate the joint angle, the formula is given below.

$$\phi(knee\ joint) = \phi(thigh) - \phi(shank)$$

3 Results and Discussion

In the project, the device measures the acceleration of the human joint. Then, using the conversion formula, the acceleration data is converted into joint angles. The normal joint range of motion in the knee is shown in Table 1.

In the accelerometer-based wearable device, the measured joint range of motion such as flexion and extension is displayed on the liquid crystal display, as shown in Figs. 6 and 7. When the range of motion is abnormal (i.e., flexion > 135° and extension > 0°), then an alert notification will be sent to the patients to monitor their mobility status and perform the exercises which are recommended by the therapists to improve their performance.

4 Conclusion

An accelerometer sensor-based wearable goniometer to measure the joint range of motion in the knee was designed and developed. This proposed device has a promising performance in monitoring a patient's movement status. The mobility parameters such as acceleration and angular velocity, and angle were collected by the sensors when the patient moves the leg upwards and downwards while sitting on the chair.

The wearable goniometer determines the range of motion, such as flexion and extension of the tibiofemoral joint. Based on the joint angle, the patients can improve their quality of walking pattern by performing the exercise such as mobilization and strengthening exercises, which are recommended by the therapists.

Fig. 6 Visualization of knee flexion angle in real time

Fig. 7 Visualization of knee extension in real time

Acknowledgements This work was performed in the Department of Biomedical Engineering at Sri Ramakrishna Engineering College, Coimbatore, India. The authors would like to express unfathomable thanks to the principal, Dr. N. R. Alamelu for the encouragement throughout the completion of the project. We would also like to express our deepest gratitude to the project coordinator Ms. V. Sri Vidhya Sakthi and project guide Ms. A. Siva Sakthi for their inspiring guidance. In addition, we would like to thank and acknowledge the contributions rendered by Dr. Seetharam and Dr. B. Bala Subramanian.

References

1. Verhoeven, F., Tordi, N., Prati, C., Demougeot, C., Mougin, F., Wendling, D.: Physical activity in patients with rheumatoid arthritis. Joint Bone Spine **83**(3) [Pubmed] (2016)
2. Heidari, B.: Rheumatoid Arthritis: Early diagnosis and treatment outcomes. Caspian J. Intern. Med. **2**(1), 161–70 [Pubmed] (2011)
3. AI- Rubaye, A.F., Kadhim, M.J., Hameed, I.H.: Rheumatoid arthritis: history, stages, epidemiology, pathogenesis, diagnosis and treatment. Int. J. Toxicol. Pharmacol. Res. **9**(2) [ncbi] [Google Scholar] (2017)
4. Yoshimoto, H., Date, N., Yonemoto, S.: Vision-based real-time motion capture systems using multiple cameras. In: IEEE International Conference on Multisensor Fusion and Integration for Intelligent Systems, pp. 247–251 [Google Scholar] (2003)
5. Donno, M., Palange, E., Di Nicola, F., Bucci, G., Ciancetta, F.: A new flexible optical fiber goniometer for dynamic angular measurements: application to human joint movement monitoring. IEEE Trans. Instrum. Meas. **57**(8) (2008)
6. Ruiz Olaya, A.F., Callejas Cuervo, M., Lara Herrera, C.N.: Wearable low-cost inertial sensor based electrogoniometer for measuring joint range of motion. Dyna 84(201), 180–185 [Google Scholar] (2017)
7. Nwaizu, H., Saatchi, R., Burke, D.: Accelerometer based human joints range of movement measurement. In: 10th International Symposium on Communication Systems, Networks and Digital Signal Processing (CSNDSP). IEEE (2016)
8. Djuric Jovicic, M.D., Jovicic, N.S., Popovic, D.B.: Kinematics of gait: new method for angle estimation based on accelerometers. Sensors **11**(11), 10571–10585 [ncbi] (2011)
9. Ilius Faisal, A., Majumder, S., Mondal, T., Cowan, D., Naseh, S., Jamel Deen, M.: Monitoring methods of human body joints: state-of-the-art and research challenges. Sensors 2019 [Pubmed] (2019)
10. A. Device: Small, low power, 3- axis $\pm$ 3 g accelerometer. ADXL335 Datasheet (2009)
11. Bonato, P.: Advances in wearable technology and applications in physical medicine and rehabilitation. J. NeuroEng. Rehab. [Google Scholar] (2005)
12. Zhou, H., Hu, H.: Human motion tracking for rehabilitation—A survey. Biomed. Signal Process. Control. **3**, 1–8 [Google Scholar] (2008)
13. Dong, W., Ming Chen, I., Lim, K.Y., Goh, Y.K.: Measuring uniaxial joint angles with a minimal accelerometer configuration. In: Proceedings in 1st International Convention on Rehabilitation Engineering & Assistive Technology: In Conjunction with 1st Tan Tock Seng Hospital Neurorehabilitation Meeting, pp. 88–91 (2007)
14. Atallah, L., Lo, B., King, R., Yang, G.Z.: Sensor placement for activity detection using wearable accelerometers. In: 2010 International Conference on Body Sensor Networks. İEEE (2010)
15. Liu, K., Liu, T., Shibata, K., Inoue, Y., Zheng, R.: Novel approach to ambulatory assessment of human segmental orientation on a wearable sensor system. J. Biomech. **42**, 2747–2752 [Pubmed] (2009)
16. Tognetti, A., Lorussi, F., Carbonaro, N., Danilo de Rossi, D.: Wearable goniometer and accelerometer sensory fusion for knee joint angle measurement in daily life. Sensors, 28435–28455 [Pubmed] (2015)

ERP Module Functionalities for the Food Supply Chain Industries

Kumar Rahul, Rohitash Kumar Banyal, and Hansika Sati

Abstract The Enterprise Resource Planning (ERP) system comprises of bundles of software that are used to coordinate all the operational, business, and other functions of various departments in the food industry and also in maintaining a smooth food supply chain [3]. It can be used in industries for their product planning, manufacturing process or service delivery, and sales. A Food-Specific ERP is designed such that it streamlines the processes which are unique to the food industries and manufacturers and to bridge gaps in certain areas which other methods cannot. The food industries have responsibilities unique to product quality, product safety, product shelf life, proper product packaging, etc. Thus, the ERP system plays a major role in optimizing the food industry's operations, increasing the flexibility, saving time, and deleting errors in the entire process. This study aims at the implementation of sales module, distribution module, and finance module in the food industry.

Keywords Enterprise resource planning · Food industry · Customer relationship management · Manufacturing

1 Introduction

ERP software, first introduced in the 1960s, started being largely implemented in the 1990s at Fortune companies. Increase of e-commerce and improvement in Food supply chain management contributed in providing shape to the ERP software industry. This new tool started providing connectivity, transparent communication, and visibility to the food industries allowing the users to keep track of the product,

K. Rahul (✉)
Department of Basic and Applied Sciences, NIFTEM, Sonipat 131028, India

R. K. Banyal
Department of Computer Science and Engineering, Rajasthan Technical University, Kota 324010, India

H. Sati
Department of Agriculture and Environmental Sciences, NIFTEM, Sonipat 131028, India

J. S. Raj et al. (eds.), *Intelligent Sustainable Systems*, Lecture Notes in Networks and Systems 213, https://doi.org/10.1007/978-981-16-2422-3_47

thus helping in a better food supply chain management. It aids in better decision-making process, by automating many tasks, thus reducing human errors and freeing the food industries to pay attention to other strategic works. Seeing this, smaller food businesses also started adopting ERP software. Now, Enterprise Resource Planning (ERP) has become a powerful tool in businesses over the past few years. All the employees can access the important details and information; thus, it serves as a shared database for operational, financial, and all other types of information that are essential for the crossflow of information essential for proper functioning of the food industries [5]. The Enterprise Resource Planning system can perform major and minor functions according to the needs of that particular food industry. Since food industries focus on high production in minimum time, the ERP system helps in saving their time that too at a low implementation cost. Many ERP modules have been designed in order to facilitate certain specific business functions, which are being followed by the food industries to perform their tasks efficiently.

In other words, we can say that if ERP is the toolbox, then its modules are the hammer, screwdriver, wrench, and other tools in the box [16]. The best part is, the organization/food industry enjoys the freedom to purchase only those ERP modules relevant to its needs, operations, and business. The beauty of the ERP software modules lies in the fact that the food industry using it can add this functionality, without changing the foundation [2]. A lengthy process of implementing a new ERP system is not needed by the food industry. For any food industry to find the space for their products in the market by overcoming the growing market competition, industries are not only focusing on the quality alone but also giving more importance to the supply chain of the product after its manufacture. Increasing the efficiency of sales and distribution process is one of the best ways to maintain their competitive edge. Utilizing the latest technologies and best human resource can help the sales module in an organization to work effectively.

1.1 Sales Module

The sales module in food industries plays an important role in implementing orders pertaining to placement, scheduling, and invoicing of the food product. Sales and distribution management is very important for any food industry in order to make profit [15]. If properly implemented, through the sales module, the food industry can optimize the efficiency of its sales and prices throughout the food supply chain.

Sales modules can be later referred by the CRM module for creating opportunities in future and lead generations. This module is considered to be an integrated execution module as it derives most of its input from other modules as well.

1.2 Order Placement

With the help of order placement, the food industry is able to manage processes related to sales processes easily, quickly, and effectively. This is involved in maintaining the price of the product, comprising the details of the discounts to be given on the particular product and its validity, and checking the availability of stock.

1.3 Order Scheduling

It is one of the important functions of sales management. As soon as the customer places his order, the resource managers in the organization begin to procure raw materials from the inventory stock. The order then moves through different stages where the product is checked for its quality shelf life, damage, etc. After this, the product consequently reaches the packaging department followed by the shipping department of the food industry where it is processed for packing with a customer label and bar code on top of it.

1.4 Shipping and Invoice

Once the product in the food industry is to be shipped, a customer invoice is generated. Invoice is the last stage in the cycle of order process sales order. Invoice screen can be accessed from sales order view. This invoice screen will display the information related to the sales order. Billing functions such as issuing of invoices, issuing credit notes followed by corresponding entries in accounts receivable, and control account of general ledger are facilitated in this process.

1.5 Sales Module Functionalities in Food Industry

It includes handling Pre-Sales and sales activities of food products, conducting market surveys to determine the various products in demand by the consumer which helps the food companies to know which of their products is selling high in the market which in turn helps them to prepare an effective market strategy; a particular target is set for the executives in all the departments of the food industry so that they work seriously and efficiently for the organization. This idea even promotes the sale of products. Once the order is scheduled over a period of time, the delivery schedule is tracked from time to time till the product reaches the consumer. Production advice is generated in such a way that more the particular food product is being sold in the market, more is its production. Order processing is based on maximum retail price

(MRP), while order placement advice for the dispatch of the product is generated by the sales module.

This is an effective way in which many products can be brought in a single order itself, which means many dispatches are made against a single order. In the food industry, it helps in the tracking of sales returns from the consumer. It is flexible to define the consumer with product prices, reference number, and other details. It provides a Facility for multiple dispatch locations for customers and dealers. Different and various types of orders can be generated to comply with the varied degree of customer needs by the sales module.

2 ERP Purchasing Module

The purchase module is a very important part of ERP as it focuses on acquiring the raw materials, which is the basic and important part of any food industry. It runs the process of buying [7]. It includes all the activities related to the purchasing of materials in the food industry. Activities involved in the purchase module involves creating the Purchase Order of food materials, entry into goods and stock receipt, creating the Purchase invoice, and matters related to the final payment.

2.1 *Analysis*

The first step in the Purchasing process focuses on the importance of identifying what order actually a food industry wants to make, and whether it is a reorder or a new order for a new product [6]. The food industries ensures product manufacturing, order received from customer ends through information system, store, payment, deliverable and feedback about services offered. These task requires a complete analysis which start from the initial phase of raw material selection for manufacturing and serve to the customer.

2.2 *Clarification*

After identifying the needed raw materials for its product, the purchasing department of the food industry aims to select the suitable brand for its product. It plans to have the raw materials of the right quality, in the right quantity, at the right time, thus playing a major role in the smooth functioning and working of that organization.

2.3 Purchase Order

The purchasing order is raised by the purchasing department of the food industry with respect to the materials needed for the processing of the final products. It acts as a reference document between the food industry and the vendor. It includes all the details pertaining to the order number of the purchase, name of the vendor, date on which the raw material is ordered, due date, name of the materials required, quantity of the materials required, shipping address along with the discount offered (if any), and the grand total. The material is received only after checking the materials against these purchase orders. The material is accepted by the purchase department of the food industry only when it complies with the purchase order or else is treated as a pending order.

2.4 Approval of Purchase Order

The purchase order generated by the food industry also needs to be approved. Nowadays, most of the food industries are using a purchase order software. This automation not only helps the food industries to track their orders but also provides a transparent communication between all the elements in the food supply chain. One can also communicate easily if any revision in the order has to be made.

2.5 Review of the Suppliers

This is a very important step in the organization as the food industry should know who the supplier is. If the purchasing order system in the food industry is automated, and if it simply is a reordering process, it is not needed as the industry has dealt with the supplier earlier, but if the food industry wishes to order a new material and that too from a new vendor then review of the supplier becomes a must. Each supplier's capability, compliance, and performance should be checked by the purchase department carefully before adding him to the system.

2.6 Supplier Selection

After proper reviewing of various suppliers, the management of the food industry decides which supplier will be fulfilling the purchase order. The selection of the supplier is mostly based on the following criteria—cost, quality, delivery, safety, convenience, risk, time, certifications, trust, and service. The purchase department finalizes the supplier for the organization. He can be either new or Pre-Scrutinized in their software.

2.7 Price and Term Negotiations

In this stage, the price quoted by the supplier is negotiated by the food industry (purchaser). In the case of an ERP software, the best prices and terms are already available from the suppliers. If the food industry wants a lesser price or better quality of the product, such terms are negotiated with the vendors in case of an existing supplier. If an order is to be placed with a new supplier, then an entire new agreement is created with the supplier and the food industry for the price, duration, quality, and terms and conditions of the order. Hence in an ERP software, this step becomes extremely simple and easy; one doesn't need to sit with the supplier personally and discuss the terms and conditions of the order, which in turn saves time and efforts.

2.8 Order Placement

The food industry places an order to the supplier on the agreed restraints. This creates an agreement between the supplier and the food industry. ERP software manages the tasks gracefully if the orders of various raw materials required by the purchaser (food industry) are to be from different suppliers. Tracking of the order and invoice, cancelations, changes in the order agreement, etc. are all available on the click of a button.

2.9 Receiving and Inspection

The goods that are getting delivered to the food industry by the supplier are checked for their quality, wholeness, quantity, etc. The inspection is done either for each item or a single item from the batch that is picked. The details of the goods are now fed in the ERP software as inbound delivery. The inventory management module is linked to this process. It automatically updates the inventory after receiving and inspection of the items to the food industry. The invoice is either sent with the goods or later; if there is any issue with the purchase, then it is updated to the supplier and the invoice is generated accordingly. The documents required like purchase agreement, invoice, receipt, etc. are all stored in the ERP and can be retrieved whenever required.

2.10 *Payments*

After receiving of the goods, if the supplier sends the invoice with the order and any issues are reported by the food industry for inaccuracy, then the invoice is changed and the payment to be done for the order is managed accordingly. The payment also depends on the agreement terms made earlier. Usually, 30-60 days are provided by the supplier to complete the payment. With the use of an ERP software, this step can be a lot easier and the changes in the invoice are updated to the food industry by the supplier, and then the payment arrangements are made accordingly which saves a lot of time.

2.11 *Purchase Module Functionalities in Food Industry*

The ERP purchasing module facilitates and automates the process of buying raw materials for the food industry from the supplier. The purchasing module is associated with the inventory management module of the ERP which updates the inventory automatically after receiving and inspection of the goods from the supplier. Multiple orders from multiple suppliers can be managed in a purchase module of the ERP. Changes in the bill/invoice can be easily updated to the supplier. The ERP purchase module takes control over the entire purchasing and procurement process.

3 Finance Module

Enterprise resource planning has an important role as it integrates functions such as accounting, inventory management, human resource management, customer relationship management, reporting, production, sales, and marketing. In finance, Enterprise Resource Planning acts as a platform for the collection of important details, information and reports related to receivables, payables, general ledgers, cash flows, fixed assets, etc. The finance module supplies useful data to other modules of ERP. Its importance in the food industry comes in, as this module supports core functions of industry such as purchasing, audit trails, and compliances to regulations. This module can give a correct idea about the state the company is in, through its profit tracking, revenues generated, losses incurred, and all the expenses features which can be automated in the ERP software [1].

This module of the ERP system takes into account all the entries related to accounts and their effect on the entire system. The total expenditures of that particular food industry and other flow of cash are reflected in the finance module. This helps in telling the exact financial position of that food industry at that particular instance. The software gives a detailed analysis report of the same which helps in planning the strategy [11]. It also reduces human error and saves time in accounting, transactions, and makes payment simpler. This is also done through automating the process and having a customized dashboard. The food industry being unique as it also includes the service sector, these features are important and help in building a relationship with customers who can directly be the end consumers or vendors, suppliers, and retailers.

3.1 Profit Tracking

One of the main functions of the financial management module in food industries is profit tracking. Tracking of profits allows us to know the exact amount to afford so as to maximize the profit in real time. It also keeps record of financial health by checking the past records of food companies [13]. This model also gives the idea that from which product by the application of particular strategy we are getting the major profit. Food companies also maintain the note of amount dissolved for the product and the amounts we receive which thereby gives the idea of profit. Future assessments can also be done on the basis of past records.

3.2 Ledger Management

Ledger management means planning, organizing, directing, and controlling the files of records which gives the total transactions with the debit and credit system. This ledger management system gives monetary balance for each account including loan taken by buyer, installment details, credit details, and debit details. Also, we can keep track on different things like assets, income, liabilities, capital amounts, and expenses. By the application of ledger management [10], financial data of food companies gets compiled in one place which makes filing tax return in more managed way. Fraudulent parties can also be tracked by this management to know if the person has not given his/her loan amount by checking the file of record of a particular person. Such frauds or odd practices can notify the company.

3.3 Accounts Payable

This module helps us to perform functions related to entry, monitoring, maintenance, and processing the payment of invoices. Companies related to the food sector deal with a plethora of suppliers and vendors on a daily basis. They heavily depend on the supply chain system. Even though we say the lesser suppliers in the chain, the more efficient the system would be, it is not practical in every case. Thus, a need arises for the company to keep a system which takes care of how much a company owes to its vendors and suppliers. Any error caused in this regard may tarnish the relationship with the suppliers or incur losses to the company. Hence, the "accounts payable" function of ERP does this job with ease obviating any human errors. The functionalities of this module lie in immediate registration and tracking of incoming invoices, entry of these invoices, and matching them automatically with the receipts [12]. An imaging feature allows converting paper documents into electronic documents. Thus, an account payable coordinates the payable data with the purchasing system in order to take control of the cash flow.

3.4 Accounts Receivable

Many companies are running the "direct to customer" model now, through their e-commerce sites. Some retailers and giants like Amazon, Jiomart, etc. also deal with grocery items. The accounts receivable feature of ERP is an important feature here as it manages the funds customers owe to them. This module tracks all the invoices that are expecting payment from the customers [9]. Companies can directly collect payments through this feature from distributors, retailers, etc. This could ensure that the payments are prompt, and any checking is also available as invoices are generated instantly and can be accessed.

There is an option to customize the portal for customers to make payment and access invoices. Many functions like "account statement", "payment reminder", and "recurring generation" can be automated through ERP system. This automation process speeds the collection process boosting customer relationship with efficient payment platform and no errors. The accounts receivable module plays an important role in classifying accounts for reconciling and controlling. It issues reminder letters with different levels of severity to send to its customers. This module helps in maintaining a credit diary containing the entire list of all unpaid invoices by its customers. The accounts receivable module aids in lowering bad debt, increasing staff productivity, maintaining cash flow, improving cash-forecasting, and reducing transaction cost.

3.5 Fixed Asset Management

Food manufacturing units have all sorts of equipment ranging from handling, processing, storing, etc. which are the fixed assets of the company. These are susceptible to damages which would have to be repaired or replaced in the future. The "fixed asset management" feature of the finance module of ERP keeps track of their life and subsequently keeps a separate budget for their repair or replacement. This is done through depreciation calculations, tax implications, and compliance requirements. It also detects tax exemption methods for this, and also prevents paying taxes for eliminated assets [4]. Tracking assets gives the company a clear view of utilization, costs, and maintenance. It helps to forecast expenditure.

3.6 Risk Management

Financial risk management takes into account aspects relating to currency risks and steps to endure the effects. The risk tool in the food industry is to detect contamination or any other type of issue in the process; it may also be any risk related to credit risk. It helps to be compliant toward ever-changing regulations which is a challenging process. From a financial point of view, it monitors cash flow in and out of the business ensuring enough cash reserve. It also manages credit risk when a customer misses a payment.

3.7 Reporting

Companies associated with the food sector have their units staggered at different locations. They may be regional offices, sub units, departments like procurement, logistics offices, etc. For evaluating their performances, revenues they create, sales, why they aren't performing as expected, expenses, and other financial components related to them, a feature called "reporting" can be used. It can be centrally monitored too. This data is visible from anywhere and helps for data-driven forecasts following decisions concerning company finances [14]. The data is also customizable, that is, it can be as data or graph, whichever we desire.

3.8 Multi-currency Management

A major percent of companies in the food industry export their products to one or many countries. So, an efficient system has to be present to receive or make payments in other currencies. The multi-currency management feature makes this a hassle-free process, making it automated. Currency conversion capabilities present in this feature help to complete transactions with ease.

3.9 Financial Module Functionalities in Food Industry

As likely we are putting the financial module in the food industry, the more will be the productivity with a reduction in time-wasting tasks. It gives real-time financial monitoring as to know what is the current financial state of the food company. It reduces the human errors which are created at the time of data entry, accounting for mistakes which can be detected, and thus avoiding these becomes possible with the help of the financial module. By the application of the financial module in food industries, frauds if any can also be detected [8]. It also integrates with other business systems in the food industry like Customer Relationship Management which gives marketing budgets. By the ledger system, we can access all the financial information of a Food Company all in one place.

4 Conclusion

Though the Enterprise Resource Planning (ERP) licensing is a costly process, and the software has a high cost of implementation and maintenance which involves complexity, we cannot neglect the fact that it can unify the food industry's IT costs and saves time along with improving the efficiency. The biggest benefit of ERP is that of the total visibility, as it makes data from every department in the food industry visible and easily accessible. Makeup of its modules adds to the benefits of ERP. Each module is designed such that it can be implemented alone. Due to this, the food industry enjoys the freedom of selecting those specific modules which suit best for the working of its organization and leaves out the rest. Along with other benefits, the most important function of ERP that a food industry enjoys is the reduction in the efforts and time that the workforce has to carry in order to fulfill the daily tasks. By reducing human errors, the ERP system offers more time to the employees of the food industry to focus on other strategic work. Thus, by improving efficiency, accurate forecasting, increasing productivity, offering flexibility and mobility, and providing integrated information, ERP software has now become a necessity for the manufacturers in food industries to perform their hectic tasks in a smooth and efficient manner in minimum time with maximum productivity.

References

1. Davenport, T.H., Brooks, J.D.: Enterprise systems and the supply chain. J. Enterprise Inf. Manag. (2004)
2. Gupta, M., Kohli, A.: Enterprise resource planning systems and its implications for operations function. Technovation **26**(5–6), 687–696 (2006)
3. Jacobs, F.R., Whybark, D.C.: Why ERP?. A Primer on SAP Implementation, McGraw-Hill Higher Education (2000)
4. Mabert, V.A., Venkataramanan, M.A.: Special research focus on supply chain linkages: challenges for design and management in the 21st century. Decis. Sci. **29**(3), 537–552 (1998)
5. Madanhire, I., Mbohwa, C.: Enterprise resource planning (ERP) in improving operational efficiency: case study. Procedia CIRP **40**, 225–229 (2016)
6. Patterson, K.A., Grimm, C.M., Corsi, T.M.: Adopting new technologies for supply chain management. Transp. Res. Part E: Logist. Transp. Rev. **39**(2), 95–121 (2003)
7. Pravin, G.: Basic Modules of ERP System. University of Pune, Retrieved on 30th January (2017)
8. Sadrzadehrafiei, S., Chofreh, A.G., Hosseini, N.K., Sulaiman, R.: The benefits of enterprise resource planning (ERP) system implementation in dry food packaging industry. Procedia Technol. **11**, 220–226 (2013)

9. Shang, S., Seddon, P.B.: A comprehensive framework for classifying the benefits of ERP systems. In: AMCIS 2000 Proceedings, vol. 39 (2000)
10. Shirazi, B.: Towards a sustainable interoperability in food industry small & medium networked enterprises: Distributed service-oriented enterprise resources planning. J. Clean. Prod. **181**, 109–122 (2018)
11. Soh, C., Kien, S.S., Tay-Yap, J.: Enterprise resource planning: cultural fits and misfits: is ERP a universal solution? Commun. ACM **43**(4), 47–51 (2000)
12. Spathis, C., Constantinides, S.: The usefulness of ERP systems for effective management. Ind. Manag. Data Syst. (2003)
13. Su, Y.F., Yang, C.: Why are enterprise resource planning systems indispensable to supply chain management? Eur. J. Oper. Res. **203**(1), 81–94 (2010)
14. Terzi, S., Cavalieri, S.: Simulation in the supply chain context: a survey. Comput. Ind. **53**(1), 3–16 (2004)
15. Wieder, B., Booth, P., Matolcsy, Z. P., Ossimitz, M.L.: The impact of ERP systems on firm and business process performance. J. Enterprise Inf. Manag. (2006)
16. Wojcik, C. M., Pretto, P. A., Courier, J., Morrow, B., Wehry Jr, J. R., Kuczynski, P.,.. & Ferguson, L. D. (1997). U.S. Patent No. 5,666,493. Washington, DC: U.S. Patent and Trademark Office

Medical Ultrasound Image Compression Using Edge Detection

T. Manivannan and A. Nagarajan

Abstract In computer era, image compression is one of the thirstiest requirements. The Compression ratio of existing algorithms is less so a new compression algorithm is needed as the urgent demand. A new image compression algorithm is proposed in this research paper using a Discrete Fourier Transform (DFT), Discrete Wavelet Transform (DWT), and Edge Detection. Daubechis wavelet family is used for DWT process. For the edge detection process, the canny edge detection method is used. A Real part of Fourier transform is used for further compression. The Wavelet method compresses the data, and the morphological process thickens the edges of detail sub-images. Canny edge detection detects edges and the information is preserved. The non-edge information is erased from the detail portion. The proposed algorithm provides a better compression ratio than the existing methods. It also produces high PSNR values.

Keywords DFT · DWT · Canny edge detection · Huffman compression · FFT · PSNR

1 Introduction

Cancer is one of the key diseases faced by mankind. There are the number of different cancers in the medical field such as lung cancer, skin cancer, and breast cancer. [1]. Cancer occurs due to the abnormal growth of cells. When it occurs, the lung cells are abnormally growing. In most of cases, affected tissues spread to other parts of the body and continue to grow and form tumor in the new place [2]. Thyroid cancer is one of the widely common cancers in the world; thyroid cancer is the fastest-growing cancer, and it affects among eighth in women and twentieth in men. The advances in the field of medical imaging should need significant quality of digital images for perfect and efficient diagnosis [3].

In this paper, boundary effects are created in the block boundary and directional information is not considered in the entropy coding. In the paper [4], compression of

T. Manivannan (✉) · A. Nagarajan
Department of Computer Applications, Alagappa University, Karaikudi, India

J. S. Raj et al. (eds.), *Intelligent Sustainable Systems*, Lecture Notes in Networks and Systems 213, https://doi.org/10.1007/978-981-16-2422-3_48

complex-valued images is handled. It does the lossy compression for the processed images. It employs radically different quantization strategies. It uses a DCT-based transformation. In the paper [5], Efficient Quadtree coding of images is applied. Here, quantization processes degrade the image quality. Liu and K. N. Ngan present a method in [6]; this uses Weighted Adaptive Lifting-Based Wavelet Transform. In the paper [7], Contourlet transform is used. But it is a high complex system to implement hardware version. In this paper [8], the quality improvement is considerably less.

2 Proposed Method

This paper proposes a compression method for Thyroid images. The compression section contains the following modules:

a. Discrete Fourier Transform

The input image is read, and the Thyroid medical Ultrasound image pixel informations are loaded in the array memory [9]. The Discrete Fourier Transform transforms the image into the frequency domain. Then, inverse DFT transform transfer them into the original form. Then, real part of the IDFT is considered the image for further processes.

b. Discrete Wavelet Transform

Two-dimensional Wavelet transform is applied on the IDFT real part Image. The Daubechis Wavelet family type is used for DWT implementation. This two-dimensional DWT leads to a decomposition of approximation coefficients and detail coefficients.

c. Morphology Process

The morphology process is used to enhance the edge information of the detail information [10]. To enhance the edges in the detail portion, the dilation process is handled. The dilation process expands the edge contents. So the dilation process thickens the edge data.

d. Canny Edge Detection

The Canny edge detection method [11] is used to detect the edges of the morphology output. Identify edges by detecting local maxima of the gradient for morphology image by using the Canny method. Gaussian filter derivation is used to calculating the gradient. Tough and weak edges are deducted when only if the week edges are connected to solid edges, the output also consolidated with week edges. So, this method is more efficient in the detection of the true week edges and other than the fool by noise.

e. Edge Data Preserving & Thresholding Operation

In the edge areas, the detailed information is preserved and quantized by 4. Then, the thresholding operation makes the possible level losses that depend on a threshold. Then, Huffman compression compresses the approximation and detail information.

f. Huffman Compression

The compressed data is stored as a file. The decompression section contains the following modules:

i. InverseHuffman Compression
ii. Inverse Quantization
iii. Inverse Wavelet Transform
iv. Discrete Fourier Transform

The decompression section involves the inverse-Huffman process. Then, the inverse quantization process is done on the decompressed data [12]. Then, the two-dimensional inverse discrete wavelet transform is applied on the inverse quantized image. The approximation part, horizontal detail part, vertical detail part, and diagonal detail part are used to make the inverse DWT transform.

3 Results and Discussions

In the proposed method, medical thyroid image is given as input and the FFT, DWT and edge detection processes are done on it and the reconstructed image is displayed. For analysis purpose, the existing method is named "Wavelet image compression". The DWT-based lossy compression is considered as an existing system for analysis propose. In the existing method [13], Discrete Wavelet Transform is used. 4 sub-images are formed by the DWT process and the first one is an approximation and the other 3 are details. Thresholding operations are also included in this system. Finally, Huffman compression is used for linear compression.

Figure 1 shows the original and reconstructed of the compressed Thyroid images at the PSNR of 30 db. It looks reasonably good with more visual quality and good compression ratio. The time taken of the proposed method is compared with the existing DWT method, and the results are shown in Table 1.

From the above analysis, it can be proved that the proposed system is taking only a reasonable time for compression and decompression.

From Table 2, it is proved that the proposed system produces a higher compression ratio than the existing DWT method.

The Quality measurement analysis for MSE and PSNR of the proposed method with the existing DWT method with the compression Ratio 1.8 is shown in Table 3 (Chart 1).

The above analysis shows that the proposed method provides less MSE value and high PSNR value than the existing method.

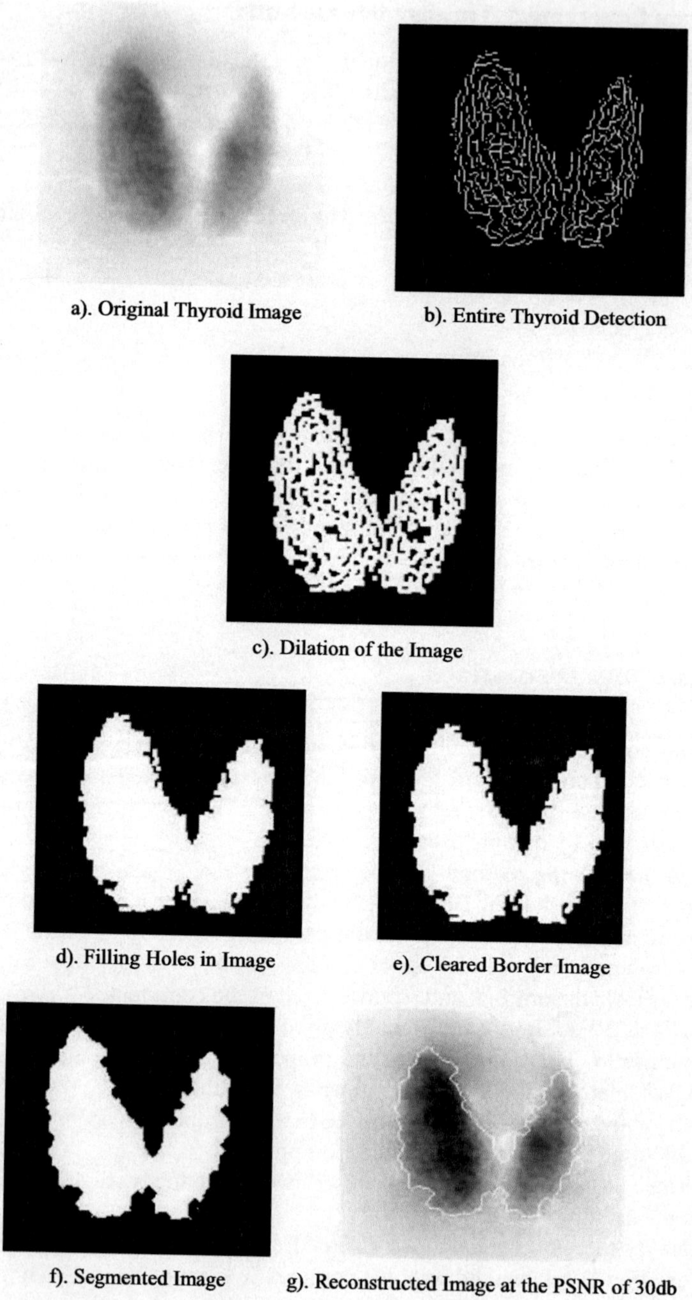

Fig. 1 **a** Original Thyroid Image. **b** Entire Thyroid Detection. **c** Dilation of the Image. **d** Filling Holes in Image. **e** Cleared Border Image

Table 1 Compression and decompression time taken analysis

Image name	Compression time taken		Decompression time taken	
	DWT [14]	Proposed Method	DWT	Proposed Method
Thyroid Image 1	0.43	3.56	0.18	3.14
Thyroid Image 2	0.46	3.46	0.18	2.48
Thyroid Image 3	0.45	3.67	0.22	2.42
Thyroid Image 4	0.48	3.61	0.15	2.65
Thyroid Image 5	0.43	3.51	0.20	2.57

Table 2 Compression ratio analysis

Image Name	Compression Ratio	
	DWT [15]	Proposed Method
Thyroid Image 1	1.78	1.89
Thyroid Image 2	1.98	2.06
Thyroid Image 3	2.09	2.27
Thyroid Image 4	2.02	2.12
Thyroid Image 5	1.99	2.06

Table 3 Quality measurement analysis for MSE and PSNR for compression Ratio 1.8

Image name	Quality measurement analysis for MSE		Quality measurement analysis for PSNR	
	DWT	Proposed method	DWT	Proposed method
Thyroid Image 1	75.23	41.91	29.34	31.93
Thyroid Image 2	49.92	36.65	31.17	32.48
Thyroid Image 3	32.94	24.93	32.95	34.14
Thyroid Image 4	47.71	32.53	31.39	33.03
Thyroid Image 5	54.53	36.70	30.72	32.41

4 Conclusion

The edge preserved thyroid medical image compression is applied in this paper. Wavelet-based method with the support of FFT and edge detection is considered. The proposed method produces a high compression ratio, higher PSNR, low MSE, and better visual quality. In future, Fuzzy concept can be adopted with this proposed system to generate more PSNR achievement.

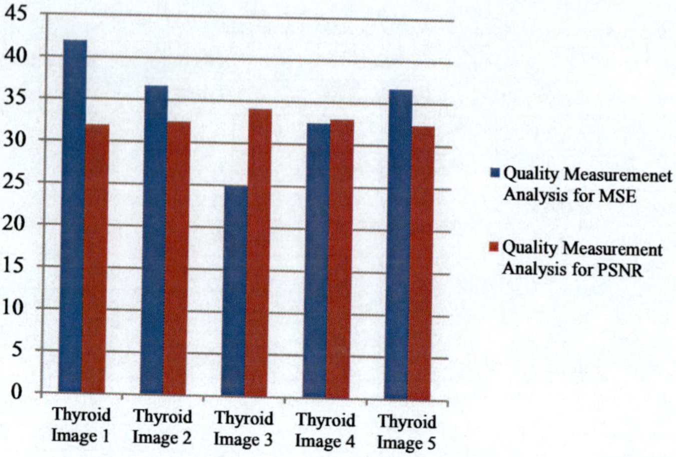

Chart 1 Quality Measurement Analysis for MSE and PSNR for Compression Ratio 1.8

References

1. Manoharan, Samuel: Improved version of graph-cut algorithm for CT ımages of lung cancer with clinical property condition. J. Artif. Intell. **2**(04), 201–206 (2020)
2. Sungheetha, Akey, Sharma, Rajesh: Real time monitoring and fire detection using internet of things and cloud based drones. J. Soft Comput. Paradigm (JSCP) **2**(03), 168–174 (2020)
3. Napoleon, D., et al.: Remote sensing image compression using 3D oriented wavelet transform. Int. J. Comput. Appl. **45**(24) (2012)
4. Ding, W., et al.: Adaptive directional lifting-based wavelet transform for image coding. IEEE Trans. Image Process. **16**(2), 416–427 (2007)
5. Eichel, P., Ives, R.W.: Compression of complex-valued images. IEEE Trans. Image Process. **8**(10), 1483–1487
6. Sullivan, G.J., Baker, R.L.: Efficient quadtree coding of images and video. IEEE Trans. Image Process. **3**(3), 327–331 (1994)
7. Do, M.N., Vetterli, M.: The contourlet transform: an efficient directional multi resolution image representation. IEEE Trans. Image Process. **14**(12), 2091–2106 (2005)
8. Liu, Y., Ngan, K.N.: Weighted adaptive lifting-based wavelet transform for image coding. IEEE Trans. Image Process. **17**(4), 500–511 (2008)
9. Georgiev et al.: Spatio-angular resolution trade-offs in integral photography. In: Proc. EGSR, 2006
10. Abdallah YMY and Abdelwahab: Application of texture analysis algorithm for data extraction in Dental X-Ray images. Int. J. Sci. Res. **3**(8), 1934–1937 (2014)
11. Kim, et al.: Application of texture analysis in the differential diagnosis of benign and malignant thyroid nodules: comparison with gray-scale ultrasound and elastography. Am. J. Roentgenol. **205**(3), 343–351 (2015)
12. Paula Branco and Ribeiro: A survey of predictive modelling under imbalanced distributions. ACM Comput. Surv. **49**(2), 31–48 (2016)
13. Raghavendra, et al.: Optimized multi-level elongated quinary patterns for the assessment of thyroid nodules in ultrasound images. Comput. Biol. Med. **95**, 55–62 (2018)

14. Sudarshan, V.K. et al.: Application of wavelet techniques for cancer diagnosis using ultrasound images: a review. Comput. Biol. Med. **69**, 97–111 (2016)
15. Xie, C., et al.: Ultrasonography of thyroid nodules: a pictorial review. Insights Imaging **7**(1). doi:https://doi.org/10.1007/s13244-015-0446-5, 77–86 (2016)

Usage of Clustering in Decision Support System

K. Khorolska, V. Lazorenko, B. Bebeshko, A. Desiatko, O. Kharchenko, and V. Yaremych

Abstract There are many ways to use cluster analysis. Mostly, it acts as a tool that allows one to look at the data as a whole. Cluster analysis can also be used for pre-processing or as an intermediate stage of other algorithms, such as classification or forecasting, or for data mining. It is necessary to mention that cluster analysis becomes vital for the development of complex AI technology, whenever one uses large amounts of data, as it allows to simplify the complexity of datasets in use. One of the areas of application of cluster analysis is the decision support system. In this paper, we present a model and algorithm based on the K-means method with dependent movements, allows one to constructively find optimal strategies and giving concrete recommendations to analytics in cybersecurity projects or other large trade projects, for example, in the field of Data Mining.

Keywords Clustering · K-means · Decision support system (DSS) · Cybersecurity · Intellectual data analysis

K. Khorolska (✉) · V. Lazorenko · B. Bebeshko · A. Desiatko · O. Kharchenko · V. Yaremych
Faculty of Information Technologies, Kyiv National University of Trade and Economics, Kyiv, Ukraine
e-mail: k.khorolska@knute.edu.ua

V. Lazorenko
e-mail: v.lazorenko@knute.edu.ua

B. Bebeshko
e-mail: b.bebeshko@knute.edu.ua

O. Kharchenko
e-mail: a.kharchenko@knute.edu.ua

V. Yaremych
e-mail: v.yaremych@knute.edu.ua

J. S. Raj et al. (eds.), *Intelligent Sustainable Systems*, Lecture Notes in Networks and Systems 213, https://doi.org/10.1007/978-981-16-2422-3_49

1 Introduction

Decision support systems are created based on certain rules that are formed as a result of data analysis. The created systems classify and analyze the data, and the methods of their processing differ in a large number of algorithms for obtaining rules and forms in which the rules are implemented. The most important thing in the decision-making process is the ability to classify rules and analyze new objects.

Cluster analysis can be one of the methods for data analysis in the decision support system. Clustering a decision support system involves allocating rules that will help influence decision-making to address structured or unstructured problems.

Clustering problems are solved by the following: a study of data; facilitation of analysis; data compression; prognostication; detection of anomalies [1].

2 Review and Analysis of Literature

There are many ways to use cluster analysis. In data mining problems with the help of cluster analysis a complex summary of data for classification is created, patterns are identified, hypotheses are formed and tested.

Cluster analysis methods are adequate for different data sets. The most common method is to combine; it is designed to combine objects into fairly large clusters, using some degree of similarity or distance between objects. A typical result of such clustering is a tree diagram. The K-means algorithm assumes such a division of objects into classes, which minimizes the differences ("distances") between objects of the same class and maximizes the distances between objects of different classes. The c-mean algorithm differs from the previous one in that the clusters are now fuzzy sets and each data element belongs to different clusters with different degrees of affiliation. Formation of a hierarchy of clusters. For their formation, it is not necessary to specify the number of clusters K, this type of clustering is more deterministic and does not require iterative refinements. Hierarchical clustering methods include two categories of algorithms. The first category is called agglomeration. The second group—divisive methods—by the method of division, a large macrocluster containing all the elements is divided into two groups, each of them also into two groups and so on.

Methods and algorithms of clustering, which are one of the fundamental principles of Data Mining, are presented quite thoroughly in the works of G. Lino ff and M. Berry [5, 7, 13], as well as Ukrainian researchers A. Zhyhir [8] and I. Strubytska [9].

Each of the authors uses its methodological approach to classification using cluster analysis methods. When constructing any classification, it is important to choose the basis and methodology of classification because these two factors can give different results.

Decision support systems are the object of scientific research of many both domestic and foreign scientists. Among them: V. Sytnyk, Yu.Ye. Petrunya, T.

Lambert, K. Hildebrant, S.M. Camomile, P.I. Bidyuk, L.O. Korshevnyuk, J. Neumann, Andrew Van Dam, W.W. Glukhov, W. Denning, S.A. Daudov, and many others.

3 The Purpose of the Study

Internet security has been one of the most important problems in the world. Anomaly detection is the basic method to defend against a new attack in intrusion detection. Network intrusion detection is the process of the events monitoring that occurs in a computing system or network and further analysis of those for signs of intrusions, defined as attempts to compromise confidentiality. A wide variety of data mining techniques have been applied to intrusion detection approaches. In data mining, clustering is the most important unsupervised learning process that is used to find the structures or patterns in a collection of unlabeled data. In this paper, we use the K-means algorithm to cluster and analyze the data. Computer simulations show that this method can detect unknown intrusions efficiently in real network connections [11].

The article aims to simplify the decision-making process in cybersecurity using a clustering algorithm.

The task of clustering is to divide the studied set of objects into groups of "similar" objects, called clusters. The solution to the classification problem is to assign each of the data objects to one (or more) of predefined classes and build, ultimately, one of the methods of classification of the data model, which determines the division of many data objects into classes note several features inherent in the problem clustering. First, the solution strongly depends on the nature of the data objects (and their attributes). Thus, on the one hand, it can be unambiguously defined, clearly quantified objects, and on the other—objects that have a probabilistic or fuzzy description. Second, the solution also depends significantly on the representation of classes (clusters) and the expected relationships of data objects and classes (clusters). Yes, it is necessary to take into account such properties as the possibility/impossibility of belonging of objects of separate classes (clusters).

The data storage module in the decision-making system should allow storing data optimized for analytical operations. It must be a database of a certain structure that provides fast execution of analytical queries. The data processing module should process the source data using statistical methods, OLAP technologies, Data Mining, expert technologies or their combination or check the consistency and reliability of expert estimates.

To solve the problem of evaluation and selection, metric input methods will be used (Euclidean distance, L1-norm, Ln-norm, supreme-norm); methods of expert evaluations (method of direct evaluations, ranking method).

The development of a decision-making system in the dialog mode will perform calculations on a quantitative assessment of efficiency. The result of such calculations will be recommendations for a reasonable choice of the best option from the alternative.

4 Methods and Models

Clustering is the method of grouping objects into meaningful subclasses so that the members from the same cluster are quite similar, and the members from different clusters are quite different from each other. Therefore, clustering methods can be useful for classifying log data and detecting intrusions [11, 15].

In modern conditions, the decision support system can be used even in preventing cyberattacks, according to the cybersecurity strategy of Ukraine, one of the priority areas is to create a system for timely detection, prevention, and neutralization of cyber threats and create conditions for modern cyber defense technologies in Ukraine. To do this, you must solve several problems, consisting of

- İmprovement of the system of storage, transmission and processing of data of state registers and databases with the use of modern information and communication technologies (including online access technologies).
- Developing new methods of preventing and disseminating information about cyberattacks, cyber incidents.
- Development of requirements (rules, guidelines) for the safe use of the Internet and the provision of electronic services by government agencies and others.

The effective solution of the set tasks directly depends on the development and implementation of the latest information technologies in the activities of cybersecurity entities, which are based on the results of scientific and scientific and technical activities of domestic and foreign scientists.

There are two main classifications of clustering algorithms:

1. Hierarchical and non-hierarchical (flat). Hierarchical algorithms build a system of nested partitions, i.e., at the output of the algorithm is a tree of clusters, with the root as the whole sample and the leaves—as the smallest clusters. Non-hierarchical algorithms build only one division of objects into clusters.
2. Clear and fuzzy. Clear algorithms give all sample objects the appropriate cluster number, which means that each object must belong to only one cluster. [10]

Intrusion detection systems compared to traditional network security measures have great advantages. Can solve the shortcomings of the original passive inspired, can also process it before the damage occurred, the appearance of the intrusion detection system has become an important part of network security. [16, 17]

The vast amount of data generated in the Internet era undoubtedly challenges the technology of large-scale data processing and data mining by using K-means clustering algorithms in data mining. It analyzes the network security problems and performance better intrusion detection system in network security analysis simulation, let more people know the network intrusion behavior produces a variety of ways and means. In this way, we can ensure the information systems leak protection in the information system itself [2].

Internet security has been one of the most important problems in the world. Anomaly detection is the basic method to defend against a new attack in intrusion detection. Network intrusion detection is the process of monitoring the events occurring in a computing system or network and analyzing them for signs of intrusions, defined as attempts to compromise confidentiality. A wide variety of data mining techniques have been applied to intrusion detections. In data mining, clustering is the most important unsupervised learning process used to find the structures or patterns in a collection of unlabeled data. We use the K-means algorithm to cluster and analyze the data in this paper. Computer simulations show that this method can detect unknown intrusions efficiently in real network connections.

Separate clustering, according to which for the data sample, which contains n records (objects), the number of clusters k is to be formed. The algorithm then divides all sample objects into k sections ($k < n$), which are clusters. The *K-means* clustering algorithm consists of four steps [3].

1. Specify the number of clusters K, which should be formed from the objects of the original sample.
2. Randomly select K records of the original sample, which will be the initial members of the clusters. The starting points from which the cluster then grows are often called « seeds » . Each such record is an « embryo » of a cluster, which consists of one element.
3. For each record of the original sample, the nearest cent of the cluster is determined.
4. The calculation of centroids—centers of gravity of clusters. This is done by determining the average for the values of each attribute of all records in the cluster.

The main point of the K-means algorithm: at each iteration, the distance between the records and the centers of the clusters is calculated to determine which of the clusters this record belongs to. The rule by which the distance in the multidimensional feature space is calculated is called a metric. In practice, the following metrics are most often used.

- Euclidean distance, the most common function of distance, represents the geometric distance in multidimensional space: [11]

$$d_e(X, Y) = \sqrt{\sum_i (x_i - y_i)^2}, \tag{1}$$

where $X = (x1, x2, \ldots, xm)$, $Y = (y1, y2, \ldots, ym)$—vectors of two records signs values of two records.

Table 1 Clustering objects

Objects	A	B	C	D	E	F	G	H	I	J
Xi	**20**	7	16	4	11	**3**	8	8	9	**17**
Yi	**10**	14	11	14	12	**15**	19	4	18	**7**

Since a set of points equidistant from some center will form a sphere (or a circle in the two-dimensional case) using Euclidean metrics, clusters obtained using the Euclidean distance will also have a shape close to spherical.

- The distance of Manhattan is calculated by the formula:

$$d_e(X, Y) = \sum_i |x_i - y_i| \tag{2}$$

In fact, this is the shortest distance between two points, drawn along lines that are parallel to the axes of the coordinate system.

5 An Example of a Solution and Computational Experiments

5.1 *Using Clustering in DSS*

Consider the example of certain rules of the decision support system, which we denote as 10 different data points in two-dimensional space, from which it is necessary to obtain three clusters, as presented in Table 1. The translation of rules into points on the plane can be performed using the method of analysis of hierarchies, determining the priority of each of them.

5.2 *Algorithm Realization*

Step 1. Determine the number of clusters into which you want to divide the original set: $k = 3$.

Step 2. The initial implementation is presented in Fig. 1. Then, we randomly select three points that will be the starting centers of the clusters. Let these be the points A, F and $J : m1 = (20; 10)$, $m2 = (3; 15)$ and $m3 = (17; 7)$.

Step 3, pass 1. For each point determine the nearest center of the cluster using the Euclidean distance. The results of calculations of distances between the centers of clusters are performed by the formula:

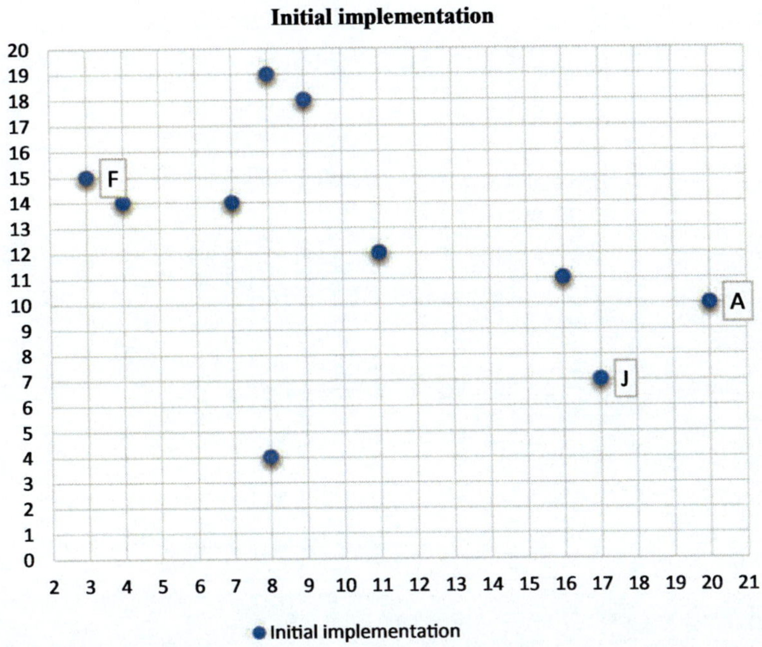

Fig. 1 Initial implementation

$$d_e(X, Y) = \sqrt{\sum_i (x_i - y_i)^2} \tag{3}$$

$$d_e(X, Y) = \sum_i |x_i - y_i| \tag{4}$$

The results are presented in the form of Table 2, which also indicates to which cluster a particular point belongs.

Thus, cluster 1 contains points A, C, cluster 2—points B, D, F, G, I, and cluster 3—points E, H, J. Once the members of the clusters are determined, you can calculate the sum of the squares of the errors:

$$E_e = \sum_{i=1}^{k} \sum_{p \in C_i} (p - m_i)^2 = 273 \tag{5}$$

$$E_m = \sum_{i=1}^{k} \sum_{p \in C_i} (p - m_i)^2 = 481 \tag{6}$$

Step 4, pass 1. For each cluster, the centroid is calculated, and the center of the cluster is moved into it (Fig. 2).

Table 2 Finding the nearest center for each point (first Pass)

			Euclidean distance				The distance of Manhattan			
Point	x	y	Distance from m1	Distance from m2	Distance from m3	Cluster independence	Distance from m1	Distance from m2	Distance from m3	Cluster independence
A	20,00	10,00	0,00	17,72	4,24	1	0,00	22,00	6,00	1
B	7,00	14,00	13,60	4,12	12,21	2	17,00	5,00	17,00	2
C	16,00	11,00	4,12	13,60	4,12	1	5,00	17,00	5,00	1
D	4,00	14,00	16,49	1,41	14,76	2	20,00	2,00	20,00	2
E	11,00	12,00	9,22	8,54	7,81	3	11,00	11,00	11,00	3
F	3,00	15,00	17,72	0,00	16,12	2	22,00	0,00	22,00	2
G	8,00	19,00	15,00	6,40	15,00	2	21,00	9,00	21,00	2
H	8,00	4,00	13,42	12,08	9,49	3	18,00	16,00	12,00	3
I	9,00	18,00	13,60	6,71	13,60	2	19,00	9,00	19,00	2
J	17,00	7,00	4,24	16,12	0,00	3	6,00	22,00	0,00	3

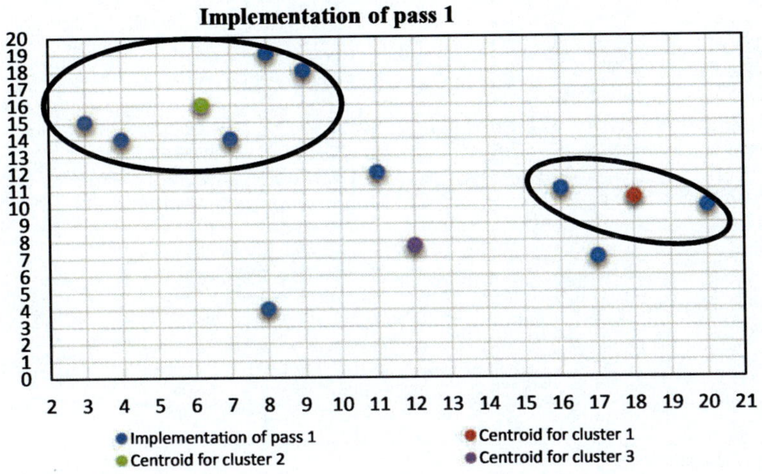

Fig. 2 Realization of 1 pass

- Centroid for cluster 1: $[(20 + 16)/2; (10 + 11)/2] = (18 : 10.5)$.
- Centroid for cluster 2: $[(7+4+3+8+9)/5; (14 + 14 + 15 + 19 + 18)/5] = (6, 2 : 16)$
- Centroid for cluster 3: $[(11+8+17)/3; (12+4+7)/3] = (12 : 7, 67)$

Step 3, pass 2. After finding new cluster centers for each point again determines the nearest center and its relationship to the corresponding cluster. To do this, the Euclidean distances between the points and centers of the clusters are calculated again. The results are shown in Table 3.

As a result of the calculations, the record J moved to cluster 1, the rest remained unchanged. The new sum of error squares:

$$E_e = \sum_{i=1}^{k} \sum_{p \in C_i} (p - m_i)^2 = 119, 77 \tag{7}$$

$$E_m E_m = \sum_{i=1}^{k} \sum_{p \in C_i} (p - m_i)^2 = 209, 17 \tag{8}$$

Step 4, pass 2. For each cluster, the centroid is recalculated, into which the center of the cluster moves. The results are presented in Fig. 3.

- New centroid for cluster 1: $[(20 + 16 + 17)/3; (10 + 11 + 7)/3] = (17.67; 9.33)$.
- New centroid for cluster 2: $\begin{bmatrix} (7+4+3+8+9)/5; \\ (14+14+15+19+18)/5 \end{bmatrix} = (6, 2; 16)$.
- New centroid for cluster 3: $[(11 + 8)/2; (12 + 4)/2] = (9.5; 8)$.

Table 3 Finding the nearest center for each point (second Pass)

Point	x	y	Distance from m1	Euclidean distance Distance from m2	Distance from m3	Cluster independence	Distance from m1	The distance of Manhattan Distance from m2	Distance from m3	Cluster independence
A	20,00	10,00	2,06	15,05	8,33	1	2,50	19,80	10,33	1
B	7,00	14,00	11,54	2,15	8,07	2	14,50	2,80	11,33	2
C	16,00	11,00	2,06	11,00	5,21	1	2,50	14,80	7,33	1
D	4,00	14,00	14,43	2,97	10,20	2	17,50	4,20	14,33	2
E	11,00	12,00	7,16	6,25	4,45	3	8,50	8,80	5,33	3
F	3,00	15,00	15,66	3,35	11,61	2	19,50	4,20	16,33	2
G	8,00	19,00	13,12	3,50	12,02	2	18,50	4,80	15,33	2
H	8,00	4,00	11,93	12,13	5,43	3	16,50	13,80	7,67	3
I	9,00	18,00	11,72	3,44	10,76	2	16,50	4,80	13,33	2
J	17,00	7,00	3,64	14,06	5,04	1	4,50	19,80	5,67	1

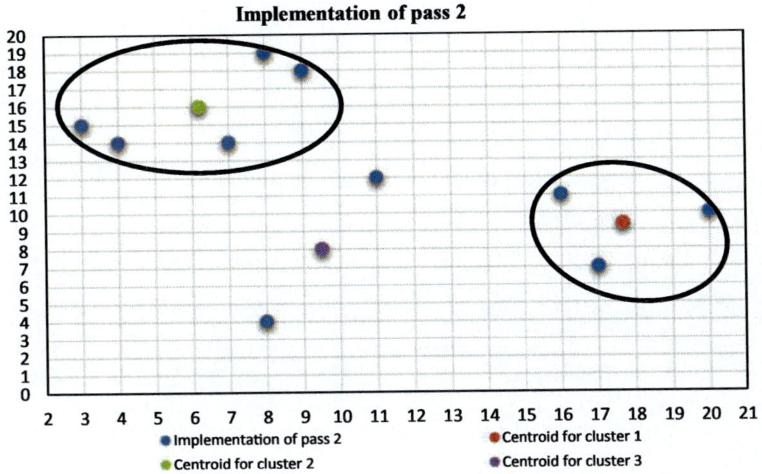

Fig. 3 Realization of 1 pass

Step 3, pass 3. For each record, the nearest center of the cluster is again. The results of the calculations are presented in Table 4.

There were no new entries that would change the cluster at this step of the algorithm. The new sum of error squares:

$$E_e = \sum_{i=1}^{k} \sum_{p \in C_i} (p - m_i)^2 = 102, 63 \tag{9}$$

$$E_m = \sum_{i=1}^{k} \sum_{p \in C_i} (p - m_i)^2 = 18, 81 \tag{10}$$

Step 4, pass 3. For each cluster, the centroid is recalculated, into which the center of the cluster moves. But since no record on this pass has changed its membership in the clusters and the position of the centroids has not changed, the algorithm completes its work (Table 5).

6 Discussion of the Results of the Computational Experiment

Using the method of cluster analysis allows you to select in the decision-making table a group of features that can be used as a starting point for the formation of a set of decision rules [16] in the decision-making system to establish the result to be provided by DSS. The clustering method combines in a given number of initial rules

Table 4 Finding the Nearest center for each point (third pass)

			Euclidean distance					The distance of Manhattan		
Point	x	y	Distance from m1	Distance from m2	Distance from m3	Cluster independence	Distance from m1	Distance from m2	Distance from m3	Cluster independence
A	20,00	10,00	2,43	15,05	10,69	1	3,00	19,80	12,50	1
B	7,00	14,00	11,64	2,15	6,50	2	15,33	2,80	8,50	2
C	16,00	11,00	2,36	11,00	7,16	1	3,33	14,80	9,50	1
D	4,00	14,00	14,44	2,97	8,14	2	18,33	4,20	11,50	2
E	11,00	12,00	7,18	6,25	4,27	3	9,33	8,80	5,50	3
F	3,00	15,00	15,72	3,35	9,55	2	20,33	4,20	13,50	2
G	8,00	19,00	13,67	3,50	11,10	2	19,33	4,80	12,50	2
H	8,00	4,00	11,04	12,13	4,27	3	15,00	13,80	5,50	3
I	9,00	18,00	12,26	3,44	10,01	2	17,33	4,80	10,50	2
J	17,00	7,00	2,43	14,06	7,57	1	3,00	19,80	8,50	1

Table 5 Clustering result

Cluster 1	Cluster 2	Cluster 3
A, C, J	B, D, F, G, I	E, H

and combines in the required number of clusters that can be derived from the general rule obtained using the decision model.

Models, algorithms, and the DSS itself, which is being developed as a part of a study based on the K-means method with dependent movements, allow one to constructively find optimal strategies and giving concrete recommendations to analytics in cybersecurity projects or other large trade projects, for example, in the field of Data Mining. The solution is based on the method of dynamic programming. This allows, in contrast to existing approaches, to find more effective solutions. To find the solution, we also used the mathematical apparatus of the K-means method with several terminal surfaces with alternate moves [12, 14].

7 Conclusions

Decision support systems are aimed primarily at supporting middle and senior management and planning and increase the likelihood of making an informed decision even in conditions of uncertainty and changing circumstances.

Cybersecurity analytics is an alternative solution to such traditional security systems, which can use big data analytics techniques to provide a faster and scalable framework to handle a large amount of cybersecurity-related data in real time. K-means clustering is one of the commonly used clustering algorithms in cybersecurity analytics aimed at dividing security-related data into groups of similar entities, which in turn can help in gaining important insights about the known and unknown attack patterns. This technique helps a security analyst to focus on the data specific to some clusters only for the analysis [4].

In the conditions of modern development, the activity of the organizations, introduction and functioning of clusters and cluster technologies is considered as one of the perspective means of maintenance of competitiveness. The effectiveness of cluster models is achieved through innovation, development of decision support systems, dissemination of information exchange, knowledge. The results of the study revealed a systemic causal relationship between the creation of cluster systems and the operation of decision-making systems. This suggests that clusters are a tool to improve the performance of systems.

Notwithstanding the above, many issues related to the development of cluster structures require more detailed research. In the future, before creating decision-making systems based on clustering methods, it is necessary to assess the possibilities of filling the system to understand the principle of the system.

References

1. Leskovec, J.: Mining of Massive Datasets. Jure Leskovec Anand Rajaraman, Jeffrey David Ullman. Stanford Univ (2010)
2. Bu, C.: Network security based on k-means clustering algorithm in data mining research. In: 8th International Conference on Social Network, Communication and Education (SNCE 2018), Advances in Computer Science Research, vol. 83. https://doi.org/10.2991/snce18.2018.130
3. Chitrakar, A.S., Petrović, S.: Efficient kmeans using triangle inequality on spark for cyber security analytics. In: Proceedings of the ACM International Workshop on Security and Privacy Analytics (IWSPA '19). Association for Computing Machinery, New York, NY, USA, pp. 37–45 (2019). doi:https://doi.org/10.1145/3309182.3309187
4. Linoff, G.S.: Data Mining Techniques: For Marketing, Sales, and Customer Relationship Management, p. 888. Wiley, Indianapolis (2011)
5. Cao L., Philip S. Yu, Zhang C, Zhang H. Data Mining for Business Applications. Springer Science; Business Media, 2008. 402 p
6. Albright, S.C., Winston, W., Zappe, Є.: Data Analysis and Decision Making, p. 948. Cengage Learning, Boston (2016)
7. Zhyhir, A.A.: Formuvannia klasteriv, yak svitova tendentsiia, poshuku shliakhiv pidvyshchennia efektyvnosti pidpryiemnytskoi diialnosti [Formation of clusters, as a global trend, finding ways to increase the efficiency of entrepreneurial activity]. Investytsii: praktyka ta dosvid – Investments: practice and experience, vol. 22, pp. 38–41 [in Ukrainian] (2015)
8. Roskladka, N.O., Roskladka, A.A., Dzygman, O.O.: Klasternyy analiz kliyent·s′koyi bazy danykh pidpryyemstv sfery posluh [Cluster analysis of the client database of enterprises in the service sector]. Tsentral′noukrayins′kyy naukovyy visnyk. Ekonomichni nauky: zb. nauk. pr. - Central Ukrainian Scientific Bulletin. Economic sciences: coll. Science. pr. - Kropyvnytskyi: CNTU - Issue 2(35), pp. 151–159 [in Ukrainian] (2019)
9. Yakymets, R.V.: Metody klasteryzatsiyi ta yikh klasyfikatsiya [Methods of clustering and their classification]. Mizhnarodnyy naukovyy zhurnal - Int. Sci. J **6**(2), 4850 [in Ukrainian] (2016)
10. Volosyuk, Yu.V.: Analiz alhorytmiv klasteryzatsiyi dlya zadach intelektual'noho analizu danykh. [Analysis of clustering algorithms for data mining problems]. Zbirnyk naukovykh prats' Viyskovoho instytutu Kyyivskoho natsional'noho universytetu imeni Taras Shevchenko - Zbirnyk naukovykh pratsiv Military Institute of Kyiv National University named after Taras Shevchenko, vol. 47, pp. 112–119 (2014)
11. Jianliang, M., Haikun, S., Ling, B.: The Application on Intrusion Detection Based on K-means Cluster Algorithm. In: International Forum on Information Technology and Applications, Chengdu **2009**, 150–152 (2009). https://doi.org/10.1109/IFITA.2009.34
12. Lakhno, V., Malyukov, V., Yerekesheva, M., Kydyralina, L., Sarsimbayeva, S., Zhumadilova, M., Buriachok, V., Sabyrbayeva, G.: Model of cybersecurity means financing with the procedure of additional data obtaining by the protection side. J. Theoret. Appl. Inf. Technol. **98**(1), 1–14. ISSN 1992-8645 (2020)
13. Mohanapriya, M., Lekha, J., Thilak, G., Meeran, M.M.: A novel method for culminating the consumption of fast food using PCA Reduction and K-means Clustering Algorithm. In: 2019 International Conference on Intelligent Sustainable Systems (ICISS), Palladam, Tamilnadu, India, 2019, pp. 549–552. https://doi.org/10.1109/iss1.2019.8908035
14. Khedr, A., Elseddawy, A., Idrees, A.: Performance tuning of k-mean clustering algorithm a step towards efficient DSS. Int. J. Innov. Res. Comput. Sci. Technol. (IJIRCST). **2**, 2347–5552 (2014)
15. Shen, H., Duan, Z.: Application research of clustering algorithm based on k-means in data mining. In: 2020 International Conference on Computer Information and Big Data Applications (CIBDA), Guiyang, China, 2020, pp. 66–69, https://doi.org/10.1109/cibda50819.2020.00023

16. Sathesh, A.: Enhanced soft computing approaches for intrusion detection schemes in social media networks. J. Soft Comput. Paradigm (JSCP) **1**(2), 69–79 (2019)
17. Suma, V.: Automatic spotting of sceptical activity with visualization using elastic cluster for network traffic in educational campus. J. Ubiquitous Comput. Commun. Technol. **2**, 88–97 (2020)

A Comparative Research of Different Classification Algorithms

Amlan Jyoti Baruah, JyotiProkash Goswami, Dibya Jyoti Bora, and Siddhartha Baruah

Abstract Classification is extensively used in Educational Data Mining (EDM). And this is very much essential as huge volume of data are generated by web-based educational applications. Among them, student enrollment records, attendance records, and examination results are some of the sensitive data, which should be handled carefully as a proper observation of this type of data may help to induce a good pattern of technological implications in educational system or may even lead to prediction results of which may help to improve the current educational system. As there exist a lot of classification algorithms that may apply in EDM, so, to find out the relevant and efficient ones with respect to EDM is a sophisticated task, which includes a rigorous review of literature study and experimental analysis. In this paper, an effort has been made to perform a comparative study on such classification techniques like common ensemble methods, e.g., bagging and boosting, J48, Naïve Bayes, LMT, ANN, REPtree, Hoeffding, Random Forest.

Keywords Classification · EDM · J48 · Naïve bayes · LMT · ANN · Reptree · Hoeffding · Random forest

A. J. Baruah (✉)
Department of Computer Science and Engineering, Assam Kaziranga University, Jorhat, Assam, India

J. Goswami
Department of Computer Applications, Assam Engineering College, Guwahati, Assam, India

D. J. Bora
School of Computing Science, Assam Kaziranga University, Jorhat, Assam, India
e-mail: dibyajyotibora@kazirangauniversity.in

S. Baruah
Department of Computer Applications, Jorhat Engineering College, Jorhat, Assam, India

J. S. Raj et al. (eds.), *Intelligent Sustainable Systems*, Lecture Notes in Networks and Systems 213, https://doi.org/10.1007/978-981-16-2422-3_50

1 Introduction

In present times, Educational Data mining is an emerging trend in computer science and engineering field. Nowadays, educational data is growing every day, so a requirement is needed for changing huge volume of educational data into useful information and knowledge. Educational data mining showcases different tools and algorithms to determine patterns by analyzing educational data. Different data mining techniques like clustering, classification, etc., can be used to extract hidden knowledge from data that come from different educational sources like web-based education, educational repositories, and educational surveys using questionnaire methods. This hidden knowledge can help the teachers or educators to improve teaching and learning methodology.

Student's academic performance is a crucial part of any educational system. The aspect of students' academic performance can be determined using predictive model based on different popular classification algorithms. There are several algorithms under classification such as J48, Naïve Bayes, LMT, ANN, REPtree, Hoeffding, and Random Forest.

In this paper, different classification techniques like common ensemble methods, e.g., bagging and boosting, J48, Naïve Bayes, LMT, ANN, REPtree, Hoeffding, Random Forest are used to perform a comparative study based on different data collected from different educational repositories considering different parameters like Accuracy, Precision, Recall, F-Measure, and Time.

2 Literature Review

Milos et. al. [1] had proposed a method where authors used WEKA tool for prediction based on attributes in two different datasets of final year student mark. Each dataset contains records of different students from one college course in the last four semesters, and final semester data are used to create the test dataset. In the paper, authors also presented experiment which gives students' grade prediction to showcase good quality of student knowledge. Using this method, educators can perform early detection of students' grade and able to find out shortcomings of students. So students are able to get more attention from educators.

Shahiri et. al. [2] had provided an overview to predict students' performance by using different data mining methods. In this paper, authors had provided a method which provides different prediction algorithm such as Artificial Neural Networks, Decision tree, Naive Bayes, and Support Vector Machine could be used to determine mostly related attributes in a students' dataset. Authors had also explained that how educational data mining techniques could improve student's achievement and success more effectively. This method can bring benefits to both students and educators.

Ramaphosa et. al [3] had showcased a comparative study of four classifiers namely Naïve Bayes, BayesNet, JRip and J48 and finds the efficient classifier based on high accuracy. In this paper, authors had compared the classifiers based on educational data that consist of city, school, grades, and Mathematics results. For this study, authors had collected 678 learner's data of Gauteng primary schools and in the experiment using WEKA tool authors had found that J48 algorithm showed the high accuracy rate.

Turbo et al. [4] *had* proposed model based on different classifiers such as K-Nearest *Neighbors* (K-NN), Convolutional Neural Network (CNN), Na¨ıve Bayes (NB), and decision trees (C4.5) to achieve predict students' performance prediction. To find out different important features from the gathered data of different sources like students, e-learning platform, etc., proposed model can be used. In this paper, *the authors* had described how *binary genetic algorithm* can improve the performance of all classifiers.

Phua et. al. [5] had presented a predictive model that applied machine learning techniques to predict students' grades of modules from past results. Based on the predicted score, this model can improve students' performance and help the instructors to identify weak students who need extra assistance. In this paper, the authors had used synthetic dataset and evaluated the performance of different machine learning techniques namely Linear Regression, K-Nearest Neighbor, and Decision Table. Here, Ensemble methods used for the study include Stacking, Bagging, and Random Forest. The experimental results showed that the Ensemble algorithms perform better than the base algorithms.

From the above review, we have found that the researchers in the field of machine learning are using different classification algorithms like K-NN, Naïve Bayes, Decision Tree, Random Forest, etc., for the classification task in Educational Data Mining (EDM). Although what they are proposing as the direct applications of the above-mentioned algorithms or a modified approach of the same, it is very difficult to merely guessing that which classification algorithm is best suitable for this problem.

By taking this problem into consideration, in this paper, we have gone through an extensive comparative analysis of these algorithms (empirical approach) in order to find out the best among them for this task of classification in EDM.

The outlines of the remaining part of the paper are as follows: Section 3 is Methods and materials, Section 4 is Experiments and result analysis, Section 5 is conclusion, and at the end, we have the reference section.

3 Methods and Materials

A. ***J48 algorithm:*** It is an advance version of ID3. It is an open-source Java implementation of C4.5 algorithm [6]. To create decision tree for classification J48 follows the concepts of greedy algorithms and reduced error pruning. In J48, training dataset can be used to build decision trees and each attribute of the data can be used to make a decision by splitting it into smaller subsets [7, 8]. Information entropy and examination of the difference in entropy are very important concepts used in J48. This difference in entropy is called normalized information gain. Attribute with highest normalized information gain is used to make the decision [9]

The Algorithm:

Stage 1: Calculate the **Entropy** of whole Dataset.

$$\mathbf{Entropy(S)} = \frac{p}{p+n}\log_2\left(\frac{p}{p+n}\right) - \frac{n}{p+n}\log_2\left(\frac{n}{p+n}\right)$$

where p and n define the values of resultant attribute.

Stage 2: For each attribute

(i) Calculate **Entropy** for each value.

$$\mathbf{Entropy(A)} = -\frac{p}{p+n}\log_2\left(\frac{p}{p+n}\right) - \frac{n}{p+n}\log_2\left(\frac{n}{p+n}\right)$$

where p and n define the values of resultant attribute for the particular value of current attribute.

(ii) For current attribute calculate **Average Information Entropy.**

$$\mathbf{I(A)} = \sum \frac{pi + ni}{p+n} Entropy(A)$$

where p_i and n_i define the values of resultant attribute for the particular value of current attribute

(iii) For current attribute **Gain** have to calculate.

$$\mathbf{Gain = Entropy(A) - I(A)}$$

Stage 3: Consider the attribute with highest **Gain** value.

Stage 4: Repeat until we get the required decision tree

B. ***Naïve Bayes*****:** The Naive Bayes Classification algorithm is based on posthumous theory of Thomas Bayes [10, 11], which uses probabilistic classification mechanisms to perform classifications. From the perspective of classification, the main goal is to find the best mapping between a piece of new data and a set of classifications within a particular problem domain [12].

Naive Bayes Classifiers rely on Bayes' Theorem, which is based on conditional probability. In this theorem, occurrence of an event depends on another event which already happened. For e.g.

$$\mathbf{P(A)} = \frac{P\left(\frac{B}{A}\right).P(A)}{P(B)} \quad \textbf{where}$$

P (A|B): the probability of occurring event A when event B has already occurred.
P (B|A): the probability of occurring event B when event A has already occurred.
P (A): The probability of event A.
P (B): The probability of event B.

Naïve Bayes classification algorithm:

***Stage 1*:** In Naïve Bayes algorithm classification parameters are independent and there is no correlation between them. Let **T** be a training set with **N** no of tuples and their associated class labels. In the training set, each tuple consists of **M** dimensional attribute vector, D = {d_1, d_2,…dn}, depicting the tuples with M attributes, respectively, A = {A_1,A_2,…,A_m} [13].
***Stage 2*:** Suppose that there are **C** no of classes, Class = {$Class_1$, $Class_2$,…, $Class_c$}. According to Naïve Bayes classifier for a tuple D, the classifier will predict the class based on the highest conditional probability. That means, the tuple D belongs to a class $Class_i$ if and only if $Class_i$ has highest conditional probability than any other class $Class_j$, where $i \neq j$.

$$P\left(\frac{Class\ i}{D}\right) > P\left(\frac{Class\ j}{D}\right) \text{and } P\left(\frac{Class\ i}{D}\right) = \frac{P(Classi/D) * P(Classi)}{P(\mathrm{D})}$$

Stage 3: Class conditional independence is considered:

$$P\left(\frac{A}{Classi}\right) = \prod_{i=1}^{m} P\left(\frac{Am}{Classi}\right) = P(A_1)^*P(A_2)^*P(A_3)\ldots^*P(A_m)$$

Stage 4: Class $Class_i$ is predicted as the output class when

$$P\left(\frac{A}{\mathrm{Classi}}\right) * P(\mathrm{Class_i}) > P\left(\frac{A}{\mathrm{Classj}}\right) * P(\mathrm{Class_j})\text{where } 1 \leq i, j \geq c \text{ and } i \neq j$$

C. ***REPTree***: Full form of REP is Reduced Error Pruning [14]. In Machine learning, pruning [15] is a technique used to reduce the size of decision trees by detaching portion of the tree that provides ability to classify records. Reduced Error Pruning is the simplest and fast pruning technique. It starts from the leaf node and replaces the node with most appropriate class. This change is accepted only if the prediction accuracy is good.
D. ***LMT:*** Full form of LMT is logistic model tree [16]. It is used to merge the idea of logistic regression and decision tree with supervised training algorithm. Logistic trees specify the arrangement of a decision tree which takes the linear regression functions at its leaves and it creates linear regression model.
E. ***ANN:*** There are so many popular educational data mining methods, one among them is neural network. Artificial Neural Network (ANN) is structured in layers. Generally, three layers are considered in ANN namely input layer, hidden layer,

and output layer. All the layers are interconnected. Each and every layer consist of one or more number of nodes and are symbolized by circles. In ANN, the links between different nodes generally define the stream of information from one node to another and normally the flow starts at the input layer and it moves towards the output layer. The input layer nodes are called submissive, as they receive data from outside world and they do not change the data. The hidden layer and output layer nodes are active, as they are involved in changing the received by these two layers. In the hidden layer whatever the data received should be multiplied by some predetermined weights. After this single-valued result can be calculated by aggregating these weights. Then to control the outputs of the nodes, a nonlinear mathematical function is applied to the value [17]. Generally, in neural networks numbers of layers and number of nodes in every layer is not limited to certain values. For analysis output layer always consider a single node [18].

F. ***Hoeffding:*** In Hoeffding tree, Hoeffding bound has to be used to create and analyze the decision tree. In order to attain certain level of confidence, how many numbers of instances have to run be decided by Hoeffding bounds [19]. Classification problem can be defined with the help of training data in the form (A, B), where "A" is defining the number of attributes and "B" is defining the final class label. The main purpose of creating a model B = f(A) is to predict the class B for different attributes A with high accuracy. In the Hoeffding algorithm, different steps have to be considered [20].

Step 1: Root has to be initialized in the tree data structure.
Step 2: Decision tree learner has to be created which can read data more efficiently. Each and every training data should be filtered down to make it a suitable leaf.
Step 3: To make decision about next step each leaf node has enough data required. At the time of splitting of any attribute, the data in a leaf node calculate the information gain.
Step 4: In this stage, using Hoeffding bound, we have to find out the best attribute which has produced better result than other attributes. Based on the result, splitting of the node for the growth of the tree can be decided. Also, memory consumption is less and delivers enhanced utilization with sampling of data

G. ***Random Forest:*** Random forest [21] is a famous ensemble technique used for classification, Regression, etc. It **is a classification algorithm that consists of subsets of different decision trees. Random forest can aggregate the prediction of the ensemble to do the final prediction.**
Steps for Random Forest [22]:

(i) Randomly select "M" data points from the training dataset.
(ii) Randomly chosen subsets are used to create decision trees
(iii) Number "K" has to be chosen to build decision trees.
(iv) Step no i & ii have to be repeated

(v) Prediction of different decision tress has to be found out so that new data points can be decided. Afterwards, new data points have to be assigned to the category that defines the majority votes.

H. ***Ensemble Method:*** Ensemble methods are based on the idea of combining the predictions of multiple classifiers to obtain best algorithmic model with better predictive performance. By combining classifiers, the variance and bias of classification can be reduced and the dependency of results on characteristics of a single training set is eliminated [23]. There are two major approaches commonly used for constructing ensemble classifiers: bagging, boosting, voting, and stacking.

Bagging:

It is a process where random sampling with replacement is used on original dataset to generate different set of sample data so that multiple models can train. After the models are trained, voting scheme has to be used to predict final results. Although bagging is a ensemble technique, it is also a strong technique that works efficiently with limited data size.

Boosting:

It is the process where predictive analysis can be amplified by expanding the concept of bagging by adding more emphasis on wrongly predicted training data. Boosting generally uses the concept of weight sampling. At the initial stage, every data point has to be weighted equally. But in every iteration, it has to be checked that data points are correctly classified or not. If there any wrong classification exists, then weight has to be increased. After that observations are based on higher weights. Again, if any observations are misclassified, then more weights have to be provided. The process will continue until all the observations come into the right class.

I. ***Evaluation Measures:*** To evaluate the classification quality, we are using five different parameters namely Accuracy, Precision, Recall, F-Measure, and Processing Time. To calculate Accuracy, Precision, Recall, and F-Measure, we take the help of confusion matrix.
Confusion matrix can be defined using the following table. (Table 1)

(i) **Accuracy:** Accuracy defines the ability of a classifier. It is the ratio of true predictions to the overall predictions.

Table 1 Confusion Matrix

		Predicted Class	
		Positive	Negative
Actual Class	Positive	True Positive (TP)	False Negative (FN)
	Negative	False Positive (FP)	True Negative (TN)

$$\textbf{Accuracy} = \frac{TP + TN}{TP + FN + FP + TN} \quad (1)$$

(ii) **Precision:** It defines the ratio of truly positive prediction to the overall positive prediction

$$\textbf{Precision} = \frac{TP}{TP + FP} \quad (2)$$

(iii) **Recall:** It defines the ratio of truly positive prediction to the overall positive actual values.

$$\textbf{Recall} = \frac{TP}{TP + FN} \quad (3)$$

(iv) **F-measure:** It is defined as the weighted harmonic mean of the test's precision and recall.

$$\textbf{F} - \textbf{measure} = 2 * \left(\frac{Precision * Recall}{Precision + Recall}\right) \quad (4)$$

Processing Time: It is defined as the time.

4 Experiments and Result Analysis

We have implemented the above-mentioned algorithms in a system of configurations 4 GB RAM, 1 TB HDD, 64-bit operating system. We have chosen WEKA for the implementation process of the algorithms. Then, the datasets upon which algorithms are tested are mentioned below with a brief introduction regarding the same. The time consumed by each of the algorithms for classification process is noted down and also the other measures are noted down to compare the results for the analysis process. So, we are mentioning the results one by one for every dataset below. The charts and the tables are given for the better understanding of the experiments.

Students' Academic Performance Dataset (xAPI-Edu-Data).

This is a data set which is gathered from Learning Management System (LMS) known as Kalboard 360. It is a LMS with multi-agent specification.

Evaluation measures for the dataset (xAPI-Edu-Data).

Traditional Classification Methods (Table 2):

From the charts and the table, it is very clear that ANN is performing better than other algorithms with respect to Accuracy, Recall, Precision and F-measure but Random

Table 2 Comparison of different traditional classification methods

Classifier	J48	REPTree	LMT	ANN	Naïve Bayers	Hoeffding	Random Forest
Accuracy	0.758333	0.677083	0.775	0.783333	0.677083	0.675	0.766
Recall	0.758	0.677	0.775	0.783	0.677	0.675	0.767
Precision	0.76	0.675	0.775	0.783	0.675	0.673	0.766
F-Measure	0.759	0.671	0.775	0.783	0.671	0.669	0.766
Time (in seconds)	0.09	0.04	1.48	9.91	0.02	0.01	0.01

Forest and Hoeffding are performing better than other algorithms with respect to Time (Fig. 1).

Bagging (Table 3):

From the charts and the table, it is very clear that ANN with bagging is performing better than other algorithms with respect to Accuracy, Recall, Precision, and F-measure but Naïve Bayes with bagging is performing better than other algorithms with respect to Time (Fig. 2).

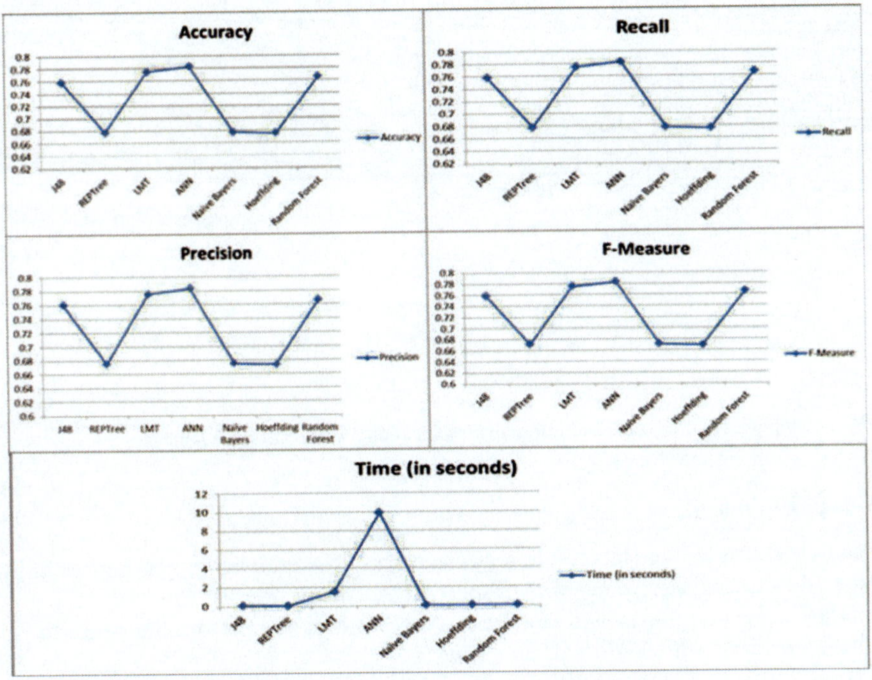

Fig. 1 Graphical representation of different traditional classification methods

Table 3 Comparison of different classification methods using Bagging

	Bagging						
Classifier	J48	REPTree	LMT	ANN	Naïve Bayers	Hoeffding	Random Forest
Accuracy	0.74735	0.704167	0.766667	0.78125	0.677083	0.677083	0.75625
Recall	0.744	0.704	0.767	0.781	0.677	0.677	0.756
Precision	0.743	0.706	0.767	0.781	0.676	0.676	0.757
F-Measure	0.743	0.704	0.767	0.781	0.672	0.672	0.756
Time (in seconds)	0.07	0.08	10.51	102.25	0.01	0.26	1.3

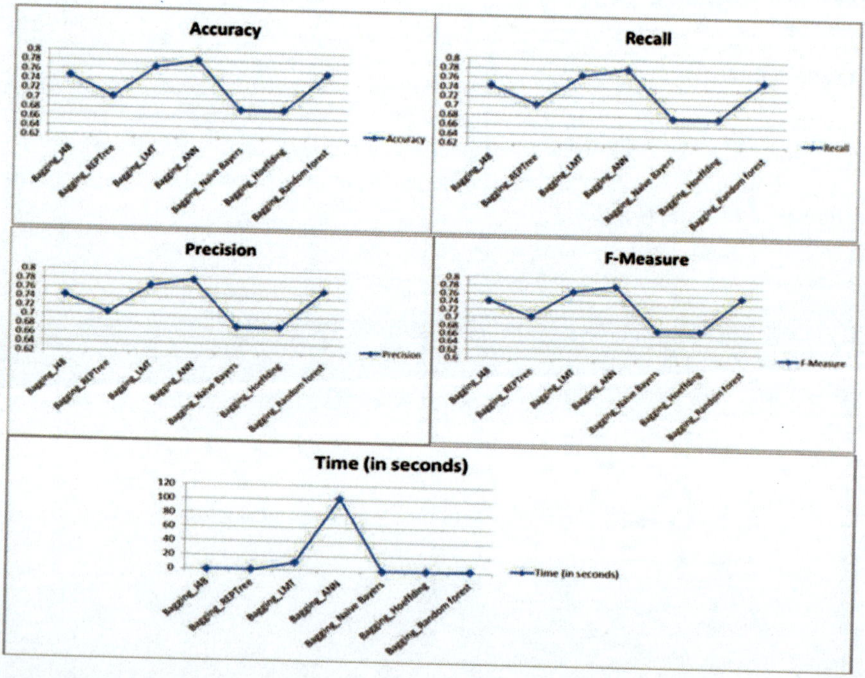

Fig. 2 Graphical representation of different classification methods using Bagging

Boosting (Table 4):

From the charts and the table, it is very clear that ANN with bagging is performing better than other algorithms with respect to Accuracy, Recall, Precision and F-measure but Naïve Bayes and Hoeffding with boosting are performing better than other algorithms with respect to Time (Fig. 3).

Table 4 Comparison of different classification methods using Boosting

			Boosting				
Classifier	J48	REPTree	LMT	ANN	Naïve Bayers	Hoeffding	Random Forest
Accuracy	0.779167	0.73125	0.75625	0.783333	0.722917	0.675	0.7625
Recall	0.779	0.731	0.756	0.783	0.723	0.675	0.763
Precision	0.779	0.732	0.756	0.783	0.724	0.673	0.763
F-Measure	0.779	0.731	0.756	0.783	0.718	0.669	0.762
Time (in seconds)	0.27	0.21	12.61	41.08	0.06	0.06	0.16

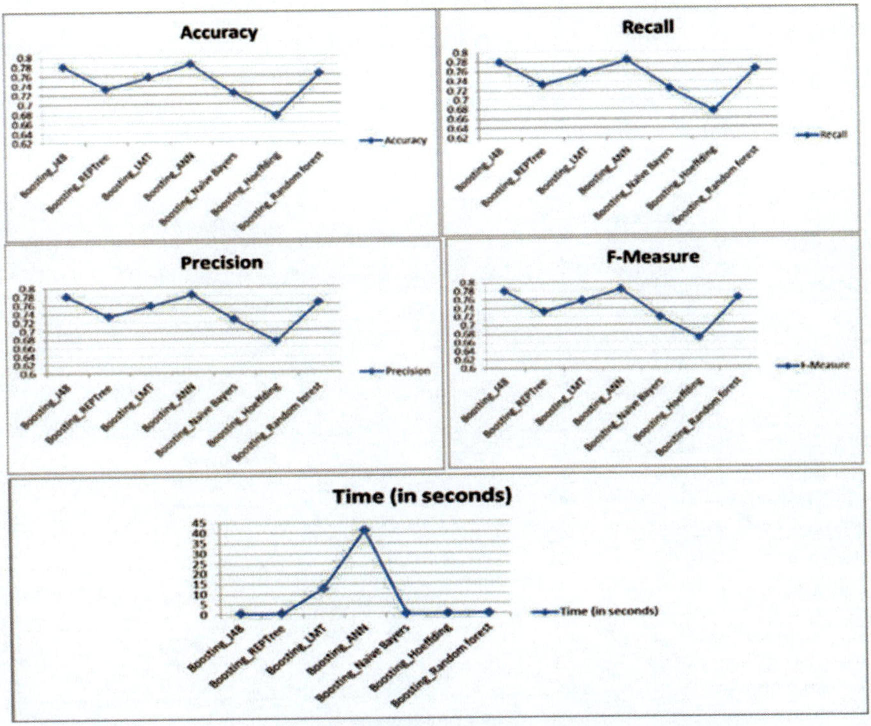

Fig. 3 Graphical representation of different classification methods using Boosting

Student Academics Performance Data Set (collegedata):

Evaluation measures for the dataset (collegedata):

This is a data set which is gathered from three engineering colleges of Assam, India. The data contain different parameters including academic details of students.

Table 5 Comparison of different classification methods for the college dataset

Classifier	J48	REPTree	LMT	ANN	Naïve Bayers	Hoeffding	Random Forest
Accuracy	0.633588	0.59542	0.648855	0.572519	0.641221	0.641221	0.656489
Recall	0.634	0.595	0.649	0.573	0.641	0.641	0.656
Precision	0.649	0.614	0.649	0.572	0.641	0.641	0.66
F-Measure	0.63	0.579	0.648	0.572	0.641	0.641	0.645
Time (in seconds)	0.03	0.01	0.51	3.7	0.01	0.02	0.04

Traditional Classification Methods (Table 5):

From the charts and the table, it is very clear that Random Forest is performing better than other algorithms with respect to Accuracy, Recall, Precision and F-measure but RepTree and Naïve Bayes are performing better than other algorithms with respect to Time (Fig. 4).

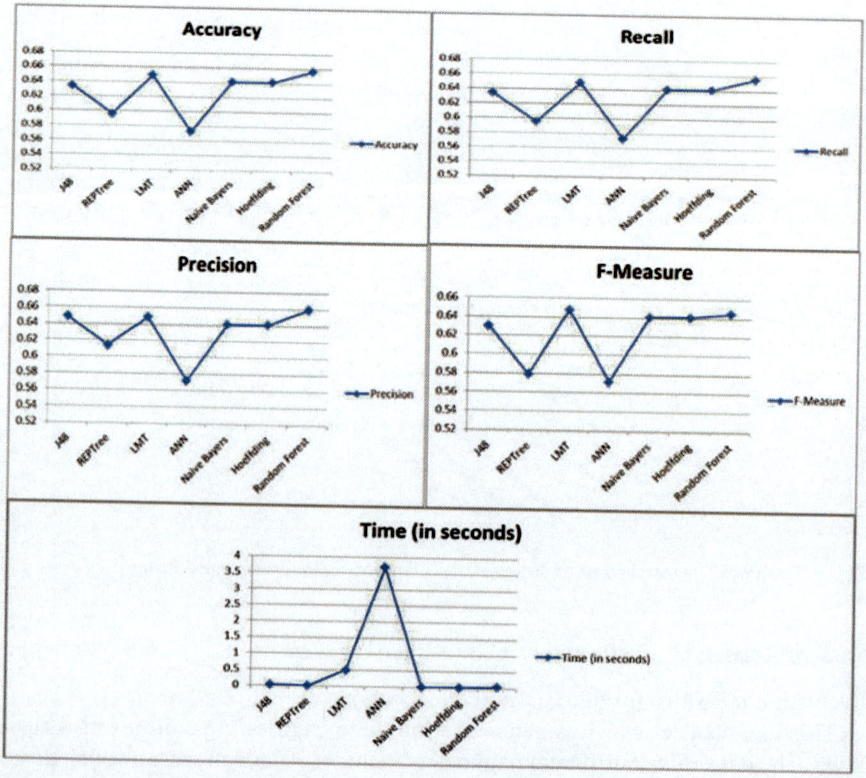

Fig. 4 Graphical representation of different classification methods for the college dataset

Table 6 Comparison of different classification methods for the college dataset using Bagging

			Bagging				
Classifier	J48	REPTree	LMT	ANN	Naïve Bayers	Hoeffding	Random Forest
Accuracy	0.648855	0.648855	0.641221	0.564885	0.625954	0.625954	0.664122
Recall	0.649	0.649	0.641	0.565	0.626	0.626	0.664
Precision	0.649	0.662	0.651	0.571	0.626	0.626	0.673
F-Measure	0.648	0.641	0.642	0.564	0.626	0.626	0.66
Time (in seconds)	0.04	0.02	2.42	35.43	0.01	0.04	0.21

Bagging (Table 6):

From the charts and the table, it is very clear that Random Forest is performing better than other algorithms with respect to Accuracy, Recall, Precision, and F-measure but Naïve Bayes with bagging is performing better than other algorithms with respect to Time (Fig. 5).

Boosting (Table 7):

From the chart and the table, it is very clear that Random Forest with boosting is performing better than other algorithms with respect to Accuracy, Recall, Precision, and F-measure but Hoeffding with boosting is performing better than other algorithms with respect to Time (Fig. 6).

5 Conclusion

After going through all the experimental results and analysis, we have found that ANN is performing better in terms of accuracy and other statistical comparison matrices. But it is taking comparatively more time than other classification algorithms and also, we have seen that for two different types of dataset, we are having different observations and these observations show us different performance by the same algorithm. So, we can conclude that based on the properties of different datasets same classification algorithm may give different accuracy levels. So, choosing a classification algorithm for a dataset is purely dependent on the properties of the dataset and also if we want accuracy at any cost then it is very clear from the experiments that ANN will be our first choice as in general it always meets the accuracy level to a standard more than average. This means there is a very less probability that its accuracy goes down or below average. Time complexity of ANN is comparatively higher. So, if we don't want to compromise with the time complexity then we have some other algorithms which are performing comparatively better than ANN and they are Naïve Bayes, Random Forest, etc. Also, we have seen that the bagging and boosting change the performance levels of these algorithms and improve the same.

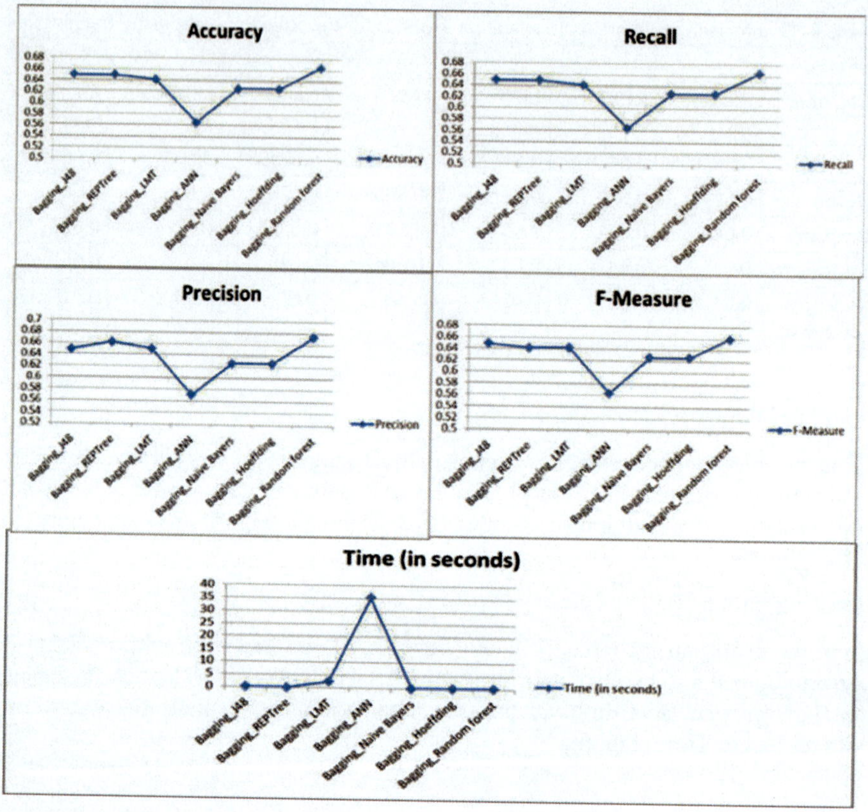

Fig. 5 Graphical representation of different classification methods for the college dataset using Bagging

Table 7 Comparison of different classification methods for the college dataset using Boosting

			Boosting				
Classifier	J48	REPTree	LMT	ANN	Naïve Bayers	Hoeffding	Random Forest
Accuracy	0.59542	0.648855	0.625954	0.572519	0.564885	0.618321	0.664122
Recall	0.595	0.649	0.626	0.573	0.565	0.618	0.664
Precision	0.597	0.65	0.632	0.572	0.567	0.619	0.681
F-Measure	0.596	0.642	0.626	0.572	0.563	0.618	0.656
Time (in seconds)	0.04	0.02	2.65	14.24	0.03	0.01	0.05

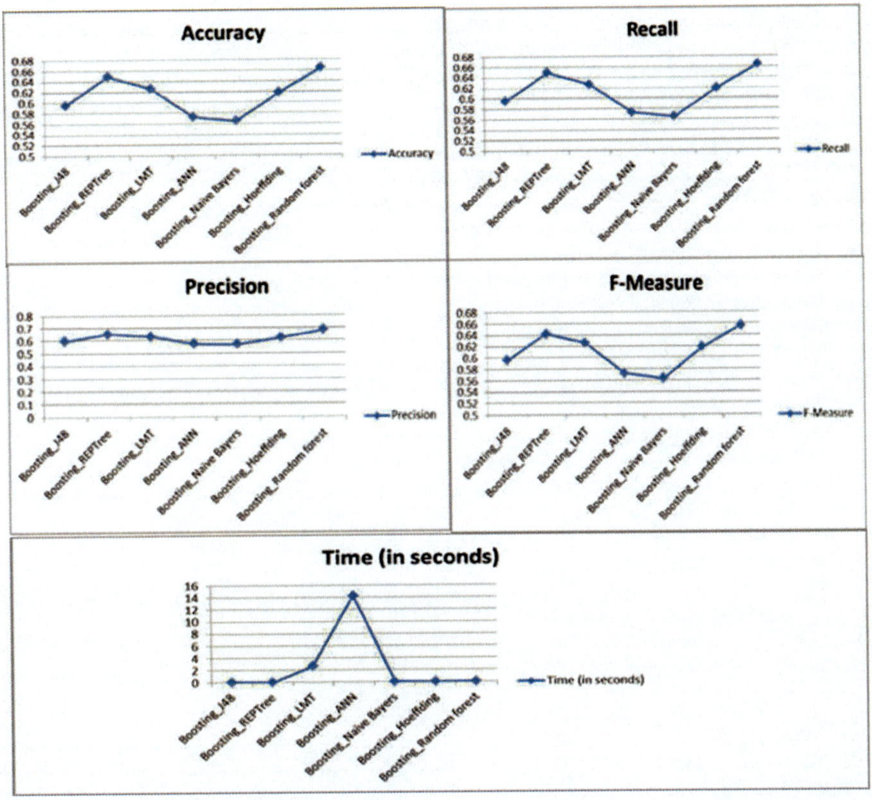

Fig. 6 Graphical representation of different classification methods for the college dataset using Bagging

So, this comparative analysis gives us a clear picture of how to go for choosing a particular classification algorithm for a particular type of dataset.

References

1. Milos, I., Petar, S., Mladen, V., Wejdan, A.: Students' success prediction using Weka tool, INFOTEH-JAHORINA, vol. 15, March 2016
2. Shahiri, A.M., Husain, W., Rashid, N.A.: A review on predicting student's performance using data mining techniques. Procedia Comput. Sci. **72**, 414–422 (2015)
3. Ramaphosa, K.I.M., Zuva, T., Kwuimi, R.: Educational data mining to improve learner performance in gauteng primary schools. In: 2018 International Conference on Advances in Big Data, Computing and Data Communication Systems (icABCD) (2018).
4. Turabieh, H.: Hybrid machine learning classifiers to predict student performance. In: 2019 2nd International Conference on New Trends in Computing Sciences (ICTCS) (2019)

5. Phua, E.U., Kadhar Batcha, N.: Comparative analysis of ensemble Algorithms'prediction accuracies in education data mining. 2020 J. Crit. Rev. (JCR) (2020)
6. Jalota, C., Agrawal, R.: Analysis of educational data mining using classification. In: 2019 International Conference on Machine Learning, Big Data, Cloud and Parallel Computing (COMITCon) (2019)
7. Kaur, G, Chhabra, A.: Improved J48 classification Algorithm for the prediction of diabetes. Int. J. Comput. Appl. **98**(22), 0975–8887 (2014)
8. Salama, G.I., Abdelhalim, M.B., Abd-elghany Zeid, M.: Breast cancer diagnosis on three different datasets using MultiClassifiers. Int. J. Comput. Inf. Technol. **1**(1), 2277–0764 (2012)
9. Dias, J.E.G.: Breast cancer diagnostic typologies by grade-of membership fuzzy modeling. In: Proceedings of the 2nd WSEAS International Conference on Multivariate Analysis and its Application in Science and Engineering
10. Bayes, T.: An essay towards solving a problem in the doctrine of chances. Philos. Trans. R. Soc. **53**(1), 1763, 370–418
11. Tabak, J.: Probability and Statistics: The Science of Uncertainty. Facts on File, Inc., NY, USA, (2004), pp. 46–50
12. Yang, F.J.: An implementation of Naive Bayes Classifier. In: 2018 International Conference on Computational Science and Computational Intelligence (CSCI), pp. 301–306 (2018)
13. Karthika, S., Sairam, N.: A Naïve Bayesian classifier for educational qualification. Indian J. Sci. Technol. **8**(16) (2015). doi: https://doi.org/10.17485/ijst/2015/v8i16/62055
14. Ali, J., Khan, R., Ahmad, N., Maqsood, I.: Random forests and decision trees. IJCSI Int. J. Comput. Sci. Issues **9**(5), 3 (2012)
15. https://e.wikipedia.org/wiki/Pruning_(decision_ trees)

An Automatic Correlated Recursive Wrapper-Based Feature Selector (ACRWFS) for Efficient Classification of Network Intrusion Features

P. Ramachandran and R. Balasubramian

Abstract Evolving technological paradigms direct information society's developments like Internet of Things (IoT), pervasive technologies. These technologies are built on networks that integrate with others for meeting end user needs. These networks are also susceptible to attacks. Technological knowledge is also used by cyber attackers for developing attacks and their numbers have increased exponentially. Hence, to safeguard networks from attackers, cybersecurity experts have become a fundamental pillar in cybersecurity and especially in Intrusion Detection Systems (IDS) which have grown into becoming the fundamental tool for cybersecurity in its provision of services on the internet. Though IDSs monitor networks for doubtful activities and send alerts on encountering such items, they are confided in real-time analytics. A new model of automated feature selections for network IDS parameters that are pre-prpocessed for efficieny of classifications is presented. This paper's proposed methodology combines multiple techniques for improving automated feature selections. The proposed technique is experimented on the KDD Cup 1999 dataset, a common source for examining IDS systems. The technique is also evaluated for efficiency in feature selection by three classifiers in terms of their test and train scores.

Keywords IDS · Feature selection · Voluinous datasets · Cyber safety · Efficient features

P. Ramachandran (✉) · R. Balasubramian
Department of Computer Science, J.J College of Arts & Science (Autonomous), J.J. Nagar, Sivapuram, Pudukkottai 622422, India
e-mail: rmc.it86mails@gmail.com

R. Balasubramian
e-mail: drrb_1951@yahoo.com

J. S. Raj et al. (eds.), *Intelligent Sustainable Systems*, Lecture Notes in Networks and Systems 213, https://doi.org/10.1007/978-981-16-2422-3_51

1 Introduction

Introduction: Current daily lives involve a lot of work on internet-based applications resulting in an explosive growth of networks and their upgrations to accommodate new evolving technologies. In spite of this growth and ehnacements including IoT, smart phone/cities and Mobile Ad-hoc NETworks (MANETs), these networks are open to cyberattacks, a persistent menace in this area. The evolving growth of technology has also resulted in the evolution of sophisticated attacks [1] which has resulted in a growth of cyber-threats. In spite of Organization safeguarding their information with strong firewalls and user authentications, they do fall a prey to cyber attacks [2] and generating a need to safeguard information with IDSs [3]. Figure 1 depicts increasing cyberattacks for a decade.

Organizational IT departments deploy IDSs for visualizing potential malicious activities within their environments. IDS as a function remains critical to enterprises. IDS is a software or a hardware appliance that monitors traffic in networks searching for malicious activities. IDSs can be distinguished based on place of detection namely Network Intrusion Detection System (NIDS) on networks and Host Intrusion Detection System (HIDS) in servers. IDSs can also be implemented using Machine Learning Techniques (MLTs) which identify signatures (misuse) or anomalies (outliers) for safeguarding the networks [4].

Anomalies are detected by assessing deviations from normal profiles, and they are found to have high False Positives (FPs) in results [5]. On the other hand, detection misuses can effectively distinguish legitimacy from adversities based on known patterns [6]. Moreover, IDS implementations use a public domain for sharing the expertise throwing it open for hackers to evade IDSs [7].

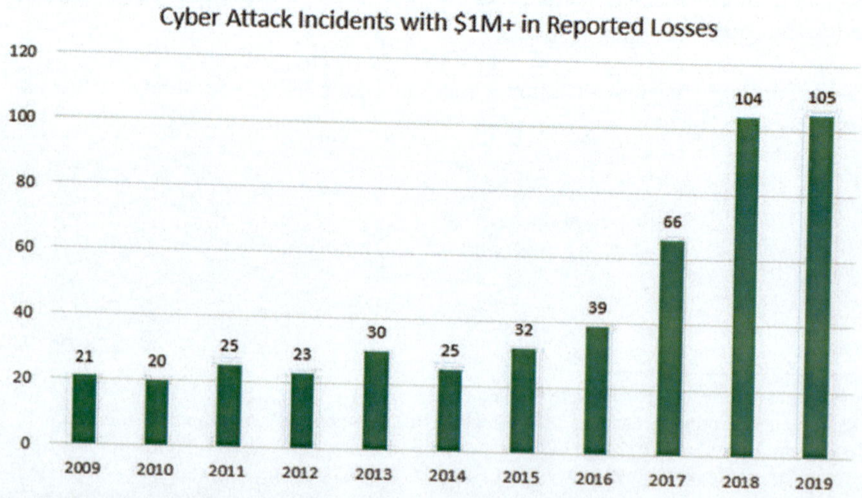

Fig. 1 Cyberattacks between 2009-201. *Source* https://sectigostore.com/blog/42-cyber-attack-statistics-by-year-a-look-at-the-last-decade/

Three main challenges arise while modeling effective IDSs for attacks. The primary problem lies in selecting apt features from a dataset as features change when attack types change. The next issue is in finding labels in real-time networks [8]. The third challenge lies in redundant information which increases network data's dimensions. Presence of redundancy in information dominates classification accuracies negatively. In data types like network information which change dynamically the complexity increases further.

Most works use unwanted features making them cumbersome in operations. MLTs like Support Vector Machines (SVMs), Decision Trees (DTs), K-Nearest Neighbors (KNN), Artificial Neural Networks (ANNs) have been widely used in IDS classifications [9]. Thus, improving methods for prevention/detection of cyberattacks acquired great importance where integration of MLTs in cybersecurity systems is preferred [10].

This paper proses an automated efficient feature selection for selecting optimal network parameters for IDSs called Automatic Correlated Recursive Wrapper-based Feature Selector (ACRWFS) for Efficient Classification of from optimally selected Network Intrusion Features. The organization of this paper is as follows. Section 2 is review of literature related to feature selections and IDS. Section 3 explains the proposed methodology and experimental results of ACRWFS executed using python language on KDD CUP 1999 dataset are displayed in Sect. 4. This paper concludes with Sect. 5.

2 Review of Related Literature

Data Mining Trechniques (DMTs) were first applied to network intrusions in 1998. DMTs and MLTs are the base for developing significant tools for cybersecurity and IDSs. The original classifications in DMTs were based on clustering due to the absence of labels and clustering-based techniques which automatically labeled them. This was then used to detect attacks. MLTs aim to find the best fitting model and currently several MLTs have been proposed for cybersecurity and IDS. Dataset clustering created classes with similar data records. Packet's behaviors were analyzed and identified as normal/abnormal based on characteristics found in data. Mini-batch K-means clustering was used for detecting IDS as it uses multiple random groups with distinct memory sizes for computations [11].

Anomalies are deviations in network traffic which can be assessed using packet flow and byte flow counters derived from device Simple Network Management Protocol (SNMP) values [12] or NetFlow [13]. The study in [14] predicted anomalous changes in traffic volumes by using wavelet analysis from NetFlow and SNMP data. Their signal analysis combined simple metrics for efficient results on high volume anomalies. The study [15] analyzed network counters by simulating TCP SYN and ICMP/UDP flooding attacks. The technique detected attacks based on bandwidth consumptions and packets distributed.

Anomaly detection algorithms proposed in [16] were based on its assessment of sudden or major traffic changes from SNMP data. Their technique was a linear parsimonious model. The model alienated legitimate from nuisance traffics. This followed by [17], introduced improvements by using Principal Component Analysis (PCA) in detecting false alarms like worm propagation which initiate substantial traffic volume changes. Practical problems connected with the use of counters were reported in the study [18] which said routers change anomaly counter when they save packets, but do not change traffic features. Numerous anomaly detection solutions are based on SNMP/NetFlow counters like NFSen [19], AKMA Labs FlowMatrix [20] and NtopNg [21] offer anomaly detections based on rule sets for reporting unwelcome behaviors

Deep Learning (DL)- based techniques help in better classifications of attack types. DL autoencoders with three layers can be used in IDS. The work in [7] advocated an analysis of detecting mechanisms of IDSs to detect further attacks easily. The study also classified types of attacks which enabled fine tuning the mechanisms for efficient detections. The study's evaluations revealed Neural Networks (NNs) followed by DTs and KNN detected attacks better in the order specified. This study resulted in the proposal of [22] which specified optimal parameters for the use of NNs in IDS. Their approach was based on a five-stage model which compared best parameters of best dataset features by normalizing them. Techniques like Apache Spark were proposed to improve pre-processing of datasets for enhanced performances [23].

Ensemble models have also been proposed in IDS for attack classifications. The study in [24] used a modified Accuracy Weighted Ensemble (AWE) for mining information. Ensemble Methods (EMs) are MLTs that combine multiple models to reduce FPs, False Positive Rates (FNRs) for producing accurate solutions and better than a single model. An ensemble using REP Tree for classifications was proposed in [25]. The scheme built its IDs model quickly while using the NSL-KDD dataset and provided higher accuracy in classification and with lowered FPs. The study [26] proposed a cluster-based ensemble classifier for IDS built on Alternating Decision Tree (ADTree) and KNN. Their proposed model outperformed others in accuracy of detections. A strong learner was built in [27] where the proposed ensemble model used C5.0, J48, Naive Bayes (NB) and Partial Decision List (PART). Their aim was to combine weaker learners for producing accurate results for detections in IDS. IoT networks were the basis for assessing malicious events like botnet attacks in the study [28]. Their technique used new statistical flow features and used AdaBoost to learn and detect attacks effectively.

Many current researches focus on hybrid approaches in feature selections followed by ensemble learnings for improved performances of IDSs. Particle Swarm Optimization (PSO) was combined with Random Forest (RF) in [29] for selecting most apposite features from classes. The model results showed elevated accuracy with reduced FPs. A comprehensive survey of IDS techniques prompted this research to concentrate on feature selections as they form the base for any classifier's accuracy.

2.1 ACRWFS Architecture

The proposed architecture uses a combination of techniques for omproving efficiency of classifiers. This scheme is an automated feature selector for network IDS parameters that are pre-pocessed where dataset information is cleaned by removing null values and non-numerical descriptions for efficient feature extractions. Pearson Correlations (PC) handle feature extractions by tracing highly correlated features. Feature selection is executed using Ordinary Least Squares (OLS) combined with Recursive and backward feature eliminations (RFE/BFE) for selecting optmal network intrusion features from KDD Cup dataset. The reliability of the systems checked by evaluating the selected features set using Gaussian Naïve Bayes (GNB), DT, and RF classifiers in terms of their training and testing accuracy scores. Figure 2 depicts the architecture of ACRWFS.

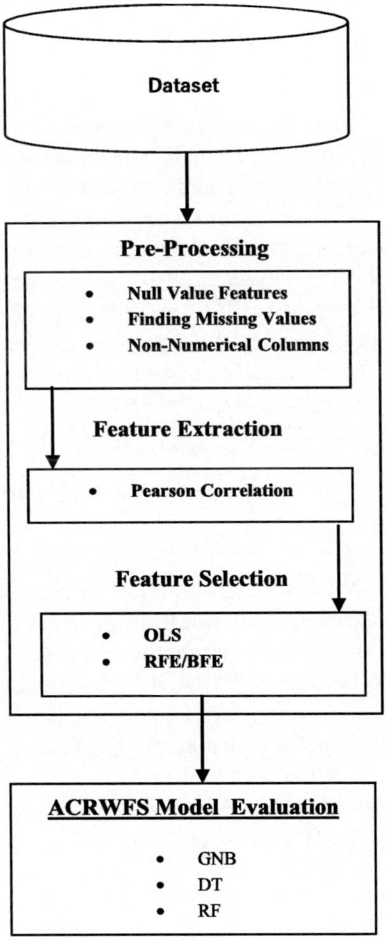

Fig. 2 Architecture of ACRWFS

Fig. 3 Missing Values Identification

Datasets are preprocessed for changing data values into numerical inputs for further processes to handle the same data like transformation/normalizations. Varying pre-processing methods lead to varied results in terms of accuracy, or unexpected results. Numeric values converted into a value of z-score gave good results while representing mean and Standard Deviation (SD) [30]. This study removes or normalizes the dataset by identifying nulls, missing values in features, and by eliminating non-numerical feature columns. The final output of the pre-processing module is then passed on for feature extractions from the dataset. Figure 3 depicts ACRWFS identification of missing values.

2.2 *ACRWFS Feature Extraction*

Feature extraction is an important for MLT processing. A feature is actually a column in a dataset. Every feature may not impact the output variable. Hence, ACRWFS addresses irrelevant features in the model by identifying correlated features for the next stage. Feature extraction in this work refers to the identification of required features and dimensionality reduction of the dataset by using filters and wrappers as execution time/accuracy is dependent on extracted features. PC investigates relationships between variables. The achieved correlation coefficients (r) in a range [-1, 1] correspond to correlation in variables and where negative values imply inverse relationships and 0 implies the absence of any relationships. PC's equation is depicted in Eq. (1)

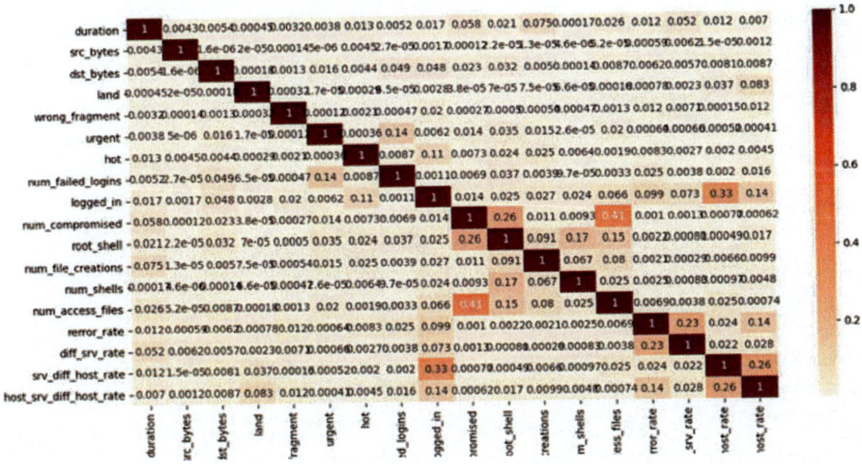

Fig. 4 ACRWFS PC Heatmap

$$Corr(X, Y) = \frac{\sum_{i=1}^{n}(xi - \bar{x})(yi - \bar{y})}{\sum_{i=1}^{n}(xi - \bar{x})^2 \sum_{i=1}^{n}(yi - \bar{y})^2} \quad (1)$$

where X, Y are coordinates of each sample point and P is the pressure at each coordinate. Thus, this study checks connection in the signature variables (x, y, p) where PC coefficient in (1) distinguish each feature. The resulting coefficients with relationships between variables results in columns that are highly correlated and whose features can be extracted for selections. Figure 4 depicts the Heatmap output of PC in python.

ACRWFS Feature Selection: It is selecting a reduced set subset of features from a bigger set. MLTs lose effectiveness due to irrelevant input features. ACRWFS uses wrappers and embedding in its feature selection process. OLS, Recursive Feature Elimination (RFE) and Backward Feature Elimination (BFE) are combined to select optimal features. OLS selects a subset of features using the sum of the squared vertical distances and finding the dividing line from which points are at a minimum distance. OLS is computed using Eq. (2)

$$\mathrm{Yi} = \beta 0 + \beta 1\ \mathrm{Xi} + \varepsilon \mathrm{i} \quad (2)$$

where Xi—Independent Variable, Yi—Dependent Variable, β0—Population Y intercept, β1- Population slope coefficient, and ε—Random Error Component. It can also be expressed as linear regression model in terms of matrix operations where predictor matrix XX of size n×(k+1)n×(k+1)), where kk is the no of predictor variables and nn the no of observations. The predictor matrix includes the values for all predictor variables and it also includes an "intercept column" (XT)0(XT)0 for which Xi0=1Xi0=1 for all 1≤i≤n1≤i≤n so that the intercept β0β0 can be treated on a par with the other

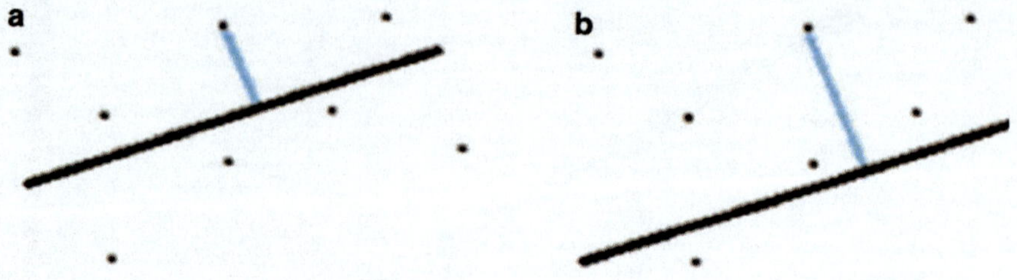

Fig. 5 OLS

regression coefficients and the linear predictor vector ξξ is ξ=Xβ. OLS is depicted in in Fig. 5.

BFE: BFE uses statistical hypotheses and P-Value in its executions. A P-value helps determine whether a feature can be accepted or rejected based on its impact on the output. ML model fits all the features according to its P-value and features having P-value >.05 are removed from the dataset. This process is repeated for all the features in the dataset.Thus, eliminating features which are not significant enough for the model.

RFE: The study uses RFE to wrap a core MLT method for filtering features prior to its selection. It ranks features in modeling with a MLT and eliminates unimportant features. Features are scored either using a MLT or by using a statistical method like regression. This study used a regression model RFE.

Results and Evaluation: This section displays stagewise experimental results of the proposed scheme executed using Python 3.9 on an AMD Athlon processor with 4 GB memory. The experiments were coded for the KDD-99 dataset which was downloaded from the UCI repository. The dataset includes intrusions in a simulated military network. Figure 6 depicts a snapshot of the fields in the dataset.

ACRWFS Pre-Processing Results: Datasets may have missing values and cause problems for MLTs. These values along with null values should be replaced with an average value or removed from the dataset. The proposed technique removed features having null values while imputing missing values statistically using the mean. ACRWFS also removes descriptive non-numerical columns from the dataset as correlations or relations can be assessed using only numerical data. Figure 7 depicts the output of pre-processing in ACRWFS.

ACRWFS Feature Extraction Results: PC correlation of independent variables is used to predict output variables. Features which had correlation of above 0.5 on their absolute values were only chosen in coefficient range of-1 to 1. All other features apart from those that were correlated were dropped or eliminated. ACRWFS creates a correlation matrix and then chooses the upper triangle of the correlation matrix as its output of Feature Extraction. The correlation matrix is depicted in Fig. 8 while Fig. 9 depicts the upper correlation matrix

ACRWFS Feature Selection Results: Feature Selection is the process of selecting relevant features manually or automatically based on certain conditions. Irrelevant features in data can decrease the accuracy of predicting models. The

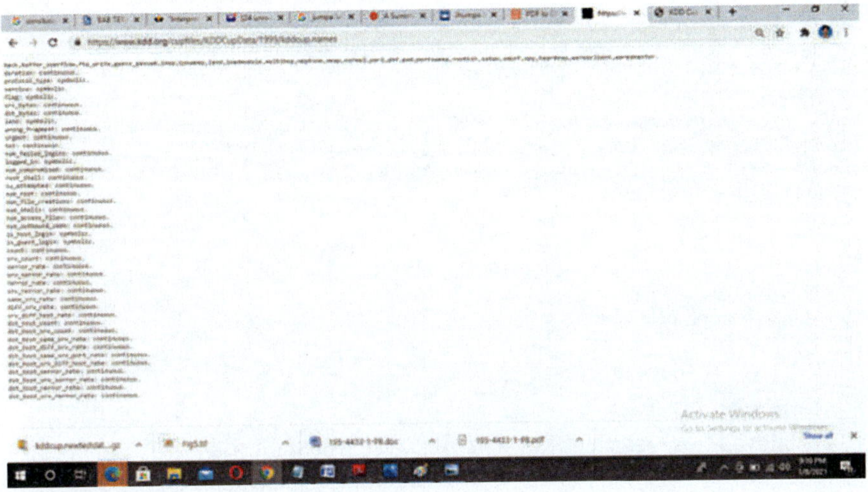

Fig. 6 KDD-99 Dataset Snapshot

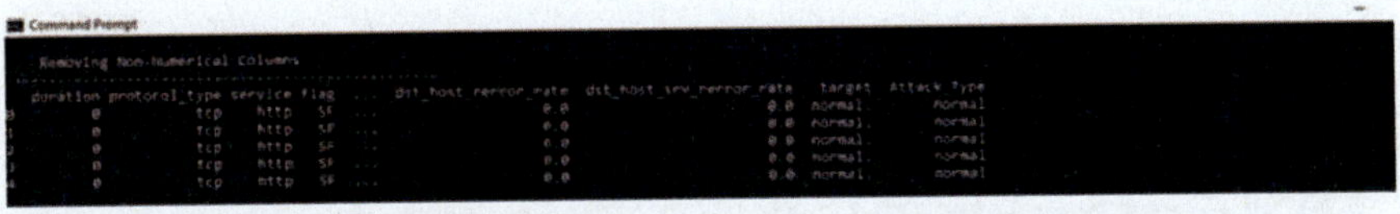

Fig. 7 Removing Non-Numerical Columns

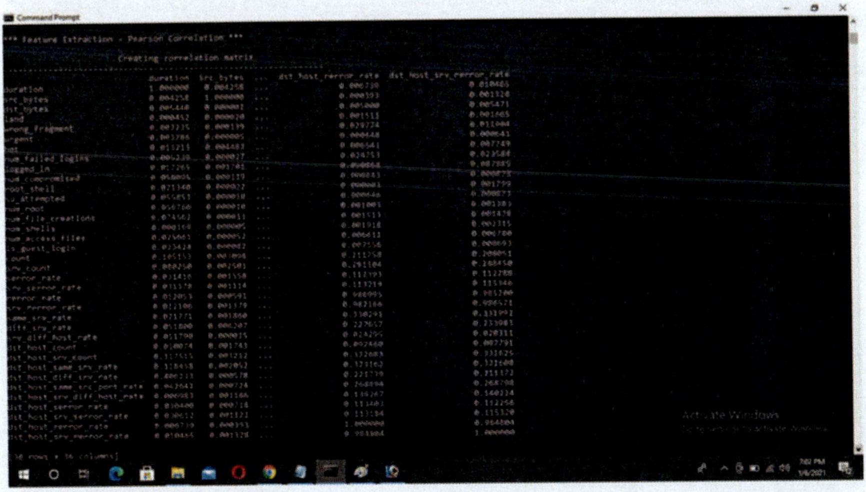

Fig. 8 PC Correlation Matrix

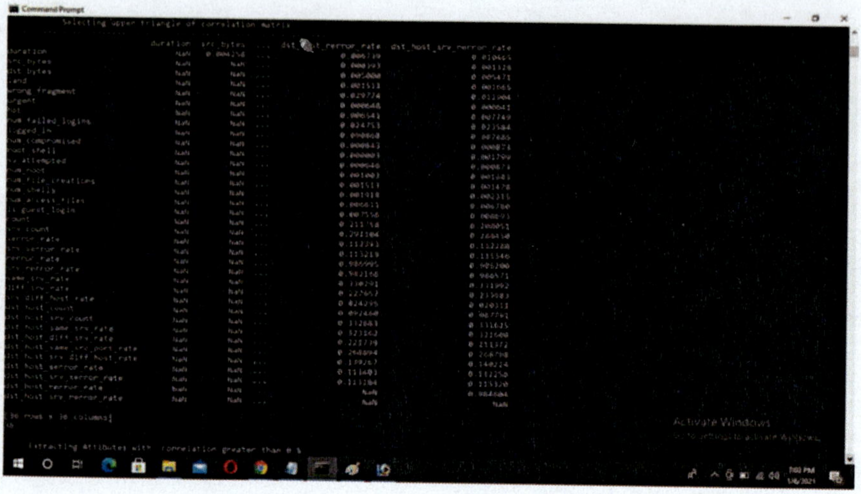

Fig. 9 PC Upper Correlation Matrix

features are mapped for their validity using the OLS technique. OLS is also a dimensionality reduction part of the ACRWFS. Figure 10 depicts the output of OLS Feature Mapping.

BFE in this work is performed with four simple steps namely (1) Select a significance level percentage (.05), (2) Fit a model with all features (variables), (3) Consider feature with highest P-Value (P > Siginificance percentage) else eliminate Feature, and (4) Refit the model with the selected feature. Figure 11 depicts the output of BFE

RFE the final step of ACRWFS's feature selection process helps in choosing the best set of features which can maximize efficiency of classifications. Figure 12 depicts the output of RFE and ACRWFS.

ACRWFS Evaluation: The proposed model's chosen feature sets from KDD-99 were classified by three known classifiers namely GNB, DT, and RF. The dataset was split into 70/30 where 70% was taken from training while the remaining for testing. Figure 13 depicts the screenshot of the classifications on the automatically derived

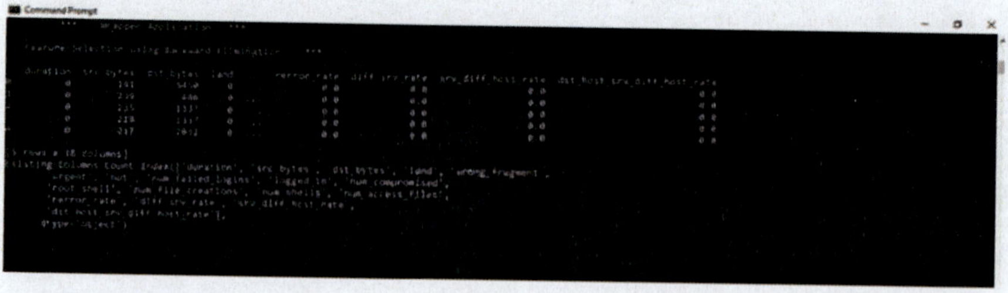

Fig. 10 OLS Output

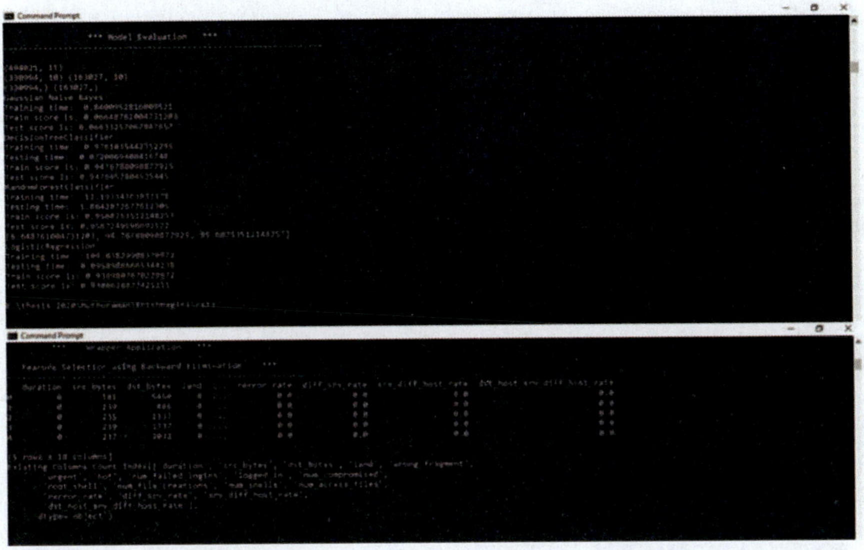

Fig. 11 BFE Output of ACRWFS

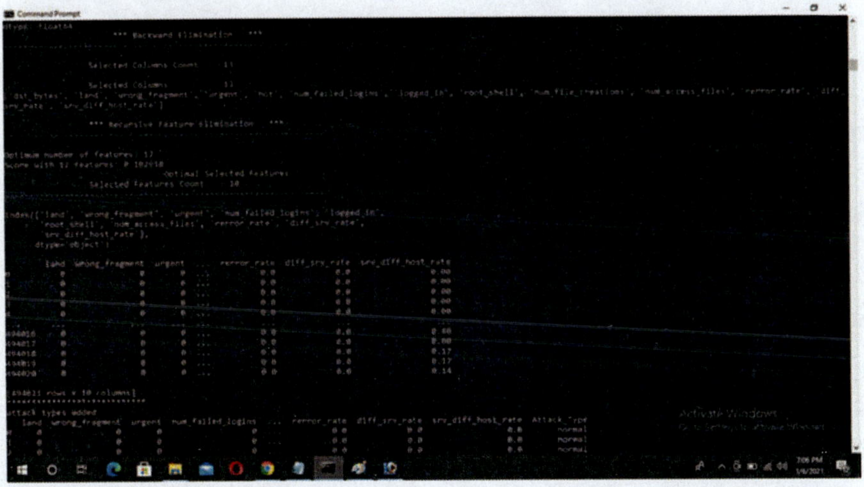

Fig. 12 ACRWFS Output

optimal feature subset of KDD-99 in terms of their training time, training score, and testing score while Fig. 14 depicts it graphically.

It can be seen from the above figure that the output of ACRWFS was best classified by RF in terms of accuracy of 95%, followed by DT with 94% and GNB with 66%. Figure 14 depicts the Training and Testing scores of the Models.

Fig. 13 ACRWFS Model Evaluation

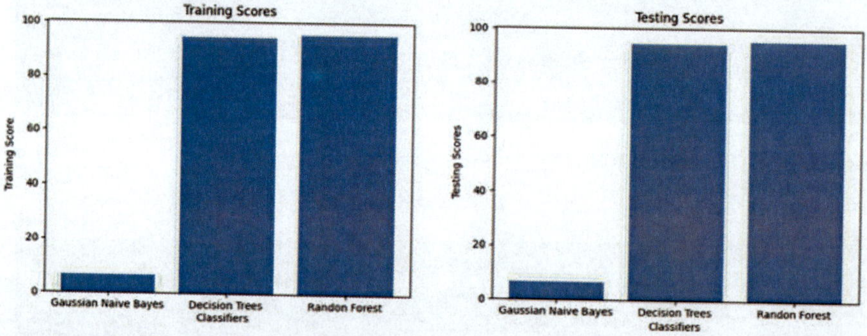

Fig. 14 Classifier Evaluation

3 Conclusion and Future Work

Network intrusions are adverse activities on networks and a kind of cybercrime. Softwares aming to protect networks including insiders belong to the IDS categoty. Researches aim to improve the capabilities through MLTs. Predictive models distinguish between bad and good connections using simulations or executions of their techniques in datasets. The datasets are also filled with noise and interferences which reduce classification accuracies. This data has to be cleaned before use. Further, datasets related to IDS data have numerous and recurring fields. The proposed technique ACRWFS created for selecting features required to classify network parameters for implementing IDS has been proposed in this work. Identifying features that are relevant, minimal and apt have to be chosen either manually or automatically. Moroover, recurrences or duplication of fields while analyzing them increase the dataset size and consume costly processing time in computers. Hence, dimensionality reduction techniques are applied for better results. OLS estimated values of PC coefficients as a part of dimensionality reduction in this work as OLS seeks to minimize features by using sum of squared residuals using linear algebra for optimal estimation of coefficients. Though it is unusual to implement OLS, it quickly reduces dimensionality which are reduced further for selection by BFE and RFE. Thus, this work's proposed technique is a novel automatic feature selection which executes pre-processing, dimensionality reduction, and feature selections for selecting the optimal set of parameters for classifications. The proposed work has suggested, implemented, and demonstrated with results of its proposed ACRWFS technique which can be implemented on IDS systems. Future work would be using an ensemble classifications model which automates the entire set of processes prior to its execution.

References

1. Key Challenges. https://www.weforum.org/centre-for-cybersecurity/home/. Accessed 15 April 2019
2. Al-Jarrah, O.Y., Alhussein, O., Yoo, P.D., Muhaidat, S., Taha, K., Kim, K.: Data randomization and cluster-based partitioning for botnet intrusion detection. IEEE Trans. Cybern. **46**, 1796–1806 (2015). https://doi.org/10.1109/TCYB.2015.2490802
3. Wang, K., Du, M., Maharjan, S., Sun, Y.: Strategic honeypot game model for distributed denial of service attacks in the smart grid. IEEE Trans. Smart Grid **8**, 2474–2482 (2017). https://doi.org/10.1109/TSG.2017.2670144
4. Joldzic, O., Djuric, Z., Vuletic, P.: A transparent and scalable anomaly-based dos detection method. Comput. Netw. **104**, 27–42 (2016). https://doi.org/10.1016/j.comnet.2016.05.004
5. Papamartzivanos, D., Mármol, F.G., Kambourakis, G.: Den-dron: Genetic trees driven rule induction for network intrusion de-tection systems. Fut. Gen. Comput. Syst. **79**, 558–574 (2018). https://doi.org/10.1016/j.future.2017.09.056
6. Kim, J., Kim, J., Thu, H.L.T., Kim, H.: Long short term memory recurrent neural network classifier for intrusion detection. In: 2016 International Conference on Platform Technology and Service (PlatCon), IEEE. pp. 1–5 (2016). https://doi.org/10.1109/platcon.2016.7456805
7. Mishra, P., Varadharajan, V., Tupakula, U., Pilli, E.S.: A detailed investigation and analysis of using machine learning techniques for intrusion detection. IEEE Commun. Surv. Tutor. **21**, 686–728 (2019)
8. Types of Intrusion Detection System. https://en.wikipedia.org/wiki/Intrusion_detection_system
9. Jianliang, M., Haikun, S., Ling, B.: The application on intrusion detection based on K-means cluster algorithm. In: International Forum on Information Technology and Application, IEEE, 15–17 May 2009, pp. 150–152
10. Geluvaraj, B., Satwik, P.M., Kumar, T.A.: The future of cybersecurity: major role of artificial intelligence, machine learning, and deep learning in cyberspace. In Proceedings of the International Conference on Computer Networks and Communication Technologies. Springer: Singapore (2019), pp. 739–747
11. Peng, K.A.I., Leung, V.C.M., Huang, Q.: Clustering approach based on mini batch kmeans for intrusion detection system over big data. In: Special Section on Cyber-Physical- Social Computing and Networking. 10.1109/ACCESS.2018.2810267
12. Harrington, D., Presuhn, R., Wijnen, B.: An Architecture for Describing Simple Network Management Protocol (SNMP) Management Frameworks. http://www.ietf.org/rfc/rfc3411.txt. Accessed 16 April 2015
13. Claise, B.: Cisco Systems NetFlow Services Export Version 9. http://tools.ietf.org/html/rfc3954. Accessed 16 April 2015

14. Barford, P., Kline, J., Plonka, D., Ron, A.: A Signal Analysis of Network Traffic Anomalies. In: Proceedings of the 2nd ACM SIGCOMM Workshop on Internet Measurement (IMW'02), Marseille, France, 6–8 November 2002; pp. 71–82
15. Kim, M.S., Kong, H.J., Hong, S.C., Chung, Hong, J.: A flow-based method for abnormal network traffic detection. Presented at IEEE/IFIP Network Operations and Management Symposium (NOMS 2004), Seoul, Korea, 19–23 April 2004, pp. 599–612
16. Casas, P., Fillatre, L., Vaton, S., Nikiforov, I.: Volume anomaly detection in data networks: an optimal detection algorithm vs. the PCA approach. In: Valadas, R., Salvador, P. (eds.) Traffic Management and Traffic Engineering for the Future Internet, vol. 5464, Lecture Notes in Computer Science. Springer, Berlin/Heidelberg, Germany (2009), pp. 96–113
17. Jingle, I., Rajsingh, E.: ColShield: An effective and collaborative protection shield for the detection and prevention of collaborative flooding of DDoS attacks in wireless mesh networks. Human-centric Comput. Inf. Sci. **4** (2014). https://doi.org/10.1186/s13673-014-0008-8
18. Zhou, W., Jia, W., Wen, S., Xiang, Y., Zhou, W.: Detection and defense of application-layer DDoS attacks in backbone web traffic. Fut. Gen. Comput. Syst. **38**, 36–46 (2014)
19. NfSen—Netflow Sensor. http://nfsen.sourceforge.net. Accessed 16 April 2015
20. AKMA Labs FlowMatrix. http://www.akmalabs.com. Accessed 16 April 2015
21. NtopNg—High-Speed Web-based Traffic Analysis and Flow Collection. http://www.ntop.org. Accessed 16 April 2015
22. Larriva-Novo, X.A., Vega-Barbas, M., Villagra, V.A., Sanz Rodrigo, M.: Evaluation of cybersecurity data set characteristics for their applicability to neural networks algorithms detecting cybersecurity anomalies. IEEE Access Appl. Sci. **8**(10), 3430 (2020)
23. Belouch, M., El Hadaj, S., Idhammad, M.: Performance evaluation of intrusion detection based on machine learning using Apache Spark. Procedia Comput. Sci. **127**, 1–6 (2018)
24. Ahmad, M., Basheri, M.J., Iqbal, Rahim, A.: Performance comparison of support vector machine, random forest, and extreme learning machine for intrusion detection. 10.1109/ACCESS.2018.2841987
25. Gaikwad, D., Thool, R.C.: Intrusion detection system using bagging ensemble method of machine learning. In: 2015 International Conference on Computing Communication Control and Automation, IEEE. pp. 291–295 (2015). https://doi.org/10.1109/iccubea.2015.61
26. Jabbar, M., Aluvalu, R., Reddy, S.S.S.:. Cluster based ensemble classification for intrusion detection system, in: Proceedings of the 9th International Conference on Machine Learning and Computing, pp. 253–257 (2017). https://doi.org/10.1145/3055635.3056595
27. Paulauskas, N., Auskalnis, J.:. Analysis of data pre-processing influence on intrusion detection using nsl-kdd dataset, in: 2017 Open Conference of Electrical, Electronic and Information Sciences (eS-tream), IEEE. pp. 1–5 (2017). https://doi.org/10.1109/estream.2017.7950325
28. Moustafa, N., Turnbull, B., Choo, K.K.R.: An ensemble intru-sion detection technique based on proposed statistical flow features for protecting network traffic of internet of things. IEEE Internet Things J. (2018). https://doi.org/10.1109/JIOT.2018.2871719
29. Malik, A.J., Shahzad, W., Khan, F.A.: Network intrusion detec-tion using hybrid binary pso and random forests algorithm. Secur. Commun. Netw. **8**, 2646–2660 (2015). https://doi.org/10.1002/sec.508
30. Larriva-Novo, X.A., Vega-Barbas, M., Villagra, V.A., Sanz Rodrigo, M.: Evaluation of cybersecurity data set characteristics for their applicability to neural networks algorithms detecting cybersecurity anomalies. IEEE Access **8**, 9005–9014 (2020)

Machine Learning-Based Early Diabetes Prediction

Deepa Elizabeth James and E. R. Vimina

Abstract There are several diseases that the world faces presently and a critical one is Diabetes mellitus. The current diagnostic practice involves various tests at a lab or a hospital and a treatment based on the outcome of the diagnosis. This study proposes a machine learning model to classify a patient as diabetic or not, utilizing the popular PIMA Indian Dataset. The dataset contains features like Pregnancy, Blood Pressure, Skin Thickness, Age and Diabetes Pedigree Function along with regular factors like Glucose, BMI and Insulin. The objective of this study is to make use of several preprocessing techniques resulting in improved accuracy over simple models. The study compares different classification models namely GaussianNB, Logistic Regression, KNN, Decision Tree Classifier, Random Forest Classifier, Gradient Boosting Classifier in several ways. Initially, missing values in the significant features are replaced by computing median of the input variables based on the outcome of whether the patient is diabetic or not. After this, feature engineering is performed by adding new features which are obtained by categorizing the existing features based on its range. Finally, Hyperparameter tuning is carried out to optimize the model. Performance metrics such as Accuracy and area under the ROC Curve (AUC) is used to validate the effectiveness of the proposed framework. Results indicate that XGBoosting Classifier is concluded as the optimum model with 88% accuracy and AUC value of 0.948. The performance of the model is evaluated using Confusion Matrix and ROC Curve.

Keywords Diabetes mellitus · Machine learning classification · Data mining · Hyperparameter tuning · XGBoosting

D. E. James (✉) · E. R. Vimina
Amrita School of Arts and Science, Amrita Vishwa Vidyapeetham, Kochi Campus, Kochi, India
e-mail: deepaelizabethjames@gmail.com

E. R. Vimina
e-mail: viminaer@asas.kh.amrita.edu

J. S. Raj et al. (eds.), *Intelligent Sustainable Systems*, Lecture Notes in Networks and Systems 213, https://doi.org/10.1007/978-981-16-2422-3_52

1 Introduction

A Mexican-American woman in her late 30s was tested for pregnancy due to a missed menstrual period and the result was positive. She earlier had six miscarriages and five normal deliveries. She had a height of 62 inches, and her weight was 198 lb. Her physical examination was normal. She did not have retinopathy nor any indication of neuropathy. However, her blood glucose reading was between 180 and 220 mg/dl. Insulin was prescribed to her to be started at the earliest.

Diabetes is a serious health issue that limits the body's ability to process blood glucose also known as blood sugar. The statistical result collected in 2017 depicted that 451 million people worldwide were living with diabetes. This is supposed to increase to 693 million by 2045 [1]. A different statistical study [2] says about the severity of diabetes that nearly half of a billion people in the world currently have diabetes. The forecast is that the number might increase to 25% of the entire population, in the year 2030 and 51% in the year 2045.

What causes Diabetes? Most of the foods that we eat are broken down into glucose and is released into your bloodstream. A hormone called insulin in the body helps to regulate the level of glucose in the blood. This is by letting the blood glucose enter into your body's cells for use as energy. The pancreas in the body releases insulin. However, when you have diabetes, pancreas either stops producing enough insulin or your body stops responding to the insulin that is being produced. As a result, too much blood sugar stays in the blood and over time this causes serious implications. Diabetes cannot be cured. However, regular exercise, healthy diet and taking medicines as needed can reduce the impact of diabetes.

Diabetes is of three types—Type 1, Type 2 and Gestational details of which are given in Table 1.

Prediabetes is another type that is considered as a phase prior to Type 2. In this type, the level of glucose of the patient is above normal but not as of Type 2. People with

Table 1 Description and characteristics of the diabetes

Diabetes type	Description	Characteristics
Type 1	This type occurs when the pancreas in the body halts to produce insulin and as a result the patients are required to take artificial insulin daily	Since this type is diagnosed mostly during childhood, it is also referred to as juvenile diabetes.
Type 2	This type occurs when the body halts to respond to the insulin produced by the pancreas in the body and this generally happens due to excess body weight and lack of physical activity	90% of the people with diabetes has Type 2 diabetes
Gestational	This usually happens during pregnancy	There is chance that this could develop into Type 2 diabetes for the mother and child

prediabetes has greater chance of getting Type 2 diabetes under specific conditions and measures.

Diabetes might lead to dangerous complications like heart disease, stroke, vision loss, amputation, etc. The methods used currently for detection of diabetes perform tests in a lab to check fasting blood glucose and oral glucose tolerance. Not only this method is time consuming but also requires multiple visits to the lab. An early prediction of diabetes would be a great advantage for the patients especially for pregnant woman and also during pandemic times as of COVID-19. This paper intends to build such a predictive model with the help of machine learning algorithms and data mining techniques. These are explained in detail in the subsequent sections.

2 Literature Review

Different researchers have used different models to study the classification of the diabetes data and few have been briefed below.

The study by Maniruzzaman et al. [3] compared the three different kernels namely radial basis function, polynomial and linear against the traditional QDA, LDA and NB. The authors also conducted extensive trials to find the best cross-validation protocol. Their experiments illustrate that the GP-based classifier with K 10 cross-validation protocol is the best performing classifier for the prediction of diabetes. The GP-based model presented in this paper gives a performance as accuracy 81.97%, sensitivity 91.79%, specificity 63.33%, positive and negative predictive values of 84.91% and 62.50% that are higher compared to other models.

Messan et al. [4] used different parameters from the dataset to explain different classification algorithms. This research used a small data sample for Diabetes prediction. However, the researchers did not incorporate the parameter of pregnancy to predict diabetes. This paper used five different algorithms namely SVM, GMM, EM, ANN and Logistic regression. And the researchers concluded that Artificial Neural Network or ANN provided highest accuracy for the prediction of Diabetes.

Francesco et al. [5] used the models Multilayer perceptron, Hoeffding Tree, Random Forest, Jrip, BayeNet and Decision Tree algorithms for prediction. In this study, the researchers used the best first and greedy stepwise feature selection algorithm for feature selection need. The study concludes that Hoeffding tree (HT) delivers high accuracy. The HoeffdingTree algorithm model used in this study utilizes real-world data and obtains a value of precision equal to 0.770 and recall value equal to 0.775.

Sisodia et al. [6], implemented three different ML classifiers namely SVM, NB and DT to predict the possibility of diabetes with maximum accuracy. They were able to establish that NB is the best performing model with an AUC of 0.819.

The study by Hassan et al. [7] selected mean value over median value because the median value has less central tendency toward the mean of the attribute distribution. They have used different ML classifiers (k-nearest Neighbour, Random Forest, Decision Trees, Naive Bayes, AdaBoost and XGBoost) and MLP. They have also used an ensemble classifier by combining the ML models for improving the diabetes prediction. The AUC of the ML model is chosen as the weight of that model for voting rather than accuracy, since AUC is unbiased to the class distribution. The classifier proposed in this study gives good performance with sensitivity value of 0.789, specificity value of 0.934, false omission rate of 0.092, diagnostic odds ratio value of 66.234 and AUC of 0.950 that outdoes other results in this area by 2% of AUC.

Alehegn et al. [8] used two different datasets namely PIDD and 130_US hospital diabetes datasets. The ML techniques used in this paper are KNN, Random Forest, J48 and Naïve Bayes. Using stacking meta classifier in case of PIDD, the proposed method provides better accuracy of 93.62%.

The proposed method of Sneha et al. [9] uses predictive analysis to determine the attributes that help to early detect Diabetes. The comparison result displays that the topmost specificity values are given by decision tree model with 98.2% and Random forest with 98% and holds best for analysis of diabetes. The best accuracy value is given by Naïve Bayesian with 82.30%.

Hina et al. [10] performed analysis of a Pima Indian dataset using various classification techniques like Zero R, Naïve Bayes, Random forest, J48, Logistic regression, MLP. Comparison is performed and diabetes is predicted. The data mining tool—WEKA, is used for the study. The study finds that for predicting diabetes, MLP is better in terms of accuracy and performance.

3 Methodology

The methodology section emphasizes on methods used for this study. The literature has various subsections that explain the Machine learning, Data mining, dataset used, proposed framework and techniques used for the evaluation of the framework.

This study utilizes machine learning and data mining techniques to build a machine learning model to predict whether a patient has diabetes or not, by classifying based on several diagnostic measurements.

Machine learning (ML) is a kind of artificial intelligence (AI) that enables software applications to gain greater accuracy at predicting outcomes without being explicitly coded to do so. Data mining is the process of arriving at inconsistencies, patterns and correlations within big data sets in order to predict outcomes. The process of grouping a given set of data into classes is called classification. As Python language has robust and efficient libraries and packages for computational needs, this language has been preferred for performing this study. The problem stated here is called binary classification with supervised learning.

Table 2 Description and characteristics of the dataset

Sr #	Attribute name	Attribute description
1	Pregnancies	The number of pregnancies of a woman
2	Glucose	Concentration of glucose in OGTT for 120 min(mg/dl)
3	Blood Pressure	Diastolic Blood Pressure value (mmHg)
4	Skin Thickness	Fold Skin Thickness value (mm)
5	Insulin	Serum Insulin for 2 h (mu U/ml)
6	BMI	Body Mass Index. This is arrived by the formula = weight/(height)^2 (kg/m2)
7	Diabetes Pedigree Function	Diabetes Pedigree Function
8	Age	Age in years
9	Outcome	Class variable for diabetes (Class value 1 for positive 0 for negative)

3.1 Dataset

The dataset adopted for this study was published basically by the National Institute of Diabetes and Digestive and Kidney Diseases. This dataset contains details about PIMA Indian females, located near Phoenix, Arizona and have been studied several times due to the high occurrence rate of diabetes in PIMA females. The dataset consists of diagnostic measurements of PIMA females who are above 20 years of age. The details of 768 females are available in the dataset, out of which 268 females had diabetes. The dataset includes 8 variables such as Age, Glucose, Blood Pressure, Number of Pregnancies and Insulin. Details about the variables are listed in Table 2.

A binary variable, Outcome, is the response variable in the dataset. It indicates whether the person was diagnosed with Diabetes or not. The data in the outcome variable depicts that 65% of the patients in the dataset are healthy and remaining 35% of the patients have been diagnosed with diabetes. Our model is proposed to predict the outcome based on the input variables and compare the results with output variable data.

3.2 Proposed Framework

Figure 1 below demonstrates the flow considered for the implementation of a machine learning supervised model used for this study, on PIMA dataset.

The proposed framework uses PIMA dataset to build a model for diabetic prediction. The various steps of the framework are listed below

Step 1: The initial step of the architecture includes different methods of pre-processing of data explained in Sect. 3.3.

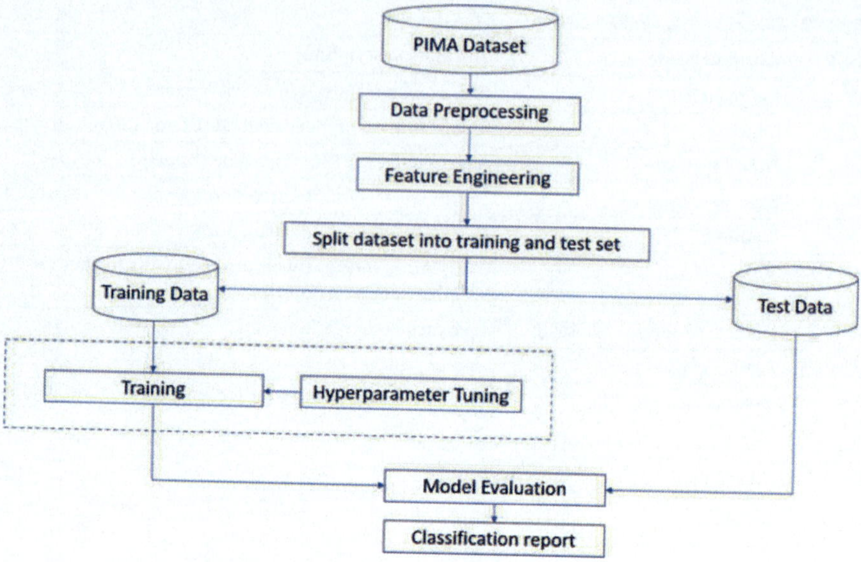

Fig. 1 The proposed framework for this study

Step 2: Feature engineering techniques are executed on the dataset by creating new features based on the range of existing features. This is explained in Sect. 3.6. Standardization of the features are conducted on the dataset by splitting the data for training and testing.

Step 3: K-Fold Cross-validation is performed to compare different classification techniques.

Step 4: In order to get an optimized model, hyperparameter tuning is also performed.

Step 5: The resultant models are set to a final comparison in order to arrive at a model with highest accuracy and AUC score.

3.3 Data Pre-processing

The chance of missing values and/or noisy and inconsistent data in real-world datasets is high. Poor data quality affects the results produced by algorithms that act on these datasets. Hence, after data collection, it is very much necessary to pre-process the data to achieve quality results. Machine learning algorithms require pre-processing of the data for better application of the context of the problem. Pre-processing helps to make the data more appropriate for data mining and analysis. The aim of data pre-processing is to deal with any data that is non-numerical in nature, excerpting significant and admissible elements, feature scaling, dealing with values that are missing and transforming features. The pre-processed stage output is the training dataset.

In the dataset used for this study, certain variables like Glucose, Blood Pressure, skin thickness, insulin, BMI, etc. have zero values. All zero values are replaced by median with respect to the outcome (diabetes or not).

3.4 *Data Visualization*

One of the best data visualization techniques more commonly used in Machine learning is the Histogram. It deals with the division of a continuous variable over a set interval or over a period of time. Each distribution represents normal distribution, outliers and skewness, etc. Skewness can be quantified to display which distribution differs from normal distribution.

Figure 2 shows Histogram Distribution for the various attributes in the diabetes dataset such as BMI, Glucose, Blood Pressure, Diabetes Pedigree Function, Age, Insulin, Pregnancies and Skin Thickness.

- Plots of certain attributes like Blood Pressure, Glucose, BMI, Skin Thickness are distributed normally except for certain outlier values.

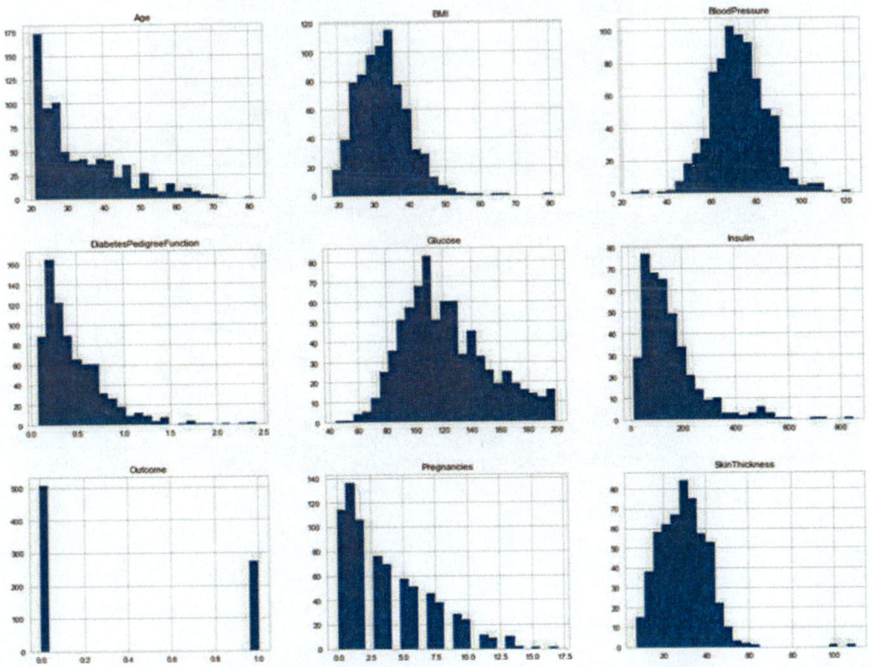

Fig. 2 Histogram of the various dataset variables

- Plots of certain other attributes like Age, Diabetes Pedigree Function, Pregnancy and Insulin are skewed.
- People with high insulin and glucose values have a higher chance of developing diabetes.
- Obese people and people with a greater number of pregnancies too are prone to developing diabetes.

3.5 *Data Correlation*

It is often easier to consider the relationship between features so as to provide a helpful dataset summary. Correlation is a statistic that calculates the degree to which two variables traverse in relation to each other. There are two types—Positive Correlation when the two variables move in same direction and Negative Correlation when it moves in the opposite direction. Figure 3 given here shows the correlation plot based on the diabetes dataset.

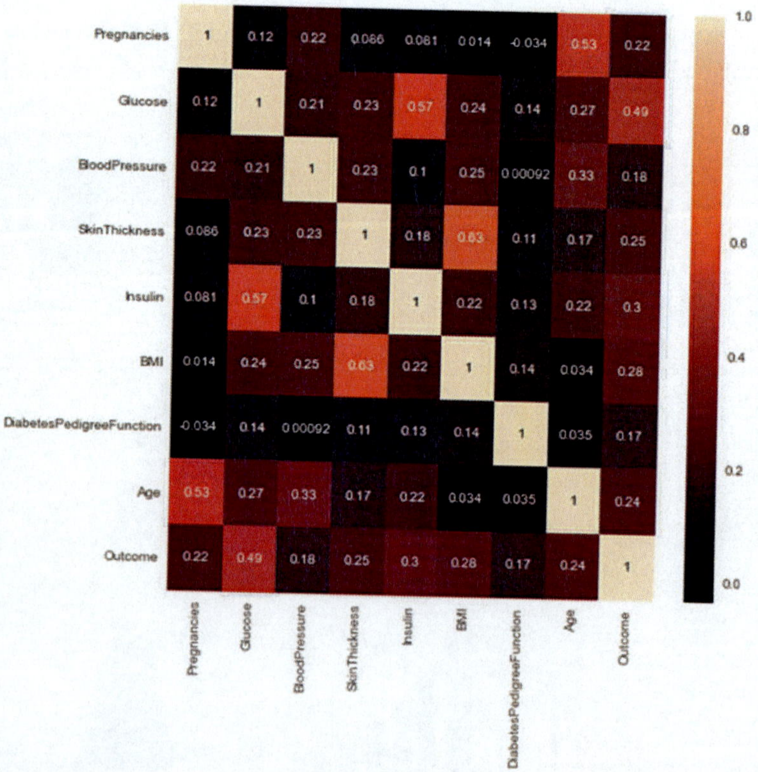

Fig. 3 Heatmap depicting the correlation between the variables

The dataset is analyzed and the correlation between the values are visualized with the help of heatmap. From this it is observed that glucose, BMI, age and insulin are the most correlated features to target variable. The correlation between 'Glucose' and 'Outcome' is 0.49 in Fig. 3, which is greater than all other attributes. That means that if 'glucose' increases, then the person can have diabetes [11].

3.6 Feature Engineering

Features or variables that define the dataset sometimes depend on each other, and this immensely influences the accuracy of the ML classification model [12]. However, creation of new features from existing dataset helps to achieve more concise and precise classifiers. Following are the new features that have been added to the dataset as part of this study.

3.6.1 Glucose Mmol Range

The unit of measurement for the 2-hour oral glucose tolerance test (OGTT) in this dataset is in milligrammes per decilitre (mg/dl). In order to apply a qualitative test result to the numeric results, this is converted to Millimoles per litre (mmol/l) by multiplying the current values by 0.0555.

A new feature has been created in this study for the glucose levels based on the glucose ranges (Glucose_Mmol_Range). The glucose target ranges proposed by the American Diabetes Association should be 3.9–7.2 mmol/l (70–130 mg/dL) on fasting, and less than 11 mmol/L (200 mg/dL) couple of hours post-meal. Following are the ranges arrived for this new feature.

- Hypoglycemia: Blood sugar level < 3.9 mmol/L
- Normal: Blood sugar level between 3.9-7.2 mmol/L
- Hyperglycemia: Blood sugar level > 7.2 mmol/L

The interpretation is that people with Hyperglycemia are at a high risk to develop diabetes.

3.6.2 Blood Pressure Range

Blood Pressure is the pressure made by the blood within the blood vessels. Blood pressure is generally articulated as systolic pressure and diastolic pressure. Systolic pressure is the maximum pressure at the time of heartbeat, and diastolic pressure is the lowest pressure between two heartbeats.

A new feature (BloodPressure_Range) has been created in this study for the BP levels based on the ranges suggested by American Heart Association. The Blood Pressure values collected in the dataset does not have information of whether it is systolic BP or diastolic BP. Since the BP values are between 40 and 120, it has been assumed that it is diastolic BP. Following are the ranges arrived for this new feature.

- BloodPressure < 80 = ''Optimal'
- Blood Pressure between 80 and 84 = 'Normal"
- Blood Pressure between 85 and 89 = 'High normal'
- Blood Pressure between 90 and 99 = 'Grade 1 hypertension'
- Blood Pressure between 100 and 109 = 'Grade 2 hypertension'
- Blood Pressure > 110 = 'Grade 3 hypertension'

The interpretation from the new feature is that people with hypertension 1, 2 or 3 has very high risk to develop diabetes.

3.6.3 Insulin Range

The hormone insulin produced in the body by pancreas allows body to use glucose in the blood for energy. In this dataset, the insulin is obtained as a two-hour serum insulin value (mu U/ml). A new feature has been created in this study for the Insulin Indicative Range (Insulin_Range). Following are the ranges arrived for this new feature:

- Insulin level between 16 and 166 = Normal
- Insulin level below 16 and above 166 = Abnormal

3.6.4 BMI Range

The measurement of a person's weight with respect to that person's height is called Body Mass Index. A new feature has been created in this study for the Body Mass Index range (BMI_Range). Following are the ranges arrived for this new feature:

- Underweight: In this range, the BMI is less than 18.5
- Normal weight: In this range, the BMI is between 18.5 and 24.9
- Overweight: In this range, the BMI is between 25 and 29.9
- Obese: In this range, the BMI is 30 or more

3.7 *Feature Transformation*

Machine learning modelling requires an appropriate choice of data representation approach for categorical data. Only then it would help in achieving desired result when doing the modelling. The effective execution of the models becomes more arduous than the algorithms itself due to system constraints. The PIMA dataset mainly comprises categorical variables—'Glucose_Mmol_Range', 'BloodPressure_Range', 'Insulin_Range', 'BMI_Range'. The need for feature translation arises when the categorical variables get transformed to its numeric representation.

Hot encoding is used to encode these features. The base for this procedure is the creation of a new variable based on value of the category variable at category level. The basic technique used in this procedure is as follows. The value of each category variable is first converted into a new variable. When the category variable is available, a value of 1 is assigned to the new variable. When the category variable is unavailable, a value of 0 is assigned to the new variable. The categorical variable is replaced by the new binary variables. This type of encoding is named dummy encoding and the new variables are referred to as dummy variables.

No numbers are unnecessarily measured in this encoding. However, more variables get added to the dataset which act as the disadvantage of this encoding. If the encoding is efficient enough, then the dataset becomes ready for applying the algorithm. Certain algorithms succeed in resizing the attributes to uniform scale. Hence, it could be said that utilizing standardization each and every feature are rescaled to have a standard deviation value of 1 and mean value of 0.

3.8 Training and Test Dataset Evaluation

Training dataset and test dataset has different purposes. The former helps with the actual classification in the model, whereas the latter helps to validate the result of the model. It is always good to have more training data so that the model has a greater number of instances to work. It is also good to have more test data to understand the generalization capacity of the model being used. Usually, 80% of the dataset is used as training dataset and remaining 20% as test dataset [13].

The PID dataset consists of an unbalanced sample of positive and negative cases. The StratifiedKCV [14] helps to conserve the percentage of all samples for each class as the original. Stratified K-Fold is used for evaluation of the stability of the models, once the models are finalized. As a first step, dataset is segregated into K different subsets or folds. The model is repetitively run K number of times. In each repetition or iteration, the dataset is trained on K-1th fold. Evaluation of the Kth fold is done with test dataset. In this study, a value of 10 is assigned to K.

Dataset is split into train and test set, after the finalization of the model. Final findings were calculated by averaging values over ten iterations from test samples. The selected top three models were refined and optimized by hyperparameter tuning. The confusion matrix reported based on the prediction from the final model is discussed in the result section.

3.9 Model Selection

Machine learning definition is provided at [15] and it says that a computer program learns from experience E in relation to class of task T and performance measure P. For this to happen, the performance of task T, measured with P, should improve with experience E. Eight different classifier models are considered as part of model comparison for this study, in order to arrive at a base model [16]. Selected algorithms are given below

- Logistic Regression: Outcome variables that are categorical in nature and predictors that are either continuous or categorical in nature are related and this is explained by the algorithm of Logistic Regression. [17].
- K-Nearest-Neighbours: KNN is an algorithm that utilizes lazy learning. Prior probability and posterior probability of each label in the KNN neighbour. A hybrid model comprising of LR and k-NN is proposed in [18].
- SVM: Support Vector Method uses nonlinear or linear process to find the best data classifying function [19]. SVM learning based on a ranking method is also available [20, 21].
- Gaussian NB: Conditional probability can be calculated using Bayes Theorem. Naive Bayes classifier is based on the Bayes theorem and is a probabilistic machine learning model that's used for classification task. A Gaussian Naive Bayes algorithm is a special type of NB algorithm that is specifically used when the features have continuous values. It's also assumed that all the features are following a Gaussian distribution, i.e. normal distribution
- Decision Tree Classifier: This model represents a tree structure where a definition is attached with each node [22].
- Random Forest Classifier: RF model is an ensemble model that constitutes of acquiring uncut decision tree. This model has a base classifier as the tree and learners containing features as the division [23]. Subsets of the dataset extracted from the parent training data are used to train the individual trees. This uses the majority of prediction to actualize the classification process. [24]. The output from various base models is joined methodically in ensemble learning in order to avoid a solo prediction. This betters the output of each base classifier [25].
- Gradient Boost Classifier: Gradient Boosting is an iterative functional gradient algorithm that reduces a loss function by iteratively selecting a function that points towards the negative gradient or a weak hypothesis.
- XGBoosting Classifier: This model is a comprehensive lifting tree system [26]. New trees are trained systematically from the errors of earlier decision trees. Top-performing base model is first finalized for the target variable. Then by opting the correct hyperparameters to further increase the accuracy, design aspects of the model get finetuned.

3.10 Hyperparameter Tuning

Each machine learning model has different design choices or input parameters to define its model architecture. These input parameters are called as Hyperparameters. The process of selecting the optimal values of the hyperparameter of any model in order to generate the best output for that model is called Hyperparameter tuning. This is also known as hyperparameter optimization.

3.11 Performance Metrics

It can be usually noticed that a single performance measure would not be able to assess a classifier from multiple perspectives. Also, there is no unified metric to define the classifier's overall performance. This ratio of accurately classified samples over the entire samples within a range is called accuracy.

Even then, since accuracy deals improperly with the inconsistent classes, it is difficult to distinguish among different sorts of misclassification. Therefore, in order to get a better understanding of the classifier's efficiency and get through with unbalanced datasets, it is suggested that certain efficiency metrics should be utilized along with accuracy.

Precision is the ability of the classifier to provide the correct positive predictions amongst all diabetic cases determined by the classifier. The proportion of these correct predictions is termed as recall or sensitivity. Optimized uniformity between recall and precision gives F1score. Receiver operating characteristic (ROC) curve is a good tool to visualize performance of a binary classifier algorithm. In this algorithm, true positive rate is plotted against false positive rate. The ROC curve has an area under it that represents a solo scalar value. This area under the curve or AUC measures the comprehensive execution of a binary classifier [27]. The AUC is a sturdy measure for the evaluation of the performance of score classifiers. The AUC contains all probable classification thresholds as its calculation is dependent on the ROC curve. The AUC is calculated by adding subsequent trapezoid areas beneath the ROC curve.

In order to account for class confusion and class overlap, and also to provide an abstract of predictions, errors, etc. made by the finalized algorithm, a confusion matrix has been used. In other words, the depiction of actual/predicted class combination for each target can be effectively done by the confusion matrix.

4 Results and Commentary

The primary interest of this study was to identify a base classifier having a target variable with high accuracy.

4.1 Classifier Comparison with Original Feature

As explained in the methodology section, the comparison of eight different algorithms is displayed here. The comparison shown in Table 3 uses the base features of the dataset which has undergone data pre-processing. Table 3 demonstrates the evaluation of these models using parameters like accuracy, F1-score, precision, AUC and recall.

Table 3 Base models comparison after data pre-processing

Sl. No.	Model	Accuracy	Precision	Recall	F1-score	AUC
0	GuassianNB	0.74	0.66	0.53	0.59	0.81
1	LogisticRegression	0.77	0.74	0.54	0.62	0.85
2	KNN	0.83	0.76	0.75	0.75	0.88
3	DecisionTree	0.80	0.76	0.71	0.72	0.80
4	RandomForest	0.85	0.80	0.77	0.79	0.92
5	GradientBoosting	0.85	0.80	0.76	0.77	0.93
6	XGBoosting	0.83	0.78	0.74	0.75	0.92
7	SVC	0.83	0.77	0.74	0.75	0.88

Table 4 Base models tenfold CV results comparison

Sl. No.	Model	Accuracy	Precision	Recall	F1-score	AUC
0	GuassianNB	0.67	0.52	0.96	0.67	0.87
1	LogisticRegression	0.83	0.77	0.75	0.75	0.89
2	KNN	0.84	0.78	0.77	0.77	0.89
3	DecisionTree	0.83	0.76	0.73	0.72	0.81
4	**RandomForest**	**0.87**	**0.82**	**0.80**	**0.79**	**0.93**
5	**GradientBoosting**	**0.87**	**0.82**	**0.79**	**0.80**	**0.94**
6	**XGBoosting**	**0.88**	**0.83**	**0.82**	**0.82**	**0.94**
7	SVC	0.84	0.78	0.76	0.76	0.90

4.2 Classifier Comparison with Newly Added Features

Table 4 shows the comparison of the 8 different classifier models, after the parameters have undergone feature engineering and feature transformation.

From Table 4, it is found that the models XGBoosting (0.88), RandomForest (0.87) and GradientBoosting (0.87) have the highest Accuracy values and hence are the top 3 models. The AUC values too are the highest for these three models in the order of GradientBoosting (0.94), XGBoosting (0.94) and RandomForest (0.93).

4.3 Classifier Comparison Post-Hyperparameter Tuning

In order to find out the optimal model among the three models with highest accuracy and AUC values (Random Forest, Gradient Boosting, XGBoosting), hyperparameter tuning is performed. Comparing the models after hyperparameter tuning gives the values as shown in Table 5.

Table 5 Results after hypertuning

Sl. No.	Model	Accuracy	Precision	Recall	F1-score	AUC
0	RandomForest	0.86	0.81	0.79	0.80	0.94
1	GradientBoosting	0.87	0.87	0.75	0.79	0.93
2	**XGBoosting**	**0.88**	**0.86**	**0.80**	**0.82**	**0.95**

From Table 5, it can be found that for RandomForest the accuracy value after hyperparameter tuning is 0.86 and AUC value is 0.94. For GradientBoosting, the accuracy value after hyperparameter tuning is 0.87 and AUC value is 0.93. For XGBoosting, the accuracy value after hyperparameter tuning is **0.88** (88%) and AUC value is **0.95**.

4.4 Confusion Matrix and Area Under ROC Curve

The above discussion and Table 4 confirm that XGBoosting is the classifier that gives the highest accuracy and AUC value for diabetes prediction on PIMA Indian dataset. A confusion matrix has been used to represent the output of the XGBoosting model.

Figure 4 depicts that 91% of healthy patients are currently classified as healthy and 83% of diabetic patients are predicted as diabetic. False prediction is minimal and hence we can confirm that XGBoosting is a good model for diabetes prediction.

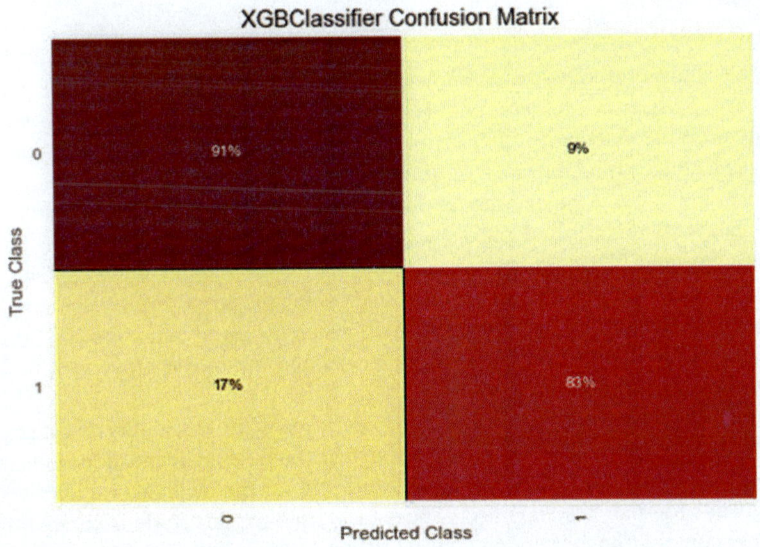

Fig. 4 Confusion Matrix of XGBoosting

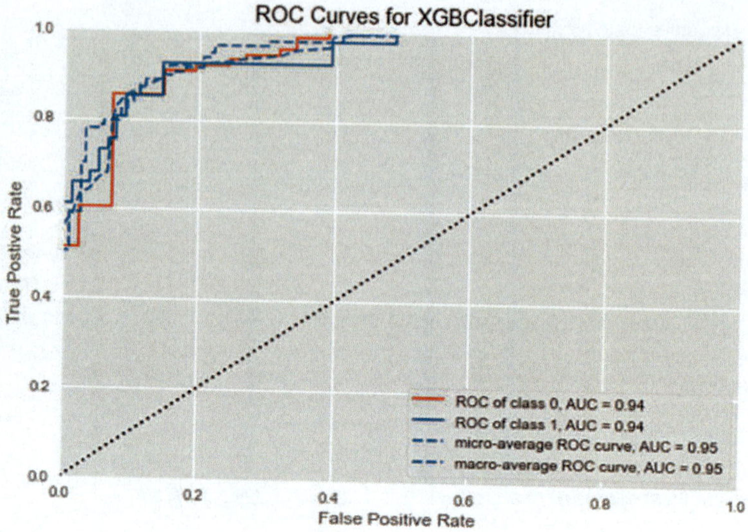

Fig. 5 ROC Curve for XGBoosting Classifier

The ROC Curve represented in Fig. 5 validates that the finalized model has better efficiency.

5 Conclusion

From this study, one could gather knowledge that higher value of factors like glucose, pregnancy, BMI, etc. in patients could cause occurrence of diabetes. Development of a highly accurate diabetes prediction model based on these factors is very significant for finding chance of occurrence of diabetes.

In this study, the proposed model does analysis of the various features of PIMA Indian Diabetes dataset and picks the best features based on correlation. Various machine learning classifier models are applied on the dataset features. The most accurate diabetes prediction has been achieved using the proposed XGBoosting classifier model. The greater accuracy and AUC value achieved using the XGBoosting Classifier shows the effectiveness of the model and the role it plays in robust and precise diabetes prediction.

Certain limitations can be taken into consideration while evaluating this study. The data used for this study was acquired several years back. And hence the results might be confined to that time. HbA1c test and urine test are the medical practices used in modern days for diabetes diagnosis. The dataset used for this study is comparatively small and contains the data collected from PIMA Indians. However, this gives a good starting point for studying diabetes prediction using machine learning algorithms.

Other similar datasets can be considered upon availability to further evaluate the model's efficiency. This study can further be utilized to predict the risk of COVID-19 in diabetes patients. Additionally, other medical contexts also could use the proposed framework to justify their generality and adaptability to predict the target classes

References

1. Cho, N.H., Shaw, J.E., Karuranga, S., Huang, Y., da Rocha Fernandes, J.D., Ohlrogge, A.W., Malanda, B.: IDF diabetes atlas: global estimates of diabetes prevalence for 2017 and projections for 2045. Diabet. Res. Clin. Pract. **138**, 271–281 (2018). https://doi.org/10.1016/j.diabres.2018.02.023
2. Saeedi, P., Petersohn, I., Salpea, P., Malanda, B., Karuranga, S., Unwin, N., Colagiuri, S., Guariguata, L., Motala, A. A., Ogurtsova, K., Shaw, J. E., Bright, D., Williams, R., IDF Diabetes Atlas Committee: Global and regional diabetes prevalence estimates for 2019 and projections for 2030 and 2045: Results from the International Diabetes Federation Diabetes Atlas, 9th edn. Diabet. Res. Clin. Pract. **157**, 107843 (2019). https://doi.org/10.1016/j.diabres.2019.107843
3. Maniruzzaman, M., Kumar, N., Menhazul Abedin, M., et al.: Comparative approaches for classification of diabetes mellitus data: Machine learning paradigm. Comput. Methods Progr. Biomed. **152**, 23–34 (2017). https://doi.org/10.1016/j.cmpb.2017.09.004
4. Komi, M., Li, J., Zhai, Y., Zhang, X.: Application of data mining methods in diabetes prediction. In: 2017 2nd International Conference on Image, Vision and Computing (ICIVC), Chengdu, pp. 1006–1010 (2017). https://doi.org/10.1109/ICIVC.2017.7984706
5. Mercaldo, F., Nardone, V., Santone, A.: Diabetes mellitus affected patients classification and diagnosis through machine learning techniques. Procedia Comput. Sci. **112**, 2519–2528 (2017). https://doi.org/10.1016/j.procs.2017.08.193
6. Sisodia, D., Sisodia, D. S.: Prediction of diabetes using classification algorithms. Procedia Comput. Sci. **132**, 1578–1585. Elsevier B.V(2018). https://doi.org/10.1016/j.procs.2018.05.122
7. Hasan Md, A., Md. Ashraful, Das, D., Hossain, E., Hasan, M.: Diabetes prediction using ensembling of different machine learning classifiers. IEEE Access, 1–1 (2020) https://doi.org/10.1109/ACCESS.2020.2989857
8. Alehegn, M., Raghvendra Joshi, R., Mulay, R.: Diabetes analysis and prediction using random forest, KNN, Naïve Bayes, And J48: an ensemble approach. Int. J. Sci. Technol. Res. **8**, 09 (2019)
9. Sneha, N., Gangil, T.: Analysis of diabetes mellitus for early prediction using optimal features selection. J. Big Data **6**, 13 (2019). https://doi.org/10.1186/s40537-019-0175-6
10. Hina, S., Shaikh, A., Sattar, S.A.: Analyzing diabetes datasets using data mining. J. Basic Appl. Sci. **13**, 466–471 (2017)
11. Asuero, A.G., Sayago, A., Gonzalez, A.: The correlation coefficient: an overview, Crit. Rev. Anal. Chem. **36**, 41–59 (2006)
12. Markovitch, S., Rosenstein, D.: Feature generation using general constructor functions. Mach. Learn. **49**, 59–98 (2002). https://doi.org/10.1023/A:1014046307775
13. Ünsal, Ö., Bulbul, H.: Comparison of classification techniques used in machine learning as applied on vocational guidance data. In: International Conference on Machine Learning and Applications, vol. 10 (2011)
14. Zeng, X., Martinez, T.R.: Distribution-balanced stratified cross validation for accuracy estimation. J. Exp. Theor. Artif. Intell. **12**, 1–12 (2000)
15. Mitchell, T.M., et al.: Machine Learning, vol. 45.37. McGraw Hill, Burr Ridge, IL, pp. 870–877 (1997)

16. Madjarov, G., Kocev, D., Gjorgjevikj, D., Džeroski, S.: An extensive experimental comparison of methods for multi-label learning. Pattern Recogn. **45**, 3084–3104, ISSN 0031-3203 (2012). https://doi.org/10.1016/j.patcog.2012.03.004
17. Peng, C.-Y.J., Lee, K.L., Ingersoll, G.M.: An introduction to logistic regression analysis and reporting. J. Educ. Res. **96**, 3–14 (2002). https://doi.org/10.1080/00220670209598786
18. Cheng, W., Hüllermeier, E.: Combining instance-based learning and logistic regression for multilabel classification. Mach. Learn. **76**, 211–225 (2009). https://doi.org/10.1007/s10994-009-5127-5
19. Özkan, Y.: Data Mining Methods. Papatya Publications, Istanbul, Turkey (2008)
20. Raj, J.S.: A novel information processing in IoT based real time health care monitoring system. J. Electron. **2**(3), 188–196 (2020)
21. Raj, J.S., Ananthi, J.V: Recurrent neural networks and nonlinear prediction in support vector machines. J. Soft Comput. Paradigm (JSCP) **1**(1), 33–40 (2019)
22. Ross Quinlan, J.: C4.5: Programs for Machine Learning. Morgan Kaufmann Publishers Inc., San Francisco, CA, USA (1993)
23. Breiman, L.: Random forests. Mach. Learn. **45**, 5–32 (2001). https://doi.org/10.1023/A:1010933404324
24. Liaw, A., Wiener, M.: Classification and regression by random forest. R news **2**, 18–22 (2002)
25. Galar, M., Fernandez, A., Barrenechea, E., Bustince, H., Herrera, F.A.: Review on ensembles for the class imbalance problem: bagging-, boosting-, and hybrid-based approaches. IEEE Trans. Syst. Man Cybern. **42**(4), 463–484 (2012). https://doi.org/10.1109/TSMCC.2011.2161285
26. Chen, T., Guestrin, C.: XGBoost: a scalable tree boosting system. In: Proceedings of the 22nd ACM SIGKDD International Conference on Knowledge Discovery and Data Mining (KDD '16). Association for Computing Machinery, New York, NY, USA, 785–794 (2016). https://doi.org/10.1145/2939672.2939785
27. Melo, F.: Area under the ROC Curve. In:Dubitzky, W., Wolkenhauer, O., Cho, K.H., Yokota, H. (eds.) Encyclopedia of Systems Biology. Springer, New York (2013). https://doi.org/10.1007/978-1-4419-9863-7_209

Acute Lymphoblastic Leukemia Detection Using Transfer Learning Techniques

K. S. Ananthu, Pambavasan Krishna Prasad, S. Nagarajan, and E. R. Vimina

Abstract Leukemia is a type of cancer that affects the body's blood forming tissues, including bone marrow. It is a dangerous illness prevalent in children under the age of five. The present diagnosis includes microscopic examination of blood cells by the hematologist. Recently, deep learning methods are extensively employed in many medical imaging applications for the diagnosis of diseases. However, one of the key issues is the limited availability of microscopic images for training the models. To overcome this difficulty, transfer learning techniques are put forward. This paper presents a comparative analysis of different transfer learning models like Xception, Inceptionv3, DenseNet201, ResNet50, and MobileNet to detect acute lymphocytic leukemia (ALL) from blood smear cells. All models were trained on ALL-IDB2 dataset and achieved an accuracy of 87.97%, 88.92%, 88.92%, 95.28%, and 97.88%, respectively.

Keywords Acute lymphoblastic leukemia · Transfer learning · Data augmentation

1 Introduction

Bone marrow is a spongy tissue within our bones that produces billions of new blood cells. There are three types of blood cells essential for the proper functioning of our body. They are red blood cells that provide oxygen to various parts of our body, white blood cells to protect against various infections, and platelets to help

K. S. Ananthu (✉) · P. Krishna Prasad · S. Nagarajan · E. R. Vimina
Department of Computer Science and IT, Amrita School of Arts and Sciences, Amrita Vishwa Vidyapeetham, Kochi, India
e-mail: KHSCI5MCA16013@asas.kh.amrita.edu

P. Krishna Prasad
e-mail: KHSCI5MCA16035@asas.kh.amrita.edu

S. Nagarajan
e-mail: KHSCI5MCA16037@asas.kh.amrita.edu

E. R. Vimina
e-mail: viminaer@asas.kh.amrita.edu

J. S. Raj et al. (eds.), *Intelligent Sustainable Systems*, Lecture Notes in Networks and Systems 213, https://doi.org/10.1007/978-981-16-2422-3_53

prevent blood from coagulating. Leukemia is malignant cancer caused by the rapid production of irregular white blood cells. Leukemia is classified based on its growth rate and where it commences (mainly in lymphoid cells or myeloid cells of the bone marrow). The key types of leukemia are chronic lymphocytic leukemia (CLL), chronic myelogenous leukemia (CML), acute lymphocytic leukemia (ALL), and acute myelogenous leukemia (AML). "Acute" means that they grow rapidly and "chronic" means moderate growth. In adults, ALL is rare but children younger than five are at high risk [1]. Leukemia is diagnosed through a physical examination, blood test, or bone marrow biopsy. The American Cancer Society estimates 60,530 new cases of blood cancer and 23,100 deaths across the USA [2]. Figure 1 [3] demonstrates leukemia incidence rates from 2000 to 2017. For all subgroups of leukemia, the five-year survival rate is 61.4%. Age, time of diagnosis, chromosome mutations, and types of leukemia are several variables influencing survival rates.

Physical microscopic analysis is very susceptible to human errors and is a time-consuming process [4]. There are several modern techniques for standardizing these examination steps and automating the detection process. The most used approaches are conventional image processing and machine learning, which include segmentation, feature extraction, and classification [5]. The classification and identification of these medical images can be achieved using convolution neural networks (CNNs) [6]. But the issue is that these CNNs need wide datasets to train (a large set of medical

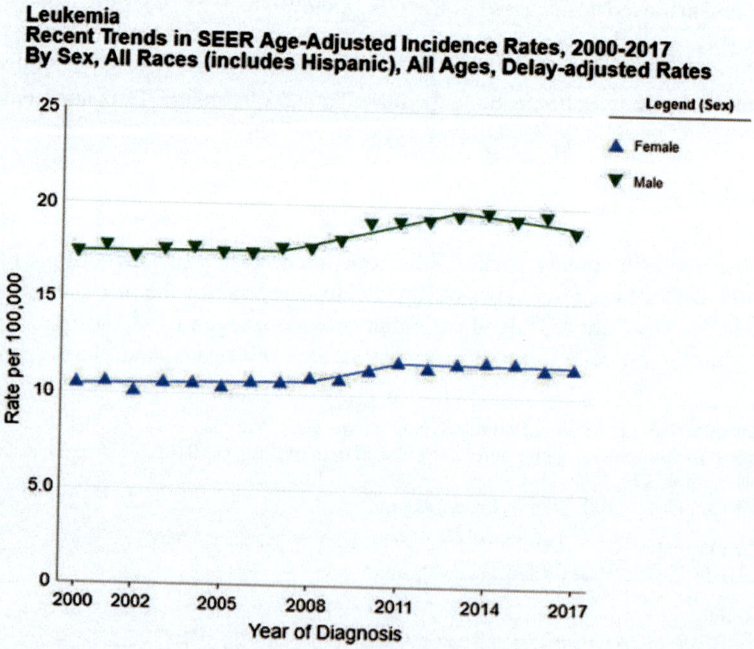

Fig. 1 Leukemia recent trends in SEER age-adjusted index rates 2000–2017. *Source* https://seer.cancer.gov/statfacts/html/leuks.html [3]

images) [7]. To train on large datasets, convolutional neural network models take substantially more time. Transfer learning is an alternative approach for both neural network training acceleration and weak datasets. Transfer learning is a machine learning method that reuses a pre-trained neural network model on a new related problem.

This study compares multiple models of transfer learning for the detection of acute lymphoblastic leukemia (ALL). Transfer learning models are pretrained models that are already trained on large datasets such as ImageNet. For improved performance, these models can be embedded with our model. Five transfer learning models Xception [8], DenseNet201 [9], ResNet50 [10], InceptionV3 [11], MobileNet [12] are evaluated on the ALL_IDB2 dataset [13] based on loss and accuracy metrics.

2 Literature Review

There are several methods for identifying leukemia. Mohammed et al. [14] employed a transfer learning approach and proposed two automated classification models for the detection of leukemia in blood smear images. In the first method, after images are preprocessed, features are extracted using AlexNet. Classification is performed using well-known classifiers. In the second model after preprocessing, fine-tuning is performed. The end result indicates that the second method provided better results when compared to the first. Rehman et al. [15] used the Alexnet model and also fine-tuned the last three layers. Various evaluation methods like Naive Bayesian, KNN, and SVM were used. LBP features were extracted and transmitted to the classifier to obtain an accuracy of 97.78%.

Roopa et al. [16] used peripheral blood smear images with color and illumination variations. An automated system was used for detecting nuclei and also to separate leukocytes. The detection of nuclei is carried out using arithmetic and morphological operations, and active contours methods are used to detect leukocytes. The final result showed that this method's sensitivity is almost around 96%.

For features extraction, several transfer learning models are available [14]. Alexnet and Ahamed et al. [17] used VGGNet to extract features from the white blood cells images. A statistically enhanced salp swarm algorithm (SESSA) is used for filtering the extracted features. The most important advantage of feature selection is to reduce training time, forestall overfitting, and thus improve accuracy. SESSA only extracts the relevant features and avoids unnecessary features. The use of CNN and SESSA for feature extraction ensures high accuracy and reduces computational complexity.

Alexandra et al. [18] proposed a method for automated identification and classification of leukemia cells from peripheral blood smear images. To identify blast cells and healthy blood cells, they selected a combination of 16 features carrying morphological and statistical information. To perform classification, artificial neural networks and support vector machines were chosen. The final results indicate that the neural network has a higher overall accuracy of 97.52%. Sarmad et al. [19] used a pretrained deep convolutional neural network for the detection of ALL and its classifications. The last layer of AlexNet has been replaced by a new layer that can classify the input images into four subcategories. Data augmentation techniques were used to reduce overfitting. The dataset was compared with different color models to verify the results. The proposed model achieved high accuracy without the need for any microscopic segmentation of the image. Prellberg and Kramer [20] used the ResNeXt50 model for the classification of leukemia from microscopic images. It is a pretrained convolutional neural network model. They applied data augmentation and fine-tuned some layers to obtain an F1 score of 88.91%.

Sarmad et al. [19] used a pretrained deep convolutional neural network for the detection of ALL and its classifications. The last layer of AlexNet has been replaced by a new layer that can classify the input images into four subcategories. Data augmentation techniques were used to reduce overfitting. The dataset was compared with different color models to verify the results. The proposed model achieved high accuracy without the need for any microscopic segmentation of images. Prellberg and Kramer [20] used the ResNeXt50 model for the classification of leukemia from microscopic images. It is a pretrained convolutional neural network model. They applied data augmentation and fine-tuned some layers to obtain an F1 score of 88.91%.

Sonali et al. [21] proposed a model for acute lymphoblastic leukemia detection from the blood smear images. For feature extraction, they used marker-based segmentation and gray level co-occurrence matrix, and for feature reduction, principal component analysis (PCA) was used. The relevant features are included in the random forest classifier. Therefore, by applying feature extraction and feature reduction, the classifier obtained higher accuracy. Sonali et al. [22] used the Y component of the CMYK image and also used a triangle method of thresholding; it preprocesses the blast cell from a microscopic image for classification of leukemia. Discrete orthonormal S-transforms were used to remove texture characteristics and the linear discriminant analysis was done to minimize dimensionality. AdaBoost algorithm with random forest classifier was used to supply reduced features. The proposed system achieved a superior accuracy of 99.66%.

3 Proposed System/Materials and Methods

The proposed system utilizes transfer learning with different profound learning models like Xception, Inceptionv3, DenseNet201, ResNet50, and MobileNet. The images from the dataset were augmented and features were collected. We then classify the images into either blast cells or normal cells. Various models are then assessed by utilizing the accuracy and loss metrics. This section provides details of the dataset, data augmentation techniques used, and feature extraction techniques (Fig. 2).

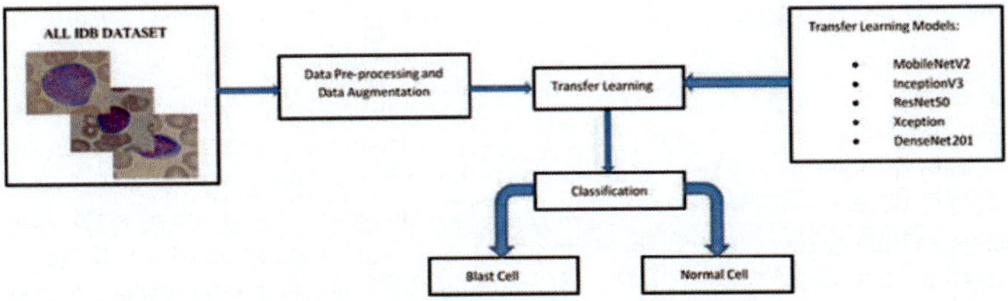

Fig. 2 Conceptual diagram

3.1 Dataset

Microscopic images of blood smear cells used in this system were acquired from ALL Image Dataset [13], a public dataset. The dataset contains 260 single-celled images from which 130 images were taken by patients with ALL and 130 were normal images. All of these images had a resolution of 2592 × 1944 with a color depth of 24-bits. Each image in the dataset was provided by an expert oncologist who collected these images using an optical laboratory microscope coupled with a Canon PowerShot G5 camera. These images are specifically designed for image classification purposes. ALL-IDB2 includes the cropped region of the significance of regular and blast cells which belong to ALL-IDB1 dataset (Fig. 3).

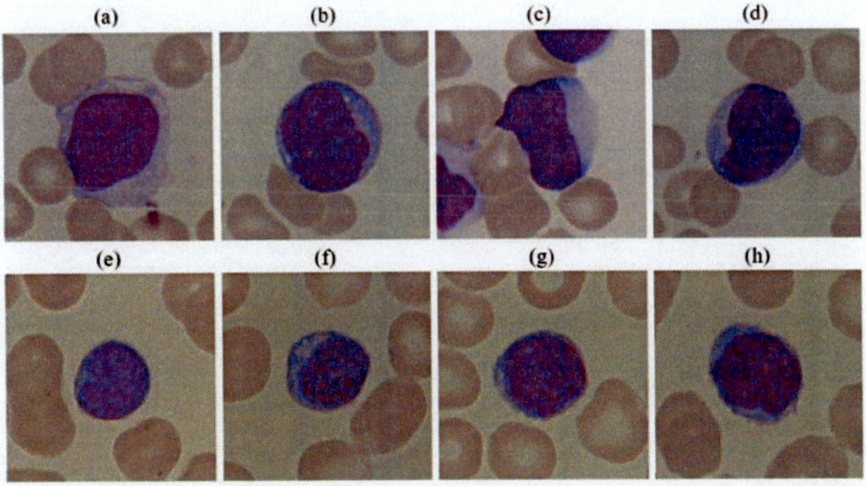

Fig. 3 Sample images from ALL_IDB2 [13] dataset. Probable blastic cells from ALL patients (**a–d**) and healthy blastic cells from non-ALL patients

3.2 Data Augmentation

Generally, the key shortcoming encountered is a limited dataset, which often can result in low precision and overfitting. Due to the limited range of images in our dataset, we made use of data augmentation techniques from Albumentations, a python library for image augmentation. Many manipulation techniques have been used, such as RandomBrightness, JpegCompression, Image rotation and Flip, Random-Contrast, and HueSaturation. The total number of images was increased to 2600 after augmentation.

Steps in Augmentation:

1. Import Albumentations
2. OpenCV is imported to read images
3. Images are passed to the augmentation pipeline
4. Augmented images are received

3.3 Feature Extraction and Classification Using Transfer Learning

We opted for transfer learning methods due to the limited dataset. Transfer learning makes use of a pretrained network with a large dataset and uses its learning to train a new network. We analyzed different transfer learning models to detect ALL. The models are:

3.4 MobileNet

Howard et al. [12] proposed the model to effectively improve accuracy considering restricted resources of embedded applications. The model contains two hyperparameters, namely width multiplier and resolution multiplier. It takes an input image of size 224 × 224 × 3 (Fig. 4).

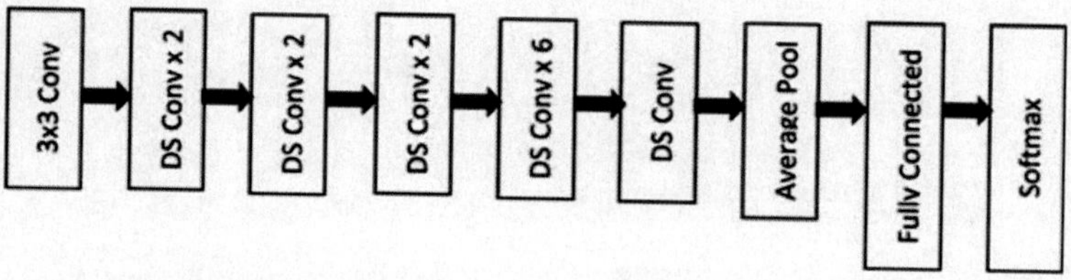

Fig. 4 Mobilenet architecture

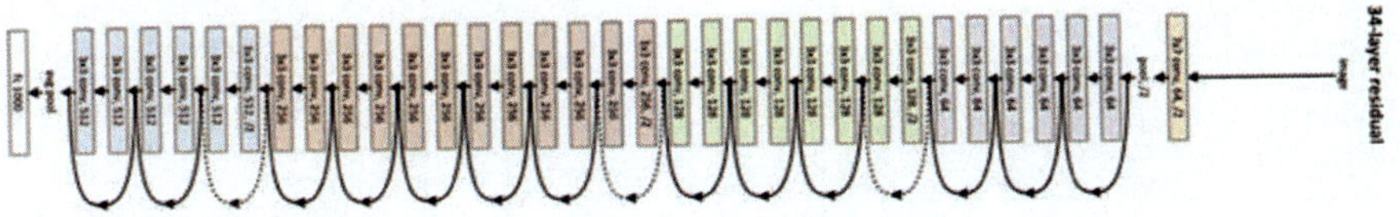

Fig. 5 Resnet50 architecture *Source* https://arxiv.org/abs/1512.03385 [24]

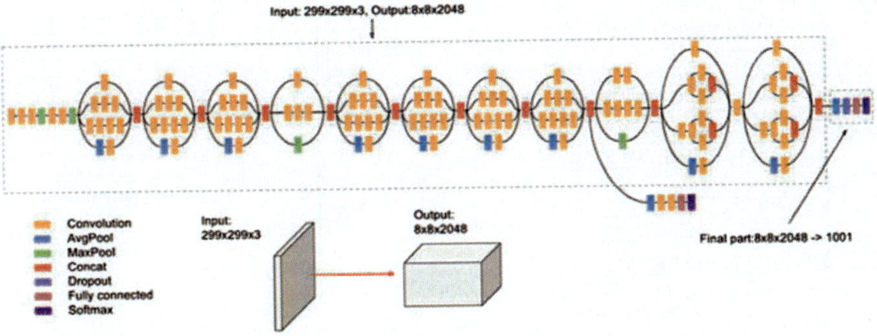

Fig. 6 InceptionV3 model architecture. *Source* https://cloud.google.com/tpu/docs/images/inceptionv3onc–oview.png [25]

3.5 ResNet50

He, Kaiming et al. [10] proposed the model which has 50 deep layers. It accepts an input image of size 224 × 224 × 3. It is trained to classify 1000 different categories. It has 48 convolutional layers, a MaxPool layer and an average pooling layer (Fig. 5).

3.6 InceptionV3

Szegedy, Christian et al. [11] developed this model and it has 48 deep layers. The model widely uses symmetric and asymmetric building blocks. It accepts an input image of size 299 × 299 × 3 and the output is 8 × 8 × 2048 (Fig. 6).

3.7 Xception

Chollet and François [8] developed this model with 71 deep layers, with an input image of size 299 × 299. It has 36 convolutional layers for feature extraction and is structured into 14 modules with a linear residual relationship between them (Fig. 7).

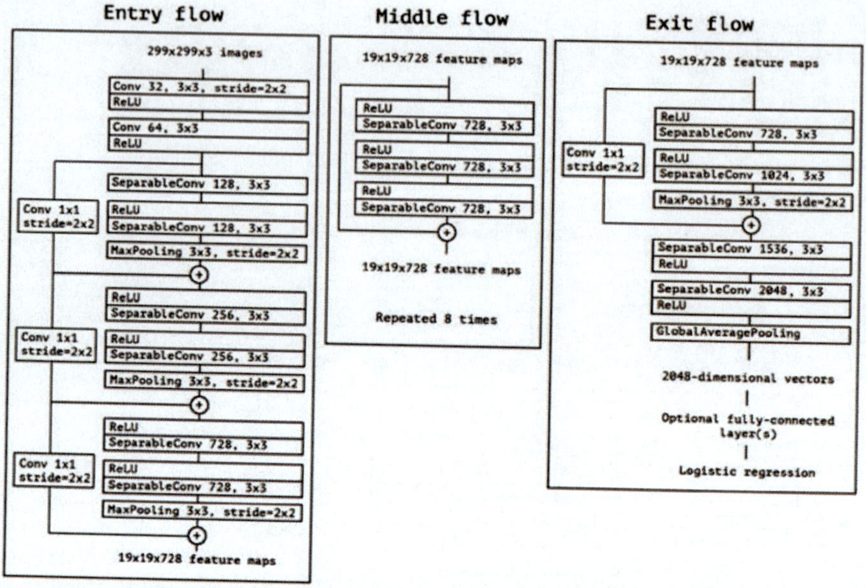

Fig. 7 Xception architecture *Source* https://arxiv.org/abs/1610.02357 [26]

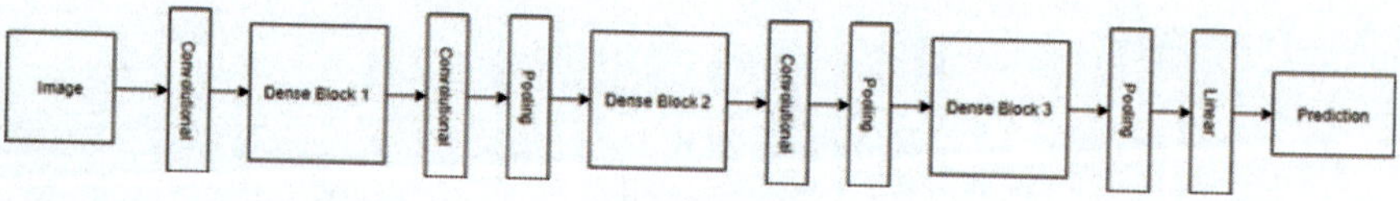

Fig. 8 Dense201 model

3.8 *Densenet201*

Huang, Gao et al. [9] developed this model with 201 deep layers and the network has an input image of shape 224 × 224. It uses dense convolutional network layers with each layer connected in a feed-forward manner (Fig. 8).

4 Experimental Results and Analysis

This section describes the implementation of our models and experiments conducted. We carried out a comparative study of the various transfer learning models to detect acute lymphocytic leukemia (ALL). The dataset contained a total of 2600 images after augmentation, including 1300 blast cell images and 1300 regular images. The data was divided into 80% as a training set and 20% as a validation set. The test batches were also produced from the validation set. Transfer learning is a pretrained model that was previously trained on a large dataset like ImageNet. In this analysis, we have experimented with five transfer learning models to find which provided better results for the detection of ALL based on loss and accuracy metrics. A batch size of 32 with 20 epochs was used to train the model. The learning rate was set at 0.0001 with Adam optimizer to compile the model. All models are compared on the basis of accuracy and loss. During each epoch, loss and accuracy are also calculated. Accuracy is the right amount of classifications made for the total number of classifications. Training accuracy is the accuracy of the model for which it was built and validation accuracy is the accuracy of the model on which it has not been seen. Loss is the difference between the predicted value and true value.

$$Loss = -\frac{1}{outputsize} \sum_{i=1}^{outputsize} y_i \log \hat{y}_i + (1 - y_i) \log\left(1 - \hat{y}_i\right) \tag{1}$$

$$Accuracy = \frac{No.of correct predictions}{Total no.of predictions} \tag{2}$$

MobileNet achieved an accuracy of 97.88%. ResNet50 has obtained superior results than Xception, Inception, and DenseNet201 in terms of accuracy and loss. Figure 9 shows the performance metrics for each transfer learning model. MobileNet network achieved better results among all the models.

DenseNet201 is a transfer learning model in which each layer collects information preceding layers. After instantiating the model, an image of input size 299*299 with a color depth of 3 is passed to the model. We freeze the convolution base before compiling to prevent weight updation while training. A GlobalAveragePooling layer was used to convert the features into a single 1920-element vector per image. Early Stopping was introduced to overcome overfitting. DenseNet201 obtained test

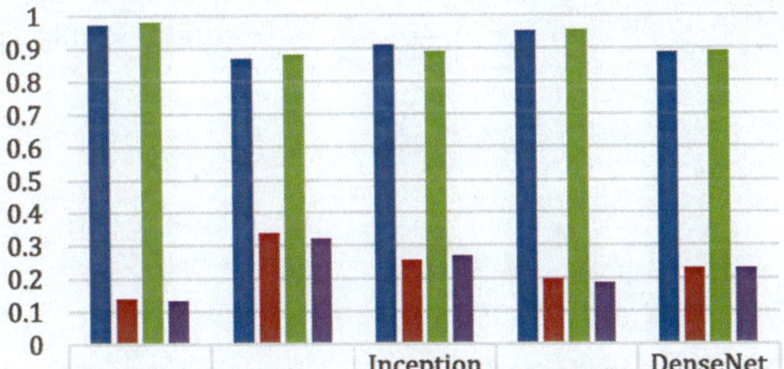

	MobileNet	Xception	Inception V3	ResNet50	DenseNet 201
Accuracy	0.9714	0.8681	0.9098	0.9512	0.8851
Loss	0.1378	0.3368	0.2529	0.1963	0.2294
Validation Accuracy	0.9788	0.8797	0.8892	0.9528	0.8892
Validation Loss	0.1322	0.3196	0.267	0.1851	0.2294

Fig. 9 Results of CNN models with transfer learning in ALL-IDB dataset

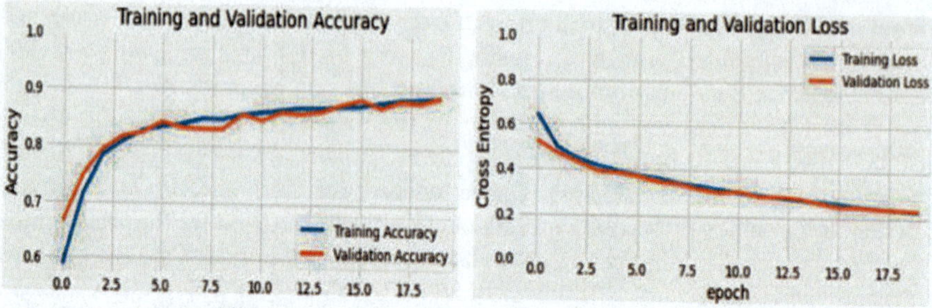

(a) (b)

Fig. 10 **a** Train and validation accuracy of DenseNet201 in each epoch **b** Train and validation loss of DenseNet201 in each epoch

accuracy of 84.37% over test batches when compared to Xception's test accuracy of 82.29%. Figure 10 demonstrates the DenseNet201 results.

Xception is a CNN that is 71 layers deep. An input image of size 299*299 with three channels is passed to the model. Similarly, we freeze the convolutional base to prevent weights being updated while training. GlobalAveragePooling layer is added to convert the features into a single 2048-element vector per image. To evade overfitting, one additional dense layer with 64 neurons along with Early Stopping was introduced. Xception achieved a test accuracy of 82.29% (Fig. 11).

InceptionV3 by Google is a convolutional neural network for image processing. After instantiating the model, an input image of size 224*224 with a color depth of 3 is passed to the model. The convolutional base was frozen to prevent weight updation. Later on, a GlobalAveragePooling layer is added to convert the features into a single 2048-element vector per image. To forestall overfitting, a dense layer with 256 neurons is added. InceptionV3 attained a test accuracy of 91.67% (Fig. 12).

ResNet50 is a transfer learning model that is 50 layers deep. The model is instantiated with an input image of size 450*450 with three channels. The convolutional

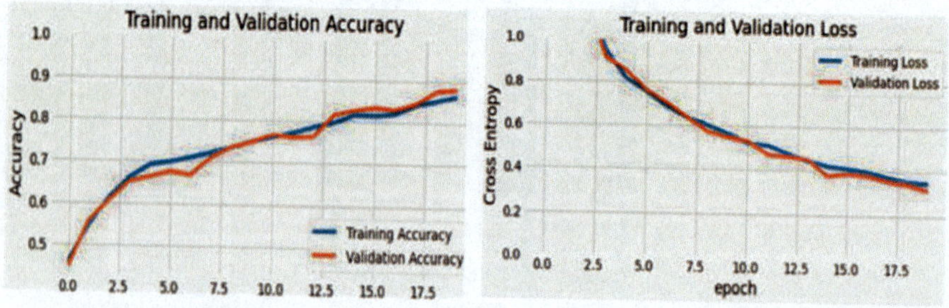

(a) (b)

Fig. 11 **a** Train and validation accuracy of Xception in each epoch **b** Train and validation loss of Xception in each epoch

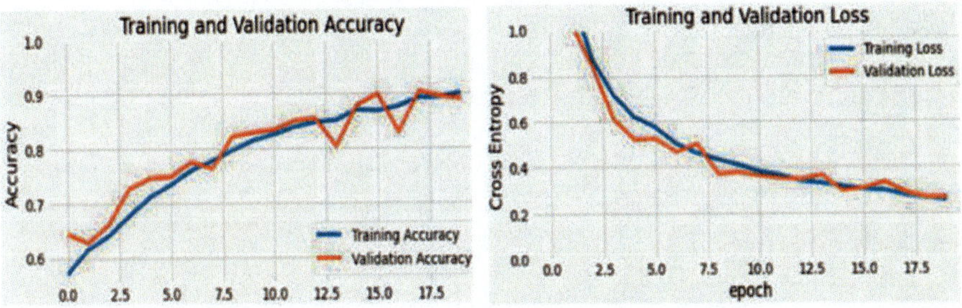

(a) (b)

Fig. 12 **a** Train and validation accuracy of InceptionV3 in each epoch **b** Train and validation loss of InceptionV3 in each epoch

base was frozen similar to all models to prevent weight updation. The GlobalAveragePooling layer is added to convert the features into a single 2048-element vector per image. To overcome overfitting, Early Stopping was implemented and the model achieved a test accuracy of 95.8%. The ResNet50 model showed steady growth in terms of accuracy and loss, and increasing the numbers of epochs may provide even better accuracy (Fig. 13).

MobileNet is a transfer learning model that is 30 layers deep. After instantiating the model, an input image of size 224*224 with a color depth of 3 is passed to the model. The convolutional base was frozen to prevent weight updation. Later on, a GlobalAveragePooling layer is added to convert the features into a single 1024-element vector per image. To forestall overfitting dropout, Early Stopping was introduced. The model obtained a test accuracy of 96.8% (Fig. 14).

After training the model, MobileNet reaches an accuracy of 97.88% on the validation set. Later on, we verified the performance of the model on new data using a test set. Figure 15 demonstrates MobileNet model prediction against test sets.

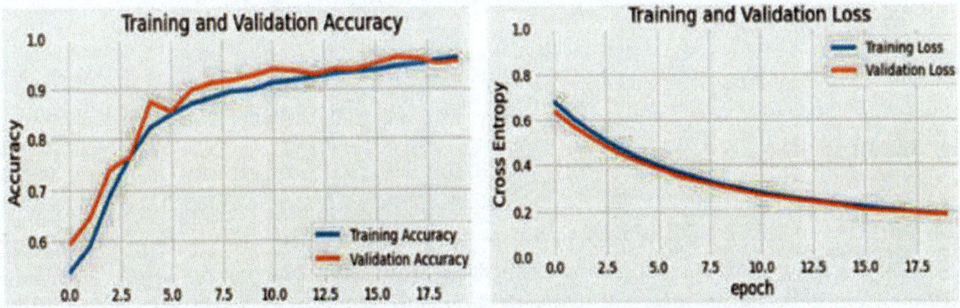

(a) (b)

Fig. 13 **a** Train and validation accuracy of ResNet50 in each epoch **b** Train and validation loss of ResNet50 in each epoch

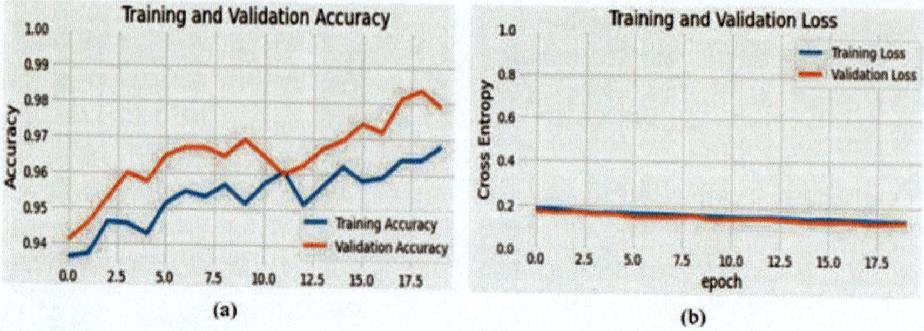

Fig. 14 **a** Train and validation accuracy of MobileNet in each epoch **b** Train and validation loss of MobileNet in each epoch

```
Predictions:
 [0 0 1 1 0 1 1 1 0 0 0 1 1 1 1 0 0 0 1 0 0 1 1 0 0 1 0 0 1 0 0 1]
Labels:
 [0 0 1 1 0 1 1 1 0 1 0 1 1 1 1 0 0 0 1 0 0 1 1 0 0 1 0 0 1 1 0 1]
```

Fig. 15 MobileNet model predicted against test batch

5 Discussion

Computer-aided diagnosing systems can assist in the early diagnosis and treatment of leukemia. Deep learning has had a significant influence on medical image processing, helping to identify, classify, and interpret images to uncover hidden patterns in them. Generally, the main shortcoming encountered in the case of medical image analysis is a limited dataset. A deep neural network can be created with a larger dataset so that it can be used on a daily basis. The limitation of the proposed work is that only images of acute lymphoblastic leukemia can be identified. Subsequent research could be done to distinguish the various forms of leukemia instead of marking images as ALL or regular images.

6 Conclusion

The purpose of this analysis was to carry out a comparative assessment of various transfer learning models for the detection of acute lymphoblastic leukemia from blood smear cells. MobileNet, ResNet50, InceptionV3, Xception, and DenseNet201 are the various types of transfer learning models used in this comparative study. MobileNet helps to minimize the network size, decreases the number of parameters, and accelerates performance but is less reliable than other models. ResNet architecture proposed residual connection, and the input to the present layer was obtained by summing up outputs from previous layers, but the complex architecture and batch normalization layers are its shortcomings. All the models were trained on the ALL-IDB2 dataset and achieved accuracy and loss of 97.88% and 13.22% for MobileNet, 95.28% and 18.51% for ResNet50, 88.92% and 26.70% for InceptionV3, 87.97% and 31.96% for Xception, and 88.92% and 23.39% for DenseNet. In this comparative study, we find that MobileNet is a superior model for the detection of acute lymphoblastic leukemia in terms of both loss and accuracy.

References

1. Vogado, L.H.S., et al.: Diagnosing leukemia in blood smear images using an ensemble of classifiers and pre-trained convolutional neural networks. In: 2017 30th SIBGRAPI Conference on Graphics, Patterns and Images (SIBGRAPI). IEEE (2017)
2. https://www.cancer.org/cancer/acute-myeloid-leukemia/about/key-statistics.html
3. https://seer.cancer.gov/statfacts/html/leuks.html
4. Inaba, H., Greaves, M., Mullighan, C.G.: Acute lymphoblastic leukaemia. The Lancet **381**(9881), 1943–1955 (2013)
5. Bodzas, A., Kodytek, P., Zidek, J.: Automated detection of acute lymphoblastic leukemia from microscopic images based on human visual perception. Front. Bioeng. Biotechnol. **8**, 1005 (2020)
6. Malon, C.D., Cosatto, E.: Classification of mitotic figures with convolutional neural networks and seeded blob features. J. Pathol. İnf. **4** (2013)
7. Vogado, L.H.S., et al.: Leukemia diagnosis in blood slides using transfer learning in CNNs and SVM for classification. Eng. Appl. Artif. Intell. **72**, 415–422 (2018)
8. Chollet, F.: Xception: Deep learning with depthwise separable convolutions. In: Proceedings of the IEEE Conference on Computer Vision and Pattern Recognition (2017)
9. Huang, G., et al.: Densely connected convolutional networks. In: Proceedings of the IEEE Conference on Computer Vision and Pattern Recognition (2017)
10. He, K., et al.: Deep residual learning for image recognition. In: Proceedings of the IEEE Conference on Computer Vision and Pattern Recognition (2016)
11. Szegedy, C., et al.: Rethinking the inception architecture for computer vision. In: Proceedings of the IEEE Conference on Computer Vision and Pattern Recognition (2016)
12. Howard, A.G., et al.: Mobilenets: Efficient convolutional neural networks for mobile vision applications. arXiv preprint arXiv:1704.04861 (2017)
13. Labati, R.D., Piuri, V., Scotti, F.: All-IDB: The acute lymphoblastic leukemia image database for image processing. In: 2011 18th IEEE International Conference on Image Processing. IEEE (2011)
14. Loey, M., Naman, M., Zayed, H.: Deep transfer learning in diagnosing leukemia in blood cells. Computers **9**(2), 29 (2020)
15. Rehman, A., et al.: Classification of acute lymphoblastic leukemia using deep learning. Microscopy Res. Technique **81**(11), 1310–1317 (2018)
16. Hegde, R.B., et al.: Image processing approach for detection of leukocytes in peripheral blood smears. J. Med. Syst. **43**(5), 114 (2019)
17. Sahlol, A.T., Kollmannsberger, P., Ewees, A.A.: Efficient classification of white blood cell leukemia with improved Swarm optimization of deep features. Sci. Rep. **10**(1), 1–11 (2020)
18. Bodzas, A., Kodytek, P., Zidek, J.: Automated detection of acute lymphoblastic leukemia from microscopic images based on human visual perception. Front.Bioeng. Biotechnol. **8**, 1005 (2020)
19. Shafique, S., Tehsin, S.: Acute lymphoblastic leukemia detection and classification of its subtypes using pretrained deep convolutional neural networks. Technol. Cancer Res. Treatment **17**, 1533033818802789 (2018)

20. Prellberg, J., Kramer, O.: Acute Lymphoblastic Leukemia Classification from Microscopic Images using Convolutional Neural Networks. ISBI, C-NMC Challenge: Classification in Cancer Cell Imaging. Springer, Singapore **2019**, 53–61 (2019)
21. Mishra, S., et al.: Gray level co-occurrence matrix and random forest based acute lymphoblastic leukemia detection. Biomed. Signal Process. Control **33**, 272–280 (2017)
22. Mishra, S., Majhi, B., Kumar Sa, P.: Texture feature based classification on microscopic blood smear for acute lymphoblastic leukemia detection. Biomed. Signal Process. Control **47**, 303–311 (2019)
23. https://arxiv.org/abs/1512.03385
24. https://cloud.google.com/tpu/docs/images/inceptionv3onc–oview.png
25. https://arxiv.org/abs/1610.02357

Hybrid Prediction Model for the Success of Bank Telemarketing

Rohan Desai and Vaishali Khairnar

Abstract Telemarketing is a technique or system of direct marketing, wherein a businessperson interacts with clients to persuade them to purchase or avail the products and facilities, either by connecting via telephone or through in-person interaction. In the present-day generation, with the humongous acceptance of cellular phones telemarketing has gained popularity as an efficient mode of marketing. In the banking domain, telemarketing is the prime support system to exchange goods and services. Banking products and services promotion to increase the business requires a comprehensive understanding of current market information and the actual client expectations. The present work has investigated traditional data mining and classification approaches which are less accurate. They could not achieve a high customer conversion rate with direct marketing campaigns. The proposed work recommends a machine learning method to foreshow the accomplishments of telemarketing requests for contracting bank term deposits. A Portuguese bank was tagged, considering the impacts of the present economic crisis. The comprehensive set of features linked with bank customer, products and services were inspected. A discussion on four machine learning (ML) models is performed along with the hybrid model, logistic regression ML model (LR), naive Bayes ML model (NB), decision trees ML model (DTs) and support vector machine ML model (SVM). The four ML models were tested and analysed with the proposed hybrid machine learning method (artificial neural network + extreme gradient boosting). The proposed hybrid machine learning method demonstrates the best results.

Keywords Machine learning · Adam · Hyper-parameter · Hyper-parameter optimisation · Deep learning

---

R. Desai (✉) · V. Khairnar
Department of Information Technology, Terna Engineering College, Navi Mumbai, India

J. S. Raj et al. (eds.), *Intelligent Sustainable Systems*, Lecture Notes in Networks and Systems 213, https://doi.org/10.1007/978-981-16-2422-3_54

1 Introduction

Marketing is the process of identifying the target customers to purchase or make a deal with a product or a service by finding a product-market fit. Product-market fit means to offer a product or a service that customers need [1]. With a proficient product-market fit understanding, businesses can put forward a product which they recognise their clients want, essentially, making it significantly easy to persuade them to purchase it. Marketing currently promotes the process to purchase the product or service. It assists in the product development process. The overall objective is to increase the sell rate of product bank term deposit and service for the industry and preserve the business repute [2]. The pressure to meet monthly targets, retain existing customers and acquire new customers has become a challenge to every organisation. Instead of vast 360° campaigns, organisations invest in direct campaigns via telemarketing.

Telemarketing has challenges like the customers do not entertain calls that are not of their interest or potential. The vast customer database available with organisations does not help in reaching out to the right customers [3]. The telemarketers often end up calling customers with no interest or the potential to subscribe to the product or services offered by organisations. Data mining is a procedure of extracting knowledge from data by the use of data analysis tools. Classification is one of the different algorithms used in data mining [4]. Traditional data mining and classification approach are less accurate. They could not achieve a high customer conversion rate with the direct telemarketing campaigns. Modernisation, exploitation and development of new technologies like net banking, direct telemarketing, mobile banking and strong market competition enforce organisations to embrace the latest techniques to expand customer database and enhance customer trust and satisfaction. Precision direct telemarketing intends to minimise costs, efficient utilisation of workforce, retain existing customers, acquire new customers and increase the conversion rate for product or service endorsed by selling campaigns. It is instrumental in maximising profits and meeting the business targets in such a competitive market and enhances sustainability in the ongoing commercial process. Artificial intelligence is any task that is performed by a machine automatically. It helps to make day-to-day tasks simpler. To thrive in a rapid evolving digital and data-driven world, financial institutions have started investing more in artificial intelligence for banking purposes [5]. The huge amount of data that the financial institutions generate on a routine basis gave rise to the necessity of machine learning [6]. Machine learning is an important segment of artificial intelligence. It can be further categorised into supervised and unsupervised. The output of the ML algorithm in supervised learning is previously known and the input data is used to predict the result. Examples of supervised training are regression and classification. In contrast, the input information of unsupervised training is available, where no same-result variables are chosen, as in the case of clustering. It helps to resolve problems that cannot be defined on a rule-based system. Machine learning allows the system to learn the data through algorithms so that it learns the patterns behind the data. These patterns allow the system to predict on future data. The work is focused on machine learning methods that can predict a certain outcome based on a specified input. Four machine learning models were used and a comparative study is produced to investigate the bank telemarketing dataset which was recorded previously to take a decision. The main contribution of this work is applying a hybrid machine learning method in a proposed framework that had the highest accuracy to classify customer's records. The ML Software development life cycle is implemented in Python.

2 Related Works

This section describes the earlier works previously made in classification using data mining and machine learning methods. An assessment is performed on the various contributions, detailed analysis methods and performances. The various application challenges, contributions, compare performances and critique methods were analysed. The taxonomy of data mining and machine learning methods is foreshowed with the knowledge of current and emerging trends in this domain as an area of focus for researchers. In [7], the research proposed a system based on classification and deep neural networks to identify customers interested in credit products. It also proposed a testing mechanism build on a moving window which allowed identification of complicated, subsequent, timed-based reliance among specific transactions. The classification algorithms applied in this study are random forest, CART and deep neural network. The achievements of these methods were examined through precision and recall metrics. Among all these methods, CART seems to be the most efficient in terms of computing power. In [8], the work aimed to foreshow the achievement of bank telemarketing. The dataset which they applied in their study consisted of 21 features. They used a visual neural network and achieved optimum network size by pruning. In [9], the author had proposed a synthetic minority oversampling technique to describe the achievement of bank telemarketing. The author got better accuracy results with methods KNN, logistic regression, SVM, ANN and naive Bayes. In [10], the authors have demonstrated that when raw data is input to a long short-term memory network, it performs well over random forest, logit machine learning techniques. In [11], the research had utilised 274 attributes to forecast the behaviour of the customer based on previous purchase information. Gradient tree boosting, logistic lasso and extreme learning machine methods were utilised for prediction. Among all these techniques, gradient tree boost outperformed the others and yielded a better score for accuracy in the ROC curve. In [12], the authors had utilised machine learning techniques like the random forest, naive Bayes and J48 decision tree for analysing bank telemarketing data and enhancing the number of subscribing customers. All the classifiers were evaluated based on the accuracy and performance metrics. Among these J48 decision tree classifier gave the best results on a refined clustered dataset. In [13], the author has utilised the deep learning method to develop the dropout prediction model for students participating in massive open online courses. A comparative study amongst different machine learning techniques such as SVM, LR, linear discriminant analysis, DT and NN is performed and decided that NN yields better precision metrics over other techniques and the performance of dropout prediction got significantly improved. In [14], the authors investigate data mining techniques to foreshow the achievements of bank telemarketing calls to promote the product bank long-term deposits. A comparative analysis of three data mining models, namely naive Bayes, J48 classifier model and random forest, is provided from which random forest outperforms the other data mining models by exhibiting better accuracy. In [15], the authors provided a comparative analysis of ML algorithms XGboost, Catboost, Adaboost, Random forest and GBM to foreshow the achievements of bank telemarketing campaigns. They utilised explainable AI methods to compute the performance of the model, wherein Catboost produced better results than older version models. In [16], the authors utilised S_Kohonen network to foretell the achievements of bank telemarketing. They enhanced the S_Kohonen network through an improved whale optimisation technique which involves optimising the weights amongst the input and competition layers. The enhanced S_Kohonen network demonstrated higher results in terms of accuracy than the existing S_Kohonen network. In [17], the authors recommend ensemble random forests over logistic regression in predicting clients likely to purchase term deposit and detecting credit card fraud as random forest presents better results than logistic regression. In [18], the authors recommend a hybrid method based on neural network and multiple criteria decision aiding (NN-MCDA). They illustrated the efficiency of NN-MCDA by assessing it through three real-world datasets. In [19], the authors utilised the data mining method to identify potential clients to offer product bank deposit through telemarketing. The experimental results of the study indicated that the radial basis function neural network demonstrated better results in accuracy and sensitivity metrics than the multilayer perceptron neural network. In [20], the authors demonstrate a method to generate cases on the basis of input from the artificial neural network (ANN) and demonstrate an integration procedure amongst ANN and case based reasoning (CBR) method to build a model to predict potential customers who will subscribe to product bank term deposit. In [21], the authors develop a decision-making model through the use of SMO, RBFN, SOM, SVM, HLVQ and MP classifiers based on neural network methods. The models are evaluated based on accuracy, specificity and support vector machine with Refief-F attribute demonstrating the highest specificity.

3 Existing System

This section describes the existing system used in the classification of bank telemarketing data. The evaluation of the existing systems was performed by assessment and analysis of earlier work done by researchers in the banking domain for direct marketing. The dataset which was used by existing systems was the customer data of a Portuguese banking organisation. The analysis of the contributions done by the previous work in this domain was performed to study the existing system. In [22], the author recommended a data mining approach with a two-stage mechanism: data pre-processing and classification processing. In the data pre-processing stage, data is filtered and incomplete records are removed; minimum, average and maximum of each feature is calculated; and each record is inserted into the database. In the classification processing stage, the dataset is bifurcated into a training and test set, and data mining techniques are applied and evaluated. In [23], the author demonstrated a data mining model framework which consisted of data pre-processing and modelling. In the data pre-processing stage raw data is pre-processed by fixing the missing values and standardisation of the data is done, and optimal variables sets are retrieved from the dataset. In the modelling stage data is bifurcated into training and testing set; data mining methods LR, SVM, NN, Bayes and DT are applied; and performance and accuracy are analysed. In [24], the authors have used the dimensionality reduction approach to classify the imbalanced dataset. The data mining techniques used were naive Bayes, J48, KNN and Bayesnet. In [25], the authors used the machine learning approach to forecast client retention. The study aimed at tagging those clients through ML algorithms who are at risk of loss. So the marketing team can contact only these clients, thereby saving time and cost for the company. The author utilised SVM, random forest, KNN and adaptive boost ML methods in which random forest gave the best results. In [26], the author has utilised the data mining method to analyse the Indian e-commerce market to predict the demand for refurbished electronics. Three random real-world e-commerce website datasets were used for analysis. The proposed system utilised data accumulation, processing and validation stages to perform the analysis in which prediction results were more accurate through the regression tree model. In [27], the authors introduced a modelling approach with three stages: pre-processing, normalisation and classification. In the pre-processing stage, codification and missing value treatment of features are performed. The normalisation stage involves eliminating the effect of an order of magnitude. In the classification stage, a comparative study on naive Bayes, DT, LR, ANN and SVM ML models with and without normalised data and performance is evaluated by accuracy, precision, recall, f-measure metrics in which all the models perform substantially competent.

4 Methodology

In this part the proposed mechanism is discussed. A hybrid machine learning technique is recommended to classify the bank telemarketing data. This technique can achieve high accuracy in classifying customer records. The block design of the recommended mechanism is presented in Fig. 1, which shows 10 stages towards developing the recommended mechanism: ML/DL business problem framing, data collection, integration, preparation and cleaning, visualisation and analysis, feature engineering, model training and parameter tuning, evaluation, deployment and prediction analysis. A separate subsection is dedicated to each stage.

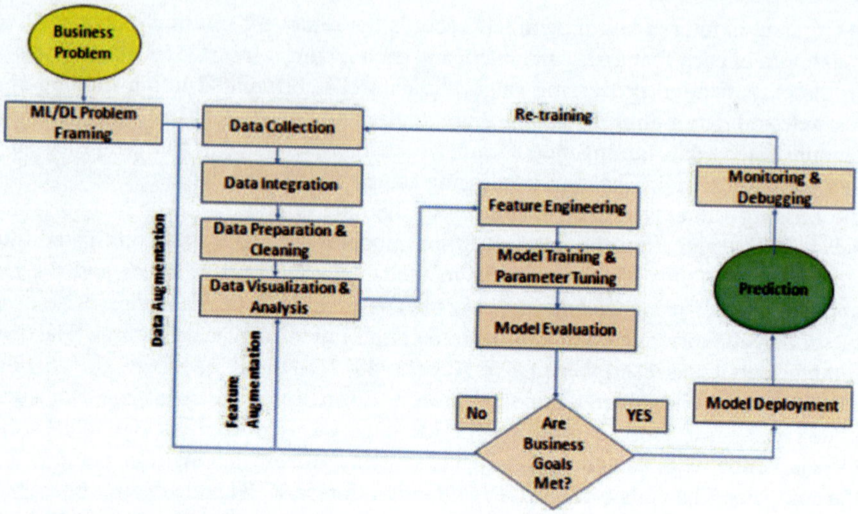

Fig. 1 Block diagram of the proposed tool

4.1 ML/DL Problem Identification

Previous research has the problem of less accuracy of prediction and could not achieve a high customer conversion rate with direct telemarketing campaigns.

4.2 Data Aggregation and Integration

The data aggregation and integration primitive in the model framework uses a Portuguese bank dataset. The Portuguese bank dataset has 41188 records and 21 attributes [28].

4.3 Data Description

The Portuguese bank dataset with comprehensive features associated with bank customer, products and social-economic characteristics was utilised.

4.4 Data Preparation and Cleaning

The numeric data is transformed by Yeo-Johnson method to treat skewness of individual features. Features like jobs, marital status, and credit default have missing values and are named as "unknown". The features in the data contain different units. It is important to normalise the data to standard scale to make it unit less. The data is scaled to a standard scale for standardised data processing.

4.5 Data Visualisation and Analysis

In this phase experimental data analysis is performed, which involves individual variable analysis as well as multivariate analysis. Fig. 2a–f shows individual variable analysis performed on the Portuguese bank dataset. In Fig. 2a, people above the age of 70 are outliers, and the median age is about 40 years. There are some customers above 90 years of age. The main targets are middle-aged people. The customer's age is positively skewed. The customer's job in Fig. 2b shows 25% are admins, 23% are blue collars and 16% are technicians. In the education analysis from Fig. 2c, 15% account for basic.9y, 23% are high.school, 29% are university.degree. Marital status in Fig. 2d shows that most of the clients are married followed by single and divorced. In Fig. 2e, the conversion rate from the previous marketing campaign was only 3%. In Fig. 2f, the output variable "y" is imbalanced, but there is no need to treat it as "yes" class, which is around 11%. Imbalanced data treatment is done when one class is very low.

4.6 Feature Engineering

In this phase attributes are drawn out from the raw details to enhance the performance of the machine learning model. The Thiel's U method is utilised to get the correlation between the categorical independent variable and target categorical variable. Figure 3 demonstrates the comprehensive set of important features determined through the hybrid machine learning method which are utilised during the hybridisation process as discussed in the further phases.

4.7 Model Training and Parameter Tuning

The dataset is split into train and test set with samples in the ratio 80:20 from the dataset. During the hybridisation process the model is containerised in a sequential mode and the network layers are stacked in linear form with activation functions to

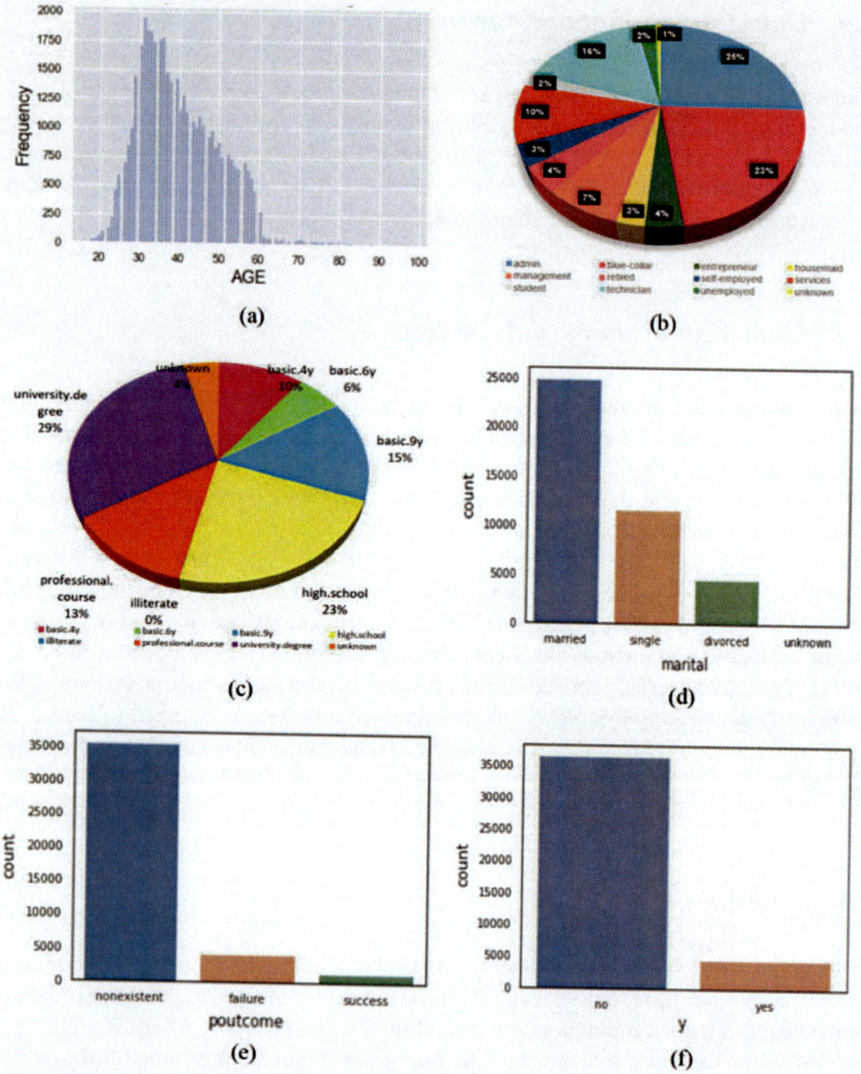

Fig. 2 Dispersion: **a** Customer age; **b** Customer job; **c** Customer education; **d** Customer marital status; **e** Poutcome, **f** Output variable 'y'

learn from the data and handle changing variables complexity. The model is optimised by using the adaptive moment estimation optimisation technique. The model is tuned to achieve the optimal batch size and validated on the validation set.

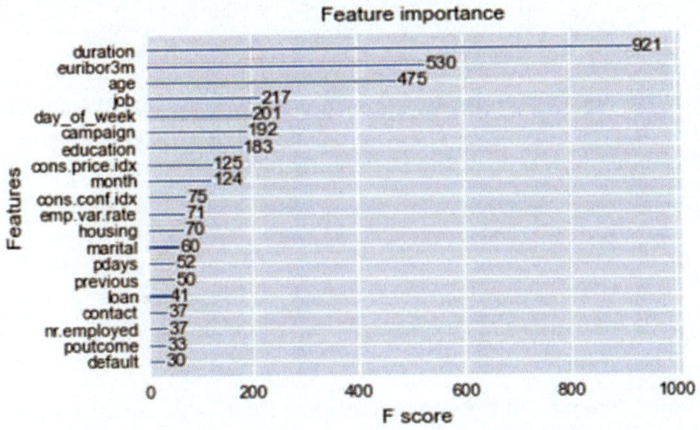

Fig. 3 Importance of the features in Portuguese bank dataset according to F score

4.8 *Model Evaluation*

In this phase the machine learning models are evaluated and the different machine learning techniques are compared. The optimum model is selected for deployment in the production environment.

Logistic Regression ML Model. It forecasts the likelihood of an end result that can be dichotomous. Numerical and categorical predictors are utilised for prediction. It develops a logistic curve, values within the range of 0 and 1 and is build by utilising the natural log of the odds of the output variable [29]. The independent variables not necessarily be normally distributed or possess uniform deviation in each group.

Figure 4 displays a logistic curve which shows the relation between the independent variable X with the moving mean of the P dependent variable. Equation 1 gives the formula.

$$P = \frac{e^{a+bX}}{1 + e^{a+bX}} \tag{1}$$

or

Fig. 4 Logistic curve

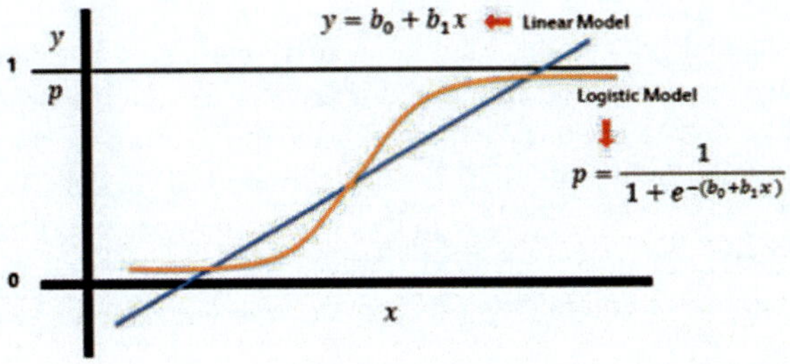

$$P = \frac{1}{1 + e^{-(a+bX)}} \tag{2}$$

where e indicates the base of natural log, whose value is 2.718 and a, b represent the variables of the model. In equation 3, P is the likelihood of a 1. Logit represents the dependent variable which happens to be natural log of odds represented as below:

$$odds = \frac{P}{1 - P} \tag{3}$$

$$\log(odds) = logit(P) = ln\left(\frac{P}{1 - P}\right) \tag{4}$$

Log of odds from equation 4 corresponds to a logit, odds are the consequence of P, the likelihood of a 1 in equation 5

$$logit(P) = a + bX \tag{5}$$

$$ln\left(\frac{P}{1 - P}\right) = a + bX \tag{6}$$

$$\frac{P}{1 - P} = e^{a+bX} \tag{7}$$

$$P = \frac{e^{a+bX}}{1 + e^{a+bX}} \tag{8}$$

Naive Bayes ML Model. In machine learning, in many cases, the area of interest is in choosing the principal hypothesis (y) given details(z). Hypothesis (y) within a classification problem possibly can be the class to allocate for the latest detail instance (z) [30]. Given the details(z) which can be utilised as preceding information regarding the difficulty, Bayes theorem imparts a technique which can evaluate the likelihood of a hypothesis by considering preceding information. Bayes theorem is demonstrated in equation 9.

$$P(y|z) = \frac{P(z|y)P(y)}{P(z)} \tag{9}$$

where P(y|z) is the likelihood of hypothesis (y) given the details(z), also known as the succeeding likelihood. P(z|y) is the likelihood of details(z), considering the hypothesis (y) was accurate. P(y) is the likelihood of hypothesis (y) being accurate, despite the details. P(z) is the likelihood of the details, despite the hypothesis. The major area of interest is in evaluating the posterior likelihood of P(y|z) out of prior likelihood p(y) with P(z), P(z|y). After computing the posterior likelihood for a number of distinct hypotheses, the one with the highest likelihood can be selected.

It's the maximum probable hypotheses and is called as maximum a posteriori (MAP) hypothesis.

SVM ML Model. Support vector machines invented in the 1990s is the most prevalent machine learning technique. It requires considerably less tuning. SVM is best described by a maximal margin classifier. An n-dimensional space is formed through input numeric parameters (X) in the data. In general, a hyperplane resembles a line that divides the input space of parameters and is selected to segregate the points present in the input parameter space based on their class values as 0 or 1. The hyperplane can be visualised being a line in two dimensions. For example, assuming all the input points can be entirely segregated through this line as given in equation 10

$$M0 + (M1 * X1) + (M2 * X2) = 0 \tag{10}$$

where M1, M2 are the coefficients that resemble the slope, and M0 is the line intercept which is determined by the ML technique. Classifications can be performed by using the line. By introducing input parameters in the line equation 10, a new point can be evaluated whether it lies over or under this line. If the point lies over the line and the equation gives a higher value than 0, the point is categorised under class 0 and for a value lower than 0 the point is categorised under class 1. The points nearer to the line can be difficult to categorise as the value could be closer to zero. More confident predictions can be achieved by the model, provided the value is of higher magnitude. Margin is the span between the line and the nearest data points. The line which segregates the two classes is the optimal one only if it has the highest margin and is known as maximal-margin hyperplane. Margin is computed as the vertical span between the line and the nearest points. These nearest points are only suitable for describing the line and in building the model. They are known as support vectors. They describe or support the hyperplane. Hyperplane learns through training data by using an optimisation technique which maximises the margin.

Decision Tree ML Model. Decision trees are the significant category of machine learning technique utilised for predictive modelling. The traditional decision tree techniques have been all around the machine learning domain for a long time and the latest developments such as random forest are amid the most influential techniques at hand. Leo Briman put forward CART—Classification and Regression Trees—to call attention to decision tree techniques that could be utilised for modelling regression problems or classification predictive modelling problems. Various platforms refer to the algorithm in different ways, like in the R programming language platform the algorithm is denoted with the latest names as in CART. But traditionally the technique is referred to as decision trees. The CART technique furnishes the base for the key machine learning techniques such as boosting decision trees, random forest and bagging decision trees.

Neural Network ML Model. Neural networks are one of the techniques of machine learning, which happens to be an application of AI. They are an illustration of a supervised learning technique and aim to estimate the function depicted by the information [31]. This is performed by evaluating the error across the predicted outputs together with expected outputs and reducing this error throughout the training

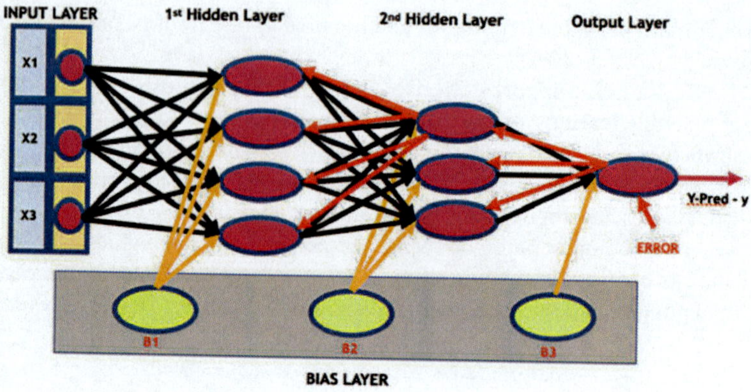

Fig. 5 Multilayer neural network

process as shown in Fig. 5. Neural networks possess at the minimum three layers consisting of neurons which represent the input neuron layer, the hidden layer(s) of neurons, as well as the output layer and the bias layer. The intermediate or the hidden layer(s) comprises several neurons in addition to connections amongst the layers. While the neural network learns the information, the weights or effectiveness of the connections amongst the aforesaid neurons are fine-tuned permitting the network to bring forward accurate predictions.

4.9 Model Deployment and Prediction Analysis

In this phase the model is integrated with an existing production environment and deployed where it can process the input information and produce the desired output. Prediction analysis is performed and the model is evaluated based on accuracy and performance metrics.

4.10 Prediction Results of Machine Learning Models

In this part the prediction results of the ML models are discussed. Fig. 6a and b shows the confusion matrix and ROC curve generated for the support vector machine ML model. Figure 7a, b shows the confusion matrix and ROC curve generated for the logistic regression ML model. Figure 8a, b shows the confusion matrix and ROC curve generated for the Bayesian ML model. Figure 9a, b shows the confusion matrix and ROC curve generated for the decision tree ML model. Figure 10a, b shows confusion matrix and ROC curve generated for hybrid ML model in which true positive (observed = 1, predicted = 1) predicted term deposit will be taken and

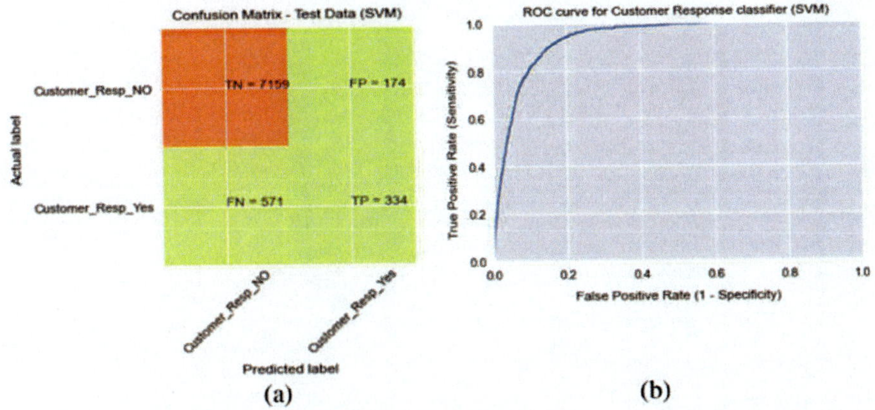

Fig. 6 SVM ML model: **a** Confusion matrix; **b** ROC curve

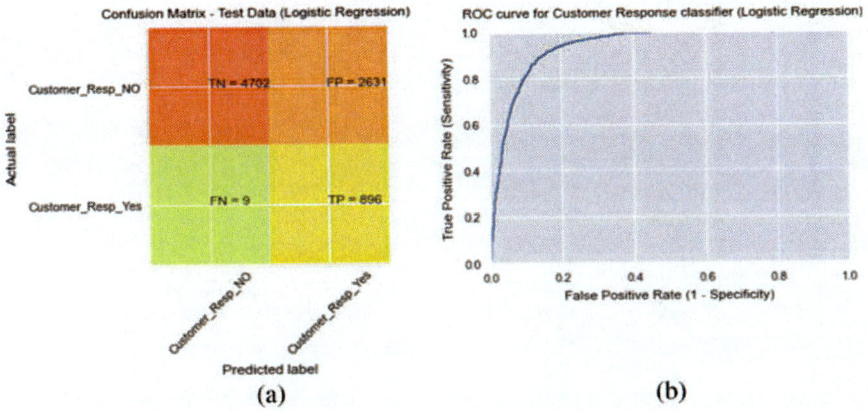

Fig. 7 Logistic regression ML Model: **a** Confusion matrix; **b** ROC curve

the customer took it; false positive (observed = 0, predicted = 1) predicted term deposit will be taken and the customer did not take it; true negative (observed = 0, predicted = 0) predicted term deposit will not be taken and the customer did not take it; false negative (observed = 1, predicted = 0) predicted term deposit will not be taken and the customer took it. The confusion matrix demonstrates the performance measure for the ML models with actual and predicted outcomes of customer responses to subscribe product bank term deposit. The ROC curve is a plot of the true positive rate on the y-axis and the false positive rate on the x-axis. It assists to choose a threshold that balances sensitivity and specificity. The area under the curve is useful as a single number summary of model performance. If randomly one positive and negative observation is selected, then the area under the

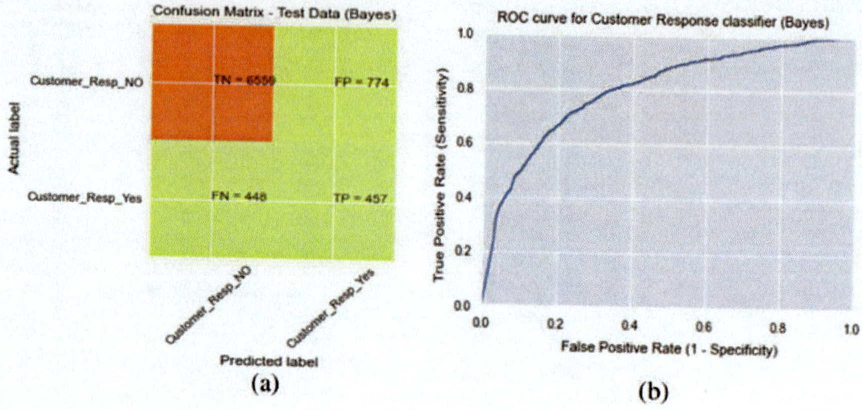

Fig. 8 Bayesian ML model: **a** Confusion matrix; **b** ROC curve

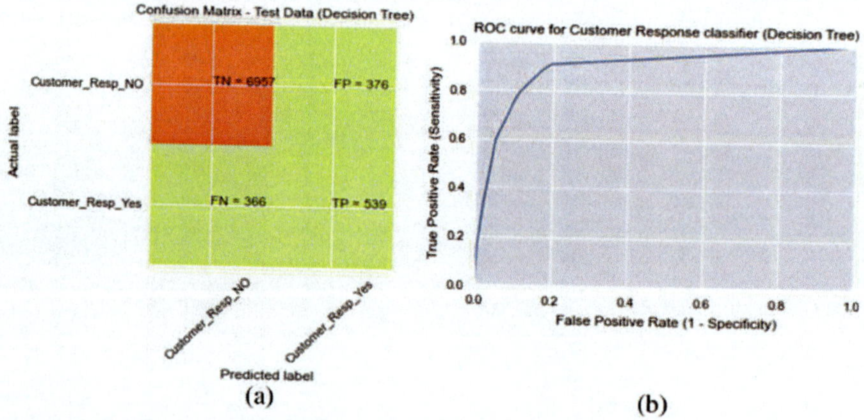

Fig. 9 Decision tree ML Model: **a** Confusion matrix; **b** ROC curve

curve represents the likelihood that the model (b) will assign a higher probability to the positive observation.

Table 1 demonstrates the performance metrics of the SVM ML model, logistic regression ML model, Bayes ML model, decision tree ML model and hybrid ML model. The accuracy of the ML models is determined by equation 11, which shows overall how often is the model accurate in predicting the customers who will subscribe for product bank term deposit. The sensitivity of the ML models is determined by equation 12, which demonstrates how sensitive is the model in detecting positive instances. Equation 13 represents the ML models specificity which states that when the actual value is negative, how often is the prediction accurate. It also demonstrates how specific or selective the model is in predicting positive instances. The preciseness of the model during the prediction of positive instances is given by equation 14,

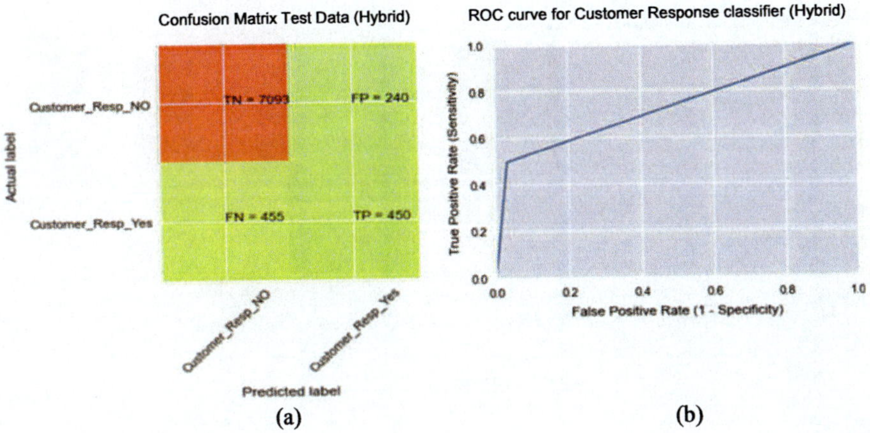

Fig. 10 Hybrid ML model: **a** Confusion matrix; **b** ROC curve

Table 1 Performance metrics—machine learning models

Performance metrics	SVM ML model	Logistic regression ML model	Bayes ML model	Decision tree ML model	Hybrid ML model
Accuracy	90.95	91.02	85.96	90.99	91.56
Sensitivity	92.37	92.37	65.63	90.93	49.72
Specificity	82.89	82.13	80.92	80.56	96.72
ROC AUC Score	93.56	93.74	80.45	90.02	73.22
True positive	334	896	457	539	450
False Negative	571	9	448	366	455
Precision score	65.74	63.92	38.94	58.90	65.21
Recall score	36.90	42.09	48.83	59.55	49.72
Cross-validation score	65.82	78.04	64.23	65.25	56.42

which demonstrates that when a positive value is predicted, how often the prediction is accurate. Equation 15 shows the recall of the ML models which demonstrates that when the actual value is positive, how often the prediction is accurate.

$$\text{ACCURACY} = \frac{TP + TN}{TP + TN + FP + FN} \tag{11}$$

$$\text{SENSITIVITY} = \frac{TP}{TP + FN} \tag{12}$$

$$\text{SPECIFICITY} = \frac{TN}{TN + FP} \tag{13}$$

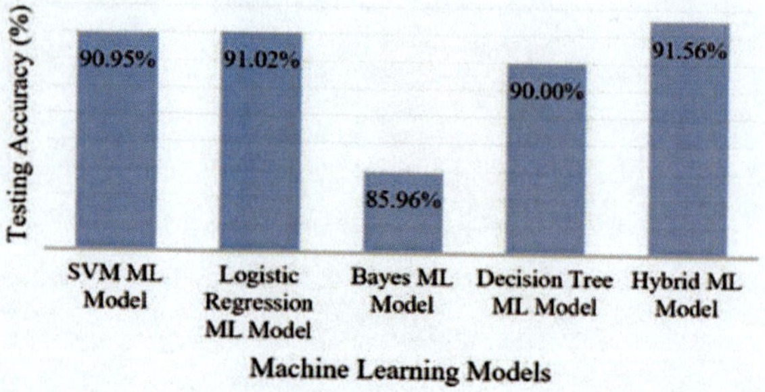

Fig. 11 Testing accuracy on Portuguese bank dataset

$$\text{PRECISION} = \frac{TP}{TP + FP} \tag{14}$$

$$\text{RECALL} = \frac{TP}{TP + FN} \tag{15}$$

Figure 11 presents the testing accuracy of the machine learning models on the Portuguese bank dataset. The support vector machine ML model gives 90.95% accuracy, the decision tree ML model gives 90% accuracy, the logistic regression ML model gives 91.02% accuracy and the Bayesian ML model gives 85.96% accuracy. The hybrid ML model gives 91.56% accuracy as compared to other models.

5 Conclusion

In the banking domain, optimisation of the targeting of telemarketing is a prevalent problem under increasing pressure to enhance the profits and reduce the loss. Taking into consideration the effects of the economic crisis, Portuguese banks were compelled to gain capital investments by acquiring term deposits. In this research, an efficient machine learning procedure is recommended for the detection of bank telemarketing customers. A current and sizeable Portuguese bank was examined and analysed through the machine learning models: logistic regression ML model, naive Bayes ML model, decision tree ML model and support vector machine ML model. These models were compared with recommended hybrid machine learning method (extreme gradient boosting + multilayer neural network) using accuracy and performance metrics. For both parameters and phases, high and valid outcomes were achieved by the hybrid technique. After achieving the desired accuracy and customer conversion rate, the model can be used for practical predictions. Future work in this domain is focused on analysing the ML models on large banking databases to meet the demand of the financial institutions to achieve the subscription targets for future banking products.

References

1. Quah, J.T.S., Chua, Y.W.: Chatbot assisted marketing in financial service industry. Int. Conf. Serv Comp. **11515**, 107–114 (2019)
2. Lessmann, S., Haupt, J., Coussement, K., et al.: Targeting customers for profit: an ensemble learning framework to support marketing decision-making. Inf. Sci. (2019)
3. Nazeeh, Ghatasheh., et al.: Business analytics in telemarketing: cost-sensitive analysis of bank campaigns using artificial neural networks. J. Appl. Sci. **10**, 2581 (2020)
4. Tricot, M.: Classification algorithms. J. Syst. Softw. **6**, 93 (1986)
5. Przegalinska, A., Ciechanowski, L., Stroz, A., Gloor, P., Mazurek, G.: In bot we trust: A new methodology of chatbot performance measures.J. Busi. Hori. **62**, 785–797 (2019)
6. Hung, P.D., Hanh, T.D., Tung, T.D.: Term deposit subscription prediction using spark MLlib and ML packages. In: Proceedings of the 2019 5th International Conference on E-Business Applications, vol. 88 (2019)
7. Ładyżyński, P., Żbikowski, K., Gawrysiak, P.: Direct marketing campaigns in retail banking with the use of deep learning and random forests. Exp. Sys. App. **134**, 28–35 (2019)
8. Patel, R., Mazumdar, H.: Prediction of bank investors using neural network in direct marketing. Int. J. Eng. App. Sci. **5**, (2018)
9. Islam, M.S., Arifuzzaman, M., Islam, M.S.: SMOTE Approach for Predicting the Success of Bank Telemarketing. 2019 TIMES-iCON (2019)
10. Sarkar, M., De Bruyn, A.: LSTM response models for direct marketing analytics: replacing feature engineering with deep learning. J. Interact. Mark. **53**, 80–95 (2021)
11. Martínez, A., Schmuck, C., Pereverzyev, S., Pirker, C., Haltmeier, M.: A machine learning framework for customer purchase prediction in the non-contractual setting. Eur. J. Oper. Res. **281**, 588–596 (2020)
12. Yadav, V., Sreelatha, M., Rajinikanth, T.V.: Classification of telemarketing data using different classifier algorithms. Int. J. Inn. Tech. Expl. Eng. **8**, 1300–1306 (2019)
13. Muthukumar Vignesh, B.N.: MOOCVERSITY - Deep Learning Based Dropout Prediction in MOOCs over Weeks. J. Sof. Comp. Para. **2**, 140–152 (2020)
14. Pradap, R., Kamaludeen, P.: Machine learning models for bank telemarketing classification and prediction. The Int. J. Ana. Exp. Mod. Analys. **11** (2019)
15. Chlebus, M., Osika, Z.: Comparison of tree-based models performance in prediction of marketing campaign results using Explainable Artificial Intelligence Tools (2020)
16. Yan, C., Li, M., Liu, W.: Prediction of bank telephone marketing results based on improved whale algorithms optimizing S_Kohonen network. Appl. Soft Comput. **92**, (2020)
17. Muppala, C., Dandu, S., Potluri, A.: Efficient predictions on asymmetrical financial data using ensemble random forests. In: Proceedings of the Third International Conference on Computational Intelligence Information, vol. 1090, (2020)
18. Guo, M., Zhang, Q., Liao, X., Chen, F.Y., Zeng, D.D.: A hybrid machine learning framework for analyzing human decision-making through learning preferences. Omega, 102263 (2020)
19. Puteri, A.N., Dewiani, T.Z.: Comparison of potential telemarketing customers predictions with a data mining approach using the MLPNN and RBFNN Methods. In: 2019 International Conference on Information and Communications Technology (ICOIACT), (2019)
20. Mustapha, S.M.F.D.S., Alsufyani, A.: Application of Artificial Neural Network and information gain in building case-based reasoning for telemarketing prediction. Int. J. Adv. Comput. Sci. Appl. **10**, 300–306 (2019)
21. Panigrahi, A., Patnaik, M.C.: Customer deposit prediction using neural network techniques. Int. J. Appl. Eng. Res. **15**, 253–258 (2020)
22. Al-shawi, A.N.F.: Hybrid datamining approaches to predict success of bank telemarketing. Int. J. Comput. Sci. Mob. Comput. **8**, 49–60 (2019)
23. Jiang, Y.: Using logistic regression model to predict the success of bank telemarketing. Int. J. Data Sci. Tech. **4**, 35–41 (2018)
24. Valarmathi, B., Chellatamilan, T., Mittal, H., Jagrit, J., Shubham, S.: Classification of imbalanced banking dataset using dimensionality reduction. In: 2019 ICCS, 1353–1357 (2019)

25. Schaeffer, S.E., Rodriguez Sanchez, S.V.: Forecasting client retention—A machine-learning approach. J. Retail. Cons. Ser. **52**, (2020)
26. Suma, V.: Data mining based prediction of demand in Indian market for refurbished electronics. J. Soft. Comp. Para. **2**, 153–159 (2020)
27. Tekouabou, S.C.K., Cherif, W., Silkan, H.: A data modeling approach for classification problems: application to bank telemarketing prediction. In: Proceedings of the 2nd International Conference on Networking on Information Systems Security, pp 1–7 (2019)
28. Dua, D., Graff, C.: UCI Machine Learning Repository [http://archive.ics.uci.edu/ml]. Irvine, CA: University of California, School of Information and Computer Science (2019)
29. De Caigny, A., Coussement, K., De Bock, K.W.: A new hybrid classification algorithm for customer churn prediction based on logistic regression and decision trees. Eur. J. Opera. Resea. **269**, 760–772 (2018)
30. Posch, K., Arbeiter, M., Pilz, J.: A novel Bayesian approach for variable selection in linear regression models. Comput. Stat. Data Anal. **144**, (2020)
31. Aggarwal, C.C.: Neural Networks and Deep Learning: A Textbook (2018)

COVID-19: Smart Shop Surveillance System

S. Kukan, S. Gokul, S. S. Vishnu Priyan, S. Barathi Kanna, and E. Prabhu

Abstract In the current situation, the spread of COVID-19 is very high and there are a lot of active COVID-19 cases everywhere around the world. This project aims at shop management systems against COVID-19. At the entrance of the shop, a contactless information transfer of the customer details is transferred by developing two Android applications: QR code generator application (for the customer) and QR code scanner application (for the shop). Scanner application is kept at the entrance of the shop in which the customer shows their QR code in which the information is present. The information is collected for ease of contact tracing. A temperature screening module is also developed to check if the customer has a fever (a COVID-19 symptom) or not. If a particular customer has a symptom, an alert message will be sent to the concerned authorities. Complete surveillance is done in the shop to check whether a customer wears a mask or not.

Keywords COVID-19 · Deep learning · CNN · MLX90614 · QR code · Flutter · Internet of things · Vgg19 · MLP

1 Introduction

Coronaviruses are a large family of viruses that may cause illness in animals or humans. COVID-19 is an infectious disease caused by a newly discovered coronavirus. COVID-19 affects different people in different ways. Most of the infected people will develop mild to moderate illness and recover without hospitalization. Some of the symptoms of COVID-19 according to the World Health Organization are given in Table 1 [1].

S. Kukan (✉) · S. Gokul · S. S. Vishnu Priyan · S. Barathi Kanna · E. Prabhu
Department of Electronics and Communication Engineering, Amrita School of Engineering, Coimbatore, Amrita Vishwa Vidyapeetham, Coimbatore, India

E. Prabhu
e-mail: e_prabhu@cb.amrita.edu

J. S. Raj et al. (eds.), *Intelligent Sustainable Systems*, Lecture Notes in Networks and Systems 213, https://doi.org/10.1007/978-981-16-2422-3_55

Table 1 Symptoms

Most common symptoms	Fever, dry cough, tiredness
Less common symptoms	Aches and pains, sore throat, diarrhea, conjunctivitis, headache, loss of taste or smell, a rash on the skin, or discoloration of fingers or toes
Serious symptoms	Difficulty of breathing or shortness of breath, chest pain or pressure, loss of speech or movement

Fig. 1 COVID-19 impact

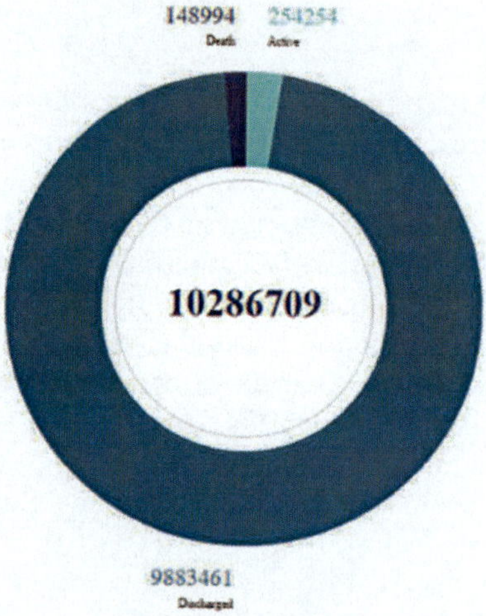

From the table, it is observed that fever is one of the main symptoms of COVID-19. So, there is a need to check the temperature of a person to confirm his/her health status [2, 3]. According to the statistics of the Ministry of Health and Family Welfare of the Government of India, there are about 254254 active COVID-19 cases in India as of 01/01/2021 which is shown in Fig. 1 [4].

Thus, COVID-19 has a greater impact on the earth in almost all sectors of society. These statistics show the level of impact of COVID-19. Washing hands and usage of alcohol-based sanitizers are the proposed method to mitigate the transmission of COVID-19. Thus, following proper respiratory etiquette is necessary. The only way to solve the spread of coronavirus is by maintaining social distancing, wearing the mask, and tracing the infected contacts to stop the spread of the disease.

2 Existing System

At present in India, there is a spread of COVID-19 for which wearing masks, social distancing, and contact tracing are some of the ways proposed and followed to stop the spread of COVID-19. In the local shops of India, customer's personal information is collected for ease of contact tracing which is done manually using paper and pen or entered by a person manually on a computer. This process is time-consuming, and not efficient enough for tracing contact as well. So a contactless system is needed. Therefore, an efficient system is needed to collect the personal information of the customers. Fever is found to be one of the main symptoms of COVID-19 [5]. All customers are screened using an IR temperature gun which measures contactless human body temperature. There is no specific system for ensuring the usage of face masks in the shop at present and it is done only by manual checking by the in-charge of the shop [6]. When the usage of the mask is checked manually, it is tedious, and continuous surveillance is nearly impossible which a machine can do efficiently than a person in charge [7]. The National Informatics Centre, a subdivision of India's Ministry of Electronics and IT (MeitY) has developed a contact tracing app called Aarogya Setu to help curb the spread of the COVID-19. Aarogya Setu app uses Bluetooth and GPS of the mobile phone to check whether the particular individual has contacted any COVID-19 patient in the database and it gives the guidelines to the individual to be followed [8]. The major limitation in the existing system is the wastage of human resources in the manual acquisition of data, temperature measurement, and face mask surveillance and few more limitations are in maintaining, searching, sorting, transfer of data, and analysis of manually stored data. There is a chance of human error in reading and recording data. Thus, there exists a system that must be more efficient and accurate. So, for these reasons, a new system is proposed which is discussed in detail in the next section.

3 Proposed System

An efficient shop surveillance system is developed for which the overall block diagram is shown in Fig. 2. The block diagram in Fig. 2 has three separate modules, where each module functions individually that together forms the smart shop surveillance system. Two applications are developed, namely a QR code generator and a QR code scanner which are used for collecting the customer's details, like name, phone number, the number of people accompanying, and address without contact. These details are saved in the local cloud of the shop [9, 10]. All the customers are screened using an IR temperature sensor and if there is any abnormality, a message is sent to the concerned authority regarding the customer's health status [2]. Face mask detection system is developed using python for continuous surveillance [6] for the usage of the mask by the customers in the shop [7]. Each model is discussed in detail below.

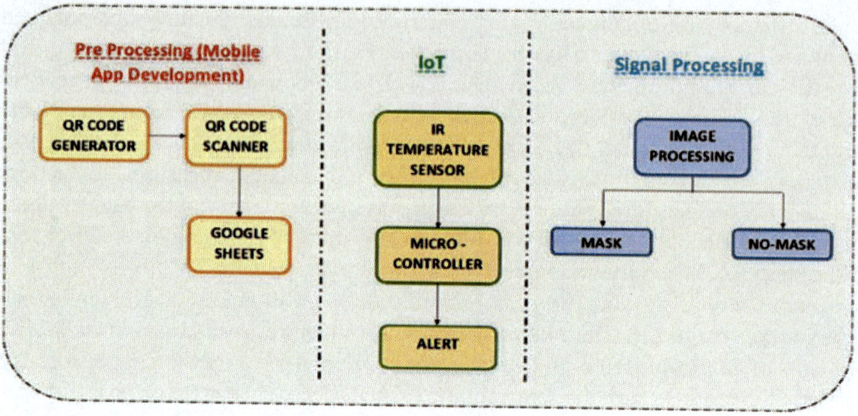

Fig. 2 Proposed block diagram of smart shop surveillance system

3.1 QR Code Generator Application

A QR code generator application is developed using the Flutter platform, android studio environment, and dart language [11]. All the customers entering the shop must have this application installed on their mobile phone and they must enter the details like name, phone number, the number of people accompanying, and address in the application and press the "generate" button to convert the customer details in the form of QR code.

3.2 QR Code Scanner Application

A QR code scanner application is developed using the Flutter platform, android studio environment, and dart language [11]. Customers must scan their generated QR code in the scanner application which decrypts all the customer details and saves them in the Google sheet along with the timestamp of the customer's arrival at the shop [12, 13]. The QR code scanner application is mainly developed to get customer details without contact. The details collected in the QR code scanner application are name, mobile number, the number of people accompanying, and address along with the date and time of scanning.

3.3 Temperature Screening Module (MLX90614)

A temperature screening module is developed using the MLX90614 temperature sensor. MLX90614 has an infrared thermopile detector and a signal-conditioning

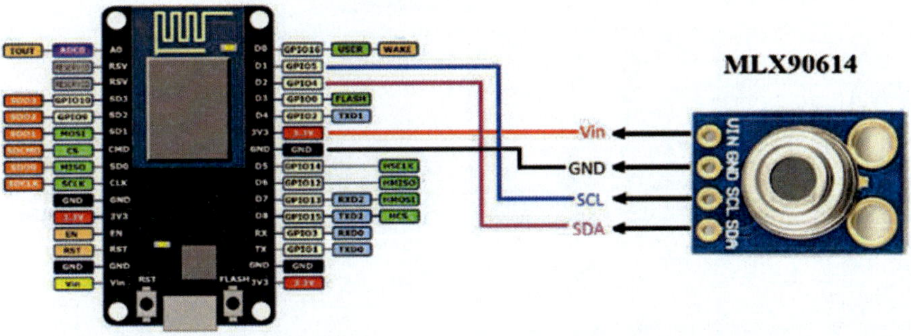

Fig. 3 Circuit diagram of IoT prototype

application processor and its main working principle is the Stefan–Boltzmann law that any object that isn't below absolute zero (0°K) emits (non-human-eye-visible) light in the infrared spectrum that is directly proportional to its temperature [5, 14].

The circuit diagram as shown in Fig. 3 describes the IoT model development process in the smart shop surveillance system. The temperature sensor collects the data and sends it to the cloud (firebase) using the microcontroller (NodeMCU) via Wi-Fi [2, 3, 7, 9]. Using MIT app inventor an application is developed to retrieve the data from firebase, whenever it is altered in the firebase. The retrieved value from the firebase is continuously monitored against the threshold set for the indication of fever in customers. When the retrieved value exceeds the threshold limit, then the application triggers the messaging service of the mobile and a message is sent to the corresponding authority [2, 3, 7, 11, 15, 20].

3.4 Facemask Detection System

A concept of deep learning is used which is a subset of machine learning which make the computation of multilayer neural network feasible. The facemask detection system is developed for continuous surveillance of customers to ensure the usage of masks in the shop using python [7]. TensorFlow is a free and open-source software library for machine learning used to train and deploy various neural networks. To build a facemask detection system, several steps must be followed which are as follows:

- Dataset collection
- Building a convolutional neural network
- Training a convolutional neural network
- Plotting the results
- Testing a convolutional neural network

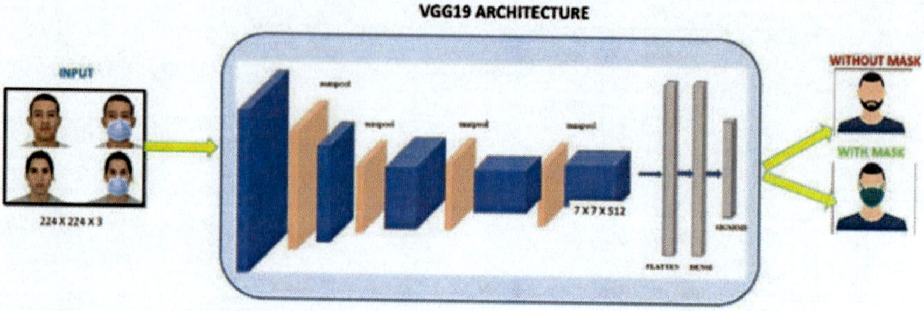

Fig. 4 Proposed model for face mask detection

Dataset plays a vital and important role in the process of building the model of the facemask detection system. A model's performance depends heavily on the dataset collected. The collected dataset must be large in terms of quantity and it must have a diverse angle of images so that all the features are extracted to good accuracy [16]. Dataset collected has approximately 6000 images of people with masks and approximately 5000 images of people without a mask. These collected images are people from different races across the globe. These images are collected from Kaggle. The collected images are used for training, testing, and validation purposes. A convolutional neural network (CNN) [17] is a type of artificial neural network that uses a machine-learning algorithm, for supervised learning and to analyze data. VGG19 is the CNN architecture used in this facemask detection system.

VGG19 is a convolutional neural network architecture which is a variant of the VGG model with 19 layers, which are input, conv3, and max-pooling, In general, a CNN has two parts: a convolutional base and a classifier. The architecture of VGG19 is shown in detail in Fig. 4. In this model, the last three fully connected layers with softmax activation function act as a classifier. This classifier is disabled and replaced by a flatten layer and the two dense layers with sigmoid as an activation function. These newly connected layers are collectively known as MLP, which is used for classification [19]. Therefore, the first 16 layers of the model are called convolutional base which is used for feature extraction purposes. A convolutional base is a stack of convolutional and pooling layers. The main purpose of the convolutional base is to extract the features from images which are then passed down to the classifier. With the help of these, the classification of features is done. Here convolutional layer extracts the features from the images such as edges, colors, orientation, etc. And they also help in reducing the dimensions. Pooling layers also do the same as a convolutional layer but these layers extract more dominant features among all extracted features and this is done to reduce the computational power required to process the data. There are two types of pooling, namely max pooling and average pooling. Max pooling returns the max value whereas average pooling returns the average of all the available values. Also, max pooling acts as a noise suppressant, and to avoid overfitting, batch normalization is done.

The MLP classifier has flatten and dense layers. Flattening is used to flatten the image we acquired from the convolutional layer into a column vector and feed it into the dense neural networks which then identify the dominating and low-level features and classifies the image using a sigmoid classifier. The results and outputs of facemask detection are discussed in detail in the next section.

4 Results and Discussions

A system is developed as shown in Fig. 2. Each result obtained is shown in detail in this section. The QR code generator application and QR code scanner application which are developed can be installed by scanning the QR code in Fig. 5. The customer's details are stored in the google sheet and is shown in Fig. 6.

The QR code generator application has the front page and info page, as shown in Fig. 7. The front page of the QR code generator application has customer details like name, mobile number, number of customers, and address. These details must be

(a) (b)

Fig. 5 QR code for generator and scanner APKs

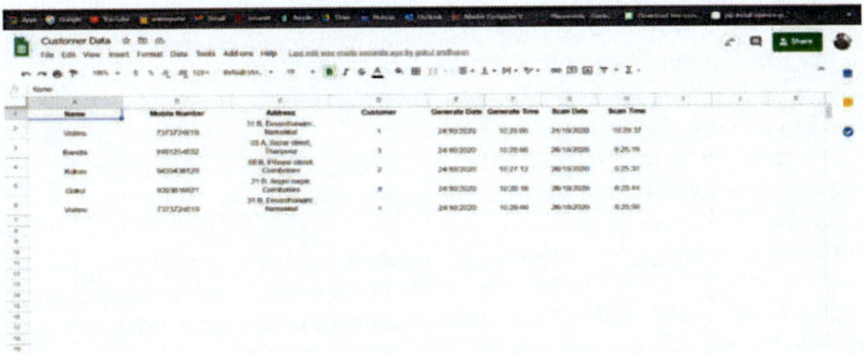

Fig. 6 Google sheet as a database for ease of contact tracing

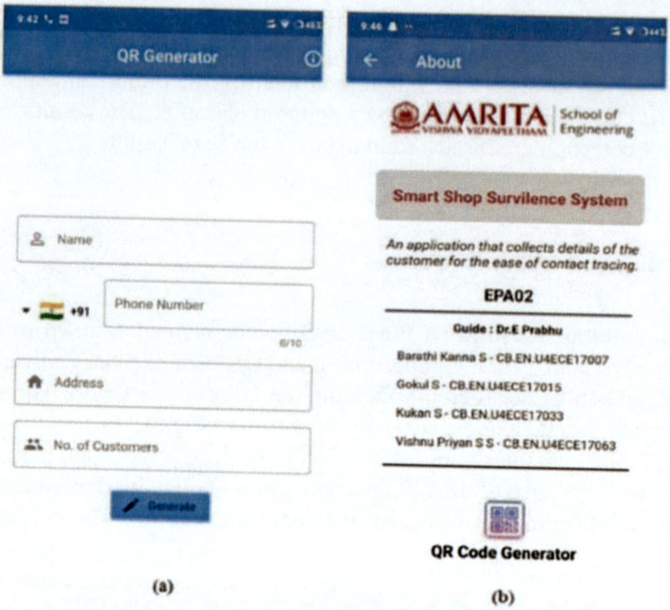

Fig. 7 QR code generator. **a** Front page, **b** About page

entered in this application after which the generate button is pressed to convert the customer's details into a QR code. The info page present in the QR code generator application has details about the application and the developer.

The QR code scanner application has the front page and the info page of the application as shown in Fig. 8. When the QR code is scanned, all the details entered in the QR code generator application are stored in Google sheets along with the timestamp of the arrival of customers at the shop. The customer's details like name, mobile number, number of customers, address, QR code generated date and time, and QR code scanned date and time are stored in Google sheets as seen in Fig. 6.

The IoT model for temperature screening is developed as seen in the circuit diagram in Fig. 3. An application developed using MIT app inventor collects temperature data of the customers and the surrounding in degree and Fahrenheit. These values in the application are retrieved from the corresponding value in firebase (cloud) as shown in Fig. 9 [3, 7, 9].

The message is sent to the concerned authority if the threshold set for the temperature is crossed indicating the customer's health status [2] which is shown in Fig. 10.

The created model for the facemask detection system's performance can be measured using accuracy and loss values. There are two types of accuracy, namely training accuracy and validation accuracy. Similarly, in loss, there are training and validation losses. The training accuracy and loss indicate the progress of the model and the validation and training loss indicate the performance of the model. The

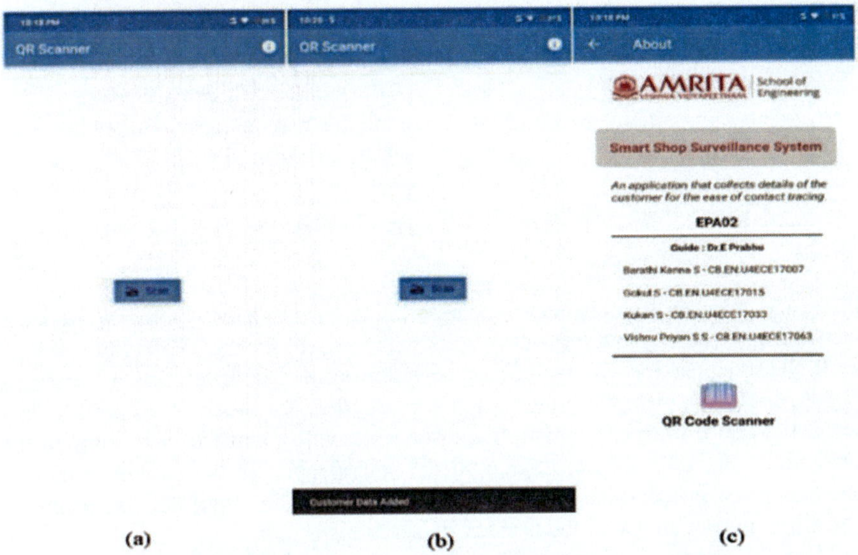

Fig. 8 QR code scanner. **a** Front page, **b** When valid QR is scanned, **c** About page

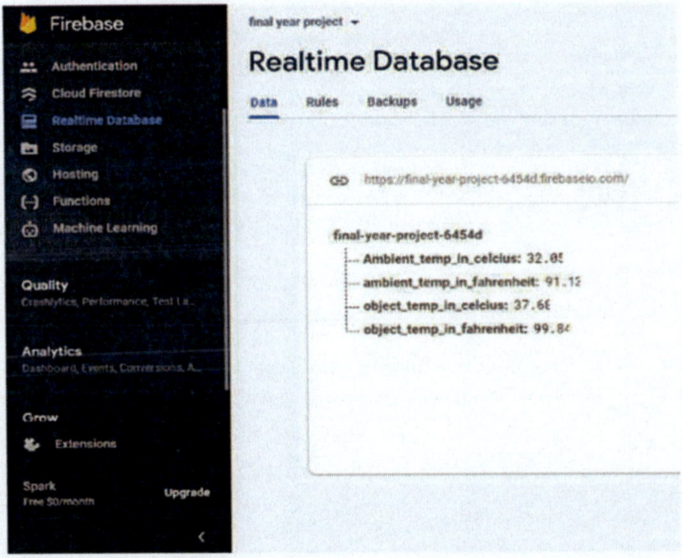

Fig. 9 Firebase as a database for automated temperature monitoring

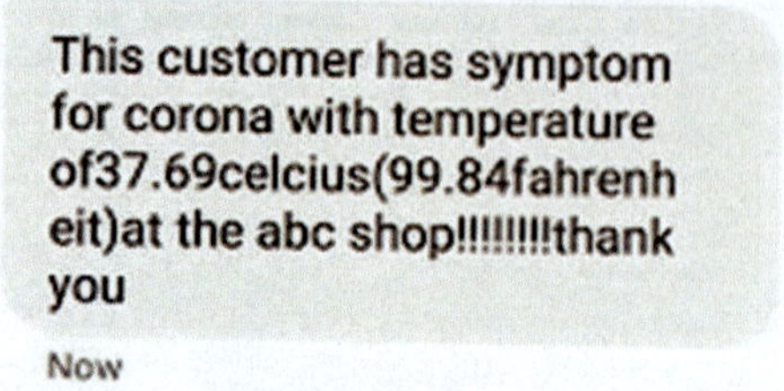

Fig. 10 Message received by officials when an abnormality is detected

difference between training and validation accuracy indicates whether a model is over-fitting. Accuracy and loss curve is shown in Fig. 11 for the VGG19 model with MLP classifier used in the face mask detection system with training accuracy of 97.58% loss of 5.75%. Here the plot is accuracy/loss vs epoch. A single epoch is where the entire training/validation dataset is passed to and fro once through the CNN model for optimization. Since all the "m" samples in the dataset cannot be passed at once, it is divided into "n" number of batches for simplification purposes. Therefore, during an epoch the number of iterations is m/n.

Thus, this graph shows the increase or decrease in performance for consecutive epochs. The Early Stopping method is used to determine the number of epochs. The

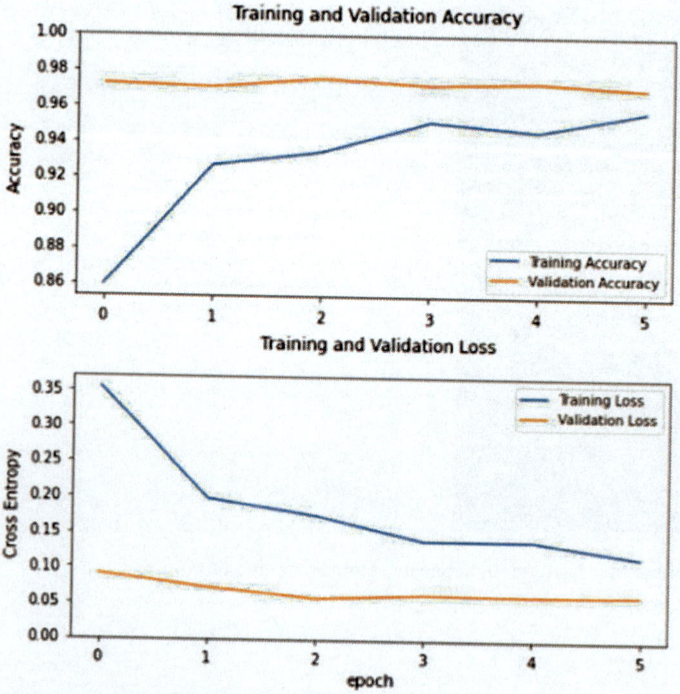

Fig. 11 Training and validation accuracy plot

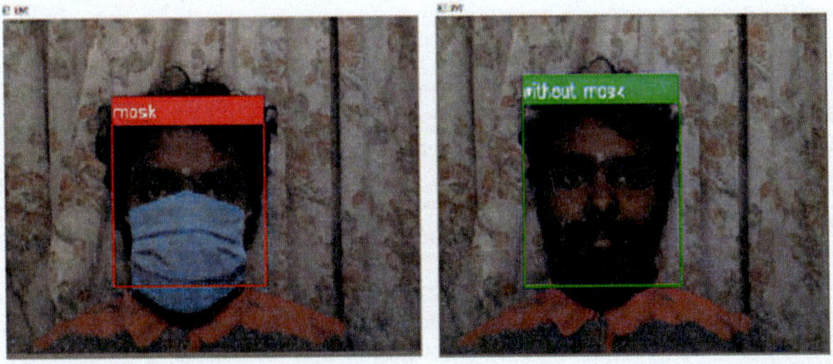

Fig. 12 Testing the model at different scenarios

epochs stop because of early stopping when the loss of the model doesn't drop further. As the trained model gives good accuracy, we get input from the video, and using the Haar cascade classifier [6], the face is detected [18] and is checked for facemask using the model created, and the output of it is shown in Fig. 12. The images of the customers without a facemask are stored separately for easy identification.

5 Conclusion

In the COVID-19 pandemic, technology is used to mitigate the effect of COVID-19. A shop is a place where all get the required items and it is a place that has a high possibility that it can turn into a COVID-19 hotspot. So it is necessary to bring a system proposed here to stop the spread of COVID-19. Thus, this system becomes an inevitable one in our lives now.

6 Further Enhancements

To mitigate the effect of COVID-19 various technologies are used and one such is proposed here. There can be further improvements done to this system for better efficiency and accuracy [2, 3]. A more accurate IR temperature sensor can be used for better accuracy and results. For the facemask detection system proposed here, various other models can be created for better results and can be used in the system.

References

1. World Health Organization, https://www.who.int/health-topics/coronavirus#tab=tab_3
2. Gu, Y., Wang, P., Duan, H., Chen, H., Ren, Y.: Design of rescue and health monitoring bracelet for the elderly based on STM32. IEEE 8th Joint International Information Technology and Artificial Intelligence Conference (ITAIC), Chongqing, China **2019**, 1259–1262 (2019)
3. Kansara, R., Bhojani, P., Chauhan, J.: Designing smart wearable to measure health parameters. In: 2018 International Conference on Smart City and Emerging Technology (ICSCET), Mumbai, 2018, pp. 1–5
4. Ministry of Health and Family Welfare of the Government of India, https://www.mohfw.gov.in/
5. Somboonkaew, A., et al.: Mobile-platform for automatic fever screening system based on infrared forehead temperature. Opto-Electronics and Communications Conference (OECC) and Photonics Global Conference (PGC), Singapore **2017**, 1–4 (2017)
6. Hoque, M.A., Islam, T., Ahmed, T., Amin, A.: Autonomous face detection system from real-time video streaming for ensuring the ıntelligence security system. In: 2020 6th International Conference on Advanced Computing and Communication Systems (ICACCS), Coimbatore, India, 2020, pp. 261–265
7. Marceline, R., Akshaya, S.R., Athul, S., Raksana, K.L., Ramesh, S.R.: Cloud storage optimization for video surveillance applications. In: 2020 Third International Conference on Smart Systems and Inventive Technology (ICSSIT), Tirunelveli, India, pp. 62–67
8. Chamola, V., Hassija, V., Gupta, V., Guizani, M.: A comprehensive review of the COVID-19 pandemic and the role of IoT, Drones, AI, Blockchain, and 5G in managing its ımpact. IEEE Access **8**, 90225–90265 (2020)
9. Li, W., Yen, C., Lin, Y., Tung, S., Huang, S.: JustIoT Internet of Things based on the firebase real-time database. In: 2018 IEEE International Conference on Smart Manufacturing, Industrial & Logistics Engineering (SMILE), Hsinchu, 2018, pp. 43–47
10. Fu, X., Feng, K., Wang, C., Zhang, J.: Improving fingerprint based access control system using quick response code. In: 2015 IEEE International Conference on Computational Intelligence and Virtual Environments for Measurement Systems and Applications (CIVEMSA), Shenzhen, 2015, pp. 1–5
11. Allison, L., Fuad, M.M.: Inter-App communication between android apps developed in app-ınventor and android studio. In: 2016 IEEE/ACM International Conference on Mobile Software Engineering and Systems (MOBILESoft), Austin, TX, 2016, pp. 17–18
12. Bani-Hani, R.M., Wahsheh, Y.A., Al-Sarhan, M.B.: Secure QR code system. In: 2014 10th International Conference on Innovations in Information Technology (IIT), Al Ain, 2014, pp. 1–6
13. Gauen, K. et al.: Comparison of visual datasets for machine learning. In: 2017 IEEE International Conference on Information Reuse and Integration (IRI), San Diego, CA, 2017
14. Melexis MLX90614 Infrared Temperature Sensor User Manual, https://www.melexis.com/media/files/documents/datasheets/mlx90614-datasheet-melexis.pdf
15. Gokul, S., Kukan, S., Meenakshi, K., Priyan, S.S.V., Gini, J.R., Harikumar, M.E.: Biometric Based Smart ATM Using RFID. In: 2020 Third International Conference on Smart Systems and Inventive Technology (ICSSIT), Tirunelveli, India, 2020, pp. 406–411
16. Liu, Y., Yang, J., Liu, M.: Recognition of QR Code with mobile phones. In: 2008 Chinese Control and Decision Conference, Yantai, Shandong, 2008, pp. 203–206
17. Baratsanjeevi, T., Deepakkrishna, S., Harrine, M.P., Sharan, S., Prabhu, E.: IoT based traffic sign detection and violation control. In: Proceedings of the 2nd International Conference on Inventive Research in Computing Applications, ICIRCA 2020, art. no. 9183081, 2020, pp. 333–339
18. Sanjay Kumar, V., Ashish, S.N., Gowtham, I.V., Ashwin Balaji, S.P., Prabhu, E.: Smart driver assistance system using raspberry pi and sensor networks. In: Microprocessors and Microsystems, 79, art. no. 103275, 2020, pp. 1–11

19. Chawan, A., Kakade, V., Jadhav, J.: Automatic Detection of Flood Using Remote Sensing Images. J. Inf. Technol. Digit. World. **2**, 11–26 (2020). https://doi.org/10.36548/jitdw.2020.1.002
20. Chandy, A.: A review on iot based medical imaging technology for healthcare applications. J. Innov. Image Process. **1**, 51–60 (2019). https://doi.org/10.36548/jiip.2019.1.006

Heart Disease Prediction Using Deep Neural Networks: A Novel Approach

Kondeth Fathima and E. R. Vimina

Abstract Cardiovascular diseases are the prime cause of global deaths. The World Health Organization statistics point cardiovascular diseases as having nearly half of the non-communicable diseases. Many researchers have designed and proposed various computerized prediction models for detecting heart diseases early. A deep neural network model is proposed in this paper, using four hidden layers to detect coronary heart diseases. Three different combinations of input layer—hidden layer—output layer were evaluated and the best model is proposed. The proposed model focuses on avoiding overfitting. The datasets used are Statlog and Cleveland datasets available in the UCI Data Repository. Accuracy, sensitivity, specificity, F1 score, and misclassification are the different metrics used to evaluate the proposed model. Further, ROC is plotted with AUC. The model gave promising figures of 98.77% (accuracy), 97.22% (sensitivity), 100.00% (specificity), 98.59% (F1 score), and 1.23% (misclassification) for the Statlog dataset. For the Cleveland dataset, the corresponding values are 96.70%, 92.86%, 100.00%, 96.30%, and 3.30%, respectively.

Keywords Heart disease prediction · Deep neural network · Imputation · Dropouts · Accuracy

1 Introduction

The member countries of the World Health Organization (WHO) agreed on a global action plan in 2013 for preventing and controlling non-communicable diseases (NCDs) [20]. NCDs account for 63% of global deaths and include mainly cardiovascular diseases (CVDs), cancers, and chronic respiratory diseases. 86% of these premature deaths are from low- and middle-income countries. These premature

K. Fathima (✉) · E. R. Vimina
Amrita School of Arts and Science, Kochi, India
e-mail: KH.SC.R1CST20013@asas.kh.amrita.edu

E. R. Vimina
e-mail: viminaer@asas.kh.amrita.edu

J. S. Raj et al. (eds.), *Intelligent Sustainable Systems*, Lecture Notes in Networks and Systems 213, https://doi.org/10.1007/978-981-16-2422-3_56

deaths can be prevented to a large extent by having efficient and responsive health systems. The goal is to have a 25% reduction in NCD-related premature deaths with nine voluntary global targets, by 2025. Two targets are directly focused on preventing and controlling CVDs, which underlines their importance. The WHO statistics point CVDs as having a 48% share of all NCDs.

Disorders of heart and blood vessels constitute CVDs. Heart attacks and strokes are critical incidents that result from blockages of blood flow to the heart and brain. The main reason for these incidents is fatty deposits build-up on the inner blood vessel walls which supplies blood to the heart or brain. The causes for these critical incidents are usually a union of risk factors like use of tobacco, hypertension, diabetes and hyperlipidemia, detrimental diet and obesity, lack of physical activity, and injurious use of alcohol [19].

The WHO risk models, including the revised model of 2019, identify age, tobacco usage indication, blood pressure, cholesterol, and body mass index as key predictor variables for predicting heart diseases [17]. With the varied nature of these and other variables, computerized heart disease prediction has become a necessity. Different authors have proposed several methods.

Enormous advancements have been taking place in the field of artificial intelligence. Computers are becoming smarter daily, and most of the humane tasks are performed by computers. Computers can learn on their own with data and can generalize results efficiently. This ability is exploited in medicine, as with any other field of expertise. The recent advancements in machine learning, especially deep learning, help physicians to make better decisions. There arises the need for efficient machine learning models for heart disease prediction.

In recent years, neural network models have proved beneficial in solving many real-world problems. Various deep learning techniques have been utilized in many fields, particularly the medical field and have gained popularity for their excellent prediction abilities. Several studies have been conducted in health care using machine learning techniques; classifying heart disease is no different. Various deep learning techniques such as CNNs, LSTMs, and RCNN have been designed, and several studies have been conducted. A neural network model is proposed in this paper with four hidden layers (HL). Various neural network models with different HLs are designed. The model with more number of HLs will be beneficial and more accurate than employing a model with fewer HLs.

The paper is given in the following order: First, Sect. 2 gives a brief overview of the earlier works. Section 3 throws light on the datasets and various pre-processing methods used. Neural network architecture and its working mechanism are also given to illustrate the three neural network models. The proposed deep neural network (DNN) has also been discussed. The experimental results with different evaluation metrics like accuracy, sensitivity, specificity, misclassification, and F1 score are given in Sect. 4. The paper ends with Sect. 5 having directions of future work.

2 Related Works

Computerized heart disease prediction became a necessity due to the varied nature of key predictor variables. The authors have proposed several methods in the literature.

Wang et al. [18] designed a stacking-based model with two levels in which L1 was the base-level and L2 was the meta-level. Meta-level accepts inputs from the base-level classifier. The classifier with the lowest correlation was found using the coefficients such as Pearson's correlation and maximum information. The best combination of classifiers with the affirmative output is found out using the enumeration algorithm. Guo et al. [7] proposed an improved linear model using a recursion enhance random forest. The model merges the features of the linear model and random forest. Support vector machine (SVM), a supervised learning model, was used to enhance the performance of the algorithm. Ali et al. [4] introduced an improved deep belief network which was configured optimally. The features not contributing to improving performance were removed using the Ruzzo-Tompa approach. Stacking two genetic algorithms using a stacked algorithm was used to develop an optimal network configuration. Khan and Algarni [11] proposed an IoMT framework using modified salp swarm optimization (MSSO) and an adaptive neuro-fuzzy inference system (ANFIS) was proposed for prediction accuracy. Using the Levy flight algorithm MSSO-ANFIS improved the search capability. Using the MSSO, the learning parameters were optimized for providing better results for ANFIS.

Fitriyani et al. [6] proposed an effective heart disease prediction model for a clinical decision support system. It included a density-based spatial clustering of applications with Noise for detecting and eliminating the outliers and a hybrid synthetic minority over-sampling technique-edited nearest neighbor for training data distribution balance. It also included XGBoost for predicting heart disease. The results were compared with other models like logistic regression, naive Bayes, SVM, multilayer perceptron, random forest, and decision tree, and also with the previous study results.

Pasha and Mohamed [15] introduced a novel feature reduction model, in which more focus was on the area under the curve (AUC) where accuracy was used for evaluation. The AUC shows the classifier model's distinguishing capability. One approach was to reduce the features and obtain a subgroup for the most contributing features in prediction. In the second approach, all individual features' accuracy and AUC are evaluated, and a subgroup is created for mostly involved features. Various combinations of these were taken for performance improvement.

Li et al. [13] developed a system based on classification algorithms that include SVM, artificial neural network (ANN), logistic regression, naïve Bayes, K-nearest neighbor, and decision tree. Classification of algorithms categorizes the input data. Algorithms such as least absolute shrinkage selection operator, relief, minimal redundancy maximal relevance were used, and for removing irrelevant and redundant features local learning was used.

Javeed et al. [8] developed a system that uses a random search algorithm (RSA) and random forest model (RFM) for selecting features and prediction of heart failure, respectively. A grid search algorithm was used to optimize the proposed model.

Ali et al. [3] introduced the stacking of two SVM models. The first model was L1 regularized, and the elimination of unimportant features was done by making their coefficients zero. The other one, L2 regularized, was used for prediction. For optimizing the two models simultaneously, a hybrid grid search algorithm was used.

Mohan et al. [14] proposed a hybrid random forest with the linear model (HRFLM) that used all features without feature selection restrictions. HRFLM made use of ANN with backpropagation. Junejo et al. [9] collected the necessary information from patients and used ANN to predict CVD. Khan [10] designed an IoT framework using a modified deep convolutional neural network (MDCNN) to predict heart disease more accurately. Blood pressure and ECG were monitored using the patient's smart watch and heart monitor. For classification, MDCNN is utilized.

Latha and Jeeva [12] concentrated on improving weak classification algorithms' weaknesses by joining them with other classification algorithms. Various ensemble techniques were used to improve accuracy.

Tama et al. [16] designed a two-tier sequential ensemble where few classifiers were used as base classifiers of the next. Gradient boosting machine, RF, and XGBoost algorithms were used. Ali et al. [1] introduced an ensemble deep learning model with feature fusion methods for better heart disease prediction. An FRF extraction module was developed for the detection and extraction of low-dimensional heart disease risk factors from unstructured EMRs. Ali et al. [2] introduced a hybrid Chi-square DNN model to eliminate irrelevant features, and the model performance is compared with conventional ANN and DNN models.

Motivated from the literature of those who used DNN, a model is proposed in this paper for improving heart disease prediction accuracy. DNN allows for robustness to data variations, generalizability for multiple applications, and scalable for more data. The optimum number of hidden layers with the best accuracy has been the focus of the proposed research.

3 Proposed Methodology

The methodology used is based on different DNN classifiers. Two datasets, Statlog and Cleveland, are used in this method. Exploratory data analysis of these datasets is done. The proposed model is chosen based on the accuracy performance of these two datasets.

3.1 Proposed Model

Three neural network models have been utilized, each with a different number of layers and neurons. The proposed model given in Fig. 1 is a DNN with one input layer (IL), four HLs, and one output layer (OL). The model was selected after experimenting with various neural network models with different HLs.

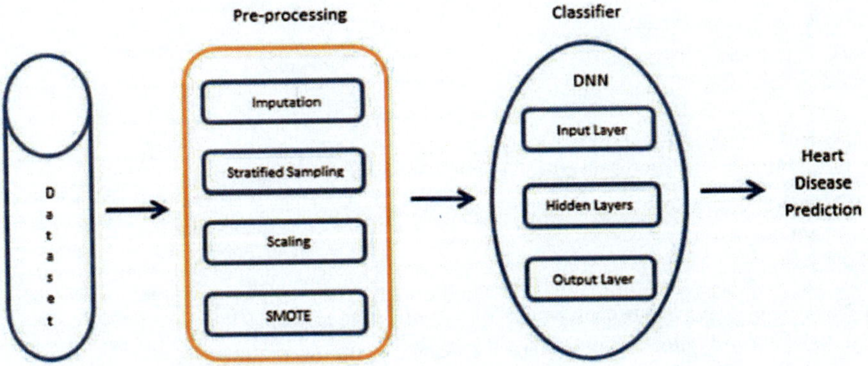

Fig. 1 Proposed model

3.2 Dataset

In this study, we employed two datasets of heart failure, Statlog and Cleveland, which are available on the machine learning repository of the University of California Irvine (UCI). These datasets (https://archive.ics.uci.edu/ml/datasets.php) have been utilized by a number of researchers due to data source reliability. The dataset consists of 303 instances, with 297 instances having complete details and six having missing details. Though it has 76 basic features, only 13 important features were used in most of the previous studies. Of these 13, 8 are categorical, and 5 are numeric, as given in Table 1. Statlog dataset also has 75 basic features, but only 13 are used for heart disease detection. 270 instances are successfully collected with no missing values.

3.3 Pre-processing

Each of the two datasets is divided in the 70:30 ratio, with 70% of the dataset is to train and 30% for test. Totally there are 14 features, including the output.

Data Imbalance. The unequal distribution of healthy and unhealthy cases in the datasets shows that the data is imbalanced. The low proportion makes it difficult for the model to learn the features of that particular class. The main causes of imbalance are biased sampling and measurement errors. The imbalances in both datasets are less, and sampling methods can correct these. Synthetic minority oversampling technique (SMOTE), a data augmentation technique, was used to address this problem [5]. In this method, a random instance from the minority class is chosen. A "k" neighbor for this instance is also chosen, and a new instance is created using a randomly selected neighbor at a randomly selected point between two instances in the features space. SMOTE is a practical approach, as the new instances created are close to the existing ones.

Table 1 Dataset attributes

No.	Symbol	Description	Data range
1	age	Subject age	29–77 years
2	sex	Subject gender	0 = female; 1 = male
3	cp	Chest pain type	1 = typical angina 2 = atypical angina 3 = non-angina pain 4 = asymptomatic
4	trestbps	Resting blood pressure	94–200 mmHg
5	chol	Serum cholesterol	126–564 mg/dl
6	fbs	Fasting blood sugar with value > 120 mg/dl	0 = false; 1 = true
7	restecg	Resting ECG	0 = normal; 1 = having ST-T wave abnormality; 2 = showing probable or definite left ventricular hypertrophy
8	thalach	Maximum heart rate	71–202
9	exang	Angina induced by exercise	0 = No; 1 = Yes
10	oldpeak	ST depression by exercise relative to rest	0–6.2
11	slope	Slope of peak exercise ST segment	1 = up sloping; 2 = flat; 3 = down sloping
12	ca	No. of major vessels colored by fluoroscopy	0–3
13	thal	Defect type	3 = normal 6 = fixed defect 7 = reversible defect

Imputation. The missing data in the Cleveland dataset are encoded as NaNs. Since the total number of instances in the dataset is limited, we cannot leave the instances with some features missing. Imputation is the replacement of the missing data with substituted values. It is usually performed using mean or median values. Mean imputation was used as its performance is better. The mean of the non-missing values is calculated, and then the missing values are replaced with it.

Stratified Splitting. The dataset is to be split for training and testing. If a random split is used, there are chances that the proportion of each class may be different and result in unsatisfactory performance at validation. In stratification, the train and test sets would be locked and the dataset is so split such that each split (train and test) has an equal proportion of healthy and unhealthy people.

Feature Scaling. The weights of the features are optimized using gradient descent in neural networks. The gradient descent is calculated as

$$\theta_j := \theta_j - \alpha\frac{1}{m}\sum_{i=1}^{m}\left(h_\theta\left(x^{(i)}\right) - y^{(i)}\right)x_j^{(i)} \tag{1}$$

To ensure a smooth transition to the minima, each gradient descent step should be updated at the same rate. The scaling of the data becomes necessary before using it further in the model. The data is scaled using standardization, a process by which the scaled values will center on the mean, and standard deviation will be unity. The formula for standardization is

$$X' = \frac{X - \mu}{\sigma} \tag{2}$$

where μ is the mean of features and σ is the standard deviation of features. X^l is the standardized value of X. Standardization is beneficial when there are outliers in data, i.e., when some observations lie at an abnormal distance when compared to other values.

3.4 Deep Neural Networks

A neural network is defined as an interconnected assembly of processing elements that act upon a function. These elements are referred to as neurons, which function as per the inter-unit strength called weights. These weights are obtained after training from a set of training patterns. Neurons in a layer get their values from activation functions. The data is fed in the forward direction, so the name forward propagation. The weights are updated during backpropagation by an optimization technique called gradient descent.

Neural networks can be classified into two. The single-layer neural network or the perceptron is used when the inputs are mapped directly to the output using activation functions. It consists of an IL and an output node. A multilayer neural network has more than one computational layer. The additional computational layers are referred to as hidden layers. Let the training instance be of the form (*X*, *y*), where $X = [x_1\ x_2 \ldots x_d]$ contains *n* feature variables and $y \in [0,1]$, where *y* is the observed value. The IL contains *d* neurons with weights $W = [w_1\ w_2 \ldots w_d]$. The output neuron's value is computed using the linear function

$$y = h(z) \tag{3}$$

where *h* is the activation function and

$$z = \sum_{j=1}^{d} w_j x_j + b \tag{4}$$

based on a threshold, depending on the activation function used. Each instance of the dataset is fed into the neural network, and the weights *W* are updated based on the error $(y - \hat{y})$

$$W = W + \alpha\left(y - \hat{y}\right)X \tag{5}$$

where y is the prediction and α is the learning rate of the neural network, to minimize the error. The above process is repeated through many cycles.

The pre-processed data is classified using three different DNN models. The first model consists of one IL, three HLs, and one OL, with the HLs having 26, 52, and 13 units, respectively. The second model has one IL, four HLs, and one OL, with the HLs having 26, 78, 52, and 13 units, respectively. The third model has one IL, two HLs, and one OL, with HLs having 52 and 13 units, respectively. The ReLU activation function is used for the layers except for the OL, and the sigmoid activation function is used for it.

Hyperparameters Setting. The proposed model uses five hyperparameters: dropout rate, number of epochs, learning rate, optimizer, and scheduler. The model was implemented in Python using the open-source machine learning library PyTorch.

Dropout. Dropout, a regularization technique is used, wherein the neural network is trained iteratively, and neuron connections that have more prediction capability are learned, and others are ignored. After experimenting with different values, a dropout rate of 0.2 was selected.

Epochs. An epoch refers to one cycle through the full dataset. As epochs increase, the model generalizes more. The number of epochs selected was 500.

Optimizer. It is used to minimize the loss function to provide the most accurate result possible. The Adam optimizer is used due to its performance on average.

Learning Rate. A small learning rate causes the weights to update slower or slower convergence. A larger learning rate causes quicker convergence, and the minima may be exceeded. An optimal learning rate of 0.001 is used.

Cosine Annealing Scheduler. In learning rate annealing, the approach starts with an initial learning rate and then gradually decreases during the training. This method is used to update the learning rate so that the best weights are obtained.

3.5 Evaluation Metrics

The proposed classification model is evaluated using different metrics, like accuracy, sensitivity, specificity, misclassification, and F1 score.

Accuracy. The ability of a model to differentiate healthy and unhealthy cases correctly is its accuracy. In the following formulae, TP, TN, FP, and FN refer to true positives, true negatives, false positives, and false negatives, respectively.

$$Accuracy = \frac{TP + TN}{TP + TN + FP + FN} \tag{6}$$

Sensitivity. The ability of a model to identify the unhealthy instances correctly is its sensitivity.

$$Sensitivity = \frac{TP}{TP + FN} \tag{7}$$

Specificity. The ability of a model to identify healthy instances correctly is its specificity.

$$Specificity = \frac{TN}{FP + FN} \tag{8}$$

Misclassification. The rate at which the data instances are misclassified.

$$Misclassification = \frac{FP + FN}{TP + TN + FP + FN} \tag{9}$$

F1 Score. Accuracy may not always be the best metric, as false negatives and false positives are also to be considered. F1 score is useful when the distribution is uneven.

$$F1Score = \frac{TP}{TP + 0.5(FP + FN)} \tag{10}$$

4 Results and Discussion

The results obtained for Statlog and Cleveland datasets have been studied. In the case of the first model for Cleveland dataset, 93.41%, 85.71%, 100.00%, 6.59%, and 92.31% are achieved for accuracy, sensitivity, specificity, misclassification, and F1 score, respectively. For the second model, the corresponding values are 96.70%, 92.86%, 100.00%, 3.30%, and 96.30%. For the third model, the values are 95.60%, 92.86%, 97.96%, 4.40%, and 95.12%, respectively. In the Statlog dataset, for the first model, 97.53%, 97.22%, 97.78%, 2.47%, and 97.22% are the corresponding values. For the second model, the values are 98.77%, 97.22%, 100.00%, 1.23%, and 98.59%. For the third model, 96.30%, 94.44%, 97.78%, 3.70%, and 95.77% are the values, respectively.

It can be inferred from the above that the second model achieved better accuracy when compared to the other two models. It is because, as the number of layers increases, the model can learn more complex features, and hence can predict more accurately.

The second model with four HLs is the proposed one (Fig. 2). The evaluation metrics for the proposed model are given in Table 2. The four HLs contain 26, 78, 52, and 13 units, respectively. Table 3 compares the performances of the proposed model with earlier cited works on Statlog and Cleveland datasets.

The evaluation was also done using the receiver operating characteristic (ROC) curve. The area under the curve (AUC) shows the classifier model's distinguishing

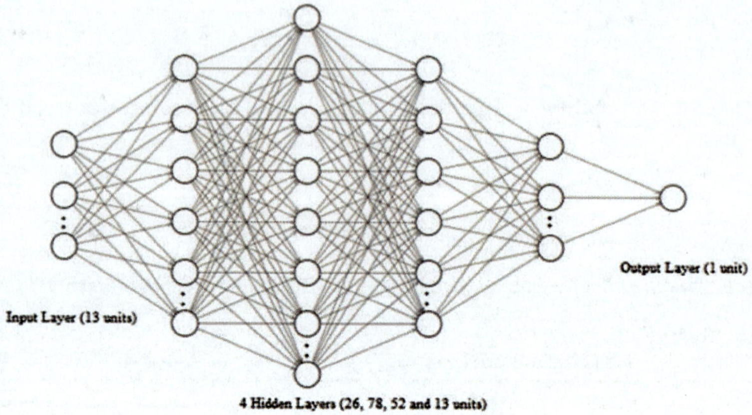

Fig. 2 Proposed DNN model

Table 2 Performace evaluation metrics

Dataset	Accuracy	Sensitivity	Specificity	F1 Score	Misclassification
Statlog	98.77	97.22	100.00	98.59	1.23
Cleveland	96.70	92.86	100.00	96.30	3.30

Table 3 Performance comparison with other approaches

Study (Year)	Approach	Accuracy
Statlog		
Pasha and Mohamed (2020) [15]	NFR Model	87.65
Tama, B.A. et al. (2020) [16]	2-Tier Ensemble	93.55
Proposed Model	***DNN***	***98.77***
Cleveland		
L. Ali et al. (2019) [2]	Chi-square DNN	93.33
A. Javeed et al. (2019) [8]	RSA-based RF	93.33
S. A. Ali et al. (2020) [4]	OCI-DBN	94.61
Latha and Jeeva (2019) [12]	Ensemble classifiers	85.40
S. Mohan et al. (2019) [14]	HRFLM	88.70
Pasha and Mohamed (2020) [15]	NFR Model	92.53
Tama, B.A. et al. (2020) [16]	2-Tier Ensemble	85.71
Proposed Model	***DNN***	***96.70***

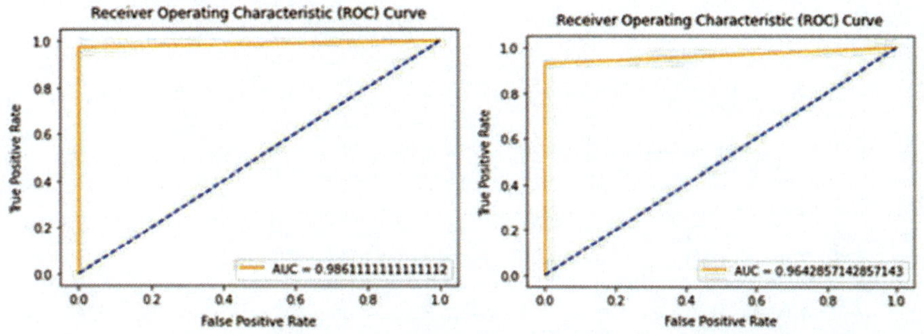

Fig. 3 ROC curves for Statlog and Cleveland datasets

capability. The false-positive rate (FPR) is on the x-axis and total positive rate (TPR) is on the y-axis. TPR is the same as sensitivity and FPR is 1—Specificity.

$$TPR = \frac{TP}{TP + FN} \tag{11}$$

$$FPR = \frac{FP}{TN + FP} \tag{12}$$

An AUC of 0.9861 is achieved for the Statlog dataset and 0.9643 for the Cleveland dataset. The higher the AUC, the better the model (Fig. 3).

5 Conclusion and Future Work

Computerized heart disease prediction mechanisms will help in the early detection of abnormalities as well as related reduction of global deaths due to cardiovascular diseases. Three neural network models are designed and tested, each with a different number of layers and neurons. A DNN is proposed in this paper, with four HLs to detect coronary heart diseases. DNN allows for robustness to data variations, generalizability for multiple applications, and scalable for more data. The datasets used are Statlog and Cleveland datasets. Different metrics used for evaluating the proposed model, such as accuracy, sensitivity, specificity, F1 score, misclassification, ROC, and AUC, gave promising results. The model gave accuracies of 98.77 for Statlog and 96.70 for Cleveland datasets. The smaller size of these datasets is a limitation of this work. Usage of datasets with more instances would help in more training and the performance improvements can be explored. As another future work, this model can be connected with IoT devices for real-time data.

References

1. Ali, F., et al.: A smart healthcare monitoring system for heart disease prediction based on ensemble deep learning and feature fusion. Inf. Fusion **63**, 208–222 (2020)
2. Ali, L., et al.: An automated diagnostic system for heart disease prediction based on Chi^2 statistical model and optimally configured deep neural network. IEEE Access. **7**, 34938–34945 (2019)
3. Ali, L., et al.: An optimized stacked support vector machines based expert system for the effective prediction of heart failure. IEEE Access. **7**, 54007–54014 (2019)
4. Ali, S.A., et al.: An optimally configured and improved deep belief network (OCI-DBN) approach for heart disease prediction based on Ruzzo-Tompa and stacked genetic algorithm. IEEE Access. **8**, 65947–65958 (2020)
5. Chawla, N.V., et al.: SMOTE: Synthetic minority over-sampling technique. jair. **16**, 321–357 (2002)
6. Fitriyani, N.L., et al.: HDPM: an effective heart disease prediction model for a clinical decision support system. IEEE Access. **8**, 133034–133050 (2020)
7. Guo, C., et al.: Recursion enhanced random forest with an improved linear model (RERF-ILM) for heart disease detection on the Internet of Medical Things platform. IEEE Access. **8**, 59247–59256 (2020)
8. Javeed, A., et al.: An intelligent learning system based on random search algorithm and optimized random forest model for improved heart disease detection. IEEE Access. **7**, 180235–180243 (2019)
9. Junejo, A., et al.: Molecular diagnostic and using deep learning techniques for predict functional recovery of patients treated of cardiovascular disease. IEEE Access. **7**, 120315–120325 (2019)
10. Khan, M.A.: An IoT framework for heart disease prediction based on MDCNN classifier. IEEE Access. **8**, 34717–34727 (2020)
11. Khan, M.A., Algarni, F.: A healthcare monitoring system for the diagnosis of heart disease in the IoMT cloud environment using MSSO-ANFIS. IEEE Access. **8**, 122259–122269 (2020)
12. Latha, C.B.C., Jeeva, S.C.: Improving the accuracy of prediction of heart disease risk based on ensemble classification techniques. Inf. Med. Unlocked. **16**, (2019)
13. Li, J.P., et al.: Heart disease identification method using machine learning classification in E-healthcare. IEEE Access. **8**, 107562–107582 (2020)
14. Mohan, S., et al.: Effective heart disease prediction using hybrid machine learning techniques. IEEE Access. **7**, 81542–81554 (2019)
15. Pasha, S.J., Mohamed, E.S.: Novel feature reduction (NFR) model with machine learning and data mining algorithms for effective disease risk prediction. IEEE Access. **8**, 184087–184108 (2020)
16. Tama, B.A., et al.: Improving an intelligent detection system for coronary heart disease using a two-tier classifier ensemble. Biomed. Res. Int. **2020**, 1–10 (2020)
17. The WHO CVD Risk Chart Working Group: World Health Organization cardiovascular disease risk charts: revised models to estimate risk in 21 global regions. Lancet Glob Health (2019)
18. Wang, J., et al.: A stacking-based model for non-invasive detection of coronary heart disease. IEEE Access. **8**, 37124–37133 (2020)
19. World Health Organization, https://www.who.int/news-room/fact-sheets/detail/cardiovascular-diseases-(cvds)
20. World Health Organization: Global action plan for the prevention and control of noncommunicable diseases: 2013-2020 (2013)

Intelligently Controlled Scheme for Integration of SMES in Wind Penetrated Power System for Load Frequency Control

S. Zahid Nabi Dar

Abstract This work envisages the integration of an energy storage device called superconducting magnetic energy storage (SMES) with a wind penetrated two-area power system. A small-sized SMES along with voltage source converter (VSC) configured on IGBTs is incorporated in the power conditioning system (PSC) with a two-quadrant chopper for bidirectional power interface. This small-sized SMES is tuned with the genetic algorithm and coupled with a doubly-fed induction generator (DFIG) for inertial support. A step disturbance of 0.01 p.u. reflecting a load outage is introduced back to back. The principle of making subsystems from the block diagram in Matlab/Simulink environment is utilized. A reduction of 66.66% in frequency and 50% in tie power deviation is observed with this arrangement.

Keywords Doubly-fed induction generator (DFIG) · Genetic algorithm (GA) · Superconducting magnetic energy storage system (SMES) · New area control error (NACE)

1 Introduction

Presently, the world is moving toward the incorporation of renewable energy sources for power production in a big way. The inertial response has a tendency to change the transient frequency reaction encountered in the power system following a disturbance [1–4]. This characteristic has restrained the renewable energy penetration to a maximum limit of 30% as stated by the French Government Ministerial Order [5].

S. Zahid Nabi Dar (✉)
Department of Electrical & Electronics Engineering, CMR Institute of Technology, Bengaluru, India

J. S. Raj et al. (eds.), *Intelligent Sustainable Systems*, Lecture Notes in Networks and Systems 213, https://doi.org/10.1007/978-981-16-2422-3_57

The main aim of this paper is to investigate the role of an energy storage device coupled with DFIG in a two-area wind penetrated power system. Research work concentrating on this arrangement has limitations and it mainly looks into maintaining electrical torque for controlling the inertial response [6–11]. In this work, the focus is on compensating rotational inertia present in the DFIG system. The majority of the research articles are dealing with stored kinetic energy in the rotor blades of the DFIG for supporting load frequency control (LFC) [12–15]. Further, in addition to virtual synthetic inertia present in the DFIG rotor and blades, a small-sized SMES of 1% rating of the total plant capacity is incorporated.

There are a number of energy storage devices available in the market such as supercapacitors, Redox flow batteries, fuel cells, flywheel, etc. SMES, in comparison, is comparatively faster in response, has high power density, and is environment-friendly as it does not release any chemical gases.

2 Designing of DFIG Model for Utilization in Simulation Studies

Generally, DFIG-configured wind turbines are modeled in order to set the turbine operational for maximum speed which leads to maximum power extraction. A reference power point is attained by the controller governed by the measured angular speed. This load power reference is given as an input to the converter that produces the power and torque by monitoring the rotor current of the generator. Thereafter, an additional control signal ΔP_f^* is added that manages the frequency deviations. The emulated inertia of installed additional control is monitored by the integral (k_{wi}) and proportional (k_{wp}) gains, respectively. The role of primary frequency control comes into play when the grid frequency violates the desired range. Under such conditions, the primary frequency control loop takes charge, by integrating the additional signal with torque equation to maintain the torque demand. As there is a decrement in the system frequency, the reference torque increments, and the set point torque is increased as kinetic energy is discharged from the rotor blades leading to a slower rotor speed as shown in Fig. 1 [12].

In Fig. 2 the integration of SMES and DFIG in a wind penetrated power system is delineated.

3 Modeling of SMES Device and Its PCS

SMES is essentially a storage device; wherein a superconducting coil behaving as a pure inductor is utilized to store energy in the form of dc circulating current. The releasing/charging of the SMES is enabled by varying the polarity of voltage (Vsm) across the SMES coil by automatically changing the duty ratio, and impressing this

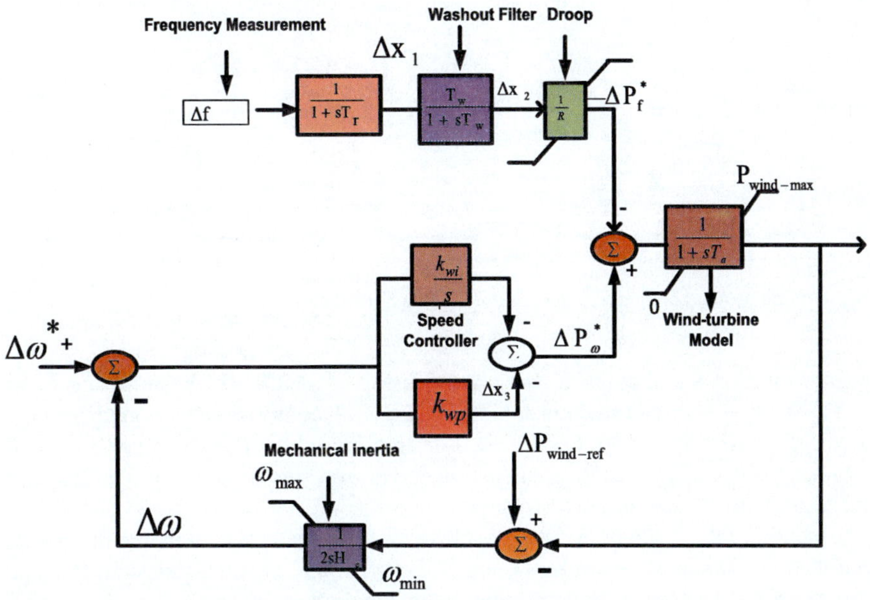

Fig. 1 Block diagram of wind turbine frequency response

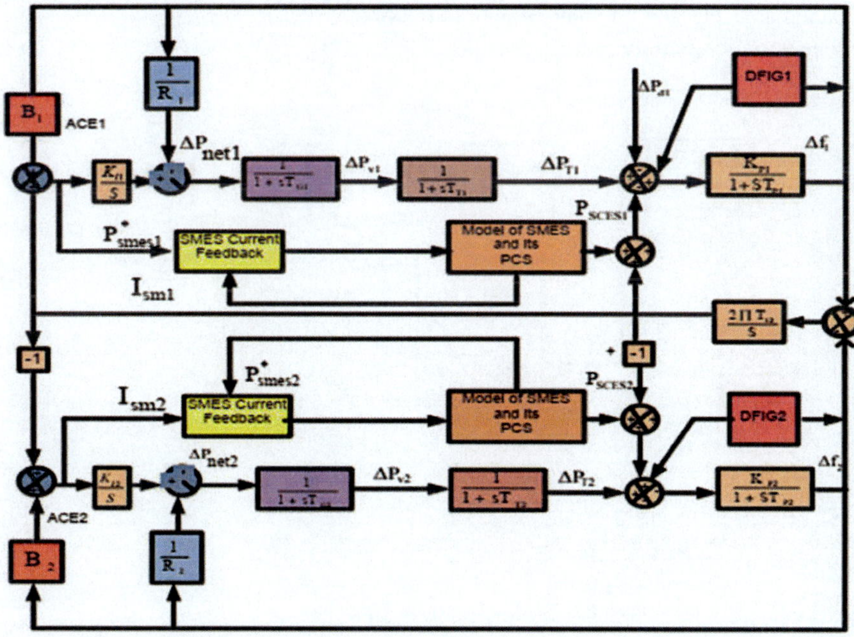

Fig. 2 Model of two-area power system with SMES and DFIG

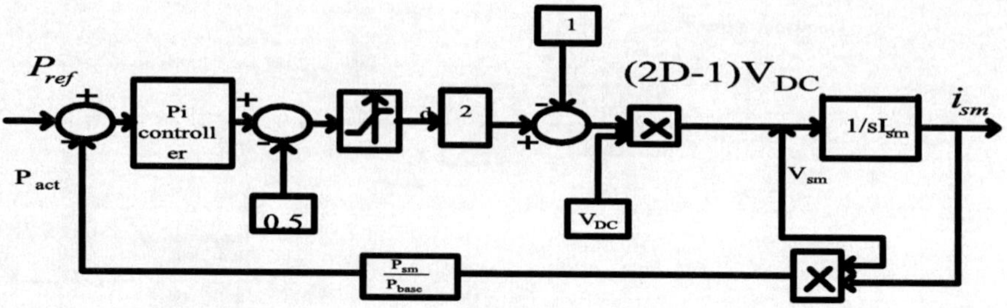

Fig. 3 PI controller configured inner control module of SMES

voltage on SMES coil helps in charging/releasing of SMES. The basic arrangement of SMES utilized in this study is depicted in Fig. 3. It comprises a two-quadrant type B chopper interlinked with a VSC via a dc-link capacitor. There is a possibility of achieving independent reactive and active power by varying voltage magnitude and phase angle in VSC-configured scheme.

Essentially three modes for duty cycle operation exists, viz., standby, discharge, and charge, which are given by $D = 0.5$, $D < 0.5$, and $D > 0.5$, respectively. The chopper for SMES coil initially presents a uniform voltage of a set value for charging the coil. Figure 4 considers the control methodology that is based on SMES genetic algorithm (ga) tuning. After GA tuning the corresponding proportional gain was obtained as 0.243 and the integral gain as 0.345.

The variation in duty ratio is given as

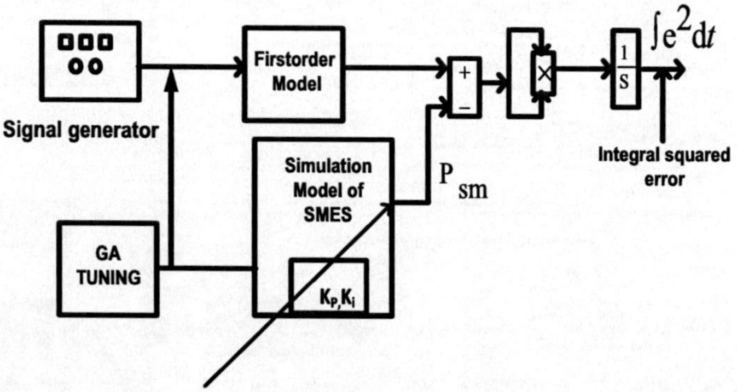

Fig. 4 GA tuning of SMES

$$m = (2D - 1) \quad (1)$$

$$V_{sm} = mV_{DC} \quad (2)$$

$$I_{cs} = V_{sm}/L_{sm}m \quad (3)$$

With the inclusion of step load disturbance of 0.01 p.u. applied back to back at time intervals 20, 100, 200, and 300 s as detailed in Fig. 5 and the rating of power converter for SMES considered 1% of the total plant capacity marks for very economical utilization of energy storage device, thus making the system reliable and less bulky.

In Fig. 6 the algorithm representing the switching states for standby, charge, and discharge mode is detailed, where $m = \{0/\pm 1\}$ indicates the case of standby, charge, or discharge mode. For standby mode, it is zero; in the case of charge $m > 0.5$ and for discharge $m < 0$. The chopper current can be established by the following equations at a certain instant:

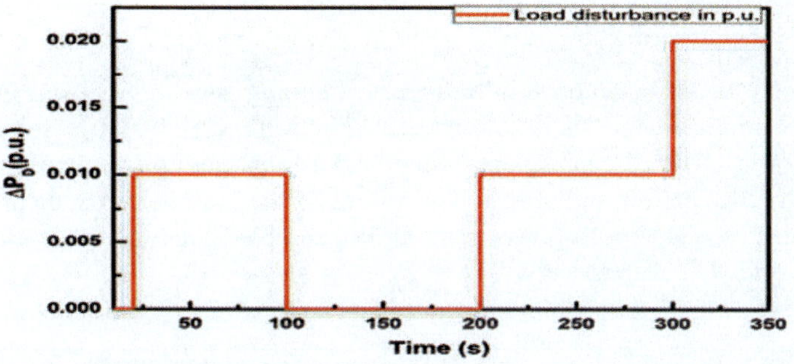

Fig. 5 Load disturbance profile

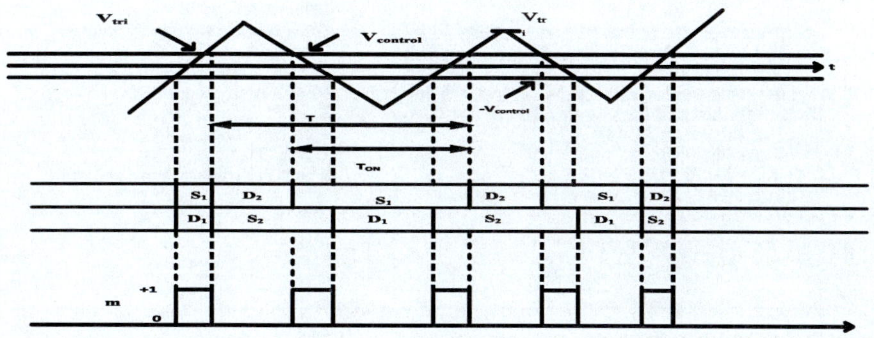

Fig. 6 Chopper operation for different switching states

$$
\begin{aligned}
I_{cs} = i_{sm} \quad & 0 < t \le dT - 0.5T \\
0 \quad & dT - 0.5T < t \le 0.5T \\
i_{sm} \quad & 0.5T < t \le dT \\
0 \quad & dT < t \le T
\end{aligned}
\tag{4}
$$

From Fig. 3 we arrive at the inner control loop block diagram of SMES representing the value m as given below:

$$
L_{sm}\frac{dI_{sm}}{dt} + R_{sm}I_{sm} = mV_{DC} \tag{5}
$$

The setting of gains by adjustment of duty cycles for bidirectional power interface is seen in this work. Moreover, the tracking of the energy storage device with a first-order reference system by employing the genetic algorithm is seen for the system to adjust automatically for different values of gain setting.

4 Simulation Results and Discussions

Improvement of about 66.66% in frequency damping is observed from Fig. 7 and 50% in tie power deviations, as observed in Fig. 8. Figures 9 and 10 represent the SMES current variation consequent to step load disturbance as given in Fig. 5. These findings are also set out in Table 1. The values of the load perturbance profile, the device data associated with power grid, DFIG, and SMES are given in Appendices "A", "B", and "C", respectively.

Appendix 'A':

Load disturbance profile for two areas

$\Delta PD1 =$ 0 .01 $0 < t \le 20$s
- 0.01 p.u. $0 < t \le 100$s
0.01p.u. $t \ge 200$s
0.01p.u. $t \ge 300$s

A two-zone power system with and without DFIG and SMES was simulated in Matlab Simulink Environment. Significant improvement was observed in the damping response of the power system with and without DFIG and SMES Appendix 'B': System Data

$H_{1m} = H_{2n} = 3.5$ p.u..MW.sec
$K_{pm} = 62$ Hz/(PU)
$K_{pn} = 62$ Hz/(PU)
$T_{pm} = 10$S
$T_{Pn} = 15$S
$T_{tm} = T_{Tn} = 1$
$T_{hm} = T_{hn} = 0.1$
$Reglation_1 = Regulation_2$ 3Hz per (p.u..MW)
$Bias_1 = Bias_2 = 0.05$(p.u.MW per radian)
$Tie_{mn} = 0.07$(p.u. .MW per Hz)

Appendix 'C': SMES and Converter

$I_{smes} = 400$A $L_{Smes} = 2.64$h,
Power conditioning system rating=0.01pu
Converter:
$V_{dc} = 800$ $P_{sm} = 100$kW(0.01 p.u.)
DFIG:$K_{wm} = 0.1$,$K_{wn} = 0.1$ $K_{wm} = 1.58$
$K_{pn} = 1.61$, $T_{rm} = T_{rn} = 0.1$ s
$T_{hm} = T_{hn} = 0.1$ Sec,$T_{tm} = T_{tn} = 1$ Sec,$T_{wm} =$
$T_{wn} = 6$s , $T_{am} = T_{an} = 0.2$ Sec
Whrein m,n represent area 1 and area 2

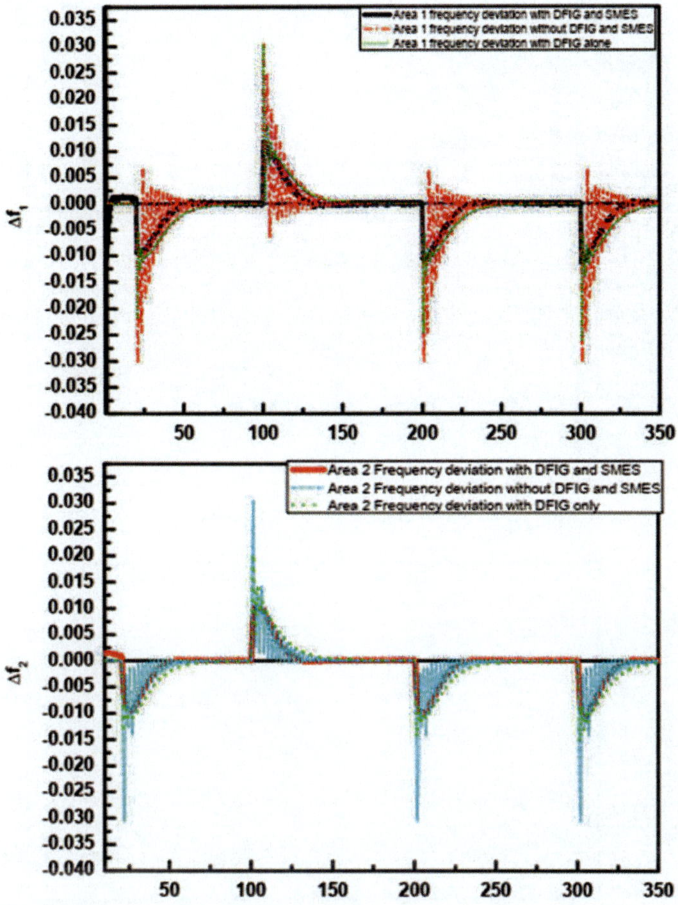

Fig. 7 Plots of frequency deviations for step load disturbance of Appendix “A”

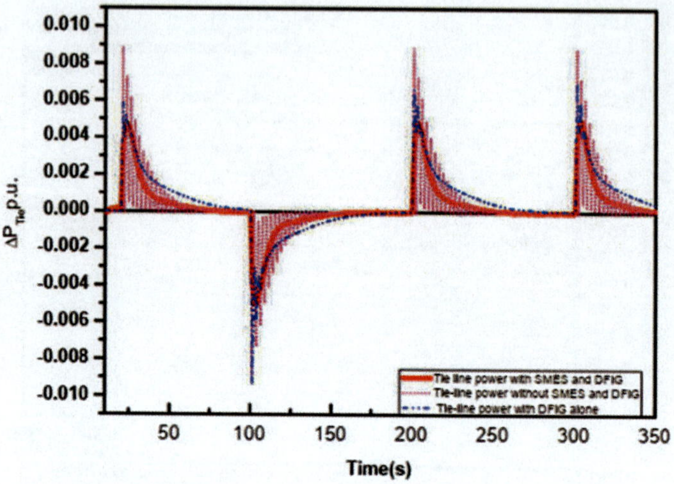

Fig. 8 Tie power deviation for step load disturbance of appendix 'A'

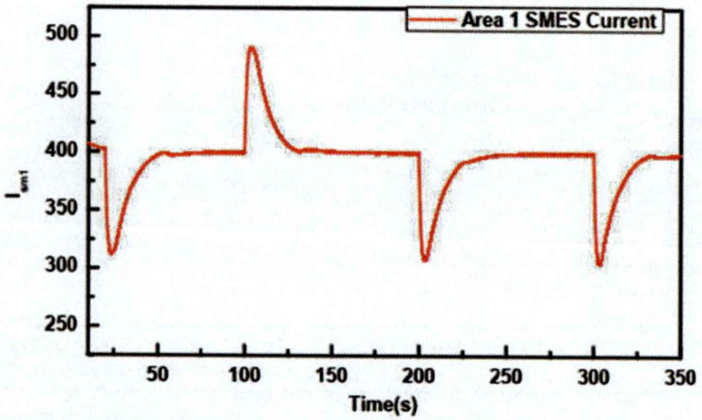

Fig. 9 Area 1 SMES current

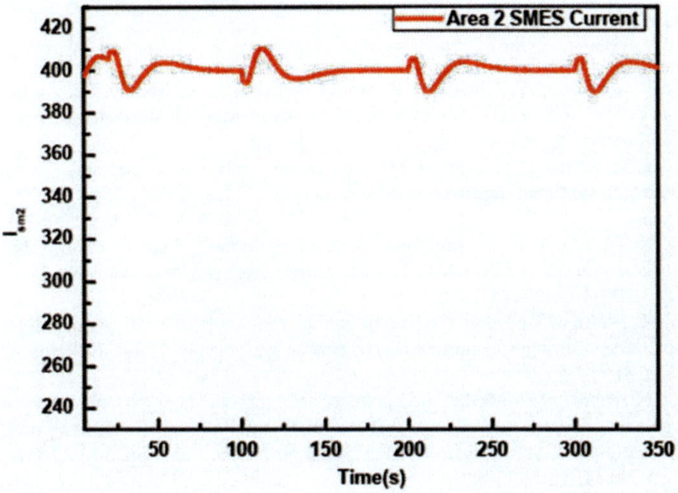

Fig. 10 Area 2 SMES current

Table 1 Comparative analysis with SMES and DFIG

Mode	ΔP_D	Deviation in peak frequency (Hz) at 20 s		Deviation (p.u.) in Tie Power
	–	Δf_1	Δf_2	ΔP_{tie1}
Without SMES and DFIG	0.01	0.03	0.03	0.009
With SMES and DFIG	0.01	0.10	0.01	0.0045
% Reduction	–	66.66	66.66	50

5 Conclusion

A two-zone power system with and without DFIG and SMES was proposed in this study. Considerable enhancement was achieved in damping power system response with the incorporation of DFIG and SMES.

It may be necessary to state herein that this work does not include generation rate constraints and governor dead band mechanism. Hence future investigations are advised with the inclusion of these parameters, and also modern methods of adaptive predictive control and data science need to be incorporated so that a very pragmatic system can be realized.

References

1. Doherty, R., Mullane, A., Nolan, G., Burke, D., Bryson, A., O'Malley, M.: An assessment of the impact of wind generation on system frequency control. IEEE Trans. Power Syst. **25**(1), 452–460 (2010)
2. Van de Jan, V., Jeroen, D.M.D., Bart, M. et al.: Droop control as an alternative inertial response strategy for the synthetic inertia on wind turbines. IEEE Trans. Power Syst. **31**(2), 1129–1138 (2016)
3. Dar, S.Z.N., Mufti, M.-U.-D.: Integrating a superconducting magnetic energy storage system for intelligent control of LFC characteristic in multi-area power system. Int. J. Ind. Electron. Drives **4**(2), 96–105 (2018)
4. Dar, S.Z.N., Mufti, M.D.: Improved load frequency characteristics with genetically tuned supercapacitor energy storage in interconnected power system. Int. J. Ind. Electron. Drives **4**(1), 16–24 (2018)
5. Dar, S.Z.N., Mufti, M.: Model predictive control of two area power system with superconducting magnetic energy storage system," 2017 International Conference on Computing, Communication and Automation (ICCCA), Greater Noida, 2017, pp. 1406–1411, https://doi.org/10.1109/ccaa.2017.8230061
6. Gevorgian, V., Zhang, Y., Ela, E.: Investigating the impacts of windgeneration participation in interconnection frequency response. IEEE Trans. Sustain. Energy **6**(3), 1004–1012 (2015)
7. Teng, F., Strbac, G.: Assessment of the role and value of frequencyresponse support from wind plants. IEEE Trans. Sustain. Energy **7**(2), 586–595 (2016)
8. Zhang, Y., Liu, X., Bin, Q.: Distributed model predictive load frequency control of multi-area power system with DFIGs. IEEE/CAA J. Automatica Sinica **4**(1), 125–135 (2017)
9. Dar, S.Z.N., Mufti, M.: Enhanced load frequency control response with integration of supervisory controlled superconducting magnetic energy storage system in wind-penetrated two-area power system. Wind Eng. **41**(5), 330–342 (2017)
10. Morren, J., Haan, W.H. de, Kling, W.L., et al.: Wind turbines Emulating Inertia and supporting primary frequency control. IEEE Trans. Power Syst. **21**(1), 433–434 (2006)
11. Heier: Grid Integration of Wind Energy: Onshore and Offshore Conversion Systems, 3rd edn. Wiley, Geremany (2014)
12. Mohamed, T.H., Morel, J., Bevrani, H., et al.: Model predictive based load frequency control-design (2012)
13. Ramtharan, G., Ekanayake, J.B., Jenkins, N.: Frequency support from doubly fed induction generator wind turbines. Renew. Power Gen., IET **1**, 3–9 (2007)
14. Ekanayake, J.B., Jenkins, N., Strbac, G.: Frequency response from wind turbines. Wind Eng. **32**(6), 573–586 (2008)
15. Jalali, M., Bhattacharya, K.: Frequency regulation and AGC in isolated systems with DFIG-based wind turbines. IEEE Power and Energy Society General Meeting (PES) (2013)

Effect of Surface Imperfections in Channel Region of an n-MOSFET on Its Vital Characteristics: A Simulation Study

Shailendra Baraniya and S. Zahid Nabi Dar

Abstract The fabrication process of silicon devices consists of a multitude of steps, like diffusion, ion implantation, epitaxial growth, etc. These steps lead to mechanical damage, fractures, surface undulations, etc. on the wafer. Some additional steps are incorporated to undo the damage done. This work attempts to study the effect of fabrication process steps on V/I characteristics of a fabricated n-MOS device. The presented work aims to study the effect of loss of vertical alignment of drain and source relative to each other. The study is conducted on a TCAD simulation platform—GTS's TCAD. The study observed a drop in I_{DS} for smaller inclinations and a rise in I_{DS} for larger inclinations. The study is conducted for 45 nm, 65 nm, and 90 nm devices.

Keywords MOSFET process · Vertically inclined channel · TCAD simulation · MOSFET characteristics · Device modeling

1 Introduction

Different steps in the semiconductor manufacturing process impact substrate mechanically and physically. The steps like ion implantation, epitaxial growth, diffusion, etc., subject the crystal to physical stresses. These steps tend to affect the variations in dimensions of features of the device being fabricated on the three axes. Considering n-MOSFET fabrication, deviations in dimensions of planner features lead to variation in channel electrical field and current density [1], influencing short channel effects [2]. Deviations in dimensions in the vertical direction impact the vertical alignment of features of devices. A vertical misalignment of drain and source manifests itself as the vertical inclination of the channel. The vertical inclination of the channel tends to alter the direction of electric field with respect to the crystal alignment, which can influence the V/I characteristics of the device.

S. Baraniya (✉) · S. Zahid Nabi Dar
Department of Electrical and Electronics Engineering, CMR Institute of Technology, Bengaluru, India
e-mail: shailendra.b@cmrit.ac.in

J. S. Raj et al. (eds.), *Intelligent Sustainable Systems*, Lecture Notes in Networks and Systems 213, https://doi.org/10.1007/978-981-16-2422-3_58

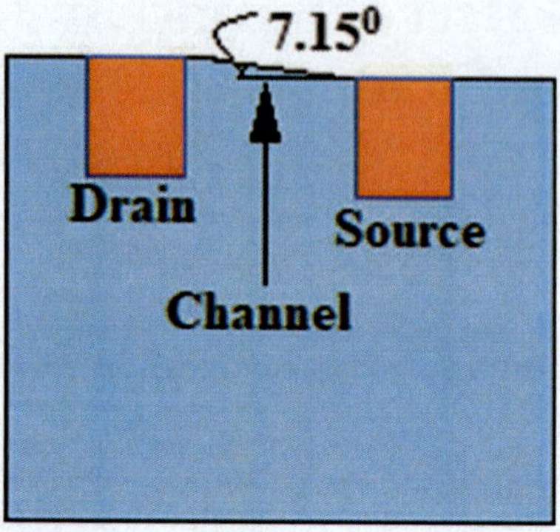

Fig. 1 A 45 nm device with D-S misalignment of 5 nm

This study focuses on characterizing n-MOSFET for various vertical slopes of the channel due to the difference in altitudes of drain and source.

2 The Problem

The undulations on crystal substrate which undergoes severe stress due to fabrication steps lead to the vertical misalignment of the drain with respect to source. The uneven surface under polysilicon at oxide–channel interface forms due to the mechanical/thermal stresses experienced due to different fabrication steps. The undulations on the interface difference in altitude of drain and source are insignificant in the case of long-channel devices. For a short-channel device, even a few nanometers difference in altitude of drain and source cannot be ignored. The model of the device (Fig. 1) in the z–y plane helps us understand the situation better. The channel inclination for a 40 nm device with a 5 nm vertical misalignment in altitudes of drain and source leads to a channel inclination of 7.15°.

3 Methodology

The device structure is modeled in silicon. The effect of surface imperfections in the channel region is modeled by displacing drain and source vertically. The length of the device (drain to source) is usually larger than the width of the channel; hence, a two-dimensional model suffices to study the effects. The device is modeled in a TCAD tool with different inclinations of the channel. The inclinations of the channel represent undulations in the channel region. The device so modeled is subjected to circuit extraction and the extracted circuit is simulated by an appropriate circuit simulator.

The modeling and simulations are performed on GTS TCAD tools. The GTS framework, from Global TCAD, is a physics-based device simulation platform. It offers three major modules: (a) device editor, (b) device viewer, and (c) simulator. The device editor, GTS Structure, is a GUI-based device structure generation and meshing tool. The GTS Structure facilitates modeling in 2-D and 3-D modes. The presented work is done in 2-D mode. The authors performed modeling and simulations on n-channel MOSFETs of channel lengths 45 nm, 65 nm, and 90 nm, on doping concentrations defined in reference design [3] (Fig. 2).

We created the device in standard planner configuration for the reference design, for the three-channel lengths. The slope of channels of these reference designs is then altered by 10°, 20°, 30°, 36.6°, 45°, and 50°. For these versions of devices, the drain and source are kept vertical and the channel under SiO_2 layer is made to incline. The device model for 65 nm design with 10° inclination is shown in Fig. 3.

Such structures are created and simulated for all inclinations from 10° to 50° for the three-channel lengths. Simulation of modeled devices in the GTS Structure requires the structure to be extracted as a circuit. This circuit is then modeled as a SPICE netlist [4]. GTS—Minimos, a module in GTS framework, performs circuit extraction from the device model and facilitates the mixed-signal simulation in the GTS framework seamlessly. The simulation settings are detailed in Table 1.

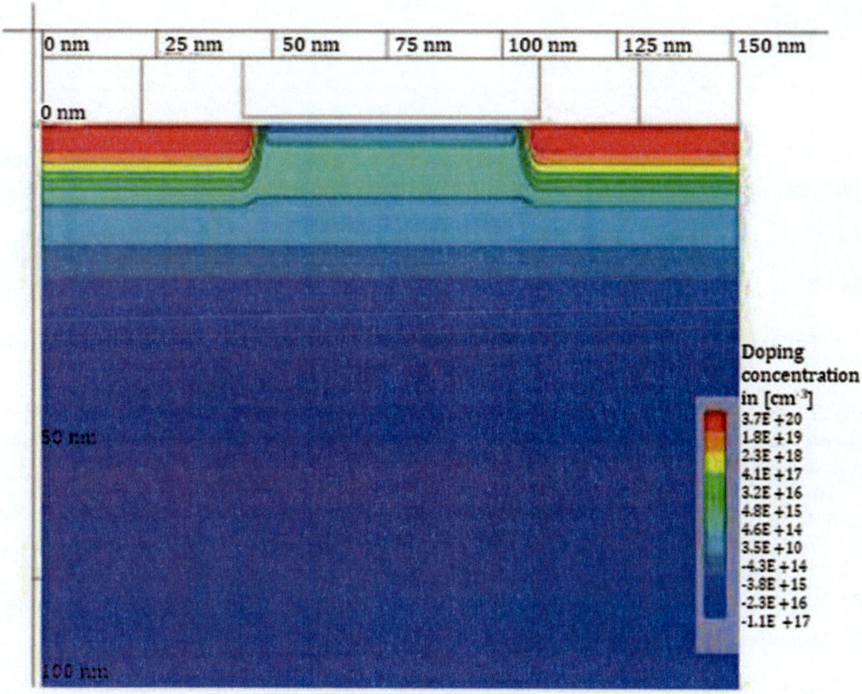

Fig. 2 An n-MOSFET reference design for 65 nm channel

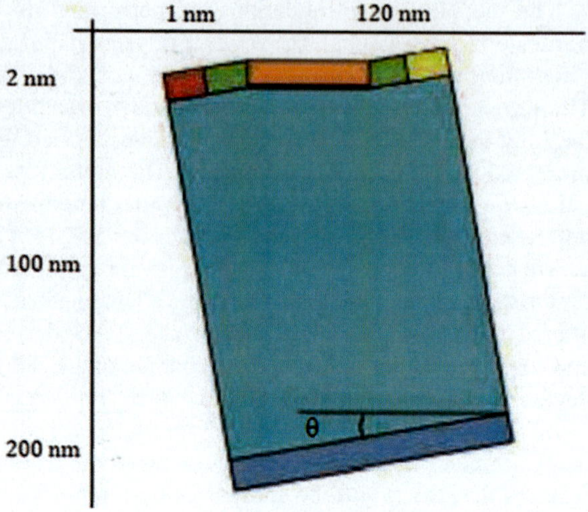

Fig. 3 Device model for channel inclination 10° (65 nm)

Table 1 Simulation settings

Simulation settings for MOSFET characteristics					
V_{GS}			V_{DS}		
First value	Last value	Increment	First value	Last value	Increment
0 V	2.4 V	0.2 V	0 V	4.0 V	0.2 V

The simulation results for one such exercise, on a 65 nm device, for reference (planner) design and that for 10° are plotted to visualize the characteristics. This is shown in Fig. 4.

4 Analysis of Simulation Output

The n-MOSFET devices are modeled for the three nodes: 45 nm, 65 nm, and 90 nm. The doping concentrations have been maintained identical. The exercise yielded 24 simulations. The simulation data so generated is analyzed using statistical procedures.

The objective of simulation data analysis is to test

1. Whether the dataset of the inclined channel device differs substantially, with respect to that of reference design?
2. If the dataset differs substantially, what is the bias for inclined channel device I_{DS} w.r.t. reference design I_{DS}?

The data analyses are carried out recalling:

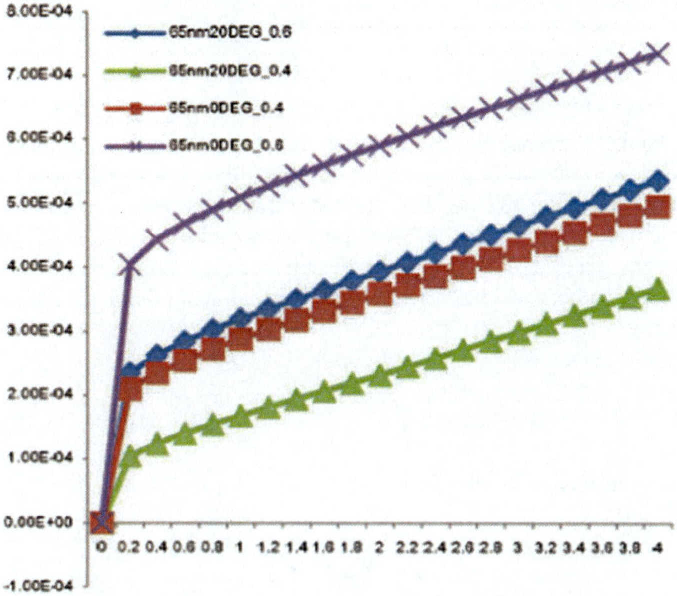

Fig. 4 V_{DS} versus I_{DS} for $V_{GS} = 0.4$, and VGS = 0.0 for $\theta = 0°$ and $\theta = 20°$

- Each simulation computes I_{DS} for a given set of values of V_{DS} and V_{GS}. A comparison of two performances is done with respective I_{DS} values.
- The simulation data need not have a Gaussian distribution.

These observations and objectives of analysis led to the selection of Wilcoxon's rank-sum procedure [5]. The Wilcoxon's procedure is applied to two datasets, to ascertain whether the datasets belong to the same original dataset, i.e., if we merge the two data sets, will the individual members of the two datasets will retain their ranks, compared to the ranks they held in their original groups.

The procedure for applying the Wilcoxon test is as follows: The two datasets are merged, and the merged dataset is ranked. The new dataset is analyzed as per the ranks of the members of the two datasets [3]. The sum of the ranks of the members of each of the two datasets in the merged dataset is computed. If the two datasets have "m" and "n" members each, the merged dataset would have (m + n) members. Let the sum of the ranks of the first dataset be W_m and that of the second dataset be W_n. Let the rank-sum W_{bf} and standard deviation σ_{bf} for best fit (i.e., the members of the two datasets consistently occupy alternate positions in the merged dataset) are, respectively, given by

$$W_{bf} = \frac{m(m+n+1)}{2}. \quad (1)$$

$$\sigma_{bf} = \sqrt{\frac{mn(m+n+1)}{12}}. \quad (2)$$

The difference between μ_m and μ_{bf} becomes zero if ($W_m - W_{bf}$) is zero. Similarly, the difference between μ_m and μ_{bf} becomes negative when ($W_m - W_{bf}$) is negative, and the difference between μ_m and μ_{bf} becomes positive when ($W_m - W_{bf}$) is positive [6]. A negative sign for ($W_m - W_{bf}$) indicates that more number of members of a dataset with m members are on the left of the median and when ($W_m - W_{bf}$) is positive more members of a dataset with m are in the right of the median. In the first case, the series with m members will have a mean smaller than that of series with n members, and vice versa.

The hypotheses are formulated as follows:

$$H_0 : \mu_m - \mu_n = 0 \; H_1 : \mu_m - \mu_n \neq 0 \quad (3)$$

Level of significance: 0.05
Test statistic:

$$Z = \frac{W_m - W_{bf}}{\sigma_{bf}}. \quad (4)$$

Acceptance region for Z, is $-1.96 \leq Z \leq 1.96$

5 Results

The results are presented in Table 2. The effect of the inclination of the channel on I_{DS} in V_{DS}–I_{DS} characteristics is summarized by an appropriate arrow.

Table 2 Summary of results

θ, the angle of inclination	Z Values					
	90 nm		65 nm		45 nm	
10°	0.321	↔	−0.16	↔	−0.31	↔
20°	−4.61	↓	−4.69	↓	−1.32	↔
30°	1.59	↔	−2.46	↓	−1.46	↔
36.6°	1.62	↔	−0.71	↔	3.82	↑
40°	2.61	↑	3.10	↑	3.35	↑
45°	3.24	↑	8.36	↑	0.82	↔
50°	3.51	↑	−0.2	↔	0.74	↔

↔—no change in I_{DS}, ↑—rise in I_{DS}, ↓—drop in I_{DS}

6 Conclusions

Correlation of Table 2 with fundamentals of device fabrication processes suggests the following:

- Small vertical inclinations have a negligible effect on device characteristics, for all three nodes of n-MOSFETs under study.
- With reference to the planner values, I_{DS} drops for channel inclinations below 36.6° and rises for inclinations above 36.6°.
- Precise control of fabrication process steps can limit deviation in performance characteristics of MOSFET devices.
- The deformations in the channel region and the internal layers manifest themselves as the vertical inclination of the channel.

References

1. Kane, T., Tenney, M.P.: Electrical Characterization of sub-30 nm Gatelength SOI MOSFETs. In: Conference Proceedings, 30th International Symposium for Testing and Failure Analysis, November, 2004 Worcester, Massachusetts
2. Khiangte, L., Kumar, K., Dhar, R.S.: TCAD Modeling and Analysis of sub-30 nm Strained Channel MOSFET. In: International Conference on Intelligent Computing and Smart Communication 2019, pp. 1383–1388. Springer, Singapore (2020)
3. Ding, J., Asenov, A.: TCAD simulations and accurate extraction of reliability-aware statistical compact models. J. Comput. Electron. **19**(1), 359–66 (2020)
4. Ghosh, A., Jain, A., Gharami, S., et al.: Small-signal model for the single-electron transistor: part I. J. Comput. Electron. **16**, 296–306 (2017). https://doi.org/10.1007/s10825-017-0971-1
5. Wilcoxon, F.: Individual comparisons by ranking methods biometrics. Bulletin **1**(6), 80–83 (1945)
6. Siegel, Sidney: Non-parametric Statistics for the Behavioral Sciences, pp. 75–83. McGraw-Hill, New York (1956)

Applications of Fuzzy Graph Structures for the Analysis of India's Growth of Smart Cities: A Systematic Review

B. Angel and D. Angel

Abstract The concept of smart cities emanated at the time once the entire globe was facing one among the worst economic recession. In 2009, the idea of smart cities had fascinated the imagination of assorted nations across the world. This mission was launched in India to enhance the quality of life of individuals and intensify their interactions with the urban environment in the year 2015. Analysis on the development of urban infrastructure has been started in the year 2016 to identify the intellectual solution for sustainable cities. The objective is to systematically review the analysis of the issues pertaining to the evolution of smart cities in India by using fuzzy graph structures (FGS). The scope of this work was to conduct a survey concentrating on studies to promote reliable and inclusive cities that provide core infrastructure and gives a satisfactory life to their citizens, clean and sustainable surroundings, and application of "Smart" solutions.

Keywords Urban infrastructure · India · Urban governance · Smart city · Graph theory · Fuzzy graph structures

1 Introduction

A graph is a mathematical illustration of a network and it describes the correspondence between vertices and edges. A graph structure allows to simultaneously study the different relationships and the related edges. A very important tool for graph theory is to represent many real-world problems. The fuzzy sets and fuzzy relations [1] concept was introduced in the year 1965 by L.A. Zadeh. In 1975, Rosenfeld developed the theory of fuzzy graphs with the help of fuzzy relations on fuzzy sets [2]. A fuzzy set [3] is a vital mathematical structure to portray a set of objects whose boundary is indefinite. When vertices and edges are uncertain then the fuzzy graph [4, 5] has a distinct significance. Ramakrishna [6] et al. explored a few concepts

B. Angel
Central Institute of Plastics Engineering and Technology, Chennai, India

D. Angel (✉)
Sathyabama Institute of Science and Technology, Chennai, India

J. S. Raj et al. (eds.), *Intelligent Sustainable Systems*, Lecture Notes in Networks and Systems 213, https://doi.org/10.1007/978-981-16-2422-3_59

based on fuzzy graph structures. Akram [7] defined some interesting results of vague graph which helps us to understand some elementary concepts of interconnected networks. Muhammed Akram [8] et al. introduced the min-product and maximal product of fuzzy graph structures and also examined their properties. The applications of FGS [9, 10] in various systems pervade to significantly interpret in many real-life scenarios.

Smart Cities Mission is an innovational and resourceful scheme launched by Prime Minister Narendra Modi on 25 June 2015 to evolve smart cities [11, 12] across India, making them a citizen-friendly and sustainable environment. This restoration communal program serves as a remedy for all the present civic challenges [13–15] that create employment and amplify incomes for all the citizens. Area-based development focused on transforming the existing areas into the sustainable green city [16, 17] and thereby improving the livability of the whole city. By eradicating the barriers [18] new sectors will be developed around cities to accommodate the enlarging population in metropolitan areas. Using TISM and fuzzy MICMAC process [19, 20], effective planning of city selection eligibility was done for the urban development of the city. Suma [21] designed the hybrid deep fuzzy hashing algorithm to improve the data efficiency in a distributed cloud environment. Hassan [22] et al. contributed to the challenges of IoT protocols to expand smart environments. The objective of this article is to design a suitable model to highlight the problems that must be rectified in achieving the expanding goals of smart cities in India.

This research article provides a qualitative methodology using fuzzy graph structures to identify the current problems of developing affordable infrastructure, adequate water supply, assured electricity, management of flood risk, proper sewerage system, etc. associated with the rapid growth of cities, and thereby degenerate quality of life for many of its individuals. This paper has many contributions in providing relevant information about the concept of fuzzy graph structures to identify the gap analysis of the current research work and for the implementation of transformative urban development. A fuzzy graph is an efficient mathematical model that helps to analyze the different issues faced by the people in different states of India. This review helps the researchers to get a detailed overview of the recent articles to further enhance the utility of this model focusing on the sustainable development of the city. The findings of this work describe how the FGS model could be a key enabler to solve many controversial issues among different states in India for shaping future smart cities. The scope of this systematic review attempts to serve as a decision-making tool for the research scholars to further examine the quantitative content analysis adopted to make up a smart city using the FGS model.

The following procedures (Fig. 1) were handled for an effective systematic review process which emphasizes planning, conducting, and documenting the reviews to find better ways of evaluating the qualitative research work.

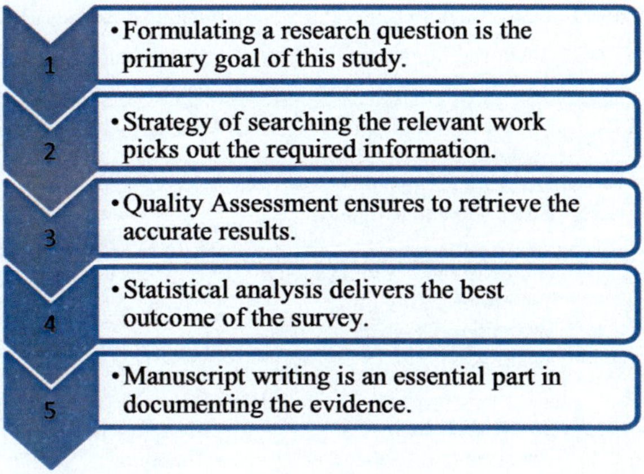

Fig. 1 Systematic review process

2 Methodology

We have searched the following databases:

- ACM Digital Library
- IEEE Xplore
- Science Direct
- Springer Link
- Research Gate
- Google Scholar
- Scopus

This review directs to explore the previous studies connected with "Smart Cities" regarding their aims, styles, and results. Therefore, a systematic analysis technique was used to review previous investigations specializing in the smart cities mission occurred within the globe. We have followed the Preferred Reporting Items for Systematic Reviews and Meta-Analysis (PRISMA) guidelines in regulating and reporting on this systematic review. The following keywords were employed to meet the eligibility criteria which retrieves the majority of the studies (which includes "Smart City" or "Smart City Mission" or "Urban Planning" or "E - Governance" or "Urbanization" or "Sustainability" or "City Development").

3 A Study on Reviewed Literature of Smart Cities in India

Table 1 presents the applications of different mathematical models which result in the identification of the particular domain in developing smart cities.

Table 2 displays the category-wise data of the reviewed papers on the identified domain of smart cities in India. The important points were gathered from the existing

Table 1 List of reviewed literature using mathematical model

S. no.	Reviewed literature	Year	Mathematical model used	Aim/objectives of the study
1	The Novel Paradigm of Economics Driven for Local Smart Sustain Cities Modeling Using Exploratory Factor Analysis and Planning Technique Using Fuzzy Evaluation Decision Making	2020	Fuzzy	This paper presents a process wherein "the Local Smart Sustain Cities Model (LSSCsM)" is combined and modeled with Exploratory Factor Analysis technique (EFA) to design a smart city
2	Henry George and Mohring–Harwitz Theorems: lessons for financing smart cities in developing countries	2019	Henry George Theorem, Mohring–Harwitz Theorem	This article suggests "beneficiaries pay" and "congesters pay" principles are combined to generate adequate revenues and to service long-tenor debt incurred for core infrastructure facilities
3	Developing a sustainable smart city framework for developing economies: an Indian context	2019	Fuzzy MICMAC	This research work will be beneficial for policy makers and practitioners in understanding of smart city development
4	Smart Sustainable Cities: Using a Fuzzy Inference System to Determine their Global Score	2019	Fuzzy Inference system	This research has collected data from four cities all around the globe using an adapted fuzzy logic model, the general sustainability score of cities in developing and developed countries
5	A Novel Method Based on the Graph Theory for Urban Energy Planning	2019	Graph Theory	In this work, a novel method to plan and optimize pathways mix of urban regional energy system is proposed
6	The Relation Between Urban Morphology and Physical Environmental Qualities: Comparing Walkability in Neighborhoods via Analyses of Spatial Statistics and Indices of Graph Theory and Space Syntax	2019	Graph Theory	This study aims to compare morphological structure and physical environmental qualities in different neighborhood of a small town and investigate whether physical environmental qualities vary by urban morphology

(continued)

Table 1 (continued)

S. no.	Reviewed literature	Year	Mathematical model used	Aim/objectives of the study
7	Barriers to the development of smart cities in Indian context.	2018	Fuzzy AHP	This research work employed fuzzy analytic hierarchy process (AHP) technique useful for eradicating the potential interferences in smart city development initiatives in India
8	A policy framework for city eligibility analysis: TISM and fuzzy MICMAC-weighted approach to select a city for smart city transformation in India	2018	TISM and fuzzy MICMAC	This article proposes a weighted criteria model to assess the city selection eligibility. The fuzzy MICMAC process is used to classify the factors based on driving power and dependence
9	Modeling Smart Cities with Hetero-functional Graph Theory	2017	Hetero-functional Graph Theory	This paper collates an hetero-functional graph theory for potential application to integrated smart city infrastructure models
10	Evaluation of Sustainability Indicators in Smart Cities for India Using MCDM Approach	2017	Fuzzy AHP	In this paper, the importance of various criteria for sustainability in a smart city is determined using fuzzy and fuzzy-AHP method
11	Application of Graph Theory in Liveable Cities	2016	Graph Theory	This study is important as it reveals the quantitative relation between livability and spatial organization of the cities
12	Examining the relation between the urban pattern and urban history using graph theory-based network indices	2016	Graph Theory	This study aims to understand the relation between development history and urban morphology in quantitative ways and the results show urban patterns measured through graph theoretic measures

Table 2 Category-wise data of the reviewed papers

Year	Mathematical model	Category	No of publications
2016	Fuzzy		1
	Graph Theory		3
	General	Statewise	0
		IOT	2
		Diverse	3
2017	Fuzzy		1
	Graph Theory		3
	General	Statewise	1
		IOT	1
		Diverse	0
2018	Fuzzy		3
	Graph Theory		1
	General	Statewise	2
		IOT	2
		Diverse	1
2019	Fuzzy		2
	Graph Theory		2
	General	Statewise	1
		IOT	0
		Diverse	2
2020	Fuzzy		1
	Graph Theory		0
	General	Statewise	2
		IOT	1
		Diverse	2

literature and it has been found that there is no study about smart cities mission using fuzzy graph structures between 2015 and 2020.

The number of publications shown in Fig. 2 was done from the year 2015 to the present after implementing the mission of smart city in India.

The analysis in Fig. 3 reveals the year-wise distribution of publications after launching the smart city mission in India. This has been done with 105 articles which facilitate in motivating the researchers for predictable planning and execution of work to get promising results. The problem of this research work is to pinpoint the factors that are more appropriate for the assessment of city "smartness" and also the alliance between "smartness" and the issues faced by the people that give way to a massive opportunity for the improved quality of affordable life of the citizens.

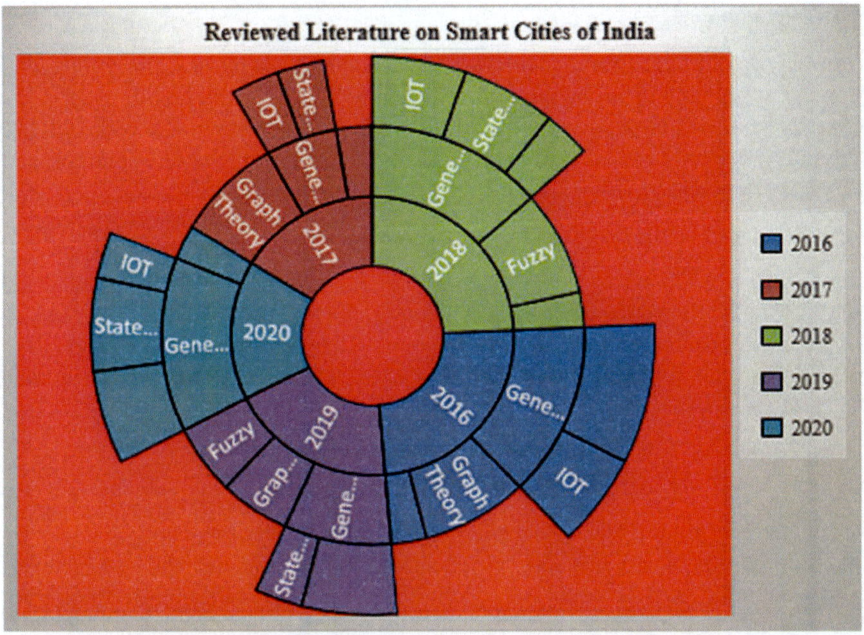

Fig. 2 Reviewed literature on smart cities of India

Fig. 3 Year-wise break-up of publications on smart cities

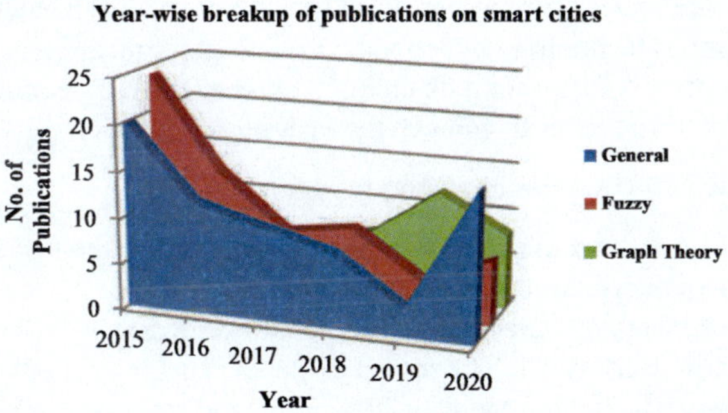

4 Results and Discussions

Based on the peer review of the recently published papers, the following research questions were framed and answered that results in some beneficial information for urban planning and development, and further studies are aimed at reaching a vision of sustainable cities in India that creates an appropriate structure which will act like a torchbearer to other endeavoring cities. A fuzzy graph uses the membership value of edges and vertices as an essential tool to model the problems. It helps us to review the various relations and also the degree of corresponding edges at the same time.

RQ1. What analysis can be done in Smart City Mission?

The analysis of Smart City Mission is to identify the issues faced by the people that promote sustainability in the community. Graph theory has the ability to analyze and arrive at a feasible solution for certain problems. A general model with vertices, paths, and network balance is established for the analysis of physical, economic, and social development of cities to optimize rapid urbanization.

RQ2. What will be the boundary of the smart city?

The key services of urban local bodies will be the boundary to make an adequate provision by funding or by any means of resources for the establishment of Greenfield development within the notified area of the city. The process of TISM and fuzzy MICMAC is an evaluation criteria model which was used in the article to categorize the factors supported on driving power and dependence to estimate eligibility of city selection.

RQ3. How does the idea of smart city work for a sustainable environment?

The target of smart cities is to benefit a clean and environmentally friendly by focusing on citizen services and cities' challenges. An imperishable environment can be achieved only through systematic deployment of innovative technologies, new techniques, and services to safeguard local and regional ecosystems.

RQ4. What are the challenges faced in Smart City Mission?

The rapid growth of population and urbanization are united to create enormous challenges for cities in India. To overcome these barriers, utilize high dependency on smart technology which may drive to a significant revolution in the existing backlog cities. A fuzzy analytic hierarchy process (AHP) technique identifies the key barriers of smart cities to achieve the vision of a smart city.

RQ5. Are smart cities required for India?

The demographic report says by the year 2050, 80% of the urban population is willing to live in cities. Hence there is going to be a heavy struggle in India due to inadequacy of housing, water, sanitation, health center, energy, transportation, employment, IT connectivity, etc. The smart city concept helps to improve the wellbeing of its citizen's welfare and to eradicate both urban and rural poverty in India.

RQ6. Why should citizen participation support the governance of the community?

Efficient governance and public participation will enhance the desirable improvements on eradicating social issues and upgrading the quality of urban infrastructure. The implication of adopting a lot of developmental activities with assured services provides environmental protection in the cities. The relation between government and society using graph theory was discussed in the article to make the cities livable.

RQ7. What are the resource pools available for the cities to acquaint with the smart solutions?

Imposing charges for land value-based taxation, property taxes, utilities, etc. are the resource pools for the cities. This set of resources can be assigned to certain cities in India using innovative planning which benefits all the stakeholders—government, business, and residents that lead to smart outcomes.

RQ8. Discuss on the futuristic vision of smart cities.

The Indian government focuses on ICT to manage every aspect of the city's operations and services that might increase the citizen's migration rate in India. Many technological platforms are involved to build the digital infrastructure of a city which helps to generate economic growth and social wellbeing that meets the people's needs in future cities in India.

RQ9. Which are the services a smart city will offer that are not available in "normal" cities?

A system of sensors should be installed to monitor the networks for urbanity and allow for real-time feedback from citizens. Smart cities promise safety to all the citizens through the implementation of centralized network security policies which reduce risks and damage toward society and ecosystems. Cybernetics is used to improve the effectiveness of citizen services and reach the standard of living. Graph parameters are tremendously important in modeling and controlling progressive interconnection networks.

RQ10. How to ensure smart cities benefit the society?

India has made rapid progress in expanding the cities and the conception of smart solutions promotes the communities to fight against the existing urban issues. Many states in India have adopted activities, like improved infrastructure, essential public utilities, safer communities, usage of digital technologies, etc., which will prevent future problems in society.

In this paper, we identify the possible solution to the smart city problems using a fuzzy graph structure whose vertices and edges are represented as common issues and different states in India, respectively. In future work, fuzzy graph structures can be designed to enhance the detailed analysis of the core-important sectors of the smart city. Cities in different states have many issues, some of which are considered to be the most controversial issues. These issues need more attention that must be solved for better employment and scope of income for the standard living of the citizens.

5 Conclusions

The accomplishment of a smart city depends on the depth and effectiveness of targeted improvement within each area or initiative and on the integrity or balance of the portfolio of initiatives across the city. The cities of the future bring together infrastructure and technology which is designed for accessibility and safety to improve the standard of living of the citizens. This work benefits the researchers to get a precise outline of the most recent articles for further investigation on the implementation of the smart city mission in India. A fuzzy graph plays a remarkable role to analyze the issues and enhance sustainable urbanization in India. Based on the scrutinization, only a few papers have been published on this particular area using FGS, so the chances are high for doing research which develops the economic growth in India. The proposed work of this paper is to develop a fuzzy graph model to identify the most ambiguous issues in different states so that prior importance can be given to that state. Therefore, the concept of smart city using fuzzy graph structures conferred in this paper needs further research, revisions, and enhancements on the analysis of the issues faced by the people due to inadequate facilities in the city. Thus, improvements in the existing conditions and schemes will increase the economy of the future city that helps India to make a place in the global modern society.

References

1. Zadeh, L.A.: Fuzzy sets. Inf. Control **8**, 338–353 (1965)
2. Rosenfeld, A.: Fuzzy graphs.In: Zadeh, L.A., Fu, K.S., Tanaka, K., Shimura, M. (eds.) Fuzzy Sets and their Applications to Cognitive and Decision Processes. Academic Press, New York, pp. 77–95 (1975)
3. Dodson, C.T.J.: A new generalization of graph theory. Fuzzy Sets Syst. **6**, 293–308 (1981)
4. Sampathkumar, E.: Generalized graph structures. Bull. Kerala Math. Assoc. **3**(2), 65–123 (2006)
5. Bhattacharya, P.: Some remarks on fuzzy graphs. Pattern Recogn. Lett. **6**(5), 297–302 (1987)
6. Ramakrishnan, R.V., Dinesh, T.: On generalized fuzzy graph structures. Appl. Math. Sci. **5**(4), 173–180 (2011)
7. Akram, M., Chen, W.J., Shum, K.P.: Some properties of vague graphs. Southeast Asian Bull. Math. **37**, 307–324 (2013)
8. Akram, M., Sitara, M.: Certain fuzzy graph structures. J. Appl. Math. Comput. **61** (2019)
9. Malik, D.S., Mathew, S., Mordeson, J.N.: Fuzzy incidence graphs: Application to human trafficking. Inf. Sci. **447**, 244–255 (2018)
10. Binu, M., Mathew, S., Mordeson, J.N.: Connectivity index of a fuzzy graph and its application to human trafficking. Fuzzy Set Syst. **360**, 117–136 (2019)
11. Hoelscher, K.: The evolution of the smart cities agenda in India. Int. Area Stud. Rev. **19**, 28–44 (2016)
12. Vadgama, C.V., Khutwad, A., Damle, M., Patil, S.: Smart funding options for developing smart cities: A proposal for India. Indian J. Sci. Technol. **8**(34) (2015)
13. Chatterjee, S., Kar, A.K.: Smart Cities in developing economies: a literature review and policy insights. In: Proceedings of the 2015 International Conference on Advances in Computing, Communications and Informatics (ICACCI), Kochi, India, pp. 2335–2340, 10–13 August (2015)
14. Yadav, G., Mangla, S.K., Luthra, S., Rai, D.P.: Developing sustainable smart city framework for developing economies: an Indian context. Sustain. Cities Soc. **47**, 101462 (2019)
15. Bhattacharya, T.R., Bhattacharya, A., Mclellan, B., Tezuka, T.: Sustainable smart city development framework for developing countries. Urban Res. Pract., 1–33 (2018)
16. Kumar, A.: Can the smart city allure meet the challenges of Indian urbanization? In: Sharma, P., Rajput, S. (eds.) Sustainable Smart Cities in India: Challenges and Future Prespectives. Springer International Publishing: Cham, Switzerland, pp. 17–40 (2017)

17. Kummitha, R.K.R., Crutzen, N.: Smart cities and the citizen-driven internet of things: A qualitative inquiry into an emerging smart city. Technol. Forecast. Soc. **140**, 44–53 (2019)
18. Rana, N.P., Luthra, S., Mangla, S.K., Islam, R., Roderick, S., Dwivedi, Y.K.: Barriers to the development of smart cities in Indian context. Inf. Syst. Front. **21**, 503–525 (2019)
19. Kumar, H., Singh, M.K., Gupta, M.P.: A policy framework for city eligibility analysis: TISM and fuzzy MICMAC-weighted approach to select a city for smart city transformation in India. Land Use Policy, vol. 82, pp. 375–390, December (2019)
20. Kumar, H., Singh, M.K., Gupta, M.P., Madaan, J.: Smart neighbourhood: a TISM approach to reduce urban polarization for the sustainable development of smart cities. J. Sci. Technol. Policy Manag. **9**, 210–226 (2018)
21. Suma, V.: A novel information retrieval system for distributed cloud using hybrid deep fuzzy hashing algorithm. JITDW 2 **3**, 151–160 (2020)
22. Hasan, M.K., Ahmed, M.M., Hashim, A.H.A., Razzaque, A., Islam, S., Pandey, B.: A novel artificial intelligence based timing synchronization scheme for smart grid applications. Wirel. Pers. Commun. **114**, 1067–1084 (2020)

Reliable Fault-Tolerance Routing Technique for Network-on-Chip Interconnect

Jayshree, Gopalakrishnan Seetharaman, and Debadatta Pati

Abstract A reliability challenge of network-on-chip (NoC)-based interconnection of storage, computational blocks, and multi-processor integrated on a single die is presented. In this paper, a fault-tolerance routing technique based on *XY* routing and dynamic *XY* (*DyXY*) routing technique has been presented to improve the reliability in the modern complex chips. Without fault-tolerance routing, there can be adverse effects on-chip communication figure-on-merits such as energy consumption and performance. Hence, the fault-tolerance routing is important for getting a high yield and reducing the negative impact of the faulty component. Moreover, the NIRGAM 2.1 simulator which is a cycle accurate SystemC simulator is used for modeling and simulation purpose. The proposed techniques are analyzed under various traffic patterns. The results are compared with the existing generic routing techniques. Finally, it is obtained that the proposed routing technique has less packet loss.

Keywords Fault-tolerance · Network-on-chip (NoC) · Reliable · Routing algorithm and system-on-chip (SoC)

1 Introduction

The ongoing miniaturization of technology to integrate more intellectual properties (IP) for satisfying customer requirements leads to a more susceptible fault in on-chip communication [1]. The design engineer of system-on-chip (SoC) uses different bus architecture IP cores. With the advancements in semiconductor technologies, the architecture of buses IP cores is not scalable to give a high performance during interaction of the processing element (PE) [2]. Apart from performance issues, reliability

Jayshree (✉) · D. Pati
Department of Electronics & Communication Engineering, National Institute of Technology, Nagaland, India
e-mail: jay.chy@nitnagaland.ac.in

G. Seetharaman
Indian Institute of Information Technology, Tiruchirappalli, India
e-mail: gsraman@iiitt.ac.in

J. S. Raj et al. (eds.), *Intelligent Sustainable Systems*, Lecture Notes in Networks and Systems 213, https://doi.org/10.1007/978-981-16-2422-3_60

is another crucial challenge in bus architecture IP cores; the fault is considered a significant issue of the complex system. The network-on-chip (NoC) has been developed to address the communication needs for future SoC architecture design [3]. The main components of NoC are router, link, network interface (NI), and PE. Each PE can be communicated through NI *via.* routers. Here, the NoC is more scalable, reliable, and reusable, to be considered as a potential candidate instead of buses. Therefore, this motivates us to opt NoC interconnect architecture for the present work.

The NoC is reliable as it enables the information transfer between the PEs *via.* multiple routes during one path's failure; the data can be traversed through the other paths. This continuous technology scalability increases the likelihood of permanent and transient faults [4]. Therefore, the fault is a vital challenge that should adequately be addressed in NoC. The permanent fault occurs by aging consequences of electromigration and inadequate technology; in general, this happens during fabrication [5, 6]. The transient involves cross talk and coupling noise problems. Primarily permanent fault issues are addressed because of the router or IP cores failures such as node and link. In node failure, full router architecture is considered a faulty part. This may happen due to the problem in any sub-component of the router data plain. Moreover, the link fault occurs due to some issues in I/P or O/P link that is connected to another router.

During node failure, the node indicates that every link related to the router is faulty. However, during the routing process, the node failure is hard to repair dynamically. The flit is hard to transmit *via.* node. In case of a link failure, flits manage to search another router through another working link. Additionally, another two more issues related to NoC are deadlock and livelock. In the literature, Asma et al. [7] mentioned that the deadlock is a cyclic dependence consequence. It can be prevented using virtual-channels introduced by Dally et al. [8] or turn model by Glass et al. [9], avoided ,and recovered [10, 11]. The livelock occurs in the network when all the adjacent link holds other packets; this can be avoided by using the minimal path or keeping restricted on non-minimal path [10]. Congestion in the network occurs due to the practice of deterministic protocol or non-uniform traffic [12, 13]. These issues can be avoided using routing policies that provide a report regarding the portion of the fabric subjected to congestion (local or not) and fault details.

The key salient features of this work are as follows.

- Design of a 2-D router which is considered as the main core of NoC-based interconnection in the SoC.
- Next, router is symmetrically connected as mesh topology in two-dimensional (2-D).
- To perform routing computation, a novel routing technique based on *XY* routing and *DyXY* routing techniques are considered.
- Implementation was performed on ZYNQ xc7z020-clg484 development on the FPGA board for the area, and its performance comparison was performed with another existing method.

- The NIRGAM 2.1 simulator, which is a cycle accurate SystemC simulator, has been used for modeling and simulation mainly for fault analysis in terms of packet loss under various synthetic traffic patterns.

The complete works are organized as follows: Sect. 2 presents the background of various routing techniques. Next, Sect. 3 discussed the proposed various routing techniques. In Sect. 4, the obtained results are analyzed and compared with the existing technique. Finally, a brief conclusion is drawn in Sect. 5.

2 Background

The NoC routing algorithms need two major figures-of-merits such as energy consumption and performance. First, one is related to NoC architecture such as the static or dynamic network; the topology is considered as 2-D/3-D regular or irregular; flow control is buffered or unbuffered. Second, that defined as the routing protocol itself, whether packets sending strategy is the source or distributed; commutation mode is a packet switching or circuit switching; and path definition is deterministic, oblivious, or adaptive. Finally, routing algorithms are to be used as any one of the storing policies such as store and forward, wormhole, or virtual cut-through.

The deterministic routing approaches can solve the deadlock issues. However, it suffers from under-utilization of channels [5], and the path is fixed. It cannot offer change paths, which lead to congestion and fault which degrade the performance [13]. Glass et al. [9] assumed the "Turn Model", which is based on permitting and prohibiting dependencies between channels. Lin et al. [10] proves that if no channels in the network are held forever, deadlock can be solved. Schwiebert et al. [11] presented the channel waiting graph, and restriction inflicted mainly for deadlock freedom to use local data for routing. Ebrahimi et al. [14] opted for the Hamiltonian path to ensure freedom from the deadlock in the NoC. To improve congestion, one uses adaptive strategies such as Q-Learning [7], which is a kind of model-free reinforcement learning. Ebrahimi et al. [15] employed routing tables for getting the information of the network. Gratz et al. [12] proposed the "Regional Congestion Awareness (RCA)" that provides information regarding the part of the network and is considered subject to congestion. Adaptive routing policies create livelock. It can be removed by imposing a restricted non-minimal path, following the minimal path, and probabilistic avoidance [5]. To improve performance and to search the optimal router in wireless networks, evolutionary algorithms have been used by Smys et al. [16]. It is necessary to either use fault tolerance, fault prevention, fault forecasting, or fault removal to solve fault challenges. Here, several fault models are presented in the literature, which attempts to use fault detection and fault tolerance [17–19].

3 Proposed Architecture and Methodology

In this paper, the 2-D virtual-channel (VC) 4-stage router is shown in Fig. 1, which modifies the architecture of proposed architecture by Lan et al. [20]. It is extended with fault-tolerance routing technique at routing computation stage as shown in Fig. 1.

It contains less number of buffer in comparison to generic wormhole flow-based router [21], generic virtual-channel flow-based router [22], and also in comparison to bidirectional link wormhole flow-based router [23]. The complete parts of the router is grouped into two different categories, the datapath, which consists of an in-out buffer unit, and $M \times M$ crossbar switch collectively for storage and transport of message payload. The rest of all sub-component of router architecture is implemented as control logic such as routing computation unit, which evaluates or identifies the output ports of message payload that to transfer. Next, request a virtual-channel input at downstream routers by the message payload through virtual-channel allocator for allocating virtual-channel. After determining the route and virtual-channel, all flits of the message payload get transferred through selected virtual-channel. This process take place within the assigned time duration of the crossbar switch by switch allocator from input/output (I/O) virtual-channel to the relevant I/O physical link. In the literature, few allocators are PIM, iSLIP, LOA, and waveform, multi-stage. Among these, the best is PIM whichhas the best performance [24] and used in the proposed architecture. These allocators are arbiters as a basic fundamental block either in fixed, variable (such as oblivious, round-robin), matrix, or queuing form to solve n number of requests for a single resource (like buffer, link, or crossbar switch port).

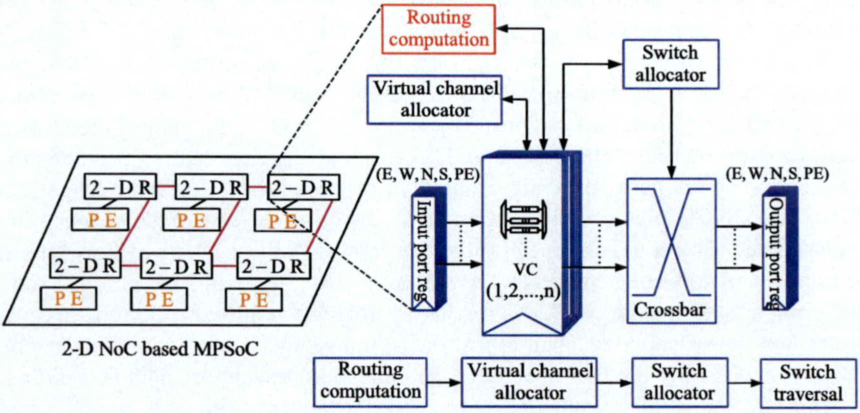

Fig. 1 The 4-stage pipelines virtual-channel router with fault-tolerance routing computation for the 2-D NoC

The topology is shown in Fig. 1, designed with less congestion network, and this is done by using a self-reconfigurable link and by proposing a fault-tolerance dynamic routing algorithm (*FTDyXY*) and its pseudo-code has been given in Algorithm 1.

Algorithm 1 Fault-tolerance routing technique.

Input: Present and destination addresses congestion detail
Output: Routing path of flits

```
for every flits of packet do
    if (CONGESTION = 0) then
        Deterministic XY routing algorithm
    else
        Minimal adaptive routing algorithm (DyXY Routing) using diagonal link
    end if
end for
```

4 Result Analysis

The fault-tolerance capacity is analyzed for the different synthetic traffic patterns such as random, bit reversal, and bit shuffle injected at the rate of 0.1 packets/node/cycle on 3×2 NoC with bidirectional link control router, and the number of faults is varied from 3 to 8. Every simulation is performed 15 times and a period of 6×10^5 cycles.

4.1 Random Traffic Pattern

The random traffic pattern is injected, and its result is illustrated in Fig. 2. Here, it is observed that *XY* [25] is having the highest packet loss in comparison to odd-even (OE), dynamic *XY*, and proposed fault-tolerance dynamic *XY* (*FTDyXY*). The *FTDyXY* has the lowest percentage packet loss.

4.2 Bit Reversal Traffic Pattern

A bit reversal traffic pattern is injected. This pattern is a more realistic pattern than a random traffic pattern shown in Fig. 3. It observed that the *FTDyXY* is still having loss percentage of packet loss in comparison to *XY*, *OE*, and *DyXY*.

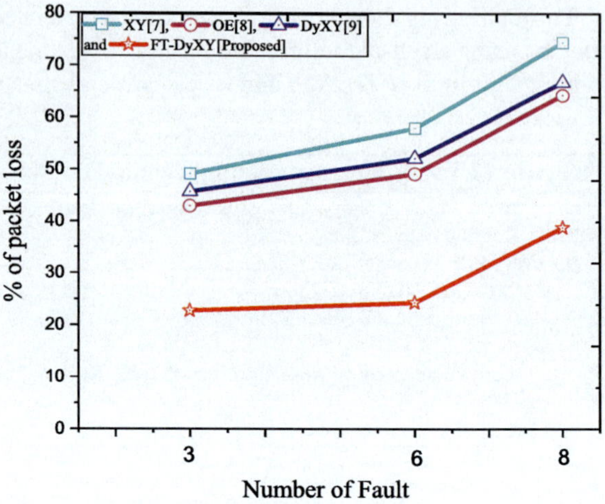

Fig. 2 The packet loss analysis under random traffic

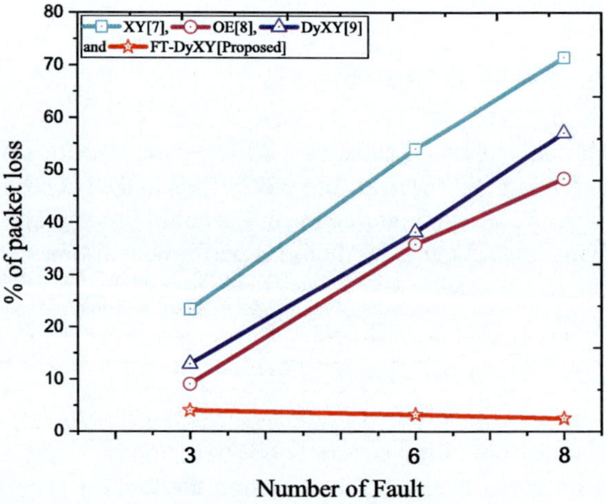

Fig. 3 The packet loss analysis under bit reversal traffic

4.3 Bit Shuffle Traffic Pattern

A bit shuffle traffic pattern is injected, which is more realistic than both previous injected patterns; its result is illustrated in Fig. 4. Under each cases, the *FTDyXY* has lower percentage of packet loss in comparison to *XY*, *OE*, and *DyXY*.

Table 1 shows that the single arrow represents the reliability of deterministic *XY* and comparative results of *XY* took as base case technique. The fault-tolerance ability of *FTDyXY* is $4 \times XY$, $2 \times OE$, and $3 \times DyXY$ for HTRPD high in comparison to *XY*, *OE*, and *DyXY*. It's all due to its property to follow non-congestion path

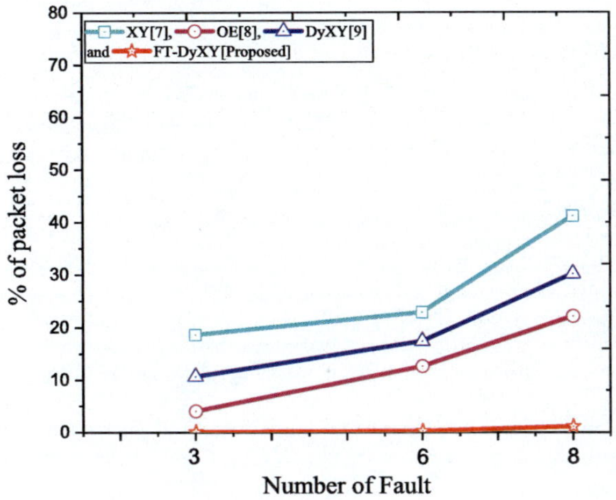

Fig. 4 The packet loss analysis under bit shuffle traffic

Table 1 Reliability comparison of proposed routing algorithm

No.	Routing algorithm	Reliability
1	Deterministic *XY*	↑
2	*OE*	↑↑
3	*DyXY*	↑↑↑
4	*FTDyXY*	↑↑↑↑

congestion and safe flow dynamic adaptive path during fault case. The comparison of reliability for the proposed *FTDyXY* routing technique $w.r.t.$ to the existing routing technique is presented.

5 Conclusion

In this work, first, we have presented a comprehensive study on prior related deadlock and livelock free, congestion-aware, and fault-tolerant routing works in NoCs. Next, we proposed the *FTDyXY* routing scheme, intending to cope with a non-static topology change caused by faults in NoCs. The complete idea relies on observing NoC elements and traffic patterns and adjusting the packets' path in response to a variance (fault occurrence) in the topology. The evaluated result on 3×2 mesh topology with various synthetic traffic models and compared outcomes shows that the proposed algorithm has less loss in terms of packets, which motivates us to extend this 2-D NoC work to 3-D NoC for mainly multi-media SoCs mainly for medical and Internet-of-Thing (IoT) application in our future work.

References

1. Sandhie, Z.T., Patel, J.A., Ahmed, F.U., Chowdhury, M.H.: Investigation of multiple-valued logic technologies for beyond-binary era. ACM Comput. Surv. **54**(1), 30 (Jan. 2021)
2. Kamal, R., Yadav, N.: NoC and bus architecture: a comparison. Int. J. Eng. Sci. Technol., vol. 4, pp. 1438–1442, 04 (2012)
3. Jayshree, S.V., Chatterjee, A., A methodology for designing LVDS interface system. Sixth Int. Symp. Embed. Comput. Syst. Des. (ISED), Dec 2016, pp. 284–288 (2016)
4. Hu, C., Conrad Meyer, M., Jiang, X., Watanabe, T.: A fault-tolerant hamiltonian-based Odd-Even routing algorithm for network-on-chip. 35th Int. Technol. Conf. Circuits/Syst, Comp. Commun. (ITC-CSCC), pp. 217–222 (2020)
5. Marri, S., Manishankar, S., Radha, D., Moharir, M.: Implementation and analysis of adaptive Odd-Even routing in Booksim 2.0 simulator. Int. Conf. Commun. El. Syst. (ICCES), pp. 76–83 (2016)
6. Jayshree, V.., Seetharaman, G.: Design and analysis of novel interconnects with network-on-chip LVDS transmitter for low delay. In: NASA/ESA Conf. Adapt. Hardw. Syst. (AHS), pp. 204–209 (2018)
7. Gabis, A.B., Koudil, M.: NoC routing protocols objective-based classification. J. Syst. Archit. **66–67**, 14–32 (2016)
8. Dally and Seitz (1987) "Deadlock-Free message routing in multiprocessor interconnection networks," IEEE Trans. Comput., vol. C-36, no. 5, pp. 547–553
9. C. J. Glass and L. M. Ni, "The turn model for adaptive routing," J. ACM, vol. 41, pp. 874–902, 1994
10. Atagoziyev, M.: "Routing algorithms for on chip networks," in Graduate School of Natural and Applied Sciences, Thesis (2008)
11. Ahmed, A.B., Ahmed, A.B., Abdallah, A.B.: "Deadlock-Recovery support for fault-tolerant routing algorithms in 3D-NoC architectures, in IEEE 7th Int. Symp. Emb. Multicore Socs **09**, 67–72 (2013)
12. R. S. Reshma Raj, C. Gayathri, S. Kalady, and P. B. Jayaraj, "Odd-Even based adaptive Two-Way routing in mesh NoCs for hotspot mitigation," in 20th Proc. Int. Conf. Distrib. Comput. Netw., ser. ICDCN '19. New York, NY, USA: Association for Computing Machinery, 2019, p. 248–252
13. Z. Sharifi, S. Mohammadi, M. Sirjani, and m. mosaffa, "Comparison of NoC routing algorithms using formal methods," 2013
14. M. Ebrahimi, M. Daneshtalab, J. Plosila, and F. Mehdipour, "MD: Minimal path-based fault-tolerant routing in on-chip networks," in 18th Asia and South Pacific Des. Autom. Conf. (ASP-DAC), Jan 2013, pp. 35–40
15. Ebrahimi, M., Daneshtalab, M., Liljeberg, P., Plosila, J., Flich, J., Tenhunen, H.: Path-based partitioning methods for 3D networks-on-chip with minimal adaptive routing. IEEE Trans. Comput. **63**(3), 718–733 (March 2014)
16. Haoxiang, D., Smys, S: Soft computing strategies for optimized route selection in wireless sensor network. J. Soft Comput. Paradigm, vol. 2, pp. 1–12, 03 (2020)
17. Xiang, D., Zhang, Y., Pan, Y.: Practical deadlock-free fault-tolerant routing in meshes based on the planar network fault model. IEEE Trans. Comput. **58**(5), 620–633 (2009)
18. Dajin Wang, "A rectilinear-monotone polygonal fault block model for fault-tolerant minimal routing in mesh," IEEE Trans. Comput., vol. 52, no. 3, pp. 310–320, 2003
19. Holsmark, R., Kumar, S.: Design issues and performance evaluation of mesh NoC with regions. in NORCHIP, pp. 40–43 (2005)
20. Y. Lan, H. Lin, S. Lo, Y. H. Hu, and S. Chen, "A bidirectional NoC (BiNoC) architecture with dynamic self-reconfigurable channel," IEEE Trans. Comput.-Aided Design Integr. Circuits Syst., vol. 30, no. 3, pp. 427–440, March 2011
21. Lu, Y., McCanny, J., Sezer, S.: "Generic low-latency NoC router architecture for FPGA computing systems," in 21st Int. Conf. Field Prog. Logic Appl., Sep. 2011, pp. 82–89 (2011)

22. Mullins, R., West, A., Moore, S.: Low-latency virtual-channel routers for on-chip networks. SIGARCH Comput. Archit. News **32**(2), 188 (Mar. 2004)
23. Lan, Y., Lo, S., Lin, Hu, Y., Che, S.: BiNoC: A bidirectional NoC architecture with dynamic self-reconfigurable channel. in 3rd ACM/IEEE Int. Symp. Netw.-on-Chip, May 2009, pp. 266–275 (2009)
24. William James Dally, B.T.: Principles and Practices of Interconnection Networks. Elsevier (2004)
25. Venkateswara Rao Musala, V. R. K. T: An efficient directional routing algorithm for network on chip. Int. J. Sci. Technol. Res., vol. 8, p. 11 (2019)

Cost-Effective and Efficient Detection of Autism from Screening Test Data Using Light Gradient Boosting Machine

Sai Pavan Kamma, Shahana Bano, Gorsa Lakshmi Niharika, Guru Sai Chilukuri, and Deepika Ghanta

Abstract Autism spectrum disorder (ASD) is a developmental disorder that affects the brain. Autism constrains a person's ability to interact and communicate with others. The cause of autism, in general, is unknown though genetics does play a role in the manifestation of the condition. In the absence of clear identifiable biomarkers, shortcomings of the available prognostic approaches create a need for a new technique that is speedy, cost-efficient, and provides an error-free diagnosis. The system should also be able to adapt to the varying characteristics of subjects with ASD. The amelioration machine learning brings to automated medical diagnosis which has inspired us to come up with a solution. An adept screening and diagnostic test for patients exhibiting known autistic symptoms is a well-compiled, specific, and approved questionnaire, which facilitates an easy and cheap diagnosis. Autistic Spectrum Disorder Screening Test data is collected from one such questionnaire. We used a combination of three publicly available datasets containing records related to ASD in children, adolescents, and adults. There are a total of 1100 instances along with 21 attributes. The proposed study uses a Light Gradient Boost (LGB) based model for classification, along with Random Search for hyperparameter optimization, which yielded a high accuracy of 95.82%.

Keywords Autism · Autism spectrum disorder (ASD) · Random search · Light gradient boosting machine (LGBM) · Pre-processing

1 Introduction

The human brain is one of the largest and most complex organs in the body, which contains roughly one hundred billion neurons. The brain is the centre for cognitive activities. The brain provides us with the ability to react to stimuli, which have emotions, and stores incidents as memories. Autism spectrum disorder is a child-development disability which causes lack of social skills, speech difficulties, and

S. P. Kamma (✉) · S. Bano · G. L. Niharika · G. S. Chilukuri · D. Ghanta
Department of Computer Science and Engineering, Koneru Lakshmaiah Education Foundation, Vaddeswaram, Andhra Pradesh, India

J. S. Raj et al. (eds.), *Intelligent Sustainable Systems*, Lecture Notes in Networks and Systems 213, https://doi.org/10.1007/978-981-16-2422-3_61

inability to grasp non-verbal cues. Autism was initially elucidated by Leo Kanner [1] in 1943 as an innate inability to create normal, biologically determined, emotional contact with others. Malfunctioning of mirror neurons and genetics are usually presented as reasons, but there is no one known cause. ASD can be hereditary [2], but the way the progeny gets the disorder is not known. People with allele alterations associated with Autism have a higher risk factor. A person suffering from Autism Spectrum Disorder may not have good people skills [1].

Fragile X Syndrome and Rett syndrome are associated as the genetic causes of autism. The general symptoms of autism may include the following [4]:

- avoiding eye contact or no awareness when others are speaking
- difficulty expressing needs
- no facial expressions or unusual facial expressions
- repeating words or phrases
- difficulty expressing needs
- hypersensitivity to sound, smell, taste, sight, or touch
- delayed speech and language skills, or lack of gesturing

ASD is found to be more common in boys as compared to girls [3]. Although on a brighter aspect, many children diagnosed with ASD live an independent, productive life. The symptoms of majority of the children assuage with age. Speech and psychological therapy can help with behavioural depression issues. In some cases, people might require continual support throughout their life. Also, autistic individuals can work and live individually [2]. The word spectrum is used to illustrate a broad range of developmental delays and symptom severity. ASD includes people who have a few mild autistic traits to those who need help with day-to-day functioning. The differences between one type and another type can be subtle and difficult to determine. A diagnosis on the spectrum means you can turn attention to assessing individual needs [4].

The diagnostic test for autism is expensive and sometimes involves extensive equipment such as Magnetic Resonance Imaging (MRI) . Various other techniques involve speech tests and observation schedules. An easier way we can quickly diagnose is through a screening tool that is built based on the eclectic phenotype and other behavioural symptoms of autism. The questionnaire consists of ten open-ended questions to which the test-taker (either parents or health care employees) can agree or disagree (example: "I often notice small sounds when others do not") based on the traits exhibited by the patient. The questionnaire also consists of general information like age, gender, and ethnicity. Two other important questions are whether any immediate family members have ASD and If the patient was born with jaundice, which drastically increases the chances of autism.

2 Related Work

Thabtah [5] has developed a quick and available mobile application for screening of ASD called ASDTests. The application design aims to screen patients of all age groups. The users are given a questionnaire of ten questions and their answers are used for the diagnosis. This work uses data collected from existing applications and machine learning (ML) techniques like Naïve Bayes classifier and Logistic Regression classifier, with the later performing better in all parameters like accuracy, sensitivity, etc. across all ages. The data collected from the app has been made available publicly for academic purposes.

Suman Raja and Sarfaraz Masood [6] have worked on a study to classily ASD using three datasets with 20 common features acquired from UCI's ML Repository. Authors of the work have implemented ML and deep learning algorithms, compared accuracy of them across all the datasets. The study reported that the CNN-based model produced the highest accuracy of 99.53%.

Fadi Thabtah [7] has proposed a screening model for autism making use of DSM-5 and Machine Learning Adaptation, the work uses datasets collected by the author. In the paper, the researcher highlights the advantages of using a DSM-5 which helps to overcome the problems with consistency of the existing tools.

M. S. Mythili et al. [8] studied autism using classification techniques. The objective of this paper was a diagnosis of autism and the severity of it. They implemented Support Vector Machine (SVM), Neural Network and Fuzzy logic techniques using WEKA workbench tools to examine student's behaviour and social interaction.

The goal of our research study is to devise a model that can handle surging data from frequent surveys conducted by health departments, parents, and hospitals on subjects who could have ASD. The data also requires a considerable amount of pre-processing and may also contain outliers, implementation regression techniques might not always produce better results. The pre-processing also includes encoding the character features whereas our proposed model negates the requirement.

3 Methodology

3.1 Importing the Dataset

The data utilized has been sourced from the machine learning Repository of University of California, Irvine (UC Irvine). Autistic Spectrum Disorder Screening Data of adult subjects [9] which consists of 704 instances with 21 attributes. ASD Screening records of kids [10] have 292 instances and 21 attributes. ASD Screening records of teenagers [11] include 104 instances with the same number of attributes as the above datasets. We used a combination of above datasets where the records are randomly shuffled to prevent the possibility of overfitting, the cumulated dataset consists of

Table 1 Overview of attributes used for the study

Attribute Id	Attributes description	Data type
1	Patient age	int32
2	Gender	String (m: Male, f: Female)
3	Ethnicity	String (White-European, Latino, Asian, Middle Eastern, etc.)
4	Whether the patient suffered from Jaundice at the time of birth	String (Yes, No)
5	If, any immediate family member suffered from autism	String (Yes, No)
6	Country of residence	String (Austria, Ireland, Jordan, etc.)
7	Test attempted Individual	String (Mother/Father, Hospital Personnel, Relative, etc.)
8	Answers to the screening test (10 questions)	String (Yes, No)
9-18	Screening Score	int32
19	Class/ASD	String (Yes, No)

1104 records. All the datasets mentioned above have the following 19 attributes common among them, which we have used for our model (Table 1).

3.2 Pre-processing the Data

The process of modifying and cleaning the raw data into feasible format for training the model is called data pre-processing. Eclectic pre-processing techniques are used to handle missing and inconsistent data. The data contains multiple columns of categorical features like ethnicity which have encoded. However, the use of Light GBM eliminates the necessity because it automatically encodes the categorical attributes. The null and missing values were imputed to increase the feasibility of the record. Dataset Standardization has been done using Standard Scalar from Scikit-Learn library.

3.3 Training and Testing of the Proposed Model

The dataset is split into a couple of parts in ratio of 80:20, one segment is for training the model, and the another is for testing data to check the model performance. Further twenty percent of the training records are used as validation data to avoid overfitting (Fig. 1).

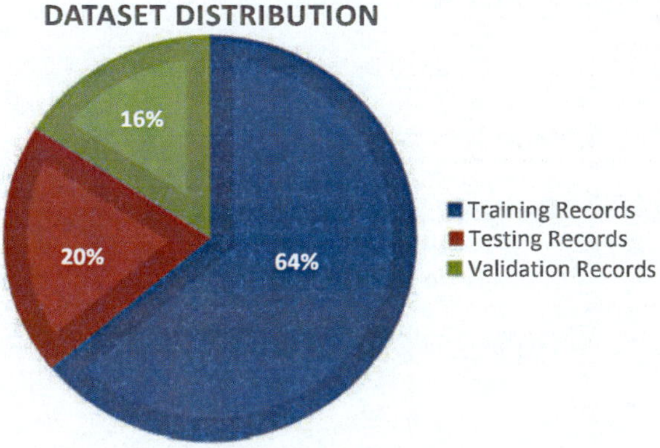

Fig. 1 Dataset Distribution

3.4 Random Search Optimization

Hyperparameter tuning is a method which is used to find the most suitable parameters of the model architecture from the available space. In simple terms, is nothing but searching for the right hyperparameters to find high precision and accuracy. Random search is a technique where random combinations of the hyperparameters are used to find the right combination for the model built. In each search pattern, random combination of parameters is considered in every iteration. Then it checks if these are the right fit for the model and the data involved, if so, the search ends otherwise it continues to another random set of parameters. Random search is a group of statistical optimization procedures that lacks the requirement of the gradient of the problem to be optimized, and Random Search can hence be used on functions that are not continuous or differentiable.

The advantage of hyperparameter tuning is that by finding the ideal combination of hyperparameters for the model we can control the performance of the model thereby reducing the loss and improving accuracy. Random search unearths the best possible values of the parameters by efficiently looking at a larger, less outcome-oriented search area in comparison to other techniques. The aim of a ML algorithm ml is to identify a mathematical relation (function) $f(n)$ that limits loss $l(n; f(n))$ over the trials n from a ground truth gaussian distribution t_n [12]. A ML algorithm ml is a relation that matches a data point $p^{(train\ data)}$ (a finite array of experiments from t_n) to a function $f(n)$ [12]. Still, the ML algorithm itself have some features named hyper-parameters λ_i, and the final algorithm for the model is available strictly after choosing λ_i, which can be denoted ml_λ, and $f(n) = ml_\lambda\ (p^{(train\ data)})$ for the training records $p^{(train\ data)}$ [12]. Random search has the benefit of finding the optimal parameters because of the randomness of the process.

```
PROCEDURE random search():

        Find x* which optimizes the function f

        Initialize x,f0

        Repeat

          Generate z, a randomly approximated point over S;

          if f(x+z)>max(f(x)) then x:=x+z;

                x*← x+z;

          RETURN(x*)

END random search;
```

3.5 Light Gradient Boosting Machine

Light GBM is an ensemble learning technique which trains a Gradient Boosted Decision Tree (GBDT), by default. LGBM works for classification and regression tasks. Microsoft is credited with the first stable release of LGBM. LightGBM is a descent-based boosting technique that employs tree-oriented learning techniques. It is developed to be distributed and efficacious as compared to other boosting algorithms. However, GBDT contains no innate set of weights, and thus the sampling methods recommended for other boosting algorithms cannot be automatically used, here the advantage of gradient sampling is observed.

The framework is fast, light and designed for distributed training. It also provides the facility of handling missing values and categorical feature support. LGBM handles categorical labels by using the data of column. It does not transform to one-hot encoding rather LGBM uses a unique technique to find the division value of categorical labels, which is quicker. Light GBM makes use of leaf-wise tree growth technique, in comparsion with depth-wise growth, this helps in faster converging [14].

The following are the advantages [14]:

- Accelerated training and superior efficacy.
- Lesser memory required for operation.
- Better performance.
- Capable of handling large-scale data.
- Enables Graphics Processing Unit (GPU) training

GBDT uses decision trees to learn a function from the input space X s to the gradient space *G* [15]. Suppose that we have a training set with n instances {x1, ... , xn}, where each xi is a vector with dimension s in space Xs . In each iteration of gradient boosting, the negative gradients of the loss function with respect to the output of the model are denoted as {g1, ... , gn}. The decision tree model splits each node at the most informative feature (with the largest information gain) [13]. For GBDT, the information gain is measured by the variance after splitting [13].

Hyperparameters which can be tuned for leaf-wise [13, 15] (LightGBM uses leaf-wise growth) Tree are

- learning_rate
- num_leaves: The value should be less than 2^{max_depth}
- max_bin: Controls the limit of bins that features will be bucketed into.

Light GBM Algorithm:
For several boosting rounds M and a differentiable loss function L:

Let $F_0(x) = arg_{\Upsilon} \min \sum_{i-1}^{n} L(y_i, \Upsilon)$

For m = 1 to M:

1. Calculate the *pseudo* residuals

$$r_m = \frac{(-\partial L(yi, Fm - 1(xi)))}{(\partial Fm - 1(xi))}$$

2. Fit decision tree $h_m(x)$ to r_{im}
3. Compute the step multiplier Υ_m for each leaf of $h_m(x)$
4. Let $F_m(x) = F_{m-1}(x) + \lambda_m \Upsilon_m h_m(x)$, where λ_m is the learning rate for iteration *m*.

3.6 Working Approach

This is a classic binary class classification problem, the classes are ASD and non-ASD. The dataset we pooled (ASD data of children, adolescents, and adults) contains 21 attributes that include test takers demographics and answers of the screening test which includes 10 questions (named A1 to A10) that test takers answered. In each of these 10 questions, the test takers were given a statement with which they had to agree or disagree. Dataset used for training consists of 1104 rows and 21 columns after pre-processing and removing less correlated features, the number of columns(features) was reduced to 19.

Pre-processing was done on this modified dataset, imputing missing values and scaling. Random search was used for hyperparameter tuning. The random search algorithm randomly searches the sample space (all possible parameter values of the Light GBM algorithm) in correspondence to the goal (find the best parameters for the data). Random search was implemented using the Sci-kit learn python library.

The best estimator of the random search is returned by the random search algorithm; these are the most appropriate hyperparameters pertaining to this task.

Light GBM is a gradient boosting ensemble method which focusses on accuracy of results and supports GPU Learning. The Light GBM also executed using Sci-kit learn. A GPU enabled version of LGBM Classifier was used which can also multiple categorical features thereby eliminating the need for encoding the values. For training of the classifier 64% of the data (earlier mentioned split) was used. The testing was done using the test set and performance metrics accuracy, precision and f1-score were used to evaluate the model. The efficiency of training the model determines the implementation result.

4 Flowchart

The flow we represented here in flowchart is started with an algorithm of importing all the packages required for building the model. These include NumPy, Pandas, Matplotlib, SkLearn, Seaborn, etc. Then the data is imported using the pandas and pre-processing is done by imputing the null values. Afterwards the data is split into train set and test set using SkLearn in 80:20 ratio, validation set is formed from the train set. Then Random search is used for finding hyperparameters of Light GBM suitable for the problem. The best set is returned for the algorithm and the training is done using the train set. Evaluation measures are used to assess the performance of the model (Fig. 2).

5 Results

The image shows a brief overview of the Autistic Spectrum Disorder Screening Data for Adult [9] (Fig. 3).

The image shows a displot (shows a histogram with a line on it) which is plotted using Seaborn library distplot plots a univariate distribution of observations (Fig. 4).

The above image is heatmap of the dataset. Heatmap is powerful way to visualize relationships between variables in high dimensional space (Fig. 5).

Bar Graph visualization of ethnicity of the subject and immediate adult relatives of the subject with ASD (Fig. 6).

Bar Graph visualization of subjects of white ethnic group and their Native Country (Fig. 7).

The image shows the classification report of the classification model, it shows well-known metrics used to calculate the effectiveness of the classification: precision, recall and f1-score for each class. The metrics are computed by using true positives (TP), true negatives (TN), false positive (FP), false negatives (FN) (Fig. 8).

The result was assessed in metrics of accuracy, precision and recall by using classification report. Classification report automatically presents the precision, recall

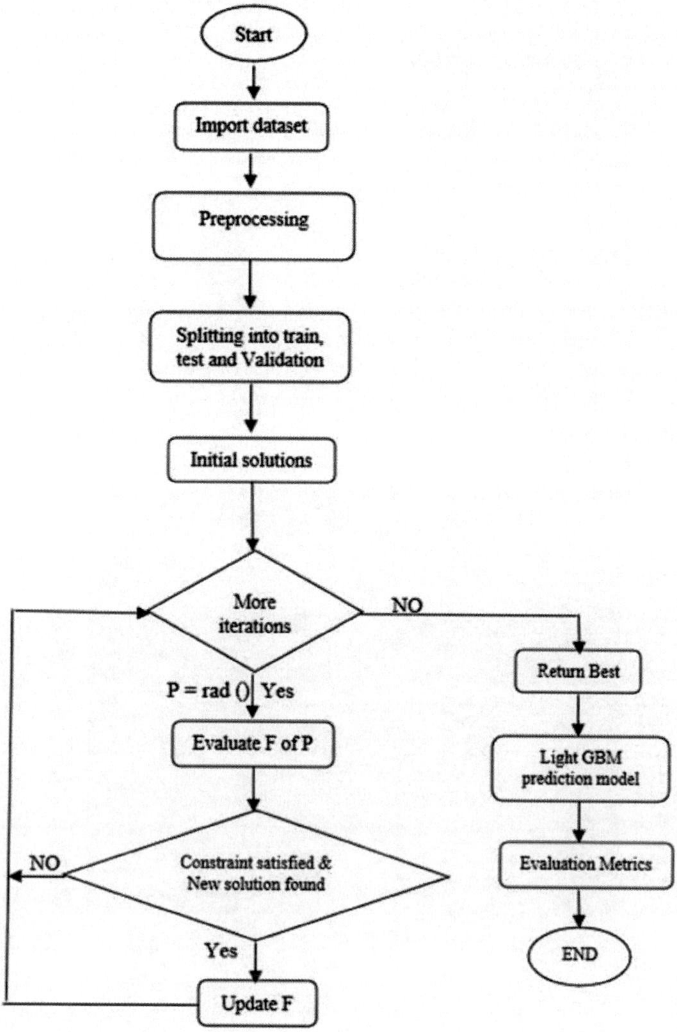

Fig. 2 Schematic of proposed work

	A1_Score	A2_Score	A3_Score	A4_Score	A5_Score	A6_Score	A7_Score	A8_Score	A9_Score	A10_Score	age	gender	ethnicity	jundice	austim	contry_of_res	result	relation	Class/ASD
0	1	1	1	1	0	0	1	1	0	0	26	0	White-European	0	0	United States	6	Self	0
1	1	1	0	1	0	0	0	1	0	1	24	1	Latino	0	1	Brazil	5	Self	0
2	1	1	0	1	1	0	1	1	1	1	27	1	Latino	1	1	Spain	8	Parent	1
3	1	1	0	1	0	0	1	1	0	1	35	0	White-European	0	1	United States	6	Self	0
4	1	0	0	0	0	0	0	1	0	0	40	0	White-European	0	0	Egypt	2	Self	0
...	...	...	...	...	...	...	...	...	...	...	...	...	...	...	...	...	...	...	...
699	0	1	0	1	1	0	1	1	1	1	25	0	White-European	0	0	Russia	7	Self	1
700	1	0	0	0	0	0	0	1	0	1	34	1	Hispanic	0	0	Mexico	3	Parent	0

Fig. 3 Snapshot of a Dataset

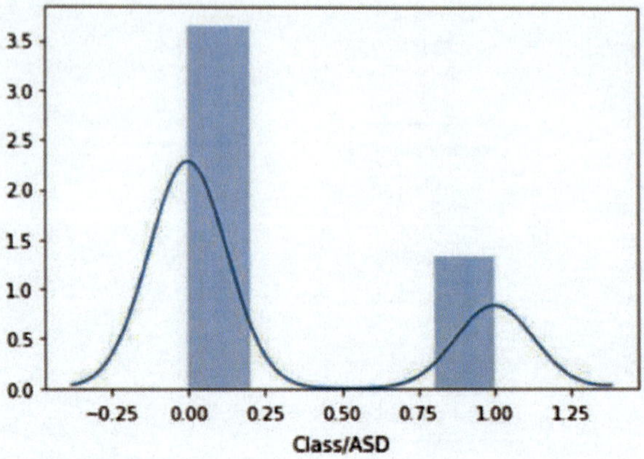

Fig. 4 Distplot of Class/ASD

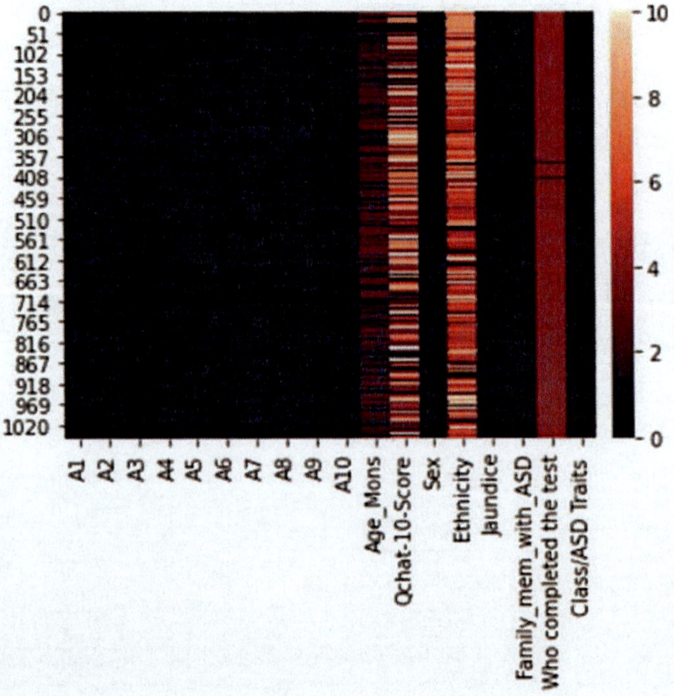

Fig. 5 HeatMap visualization of the dataset

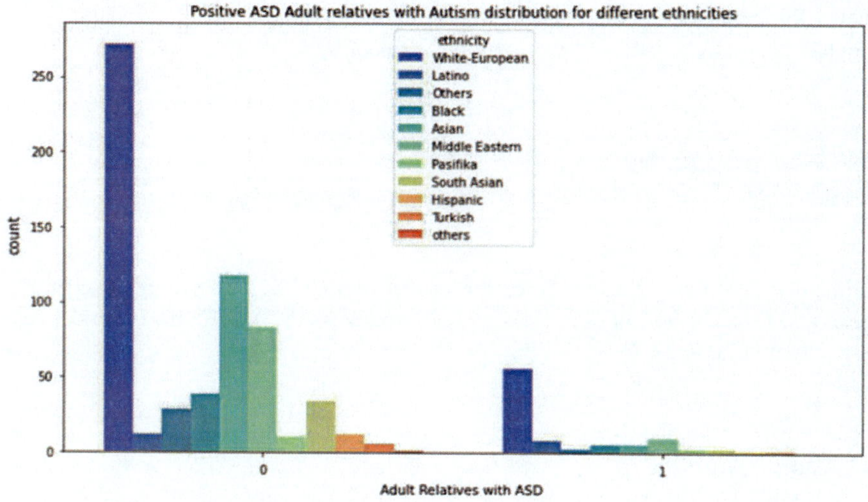

Fig. 6 Positive ASD Adult relatives vs different ethnicities

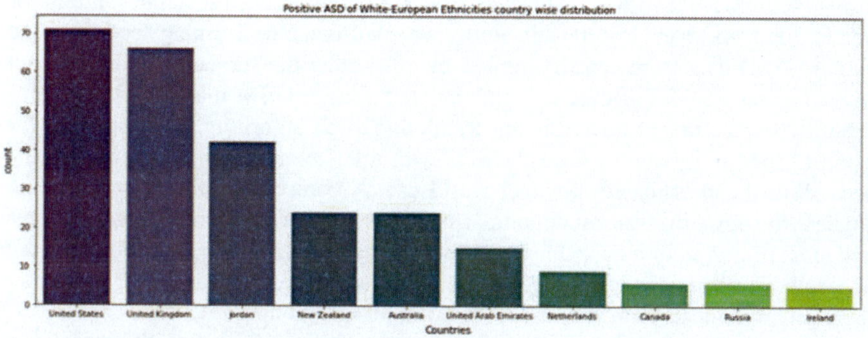

Fig. 7 Positive ASD in subjects belonging to white ethnicity versus Country of residence

Fig. 8 Classification report of the model

	precision	recall	f1-score
0	0.94	1.00	0.97
1	1.00	0.81	0.90
accuracy			0.95
macro avg	0.97	0.91	0.93
weighted avg	0.96	0.95	0.95

and f1-score of the experimentation with the help of confusion matrix. The formulae for the above-mentioned metrics can be seen in the below given equations. The best fit hyperparameters (learning_rate = 0.05,num_leaves = 11, n_jobs = 3) returned by random search were used for the LGBM Classifier which was trained using 300

Table 2 Elements of a Confusion Matrix

Original data values	Prognostic data values	
	TP	FP
	FN	TN
Accuracy $= \frac{TP+TN}{(TN+TP+FP+FN)}$	Precision $= \frac{TP}{TP+FP}$	
Recall $= \frac{TP}{TP+FN}$	f-1 Score $= 2 \times \frac{Precision*Recall}{Precision+Recall}$	

estimators. The overall accuracy of the model is 95.82%. The average precision, recall and f1-score for both the classes (ASD and non-ASD) is 97%, 91% and 94%, respectively (Table 2).

6 Conclusion

Around One in fifty-four children are diagnosed with autism according to the reports from the Centre for Disease Control's Autism and Developmental Disabilities Monitoring (ADDM) Network [3]. This work focuses on providing a scalable solution for efficient autism screening. In this study, we used machine learning techniques to classify ASD data from a screening test as to whether the subject is autistic or not (ASD or non-ASD). For this, we implemented a Light GBM model with the algorithm hyperparameters tunned using Random Search algorithm. The rationale for using LGBM is it handles huge datasets with relative ease thereby improving the accessibility and efficiency for real-world use. A comparison with some existing studies [6] shows that our model outperforms even while using three fewer features and most importantly required less pre-processing. This work could be further used to create an automatic screening tool using a mobile application for several other ASD screening tests with subjects from different age groups.

References

1. Eisenberg, L.: Leo Kanner, MD 1894–1981. Am. J. Psychiatry **138**(8), 1122–1125 (1981)
2. https://www.ninds.nih.gov/Disorders/Patient-Caregiver-Education/Fact-Sheets/Autism-Spectrum-Disorder-Fact-Sheet
3. Baio, J., et al.: Prevalence of autism spectrum disorder among children aged 8 years—autism and developmental disabilities monitoring network, 11 sites, United States, 2014. MMWR Surveillance Summaries 67.6 (2018), p. 1
4. https://www.healthline.com/health/types-of-autism#autism-symptoms
5. Thabtah, F.: An accessible and efficient autism screening method for behavioural data and predictive analyses. Health Inf. J. **25**(4), 1739–1755 (2019)
6. Raj, S., Masood, S.: Analysis and detection of autism spectrum disorder using machine learning techniques. Procedia Comput. Sci. **167**, 994–1004 (2020)

7. Thabtah, F.: Autism spectrum disorder screening: machine learning adaptation and DSM-5 fulfillment. In: Proceedings of the 1st International Conference on Medical and Health Informatics (2017)
8. Mythili, M.S., Mohamed Shanavas, A.R.: A study on Autism spectrum disorders using classification techniques. Int. J. Soft Comput. Eng. **4.5**, 88–91 (2014)
9. Thabtah, F.F.: Autistic Spectrum Disorder Screening Data for Adult (2017). https://archive.ics.uci.edu/ml/machine-learningdatabases/00426/
10. Thabtah, F.F.: Autistic Spectrum Disorder Screening Data for Children (2017). https://archive.ics.uci.edu/ml/machine-learningdatabases/00419/
11. Thabtah, F.F.: Autistic Spectrum Disorder Screening Data for Adolescent (2017), https://archive.ics.uci.edu/ml/machine-learningdatabases/00420/
12. Bergstra, J., Bengio, Y.: Random search for hyper-parameter optimization. J. Mach. Learn. Res. **13**(1), 281–305 (2012)
13. Ke, G., et al.: Lightgbm: a highly efficient gradient boosting decision tree. Advances in Neural Information Processing Systems (2017)
14. https://lightgbm.readthedocs.io/en/latest/index.html
15. Friedman, Jerome, H.: Greedy function approximation: a gradient boosting machine. Annals Stat., 1189–1232 (2001)

Qualitative Analysis of Edible Oils Using Low Field ^{1}H NMR Spectroscopy and Multivariate Statistical Methods

J. Aswathy, Patel Surendra Singh, V Sai Krishna, Navjot Kumar, and P. C. Panchariya

Abstract Low field ^{1}H Nuclear Magnetic Resonance (LF NMR) Spectroscopy is an efficient tool to capture a research sample's content and purity. LF NMR is employed for the qualitative analysis of Edible oils that are available in Indian Market. Edible oils used for the study include Coconut, Groundnut, Olive, Mustard, Rice bran, and Soyabean oil. Principal Component Analysis (PCA) is used to build a model initially that could classify the oils based on their chemical composition. This model built using PCA could capture 96% of total variance in data. Linear Discriminant Analysis (LDA) was used to build a model that could classify with 100% accuracy. The results provide successful proof for detecting adulteration and further classification of edible oils using Low field ^{1}H NMR Spectroscopy in conjunction with Multivariate Statistical methods such as PCA and LDA.

Keywords Low field ^{1}H NMR Spectroscopy · Multivariate statistical methods · Principal Component Analysis (PCA) · Linear Discriminant Analysis (LDA) · Adulteration detection · Classification · Edible oils

1 Introduction

Edible vegetable oils, due to their high nutritional value and substantial health benefits, are an integral part of cuisine, both in domestic and food manufacturing sectors. Due to lesser profit margins of selling the edible oils in pure form, manufacturers of the oil resort to adulterating the pure oil with less expensive oils. Some of the less expensive oils include refined oils that are hazardous to health. Due to increased

J. Aswathy (✉) · P. Surendra Singh · V. Sai Krishna · N. Kumar · P. C. Panchariya
Academy of Scientific and Innovative Research (AcSIR), Ghaziabad 201002, India

CSIR-Central Electronics Engineering Research Institute, Pilani 333031, India

J. S. Raj et al. (eds.), *Intelligent Sustainable Systems*, Lecture Notes in Networks and Systems 213, https://doi.org/10.1007/978-981-16-2422-3_62

awareness from consumers, producers, and policymakers, food authentication has gained importance in the recent years [8].

Different analytical techniques are used to detect the presence of adulterants in edible oils. Some of the methods used include Near Infrared (NIR) spectroscopy [5], Mass spectroscopy, High Performance Liquid Chromatography (HPLC) [16], Synchronous fluorescence [15], Fourier Transform Infrared (FTIR) spectroscopy [24],Nuclear Magnetic Resonance (NMR) spectroscopy [23], Gas chromatography[2, 16], and Vibrational spectroscopy [6]. These analytical techniques have their own pros and cons based on their performance parameters such as sensitivity and their expenditure. Low field Nuclear Magnetic Resonance (LF NMR) spectroscopy forms a preferred option that is used in industry for food quality assessment, because of its low cost and rapid detection technique.

Research was carried out to assess the oil content, moisture levels as well as microstructure changes in french fries, using LF NMR spectroscopy. The study comments on the influence of fatty acid composition, on the behavior of oil absorption and surface morphology [25]. LF NMR as a powerful tool was used in the detection of adulteration of peanut oil. Quality changes that occur in peanut oil upon adulterating it with various oils were studied using LF NMR spectroscopy [27]. A comparative study between low field and high-field NMR was done to detect the adulteration of cold pressed rapeseed oil with refined oils, and to quantitatively detect adulteration percentage. Though high-field NMR spectra produced better classification results, LF NMR still produced competitive classification results in detecting the presence of adulterants [18].

Most of the work done in the analysis of edible oils are done on foreign oil brands and there is less work done in the analysis of edible oils belonging to Indian Brands. Hence the main aim of this study is to classify edible oils that are available in Indian Market and to detect the presence of adulterants in them using Low Field -Nuclear Magnetic Resonance spectroscopy (LF NMR). The oil samples used for study include Coconut, Groundnut, Olive, Mustard, Ricebran , and Soyabean oil. Multivariate Statistical data analysis techniques are applied on the acquired spectra for qualitative analysis of expensive oils adulterated with less expensive oils. For example, Mustard oil being adulterated with Rice bran oil, Virgin Olive oil being adulterated with Soyabean oil, etc.

Introduction section covers the issue of adulteration and the need for detecting adulteration using various techniques. NMR analysis section defines the molecular process involved and the advantages of LF NMR spectroscopy along with literature review. Materials and methods section provide insight into the sample preparation, processing, and acquisition of LF NMR spectra. Results of the experiment by applying Principal Component Analysis (PCA) and Linear Discriminant Analysis (LDA) and their optimization for better results are discussed in the results and discussion section.

2 NMR Analysis

NMR spectroscopy helps in determining the chemical composition of a sample under study by obtaining the overall structure of the molecule. It does so by examining the nucleus of the molecule that we wish to study by placing it in a magnetic field. The nucleus which is dense and positively charged when subjected to a strong magnetic field, the protons in them start spinning.

Different nuclei have different spins and those nuclei with spin equals $\frac{1}{2}$ can exist in one of the two possible states. This nature is possessed by the following isotopes: ^{1}H, ^{13}C, ^{15}N, ^{19}F, ^{31}P. The hydrogen nuclei are the most commonly found nuclei in all organic compounds and thus the most frequently analyzed nuclei. Hence the name ^{1}H NMR Spectroscopy.

Let us consider the magnetic field B_0 is applied along the *z axis* at equilibrium. A short Radio Frequency (RF) pulse is applied at right angle to the magnetic field. This RF pulse flips down the protons to *xy plane* and continue to precess. This process continues until RF pulse is present and the protons comeback to original position upon stopping the RF pulse. This process is known as Free Induction Decay (FID). The precession happens at a particular frequency known as Larmor frequency or precessional frequency described by the equation,

$$\nu = \frac{\gamma B_0}{2\pi} \tag{1}$$

where

- ν refers to Larmor or precessional frequency
- γ refers to gyromagnetic ratio, constant for a given isotope
- B_0 refers to magnetic field strength

Based on the electron density distribution, the electron creates its own magnetic field which opposes the applied magnetic field B_0. A proton with higher electron density around it experiences a lower magnetic field and hence lower Larmor frequency and vice versa. This flipping down of magnetization vector from *z axis* to *xy plane* and continued precession induces a current in the receiver coil. This signal when detected and amplified results in a sine wave of decreasing magnitude, explaining the re-alignment of protons in the nucleus to *z axis*. This process is known as Free Induction Decay (FID). The FID is representation in time domain. On application of Fourier Transform, the time domain is converted to frequency domain and results in NMR spectrum. The Fig. 1 shows an example of the ^{1}H LF NMR Spectra obtained from NMReady-60 Pro (Nanalysis, Canada), for six different types of edible oils used in study.

To describe the returning back of H proton from excited magnetic state to equilibrium, the terms longitudinal relaxation time (T_1) and transverse relaxation time (T_2) are used.T_1 relaxation, also known as relaxation of the spin-lattice, involves the transfer of energy in between excited nuclear spins and lattice. It results in decrease

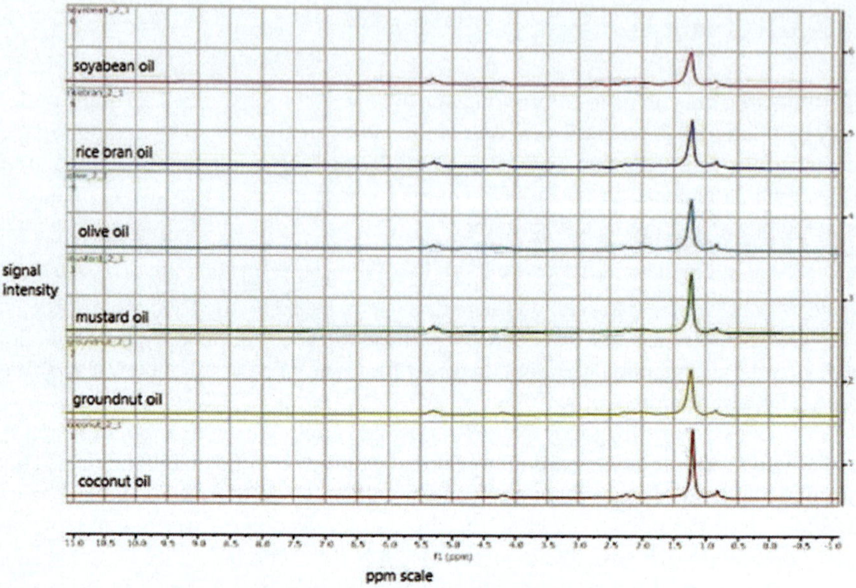

Fig. 1 ^{1}H LF NMR Spectra of edible oils used in the study

of overall energy. Spin-spin relaxation, otherwise known as (T_2) relaxation, refers to the transactions amid nuclear magnetic moments (spins) [7]

Nuclear Magnetic Resonance (NMR) Spectroscopy is useful technique to analyze food matrices, which are quite complex in nature. It is a non-destructive and rapid method that can be used for sensing frauds of foods and drinks [14]. In a single step, it can provide details on a broad range of features and has fast data acquisition rate [11]. Nuclear Magnetic Resonance has emerged as an effective tool for qualitatively detecting the adulterants present in the oil samples. Based on the chemical environment, NMR helps to decide the molecular structure through ppm scale resonance frequency shifts [22]. NMR in combination with chemometrics is a useful tool in controlling the quality of food products such as edible oils, milk, honey, and paprika powder [1, 3, 11, 20].

2.1 *Low Field NMR Spectroscopy*

However, NMR spectroscopy instrumentation requires high-field homogeneous magnetic fields [12] with superconducting magnets that require cryogenic cooling for probes, which makes them quite expensive [19]. Low field Nuclear Magnetic Resonance (LF NMR) makes use of magnetic fields of lower strength to capture the spectra. But they do have an intrinsic drawback of limited sensitivity due to minor

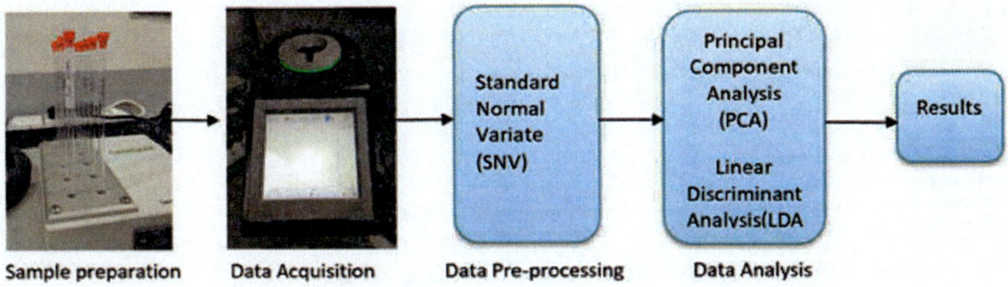

Fig. 2 A Complete workflow of the steps involved

variations in thermodynamic equilibrium concentrations of the spin states [4]. The low sensitivity can cause measurement of small shifts in resonance frequency difficult. Gradient-based solvent suppression is one of the methods applied to overcome sensitivity issue [9]. Shim pulses are used in LF NMR spectrometers to overcome the inhomogeneity in permanent magnet [22]. Thus it makes LF NMR Spectroscopy suitable for routine analysis that is low cost and suitable in industrial settings. They provide smaller footprint when compared with High-field NMR spectroscopy. The benchtop spectrometer that is used for the study is cryogen free, having a permanent magnet of strength 1.4T, from NMReady-60 Pro (Nanalysis, Canada). The standard NMR tubes used has a dimension of 5 mm.

3 Materials and Methods

3.1 Work Flow

Preparation of the sample is the foremost step involved in performing experimentation. The next step is acquiring the data using Nanalysis Benchtop LF NMR Instrument, followed by pre-processing the acquired data to remove various artifacts. The final step involved is analyzing the pre-processed data by applying algorithms and arriving at the results. A complete workflow involved in performing experiment and arriving at the results using blockdiagram approach is described as shown in the Fig. 2

3.2 Sample Preparation and Processing

The oil samples used for the study include Coconut, Groundnut, Olive, Mustard, Ricebran, and Soyabean oil. The oils were purchased from IICT Hyderabad and stored at room temperature. The samples were diluted using deuterated standard chloroform ($CDCl_3$) and a standard solution of deuterated Dimethyl Sulfoxide (DMSO d6). The purpose of using solvents is to produce a more defined spectrum (narrow line width) when compared with using pure oil samples as such for testing. A fraction of 480 μl of oil, succeeded by 120 μl of $CDCl_3$, was titrated directly into Nanalysis standard 5 mm tubes.

3.3 Acquisition of LF ^1H NMR Spectra

Proton 1D spectra were recorded at 20° C using NMReady-60 Pro (Nanalysis, Canada) Benchtop Spectrometer operating at 60 MHz. The spectral acquisition parameters include Time domain 65536 points, 16 scans, spectral width 12 ppm, acquisition time 190 s. All the files were stored in a separate folder in "comma separated values" (.csv). The analysis of data was performed using The Unscarmbler Software (Version 10.5, CAMO Software Inc.,).

4 Results and Discussion

4.1 Pre-processing Methods

Need for Pre-processing The pre-processing techniques are used to remove scattering and other distrubances in the spectra captured. Other disturbances include presence of outliers, difference due to measurement geometry, etc. [10]. The resultant spectrum after applying pre-processing technique can be used for building models with better classification ability. Since carrying out the analysis using, the Unscarmbler software was difficult with time domain 65536 data points; the initial step was to extrapolate to time domain 4096 data points.Figure 3 shows the spectra obtained before pre-processing on extrapolated data points using The Unscrambler Software.

It is preferred to have a NMR spectrum with consistent baseline of 0. But from the raw spectrum it could be inferred that spectral baseline is not to the scale of 0. Hence baseline correction need to be performed. If it is not performed then upon integration data shows non-uniform signal area and may be misintepreted as variation corresponding to the concentration of sample under study.

SNV The Standard Normal Variate (SNV) is applied as pre-processing technique to the extrapolated data points to perform baseline correction. In SNV, each spectrum is transformed by subtracting the mean of the spectrum from corresponding individual spectrum and dividing by the standard deviation of the spectrum. It helps in removing the multiplicative interferences of light scatter and particle size [17] and brings all spectra to the same scale [10]. The SNV is calculated as follows:

$$\bar{x_i} = \frac{\sum_{j=1}^{m} x_{ij}}{m} \quad (2)$$

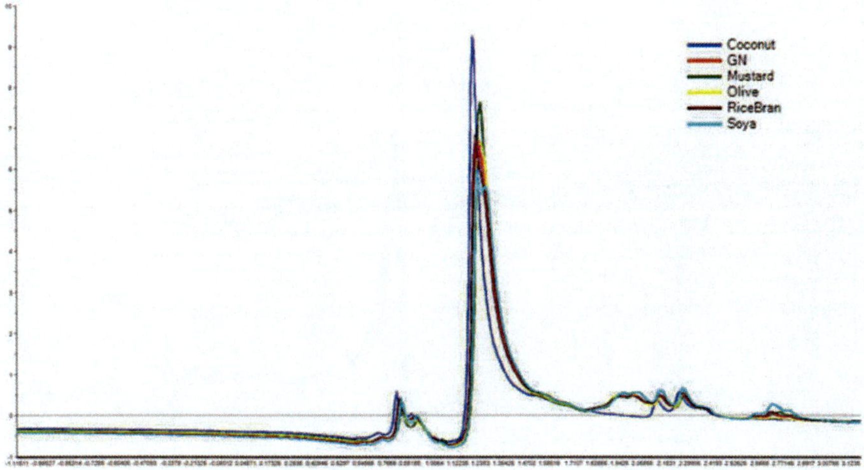

Fig. 3 Raw spectra before pre-processing

$$\sigma = \sqrt{\frac{\sum_{j=1}^{m}(x_{ij} - \bar{x_i})^2}{m - 1}} \tag{3}$$

$$x_{ij}(SNV) = \frac{x_{ij} - \bar{x_i}}{\sigma}, \tag{4}$$

where

- $\bar{x_i}$ refers to the mean spectrum,
- σ refers to the spectrum standard deviation,
- m refers to the number of wavelengths,
- $x_{ij}(SNV)$ refers to the corrected reflectance for sample i.

Figure 4 shows spectra obtained after applying SNV pre-processing method on raw extrapolated spectra. On comparison of Fig. 3 with Fig. 4, it could be observed that multiplicative interference of light and offset is removed.

4.2 *Principal Component Analysis (PCA)*

PCA is a multivariate statistical method. It modifies data from higher dimension to lower dimension. It uses orthogonal transformation to project the data to a new coordinate system. The first axis of the new coordinate captures maximum variance called First Principal Component (PC-1), followed by second axis called Second Principal Component (PC-2) capturing the second highest amount of variance [26] and so on. In doing so it helps in capturing the similarities and differences in data which are

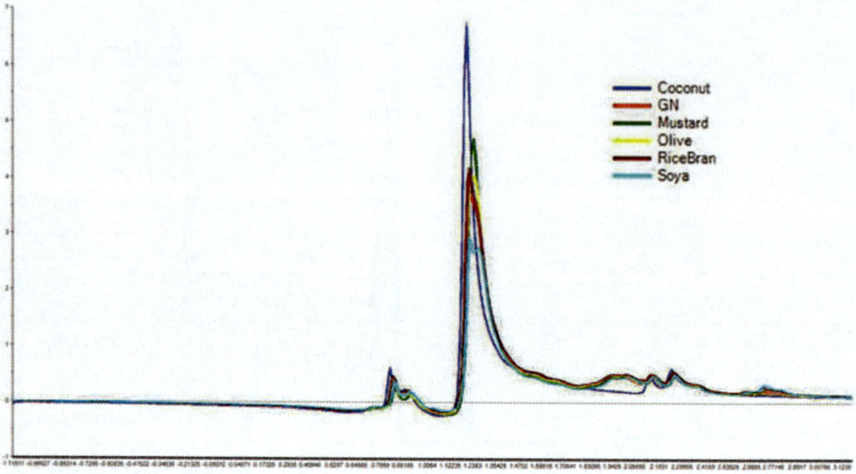

Fig. 4 After applying SNV

otherwise difficult to observe in higher dimensions. Also the transformation happens without much loss of data, which is an added advantage. PCA finds application in various fields such as image compression and face recognition.

Mathematically PCA finds the eigenvectors and eigenvalues from the given data. For a p dimensional data there will be p eigenvectors and p eigenvalues. The eigenvectors and eigenvalues are also known as scores and loadings respectively. The eigenvalues are arranged in descending order. Correspondingly eigenvectors are also arranged to get the components of order of significance. The lesser significant component can be ignored to produce a new set of coordinate system with lesser dimension than the original higher dimension.

PCA was applied to the extrapolated and pre-processed data. All the loadings were taken into account. The resulting PC-1 could capture 91% of the variance in data and PC-2 could capture 6% of the variance in data. Figure 5 shows PCA plot obtained from extrapolated and pre-processed data, which captures a total of 97% of variance from the two principal components.

It could be observed from Fig. 5 that classification of various types of edible oils is achieved with the help of PCA. But the samples representing edible oils belonging the same group are not close enough. Hence in order to improve the grouping of data points belonging to the same group, optimization of the loading variables is done. This optimization is achieved by neglecting the loading variable corresponding to the least significant eigenvalue. Thus it results in a new coordinate system with number of axes one less than the original.

The result of optimization of loading variables in terms of grouping is well evident from Fig. 6, when compared with Fig. 5, where the performance of PCA in grouping similar data is less. Each separated group represented samples belonging to specific kind of edible oils such as Coconut, Groundnut, Mustard, Olive, Ricebran, and Soy-

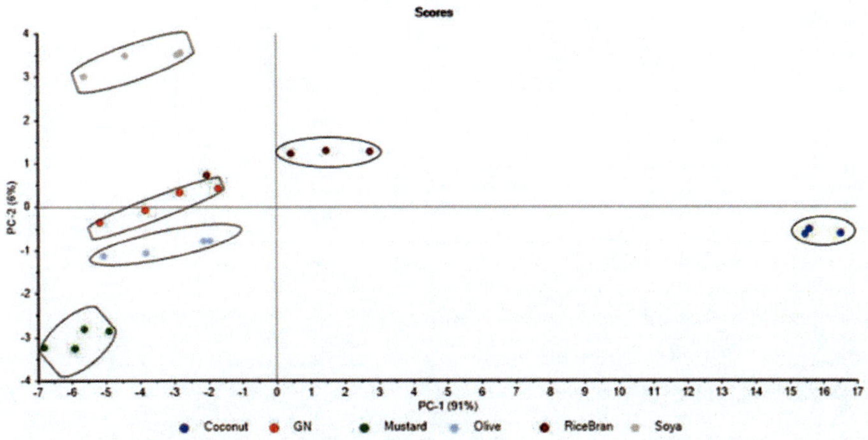

Fig. 5 PCA applied on extrapolated pre-processed data

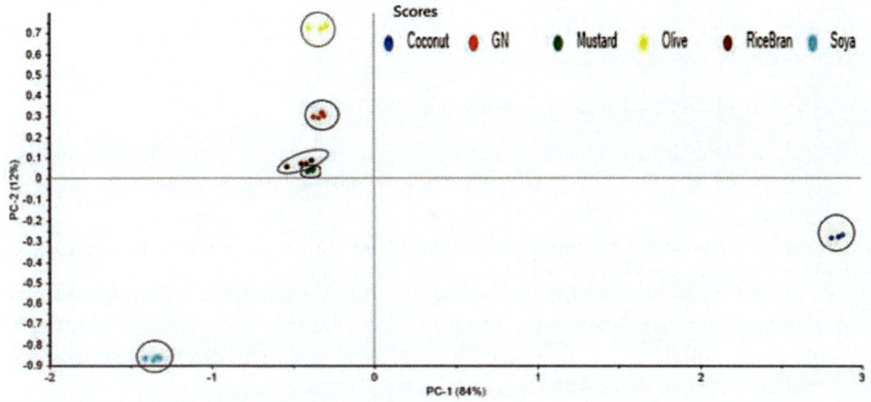

Fig. 6 PCA plot on optimized loading variables

abean oil. Thus PCA classifies and helps in qualitative analysis of different edible oils.

4.3 *Linear Discriminant Analysis (LDA)*

Linear Discriminant Analysis is a supervised classification method. It works on the principle of maximizing the separation among the classes. It projects features from higher dimension to lower dimension[13]. LDA assumes that our data follows Guassian distribution and it uses two criteria for classification:

– Maximize the distance between the mean of two classes.

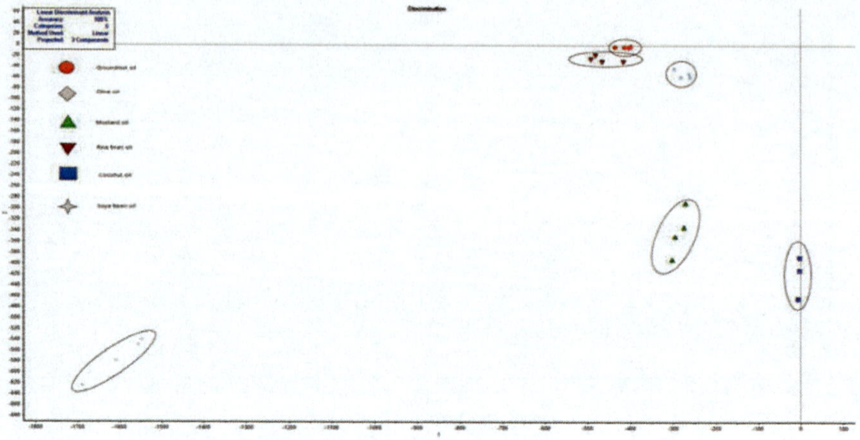

Fig. 7 LDA applied on extrapolated pre-processed data

– Minimize the within class variance.

To realize this LDA performs the following three steps:

– Calculate the distance between mean of different classes or between class matrix.
– Calculate the distance between mean and samples present in each class or within class matrix.
– Project to a lower dimension which satisfies the above said criteria of LDA. [21]

On the pre-processed and extrapolated data LDA was applied. The model was able to predict with 100% accuracy. Figure 7 shows the result of LDA model fitted to a extrapolated and SNV pre-processed data. LDA was able to classify the data into six classes and could project in terms of three principal components.

Based on the fitting obtained, a confusion matrix was drawn which helps to identify the performance measurement of classification. Table 1 shows the effectiveness of model through confusion matrix.

Recall: It refers to the total number of correctly predicted classes out of all positive classes.

$$Recall = \frac{TP}{TP + FN} = 1 \tag{5}$$

Precision: Out of correctly predicted positive classes, precision refers to how many are actually positive.

$$Precision = \frac{TP}{TP + FP} = 1 \tag{6}$$

Table 1 Confusion Matrix of the result obtained from LDA

Confusion Matrix	PREDICTED CLASS					
	Soyabean oil	Ricebran oil	Olive oil	Mustard oil	Groundnut oil	Coconut oil
Actual class						
Soyabean oil	4	0	0	0	0	0
Ricebran oil	0	4	0	0	0	0
Olive oil	0	0	4	0	0	0
Mustard oil	0	0	0	4	0	0
Groundnut oil	0	0	0	0	4	0
Coconut oil	0	0	0	0	0	4

F-measure: It is a parameter to make two models cmparable.

$$F - measure = \frac{2 * Recall * Precision}{Recall + Precision} = 1 \quad (7)$$

Accuracy:

$$Accuracy = \frac{TP + TN}{TP + TN + FP + FN} = 1 \quad (8)$$

5 Conclusion and Future Scope

This study provides proof that Low field ^{1}H NMR Spectroscopy in combination with Multivariate Statistical methods such as PCA and LDA can be used to classify different types of food products such as Edible oils present in the Indian market and thereby paves way for detecting adulteration. This study also gives an insight that low field NMR spectroscopy also provides competitive results, when compared with High-field NMR spectrometers, when factors such as size and cost of equipment are considered. The PCA model built by optimizing the loading variables could capture 96% of the total variations in data and LDA model could predict with 100% accuracy. Quantitative analysis of edible oils to detect the percentage of adulterants present in each edible oil can be the future scope of the study.

Acknowledgements We thank our Director for his continuous support and encouragement. We also thank the Signal Analytics team consisting of the members of Central Electronics Engineering Research Institute (CSIR-CEERI) for their excellent support in carrying out the experimental work. The work is supported by a grant from CSIR Mission Mode project HCP-0016.

References

1. Agiomyrgianaki, A., Petrakis, P.V., Dais, P.: Detection of refined olive oil adulteration with refined hazelnut oil by employing nmr spectroscopy and multivariate statistical analysis. Talanta **80**(5), 2165–2171 (2010)
2. Andrikopoulos, N., Giannakis, I., Tzamtzis, V.: Analysis of olive oil and seed oil triglycerides by capillary gas chromatography as a tool for the detection of the adulteration of olive oil. Journal of chromatographic science **39**(4), 137–145 (2001)
3. Bergana, M.M., Adams, K.M., Harnly, J., Moore, J.C., Xie, Z.: Non-targeted detection of milk powder adulteration by 1h nmr spectroscopy and conformity index analysis. Journal of Food Composition and Analysis **78**, 49–58 (2019)
4. Blümich, B., Casanova, F., Appelt, S.: Nmr at low magnetic fields. Chemical Physics Letters **477**(4–6), 231–240 (2009)
5. Christy, A.A., Kasemsumran, S., Du, Y., OZAKI, Y.: The detection and quantification of adulteration in olive oil by near-infrared spectroscopy and chemometrics. Analytical Sciences **20**(6), 935–940 (2004)
6. Domingo, E., Tirelli, A.A., Nunes, C.A., Guerreiro, M.C., Pinto, S.M.: Melamine detection in milk using vibrational spectroscopy and chemometrics analysis: A review. Food Research International **60**, 131–139 (2014)
7. Dybowski, C.: Practical nmr relaxation for chemists by vladimir i. bakhmutov (texas a&m university). john wiley & sons, ltd: Chichester, uk 2004. xii+ 202 pp. isbn 0-470-09445-1. (2005)
8. Fadzillah, N.A., bin Che Man, Y., Rohman, A., Rosman, A.S., Ismail, A., Mustafa, S., et al.: Detection of butter adulteration with lard by employing 1h-nmr spectroscopy and multivariate data analysis. Journal of Oleo Science p. ess14255 (2015)
9. Gouilleux, B., Charrier, B., Akoka, S., Giraudeau, P.: Gradient-based solvent suppression methods on a benchtop spectrometer. Magnetic Resonance in Chemistry **55**(2), 91–98 (2017)
10. Grisanti, E., Totska, M., Huber, S., Krick Calderon, C., Hohmann, M., Lingenfelser, D., Otto, M.: Dynamic localized snv, peak snv, and partial peak snv: Novel standardization methods for preprocessing of spectroscopic data used in predictive modeling. Journal of Spectroscopy **2018**, (2018)
11. Hatzakis, E.: Nuclear magnetic resonance (nmr) spectroscopy in food science: A comprehensive review. Comprehensive Reviews in Food Science and Food Safety **18**(1), 189–220 (2019)
12. Hills, B.: Applications of low-field nmr to food science. In: Annual reports on NMR spectroscopy, vol. 58, pp. 177–230. Elsevier (2006)
13. Hou, X., Wang, G., Wang, X., Ge, X., Fan, Y., Jiang, R., Nie, S.: Rapid screening for hazelnut oil and high-oleic sunflower oil in extra virgin olive oil using low-field nuclear magnetic resonance relaxometry and machine learning. Journal of the Science of Food and Agriculture (2020)
14. Hu, Y., Wang, S., Wang, S., Lu, X.: Application of nuclear magnetic resonance spectroscopy in food adulteration determination: the example of sudan dye i in paprika powder. Scientific reports **7**(1), 1–9 (2017)
15. Huyan, Z., Ding, S., Liu, X., Yu, X.: Authentication and adulteration detection of peanut oils of three flavor types using synchronous fluorescence spectroscopy. Analytical Methods **10**(26), 3207–3214 (2018)
16. Jabeur, H., Zribi, A., Makni, J., Rebai, A., Abdelhedi, R., Bouaziz, M.: Detection of chemlali extra-virgin olive oil adulteration mixed with soybean oil, corn oil, and sunflower oil by using gc and hplc. Journal of Agricultural and Food Chemistry **62**(21), 4893–4904 (2014)
17. Liu, Y., Liu, Y., Chen, Y., Zhang, Y., Shi, T., Wang, J., Hong, Y., Fei, T.: The influence of spectral pretreatment on the selection of representative calibration samples for soil organic matter estimation using vis-nir reflectance spectroscopy. Remote Sensing **11**(4), 450 (2019)
18. McDowell, D., Defernez, M., Kemsley, E.K., Elliott, C.T., Koidis, A.: Low vs high field 1h nmr spectroscopy for the detection of adulteration of cold pressed rapeseed oil with refined oils. LWT **111**, 490–499 (2019)

19. Parker, T., Limer, E., Watson, A., Defernez, M., Williamson, D., Kemsley, E.K.: 60 mhz 1h nmr spectroscopy for the analysis of edible oils. TrAC Trends in Analytical Chemistry **57**, 147–158 (2014)
20. Song, X., She, S., Xin, M., Chen, L., Li, Y., Vander Heyden, Y., Rogers, K.M., Chen, L.: Detection of adulteration in chinese monofloral honey using 1h nuclear magnetic resonance and chemometrics. Journal of Food Composition and Analysis **86**, 103390 (2020)
21. Tharwat, A., Gaber, T., Ibrahim, A., Hassanien, A.E.: Linear discriminant analysis: A detailed tutorial. AI communications **30**(2), 169–190 (2017)
22. Topgaard, D., Martin, R.W., Sakellariou, D., Meriles, C.A., Pines, A.: "shim pulses" for nmr spectroscopy and imaging. Proceedings of the National Academy of Sciences **101**(51), 17576–17581 (2004)
23. Wang, R., Liu, K., Wang, X., Tan, M.: Detection of sesame oil adulteration using low-field nuclear magnetic resonance and chemometrics. International Journal of Food Engineering **15**(7), (2019)
24. Wu, Z., Li, H., Tu, D.: Application of fourier transform infrared (ft-ir) spectroscopy combined with chemometrics for analysis of rapeseed oil adulterated with refining and purificating waste cooking oil. Food analytical methods **8**(10), 2581–2587 (2015)
25. Yang, D., Wu, G., Li, P., Qi, X., Zhang, H., Wang, X., Jin, Q.: The effect of fatty acid composition on the oil absorption behavior and surface morphology of fried potato sticks via lf-nmr, mri, and sem. Food Chemistry: X **7**, 100095 (2020)
26. Zhang, X., Qi, X., Zou, M., Liu, F.: Rapid authentication of olive oil by raman spectroscopy using principal component analysis. Analytical letters **44**(12), 2209–2220 (2011)
27. Zhu, W., Wang, X., Chen, L.: Rapid detection of peanut oil adulteration using low-field nuclear magnetic resonance and chemometrics. Food chemistry **216**, 268–274 (2017)

K-Means Algorithm: An Unsupervised Clustering Approach Using Various Similarity/Dissimilarity Measures

Surendra Singh Patel, Navjot Kumar, J. Aswathy, Sai Krishna Vaddadi, S. A. Akbar, and P. C. Panchariya

Abstract Clustering is an unsupervised method of classifying data objects into similar groups based on some features or properties usually known as similarity or dissimilarity measures. K-Means is one of the most popular clustering methods that come under the hard clustering group. In this clustering method, any data object can belong to a single cluster. On the other hand, in soft clustering methods (e.g., fuzzy c-means clustering), the data object can be clustered in more than one cluster with some degree which is specified by the membership value with the limitation imposed as the summation of these membership values should be equal to individual. While the clustering method of K-Means is a comparatively old technique, it still has tremendous popularity in terms of being used in applications for data grouping and machine learning. In this article, K-Means approach with five different distance measures such as Euclidean, Squared Euclidean, Half Squared Euclidean, Cosine, and City Block distance has been explored. A comparative study is made based on the performance of these similarity criteria on real-time Edible oil dataset acquired using MIR spectroscopy. In addition, it attempts to investigate the measure of similarity for a specific collection of unique patterns carrying data. In the MATLAB R2015b environment given by Mathworks, the K-Means algorithm with different similarity and dissimilarity measures were formulated and implemented.

Keywords Supervised/unsupervised learning · Clustering · Classification · K-Means clustering · Similarity and dissimilarity measures

1 Introduction

Clustering is an unsupervised form of classification that has similar characteristics and divides data samples into similar categories. It is possible to divide clustering into two subgroups, i.e., hard and soft clustering. Hard clustering means that a single data point may belong to a single community, while soft clustering enables data points in

S. S. Patel (✉) · N. Kumar · J. Aswathy · S. K. Vaddadi · S. A. Akbar · P. C. Panchariya
Academy of Scientific and Innovative Research (AcSIR), Ghaziabad, UP, India

CSIR-Central Electronics Engineering Research Institute, Pilani, Rajasthan, India

J. S. Raj et al. (eds.), *Intelligent Sustainable Systems*, Lecture Notes in Networks and Systems 213, https://doi.org/10.1007/978-981-16-2422-3_63

more than one cluster with a certain amount of data. Sensible grouping of data objects is to be considered an important part of data analysis to realize the underlying pattern of unlabeled data. Data clustering is the means of understanding the characteristics or features of data objects by assigning them into similar groups based on these features. The unsupervised methods of data mining are identified as a bit complex as compared to supervised methods. Some of the reasons could be identified as collecting the samples, labeling may surprisingly be an expensive approach, and in many data mining applications properties of the data might vary with time and subsequently the pattern may be altered specifically in food applications where the quality may vary due to change. Hence, these changes in the characteristics of the pattern can be tracked in unsupervised learning methods and eventually, improved performance can be observed [1]. Measuring the distance or similarity between two data objects is considered an essential step in data mining applications [2]. Without being affected by the fact that complexity in cluster analysis is greater as compared to classification, there is an enormous scope of clustering techniques in any type of multivariate data analysis that has been reported [3]. With the evolution of high-end storage and computational capability instruments and modern mathematical approaches, it has become possible to handle an unprecedented amount of data. The goal of data mining is to extract the dominant features of the data so that the underlying structure of the data can be identified and a natural cluster can be formed [4]. The methods like segmentation, classification, and clustering seem to be similar. Since classification is a supervised method of grouping the data by assigning them into previously defined groups, while the clustering methods split the data objects which are not previously defined, and segmentation on the other hand is the method of splitting the data objects based on similar characteristics [5]. Preprocessing is considered one of the major steps for extracting meaningful information from a huge amount of data from spectroscopic analytical instruments. Navjot et al. used the k-means algorithm for comparing the performance of various preprocessing techniques that are often used in multivariate data analysis [6]. When using the k-means algorithms, the selection of the initial cluster center is a sensitive step as it ultimately affects the liability of data points in the desired groupings. To understand the impact of initial cluster center assignment [7], four initialization methods have been studied. In addition to it, some optimization-based methods are also proposed to initialize the cluster center in [8] and refining initial starting conditions for the K-means algorithm to avoid convergence to numerous local minima [9]. Khan et al. proposed an algorithm to initialize the cluster center to make the convergence to a better local minimum [10]. There are several methods proposed in the literature to improve the efficacy of traditional k-means algorithms, e.g., fuzzy entropy to improve the performance of traditional k-means [11], hybrid genetic algorithm [12] [13]. Image segmentation is an important step in digital image processing. Clustering can also help in identifying the best segmentation technique to get better insights into image [14].

2 K-Means Clustering Algorithm

K-Means is one of the hard clustering methods of classification. It splits the whole data samples into similar groups based on their similarity measure. Euclidean distance-based similarity measure is the most commonly used method in these techniques. The basic algorithm has the following steps:

Step 1: Choose the initial cluster centroids $C_1, C_2 \ldots C_k$ (where k is the number of clusters) from the given data set $X_1, X_2, X_3 \ldots X_n$ either randomly or using any analytical method.

Step 2: For each point X_i, find the nearest centroid C_j

$$\arg\min_j D(x_i, c_j)$$

where D: Distance measure between the data point x_i and cluster center c_j.

- Assign the point X_i to the cluster j.

Step 3: Compute new cluster center as

$$C_j^* = \frac{1}{n_j} \sum_{x_i \to C_j} x_i$$

Step 4: If $C_j^* = C_j$ then the algorithm converges otherwise repeat from Step 2.

Step 5: If the algorithm does not converge in Step 4 then it indicates that the algorithm executed the maximum number of iterations.

The purpose of the calculating distance between any two points is to check whether these two-point are nearer to each other then it can be inferred that these two points are similar and vice versa. In the following section, the definition and mathematical expression of five different types of similarity/dissimilarity measures are presented.

2.1 *Euclidean Distance*

The Euclidean distance is also known as the L2 norm. The Euclidean distance between two points p1 and p2 having 'n' dimensions can be mathematically expressed as

$$D_{Eucl.} = \sqrt{\sum_{i=1}^{n} (p1_i - p2_i)^2}$$

2.2 City Block or Manhattan Distance

The city block distance is also termed as L1 norm and this is the special case of Minkowski (Lp) distance with p = 1. The distance can be calculated between two points p1 and p2 as

$$D_{city} = \sum_{i}^{n} |(p1_i - p2_i)|$$

2.3 Squared Euclidean Distance

In K-Means clustering, the results may not differ from the squared Euclidean distance, since in this distance metric the square root does not have to be calculated, the clustering process appears to be faster. The distance metric can be calculated with the following mathematical expression:

$$D_{Sq\_Eucl.} = \sum_{i=1}^{n} (p1_i - p2_i)^2$$

2.4 Half Squared Euclidean Distance

In half a square Euclidean distance, not everyone has to calculate a square root instead of a square Euclidean distance. So, mathematically, it can be expressed as follows:

$$D_{HSq\_Eucl.} = \frac{1}{2} \sum_{i=1}^{n} (p1_i - p2_i)^2$$

2.5 Cosine Similarity and Cosine Distance

Cosine similarity calculates the cosine angle between two points instead of distance calculation that can mathematically expressed as

$$\text{Cosine similarity} = \frac{\sum_i^n p1_i * p2_i}{\sqrt{\sum_i^n p1_i^2}\sqrt{\sum_i^n p2_i^2}}$$

And eventually,

$$\text{Cosine distance} = 1 - \text{Cosine similarity}$$

3 Results and Discussion

In this research, seven edible oils (Mustard, Ricebran, Sunflower, Soya bean, Groundnut, Cottonseed, and Olive oil) data are being used to compute the efficacy of various similarity/dissimilarity measures in the K-means algorithm. The edible oils data was acquired using PYREOS make Attenuated Total Reflection (ATR with ZnSe crystal surface) based MIR spectrophotometer in the spectral range of 1818–909 cm^{-1}. A raw MIR spectrum of seven different oils is depicted in Fig. 1.

The MIR spectra of oil data contain 128 variables, which is quite a large number in the view of computational complexity. Hence, Principal Component Analysis (PCA) is one of the most commonly used variable reduction techniques [15, 16] utilized to improve the computational cost when dealing with multivariate data. In addition to this PCA is commonly used to visualize the high dimensional data for further data

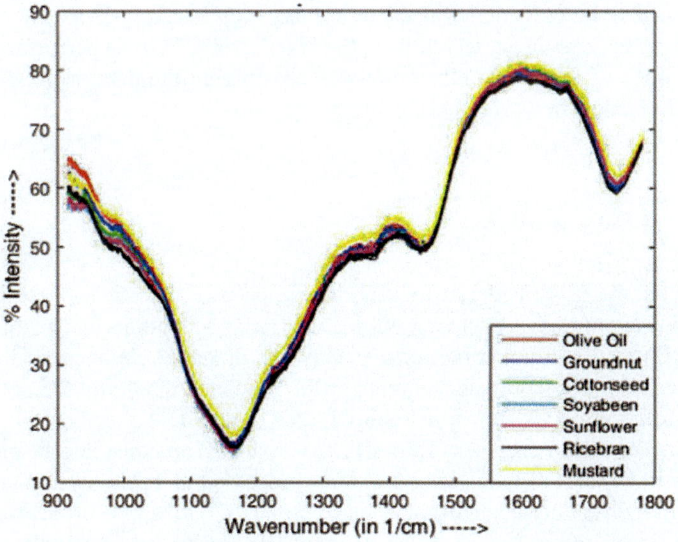

Fig. 1 Raw MIR spectra of different edible oils

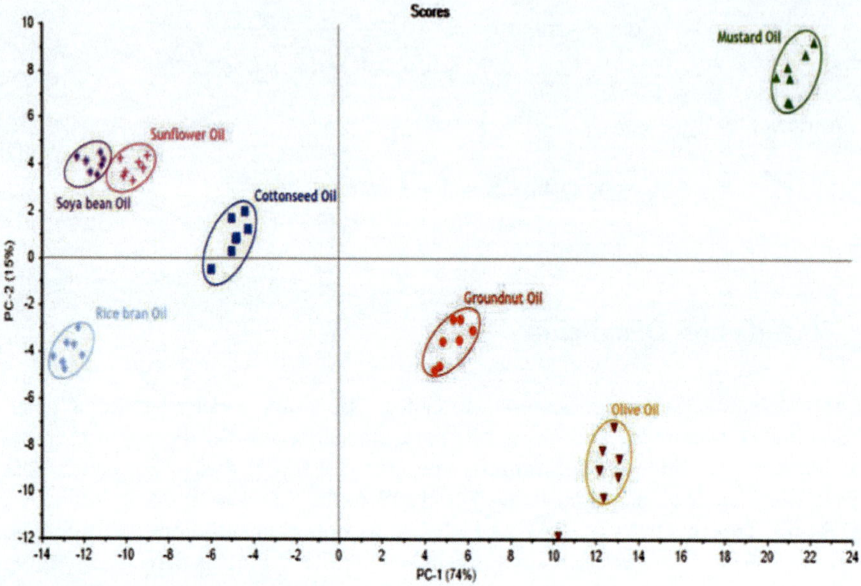

Fig. 2 PCA Plot of different edible oils

exploration [17]. The first seven principal components explaining approximately 99% of the information were used in the calculation of K-means to identify the efficient clusters. The PCA score plot for seven different edible oils is shown in Fig. 2. PCA method can also be considered as a preliminary step in exploratory data analysis while performing clustering techniques. So in this work, PCA is performed on raw data to identify whether data clustering was possible on the data or not. The PCA score plot indicates the grouping of different edible oils.

3.1 Confusion Matrix

Confusion matrix is considered as a performance measure for any clustering or classification technique. A perfect grouping of similar data points can be understood with the help of a confusion matrix that only has diagonal elements and all other elements as zero. The confusion matrix for the predicted class and the actual class for seven different oils is shown in Tables 1, 2, 3, 4, and 5.

From the confusion matrices (Tables 1, 2, 3, 4, and 5) and the classification accuracy of the individual distance measurements presented in Table 6 can be observed that the half-square Euclidean distance measurement produced the best classification accuracy in terms of grouping similar category oils into one cluster as desired.

Table 1 Confusion matrix for Euclidean distance

Predicted → / True ↓	RB	GN	CS	SB	SF	RB	MU
RB	7	0	0	0	0	0	0
GN	0	7	0	0	0	0	0
CS	0	7	0	0	0	0	0
SB	0	0	4	3	0	0	0
SF	0	0	0	0	7	0	0
RB	0	0	0	0	0	7	0
MU	0	0	0	0	0	0	7

Table 2 Confusion matrix for city block distance

Predicted → / True ↓	RB	GN	CS	SB	SF	RB	MU
RB	7	0	0	0	0	0	0
GN	3	4	0	0	0	0	0
CS	0	3	4	0	0	0	0
SB	0	0	4	5	0	0	0
SF	0	0	0	4	5	0	0
RB	0	0	0	0	0	7	0
MU	0	0	0	0	0	5	2

Table 3 Confusion matrix for squared Euclidean distance

Predicted → / True ↓	RB	GN	CS	SB	SF	RB	MU
RB	7	0	0	0	0	0	0
GN	2	5	0	0	0	0	0
CS	0	2	5	0	0	0	0
SB	0	0	2	5	0	0	0
SF	0	0	0	0	7	0	0
RB	0	0	0	0	0	7	0
MU	0	0	0	0	0	0	7

Table 4 Confusion matrix for half squared Euclidean distance

Predicted → / True ↓	RB	GN	CS	SB	SF	RB	MU
RB	7	0	0	0	0	0	0
GN	0	3	4	0	0	0	0
CS	0	0	0	7	0	0	0
SB	0	0	0	7	0	0	0
SF	0	0	0	0	7	0	0
RB	0	0	0	0	0	7	0
MU	0	0	0	0	0	0	7

Table 5 Confusion matrix for Cosine distance

Predicted → / True ↓	RB	GN	CS	SB	SF	RB	MU
RB	7	0	0	0	0	0	0
GN	0	7	0	0	0	0	0
CS	0	0	5	2	0	0	0
SB	0	0	0	5	2	0	0
SF	0	0	0	0	5	2	0
RB	0	0	0	0	0	7	0
MU	0	0	0	0	0	5	2

Table 6 Classification accuracy

Distance measures	Classification accuracy
Euclidean	77.55
City Block	69.39
Squared Euclidean	**87.76**
Half Squared Euclidean	77.55
Cosine Distance	77.55

4 Conclusion

Various distance measurements are reported in the literature and can be found in much standard statistical analysis software, but in this article, the commonly used five-distance measurements are implemented and tested on seven different categories of edible oil data.

While running the algorithm several times, the inconsistent cluster centers and subsequently uncertain clusters are being formed and it can be concluded that the K-means algorithm is not deterministic. This non-deterministic behavior of the K-means algorithm resulted due to the random initialization of cluster centers from one run to another. Besides, the K-means clustering is affected by the outliers. This effect of outliers can be further minimized by using the average calculation of the data points while calculating the distance. The algorithm was tested several times for each distance measurement to obtain the best-optimized cluster center and classification accuracy. The average of values obtained in every test was presented in the form of the confusion matrix along with the classification accuracy. It can also be observed from the outcomes of this study is that unsupervised learning method in combination with spectral information is capable in extracting the underlined pattern associated with different edible oils.

Acknowledgements We are thankful to the Director, CSIR-CEERI, Pilani for his continuous encouragement to carry out this work. We would also like to express gratitude to all members of Signal Analytics Group (SAG) and CEERI Jaipur center for their cooperation and help.

References

1. Duda, R.O., Hart, P.E.: DGS Pattern Classification, 738 (2001)
2. Boriah, S.: Similarity Measures for Categorical Data : A Comparative Evaluation
3. Jain, A.K.: Data clustering: 50 years beyond K-means. Pattern Recogn. Lett. **31**, 651–666 (2010). https://doi.org/10.1016/j.patrec.2009.09.011
4. Brunton, S.L., Kutz, J.N.: Data Driven Science & Engineering - Machine Learning, Dynamical Systems, and Control. 572 (2017)
5. Sakthivel, E.: Application of Fuzzy Logic in Data Mining (2015)
6. Kumar, N., Panchariya, P.C., Patel, S.S., et al.: Application of Various Pre-Processing Techniques on Infrared (IR) Spectroscopy Data for Classification of Different Ghee Samples. In: 2018 Fourth International Conference on Computing Communication Control and Automation (ICCUBEA), pp 1–6 (2018)
7. Lozano, J.A., Larra, P.: An empirical comparison of four initialization methods for the K -Means algorithm. Pattern Recogn. Lett. **20**, 1027–1040 (1999)
8. Babu, G.P., Murty, M.N.: Simulated annealing for selecting initial seeds in the k-means algorithm, pp. 85–94 (1994)
9. Bradley, P.S., Bradley, P.S.: Refining Initial Points for K-Means Clustering One Microsoft Way Refining Initial Points for K-Means Clustering, pp. 91–99 (1998)
10. Khan, S.S., Ahmad, A.: Cluster center initialization algorithm for K -means clustering. Expert Syst. Appl. **25**, 1293–1302 (2004). https://doi.org/10.1016/j.patrec.2004.04.007
11. Sharma, J., Panchariya, P.C., Purohit, G.N.: Clustering Algorithm Based on K-means and Fuzzy Entropy for E-nose Applications, pp. 3–5 (2013)
12. Krishna, K., Murty, M.N.: Genetic K-Means Algorithm. IEEE Trans. Syst. Man Cybern. Part B: Cybern. **29**, 433–439 (1999)
13. Babu, G.P., Murty, M.N.: A near-optimal initial seed value selection in K- means algorithm using a genetic algorithm. Pattern Recognit. Lett. **14**, 763–769 (1993)
14. Manoharan, S.: Performance analysis of clustering based image segmentation techniques. J. Innov. Image Process. **02**, 14–24 (2020)
15. Kumar, N., Panchariya, P.C., Kiranmayee, A.H., et al.: Rapid classification of different types of ghee using mid infrared spectroscopy. In: Proceedings - TIMA 2017: 9th International Conference on Trends in Industrial Measurement and Automation (2017). https://doi.org/10.1109/TIMA.2017.8064802
16. Preview A: Principal Components Analysis, pp. 374–377 (2010)
17. Gewers, F., Ferreira, G.R., Arruda, H.F. De, Silva, F.N.: Principal Component Analysis : A Natural Approach to Data Exploration (2018)

Blockchain-Based User Authentication in Cloud Governance Model

Ankur Biswas and Abhishek Roy

Abstract Evolution of digital communication has facilitated human beings to avail electronic services at their doorsteps. User can overcome the compulsion of their physical presence for accessing any type of service and hence can perform multiple tasks while availing electronic services within the same time frame. Furthermore, the present situation of global pandemic of corona virus (COVID19) has shown the necessity of electronic communication, so as to maintain the rules of social distancing to prevent transmission of virus within our society. In this entire process, utmost focus should be given over security of sensitive information which are transmitted through open digital medium (like Cloud). Advanced security schemes like Blockchains may be used to ensure digital identity of users. This task becomes more compact if an integrated environment can be designed for deployment, monitoring, and maintenance of multifaceted electronic services in a secured and Citizen centric manner. In this case, a multivariate smart card will act as a single window interface between the Citizen and Government for transmission of **SERVICE REQUEST** and **SERVICE RESPONSE**, respectively. The application of Blockchain within this integrated electronic service delivery model will build trust among the end users. To achieve this objective, authors have proposed Blockchain-based user authentication scheme during Citizen to Government (C2G) type of Cloud Governance transaction.

Keywords Cloud computing · User authentication · Blockchain · Smart card

1 Introduction

With the gradual evolution of Information and Communication Technology (ICT), human beings have learned to communicate over virtual mediums like Internet, Cloud, etc. The application of these virtual medium has facilitated us to communicate in a faster, accurate, and budget friendly manner through open digital space, i.e., Internet. Due to this approach, the compulsion of physical presence of user (while

A. Biswas (✉) · A. Roy
Department of Computer Science & Engineering, Adamas University, Kolkata, India

J. S. Raj et al. (eds.), *Intelligent Sustainable Systems*, Lecture Notes in Networks and Systems 213, https://doi.org/10.1007/978-981-16-2422-3_64

availing any type of service) has been reduced to a significant level thereby exploring scope of performing multiple task simultaneously for optimum utilization of time. Moreover, the present situation of global pandemic of corona virus (COVID19) has compelled us to avail electronic mode of message communication to maintain the rules of social distancing to break the chain of virus transmission within society. However in case of electronic communication, as the communicating parties are unable to meet with each other over physical mode, security of sensitive information that are being transmitted through open digital medium (i.e., Internet) should be maintained at highest level. The entire success of electronic communication is dependent on the security of electronic messages [18]. If these sensitive information are divulged to any unauthorized user [3], the integrity of electronic message communication will be highly compromised. To prevent such disasters, advanced security concept like Blockchain may be used for authentication [8–11] of user and detection of invalid user to maintain the integrity of any electronic transaction like Cloud Governance, Cloud Banking, Cloud Healthcare system, Cloud Transport management, etc. This entire task becomes more compact if an integrated environment can be designed for deployment, monitoring, and maintenance of multifaceted electronic services in a secured and Citizen centric manner. Citizen will use this smart card [26], namely, Multipurpose Electronic Card (MEC) to send **SERVICE REQUEST** to the Service Provider and receive corresponding **SERVICE RESPONSE** as reply from the Service Provider. To build trust of end user, in this paper authors have shown the application of Blockchain for authentication of end user during Cloud Governance transaction.

Section 2 of this paper describes the basic concept of Blockchain. Literature review describes in the Sect. 3. The origin of our research work, i.e., multivariate smart card-based Cloud Governance is briefly mentioned in Sect. 4 of this paper. Section 5 describes our Blockchain-based user authentication scheme over Cloud Governance model. Finally, the conclusion of this research work along with its future dimension is explored in Sect. 6 of this paper.

2 Basics of Blockchain

Blockchain is a distributed and decentralized ledger [4, 30, 31, 33]that stores data of any electronic transaction and share them publicly with all nodes across its network. As shown in Fig. 1, Blockchain consists of collection of blocks [12], where each block acts as a data storage unit to store data within its data field. Each block is uniquely identified using an unique value [5], which actually looks like a random string of characters. This unique value is called Hash[29], because it is generated from a cryptographic hash function [21] which generates an unique output for every unique input. The input of block hash includes the data it needs to store and metadata about that block, like the information of lastly created hash, etc. The last hash of hash value is set for every new block to hash of block that came before it. In this way a chain of block is created using the subsequent blocks. Each new block reference

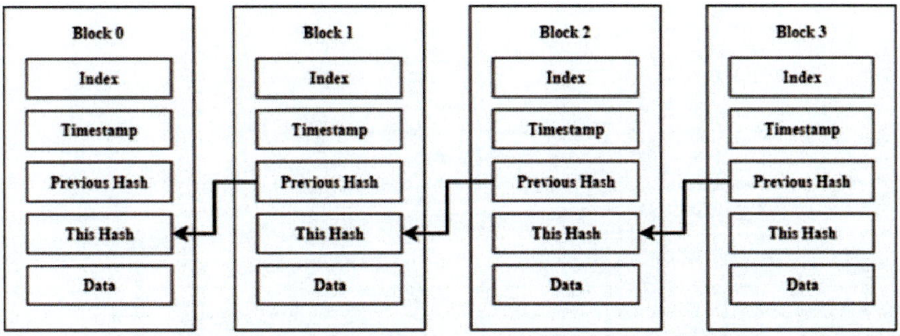

Fig. 1 Conceptual diagram of blockchain

is a hash of the last block in the current chain. We are using Blockchain because it is fair, democratized, and completely public [20]. Due to its compactness and inter-dependencies among the blocks within the same chain, it provides a highly secure [22, 32, 34] data storage mechanism. To breach the security protocol of Blockchain-based electronic data storage system, hacker should take control of very large number of nodes and computers, which is not an easy task to achieve. However, as the blockchain is public in nature, attacker could attempt to intercept the electronic transaction. To counter such malicious attempts, crypto-currency uses cryptography to protect Blockchain using algorithms to obscure data through cryptic messages. In this paper, we have discussed Blockchain-based user authentication within Cloud Governance model during Citizen to Government (C2G) type of transaction.

3 Literature Review

In paper [2] author have proposed a Cloud Governance model for availing multiple services using a single smart card=based interface, namely, Multipurpose Electronic Card (MEC) [27]. To maintain the security perspective of this proposed model, its Intrusion Detection System during Cloud Governance transaction is discussed in paper [25]. As this model is projected as a multifaceted electronic service delivery model, in paper [26] authors have discussed the database structure for Electronic Voting (i.e., E-Voting) using this smart model. To explore its research scope, object-oriented modeling during Cloud Healthcare system is discussed in paper [28]. In this paper we have used this smart model as origin of our research work, which is discussed in the subsequent sections.

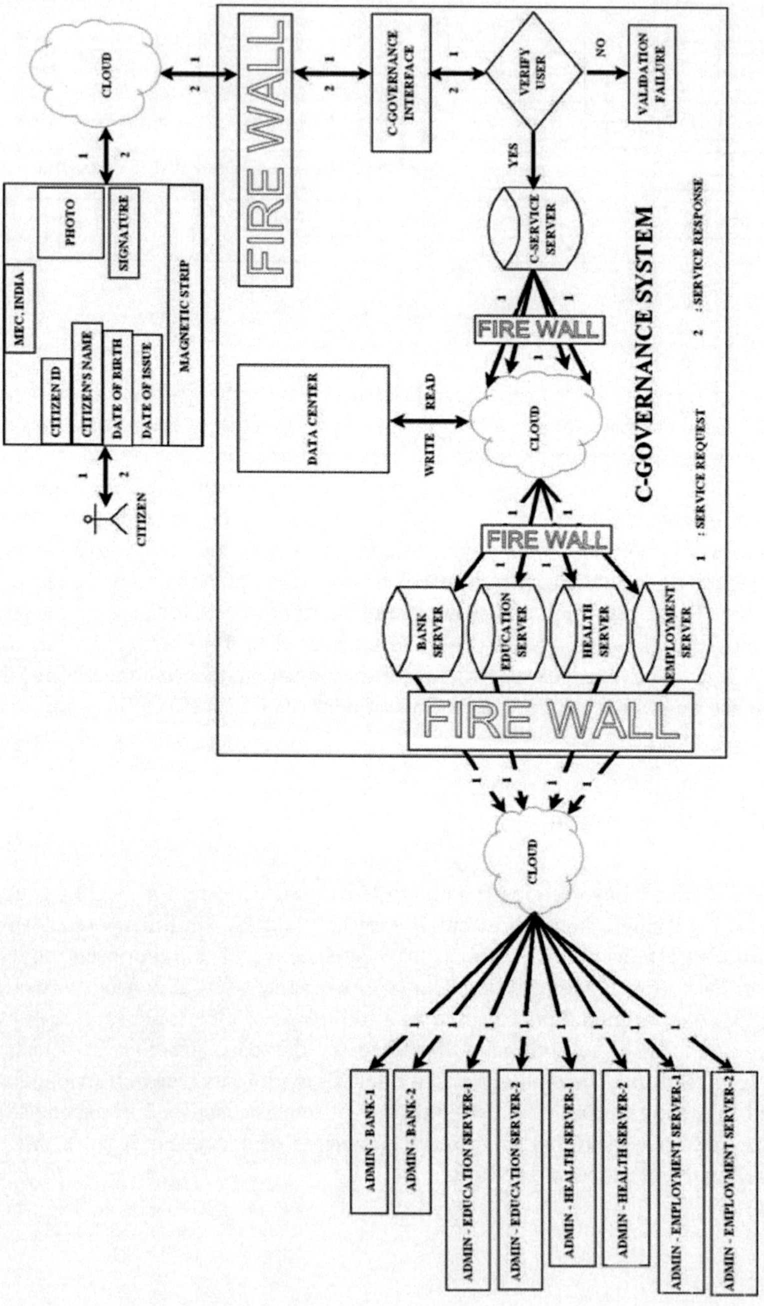

Fig. 2 Conceptual diagram of cloud governance model

4 Origin of Work

Cloud Governance model [1, 2, 25–28] is the origin of our research work and we have implemented user authentication using Blockchain during Cloud Governance transaction. This model facilitates delivery of multivariate electronic services to Citizen, like Banking, Healthcare, Voting, Education, etc., in a Citizen centric manner using a Multipurpose Electronic Card [1]. The primary components of this Cloud Governance model are mentioned below:

1. Citizen (i.e., Service Seeker).
2. Government (i.e., Service Provider / Administrator).
3. Third Party Service Provider (for example, Bank, Hospital, College, University, Transport Services, Employer, etc.).

As shown in Fig. 2, Citizen performs electronic transaction with Government [2] to send and receive **SERVICE REQUEST** and **SERVICE RESPONSE** through Path-1 and Path-2, respectively. The sequence of operations during Citizen to Government (C2G) type of Cloud Governance transaction is discussed below:

1. **Citizen side**: Citizen initiates electronic transaction and transmit SERVICE REQUEST to Government using Multipurpose Electronic Card (MEC) through public cloud medium.
2. **Government side**: Government receives SERVICE REQUEST from the Citizen, which passes through the Firewall of Cloud Governance model.
3. **Government side**: Government verifies the identity of Citizen.

Scenario 1 (Verification Failure):

   i. Transaction is aborted.
   ii. Citizen is informed about Failure through System Timeout.

Scenario 2 (Verification Success):

   i. SERVICE REQUEST is allowed to proceed further.

4. **Government side:** SERVICE REQUEST arrives at Cloud Service Server for identification of specific type of service requested by Citizen.
5. **Government side:** Cloud Service Server performs READ and WRITE operations over Data Center connected to its Private Cloud and sends SERVICE REQUEST to the service specific server like Bank, Education, Healthcare, Employment, Voting, etc.
6. **Government side:** Service specific server receives SERVICE REQUEST through Private Cloud of Cloud Governance model and transmit it to Third-Party Service Provider through Public Cloud for generation of corresponding SERVICE RESPONSE.

The success rate of this entire electronic service delivery is dependent over the identity verification of Citizen, who is the ultimate beneficiary of this Cloud Governance model. To ensure the identity of user through strict cryptographic security protocol, we have applied Blockchain during electronic transaction which is discussed in subsequent section of this paper.

5 Application of Blockchain in Cloud Governance Model

In our proposed Cloud Governance model we have assigned individual Blockchain for updated record maintenance of respective user. Here we have used Permissioned Blockchain [6, 15] based network for authentication of user during Citizen to Government (C2G) type of Cloud Governance transaction. In this case a valid member of Permissioned Blockchain network holds the Distributed Ledger [7, 13] and Citizen sends its credentials to the registration portal for its verification. In case of successful verification, registration portal send Citizen information to Blockchain as the part of Citizen-Registration-Request (Service Request) to perform Blockchain Mining [14] for updation of Blockchain Ledger. Only after updation of Citizen information, Registration Portal issues credential and unique Multipurpose Electronic Card (MEC) ID [16, 17] to the Citizen. Thus, Citizen can use this unique Multipurpose Electronic Card (MEC) ID for its verification while availing the electronic services of our Cloud Governance model. The user registration of our concept is explain in subsequent section of this paper.

5.1 User Registration Using Blockchain

The essential players of our proposed framework are Client (i.e., Citizen), Document Verifier, Government (i.e., State and its agencies) and Third-Party Service Provider which are connected through Public Cloud [23]. As shown in Fig. 3 the step-wise explanation of our proposed Blockchain Based User Registration System is described below:

CLIENT Side:

Step 1: Citizen input data and uploads it into the system to register through Public Kiosk [19] (i.e., Public Cloud).

DOCUMENT VERIFIER Side:

Step 2: The physical documents play a vital role for verification of user (i.e., Citizen). Document Verifier verifies the uploaded document with its original counterpart.

Step 3: Document Verifier transmit SERVICE REQUEST to Government for identification of Citizen mentioned within it.

GOVERNMENT Side:

Step 4: Government uses SERVICE REQUEST to store Citizen data securely into Blockchain. This step is shown as User Registration Process in Fig. 2.

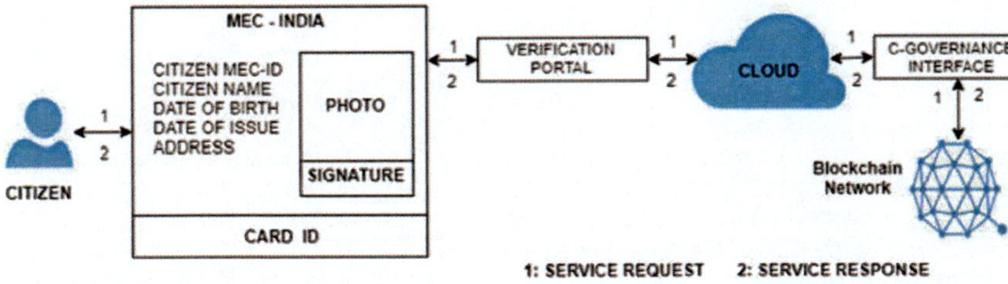

Fig. 3 Blockchain-based user registration in basic cloud c-governance model

Step 5: After storing Citizen data into Blockchain, Blockchain returns Hash as SERVICE RESPONSE.

Step 6: Cloud Governance Interface returns unique Multipurpose Electronic Card (MEC) ID to the Citizen and Document Verifier as SERVICE RESPONSE through Public Cloud [24].

DOCUMENT VERIFIER Side:

Step 7: Finally, Document Verifier Issues Multipurpose Electronic Card (MEC) Card to the Citizen after receiving unique Multipurpose Electronic Card (MEC) ID through SERVICE RESPONSE.

5.2 *User Authentication Using Blockchain*

The authentication of user of our proposed Cloud Governance model is discussed thoroughly in this section. As shown in Fig. 4, our proposed Blockchain-based user authentication is further explained below:

CITIZEN Side:

Step 1: Citizen uses Multipurpose Electronic Card (MEC) to transmit SERVICE REQUEST during Citizen to Government (C2G) type of Cloud Governance transaction using Public Kiosk (i.e., Public Cloud).

CLOUD Communication:

Step 2: Public Kiosk (i.e., Public Cloud) transmit SERVICE REQUEST to Cloud Governance model.

GOVERNMENT Side:

Step 3: Cloud Governance Interface send SERVICE REQUEST to Blockchain for verification.

Step 4: Blockchain returns Blockchain Hash to Cloud Governance Interface.

Step 5: Cloud Governance System verifies Citizen with its Blockchain Hash value.

Step 6: Anyone of the following situation will prevail.

Situation 1 (Success): In case of successful verification of Citizen, SERVICE REQUEST is sent to Cloud Service Server.

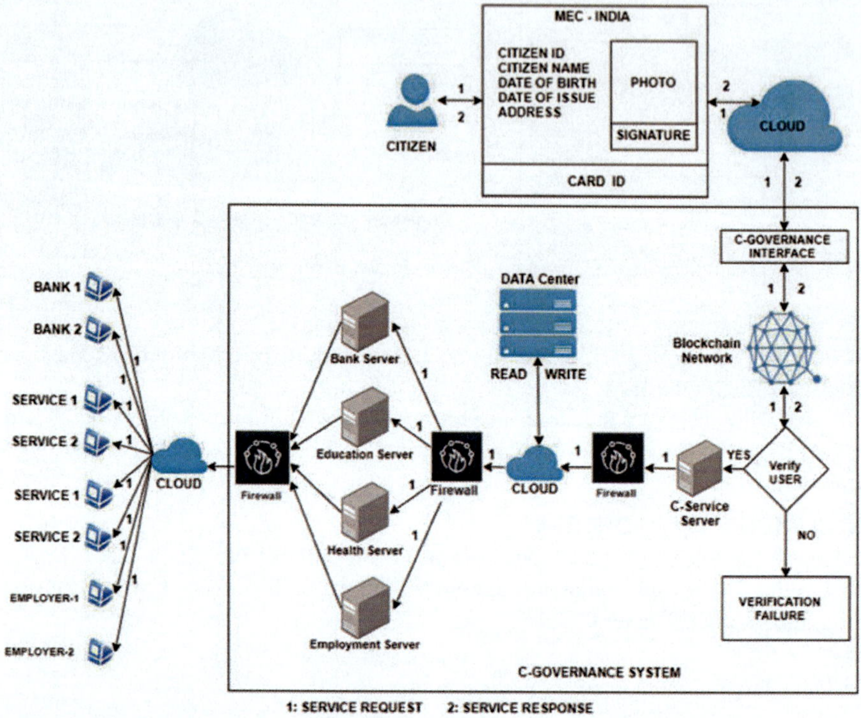

Fig. 4 Blockchain-based user authentication in cloud governance model

Situation 2 (Failure): In case of verification failure, SERVICE REQUEST is aborted and Citizen is notified through System Timeout.

Step 7: SERVICE REQUEST proceeds from Cloud Service (i.e., C-Service) Server toward Private Cloud through Firewall.

Step 8: The Data Center attached to the Private Cloud performs essential Read and Write operations about the SERVICE REQUEST.

Step 9: Depending upon the SERVICE REQUEST of Citizen, respective service server is selected.

Step 10: This step denotes the last phase of Government interference over the entire electronic communication, where SERVICE REQUEST is transmitted to Third-Party Service Provider through Public Cloud for its final execution.

THIRD-PARTY SERVICE PROVIDER Side:

Step 11: The Third-Party Service Provider will receive SERVICE REQUEST of Citizen and execute it to generate SERVICE RESPONSE. Step 12: Third-Party Service Provider transmits SERVICE RESPONSE to Citizen through Public Cloud.

Depending upon the type of service requested by Citizen, sometimes Government may also act as Service Provider and in that case SERVICE RESPONSE will be generated by Government (instead of Third-Party Service Provider), which will be communicated to Citizen through Public Cloud. Since we performed Blockchain-based user verification of our proposed Cloud Governance model, there exists significant amount of similarities between Fig. 2 and 4 of this paper. The presence of Blockchain Network in Fig. 4 is its primary difference with Fig. 2, which is in fact the main model of this Blockchain-based concept.

In this paper we have conceptualized the application of Blockchain for authentication of user during C2G type of cloud for transaction. The implementation of this concept along with relevant result analysis will be considered as future scope of the work.

6 Conclusion

In this paper authors have described application of Permissioned Blockchain for user authentication during Citizen to Government (C2G) type of Cloud Governance transaction. This Cloud Governance model will attempt to deliver electronic services to the Citizen through the cloud platform. Citizen will use a single smart card namely Multipurpose Electronic Card (MEC) to send multifaceted **SERVICE REQUEST** to the Service Provider and receive corresponding **SERVICE RESPONSE** in reply from the Service Provider. It is utmost necessary to build trust about the integrity of this electronic service delivery model among its ultimate beneficiaries. As mentioned above, this trust can be developed using advanced technologies like Blockchian for verification of end user during the electronic transactions. This Blockchain-based verification scheme can be applied over any type of electronic message communication like Cloud Banking, Cloud Healthcare system, Cloud Transport management, etc. We have used a multivariate smart card-based interface between the Citizen and Government (i.e., Service Provider) for exchange of **SERVICE REQUEST** and **SERVICE RESPONSE** in a secure manner through Cloud.

References

1. Roy, A.: Object-Oriented Modeling of Multifaceted Service Delivery System Using Connected Governance. In: Jena A., Das H., Mohapatra D. (eds) Automated Software Testing. ICDCIT 2019. Services and Business Process Reengineering. Springer, Singapore. pp. 1–25. (2020). https://doi.org/10.1007/978-981-15-2455-4-1

2. Roy, A.: Smart delivery of multifaceted services through connected governance. In: 3rd International conference on computing methodologies and communication (ICCMC). IEEE. India. pp 476—482. (2019). https://doi.org/10.1109/ICCMC.2019.8819851
3. Biswas, A., Roy, A.: A study on Dynamic ID based user authentication system using smart card.: ajct [Internet]. (2019). [cited 26Feb.2020];5(2). http://www.asianssr.org/index.php/ajct/article/view/871
4. Christidis, K., Devetsikiotis, M.: Blockchains and smart contracts for the Internet of Things. IEEE Access **4**, 2292–2303 (2016). https://doi.org/10.1109/ACCESS.2016.2566339
5. Pinno, O.J.A., Gregio, A.R.A., De Bona, L.C.E.: ControlChain: blockchain as a central enabler for access control authorizations in the IoT. In: GLOBECOM 2017–2017 IEEE Global Communications Conference, Singapore (2017)
6. Tapas, N., Merlino, G., Longo, F.: Blockchain-based IoT-cloud authorization and delegation. In: 2018 IEEE International Conference on Smart Computing (SMARTCOMP), Taormina, pp. 411–416 (2018)
7. Benhamouda, F., Halevi, S., Halevi, T.: Supporting private data on hyperledger fabric with secure multiparty computation. In: IEEE International Conference on Cloud Engineering (IC2E), April (2018)
8. Ying, N., Yao, Z., Hua, Z.: The study of multi-level authentication-based single sign-on system. In: 2009 2nd IEEE International Conference on Broadband Network and Multimedia Technology, pp. 448–452 (2009)
9. Kansal, S., Kaur, N.: Multi-level Authentication for Internet of Things to establish secure healthcare network. Int. J. Adv. Res. Ideas Innov, Technol (2016)
10. Peter, S., Gopal, R.K.: Multi-level authentication system for smart home-security analysis and implementation. In: 2016 International Conference on Inventive Computation Technologies(ICICT), Coimbatore (2016)
11. Gupta, S., Gabrani, G.: A dynamic two-level priority based authentication system for job scheduling in a heterogeneous grid environment. In: 2016 SAI Computing Conference (SAI), London, pp. 1100–1106 (2016)
12. Odelu V. (2020) IMBUA: Identity Management on Blockchain for Biometrics-Based User Authentication. In: Prieto J., Das A., Ferretti S., Pinto A., Corchado J. (eds) Blockchain and Applications. BLOCKCHAIN 2019. Advances in Intelligent Systems and Computing, vol 1010. Springer, Cham. https://doi.org/10.1007/978-3-030-23813-1-1
13. Yu, Y., Zhao, Y., Li, Y., Du, X., Wang, L., Guizani, M.: Blockchain-Based Anonymous Authentication With Selective Revocation for Smart Industrial Applications. IEEE Transactions on Industrial Informatics. **16**, 3290–3300 (2020). https://doi.org/110.1109/TII.2019.2944678
14. Almadhoun, R., Kadadha, M., Alhemeiri, M., Alshehhi, M., Salah, K.: A User Authentication Scheme of IoT Devices using Blockchain-Enabled Fog Nodes. 2018 IEEE/ACS 15th International Conference on Computer Systems and Applications (AICCSA). (2018). https://doi.org/10.1109/AICCSA.2018.8612856
15. Park, B., Lee, T., Kwak, J.: Blockchain-Based IoT Device Authentication Scheme. Journal of the Korea Institute of Information Security and Cryptology. **27**, 343–351 (2017). https://doi.org/10.13089/JKIISC.2017.27.2.343
16. Ghosh, A., Das, T., Majumder, S., Roy, A.: Authentication of User in Connected Governance Model. Data Science and Analytics. **110–122**, (2020). https://doi.org/10.1007/978-981-15-5830-6_10
17. Khatun, R., Bandopadhyay, T., Roy, A.: Data modelling for e-voting system using smart card based e-governance system. Int. J. Inf. Eng. Electron. Bus. **9**, 45–52 (2017). https://doi.org/10.5815/ijieeb.2017.02.06
18. Applebaum, B.: Key-Dependent Message Security: Generic Amplification and Completeness. Advances in Cryptology - EUROCRYPT **2011**, 527–546 (2011). https://doi.org/10.1007/978-3-642-20465-4_29
19. MAGUIRE, M.: A review of user-interface design guidelines for public information kiosk systems. International Journal of Human-Computer Studies. 50, 263-286 (1999). https://doi.org/10.1006/ijhc.1998.0243

20. Judmayer, A., Stifter, N., Schindler, P., Weippl, E.: Blockchain: Basics. Business Transformation through Blockchain. **339–355**, (2018). https://doi.org/10.1007/978-3-319-99058-3_13
21. Treleaven, P., Gendal Brown, R., Yang, D.: Blockchain Technology in Finance. Computer. **50**, 14–17 (2017). https://doi.org/10.1109/MC.2017.3571047
22. Karame, G., Capkun, S.: Blockchain Security and Privacy. IEEE Security & Privacy. **16**, 11–12 (2018). https://doi.org/10.1109/MSP.2018.3111241
23. Shakya, S.: AN EFFICIENT SECURITY FRAMEWORK FOR DATA MIGRATION IN A CLOUD COMPUTING ENVIRONMENT. Journal of Artificial Intelligence and Capsule Networks. **01**, 45–53 (2019)
24. Haoxiang, W., Smys, S.: Secure and Optimized Cloud-Based Cyber-Physical Systems with Memory-Aware Scheduling Scheme. Journal of Trends in Computer Science and Smart Technology. **2**, 141–147 (2020)
25. Biswas, S., Roy, A.: An Intrusion Detection System Based Secured Electronic Service Delivery Model. In: 2019 3rd International conference on Electronics, Communication and Aerospace Technology (ICECA), Coimbatore, India, pp. 1316–1321. (2019). https://doi.org/10.1109/ICECA.2019.8822016
26. Khatun, R., Bandopadhyay, T., Roy, A.: Data Modeling for E-Voting System Using Smart Card based E-Governance System. International Journal of Information Engineering and Electronic Business. **9**, 45–52 (2017). https://doi.org/10.5815/ijieeb.2017.02.06
27. Roy, A., Karforma, S.: Uml Based Modeling of ECDSA for Secured and Smart E-Governance System. Computer Science and Information Technology (CS and IT). (2013). https://doi.org/10.5121/csit.2013.3219
28. Mohapatra, S., Paul, K., Roy, A.: Object-Oriented Modeling of Cloud Healthcare System Through Connected Environment. Applications of Internet of Things. **151–164**, (2020). https://doi.org/10.1007/978-981-15-6198-6_14
29. Nofer, M., Gomber, P., Hinz, O., Schiereck, D.: Blockchain. Business & Information. Systems Engineering **59**(3), 183–187 (2017). https://doi.org/10.1007/s12599-017-0467-3
30. Risius, M., Spohrer, K.: A Blockchain Research Framework. Business & Information Systems Engineering **59**(6), 385–409 (2017). https://doi.org/10.1007/s12599-017-0506-0
31. Beinke, J., Fitte, C., Teuteberg, F.: Towards a Stakeholder-Oriented Blockchain-Based Architecture for Electronic Health Records: Design Science Research Study. Journal of Medical Internet Research **21**(10), p.e13585 (2019). https://doi.org/10.2196/13585
32. Batubara, F., Ubacht, J., Janssen, M.: Challenges of blockchain technology adoption for e-government. Proceedings of the 19th Annual International Conference on Digital Government Research: Governance in the Data Age. (2018). https://doi.org/10.1145/3209281.3209317
33. Motta, G., Tekinerdogan, B., Athanasiadis, I.: Blockchain Applications in the Agri-Food Domain: The First Wave. Frontiers in Blockchain. **3**, (2020). https://doi.org/10.3389/fbloc.2020.00006
34. Xie, J., Tang, H., Huang, T., Yu, F., Xie, R., Liu, J., Liu, Y.: A Survey of Blockchain Technology Applied to Smart Cities: Research Issues and Challenges. IEEE Communications Surveys & Tutorials. **21**, 2794–2830 (2019). https://doi.org/10.1109/COMST.2019.2899617

Balanced Cluster Head (CH) Selection in Four-Level Heterogeneous Wireless Sensor Networks (WSN's)

V. Baby Shalini

Abstract WSN is usually made up of a hefty amount of low priced, moderate power-driven multifunctional sensors, which are positioned in a target region. Basically, these nodes are small in magnitude however they are fortified with one or more sensors, microprocessors as well as radio transmitter and receiver. In innumerable applications, arrangement of sensors is carried out in an ad hoc manner without taking into account of cautious preplanning as well as engineering. As soon as the sensors are scattered, they must be talented to independently unify it selves into a wireless network. In specific, sensors are generally battery powered and should drive without presence for a moderately extended time period. In utmost circumstances, it is very strenuous and even impractical to revitalize the batteries of sensors. To protract the network existence, clustering is a crucial technique with more proficient the cluster leader selection likelihood. The sensors with unexpended energy and network average energy will have lot of possibilities to be the cluster heads, when contrasted with nodes that have minimal energy. This paper presents four-level Balanced Energy-Efficient Network Integrated Super Heterogeneous (FL-BEENISH) protocol and simulation result shows that the proposed method performs better than the DDEEC and EDDEEC protocol.

Keywords Target · Microprocessor · Transmitter · Receiver · Ad hoc · Lifespan · Clustering · Cluster head

1 Introduction

WSN's have been comprehensively considered as one of the vital innovations for the twenty-first century [1]. Facilitated by contemporary progresses in microelectronic mechanical systems (MEMS) as well as radio communique techniques, miniscule, inexpensive, and ingenious sensors are dispersed in a physical zone and interacted wirelessly and the Internet affords unprecedented prospects for an assortment of

V. Baby Shalini (✉)
Information Technology, Kalasalingam Academy of Research and Education, Krishnankoil, Srivilliputhur, India
e-mail: v.babyshalini@klu.ac.in

J. S. Raj et al. (eds.), *Intelligent Sustainable Systems*, Lecture Notes in Networks and Systems 213, https://doi.org/10.1007/978-981-16-2422-3_65

civilian as well as military applications [2, 3]. In the ancient decade, WSN's have acquired tremendous attention from industry as well as academia over the world. An enormous quantity of research events are performed done to examine and resolve innumerable strategies and application disputes, besides notable advances have been made in the growth and arrangement of WSNs. Recognized from customary wireless networks, WSN's acquire unique individualities, i.e., denser degree of node positioning, higher inconsistency of sensors, harsh energy, calculation as well as capacity restraints [4], which present numerous novel challenges in the growth of WSN's. Because of stern energy restraint of sensors, structural design of a network hugy affects the energy consumption and in this way lifespan also gets grieved. In general, sensor network comprises of a hefty quantity of sensors that are thickly dispersed in an area as well as single or extra sink are situated nearby to or within the sensing zone, as exposed in Fig. 1. The sink nodes enquire the sensors in the recognizing area to ensure the perceiving task and propel it to sink(s) [5]. In the meantime, the sink(s) additionally act as a passage for open air systems, For instance, the Internet. It accumulates information from the sensors, executes the handling on gathered data, and afterward advances the appropriate information through Internet to the clients who demanded it or usage the information.

To propel the data to the base station(BS), each and every sensor can utilize sole hop long distance communication leading to a network called single hop network organization illustrated in Fig. 2.

Nonetheless, this is pricy with regard to energy utilization. To resolve this issue, multihop petite distance communique as delineated in Fig. 3 is greatly preferred. A node can confer the recognized information toward the BS by means of intermediary sensors, which in turn the energy exploitation is diminished for communique purpose.

The structure of a multihop network can be of dual categories: flat as well as hierarchical or cluster-based network. In a former network as shown in Fig. 4, each one node does similar role in accomplishing a sensing task. In a later network as exposed in Fig. 5, sensors are systematized into groups [6], where the cluster members (CM's) forward their perceived data to the CH's which in turn dispatch it to BS.

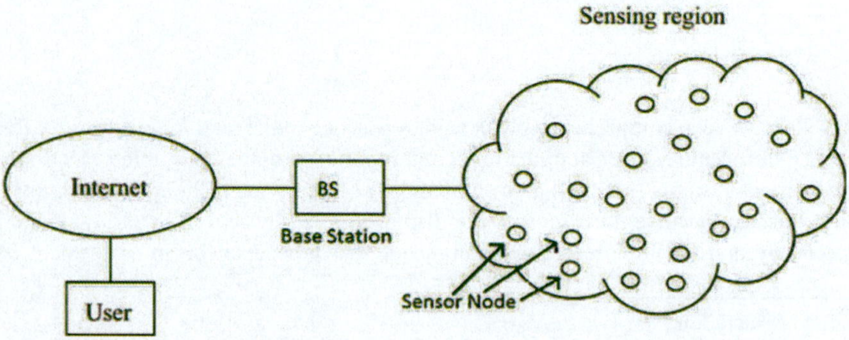

Fig. 1 Sensor network architecture

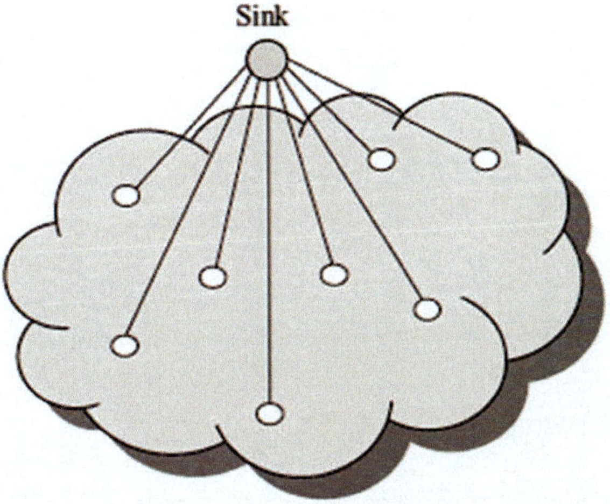

Fig. 2 Single hop structure

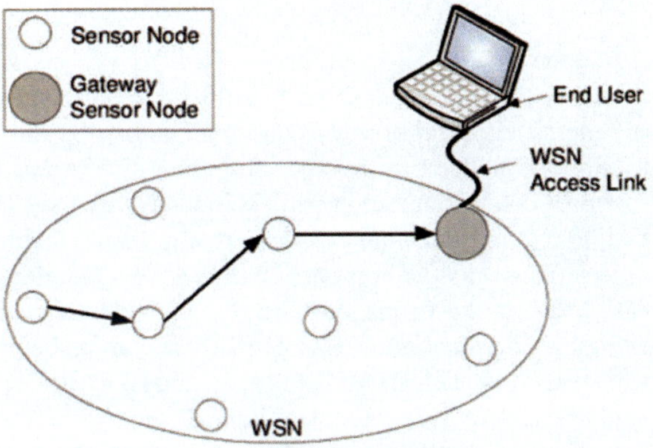

Fig. 3 Typical multi HopWSN architecture

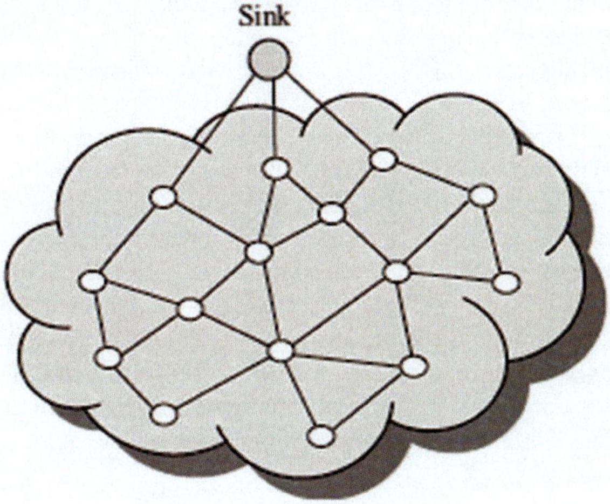

Fig. 4 Flat network architecture

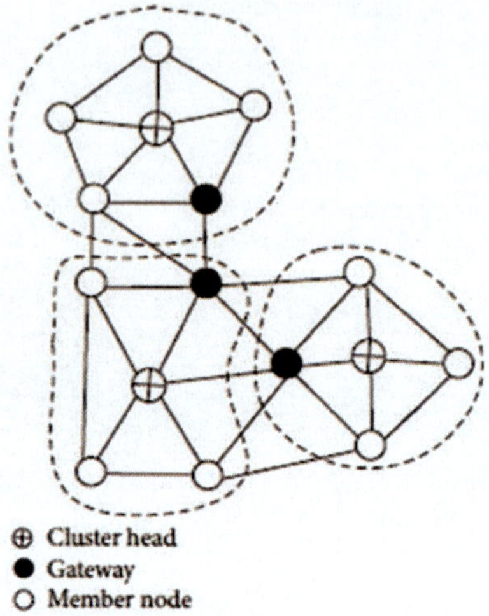

Fig. 5 Cluster-based WSN

A node with not as much energy can do the perceiving task and send the recognized data to CH, as a result it will diffuse the aggregated data to the sink. By doing so, energy consumption for communique can be diminished, but also balance the traffic load as well as enhance the scalability when the network size grows. The foremost dispute with clustering is in what way the CH's are selected, in addition to that how to organize the clusters [7]. Clustering techniques can be enforced in two diverse networks (1) homogeneous WSN'S (2) heterogeneous WSNs. In erstwhile case, sensors are furnished with the similar energy level. In latter network, these energy levels fluctuate. LEACH [8] is accomplished for homogenous WSNs. SEP [9], DEEC [10–12], DDEEC [13, 14], and EDDEEC [15, 16] are certain examples of heterogenous WSN protocols. The real issue with clustering is the manner in which how the CH's are chosen and the arrangement of the clusters [7]. As indicated by the distance among CM's and their respective CH, a sensor network can be split into two styles, namely, single and multi hop clustering architecture. In single hop clustering architecture, CM's pass on the sensed information direct to their respective CH as shown in Fig. 6.

In multihop clustering network, CM's propel their detected information to the sink through intermediary nodes as shown in Fig. 7.

The proposed FL-BEENISH Protocol is equipped with four categories of sensors: normal, advanced, super, and ultra-super nodes. Here last triple sensors possess extra energy in comparison to ordinary ones and CH's is elected with reference to nodes' left over energy. So, the last triple nodes are principally chosen as CH's for the principal rounds, after time elapses if the energy of a particular senor's decreases as same as energy level of normal sensors will have similar CH determination probability like the normal sensors. Thusly, energy is disseminated proficiently within the network and elongates the sensor network duration.

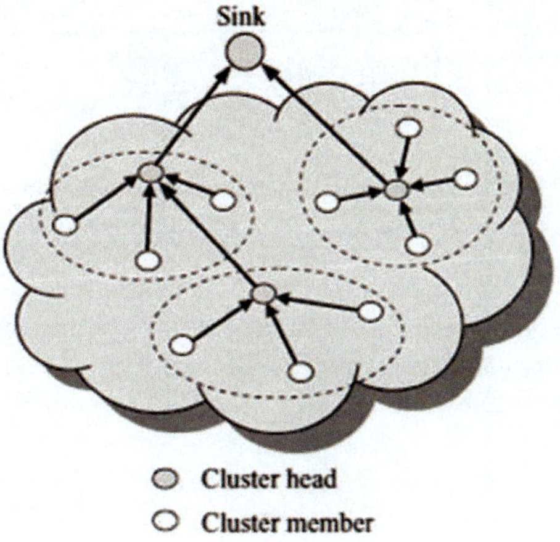

Fig. 6 Single hop clustering architecture

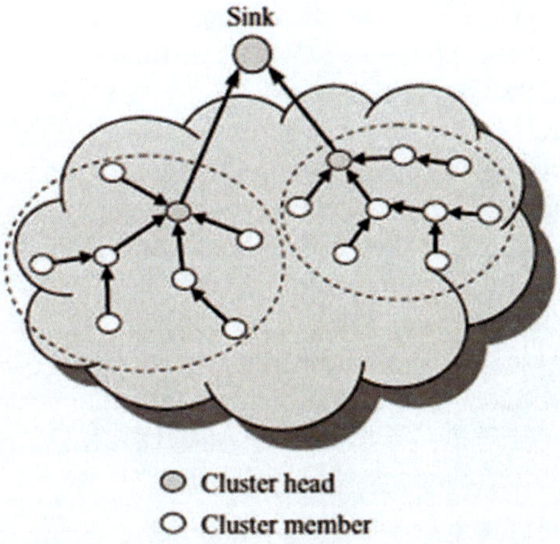

Fig. 7 Multi hop clustering architecture

The leftovers of this research examination are systematized in the ensuing style. Section 2 incorporates the interrelated work, Sect. 3 illustrates the radio and network model, Sect. 4 illustrates the proposed protocol, Sect. 5 deals with result and discussion, Sect. 6 concludes the paper.

2 Related Work

Heinzelman et al. [8] have explained the details of cluster-based scheme in which the network is apportioned into groups, a dedicated node acting as CH in each cluster. LEACH process is segregated into many rounds with dual phases, namely, (i) setup phase (ii) steady-state phase as shown in Fig. 8 for carrying out in each round.

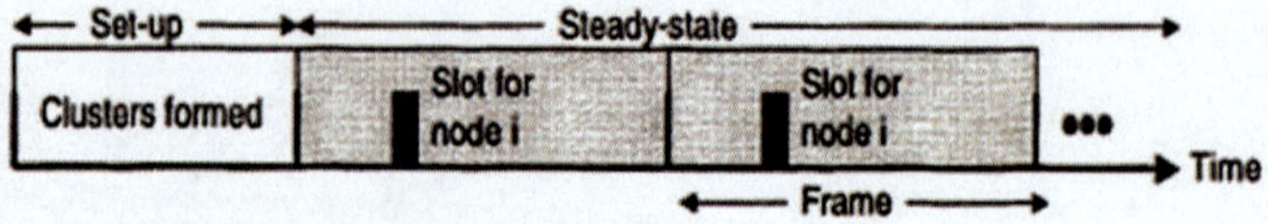

Fig. 8 LEACH operation delineated by time slots

During setup phase, all sensor nodes produce a number in a haphazard manner somewhere in the range of 0–1. If this number is not as much as the following threshold presented in Eq. 1 for a node, the node becomes active as CH for the existent round(r):

$$T(n) = \begin{cases} \frac{p}{1-p\left(r* \bmod \frac{1}{p}\right)} & if\ n\varepsilon G \\ 0 & otherwise \end{cases} \tag{1}$$

Here P shows the desired proportion of CH in sensor population and G as the faction of nodes that do not function as CH's in the most recent 1/P rounds. A certain node chosen as CH effectively notifies the announcement information to the rest of the nodes. The remaining nodes decide on the cluster for accompaniment for this extant round on the basis of the potency of the announcement received. In addition, it dispatches a membership message to its relevant CH. During the steady-state stage, data is delivered from CM's to the corresponding CH to their allocated Time Division Multiple Access (TDMA) slot only and at the same time, radio of other clusters MN's can be turned off, thereby minimizing energy dissipation. The entire data collected by CH from CM's is accumulated and forwarded to BS. After some moments, the network falls into setup phase again and gets directed into one more round for selection of CH [17].

Advantages of LEACH [18]:

(1) LEACH uses a randomized selection of CH, with the goal of the ability if sensor nodes getting the same opportunity to act as CH and, once selected, it cannot become CH again in the succeeding rounds, thus providing load sharing among sensor nodes.
(2) Use of a TDMA slot inhibits CH's from unnecessary conflicts.

Limitations of LEACH:

(1) LEACH cannot be utilized for large networks as it carries out single hop inter-cluster communication for dispatching the data from CH to BS. Thus, the farthest node to the BS expires earlier than others.
(2) Since preference for CH's is haphazard, this protocol does not warrant a fine consistent circulation of CH and ceases in a finite amount of rounds.
(3) Transmission distance and the node's energy are not considered as a CH criteria for the selection of CH, leading to uneven energy consumption for different nodes.

(4) Sometimes sensors with bad preliminary energy can be selected as CH for the first round itself. In such a situation, the node will demise too soon. This leads to a problem termed as energy hole and becomes a coverage problem.

Hosseinirad et al. [19] clearly explained that Low Energy Adaptive Clustering Hierarchy-Imperialist Competitive Algorithm(LEACH-ICA) is an improvement over Low Energy Adaptive Clustering Hierarchy (LEACH) protocol. Generally a sensor network consists of static sensor nodes that are randomly distributed in an grid-based architecture. Once the clusters are organized, each node in the network will adjudge whether or not it acts as a CH for an extant round based on the appropriate location, so that CM's can transfer their packets with minimum amount of energy. This will significantly conserve the energy, increase the network lifetime, and maintain the connectivity. The disadvantage of LEACH-ICA is that this protocol is suitable only for homogeneous networks and not considered the energy of CH.

Loscri et al. [20] showed that two-level hierarchy LEACH(TL-LEACH) is an extension of LEACH protocol. The author primarily focuses on selecting the two hierarchies of CH's (i.e., primary and secondary CH), energy efficiency is improved when compared to LEACH protocol. Here, data transferred from the sensors to the BS takes place in four stages. In the advertisement stage, individual sensor randomly adjudges itself as a primary CH, secondary CH or an ordinary node (ON) in every round. In cluster setup phase, a node opted as a primary CH must publicize to secondary CH which as a consequence convey it to ON's. Then, each secondary CH picks the appropriate primary CH to which it appertains and transmits an advertisement message. Similarly, each ON should pick the pertinent secondary CH and notifies it through an ad message. In schedule creation stage, TDMA plan is formed by primary CH and diverse time slot is ascribed for every node in its group for transmission. Code Division Multiple Access (CDMA) code is used by all primary CH and instructs all sensors in its group to make usage of code. In like manner, individual secondary CH can instruct about code and TDMA plan to ON's in its clusters. In data transmission phase, each node can transmit their data-based on TDMA schedule decided by its primary CH and secondary CH. Each and every ordinary node (ON) present in the network will monitor the physical environment and then the sensed data is passed from all the ON's to the secondary CH's. Data fusion can be performed by the secondary CH's and then transmitted to the primary CH's. Data fusion can also be performed by primary CH's and forward the accumulated data to the BS. The main disadvantage is two hops for data dispatch from ON's to BS can breed much more energy utilization and is also not befitting for voluminous networks. Here, the author does not consider any energy level for selecting the CH, it cannot guarantee real load balancing if there should be an occurrence of nodes with different amounts of preliminary energy.

Smaragdakis et al. [9] are known for a Stable Election Protocol (SEP) which is a variation of LEACH. It is designed mainly for dual-level heterogeneous networks composed of doublet types of sensors (particularly Advance and normal nodes). In general, advance sensors retain extra energy and this has to be turned CH's more frequently compared to normal nodes. Here CH's chosen pertain to the initial energy of sensors. The main disadvantage is that this decision of CH's is not dynamic. As a consequence, advanced nodes that are far-flung with reference to the sink drain their energy faster and depart from their existence soon. SEP is not suitable for multi-level heterogeneous networks.

Malluh et al. [21] have suggested an improved form of SEP. Here also, advanced nodes are selected as CH's more often than the normal sensors. In addition, the number of nodes connected with each CH is considered. Thus enables equtable distribution for the sensor nodes between the CH's. Also, in the event of an excess of one sensor accessible to be a CH at certain round, EM-SEP chooses the higher energy sensor as a CH. Those two factors protract the stable period of the sensor network. The main drawback in this method is that inter-cluster communication cost is not considered.

Rehman et al. [22] have provided a description of a stability-based clustering protocol which is an extension of SEP. Here, heterogeneous sensors are randomly dispersed in WSN. A node with a minutest Energy Consumption Rate (ECR) in the proceeding round is preferred as CH. ECR implies the existence of a difference between initial energy and residual energy of a particular node. The main shortcoming is that advanced nodes always get punished by being selected as a CH.

Kumar et al. [23] have explicated an energy proficient heterogeneous clustered method in which CH's for a cluster are elected on the basis of a weighted selection prospects of every node in regard to their residual energy in every round. The optimal likelihood of a node the choice of as a CH is based on spatial density. The main disadvantages are poor stability, and difficulty in calculating the weight probability and spatial density.

Younis et al. [24] explain about Hybrid Energy-Efficient Distributed Clustering (HEED) approach for ad hoc networks which utilizes sensor with minimal residuary energy as principal restraint as well as least distance to communique the perceived data with neighbors is taken as auxiliary requirement for picking the CH. This scheme has a minimal processing time complexity per node and moreover finishes after a stable number of rounds.

Duan et al. [25] have indicated a bi-level energy sensible heterogeneous clustering method in which CH for a particular cluster is selected on the basis of the proportion between sensor nodes residual energy and average energy of the network. The main disadvantage is energy efficiency is limited.

Zhu et al. [12] presented a protocol Distributed Energy-Efficient Clustering (DEEC) in which CH election depends on likelihood proportion of remnant energy and network average energy. Accordingly DEEC doesn't require any global knowledge of energy at every round. This algorithm constantly punishes the advanced nodes, especially when their left over energy diminishes and becomes in the range of the normal sensor. In this circumstances, the advanced nodes will perish rapidly than the others.

Elbhiri et al. [26] demonstrated that Stochastic DEEC (SDEEC) is a self organized network with dynamic clustering concept in which entire network is first divided into dynamic clusters. In each cluster, every sensor autonomously opts themselves as a CH based on its elementary and residual energy. After electing CH, the non-cluster member nodes will directly communicate with their respective CH's in their allotted TDMA time slot, in order to acquire all the perceived data from the CM's in the cluster, CH node must keep its receiver on. After that, a signal processing action is done by CH for compressing the sensed data into a single valuable information. This information is forwarded by the CH to the BS. Each non-CH can deactivate their radio to snooze mode to rescue their energy. The obstacle is that if member nodes modify their state to snooze mode when CH is carrying out the agglomeration activity, they don't recognize the succeeding round of CH election.

Elbhiri et al. [14] elucidate about DDEEC (Developed DEEC) which utilizes dual-level heterogeneous nodes in the network to solve the energy dispersal issue in DEEC by making amendment in likelihood function for sustaining the life expectancy of advanced sensors.

Javaid et al. [16] elaborated about Enhanced DDEEC in which three sorts of sensors with different energy are set out and CH is chosen with amended prospect function. At the preliminary rounds, super, and advanced sensors have the occasion to pretend to be as CH, later on a statuses ensues at which all sensors will have the similar CH election prospect.

Saini et al. [27] made clear about a CH election protocol termed it as Threshold DEEC (TDEEC) in which the network encompasses three diverse sensors with three dissimilar energy. Based on fraction of left over energy of a sensor to average network energy of that round relating to optimal quantity of CH is deliberated as criterium for electing CH.

Javaid et al. [28] elucidate about Balanced Energy-Efficient Network Integrated Super Heterogeneous (BEENISH) protocol in which perception zone encompasses four sensor varieties. In this scheme, ultra-super, super, and advanced nodes have surplus energy, when compared to super, advanced, and normal nodes, respectively. The ultra-super nodes are nominated as CH for more times which in turn diminishes their energy expeditiously.

3 Radio Model and Heterogeneous WSN Model

3.1 Radio Model

Network Existence is momentously afflicted by radio energy model which is shown in Fig. 9 and the parameters are shown in Table 1.

E_{TX} represents energy expenditure for despatching K-bit of packet up to d meters distance by sensor node transmitter is less or greater than threshold distance thd_0 is given as

$$E_{TX}(K, d) = K^*\left(E_{elec} + d^2 * \varepsilon d_{fs}\right) \quad for \quad d < thd_0$$

$$E_{TX}(K, d) = K^*\left(E_{elec} + d^4 * \varepsilon d_{mp}\right) \quad for \quad d \geq thd_0 \tag{2}$$

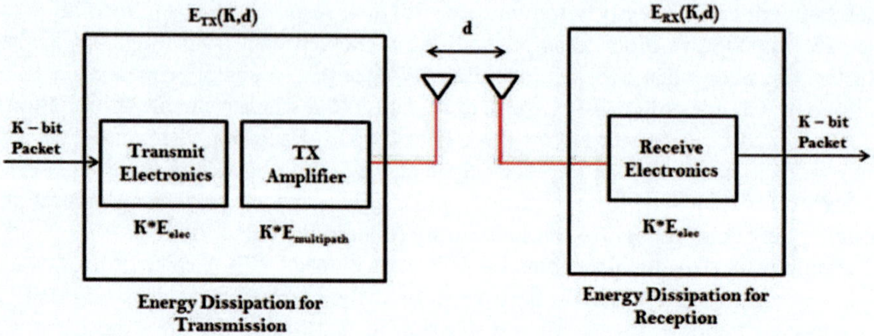

Fig. 9 Energy dissipation model

Table 1 Distinct presumptions with reference to the radio characteristics

Operation	Energy dissipated
E_{elec}	50 nJ/bit
εd_{fs}	10 pJ/bit/m^2
εd_{mp}	0.0013 nJ/bit/m^4
Energy Dissipation intended for Data Fusion (E_{DA})	5 nJ/bit/message

Here, E_{elec} signifies energy dispersal by radio for operating the circuit of transceiver, εd_{fs} and εd_{mp} implies energy exhausted in free space and multipath milieu.

Getting a K message requires the following energy that a node dissipates for the radio reception at a distance of d meters.

$$E_{RX}(K, d) = K * E_{elec} \tag{3}$$

Threshold distance thd$_o$ value is calculated by using the formula

$$thd_0 = \frac{\varepsilon d_{fs}}{\varepsilon d_{mp}} \tag{4}$$

3.2 Heterogeneous WSN Network Model

It may comprise of two, three, or multi varieties of sensors based on their energy hierarchy and named it as two, three, or multi-level heterogeneous WSN's, respectively. Here, FL-BEENISH embraces four diverse energy hierarchies of nodes: Normal sensors are embedded with Ei energy. Advanced, super, and ultra-super nodes of fraction m, m0, and m1 have a1, b1, and c1 times extra energy when contrasted to normal nodes.

The total initial energy of ultra-super, super, advanced, and normal nodes in WSN are as follows

$$E_{\text{ultra super}} = TS * m * m_0 * m_1 * E_i * (1 + c1) \tag{5}$$

$$E_{\text{super}} = TS^* m^* m_0 * (1 - m_1)^* E_i * (1 + b1) \tag{6}$$

$$E_{\text{advanced}} = TS^* m * (1 - m_0)^* E_i * (1 + a1) \tag{7}$$

$$E_{\text{normal}} = TS^*(1 - m) * E_i \tag{8}$$

where E_i represent initial energy of a sensor node.

Accordingly, aggregate energy of entire sensors will be given as

$$\begin{aligned} E_{\text{TotEneg}} &= E_{\text{ultra super}} + E_{\text{super}} + E_{\text{advanced}} + E_{\text{normal}} \\ &= TS^* E_i * \left(1 + m^* a_1 - m^* m_0 * a_1 + m^* m_0 * b_1 - m^* m_0 * m_1 * b1 + m^* m_0 * m_1 * cl\right) \\ &= TS^* E_i * \left(1 + m^* a_1 * (1 - m_0) + m^* m_0 * b1 * (1 - m_1) + m^* m_0 * m_1 * cl\right) \end{aligned} \tag{9}$$

4 The Four-Level BEENISH Protocol

This protocol practice the notion of four-level hierarchies of sensors such as normal, advance, super, and ultra-super nodes.

Let ni be total rounds for a sensor to opt as CH known as rotating epoch. If rotating epoch is alike for all the sensors present in the WSN implies, then the energy won't around dispersed and same time minimal energy nodes will decrease more rapidly than the high energy nodes. In general, high energy nodes are more times preferred as CH as compared to minimal energy nodes, therefore, FL-BEENISH pick dissimilar epoch for different nodes concerning their left over energy E_{loe}. In FL-BEENISH, ultra-super nodes are to a great extent chosen as CH when contrasted with left over nodes. Thereby, energy consumption by all nodes are evenly apportioned. Here, selection of CH is entirely based on left over energy of the sensor (E_{loe}) with respect to average network energy (E_{avg}).

E_{avg} is the energy that individual sensor possess in extant round to retain the network survive as far as possible and also utilized as reference energy for every individual node.

E_{avg} of rth round can be estimated by using the formula

$$E_{avg} = \frac{1}{TS} * E_{TotEneg} * \left(1 - \frac{r}{NLS}\right) \tag{10}$$

TS represents total number of sensors initially dispersed in the network, $E_{TotEneg}$ represent whole preliminary energy of the entire network, NLS is the sensor network life duration and is found by means of

$$NLS = \frac{E_{TotEneg}}{E_{EnergyDisperse}} \tag{11}$$

$E_{EnergyDisperse}$ is the entire energy exhausted by the network is calculated as

$$L^*\left(TS^*\left(2^*E_{elec} + E_{DA}\right) + g^*\varepsilon d_{mp} * dis^4_{to\ BS} + TS^*\varepsilon d_{fs} * dis^2_{to\ CH}\right) \tag{12}$$

g is the quantity of cluster $dis^4_{to\ BS}$ as well as $dis^2_{to\ CH}$ represent communique separation between CH and BS and CM and CH and is appraised with the formula,

$$dis_{\text{to } BS} = 0.765 * \frac{X_M}{2} \quad \text{and} \quad dis_{\text{to } CH} \frac{X_M}{\sqrt{2\Pi g}} \tag{13}$$

By taking the derivative of $E_{EnergyDisperse}$ with respect to g and then equate to 0, optimum quantity of clusters is estimated as

$$g_{opt} = \sqrt{\frac{TS^*\varepsilon d_{fs}}{2\Pi^*\varepsilon d_{mp}} * \frac{X_M}{dis_{toCH}}} \tag{14}$$

The prospect of sensors are

$$p_i = \begin{cases} \frac{p_{opt} * E_{loe}}{(1 + a1m(1 - m_0) + b1mm_0(1 - m_1) + mm_0m_1c1)^* E_{avg}} & for\ normal\ nodes\ if\ E_{loe} > T_{absolute} \\ \frac{p_{opt} * (1 + a1) * E_{loe}}{(1 + a1m(1 - m_0) + b1mm_0(1 - m_1) + mm_0m_1c1)^* E_{avg}} & for\ advanced\ nodes\ if\ E_{loe} > T_{absolute} \\ \frac{p_{opt} * (1 + b1)^* E_{loe}}{(1 + a1m(1 - m_0) + b1mm_0(1 - m_1) + mm_0m_1c1) * E_{avg}} & for\ super\ nodes\ if\ E_{loe} > T_{absolute} \\ \frac{p_{opt} * (1 + c1) * E_{loe}}{(1 + a1m(1 - m_0) + b1mm_0(1 - m_1) + mm_0m_1c1)^* E_{avg}} & for\ ultrasuper\ nodes\ if\ E_{loe} \le T_{absolute} \end{cases} \tag{15}$$

Sensor with high left over energy will be opted as CH more habitually when compared to rest of the nodes with lower energy.

The absolute residual energy $T_{absolute}$ is written as

$$T_{absolute} = z * E_i \tag{16}$$

where z = 0.7 is the appropriate value.

A point comes in a network where unexpended energy of ultra-super, super, and advanced sensors remain same as normal sensors' energy. At that moment, ultra-super nodes get penalized frequently and they decrease more rapidly than normal nodes. To overcome this problem, proposed protocol makes changes in the probability function based on threshold residual energy value. The modified possibilities for the CH methodical election are given as follows:

$$p_i = \frac{c^* p_{opt} * (1 + c1)^* E_{loe}}{(1 + a1m(1 - m_0) + b1mm_0(1 - m_1) + mm_0m_1c])^* E_{avg}} \quad for\ all\ node \tag{17}$$

where c is a variable which controls the clusters in the network. On the off chance that c value is greater, at that point there are more CH's conveying straightforwardly to the BS. As a solution, c value is fixed to 0.025 for enhanced network proficiency. Consequently, the CH choice will be balanced and more competent.

At each round, node concludes whether to opt as CH dependent on threshold computed by the suggested proportion of CH's and similarly how many times the sensors are pretended as CH up until this point. This conclusion is done by the nodes by taking the random values somewhere in the range of 0 and 1. If the number is not as much as threshold value, the node turns into a CH for the extant round.

The threshold is

$$Threshold = \frac{p_i}{1 - p_i\left(r \bmod \left(\frac{1}{p_i}\right)\right)} \quad if\ S\varepsilon G \tag{18}$$

where pi, r, and G signify the desired fraction of CH, the extant round number and the set of sensors that had not yet act as CH in the preceding 1/pi rounds. The flow chart of the proposed protocol is shown in Fig. 10.

5 Result and Discussion

In this section, the performance of FL-BEENISH protocol was evaluated using MATLAB and compared with previous algorithms like DEEC and EDDEEC. Here, WSN arrange model is made out of static sensors embrace four divergent energy in MXM field alongside base station is set midway. Here, 40 ordinary nodes have preliminary energy, 30 advanced, 18 super, and 12 ultra-super sensors hold 1.5, 2, and 2.5 times surplus energy. The parameters are exploited for accomplishing the experimentations are portrayed in Table 2.

The metrics used to appraise the FL-BEENISH method are

(1) Network Longevity is the meticulous round number at which entire nodes die.
(2) Quantity of packets which is acquired by BS from CH.

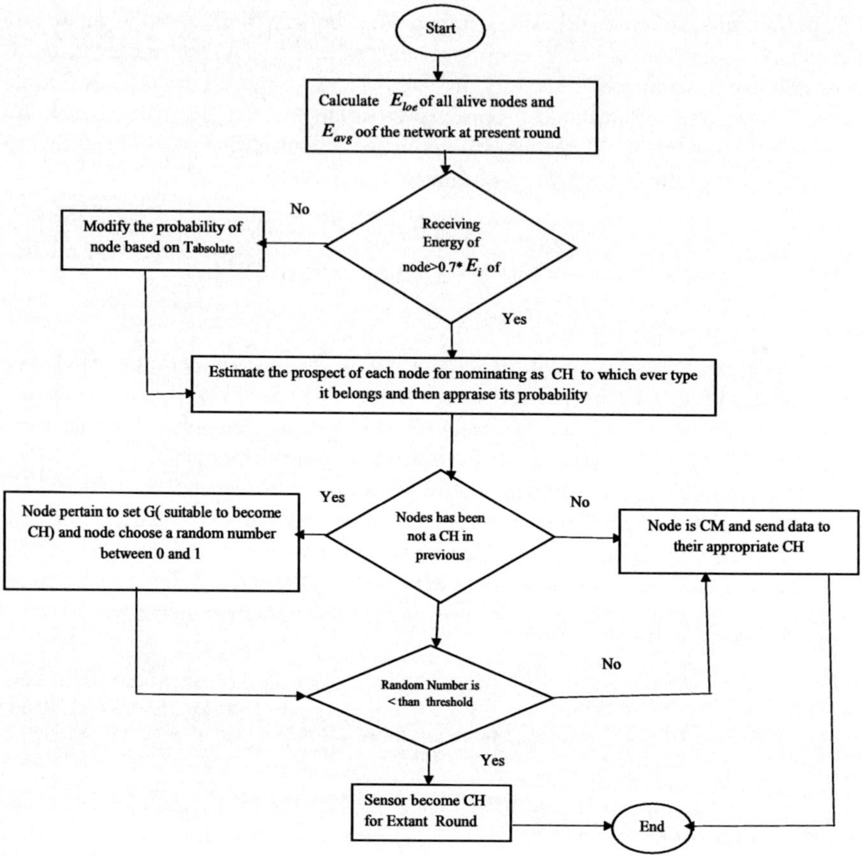

Fig. 10 Flow chart of proposed protocol

Table 2 Experimental parameters

No	Parameter	Description	Value
1	TS	Total Quantity of Sensors in the Perceiving Zone	100
2	X_M, Y_M	X and Y Dimensions in the Zone	100
3	p_{opt}	CH Likelihood	0.1
4	thd_0	Threshold Distance	80 m
5	E_i	Principal Energy	0.5 J
6	L	Data Size	4000 bits
7	m, m_0, m_1	Proportion of advanced, super, and ultra-super nodes	0.6,0.5,0.4

Altogether sensor deceased vs number of rounds is exposed in Fig. 11. At this point, the entirety node loses their lifespan at 5106, 5859, 6891 round number for DDEEC, EDDEEC, and FL-BEENISH protocols, respectively. It clearly shows that by introducing ultra-super nodes in network, the lifespan increases when compared to existing protocols. The result shows that FL-BEENISH increases 15% more rounds of network lifetime than DDEEC and EDDEEC (Table 3).

The Packets received by the base station are revealed in Fig. 12. The total number of packets gotten by the base station is 227479,321573,335077bits, respectively (Table 4).

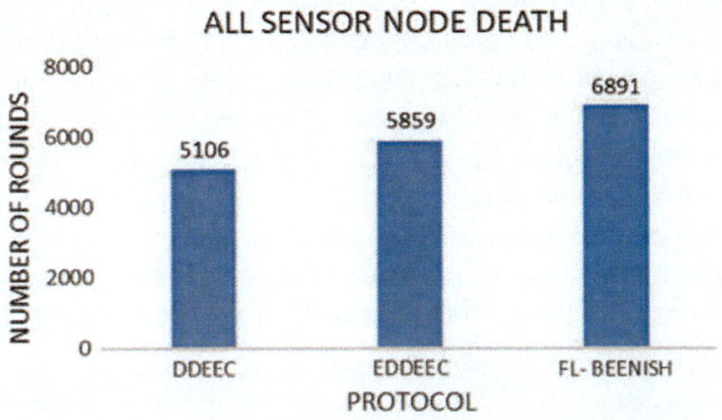

Fig. 11 All sensor node death

Table 3 All node dead vs number of rounds

Protocol	Altogether sensor death (In Rounds)
DDEEC	5106
EDDEEC	5859
FL- BEENISH	6891

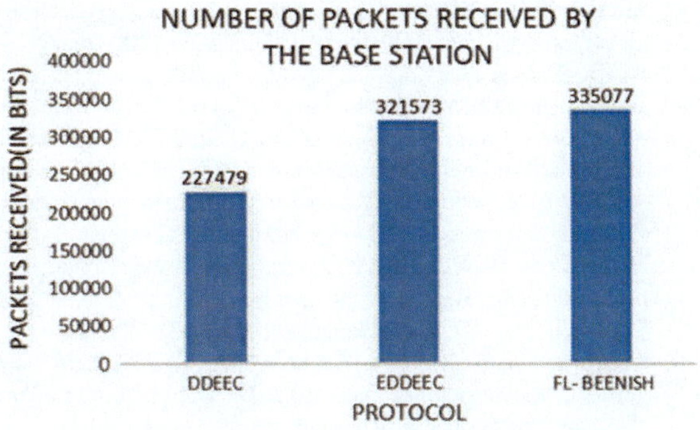

Fig. 12 Packets received by the base station

Table 4 Packets received by the base station

Protocol	Packets received by the base station (In bits)
DDEEC	227479
EDDEEC	321573
FL-BEENISH	335077

6 Conclusion

FL-BEENISH approach for WSN is elucidated in this paper by scrutinizing the network energy utilization dependent on energy heterogeneity. The experimental investigation is conducted based on dual facets, i.e., network existence and packets transported to the BS. Here, heterogeneity in the network is added by embedding ultra-super sensors with additional energy than remaining nodes. The proposed strategy utilizes the portion of specific sensor left over energy as well as sensor network average energy as CH choice criterium and progresses the data communique, diminishes energy utilization, and prolongs the duration of the network. The future scope of this work would embrace more than four energy levels of nodes and more work should be performed in future to extend the network lifespan with mobile sink node.

References

1. Ideas for the 21st Century, Business Week, pp. 78–67 (1999)
2. Chong, C.Y., Kumar, S.P.: Sensor networks: evolution, opportunities and challenges. Proc. IEEE **91**(8), 1247–1256 (2003)
3. Akyildiz, I.F., Su, W., Sankarasubramaniam, Y., Cayirci, E.: A survey on sensor networks. IEEE Commun. Mag. **140**(8), 102–114 (2002)
4. Mhatre, V., Rosenberg, C.: Design guidelines for wireless sensor networks: communication. Cluster Aggregat Ad Hoc Netw **2**(1), 45–63 (2004)
5. Al Karaki, J.N., Kamal, A.E.: Routing techniques in wireless sensor networks: a survey. IEEE Wirel. Commun. **l**(11, 6), 6–28 (2004)
6. Ahmad. A., Javaid, N., Khan, Z.A., Qasim, U., Alghamdi. T.A.: $(ACH)^2$: Routing scheme to maximize lifetime and throughput of WSN's, IEEE Sens. J. **14**(10), 3516–3532 (2014)
7. Abbasi, A.A., Younis, M.: A survey on clustering algorithms for wireless sensor networks. Comput. Commun. **30**(14–15), 2826–2841 (2007)
8. Heinzelman, W.R., Chandrakasan, A.P., Balakrishnan, H.: Energy efficient communication protocol for wireless micro sensor networks. In: Proceedings of the 33rd Hawaii International Conference on System Sciences (HICSS-33) (2000)
9. Smaragdakis, G., Matta, I., Bestavros, A.: SEP: A stable election protocol for clustered heterogeneous wireless sensor networks. In: Second International Workshop on Sensor and Actor Network Protocols and Applications, pp. 1–11 (2004)
10. Chamam, A., Pierr, S.: A Distributed energy efficient clustering protocol for wireless sensor networks. J. Comput. Electr. Eng. **36**(2), 303–312 (2010)
11. Singh. S., Malik, A., Kumar, R.: Energy efficient heterogeneous DEEC protocol for enhancing lifetime in WSN's. Int. J. Eng. Sci. Technol. **20**(1), 345–353 (2017)

12. Qing, L., Zhu, Q., Wang, M.: Design of a distributed energy efficient clustering algorithm for heterogeneous wireless sensor network. Elsevier, Comput. Commun. **29**, 2230–2237 (2006)
13. Belbhiri, R.S., El Fkihi, S., Aboutajdine, D.: Developed distributed energy efficient clustering (DDEEC) for heterogeneous wireless sensor networks. In: 5th International Symposium on Communications and Mobile Network (ISVC) (2010)
14. Elbhiri, B., Rachid, S., El Fkihi, S., Aboutajdine, D.: Developed distributed energy efficient clustering (DDEEC) for heterogeneous WSN. In: 5th International Symposium on I/V Communications and Mobile Networks (2010)
15. Javaid, N., Babar Rasheed, M., Imran, M., Guizani, M., Ali Khan, Z., Ali Alghamdi, T., Ilahi, M.: An energy efficient distributed clustering algorithm for heterogeneous WSNs. EURASIP J. Wirel. Commun. Network. (2015)
16. Javaid, N., Qureshi, T.N., Khan, A.H., Iqbal, A., Akhtar, E., Ishfaq, M.: EDDEEC: enhanced developed distributed energy efficient clustering for heterogeneous wireless sensor networks. Procedia Comput. Sci. **19**, 914–919 (2013)
17. Tyagi, S., Kuma, N.: A systematic review on clustering and routing techniques based upon LEACH protocol for wireless sensor networks. J. Netw. Comput. Appl. **36**(2), 623–645 (2013)
18. Lai. W.K, Fan. C.S., Lin, L.Y.: Arranging cluster sizes and transmission ranges for wireless sensor networks. Inf. Sci. **183**(1), 117–131 (2012)
19. Hosseinirad, S.M., Mohammadi, M.A., Basu, S.K., Pouyan, A.A.: LEACH routing algorithm optimization through imperialist approach. Int. J. Eng. Trans. A, Basics **27**(1), 39–50 (2014)
20. Loscri, V., Morabito, G., Marano. S.: A two level hierarchy for low energy adaptive clustering hierarchy. In: Proceedings of the 2nd IEEE Semi Annual Vehicular Technology Conference, pp. 1809–1813 (2005)
21. Malluh. A.A., Elleithy. K.M., Qawaqneh. Z., Mstafa. R.J., Alanazi, A.: EM-SEP: an efficient modified stable election protocol. In: Proceedings of the 2014 Zone 1 Conference of the American Society for the Engineering Education (2014)
22. Rehman, O., Javaid, N., Manzoor, B., Hafeez, A., Iqbal, A., Ishfaq, M.: Energy consumption rate based stable election protocol (ECRSEP) for WSN's. Procedia Comput. Sci. **19**, 932–937 (2013)
23. Kumar, D., Aseri, T.C., Patel, R.B.: EEHC: energy efficient heterogeneous clustered scheme for wireless sensor networks. Comput. Commun. **32**(4), 662–667 (2009)
24. Younis, O., Fahmy, S.: HEED: a hybrid, energy efficient, distributed clustering approach for ad hoc sensor networks. IEEE Trans. Mob. Comput. **3**(4), 366–379 (2004)
25. Duan, C., Fan, H.: A distributed energy balanced clustering protocol for heterogeneous wireless sensor networks. In: Proceedings of International Conference on Wireless Communication, Networking and Mobile Computing, pp. 2469–2473 (2007)
26. Elbhiri, B., Saadane, R., Aboutajdine, D.: Stochastic distributed energy efficient clustering (SDEEC) for heterogeneous wireless sensor networks. ICGST CNIR J. **9**(2), 11–17 (2009)
27. Saini, P., Sharma, A.K.: Energy efficient scheme for clustering protocol prolonging the lifetime of heterogeneous wireless sensor networks. Int. J. Comput. Appl. **6**(2) (2010)
28. Qureshi, T.N., Javaid, N., Khan, A.H., Iqbal, A., Akhtar, E., Ishfaq, M.: Balanced energy efficient network integrated super heterogenous protocol for wireless sensor networks. Procedia Comput. Sci. **19**, 920–925 (2013)

Author Index

J. S. Raj et al. (eds.), *Intelligent Sustainable Systems*, Lecture Notes
in Networks and Systems 213, https://doi.org/10.1007/978-981-16-2422-3